Springer Proceedings in Mathematics & Statistics

Volume 446

This book series features volumes composed of selected contributions from workshops and conferences in all areas of current research in mathematics and statistics, including data science, operations research and optimization. In addition to an overall evaluation of the interest, scientific quality, and timeliness of each proposal at the hands of the publisher, individual contributions are all refereed to the high quality standards of leading journals in the field. Thus, this series provides the research community with well-edited, authoritative reports on developments in the most exciting areas of mathematical and statistical research today.

Dimitrios Vlachos
Editor

Mathematical Modeling in Physical Sciences

12th IC-MSQUARE, Belgrade, Serbia, August 28–31, 2023

Springer

Editor
Dimitrios Vlachos
Department of Economics
University of Peloponnese
Tripoli, Greece

ISSN 2194-1009 ISSN 2194-1017 (electronic)
Springer Proceedings in Mathematics & Statistics
ISBN 978-3-031-52964-1 ISBN 978-3-031-52965-8 (eBook)
https://doi.org/10.1007/978-3-031-52965-8

Mathematics Subject Classification: 74P15, 60G35, 62M10, 68W20, 92D25, 65C05, 03E72

This Springer imprint is published by the registered company Springer Nature Switzerland AG
The registered company address is: Gewerbestrasse 11, 6330 Cham, Switzerland

Paper in this product is recyclable.

International Scientific Committee

Morten Hjorth-Jensen, University of Oslo, Norway
Jacob Barhen, Oak Ridge Natl. Lab, Oak Ridge, USA
Georgios Loukes-Gerakopoulos, TPI, FSU Jena, Germany
Javier Hernández-Andrés, University of Granada, Spain
Maia Martcheva, University of Florida, USA
Theocharis Kosmas, University of Ioannina
Angelo Tartaglia, Politecnico di Torino, Italy
Yoshiharu Yamamoto, The University of Tokyo, Japan
Amaury Mouchet, Université de Tours, France
Mairi Sakellariadou, King's College of London, UK
Kostas Skenderis, University of Amsterdam, The Netherlands
Orfeu Bertolami, University of Porto, Portugal
Saurya Das, University of Lethbridge, Canada
Michel A. Van Hove, City University of Hong Kong, Hong Kong
Rho Shin Myong, Gyeongsang National University, Republic of Korea
Sergey Nazarenko, University of Warwick, UK
Emilio Elizalde, CSIC, Barcelona, Spain
Theodosios Christodoulakis, University of Athens, Greece
Bari Ferenc, University of Szeged, Hungary
Mark Cross, Swansea University, UK
Kostas Kokkotas, Eberhard Karls University of Tübingen, Germany
Shiyi Chen, Peking University, People's Republic of China
Heinrich Begehr, Germany
Gerard R. Lemaitre, France
Robert Finn, Stanford University, USA

Organizing Committee

Chair: Evangelia D. Vlachou, Ionian University, Greece
Amanda Mavrogianni, Sivitanideios School, Greece
Ioannis Torakis, National and Kapodistrian University of Athens
Sotirios-Neilos D. Vlachos, National Technical University of Athens, Greece

Preface

The International Conference on Mathematical Modeling in Physical Sciences (IC-MSQUARE) is an interdisciplinary scientific event which aims to promote the knowledge and the development of high-quality research in mathematical fields that have to do with the applications of other scientific fields and the modern technological trends that appear in them, these fields being those of Physics, Chemistry, Biology, Medicine, Economics, Sociology, Environmental Sciences, etc.

The 12th IC-MSQUARE took place in Belgrade, Serbia, between August 28 and 31, 2023. In addition, (because in previous years the conference, due to the pandemic, took place only with online presentations), the possibility of participation via the internet was also adopted this time. Thus, apart from the talks and workshops held on site, we also had a large collection of pre-recorded presentations that were available to the participants through the conference website.

The Conference was attended by more than 350 participants and hosted about 320 oral and virtual presentations while counted more than 1000 pre-registered authors. The 12th IC-MSQUARE consisted of different and diverging workshops and thus covered various research fields where Mathematical Modeling is used, such as Theoretical and Mathematical Physics, Neutrino Physics, Non-integrable Systems, Dynamical Systems, Computational Nanoscience, Biological Physics, Computational Biomechanics, Complex Networks, Stochastic Modeling, Fractional Statistics, DNA Dynamics, Macroeconomics, Social Modeling, etc.

The scientific program was rather heavy, however, according to all attendees, the program was excellent with high level of talks and the scientific environment was fruitful, thus all attendees had a creative time.

All papers submitted to the Conference have been reviewed using a single-blind peer-review process. The review process adopted a scoring system (0–5 points) who rated the Relevance, Scientific Quality, Contribution, Title and Paper Presentation, Tables and Figures, Language, and the References of each paper. The threshold for an article to be accepted was a score above 4. Additionally, each paper was tested for similarity with already published content using the Turnitin software. All sources of similarities were checked and only original content was accepted. In this way, out of more than 120 papers that were evaluated for this volume, only 51 were accepted for publication.

We would like to thank the Reviewers, the Members of the International Scientific Committees as well as the Members of the Organizing Committee.

Tripoli, Greece

Dimitrios Vlachos
The Conference Chairman

Contents

Mathematical Methods in Physical and Engineering Applications

Single and Multi-material Topology Optimization of Continuum Structures: ABAQUS Plugin

Dhaval Patel, Thomas Rockenbauer, Sandra Schlögl, and Margit Lang

Abstract This research addresses the need for versatile topology optimization techniques capable of optimizing both single and multi-material designs. The extended Python code incorporates the modified Bi directional Evolutionary Structural Optimization (BESO) algorithm and a material interpolation scheme to enhance its capabilities. Engineers and designers can utilize this improved approach to optimize the performance of structures, especially in the context of multi-material configurations. To increase its practicality, the code is converted into ABAQUS plugin, seamlessly integrating it with the widely used finite element analysis software. Validation examples, conducted in the ABAQUS environment, demonstrate the compatibility and accuracy of the code. This research provides an efficient and accurate solution for topology optimization, addressing the demands of multi-material designs in various engineering applications. Examples of such applications include lightweight design in automotive and aerospace industries, customized implants in biomedical engineering, and optimal material distribution in architectural structures.

Keywords Topology optimization · ABAQUS plugin · Bi-directional evolutionary structural optimization (BESO) · Multi-materials optimization

1 Introduction

Structural optimization is a method used in engineering and design to determine the most efficient material layout and shape of a structure, considering specific design criteria and constraints. This technique originated in the 1980s and has progressed as technology has advanced. The three primary techniques for structural optimization are: (1) topology optimization, which optimizes the material distribution for a desired structural performance, (2) shape optimization, which enhances the design geometry

D. Patel (✉) · S. Schlögl · M. Lang
Polymer Competance Center Leoben GmbH, Leoben, Austria
e-mail: dhaval.rasheshkumar.patel@pccl.at

T. Rockenbauer
Luxinergy GmbH, Leoben, Austria

D. Vlachos (ed.), *Mathematical Modeling in Physical Sciences*, Springer Proceedings in Mathematics & Statistics 446, https://doi.org/10.1007/978-3-031-52965-8_1

to improve performance (e.g. to avoid stress concentration), and (3) size optimization, which determines the optimal dimensions (e.g. thickness) of a component. By utilizing topology optimization, engineers can design structures that are stronger, lighter, and more efficient, while meeting performance requirements such as durability and stability. This results in improved structural performance. The ability to remove redundant material from areas that do not carry significant loads, furthermore increases material efficiency by offering the possibility to use resources in a more sustainable way. Consequently, the use of topology optimization is expanding in various fields, including aerospace, automotive, civil and biomedical engineering. As computational power and optimization algorithms continue to improve, it is expected that topology optimization will become even more prevalent in engineering design.

Topology optimization encompasses various techniques that have unique strengths and limitations. Some of the most commonly used methods include Solid Isotropic Material with Penalization (SIMP) [1], Evolutionary Structural Optimization (ESO) [2], and Level Set Method (LSM) [3]. Each method offers a different approach to solving topology optimization problems and is selected based on the specific design objectives and constraints. SIMP is a commonly used topology optimization technique that assigns material density to each element in the design domain and penalizes intermediate densities to obtain a clear "black-white" design. However, it can lead to so-called checkerboard patterns [4] and numerical instabilities [5]. To address these issues, various restriction methods have been developed, such as perimeter control, gradient constraints, or blurring filters [6–8]. ESO is a topology optimization technique that mimics the process of natural selection to obtain an optimal topology by gradually removing "inefficient" material and redistributing loads. It is useful in conceptual design and can produce organic-looking shapes. Bi-directional evolutionary structural optimization (BESO) is an extension of ESO that enables the addition of "efficient" material while removing "inefficient" material. However, both methods suffer from numerical instability issues that can be addressed by incorporating perimeter control or filters [6–8]. The most distinct difference with respect to SIMP is that with ESO or BESO no intermediate densities are produced. For ESO and BESO basically only two "material states" exist—"0 = void" and "1 = full material". Therefore, no interpretation of intermediate densities is necessary for the design engineer. LSM is a topology optimization technique that utilizes a level set function to represent the design domain, enabling straightforward manipulation of the topology. LSM is beneficial for problems with multiple load cases or design objectives, and it can create smooth and continuous designs. However, it is computationally costly and may be mesh-dependent [9].

Over the past four decades, based on these methods, numerous educational articles have been published with the objective of introducing fundamentals of various topology optimization algorithms through computer program implementations. For instance, Sigmund developed a 99-line MATLAB code for the SIMP method [1], while Andreessen et al. developed an 88-line MATLAB code that is an extension of Sigmund's code with improved efficiency [10]. Yongsheng et al. presented an efficient 137-line MATLAB code [11] for 2D compliance minimization utilizing the BESO method presented by Huang and Xie [12], and Liu and Tovar presented a modified SIMP model for 3D topology optimization for linear structures with regular eight-noded elements [13]. Moreover, Zhi Hao Zuo et al. [14] developed a compact and simple Python code for complex 3D topology optimization utilizing the BESO method. Additionally, several commercial and open-source tools for topology optimization are available, as stated in the comparative study published in article [9].

As mentioned before, most studies and applications of topology optimization have been limited to single material structures. As a result, there is a noticeable gap in research regarding topology optimization for structures with multiple materials. Several researchers have attempted to address this issue through different approaches. For instance, Zhou and Wang [15] developed a phase-field method with a generalized Cahn-Hilliard model for the optimization of multi-material structural topology. An algorithm was introduced by Huang and Xie [16] that utilizes the BESO method for topology optimization of continuum structures. The algorithm is capable of handling both single and multiple materials and incorporates the material interpolation approach from the SIMP method. Hvejsel and Lund [17] proposed two multi-material interpolation schemes based on SIMP and (RAMP) Rational Approximation of Material Properties material interpolation schemes to enable topology and multi-material optimization formulated within a unified parametrization. A novel problem formulation was proposed by Gao and Zhang [18] to address the topology optimization of lightweight structures comprising multiple phases of materials. The formulation introduces a mass constraint to guide the optimization process. Gaynor and Yang [19] combined the force flow path with topology optimization to provide topology optimization methods for two- and three-dimensional reinforced concrete structures. Wang et al. [20] proposed a new Multi-Material Level Set (MM-LS) topology description model for topology and shape optimization of structures involving multiple materials. A multi-material topology optimization method utilizing a reaction-diffusion equation based on the level set method was proposed by Cui et al. [21]. Recently, a multi-material topology optimization method based on BESO has been introduced. This method specifically addresses structures with multiple materials that possess notably different mechanical properties in tension and compression [22, 23].

Despite many research efforts, there is still a lack of open-source codes available for multi-material topology optimization. To overcome this issue, the author is motivated by the works of Zuo and Xie in [14] and Huang and Xie in [16] and intends to extend the Python code available from [14]. The objective is to make it more computationally efficient and to enable multi-material topology optimization using the formulation of the modified BESO method introduced in the research article [16]. This endeavour aims to contribute to the advancement of multi-material topology optimization research and provide an efficient tool for the scientific community. The code can be found in this GitHub repository: https://github.com/dhavalrpatel2511/Topology_Optimization.git.

2 Methodology

This research article focuses on the implementation of the modified BESO method for single and multi-material topology optimization, which is based on the previously published work by X. Huang et al. [16]. While the formulation of the topology optimization problem is not extensively discussed in this article, readers can refer to the original work [14, 16] for a more detailed explanation. Nonetheless, the essential information regarding the topology optimization problem shall be provided here.

2.1 Problem Statements

The objective of the topology optimization problem is to minimize the compliance of a structure while satisfying a constraint on the volume of the material to be used in the final design. Mathematically, this approach to single-material optimization can be expressed as:

$$\begin{aligned} &\min_{X} : C_m(\mathbf{X}) = \frac{1}{2}\mathbf{F}^{\mathbf{T}}\mathbf{U} = \frac{1}{2}\mathbf{U}^{\mathbf{T}}\mathbf{K}\mathbf{U} \\ &\text{subject to: } \mathbf{X} = [x_e];\; x_e = 1 \; or \; x_{min},\; \forall e = 1, ..., N \\ &\mathbf{F} = \mathbf{K}\mathbf{U} \\ &V(\mathbf{X}) = \sum_{X} x_e V_e = V^* \end{aligned} \tag{1}$$

The mean compliance $C_m(\mathbf{X})$, which is the objective function, represents the energy required to deform the structure under a given set of loads, and it is proportional to the strain energy stored in the material. **F** represents the global force vector, **U** is the displacement vector, **K** is the stiffness matrix and V(**X**) refers to the volume of the design domain occupied by the material. The binary design variable x_e denotes the density of *e*th element. To avoid the singularity of the stiffness matrix, a small value of x_{min}, for example, 0.001 is used to denote the void elements.

With respect to multi-material optimization the objective is to find the optimal distribution of multiple materials with differing modulus of elasticity $E_1, E_2, ..., E_n$ (where $E_1 > E_2 > ... > E_n$). The corresponding optimization problem can be formulated as follows:

$$\begin{aligned} &\min_{X} : C_m(\mathbf{X}) = \frac{1}{2}\mathbf{F}^{\mathbf{T}}\mathbf{U} = \frac{1}{2}\mathbf{U}^{\mathbf{T}}\mathbf{K}\mathbf{U} \\ &\text{subject to: } \mathbf{X} = [x_{ij}];\; x_{ij} = 1 \text{ or } x_{min},\; (j = 1,\; 2,\; ...,\; n-1) \\ &\mathbf{F} = \mathbf{K}\mathbf{U} \\ &V_j^* - \sum_{i=1}^{n} V_i x_{ij} - \sum_{i=1}^{j-1} V_i^* = 0 \end{aligned} \tag{2}$$

where x_{ij} denotes the density of the ith element for the jth material, and V_j* refers to the prescribed volume of the jth material.

2.2 Sensitivity Numbers

In the work of Huang and Xie [16], the material interpolation scheme from the SIMP method and sensitivity numbers for various topology optimization variants (such as single material void topology optimization, two non-zero material, or multi-material topology optimization) are briefly described. However, for the purpose of implementation, the equations for computing the sensitivity numbers for all these different variants of topology optimizations are provided here.

The sensitivity number in the BESO algorithm is a measure of how the overall structural performance is influenced by a change in "design" of a specific element. Based on the sensitivity number the algorithm decides which elements to modify or remove during the optimization process to achieve an optimal material distribution for the given objective. The sensitivity number for topology optimization with a single material and void is defined as follows:

$$\alpha_i = -\frac{1}{p}\frac{\partial C}{\partial x_i} = \begin{cases} \frac{1}{2}\mathbf{u}_i^{\mathbf{T}}\mathbf{K}_i\mathbf{u}_i & \text{when } x_i = 1 \\ \frac{x_{min}^{p-1}}{2}\mathbf{u}_i^{\mathbf{T}}\mathbf{K}_i\mathbf{u}_i & \text{when } x_i = x_{min} \end{cases} \tag{3}$$

The sensitivity numbers for topology optimization with two non-zero materials and with multi-materials are similar in form and can be defined as follows:

$$\alpha_i = \begin{cases} \frac{1}{2}\left[1 - \frac{E_2}{E_1}\right]\mathbf{u}_i^{\mathbf{T}}\mathbf{K}_i^1\mathbf{u}_i & \text{formaterial1} \\ \frac{1}{2}\frac{x_{min}^{p-1}(E_1 - E_2)}{x_{min}^{p}E_1 + (1 - x_{min}^{p})E_2}\mathbf{u}_i^{\mathbf{T}}\mathbf{K}_i^2\mathbf{u}_i & \text{formaterial2} \end{cases} \tag{4}$$

$$\alpha_{ij} = \begin{cases} \frac{1}{2}\left[1 - \frac{E_{j+1}}{E_j}\right]\mathbf{u}_i^{\mathbf{T}}\mathbf{K}_i^j\mathbf{u}_i & \text{for material } 1, \ldots, \text{j} \\ \frac{1}{2}\frac{x_{min}^{p-1}(E_j - E_{j+1})}{x_{min}^{p}E_j + (1 - x_{min}^{p})E_{j+1}}\mathbf{u}_i^{\mathbf{T}}\mathbf{K}_i^{j+1}\mathbf{u}_i & \text{for material j+1}, \ldots, \text{n} \end{cases} \tag{5}$$

The equations described above indicate that the sensitivity of voids in single material topology optimization, as well as other materials in various topology optimization methods, except for the first material, is affected by the selection of the penalty exponent. In particular, as the penalty exponent approaches infinity, the sensitivity values of these materials or voids tend to approach zero.

2.3 Filter Scheme and Convergence Criterion

The sensitivity numbers defined in the previous section are often adapted to obtain a mesh-independent solution. In order to achieve this, a blurring filter can be utilized, such as the one proposed by Huang and Xie [6, 8]. In this paper, a simplified elemental sensitivity filtering scheme [14] is employed to maintain the solution accuracy while avoiding the need for a newly developed computationally slower nodal sensitivity filtering scheme presented in the study by Huang et al. [16]. Nodal sensitivity analyzes localized information, while elemental sensitivity provides a broader understanding of design impacts for large-scale optimization. The simplified elemental sensitivity filtering scheme is used in this paper as follows:

$$\alpha_e = \frac{\sum_j w(r_{ej})\alpha_j}{\sum_j w(r_{ej})} = \sum_j (\frac{w(r_{ej})}{\sum_j w(r_{ej})}\alpha_j) = \sum_j \eta_j \alpha_j \tag{6}$$

$$w(r_{ej}) = max(0, r_{min} - r_{ej}) \tag{7}$$

Equation (6) uses the distance between the centers of elements e and j (r_{ej}), a weight function (w) to average the raw sensitivities, and a filter radius (r_{min}). It is important to note that the weight factor $w(r_{ej})$ is predetermined and does not depend on the sensitivity values, see Eq. (7).

The element sensitivity filter scheme requires the computation of the distance between each element that lies within the filter radius, which is used to calculate the weighting factor based on the relative distances. However, calculating the distance between each individual element and all other elements is computationally inefficient and time-consuming. Therefore, to improve efficiency, the *getByBoundingBox(...)* function, available in ABAQUS, is used in this study. Thereby, only the distances between elements that lie within the filter radius, instead of the distances of the individual element to all other elements, as described in the original publication [14], are considered.

To ensure convergence of the solution, it is advised to incorporate the historical information of the sensitivities when using discrete methods like BESO [12]. This can be achieved by averaging the sensitivity of the current iteration with that of the previous iteration as shown in Eq. (8), where k is the current iteration number.

$$\alpha_e = \frac{\alpha_e^k + \alpha_e^{k-1}}{2} \tag{8}$$

Convergence criteria are needed in the BESO algorithm to determine when the optimization process has reached a satisfactory solution. They provide a stopping criterion, ensuring the algorithm doesn't run infinitely. Convergence criteria monitor specific parameters or metrics, such as changes in the objective function or design variables, to assess when the optimization has converged to an acceptable level of accuracy. This helps save computational resources and ensures the reliability and quality of the optimized topology.

$$Error = \frac{\left|\sum_{i=1}^{N}(C_{k-i+1} - C_{k-N-i+1})\right|}{\sum_{i=1}^{N}(C_{k-N-i+1})} \leq \tau \tag{9}$$

The criterion shown in Eq. (9) uses a tolerance value τ and an integer value N (fixed at 0.01% and 5 respectively in this paper) to ensure that subsequent iterations do not significantly improve the design. Specifically, the convergence criterion requires a stable compliance over a minimum of ten consecutive iterations, which reduces the impact of above equation on the final solution. The simplified form of Eq. (9) can be represent as follows:

$$Error = \frac{C[k-4:k+1] - C[k-9:k-4]}{C[k-9:k-4]} \leq \tau \tag{10}$$

2.4 Optimization Procedure

A step-by-step explanation of topology optimization procedure is given below.

Step 1: Mesh the design domain with finite elements and apply appropriate boundary conditions and loads.
Step 2: Determine the coordinates of the center of each element and compute the weight factor for each element based on the distance to neighboring elements within the filter radius.
Step 3: Provide the inputs such as Volume fraction, Evolution rate, Young's modulus, Poisson's ratio and most important the Design-Case, which refers to the variants of topology optimization (1. solid-void, 2. two non-zero materials and 3. Multi-Material topology optimization).
Step 4: Perform finite element analysis (FEA) to determine the compliance of the entire model and the strain energy of each element.
Step 5: Compute the sensitivity number for each element based on the respective design domain variation.
Step 6: Bi-directional Evolutionary Structural Optimization (BESO) procedure.
Step 7: Update material properties and design domain for next optimization iteration based on current sensitivity numbers and material distribution.
Step 8: Perform steps 4–7 iteratively until the desired volume is reached and the convergence criteria are met.

The flowchart (Fig. 1) shows a graphical representation of the step-by-step procedure of this topology optimization process. The BESO algorithm exhibits similarities

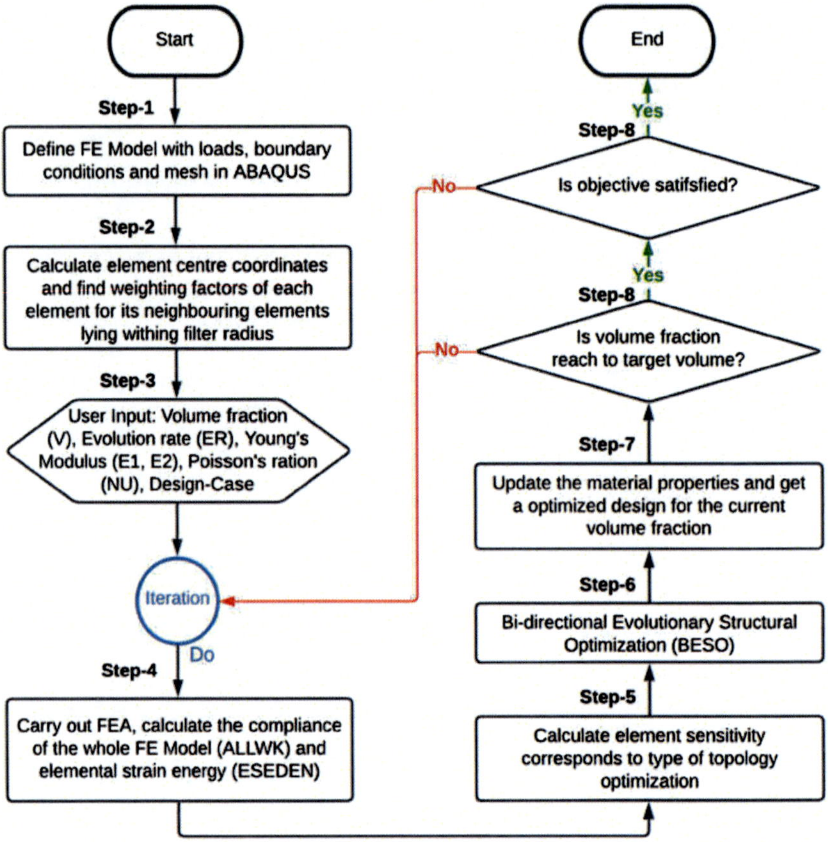

Fig. 1 Flowchart illustrating the topology optimization procedure

between the solid-void design and the two non-zero materials design variants. However, the BESO algorithm for multi-materials design variants has some differences. A detailed explanation of this BESO variant is provided in the pseudo-code (1). The multi-material BESO procedure starts with a full design containing material 1. It utilizes the evolutionary ratio (ER) to trigger the volume reduction of material 1. Simultaneously, the volume of material 3 gradually increases until the predefined volume fraction is reached. Afterwards, the volume of material 3 remains constant and the volume of material 2 increases until the target volume is achieved. The step-wise procedure is outlined in the pseudo-code, where lines 1–26 handle the optimization for material 1, lines 27–52 for material 2.

Algorithm 1 Modified BESO Algorithm for Multi-Materials Topology Optimization

```
Find the minimum (min1) and maximum (max1) values of elemental sensitivity numbers based on stiff and void material properties (Ae1_i).
Calculate the target volume of the void material in the next optimal design (T_v1 = V_f1 * Total number of elements).
repeat
  Update the threshold (ε1_k = (max1+min1)/2.0)
  for each element i in the design domain do
    if Ae1_i > ε1_k then
      Set x1_i = 1.0 (fully dense)
    else
      Set x1_i = 0.00001 (fully porous)
    end if
  end for
  if sum(x1_i) - T_v1 > 0 then
    Set min1 = ε1_k
  else
    Set max1 = ε1_k
  end if
until (max1-min1)/max1 > 0.00001
Initialization of empty lists for Void group of elements.
Initialize the temporary dictionary of elemental sensitivity numbers based on stiff and soft material properties (Ae2_i).
for each element i in the design domain do
  if x1_i = 0.00001 then
    Void.append(i)
  else
    TAe2(i) = Ae2(i)
  end if
end for
Find the minimum (min2) and maximum (max2) values of elemental sensitivity from temporary dictionary (Ae2_i).
Calculate the target volume of the stiff material in the next optimal design (T_v2 = V_f2 * Total number of elements)
repeat
  Update the threshold (ε2_k = (max2+min2)/2.0)
  for each element i in the design domain do
    if TAe2_i > ε2_k then
      Set x2_i = 1.0 (Stiff Material)
    else
      Set x2_i = 0.00001 (Soft Material)
    end if
  end for
  if sum(x2_i) - T_v2 > 0 then
    Set min2 = ε2_k
  else
    Set max2 = ε2_k
  end if
until (max2-min2)/max2 > 0.00001
Initialization of empty lists for Stiff and Soft groups of elements.
for each element i in the design domain do
  if x2_i = 1.0 then
    Stiff.append(i)
  else
    Soft.append(i)
  end if
end for
Section assignment for both Solid and Void groups of elements.
```

3 Results

Within this section, the focus lies on investigating the various design optimization variants as listed in Table 1. Specifically, the topology optimization problem

Table 1 Various design optimization variants

Variants	Stiff material volume (%)	Soft material volume (%)	Void volume (%)
3.1	30	–	70
3.2	50	50	–
3.3	15	25	60

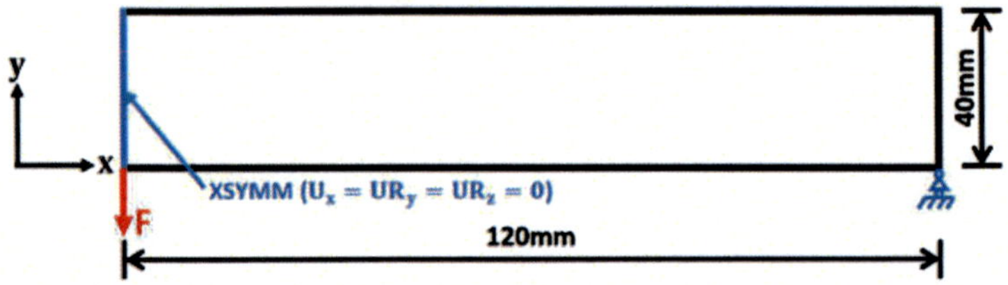

Fig. 2 Design domain of the right half of the full beam model

Table 2 Material properties

	Stiff material	Soft material	Void
Young's modulus	1000 MPa	100 MPa	1e-9 MPa
Poisson's ratio	0.3	0.3	0.3

is analyzed, concerning a beam structure with dimensions of 240×40 mm. However, exploiting the beam's symmetry, we shall solely examine the right half, which possesses a domain size of 120×40 mm, see Fig. 2. The bottom right corner of the right half is subjected to support, while the left edge adheres to X-symmetry boundary conditions. Additionally, a 10N load (F) is applied at the bottom left corner, as illustrated in Fig. 2. The application of X-symmetry boundary conditions, in this context, serves to accurately replicate the overall behavior of the entire beam structure under the specified load. These boundary conditions effectively constrain three degrees of freedom at the nodes along the left edge of the beam: the displacement along the X-axis (U_x), the rotation along the Y-axis (UR_y), and the rotation along the Z-axis (UR_z). The domain is discretized using four-node plane stress elements (CPS4) with a size of 1 mm.

As mentioned earlier, the topology optimization is performed using specific parameters. These parameters include a minimum element density (xmin) of 0.0001, a filter radius (rmin) of 3.0, a penalty exponent (p) set to infinity, and an Evolutionary Ratio (ER) of 2.0%. Furthermore, Table 2 includes the material properties of the stiff material (Material 1), soft material (Material 2), and void. The properties include the respective values for Young's modulus and Poisson's ratio assuming liner elastic material behavior.

3.1 Optimized Solid-Void Design

This section focuses on optimizing the beam structure, shown in Fig. 2, using a single material. The objective is to design the structure with maximum stiffness by minimizing its mean compliance while ensuring that only 30% of the material is present in the final design domain. The topology parameters mentioned earlier are used for this optimization process. The material properties for the stiff material and void can be referred to in Table 2.

Figure 3a shows the optimal design achieved using the newly developed BESO algorithm as described in the reference paper [16]. The mean compliance of this optimal design is measured to be 7.26 Nmm. On the other hand, Fig. 3b shows the optimal design obtained using the author's extended algorithm, which exhibits a mean compliance of 7.25 Nmm. Although both optimal designs differ slightly in appearance, the design from this paper fulfills the objective function and demonstrates slightly lower mean compliance, indicating increased stiffness. Moreover, this design bears some resemblance to the optimal design obtained through the continuation method described in Fig. 3c from the reference paper [16], which achieves a mean compliance value of 7.18 Nmm.

3.2 Optimized Design with Non-zero Materials

In this section, the topology optimization problem, illustrated in Fig. 2, is investigated, considering two non-zero materials in the optimal design, excluding the void. Again, the objective is to maximize the stiffness of the structure by minimizing its mean compliance while ensuring an optimal distribution of the two materials. Each material

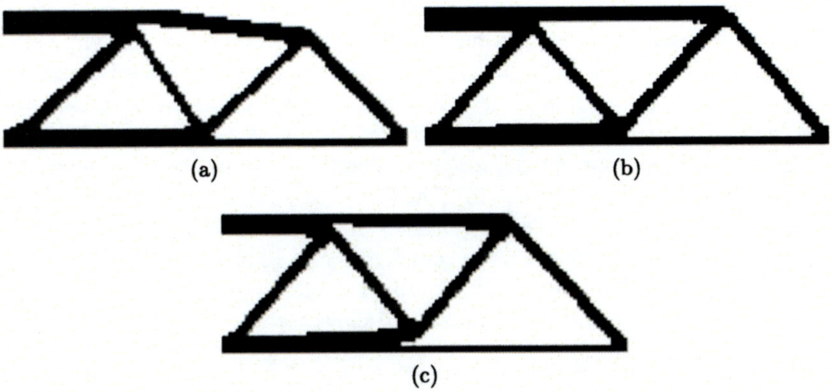

Fig. 3 Comparative optimal designs obtained from: **a** reference paper [16], **b** current research paper, and **c** continuation method [16]

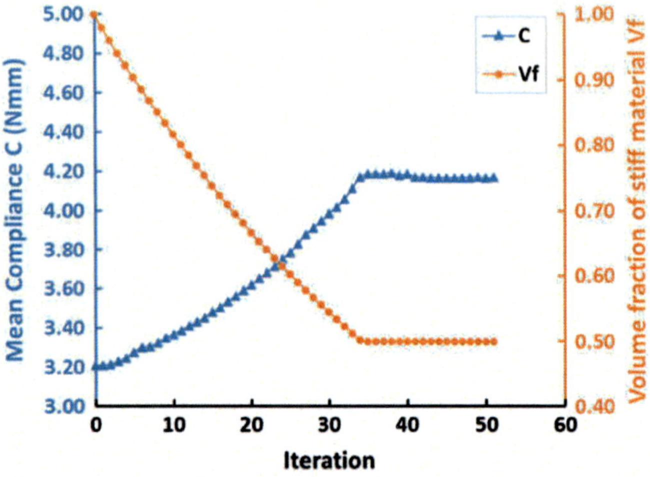

Fig. 4 Evolutionary histories of the mean compliance and volume fraction

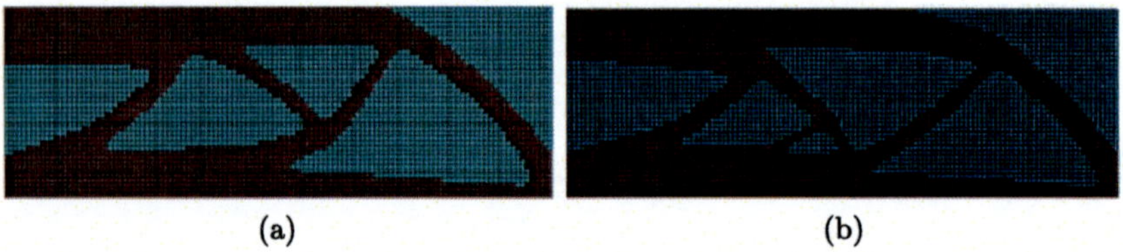

Fig. 5 Comparative two material optimal designs obtained from: **a** reference paper [16], **b** current research paper

is allocated an equal contribution of 50% in the final design domain. Figure 4 shows that the mean compliance of the whole model increases initially as the volume fraction of the stiff material decreases, and then it converges to an almost constant value after the objective volume is achieved. The topology parameters mentioned earlier are utilized for this optimization process, and the properties of the non-zero materials (stiff and soft materials) can be found in Table 2.

Figure 5a) presents the optimal design obtained using the two non-zero materials, as outlined in the reference paper [16]. The mean compliance of this design is determined to be 4.21 Nmm. Figure 5b) illustrates the optimal design attained using the author's code, which exhibits a mean compliance of 4.10 Nmm. While the material distribution in both optimal designs varies slightly, the design derived from this paper successfully meets the objective function and showcases a lower mean compliance, indicative of enhanced stiffness.

3.3 Optimized Design with Void and Two Materials

The author's code has demonstrated its effectiveness in addressing the optimization problem for designs involving two materials as well as solid-void configurations. It successfully integrates two non-zero materials, optimizes their distribution, and achieves the desired objectives, resulting in enhanced structural performance. While numerous open-source codes and software exist for similar purposes, the primary focus lies in achieving the same level of success for multi-material designs. This section aims to investigate and explore the optimization of multi-material designs including void, which presents unique challenges and requires specialized techniques.

In this section, the topology optimization problem, depicted in Fig. 2, considering two non-zero materials and the void is analyzed. The objective is to maximize the stiffness of the structure by minimizing its mean compliance, while ensuring an optimal distribution of the materials. The stiff and soft materials are allocated contributions of 15% and 25%, respectively, resulting in a final design domain consisting of 40% material and 60% void. The previously mentioned topology parameters are employed for this optimization process, and the properties of all the materials can be found in Table 2.

Figure 6a illustrates the optimal design obtained from the reference paper [16], with a mean compliance of 13 Nmm. Conversely, Fig. 6b displays the optimal design achieved using the author's code, which exhibits a mean compliance of 12.6 Nmm. Despite slight variations in material distribution, due to slightly different implementation (e.g. different sensitivity filter), the design presented in this paper meets the objective function and exhibits a lower mean compliance, indicating enhanced stiffness.

(a) (b)

Fig. 6 Comparative multi-materials optimal designs obtained from: **a** reference paper [16], **b** current research paper

4 Conclusions

To conclude, this research paper introduces a comprehensive code that has been developed and implemented based on the formulation proposed in the reference paper authored by Huang and Xie [16]. The code effectively addresses three different variants of design optimization, namely solid-void, two materials, and multi-materials optimization. The primary objective of the code is to generate optimal designs that maximize stiffness by minimizing mean compliance while satisfying volume constraints. The results obtained from the code consistently show designs that exhibit slightly higher stiffness compared to those reported in the reference paper. This demonstrates the effectiveness of the code in achieving the desired objective of improved structural performance.

However, it is worth mentioning that the code does face certain challenges, particularly in terms of computational time. It requires more time to converge compared to the demonstrated results in the reference paper. This indicates an area for potential future improvements, with opportunities to enhance the code's efficiency and reduce the computational time required for optimization.

Furthermore, the code has been successfully converted into an ABAQUS plugin, enhancing its usability and convenience for users. This enables researchers and engineers to easily apply the code to their specific design problems within the ABAQUS framework. In the future, the code can be enhanced to optimize structures with complex geometries, multiple materials, and diverse design objectives and constraints. These improvements will broaden its application in structural engineering, addressing a wider range of optimization challenges.

Overall, this research paper contributes to the field of topology optimization by providing a comprehensive code that can be applied to different design variants, yielding stiffer designs and showcasing its potential for future advancements in the field.

Acknowledgements The research work was performed within the PdZ-project "3DFit4Wear - Neue Wege zur wirtschaftlichen Herstellung von adaptierbaren und wiederverwertbaren Orthesen über 3D Druck" at the Polymer Competence Center Leoben GmbH (PCCL, Austria) within the framework of the "Produktion der Zukunft"-program of the Federal Ministry for Climate Action, Environment, Energy, Mobility, Innovation and Technology and the Federal Ministry for Digital and Economic Affairs with contributions by Montan University Leoben (Chair of Chemistry of Polymeric Materials & Chair of Science and Testing of Polymer), In-Vision Technologies AG, and Luxinergy GmbH. The PCCL is funded by the Austrian Government and the State Governments of Styria, Lower Austria and Upper Austria.

References

1. Sigmund, O.: A 99 line topology optimization code written in matlab. Struct. Multidiscip. Optim. **21**, 120–127 (2001)
2. Nha Chu, D., Xie, Y.M., Hira, A., Steven, G.P.: Evolutionary structural optimization for problems with stiffness constraints. Finite Elements Anal. Des. **21**(4), 239–251 (1996)

3. Wang, M.Y., Wang, X., Guo, D.: A level set method for structural topology optimization. Comput. Methods Appl. Mech. Eng. **192**(1–2), 227–246 (2003)
4. Philip Bendsoe, M., Sigmund, O.: Topology Optimization: Theory, Methods, and Applications. Springer Science & Business Media (2003)
5. Sigmund, O., Petersson, J.: Numerical instabilities in topology optimization: a survey on procedures dealing with checkerboards, mesh-dependencies and local minima. Struct. Optim. **16**, 68–75 (1998)
6. Bourdin, B.: Filters in topology optimization. Int. J. Numer. Meth. Eng. **50**(9), 2143–2158 (2001)
7. Sigmund, O.: Design of material structures using topology optimization. Ph.D. thesis, Technical University of Denmark Lyngby (1994)
8. Sigmund, O.: On the design of compliant mechanisms using topology optimization. J. Struct. Mech. **25**(4), 493–524 (1997)
9. Yago, D., Cante, J., Lloberas-Valls, O., Oliver, J.: Topology optimization methods for 3d structural problems: a comparative study. Arch. Comput. Methods Eng. **29**(3), 1525–1567 (2022)
10. Andreassen, E., Clausen, A., Schevenels, M., Lazarov, B.S., Sigmund, O.: Efficient topology optimization in matlab using 88 lines of code. Struct. Multidiscip. Optim. **43**, 1–16 (2011)
11. Han, Y., Bin, X., Liu, Y.: An efficient 137-line matlab code for geometrically nonlinear topology optimization using bi-directional evolutionary structural optimization method. Struct. Multidiscip. Optim. **63**, 2571–2588 (2021)
12. Huang, X., Xie, Y.M.: Convergent and mesh-independent solutions for the bi-directional evolutionary structural optimization method. Finite Elem. Anal. Des. **43**(14), 1039–1049 (2007)
13. Liu, K., Tovar, A.: An efficient 3d topology optimization code written in matlab. Struct. Multidiscip. Optim. **50**, 1175–1196 (2014)
14. Zuo, Z.H., Xie, Y.M.: A simple and compact python code for complex 3d topology optimization. Adv. Eng. Softw. **85**, 1–11 (2015)
15. Michael Yu Wang and Shiwei Zhou: Synthesis of shape and topology of multi-material structures with a phase-field method. J. Comput. Aided Mater. Des. **11**, 117–138 (2004)
16. Xiaodong Huang and Yi Min Xie: Bi-directional evolutionary topology optimization of continuum structures with one or multiple materials. Comput. Mech. **43**, 393–401 (2009)
17. Christian Frier Hvejsel and Erik Lund: Material interpolation schemes for unified topology and multi-material optimization. Struct. Multidiscip. Optim. **43**, 811–825 (2011)
18. Gao, T., Zhang, W.: A mass constraint formulation for structural topology optimization with multiphase materials. Int. J. Numer. Meth. Eng. **88**(8), 774–796 (2011)
19. Yang, Y., Moen, C.D., Guest, J.K.: Three-dimensional force flow paths and reinforcement design in concrete via stress-dependent truss-continuum topology optimization. J. Eng. Mech. **141**(1), 04014106 (2015)
20. Wang, Y., Luo, Z., Kang, Z., Zhang, N.: A multi-material level set-based topology and shape optimization method. Comput. Methods Appl. Mech. Eng. **283**, 1570–1586 (2015)
21. Cui, M., Chen, H., Zhou, J.: A level-set based multi-material topology optimization method using a reaction diffusion equation. Comput. Aided Des. **73**, 41–52 (2016)
22. Li, Y., Min Xie, Y.: Evolutionary topology optimization for structures made of multiple materials with different properties in tension and compression. Composite Struct. **259**, 113497 (2021)
23. Li, Y., Yuan, P.F., Xie, Y.M.: Topology optimization of structures composed of more than two materials with different tensile and compressive properties. Composite Struct. **306**, 116609 (2023)

Adaptive Filtering of Distributed Data Based on Modeling the Perception Mechanisms of Living Sensory Systems

Viacheslav E. Antsiperov

Abstract The article is devoted to the results of the adaptive filtering synthesis in the frames of neuromorphic model inspired by biological sensory system. To adequately model the perception mechanisms the analysis and synthesis are carried out on the base of the most realistic representation of input data—external stimuli—in the form of the stream of receptor registration events. Statistical model of that stream is chosen in the form of Poisson two-dimensional point processes. On this basis, a statistical description of the input data in the form of a sampling representation is proposed. The model of neural encoding of input data is considered in the article within the framework of the receptive fields concept. A number of well-known neuromechanisms are implemented in the encoding model, including in particular, center/surround inhibition. Decoding issues are considered in the context of stimuli spatial contrasts restoration, which partially models the responses of so-called simple cells in the visual cortex. It is shown that the model of coupled ON-OFF-decoding admits the sharp image details restoring in the form of local edges. At the end of the article, to justify the adequacy of the synthesized encoding procedure—adaptive filtering—we demonstrate an example of image edge directed interpolation and discus the results in the terms of perceptual quality.

Keywords Adaptive filtering · Linear-nonlinear Poisson model · Neuromorphic systems · Receptive fields · Center/surround inhibition · Perception

1 Introduction

In the first year of the millennium (2001) was published the Haykin's seminal editorial preface to the summer issue of the IEEE Signal Processing Magazine [1]. It can be considered, without exaggeration, as a manifesto for a new era of signal processing, image enhancement, data streaming (in artificial or real nets), etc. Being an authoritative researcher in the field of adaptive signal processing and knowing the

V. E. Antsiperov (✉)
Kotelnikov Institute of Radioengineering and Electronics RAS, Moscow, Russian Federation
e-mail: antciperov@cplire.ru

D. Vlachos (ed.), *Mathematical Modeling in Physical Sciences*, Springer Proceedings in Mathematics & Statistics 446, https://doi.org/10.1007/978-3-031-52965-8_2

real problems firsthand, Haykin formulated the main paradigm of data processing methods in the new 21st century as the maximum possible fusion of mathematics and physics of the subject area, ensuring, on the one hand, high performance of algorithms on real data, and on the other hand, the possibility of learning from data.

A significant part of the statistical methods for processing digital data developed in the past, the 20th century, is positioned today as classical signal processing (DSP). Most classical methods are based on two main stochastic assumptions about signals—Gaussianity and Stationarity. The widespread use of these assumptions is associated, firstly, with the compactness of the data description within the framework of the Gaussian model due to the use of only the first two statistical moments and, secondly, with a noticeable simplification of analysis under the assumption of translational invariance—in the case of Gaussian processes—stationarity. Thus a major part of the classical DSP is the theory of linear filtering, discovered almost simultaneously in the works of Wiener and Kolmogorov [2]. Today it is a rich and well-established theory of analysis and engineering design for a wide range of applications.

Unfortunately, today the existing limitations of classical methods have become an obvious fact. This is mainly due to the general fact that many, if not all, physical phenomena that provide data in practice are in fact both non-stationary and non-Gaussian (and, as a rule, nonlinear) [3]. In particular, almost all real probability distributions, unlike Gaussian ones, have long tails and require more *a priory* information than second-order statistics. The problem is often complicated by the fact that, knowing that the underlying statistics is not Gaussian, we don't know what it really is. Thus, practical experience accumulated by the beginning of the millennium showed that Gaussian distributions constitute only a rather small part of the set of all realistic distributions. This fact has been confirmed for a wide range of applications including imaging, teletraffic, communications, hydrology, geology, and economics. As a result, there was a strong belief that in the development of specific applications, it is necessary, not excluding universal classical methods, to give priority to the "physical" specifics of the corresponding subject area. In other words, signal processing achieves its best results when it successfully combines the unique ability of mathematics to generalize with a deep understanding of the specific information that can be extracted from the underlying physics. Haykin declares the following directions along which this fusion should be carried out: *prior information, regularization* and *adaptivity* [1].

Today, two decades after the publication of Haykin's preface, several more important trends can be added to the manifesto. In particular, there is a tendency to use in DSP not only the physical specifics of data, but also well-known physical principles. We note here a wide range of recurrent methods using nonlinear diffusion equations [4] or, for example, using simulated annealing [5]. And, of course, one should especially note the unprecedentedly wide distribution of methods based on the principles of modeling biological neural networks—various information systems of artificial neural nets [6]. It should be noted that recently a kind of renaissance is observed in the study of neuromorphic systems, the development of which is associated with the name of Carver Mead and dates back to the last decades of the last century [7]. Neuromorphic systems such as the electronic retina and cochlea [8] in addition to modeling neural networks also seek to model the biological principles of sensory perception—to adequately represent input data. The increased interest in such systems is associated, in particular, with the search for ways to solve the problem of efficient transfer of recorded data to the processing center—the so-called bottleneck problem in von Neumann-type computing systems.

Considering the listed above trends in data processing, we present in this article the results of the developing adaptive filtering mechanisms/procedures inspired by neuromorphic model of biological sensory system. Since the features of adaptive filtering in the chosen model are largely motivated by research in the field of neuromorphic systems [8], its internal content relies heavily on a number of mechanisms of neural perception of external stimuli by living sensory systems and, to a certain extent, it models them. This also applies to the neural representation of the recorded signals in the form of Poisson events associated with the triggered receptors, and to the aggregation of registered data through a system of receptive fields, and as well as to the method of neuro-coding data based on centre/surround lateral inhibition.

In this regard the formal receptive fields are equipped with a concentric structure of the center and antagonistic surround, and in terms of activation features they are divided into ON- and OFF-types. Accordingly, the optimal encoding (adaptive filtering) based on receptive fields is synthesized in the form similar to linear-nonlinear Poisson model, which has been widely used recently in the field of the neuronal perception [9].

Decoding issues are also inspired by well-known biological facts about registered external stimuli internal neural representation. In particular, it concerns contrast restoration, which partially models the responses of simple cells in the primary visual cortex. We show that the model of coupled ON- OFF-decoding admits the restoring of sharp image details in the form of local edges. At the end of the article, to justify the adequacy of the synthesized encoding procedure—adaptive filtering—we demonstrate the results of decoded edge directed image interpolation.

2 Data Sampling Representation and NLN Encoding Model

As noted in the introduction, the methods for modeling neuromorphic systems are a popular topic today. This is especially true of the Human Visual System (HVS), where the number of publications devoted to various aspects of modeling is growing rapidly. In particular, this applies to the periphery of the HVS—retina, where a significant amount of neurophysiological data has been accumulated both on its structure and on its functions. Recently, a number of works have also appeared devoted to the interpretation of processes in the retina as (nonlinear) filtering of input spatial stimuli—optical radiation detected by the eye. For example, the article [10] is devoted to such an interpretation; it also contains a detailed description in DSP context of layered structure of the retina and the functions of its individual layers. We also note a good bibliography on the problem under discussion contained in this paper.

In order to outline the main scope of the subject and fix the terminology for further discussion, let us recall some facts about the cellular structure and neural connectivity in the retina [10, 11]. The retinal tissue has a pronounced layered structure. The outer layer is represented by photoreceptor cells (rods, cones), which register the radiation formed by the optical system of the eye. The inner layer is made up of retinal output neurons—ganglion cells (RGC)—the only retinal cells with axons that make up together the optic nerve. It is along the axons of the optic nerve that the action potential pulses (spike trains) are transmitted from the retina to the primary visual cortex. It should be noted that spike trains are not a direct result of the registering the input radiation by receptors, but the product of successive transformations in the cells of the intermediate layers—interneurons. In terms of functional blocks, the retina can also be represented as a series of intersecting layers: the outer plexiform layer (OPL), which includes photoreceptors, horizontal and bipolar cells, the inner plexiform layer (IPL), which contains bipolar and amacrine cells, and ganglionic layer (GL), which consists of amacrine and ganglion cells. It is believed that the OPL layer approximately linearly filters the detected incident light (photons) into electric signal, the IPL layer is responsible for the nonlinear rectification of the filtered signal, and the GL generates spike trains, which are a kind of pulse code [10]. If we take into account that the number of retinal receptors reaches ~10^8, and the number of optic nerve axons is only 10^6 [12], then the degree of input data compression of the retina's encoding is ~ one hundred times. The need for such encoding is related to the limited capacity of the data channel—the optical nerve in accordance with Shannon's information theory.

From the above qualitative description, it follows that the input data registered in the outer layer of the retina undergo complex changes as the output data of its inner layer is formed. Therefore, for an adequate quantitative description of the encoded output data, which is in turn the data at the input of the optic nerve (channel coding), an appropriate statistical model is needed. As a result of long research in this area, the models based on the concept of center-surround receptive fields have been recognized as the most successful, see Fig. 1b). Here, the receptive field (RF) refers to that area of the retina, that contains in its outer layer those receptors whose activation affects the excitation of a particular output neuron (see [11]). In its best-known version, the model is a linear filtering with an RF-window of the registered stimulus, followed by a non-linear transformation of the filtered signal. The purpose of the latter is to prevent negative predicted firing rates as well as implement the thresholding and saturation of responses. This model is known as the linear-nonlinear (LN) model [13] (see Fig. 1c) and has become the de facto standard for the RGC response model and underpins many approaches to data analysis and neural computing.

Formally, the components of the LN model are defined as follows. If we denote by $f(\vec{x})$ the activation of the retina receptors, which is proportional to the intensity of the detected radiation $I(\vec{x})$, $\vec{x} \in \Omega$ on its outer surface Ω, then the output of OPL—the linear L component of the model—$A(\vec{x})$ has the form of filtered $f(\vec{x})$ [10]:

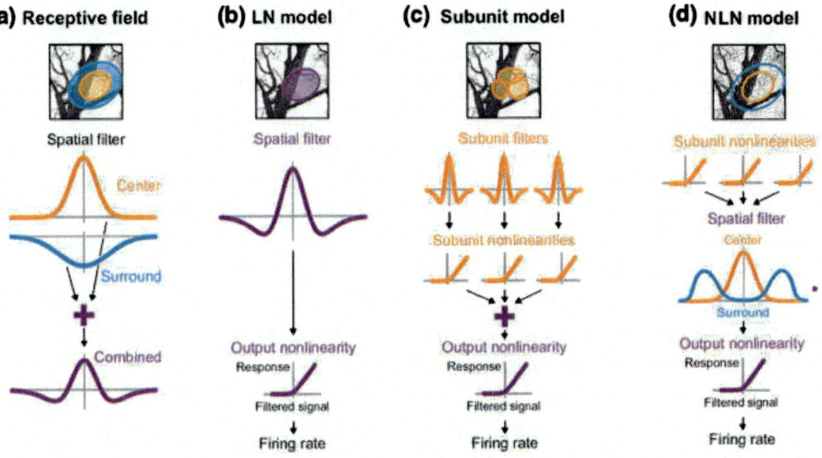

Fig. 1 Several popular models of receptive fields, adapted from [13]

$$A(\vec{x}_i) = \int_\Omega K(\vec{x}_i - \vec{x}) f(\vec{x}) d\vec{x}, \tag{1}$$

where $\vec{x}_i$ is the coordinate of a bipolar cell related to the center of i-th RF (and to the corresponding RGC output). Typically, the filter kernel $K(\vec{x})$ is approximated by the difference of a pair of windows in the form of Gaussian distributions—one is narrow, with a width σ_c, the other is 1.5–2 times wider, with a width $\sigma_s \approx 2\sigma_c$ (see Fig. 1b). Sometimes, instead of the difference of a pair of Gaussian distributions, in so-called LoG model, the Laplacian of one of them is used. The linear component L of the model is usually associated with the OPL layer of the retina. The non-linear component N is associated with the IPL layer of the retina and is represented as a static non-linearity that generates random (Poisson) spike counts, modeling the RGC at the output of the retina, see Fig. 1c.

The main advantage of the LN model described above lies in its remarkable simplicity and interpretability of the functions performed. Unfortunately, as recent neurophysiological studies have shown, the LN model is often not a good predictor of responses for RGCs in the case of realistic stimuli, natural images.

The latter is clearly manifested in the presence of spatial contrasts in images of the scales of the receptive field. The fine spatial structure of natural stimuli is not captured by the LN model [13]. For this reason, a number of attempts have been madc to extend the standard LN model by including in the concept of receptive fields a number of elements sensitive to spatial structure of the signal in the RF region by subdividing the center into smaller subunits.

This led to the so-called LN-LN or "subunit" model [13], see Fig. 1d. In this model, each of the subunits is itself represented by an LN model, which includes its own independent spatial filter, whose signal is non-linearly transformed before summation into an integrated ganglion cell excitation (Fig. 1d). The subunits are thought to correspond to bipolar cells that provide excitatory input to ganglion cells, so excitation in this model already exhibits non-linear spatial integration.

In addition, non-linearity at the bipolar cell input may also contribute to non-linear spatial integration upstream of bipolar cell subunits. The subunit structure of the receptive fields makes it possible to explain some functional properties and specific features of the retina, such as sensitivity to high spatial frequencies and to small objects below the scale of receptive field centers. This can be characterized as ability to encoding local changes in illumination even without total changes in brightness throughout the receptive field.

Some time ago, in order to overcome the disadvantages of the LN model we also proposed a new approach to modeling retinal functions [11, 14]. It differs from the LN-LN model mentioned above mainly in terms of input data representation. Namely, instead of considering the values of the registered intensity $I(\vec{x})$, or of the function of receptors activation $f(\vec{x})$ that is proportion to it, as input data, we proposed to take as the input of the model the actual result of registration, which is a set of photo counts $X = \{\vec{x}_j\}$ formed by the receptors. From a physical point of view, such an approach seems to be more adequate and universal, since from the very beginning it takes into account the uniform representation of the input data, regardless, for example, of whether the detected radiation is weak, medium, or very bright [14]. A number of arguments in favor of the approach proposed in terms of the quality of images formed from counts can be found in [15]. In general, from a formal point of view, the model proposed occupies an intermediate position between the LN and LN-LN models, which is why it is reasonable denoted as the NLN model shown in Fig. 1e.

Representation of the input data in the NLN model—the input image—is based on the concept of an ideal image [14]. The latter, compared with the classical representation in the form of registered intensity $I(\vec{x})$, is given by a set of random events—counts (sometimes called photocounts) [14, 15]. The counts of an ideal image are in turn the output of an ideal imaging device [14], whose input is the intensity $I(\vec{x})$—generally speaking, a continuous function of $\vec{x} \in \Omega$. The result of $I(\vec{x})$ registration—the ideal imaging device output—is the discrete set of random counts $X = \{\vec{x}_j\}$, where $\vec{x}_j$, $j = 1, ..., N$ are the coordinates of the registered events inside the light-sensitive two-dimensional surface Ω. It should be noted that the number of counts N is also a random variable, whose statistics is given by the Poisson distribution with mean $\overline{N} = \int_\Omega \alpha I(\vec{x}) d\vec{x}$, where the coefficient $\alpha = \eta(h\overline{\nu})^{-1}T$ is determined by the average energy $h\overline{\nu}$ of photons registered, by dimensionless quantum efficiency η of the ideal imaging device photosensitive material and by the registration time T. Noteworthy is the fact that for any other region $\Omega_i \subset \Omega$, such as RF, the number of counts will also be Poisson random with an average $\overline{N}_i = \int_{\Omega_i} \alpha I(\vec{x}) d\vec{x}$ (compare it with (1)).

Unfortunately, in many practical situations the number of counts $\overline{N}$ is too large (the flux of photons from the Sun through $\Omega \sim 1\ \text{cm}^2$ per 1 ms is $\sim 10^{15} - 10^{16}$, and for typical $\eta = 0.10$ we will have $\overline{N} \sim 10^{14} - 10^{15}$), so the direct use of a theoretical ideal image model is difficult. To solve this problem, we proposed the following decision [11, 14]. From the very beginning, we fix some acceptable value k of the representation size and, considering the ideal image as some general population X, we produce a random sample $X_k = \{\vec{x}_{j_1}, \ldots, \vec{x}_{j_k}\}$ from it in k counts. Obviously, such a sample, in full accordance with the principles of classical statistical theory, will still represent the input image. Such an X_k we call a sample of random counts or, in short, a sampling representation. The statistical description of the sampling representation X_k easily follows from the Poisson statistics and has the following simple form [14]:

$$\begin{gathered} \rho\left(X_k = \{\vec{x}_j\}, |I(\vec{x})\right) = \prod_{j=1}^{k} \rho\left(\vec{x}_j | I(\vec{x})\right), \\ \rho\left(\vec{x} | I(\vec{x})\right) = I\left(\vec{x}\right) / \int_{\Omega} I\left(\vec{x}\right) d\vec{x}. \end{gathered} \quad (2)$$

Note, that sometimes representations like (2) are called preset counts model [15].

To illustrate the representation of images by samples of counts and to prepare sampling representation example for further discussion, we generated, in accordance with (2), the set X_k of $k = 10.000.000$ counts, for simple artificial image "rat-06" from the standard data set MPEG7 Core Experiment CEShape-1 [16], which is presented in Fig. 2. The GIF image "rat-06" of size 352 × 288 pixels, color depth $\upsilon = 8$ bits was converted to PNG format with the same color depth and an image size of $s \times s = 1.000 \times 1.000$ pixels. Wherein only two shades 100 and 255 of grey were used for the image. $k = 10.000.000$ counts were generated by the Monte-Carlo accept-reject method with a uniform auxiliary distribution $g\left(\vec{x}\right) = (s \times s)^{-1} = 10^{-6}$ and an auxiliary constant M equal to the largest value of pixels $2^{\upsilon} = 256$ (see details in [11]).

The sampling representation (2) is very convenient in many respects [14, 15]. With its help, for example, it is easy to find the probability of detecting any count in some finite (not infinitely small $d\vec{x} \rightarrow 0$) region $\Omega_i \subset \Omega$:

Fig. 2 Image representation by sample of counts (sampling representation). Left fragment is the original "rat-06" image [16]. Right—sampling representation of size 10 000 000 counts

$$P_{\Omega_i|I(\vec{x})} = \frac{\int_{\Omega_i} I(\vec{x})d\vec{x}}{\int_{\Omega} I(\vec{x})d\vec{x}} = \frac{\sigma_i \overline{\mathcal{I}}_i}{\mathcal{W}}, \overline{\mathcal{I}}_i = \frac{1}{\sigma_i}\int_{\Omega_i} I(\vec{x})d\vec{x}. \tag{3}$$

where σ_i is the finite area of the region, $\overline{\mathcal{I}}_i$ is the average intensity on Ω_i and $\mathcal{W}$ is the total power of the radiation registered. Based on (3), we can also find the probability of detecting exactly k_i counts from X_k in Ω_i as the probability of k_i successes in a binomial experiment with a series of k Bernoulli trials and with (3) as probability of success. In addition, in the asymptotic case $k \to \infty$, $\mathcal{W} \to \infty$, $\mathcal{W}/k \to h = const$ the binomial distribution again transforms into a Poisson one with mean $\overline{k}_i = \sigma_i \overline{\mathcal{I}}_i / h$:

$$P_{k_i|\Omega_i} = \frac{k!}{k_i!(k-k_i)!}\left(\frac{\sigma_i \overline{\mathcal{I}}_i}{\mathcal{W}}\right)^{k_i}\left(1 - \frac{\sigma_i \overline{\mathcal{I}}_i}{\mathcal{W}}\right)^{k-k_i} \approx \frac{\left(\overline{k}_i\right)^{k_i}}{k_i!}\exp\left(-\overline{k}_i\right). \tag{4}$$

Further, with the help of (4) it is possible to refine the division of counts k_i within a certain receptive field Ω_i by the number of counts k_i^c in its center Ω_i^c and k_i^s in its surround Ω_i^s, $\Omega_i^c \cup \Omega_i^s = \Omega_i$. Assuming in this connection that the center and surround of RF do not intersect, $\Omega_i^c \cap \Omega_i^s = \varnothing$, we can obtain according to the previous discussion, that the numbers k_i^c and k_i^s are the independent Poisson random variables and their joint distribution in accordance with (4) has the form:

$$P_{k_i^c,k_i^s|\Omega_i} = \frac{\left(\sigma_c \overline{\mathcal{I}}_c/h\right)^{k_i^c}}{k_i^c!}\frac{\left(\sigma_s \overline{\mathcal{I}}_s/h\right)^{k_i^s}}{k_i^s!}\exp\left(-\frac{\left(\sigma_c \overline{\mathcal{I}}_c + \sigma_s \overline{\mathcal{I}}_s\right)}{h}\right), \tag{5}$$

where σ_c and σ_s are the areas of center and surround, $\overline{\mathcal{I}}_c$ and $\overline{\mathcal{I}}_s$ are average intensities on Ω_i^c and Ω_i^s of i-th RF (index i in $\overline{\mathcal{I}}_c$ and $\overline{\mathcal{I}}_s$ is implied). If we pass from variables k_i^c and k_i^s back to random k_i^c and $k_i = k_i^c + k_i^s$, then, recalling that $\sigma_i = \sigma_c + \sigma_s$, $\overline{\mathcal{I}}_i = \left(\sigma_c \overline{\mathcal{I}}_c + \sigma_s \overline{\mathcal{I}}_s\right)/\sigma_i$ and $\overline{k}_i = \sigma_i \overline{\mathcal{I}}_i / h$, we will obtain the following joint distribution:

$$\begin{gathered} P_{k_i^c,k_i|\Omega_i} = B\left(k_i^c|k_i, p_i\right)P_{k_i|\Omega_i}, \\ B\left(k_i^c|k_i, p_i\right) = \frac{k_i!}{k_i^c!\left(k_i - k_i^c\right)!}p_i^{k_i^c}(1-p_i)^{k_i-k_i^c}, p_i = \frac{\sigma_c \overline{\mathcal{I}}_c}{\sigma_i \overline{\mathcal{I}}_i}. \end{gathered} \tag{6}$$

where $B\left(k_i^c|k_i, p_i\right)$ is binomial distribution and $P_{k_i|\Omega_i}$ is Poisson one (4). The last modification of (5), (6) is motivated by the approximation of binomial distribution in the neighborhood of its mean $p_i k_i$ by Gaussian one. So, introducing random $\Delta_i = k_i^c - (\sigma_c/\sigma_i)k_i$ and denoting its mean by $\delta_i = p_i k_i - (\sigma_c/\sigma_i)k_i$ we finally get:

$$\begin{gathered} P_{\Delta_i,k_i|\Omega_i} = \frac{1}{\sqrt{2\pi\gamma_i^2}}\exp\left(-\frac{(\Delta_i-\delta_i)^2}{2\gamma_i^2}\right)P_{k_i|\Omega_i}, \\ \gamma_i^2 = \frac{\sigma_c\sigma_s}{\sigma_i^2}\left(\frac{\overline{\mathcal{I}}_c\overline{\mathcal{I}}_s}{\overline{\mathcal{I}}_i^2}\right)k_i, \delta_i = \frac{\sigma_c\sigma_s}{\sigma_i^2}\left(\frac{\overline{\mathcal{I}}_c - \overline{\mathcal{I}}_s}{\overline{\mathcal{I}}_i}\right)k_i. \end{gathered} \tag{7}$$

Let us make some remarks regarding the joint probability distribution (7) associated with i-th RF. First, the Gaussian distribution in front of $P_{k_i|\Omega_i}$ is the conditional distribution of the random Δ_i, which is the difference between the number of counts k_i^c in the RF center and its predictor $(\sigma_c/\sigma_i)k_i$ by k_i. If $\overline{\mathcal{I}}_c$ is equal to $\overline{\mathcal{I}}_s$, Δ_i has a zero mean $\delta_i = 0$, which can be interpreted as lateral inhibition of the center in the case of uniform illumination of the field. On the contrary, in the case of $\overline{\mathcal{I}}_c \gtrless \overline{\mathcal{I}}_s$ the mean value δ_i is not zero, and at large intensity changes on RF Δ_i will probably have large absolute ON- ($\delta_i > 0$) or OFF- ($\delta_i < 0$) values. Secondly, the conditional dependence of Δ_i on k_i is carried out through both parameters δ_i and γ_i^2, which implies the dependence on k_i not only of the intensity change forecast, but also of the forecast errors. Therefore, thirdly, the assessment of non-uniformity of the intensity $I(\overrightarrow{x})$ on RF must be based not only on the value of δ_i and a priori statistics on the intensities $\overline{\mathcal{I}}_c$ or $\overline{\mathcal{I}}_s$, but also on the registered $\overline{\mathcal{I}}_i \sim k_i$ (3) on Ω_i.

The above qualitative reasoning about the use of RF data k_i and Δ_i to estimate the average intensity $\overline{\mathcal{I}}_i$ on the field and the presence of a contrast in its behavior δ_i (including the estimate of the change value) can be strictly statistically justified if the input data model (7) is supplemented with some a priori statistical intensity $I(\overrightarrow{x})$ model on the retina Ω. In particular, in [11] it is shown that for the simplest model $\rho(\overline{\mathcal{I}}_c, \overline{\mathcal{I}}_s) = \omega\delta(\overline{\mathcal{I}}_c - \overline{\mathcal{I}}_s)\wp(\overline{\mathcal{I}}_s) + (1-\omega)\wp(\overline{\mathcal{I}}_c)\wp(\overline{\mathcal{I}}_s)$, statistical inference, based on the Bayesian approach, can be carried out to the end and explicit expressions for the *posterior* intensities $\widehat{\mathcal{I}}_c$ and $\widehat{\mathcal{I}}_s$ can be obtained. In *a priory* model $\rho(\overline{\mathcal{I}}_c, \overline{\mathcal{I}}_s)$ weights ω and $1-\omega$ can be interpreted as the probability of the 0–hypothesis H_0 that $\overline{\mathcal{I}}_c$ and $\overline{\mathcal{I}}_s$ are equal, and as the probability of the alternative $\overline{H_0}$ that $\overline{\mathcal{I}}_c$ and $\overline{\mathcal{I}}_s$ are independent, distribution $\wp(\overline{\mathcal{I}}_c)/\wp(\overline{\mathcal{I}}_s)$ is the a priori marginal probability of intensity $\overline{\mathcal{I}}_c/\overline{\mathcal{I}}_s$. Referring interested specialists to [11], we present here only the final result:

$$\frac{1}{h}\widehat{\mathcal{I}}_c(\Delta_i, k_i) = \frac{k_i}{\sigma_i} + \frac{1}{\frac{\omega}{1-\omega}\Lambda_{k_i}(\Delta_i) + 1}\frac{\Delta_i}{\sigma_c}, \tag{8}$$

where the likelihood ratio $\Lambda_{k_i}(\Delta_i)$ of the hypotheses H_0 and $\overline{H_0}$ for given Δ_i and k_i is

$$\Lambda_{k_i}(\Delta_i) = \frac{P_{\Delta_i,k_i|\Omega_i,H_0}}{P_{\Delta_i,k_i|\Omega_i,\overline{H_0}}} \approx \sqrt{\frac{\sigma_c\sigma_s}{2\pi\sigma_i^2}}\lambda\sigma_i \exp\left\{-\frac{\Delta_i^2}{2n\sigma_c\sigma_s/\sigma_i^2}\right\}, \tag{9}$$

and λ is the a priori mean of intensity $\overline{\mathcal{I}}$, characteristic scale of $\wp(\overline{\mathcal{I}})$, which corresponds to a rough approximation $\wp(\overline{\mathcal{I}}) \approx \lambda^{-1}$. For $\widehat{\mathcal{I}}_s$ we can write a similar result, but it follows directly from (8) and the identity $\sigma_c\widehat{\mathcal{I}}_c + \sigma_s\widehat{\mathcal{I}}_s = hk_i$.

Considering *posterior* intensities $\widehat{\mathcal{I}}_c$ and $\widehat{\mathcal{I}}_s$ as optimal estimates of real, unknown $\overline{\mathcal{I}}_c$ and $\overline{\mathcal{I}}_s$, we arrive at an expli cit form of the coding procedure—the main function of the retina. The dependences of the likelihood ratio $\Lambda_{k_i}(\Delta_i)$ (9) and *posterior* biased intensity estimate $\delta\widehat{\mathcal{I}}_c(\Delta_i) = \widehat{\mathcal{I}}_c(\Delta_i, k_i)/h - k_i/\sigma_i$ (8) on $\Delta_i = k_i^c - (\sigma_c/\sigma_i)k_i$ are shown in Fig. 3 . The encoding procedure (8) can be essentially simplified if we

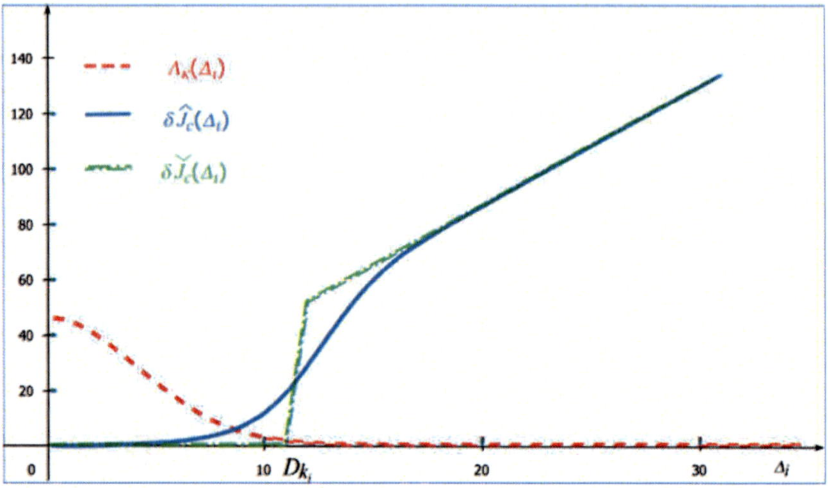

Fig. 3 The dependences of the ON-RF ($\Delta_i \geq 0$) likelihood ratio $\Lambda_{k_i}(\Delta_i)$ (9), *posterior* biased intensity estimate $\delta\hat{\mathcal{I}}_c(\Delta_i) = \hat{\mathcal{I}}_c(\Delta_i, k_i)/h - k_i/\sigma_i$ (8) and its encoding piecewise linear approximation $\delta\check{\mathcal{I}}_c(\Delta_i) = 0$, when $\Delta_i < D_{k_i}$ and $\frac{\Delta_i}{\sigma_c}$, when $\Delta_i > D_{k_i} = 0$(10) on Δ_i; $D_{k_i} = 12, k_i = 100$

roughly approximate $\Lambda_{k_i}(\Delta_i)$ (9) by zero for large Δ_i and by infinity $+\infty$ for small ones, that is, performing enco ding by deciding whether the hypothesis H_0 is true or its alternativ e H_0 is true. Thus, choosing some threshold D_{k_i} separating small and large values Δ_i, we arrive at the f ollowing coding procedure:

$$\frac{1}{h}\check{\mathcal{I}}_c(\Delta_i, k_i) = \begin{cases} \frac{k_i}{\sigma_i}, \ \Delta_i < D_{k_i}; \\ \frac{k_i}{\sigma_c}, \ \Delta_i > D_{k_i}; \end{cases} \tag{10}$$

The value D_{k_i} in (10), defining the boundary between the small and large Δ_i, can be found from the condition $\Lambda_{k_i}\left(D_{k_i}\right)\omega/(1-\omega) = 1$ (see (9)):

$$D_{k_i} = \sqrt{\frac{\sigma_c\sigma_s}{\sigma_i^2}\left[2\ln\left(\frac{\omega}{1-\omega}\right) + \ln\left(\frac{\sigma_c\sigma_s}{2\pi\sigma_i^2}\right) + \ln(\lambda\sigma_i)\right]}\sqrt{k_i} \tag{11}$$

Here we should make an important remark about the threshold D_{k_i} (11)—as it was assumed above, it turns out to be dependent on the detected intensity $\overline{\mathcal{I}}_i \sim k_i$ (3) on Ω_i.

The dependence of the encoding procedure—shrinking biased intensity $\delta\hat{\mathcal{I}}_c(\Delta_i) = \hat{\mathcal{I}}_c(\Delta_i, k_i)/h - k_i/\sigma_i$ (10) on $\Delta_i = k_i^c - (\sigma_c/\sigma_i)k_i$ is shown in Fig. 3. It is interesting to note that, on the one hand, the plot shape resembles popular artificial neural net activation functions, such as ReLU and, to an even greater extent, Leaky

ReLU, and on the other hand, it is similar to a generic compact shrinkage functions design that incorporates the filtering adaptivity [17]. Thus, there are deeper connections between these facts than it seems at first glance.

3 Conclusions

In order to evaluate the effectiveness of the proposed coding procedure (based on adaptive filtering (8), (10), it is useful to assess the degree of image distortion after decoding (reconstruction). Without resorting to formal statistical synthesis of decoders, it is possible, at least in the first approximation, to use well-known interpolation methods for low-resolution images. Considering that the RF count numbers $\{k_i\}$ are essentially the output of a smoothing filter with a sliding window of RF sizes, we can immediately say that the distortions associated with that part of code have the form of a blur of the original image (see Fig. 4 left). The filtering associated with the $\{\Delta_i\}$ codes is performed by constant positive center embedded within a constant negative surround, zero DC response filter, which exactly coincides with center-on-surround-off (COSO) filter proposed in [18]. The latter was proposed to mimic the point spread function for the Laplacian-of-Gaussian (LoG) filter, used by Marr and Hildreth in the theory of edge detection [19]. So, as the $\{\Delta_i\}$ codes are associated with the Laplacian-operator and through its zero-crossing points with the edges of the images, one can immediately use numerous edge interpolation methods [20], on the basis of which the perceptual quality of the reconstructed images can be estimated.

To illustrate the reconstruction (decoding) of the images, we used the simplest bilinear interpolation for the encoded sampling representation "rat-06", shown in Fig. 1 on the right. The area Ω of the representation was covered by a set of 2500 receptive fields located at the nodes of a 50×50 square grid (see details in [11]).

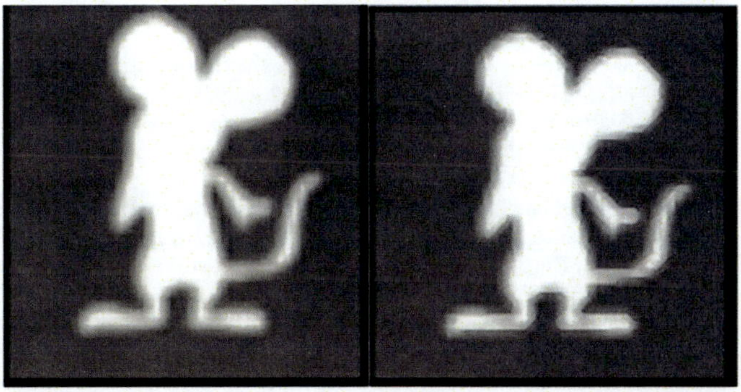

Fig. 4 The reconstructed (decoded) image "rat-06" from the sampling representation Fig. 1, encoded by proc. (10) on a 50×50 RFs grid. Left—the smoothed image decoded with RF count numbers $\{k_i\}$ only, right—edge-directed interpolation with edges detected by the $\{\Delta_i\}$

In each i-th node the values of RF k_i and Δ_i were calculated, after which Δ_i was censored with a threshold $D_{k_i} = \sqrt{k_i}$. The classical bilinear interpolation, shown in Fig. 4 left, demonstrates reconstruction based only on the $\{k_i\}$ data, interpolating them first linearly along the vertical grid edges, and then linearly over all cell rows basing on the already interpolated values of the vertical edges. The edge-directed bilinear interpolation, shown in Fig. 4 right, also uses a two-pass reconstruction. On the first pass the values $\{k_i\}$ are also interpolated along all vertical and horizontal grid edges. But in the case, when the nodes of the given grid edge have non-zero values Δ_i and Δ_j of different signs this grid edge is considered as intersecting the image edge and, basing on the $\Delta_i - \Delta_j$ linear interpolation zero-crossing, the intersection point can be found. As a result, if an edge intersects an image edge, the values of k_i and k_j at its nodes piecewise-constantly continue along the edge until they meet at the intersection point. If the edge does not intersect the image edge, a linear interpolation $k_i - k_j$ is performed on it, as in the classical case. At the second pass the values in the grid cells are linearly interpolated according to the values on their edges. If no edge of the cell intersects the image edge, or there is only one such vertical edge, or there are more than two such edges, the interpolation is performed along the cell rows, as in classical interpolation. If in the cell exists only one horizontal edge that intersects the image edge, the interpolation is performed over the columns of the cell. If there are exactly two grid edges intersecting the image edge, then connecting their intersection points by the line we reconstruct the fragment of the image edge in a cell and the linear interpolation is carried out along straight lines in the cell parallel to this image edge. The result (see Fig. 4 right) is a better image perceptual quality compared to classical interpolation (see Fig. 4 left).

Funding Present research was carried out at the expense of budgetary financing within the framework of the State Order at the Kotelnikov Institute of Radio-Engineering and Electronics of the Russian Academy of Sciences (State Contract "RELDIS").

References

1. Haykin, S.: Signal processing: where physics and mathematics meet. IEEE Signal Process. Mag. **18**(4), 6–7 (2001). https://doi.org/10.1109/msp.2001.939832
2. Pollock, D.S.G.: Statistical signal extraction and filtering: a partial survey. In Belsley, D.A., Kontoghiorghes, E.J. (eds.) Handbook of Computational Econometrics, pp. 321–376 (2009). https://doi.org/10.1002/9780470748916.ch9
3. Haykin, S.: Signal Processing in a Nonlinear, NonGaussian, and Nonstationary World. Lecture Notes in Computer Science, pp. 43–53 (2005). https://doi.org/10.1007/11520153_3
4. Didas, S., Weickert, J., Burgeth, B.: Properties of higher order nonlinear diffusion filtering. J Math Imaging Vis **35**(3), 208–226 (2009). https://doi.org/10.1007/s10851-009-0166-x
5. Chen, S., Istepanian, R., Luk, B.L.: Digital IIR filter design using adaptive simulated annealing. Digit Signal Process **11**(3), 241–251 (2001). https://doi.org/10.1006/dspr.2000.0384
6. Feldkamp, L.A., Puskorius, G.V.: A signal processing framework based on dynamic neural networks with application to problems in adaptation, filtering, and classification. Proc. IEEE **86**(11), 2259–2277 (1998). https://doi.org/10.1109/5.726790
7. Mead, C.: How we created neuromorphic engineering. Nat Electron **3**(7), 434–435 (2020). https://doi.org/10.1038/s41928-020-0448-2

8. Lande, S.: Neuromorphic Systems Engineering: Neural Networks in Silicon, 1st edn. Kluwer Academic Publishers, Boston (1998)
9. Triplett, M.A., Goodhill, G.J.: Probabilistic encoding models for multivariate neural data. Front Neural Circuits **13**, 1–1 (2019). https://doi.org/10.3389/fncir.2019.00001
10. Doutsi, E., Fillatre, L., Antonini, M., Gaulmin, J.: Retina-inspired filter. IEEE Trans. Image Process. **27**(7), 3484–3499 (2018). https://doi.org/10.1109/tip.2018.2812079
11. Antsiperov, V.: New centre/surround retinex-like method for low-count image reconstruction. In: Proceedings of the 12th International Conference on Pattern Recognition Applications and Methods (ICPRAM 2023), pp 517–528. SCITEPRESS – Science and Technology Publications (2023). https://doi.org/10.5220/0011792800003411
12. Schiller, P.H., Tehovnik, E.J.: Vision and the Visual System. Oxford University Press, Oxford (2015). https://doi.org/10.1093/acprof:oso/9780199936533.001.0001
13. Zapp, S.J., Nitsche, S., Gollisch, T.: Retinal receptive-field substructure: scaffolding for coding and computation. Trends Neurosci (Regular ed.) **45**(6), 430–445 (2022). https://doi.org/10.1016/j.tins.2022.03.005
14. Antsiperov, V., Kershner, V.: Retinotopic image encoding by samples of counts. In: De Marsico, M., Sanniti di Baja, G., Fred, A. (eds.) Pattern Recognition Applications and Methods, ICPRAM 2021–2022. Lecture Notes in Computer Science, vol. 13822. Springer, Cham (2023). https://doi.org/10.1007/978-3-031-24538-1_3
15. Caucci, L., Barrett, H.H.: Objective assessment of image quality, V Photon-counting detectors, and list-mode data. J. Opt. Soc. Am. A **29**(6), 1003 (2012). https://doi.org/10.1364/josaa.29.001003
16. Latecki, L.J., Lakamper, R., Eckhardt, T.: Shape descriptors for non-rigid shapes with a single closed contour. In: Proceedings IEEE Conference on Computer Vision and Pattern Recognition CVPR 2000 (Cat.No.PR00662), pp. 424–429. https://doi.org/10.1109/CVPR.2000.855850
17. Alt, T., Weickert, J.: Learning a generic adaptive wavelet shrinkage function for denoising. In: ICASSP 2020, pp. 2018–2022. https://doi.org/10.1109/ICASSP40776.2020.9054644
18. Allebach, J., Wong., P.W.: Edge-directed interpolation. In: Proceeding of the 3rd IEEE International Conference on Image Processing, vol. 2, pp. 707–710 (1996). https://doi.org/10.1109/icip.1996.560768
19. Marr, D., Hildreth, E.: Theory of edge detection. Proc. R. Soc. Lond. Ser. B Biol. Sci. **207**(1167), 187–217 (1980). https://doi.org/10.1098/rspb.1980.0020. PMID 6102765
20. Kok, C.W., Tam, W.-S.: Edge-directed interpolation. In: Digital Image Interpolation in MatLab, 1st edn, pp. 191–209. IEEE Press, Wiley (2019). https://doi.org/10.1002/9781119119623.ch7

A Method for Fast and Robust Sungular Spectrum Analysis

Adamantia Mavrogianni

Abstract One of the main problems encountered in the analysis of real-world time series, is the existence of outliers originating from various sources other than the dynamics of the series. Such sources can be a momentary noise or an imperfection in the recording devices. This problem has led to the need to construct robust methods using metrics other than the usual mean squared error. In combination with non-parametric techniques such as the very popular Singular Value Decomposition (SSA), it can give iterative algorithms that deal with this problem but create huge needs in computing power, especially in problems where we need fast response of the system (for example in high frequency trading). In this paper, a technique addressing both problems is presented which exploits the peculiarity of the Singular Value Decomposition (SVD) analysis of square symmetric matrices.

Keywords Robust singular spectrum analysis

1 Introduction

Sungular Spectrum Analysis (SSA) is a non-parametric technique that is used to analyze time series as well as to forecast them. Its main advantages are:

1. It can be used on a wide range of time series without needing any strong assumption about them, such as stationarity for example.
2. It is not based on any model nor dependent on some hyper-parameters.
3. It can be used for long-term forecasts since the analysis of the time series into components also includes different time scales.

An excellent presentation of the theory behind the SSA technique can be found in the Refs. [8, 31]. Also, a multitude of applications of SSA techniques and others inspired by Singular Value Decomposition (SVD) analysis can be found in industry [14, 18, 19], demography [9, 14], business, economics and finance [1, 5, 6, 9, 16,

A. Mavrogianni (✉)
Sivitanidios Public School of Arts and Professions, Kallithea, Greece
e-mail: amanda.mavrogianni@gmail.com

D. Vlachos (ed.), *Mathematical Modeling in Physical Sciences*, Springer Proceedings in Mathematics & Statistics 446, https://doi.org/10.1007/978-3-031-52965-8_3

17, 20], materials science [23, 26, 28, 29], optimization and modeling [12, 13, 24, 27] hardware and signal processing [15, 21], climatology [2, 22], neural networks [30] even in nuclear science [25].

The main feature of the SSA method is the analysis of a matrix using the SVD technique. To implement this we need first of all to create a matrix. Here are different approaches:

1. In its simplest form we can create a trajectory matrix where we use segments of the time series as lines where each line has a lag with respect to the previous one. In this way a Hankel matrix is constructed. A variation of the method is to construct a pivot table where each column of the table is a separate (but contiguous) segment of the time series.
2. the second approach, we construct from the trajectory matrix the autocorrelation matrix, a technique we will exploit later because it produces square symmetric matrices.

But let's look in detail at the process, which we can easily extend to multiple series.

1.1 Embedding

Consider that we have a time series with N samples $\{x_0, x_1, \ldots, x_{N-1}\}$. We can create 2-kind of matrices, the trajectory matrix X

$$X = \begin{pmatrix} x_0 & x_1 & x_2 & \ldots & x_{N-L} \\ x_1 & x_2 & x_3 & \ldots & x_{N-L+1} \\ x_2 & x_3 & x_4 & \ldots & x_{N-L+2} \\ \vdots & \vdots & \vdots & \vdots & \vdots \\ X_{L-1} & x_L & x_{L+1} & \ldots & x_{N-1} \end{pmatrix} \tag{1}$$

and the pivot matrix X_p

$$X_p = \begin{pmatrix} x_0 & x_L & x_{2L} & \ldots & x_{N-L} \\ x_1 & x_{L+1} & x_{2L+1} & \ldots & x_{N-L+1} \\ x_2 & x_{L+2} & x_{2L+2} & \ldots & x_{N-L+2} \\ \vdots & \vdots & \vdots & \vdots & \vdots \\ X_{L-1} & x_{2L-1} & x_{3L-1} & \ldots & x_{N-1} \end{pmatrix} \tag{2}$$

The trajectory matrix X has the Hankel property

$$X_j^{n-j} = X_i^{n-i} \tag{3}$$

$\forall n = 0, 1, \ldots, 2N-2$ and $\forall i, j = 0, 1, \ldots, n$ and in the case of the pivot matrix, L must be a divisor of N.

1.2 Decomposition

Consider m series with transition matrices $X_i, i = 0, 1, \ldots, m-1$ with $X_i \in \mathbb{M}_{L\times K}$. Thus, foreach series we have $N = L + K - 1$ samples. Let

$$\mathbf{X} = [\mathbf{X}_0, \mathbf{X}_1, \ldots, \mathbf{X}_{m-1}] \tag{4}$$

and perform SVD decomposition to $\mathbf{X}$:

$$\mathbf{X} = \mathbf{U} \cdot \mathbf{S} \cdot \mathbf{V}^T \tag{5}$$

where $\mathbf{U}^T \cdot \mathbf{U} = \mathbf{U} \cdot \mathbf{U}^T = \mathbb{I}_{L\times L}$. Now set:

$$V_{(\mu),j} = \mathbf{X}_\mu^T \cdot U_j \tag{6}$$

Moreover:

$$\begin{aligned} \mathbf{X}^T \cdot \mathbf{U}_j &= \mathbf{V} \cdot \mathbf{S} \cdot \mathbf{U}^T \cdot U_j && (7)\\ &= \mathbf{V} \cdot \mathbf{S} \cdot e_j && (8)\\ &= s_j V_j && (9) \end{aligned}$$

where V_j is the jth column of $\mathbf{V}$. Since

$$\mathbf{X}^T \cdot \mathbf{U}_j = [\mathbf{X}_0 \cdot U_j, \mathbf{X}_1 \cdot U_j, \ldots, \mathbf{X}_{m-1} \cdot U_j] \tag{10}$$

we have:

$$V_{(\mu),j} = s_j V_j^T [\mu k : (\mu + 1)k] \tag{11}$$

Notice here that although the vectors V_j are orthonormal, the vectors $V_{(\mu),j}$ are not. Thus:

$$s_{(\mu),j} = \| V_{(\mu),j} \| \tag{12}$$

1.3 Reconstruction

We only take the first d vectors U_j that correspond to $s_j > \epsilon$ where ϵ is a custom threshold (in the ideal case we take $\epsilon = 0$). Then, each series can be reconstructed as:

$$\hat{\mathbf{X}}_\mu = \sum_{j=0}^{d-1} U_j \cdot V_{(\mu),j}^T \tag{13}$$

The contribution of each component $\hat{\mathbf{X}}_{(\mu),j} = U_j \cdot V_{(\mu),j}^T$ is

$$c_{(\mu),j} = \frac{s_{(\mu),j}}{\sum_{i=0}^{d-1} s_{(\mu),i}} \tag{14}$$

The reconstructed time series $\hat{F}_\mu$ is the time series that corresponds to the transition matrix

$$\mathbb{H}[\hat{\mathbf{X}}_\mu] \tag{15}$$

where the operator $\mathbb{H}$ when applied to a matrix $\mathbf{A}$, produces the closest Hankel matrix to $\mathbf{A}$.

1.4 Forecast

Consider a vector Y and let us calculate the projection of Y to the space $\mathrm{Span}[U_i]$ which spanned by the linearly independent vectors U_i. Then:

$$\begin{aligned} Y' &= \Pi \cdot Y & (16)\\ &= \mathbf{U} \cdot \left[\mathbf{U}^T \cdot \mathbf{U}\right]^{-1} \cdot \mathbf{U}^T \cdot Y & (17)\\ &= \mathbf{U} \cdot \mathbf{U}^T \cdot Y \implies & (18)\\ \Pi &= \mathbf{U} \cdot \mathbf{U}^T & (19) \end{aligned}$$

We have 2-ways to forecast the time series:

1. We can perform a linear regression by letting the $L-1$ rows of the transition matrix to predict the last row. Thus, if $\mathbf{X}^\nabla$ is taken from $\mathbf{X}$ by removing the last row and Y^T is the last row of $\mathbf{X}$ we have:

$$b^T \cdot \mathbf{X}^\nabla = Y^T \implies b = \left(\mathbf{X}^\nabla \cdot (\mathbf{X}^\nabla)^T\right)^{-1} \mathbf{X}^\nabla \cdot Y \tag{20}$$

Then the forecast y is just

$$y = b^T \cdot F_\Delta \tag{21}$$

where F_Δ is the last column of the transition matrix without the first (row) element.
2. If π_j is the last element (row) of the vector U_j, then we build the vector R such that

$$R = \frac{1}{1-\nu^2} \sum_i \pi_i U_i^\nabla \tag{22}$$

where $\nu^2 = \sum_i \pi_i^2$. Then the forecast y is

$$y = R^T \cdot F_\Delta \tag{23}$$

where F_Δ is the last column of the transition matrix without the first (row) element.

By applying the forecast method to known values of the time series, we can calculate the variance of the forecast.

2 Robust SSA

Since the SSA algorithm is based on SVD analysis, this means that each eigenvector corresponds to that direction that maximises the variance of the points that correspond to the columns of the table. Therefore, instead of using the L_2 norm to calculate the spread of points, one can use another norm, such as L_1, which is more robust to the existence of outliers (or even a more specialised norm such as Huber's one). In the literature, one can find many robust alternatives to classical SVD [3, 4, 11], with a very useful overview here [7]. One should be careful when choosing a norm other than the L_2 norm, because then the decreasing order of the calculated eigenvalues is not guaranteed as well as the orthogonality of the calculated eigenvectors. Here we will use the technique of Hawkins [10] which is based on an iterative estimation of the vectors U and V alternatively. The technique is shown in Algorithm 1.

Algorithm 1 Robust SSA

Require: Initial estimate for U ▷ Usually $U_{(0)}$ is the median of the rows
  $k \leftarrow 0$
  while Correction > threshold **do**
    $\mathbf{c} \leftarrow \min_{\mathbf{c}} \left\{ X - \mathbf{c} \cdot U_{(k)}^T \right\}$
    $V_{(k)} \leftarrow \frac{\mathbf{c}}{||\mathbf{c}||}$
    $\mathbf{d} \leftarrow \min_{d} \left\{ X - \mathbf{d} \cdot V_{(k)}^T \right\}$
    $U_{(k+1)} \leftarrow \frac{\mathbf{d}}{||\mathbf{d}||}$
    Correction $\leftarrow ||U_{(k+1)} - U_{(k)}||$
    $k \leftarrow k+1$
  end while
  $\lambda \leftarrow \min_{\lambda} \left\{ X - \lambda U \cdot V^T \right\}$

After calculating the eigenvector U and the corresponding eigenvalue, the algorithm proceeds by subtracting the contribution of the calculated eigenvector and applying again the algorithm to find the next eigenvector. Thus, the new trajectory matrix becomes

$$X' = X - \lambda U V^T \tag{24}$$

It is clear from Algorithm 1 that in order to find one eigenvector we have to perform $2 \times l + 1$ minimizations (where l is the number of repetitions that are needed to obtain the predefined threshold). This can be proved prohibitive especially in cases where we need fast iterations of the algorithm, which happens in the case of rapidly changing signals.

3 Fast Robust SSA

In order to face the problem that arise from minimisation tasks, we return to the autocorrelation matrix. Let

$$C = X \cdot X^T \tag{25}$$

then if X is an $L \times K$ matrix, then C is a symmetric square $L \times L$ matrix. In practice, usually $L << K$. Then

$$X = \sum_{i=1}^{d} \sqrt{\lambda_i} U_i \cdot V_i^T \implies \tag{26}$$

$$C = \sum_{i=1}^{d} \sqrt{\lambda_i} U_i \cdot V_i^T \left(\sum_{j=1}^{d} \sqrt{\lambda_j} U_i \cdot V_j^T \right)^T \tag{27}$$

$$= \sum_{i,j=1}^{d} \sqrt{\lambda_i \lambda_j} U_i \cdot V_i^T \cdot V_j \cdot U_j^T \tag{28}$$

$$= \sum_{i=1}^{d} \lambda_i U_i \cdot U_i^T \tag{29}$$

since $V_i^T \cdot V_j = \delta_{i,j}$. One then can calculate the eigenvectors U_i and then the corresponding V_i by the equation

$$V_i = \frac{1}{\sqrt{\lambda_i}} X^T \cdot U_i \tag{30}$$

The calculation algorithm now is shown in Algorithm 2.

Algorithm 2 Fast and Robust SSA

while $\lambda >$ threshold **do**
  $U \leftarrow \min_{U} |X - U \cdot U^T|$
  $\lambda \leftarrow ||U||^2$
  $U \leftarrow \frac{U}{\sqrt{\lambda}}$
  return λ, U
  $X \leftarrow X - \lambda U \cdot U^T$
end while

4 Discussion

The use of the robust SSA technique can give significant advantages over the traditional one. Primary results on time series involving currency exchange rates have provided strong evidence for the use of this technique. As an example we mention here the use of the technique to construct input data in a simple neural network. The input data is generated by the following process:

1. Consider the time series (in reverse time order)

$$X_n = X_0, X_{-n}, X_{-2n}, X_{-3n}, \ldots, X_{-kn} \tag{31}$$

   where the values for n ranges from 1 to d. In this way we build d-time series
2. Each X_n is a part of the initial time series in different time scales. We treat these series as a multivariate time series and we apply the both the traditional and the robust SSA.
3. In signals like exchange rates, usually we have a strong component (the largest eigenvalues) while the others are at least 2 orders of magnitude smaller. As a result, we keep only the first eigenvalue and we reconstruct the signals.
4. We apply the forecast technique for each signal and we construct a vector $\mathbf{y} = (y_1, y_2, \ldots, y_d)^T$. Each component y_j of the vector is now a forecast for the next j time units.
5. A simple neural network is trained with the vectors $\mathbf{y}$ in order to predict if a certain trading action is profitable (for example ia buy order with a specific take profit and stop loss range).

In the Table 1 we summarise the results in a backtest session for several pairs of currencies for both methods.

Table 1 Successful percentage of orders

Pair	SSA (%)	Robust SSA (%)
EUR/USD	72	77
EUR/GBP	69	79
EUR/JPY	73	75
EUR/CAD	71	76
EUR/CHF	67	78

References

1. Alexandros, N.K., Sakas, D.P., Vlachos, D.S., Dimitrios, N.K.: Comparing scrum and xp agile methodologies using dynamic simulation modeling. In: Strategic Innovative Marketing: 5th IC-SIM, Athens, Greece 2016, pp. 391–397. Springer International Publishing (2017)
2. Bond, G., Kromer, B., Beer, J., Muscheler, R., Evans, M.N., Showers, W., Hoffmann, S., Lotti-Bond, R., Hajdas, I., Bonani, G.: Persistent solar influence on north Atlantic climate during the Holocene. Science **294**(5549), 2130–2136 (2001)
3. Croux, C., Filzmoser, P., Rosario Oliveira, M.: Algorithms for projection–pursuit robust principal component analysis. Chem. Intell. Laboratory Syst. **87**(2), 218–225 (2007)
4. Croux, Christophe, Ruiz-Gazen, Anne: High breakdown estimators for principal components: the projection-pursuit approach revisited. J. Multivariate Anal. **95**(1), 206–226 (2005)
5. Dimitrios, N.K., Sakas, D.P., Vlachos, D.S.: The contribution of dynamic simulation model of depiction of knowledge, in the leading process of high technology companies. Key Engineering Mater. **543**, 406–409 (2013)
6. Dimitrios, N.K., Sakas, D.P., Vlachos, D.S.: Modeling publications in academic conferences. Procedia-Soc. Behav. Sci. **147**, 467–477 (2014)
7. Filzmoser, Peter, Todorov, Valentin: Robust tools for the imperfect world. Inf. Sci. **245**, 4–20 (2013)
8. Golyandina, N., Nekrutkin V., Zhigljavsky, A.A.: Analysis of time series structure: SSA and related techniques. CRC Press (2001)
9. Hassani, Hossein, Thomakos, Dimitrios: A review on singular spectrum analysis for economic and financial time series. Stat. Interface **3**(3), 377–397 (2010)
10. Hawkins, D., Liu, L., Stanley Young, S.: Robust singular value decomposition technical report number 122 (2001). National Institute of Statistical Sciences 19 (2002)
11. Hubert, M., Rousseeuw, P.J., Branden, K.V.: Robpca: a new approach to robust principal component analysis. Technometrics **47**(1), 64–79 (2005)
12. Kosmas, O.T., Vlachos, D.S.: Phase-fitted discrete lagrangian integrators. Comput. Phys. Commun. **181**(3), 562–568 (2010)
13. Kosmas, O.T., Vlachos, D.S.: Local path fitting: a new approach to variational integrators. J. Comput. Appl. Math. **236**(10), 2632–2642 (2012)
14. Mahmoudvand, R., Konstantinides, D., Rodrigues, P.C.: Forecasting mortality rate by multivariate singular spectrum analysis. Appl. Stoch. Models Bus. Ind. **33**(6), 717–732 (2017)
15. Muruganatham, B., Sanjith, M.A., Krishnakumar, B., Murty, S.S.: Roller element bearing fault diagnosis using singular spectrum analysis. Mech. Syst. Signal Process. **35**(1-2), 150–166 (2013)
16. Nasiopoulos, D.K., Sakas, D.P., Vlachos, D.S., Mavrogianni, A.: Modeling of market segmentation for new it product development. In: AIP Conference Proceedings, vol. 1644, pp. 51–59. American Institute of Physics (2015)
17. Nasiopoulos, D.K., Sakas, D.P., Vlachos, D.S., Mavrogianni, A.: Simulation of generation of new ideas for new product development and it services. In: AIP Conference Proceedings, vol. 1644, pp. 60–68. American Institute of Physics (2015)

18. Rodrigues, P.C., De Carvalho, M.: Spectral modeling of time series with missing data. Appl. Math. Model. **37**(7), 4676–4684 (2013)
19. Paulo Canas Rodrigues and Rahim Mahmoudvand: Correlation analysis in contaminated data by singular spectrum analysis. Quality Reliab. Eng. Int. **32**(6), 2127–2137 (2016)
20. Sakas, D.P., Vlachos, D.S., Nasiopoulos, D.K.: Modeling the development of the online conference's services. Library Rev. **65**(3), 160–184 (2016)
21. Salgado, D.R., Alonso, F.J.: Tool wear detection in turning operations using singular spectrum analysis. J. Mater. Process. Technol. **171**(3), 451–458 (2006)
22. Schlesinger, M.E., Ramankutty, N.: An oscillation in the global climate system of period 65–70 years. Nature **367**(6465), 723–726 (1994)
23. Skafidas, P.D., Vlachos, D.S., Avaritsiotis, J.N.: Modelling and simulation of abnormal behaviour of thick-film tin oxide gas sensors in co. Sensors Actuators B: Chem. **21**(2), 109–121 (1994)
24. Vlachos, D.S.: Optimal ship routing based on wind and wave forecasts. Appl. Numer. Anal. Comput. Math. **1**(2), 547–551 (2004)
25. Vlachos, D.S.: Self-calibration techniques of underwater gamma ray spectrometers. J. Environ. Radioact. **82**(1), 21–32 (2005)
26. Vlachos, D.S., Papadopoulos, C.A., Avaritsiotis, J.N.: A technique for suppressing ethanol interference employing seebeck effect devices with carrier concentration modulation. Sensors and Actuators B: Chem. **44**(1–3), 239–242 (1997)
27. Vlachos, D.S., Simos, T.E.: Partitioned linear multistep method for long term integration of the n-body problem. Appl. Numer. Anal. Comput. Math. **1**(2), 540–546 (2004)
28. Vlachos, D.S., Skafidas, P.D., Avaritsiotis, J.N.: Transient effects of tin oxide co sensors in the presence of water vapor. Appl. Phys. Lett. **63**(13), 1760–1761 (1993)
29. Vlachos, D.S., Xenoulis, A.C.: Gas detection sensitivity and cluster size. Nanostruct. Mater. **10**(8), 1355–1361 (1998)
30. Wu, A., Hsieh, W.W., Tang, B.: Neural network forecasts of the tropical pacific sea surface temperatures. Neural Netw. **19**(2), 145–154 (2006)
31. Zhigljavsky, Anatoly: Singular spectrum analysis for time series: introduction to this special issue. Stat. Interface **3**(3), 255–258 (2010)

An Algorithm for Non-negative Leslie Matrices

Jésica Pantáz, Jonnathan Rodríguez, and Luis Medina

Abstract The Leslie matrix plays a crucial role in analyzing the changes in survival and birth rates, thus determining the population's evolution. However, when it comes to evaluating continuous solutions of population growth models, discrete solutions are necessary. The Leslie matrix model, being a discrete model based on the age stage of the population, becomes particularly relevant in this context. In this work, we present a computational mathematical tool designed be used in the Leslie matrix model, which consists of reconstructing a Leslie matrix of any order from a given set of real numbers.

Keywords Randomized algorithms · Recursive methods · Leslie matrix · Mathematica algorithms

1 Introduction and Preliminaries

In 1900 in Scotland, Patrick Holt Leslie was born. Leslie studied at Christ Church College of Oxford University and obtained a bachelor's degree in physiology in 1921. After a few years working as an assistant in bacteriology in the pathology department, he turned to statistics and joined the Animal Population Bureau in 1935 [1]. The Leslie model, known as the Leslie matrix, is a discrete, age-structured model of population growth and is very popular in biology. This model considers the age for the growth of the number of females in a population (animals or plants).

In [4], Donovan et al. give an illustration map about the Leslie life cycle of a population, see Fig. 1, where:

J. Pantáz
Departamento de Física, Facultad de Ciencias Básicas, Universidad de Antofagasta, Av Angamos 601, Antofagasta, Chile

J. Rodríguez · L. Medina (✉)
Departamento de Matemáticas, Facultad de Ciencias Básicas, Universidad de Antofagasta, Av Angamos 601, Antofagasta, Chile
e-mail: luis.medina@uantof.cl

D. Vlachos (ed.), *Mathematical Modeling in Physical Sciences*, Springer Proceedings in Mathematics & Statistics 446, https://doi.org/10.1007/978-3-031-52965-8_4

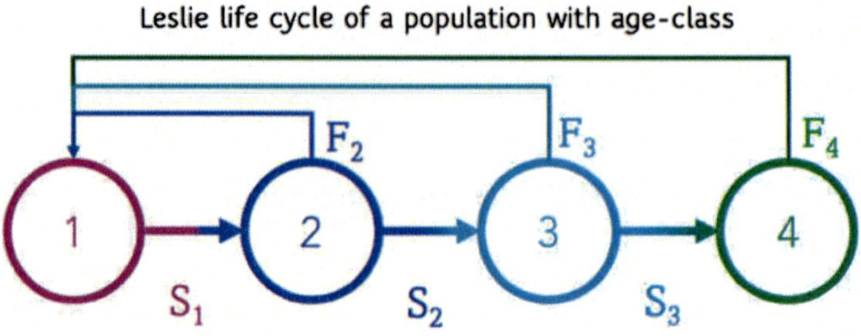

Fig. 1 Leslie model

(a) The circles represent the age.
(b) The arrows between circles represent survival probabilities.
(c) The block arrows at the top of the diagram represent births.

The Leslie model in Fig. 1 can be represented as a Leslie matrix of the form

$$L = \begin{pmatrix} a_1 & a_2 & a_3 & \dots & a_{n-1} & a_n \\ b_1 & 0 & 0 & \dots & 0 & 0 \\ 0 & b_2 & 0 & \dots & 0 & 0 \\ \vdots & \vdots & \ddots & \ddots & \vdots & \vdots \\ 0 & 0 & \dots & b_{n-2} & 0 & 0 \\ 0 & 0 & \dots & 0 & b_{n-1} & 0 \end{pmatrix}, \tag{1}$$

where for $i = 1, \dots, n$, the a_i are real positive numbers that represent fertilization and $b_j \in (0, 1]$ $j = 1, 2, \dots, n-1$, represent the fraction of individuals who survive from the age class x and pass to the class x + 1, since both indexes are non-negative and, considering that not all entries are zero, the Leslie matrix is a non-negative matrix. If $b_i = 1$, $i = 1, 2, \dots, n-1$, L is called the companion matrix and is usually denoted by C, that is,

$$C = \begin{pmatrix} a_1 & a_2 & a_3 & \dots & a_{n-1} & a_n \\ 1 & 0 & 0 & \dots & 0 & 0 \\ 0 & 1 & 0 & \dots & 0 & 0 \\ \vdots & \vdots & \ddots & \ddots & \vdots & \vdots \\ 0 & 0 & \dots & 1 & 0 & 0 \\ 0 & 0 & \dots & 0 & 1 & 0 \end{pmatrix}.$$

Double Leslie matrices were defined in [13]. The generalization of these matrices was made by Butcher and Wright in [3] in 2006; subsequently, the Leslie, doubly Leslie, companion, and doubly companion matrices were univocally reconstructed from the maximal eigenvalues of its all leading principal submatrices and one or more of its eigenpairs [8], also in this work, a doubly Leslie matrix was defined as follows:

Definition 1 ([8]) A doubly Leslie matrix is an $n \times n$ matrix of the form

$$\mathcal{L} = \begin{pmatrix} a_1 & a_2 & a_3 & \dots & a_{n-1} & a_n + c_n \\ b_1 & 0 & 0 & \dots & 0 & c_{n-1} \\ 0 & b_2 & 0 & \dots & 0 & c_{n-2} \\ \vdots & \vdots & \ddots & \ddots & \vdots & \vdots \\ 0 & 0 & \dots & b_{n-2} & 0 & c_2 \\ 0 & 0 & \dots & 0 & b_{n-1} & c_1 \end{pmatrix} \qquad (2)$$

where $0 < b_i \leq 1,\ i = 1, 2, \dots, n-1$, and $a_j \geq 0,\ j = 1, 2, \dots, n$. The double Leslie matrices were defined as a generalization of a doubly companion matrix introduced in [3]. The doubly Leslie $\mathcal{L}$ previously defined is a non-negative matrix. Furthermore, if $c_j = 0,\ j = 1, 2, \dots, n$, a Leslie matrix is obtained.

The structure of doubly companion, Leslie, and companion matrices is a particular case of a doubly Leslie matrix $\mathcal{L}$ of the form (2). Since the entries $a_j > 0,\ j = 2, 3, \dots, n$, all the matrices are irreducible non-negative matrices, and, as a consequence of that, these matrices have a unique positive real eigenvalue whose module is greater than its other eigenvalues. This eigenvalue is usually called the Perron eigenvalue [5] and has an associated eigenvector whose components are all positive.

In 2006, Peng et al. in [7] considered the problem of constructing a matrix from spectral information, and then Pickmann et al. in [8–11], obtained significant and relevant results regarding the inverse eigenvalue problem for Leslie and doubly Leslie matrices. In [2, 6], the authors present a complete and detailed work on constructing such matrices from their eigenvalues and the construction of Leslie stochastic matrices with zero traces from the coefficients of their characteristic polynomial, respectively.

Remark 1 Throughout this paper, we use the following notation, given a matrix $n \times n$ A, for $j = 1, 2, \dots, n$,

(1) A_j the $j \times j$ principal sub-matrix $j \times j$ of A.
(2) $\sigma(A_j)$ the spectrum of A_j.
(3) $P_j(\lambda)$ the characteristic polynomial of A_j.
(4) $\lambda_j^{(j)}$ the maximal eigenvalue of A_j.

A relevant result on the previous remark is the following [5].

Theorem 1 ([5]) *The maximal eigenvalue of a principal submatrix of a non-negative matrix A cannot exceed the maximal eigenvalue of A.*

Remark 2 The maximal eigenvalue of a non-negative matrix A is greater than or equal to the maximal eigenvalue of any of its leading principal submatrices A_j. The set of non-negative matrices has a special subset. Matrices from that subset, called irreducible matrices.

In this paper, considering the problem given by Pickman et al. in [11], we present computational algorithms (executed in Mathematica software) that reconstruct four matrices: Leslie matrices, doubly Leslie matrices and companion and doubly companion matrices, from a random set of numbers and a specific set of numbers, considering in the examples presented four decimal point.

2 An Algorithms for Leslie and Companion Matrices from Special Spectral Data and Specific Set of Data

This section presents the efficiency of the algorithm which constructs a Leslie matrix L from the maximum eigenvalues of all its principal submatrices and one of its eigenpairs. Furthermore, we present a synthetic example with random values ($n = 5$) and an example considering the population of collared lizards, where its population is modeled using a Leslie matrix. In addition, we present a construction of a companion matrix C only from the maximum eigenvalues of all its principal submatrix principals.

Our algorithm uses recursive methods and works in two ways:

- The first mode is random sets. The algorithms will take a random set of numbers and will return a Leslie matrix (as a well as Companion Matrix, doubly Leslie, and doubly companion matrix); with this mode, you also have previously significant information about your experiment.
- In the second mode, you can place the set of numbers, and the algorithm will return the Leslie matrix (as well as the companion matrix, doubly Leslie, and doubly companion matrix). In particular, using this configuration allows us to reduce significant experimental time because we will beforehand have an approximation of the rate of fertilization and the rate of survivors.

The algorithms in this section reconstruct a Leslie matrix and a companion matrix, respectively. The algorithm codes were programmed in Mathematica software. We use the following theorem [11] to prove the algorithms.

Theorem 2 ([11]) *Consider giving the set of n real numbers*

$$\left\{\lambda_1^{(1)}, \lambda_2^{(2)}, \ldots, \lambda_{n-1}^{(n-1)}, \lambda_n^{(n)}\right\}$$

and the positive vector $\mathbf{x} = (x_1, x_2, \ldots, x_n)^T$. *Then there exists a* $n \times n$ *Leslie matrix* L *of the form (1), such that* $\lambda_j^{(j)}$ *is the maximal eigenvalue of its leading principal submatrix* L_j, $j = 1, 2, \ldots, n$, *and* $\left(\mathbf{x}, \lambda_n^{(n)}\right)$ *is an eigenpair of L, if and only if*

$$0 \leq \lambda_1^{(1)} < \lambda_2^{(2)} < \cdots < \lambda_{n-1}^{(n-1)} < \lambda_n^{(n)} \tag{3}$$

and

$$0 < \lambda_n^{(n)} \frac{x_{j+1}}{x_j} \leq 1, \ \ j = 1, 2, \ldots, n-1. \tag{4}$$

For Theorem 2, the first Algorithm shows the construction of a random order Leslie matrix.

Algorithm 1: Matrix Leslie (Random Order)

```
ClearAll["Global`*"]
k = 0;
Do[
 n = 5;
 \[Lambda]1[k + 1] =
{\[Lambda][1],
 \[Lambda][2],
 \[Lambda][3],
 \[Lambda][4],
 \[Lambda][5]} =
Sort[{RandomReal[{0, 1}],
RandomReal[{0,1}],
RandomReal[{0,1}],
RandomReal[{0,1}],
RandomReal[{0,1}]}];

x1[k + 1] = {x[1], x[2], x[3], x[4], x[5]} =
Sort[{RandomReal[{0, 1}],
RandomReal[{0,1}],
RandomReal[{0,1}],
RandomReal[{0,1}],
RandomReal[{0,1}]}];

x1[k + 1] = x1[k + 1]/Norm[x1[k + 1]];
b[i_] := \[Lambda] [n]*(x[i + 1]/x[i]);

p[j_, \[Lambda]_] :=
\[Lambda]*p[j - 1, \[Lambda]] - a[j]
*\!\(\*UnderoverscriptBox[\(\[Product]\),
\(i = 1\), \(j - 1\)]\(b[i]\)\);

 p[0, \[Lambda]_] := 1;
 b[0] = 0;

 a[j_] := (\[Lambda] [j]*p[j - 1, \[Lambda][j]])/
 (\!\(\*UnderoverscriptBox[\(\[Product]\), \(i = 1\), \(j - 1\)]\(b[i]\)\));

 a[1] = \[Lambda][1];
 If[0 < b[1] <= 1 && 0 < b[2] <= 1 && 0 < b[3] <= 1 && 0 < b[4] <= 1,
  k = k + 1];
 If[0 < b[1] <= 1 && 0 < b[2] <= 1 && 0 < b[3] <= 1 && 0 < b[4] <= 1,
  A[k] = Sum[SparseArray[{{1, k} -> a[k]}, {n, n}], {k, 1, n}] +
     Sum[SparseArray[{{k + 1, k} -> b[k]}, {n, n}], {k, 1, n - 1}]];
 Clear[A1, A2, A3, A4];

 If[0 < b[1] <= 1 && 0 < b[2] <= 1 && 0 < b[3] <= 1 && 0 < b[4] <= 1,
  A1 = Drop[A[k], -1]];
 If[0 < b[1] <= 1 && 0 < b[2] <= 1 && 0 < b[3] <= 1 && 0 < b[4] <= 1,
  A1 = Transpose[Drop[Transpose[A1], -1]]];
 If[0 < b[1] <= 1 && 0 < b[2] <= 1 && 0 < b[3] <= 1 && 0 < b[4] <= 1,
  A2 = Drop[A1, -1]];
 If[0 < b[1] <= 1 && 0 < b[2] <= 1 && 0 < b[3] <= 1 && 0 < b[4] <= 1,
  A2 = Transpose[Drop[Transpose[A2], -1]]];
 If[0 < b[1] <= 1 && 0 < b[2] <= 1 && 0 < b[3] <= 1 && 0 < b[4] <= 1,
  A3 = Drop[A2, -1]];
 If[0 < b[1] <= 1 && 0 < b[2] <= 1 && 0 < b[3] <= 1 && 0 < b[4] <= 1,
```

```
 A3 = Transpose[Drop[Transpose[A3], -1]]];
If[0 < b[1] <= 1 && 0 < b[2] <= 1 && 0 < b[3] <= 1 && 0 < b[4] <= 1,
 A4 = Drop[A3, -1]];
If[0 < b[1] <= 1 && 0 < b[2] <= 1 && 0 < b[3] <= 1 && 0 < b[4] <= 1,
 A4 = Transpose[Drop[Transpose[A4], -1]]];
 If[0 < b[1] <= 1 && 0 < b[2] <= 1 && 0 < b[3] <= 1 &&
  0 < b[4] <= 1, \[Lambda]1[
    k] = {\[Lambda][1], \[Lambda][2], \[Lambda][3], \[Lambda][
     4], \[Lambda][5]}];
If[0 < b[1] <= 1 && 0 < b[2] <= 1 && 0 < b[3] <= 1 &&
  0 < b[4] <= 1, \[Lambda]2[k] = {Re[Eigenvalues[A4][[1]]],
   Re[Eigenvalues[A3][[1]]], Re[Eigenvalues[A2][[1]]],
   Re[Eigenvalues[A1][[1]]], Re[Eigenvalues[A[k]][[1]]]}];
x2[k] = First[Re[Abs[Eigenvectors[A[k]]]]];, 10000]

In: k
out: 439
A[k] // MatrixForm
```

In Algorithm 1, a random order Leslie Matrix is reconstructed, for example, here from 10,000 matrices, the algorithm found 439 Leslie Matrices, this number can change each time the algorithms run, to display the matrix, just replace k with a number, i.e.

If in algorithm 1, $k = 1$ then A[1]// MatrixForm we get

$$\begin{pmatrix} 0.1838 & 0.02699 & 0.03566 & 0.0582 & 0.407 \\ 0.8315 & 0 & 0 & 0 & 0 \\ 0 & 0.8878 & 0 & 0 & 0 \\ 0 & 0 & 0.8163 & 0 & 0 \\ 0 & 0 & 0 & 0.9029 & 0 \end{pmatrix}$$

and, for A[339]// MatrixForm, the following matrix is obtained,

$$\begin{pmatrix} 0.03681 & 0.006181 & 0.0004528 & 0.0258 & 0.07778 \\ 0.7328 & 0 & 0 & 0 & 0 \\ 0 & 0.3237 & 0 & 0 & 0 \\ 0 & 0 & 0.3388 & 0 & 0 \\ 0 & 0 & 0 & 0.3389 & 0 \end{pmatrix}.$$

In particular, in Theorem 2, if a specific set of real numbers is considered, we propose Algorithm 2 which immediately returns a Leslie matrix of a specified order.

The second algorithm shows the construction of a specific order Leslie matrix.

Algorithm 2: Matrix Leslie (Specific Order)

```
Clear["Global\[OpenCurlyQuote]*"]
n = 5;
\[Lambda][1] = 0.1177;
\[Lambda][2] = 0.2000;
\[Lambda][3] = 0.3963;
\[Lambda][4] = 0.5967;
\[Lambda][5] = 1.0063;

x[1] = 0.0965;
x[2] = 0.0836;
x[3] = 0.0725;
x[4] = 0.0273;
x[5] = 0.0112;

b[i_] := \[Lambda] [n]*(x[i + 1]/x[i]);
p[j_, \[Lambda]_] := \[Lambda]*p[j - 1, \[Lambda]] - a[j] *\!\(
\*UnderoverscriptBox[\(\[Product]\), \(i = 1\), \(j - 1\)]\(b[i]\)\)

p[0, \[Lambda]_] := 1;
b[0] = 0;
a[j_] := (\[Lambda] [j]*p[j - 1, \[Lambda][j]])/(\!\(
\*UnderoverscriptBox[\(\[Product]\), \(i = 1\), \(j - 1\)]\(b[i]\)\));
a[1] = \[Lambda][1];

A := Sum[SparseArray[{{1, k} -> a[k]}, {n, n}], {k, 1, n}] +
   Sum[SparseArray[{{k + 1, k} -> b[k]}, {n, n}], {k, 1, n - 1}];
A // MatrixForm
```

Remark 3 In this case from specific numbers, we say

$$\{0.1177; 0.2000; 0.3963; 0.5967; 1.0063\}$$

(these numbers can be modified, but **must** satisfies the condition of Theorem 2), the Algorithm 2 returns the following Leslie matrix.

$$\begin{pmatrix} 0.2006 & 0.01658 & 0.1045 & 0.5672 & 3.299 \\ 0.725 & 0 & 0 & 0 & 0 \\ 0 & 0.7258 & 0 & 0 & 0 \\ 0 & 0 & 0.3151 & 0 & 0 \\ 0 & 0 & 0 & 0.3433 & 0 \end{pmatrix}.$$

We will verify our code, we use an example given in [12], in this work the authors present the equation of the Leslie matrix associated with the population of the cabbage lizard, we are going to reconstruct the following matrix (Table 1)

Table 1 Spectral information

j	1	2	3	4	5	6	7	8
$\lambda_j^{(j)}$	0	0.4118	0.4118	1.092	1.092	1.14	1.14	1.153
$x_j^{(n)}$	0.7438	0.0547	0.3221	0.1975	0.0572	0.031	0.02151	0.009326

$$\begin{pmatrix} 0 & 0.2 & 0 & 3 & 0 & 3.75 & 0 & 4.25 \\ 0.848 & 0 & 0 & 0 & 0 & 0 & 0 & 0 \\ 0 & 0.679 & 0 & 0 & 0 & 0 & 0 & 0 \\ 0 & 0 & 0.707 & 0 & 0 & 0 & 0 & 0 \\ 0 & 0 & 0 & 0.334 & 0 & 0 & 0 & 0 \\ 0 & 0 & 0 & 0 & 0.625 & 0 & 0 & 0 \\ 0 & 0 & 0 & 0 & 0 & 0.8 & 0 & 0 \\ 0 & 0 & 0 & 0 & 0 & 0 & 0.5 & 0 \end{pmatrix}.$$

The following table has the spectre information the above Leslie matrix.

Remark 4 The $\lambda_j^{(j)}$ is the maximal eigenvalue of its leading principal submatrix, in our case its $\lambda_8^{(8)} = 1.153$ the same value that authors found in [12], and since we take eigenvector associated to the maximal eigenvalue we obtain the following vector; $\mathbf{x} = (0.7438, 0.0547, 0.3221, 0.1975, 0.0572, 0.031, 0.02151, 0.009326)^T$

The Leslie matrix reconstructed is the following

$$\begin{pmatrix} 0 & 0.2 & 0 & 3 & 0 & 3.81 & 0 & 4.1 \\ 0.848 & 0 & 0 & 0 & 0 & 0 & 0 & 0 \\ 0 & 0.679 & 0 & 0 & 0 & 0 & 0 & 0 \\ 0 & 0 & 0.707 & 0 & 0 & 0 & 0 & 0 \\ 0 & 0 & 0 & 0.334 & 0 & 0 & 0 & 0 \\ 0 & 0 & 0 & 0 & 0.625 & 0 & 0 & 0 \\ 0 & 0 & 0 & 0 & 0 & 0.8 & 0 & 0 \\ 0 & 0 & 0 & 0 & 0 & 0 & 0.5 & 0 \end{pmatrix}$$

Remark 5 There is a slight difference about both matrix, the main reason it that our algorithm its more precised.

Now, from only the maximal eigenvalue $\lambda_j^{(j)}$ of its leading principal submatrix C_j, $j = 1, 2, \ldots, n$, we present a result for the construction of a companion matrix C of the form (1) with $b_i = 1$, $i = 1, 2, \ldots, n-1$.

Theorem 3 ([11]) *Let the set of n real numbers* $\left\{\lambda_1^{(1)}, \lambda_2^{(2)}, \ldots, \lambda_{n-1}^{(n-1)}, \lambda_n^{(n)}\right\}$ *be given. Then there exists a companion matrix* $n \times n$ C *of the form* (1) *with* $b_i = 1$, $i = 1, 2, \ldots, n-1$, *such that* $\lambda_j^{(j)}$ *is the maximal eigenvalue of its leading principal submatrix* C_j, $j = 1, 2, \ldots, n$, *if and only if*

$$0 \leq \lambda_1^{(1)} < \lambda_2^{(2)} < \cdots < \lambda_{n-1}^{(n-1)} < \lambda_n^{(n)}. \quad (5)$$

To solve the Theorem 3, we show the following algorithm for the construction of a Companion matrix of specific order.

Algorithm 3: Companion matrix (Specific Order)

```
Clear["Global`*"]
n = 5;
\[Lambda][1] = 0.0785;
\[Lambda][2] = 0.9365;
\[Lambda][3] = 1.1009;
\[Lambda][4] = 1.2365;
\[Lambda][5] = 1.3000;

p[j_, \[Lambda]_] := \[Lambda]*p[j - 1, \[Lambda]] - a[j];
p[0, \[Lambda]_] := 1;

a[j_] := (\[Lambda] [j]*p[j - 1, \[Lambda][j]]);
a[1] = \[Lambda][1];
b[i_] = 1;

A := Sum[SparseArray[{{1, k} -> a[k]}, {n, n}], {k, 1, n}]
+ Sum[SparseArray[{{k + 1, k} -> b[k]}, {n, n}], {k, 1, n - 1}];

A // MatrixForm
```

Remark 6 The same way that above algorithm, from specific numbers

$$\{0.0785; 0.9365; 1.1009; 1.2365; 1.3000\}$$

(these numbers can be modified, but **must** satisfies the condition of Theorem 3), the Algorithm 3 returns the following companion matrix.

$$\begin{pmatrix} 0.0785 & 0.8035 & 0.3545 & 0.5223 & 0.4452 \\ 1 & 0 & 0 & 0 & 0 \\ 0 & 1 & 0 & 0 & 0 \\ 0 & 0 & 1 & 0 & 0 \\ 0 & 0 & 0 & 1 & 0 \end{pmatrix}.$$

3 Doubly Leslie and Doubly Companion Matrices from Special Spectral Data

The algorithms in this section reconstruct a doubly Leslie matrix and a doubly companion matrix, respectively. The algorithm codes were programmed in Mathematica software. For this case, we use the following theorem; see [11].

Theorem 4 ([11]) *Let the set of n real numbers* $\left\{\lambda_1^{(1)}, \lambda_2^{(2)}, \ldots, \lambda_{n-1}^{(n-1)}, \lambda_n^{(n)}\right\}$ *and the positive vectors* $\mathbf{y} = (y_1, y_2, \ldots, y_{n-1})^T$ *and* $\mathbf{x} = (x_1, x_2, \ldots, x_n)^T$ *be satisfying*

$$0 \leq \lambda_1^{(1)} < \lambda_2^{(2)} < \cdots < \lambda_n^{(n)}, \tag{6}$$

$$0 < \lambda_{n-1}^{(n-1)} \frac{y_{i+1}}{y_i} \leq 1, \; i = 1, 2, \ldots, n-2, \tag{7}$$

$$\frac{x_j}{x_{j+1}} = \frac{y_j}{y_{j+1}}, \; j = 1, 2, \ldots, n-1, \tag{8}$$

and

$$0 < \lambda_n^{(n)} \frac{x_n}{x_{n-1}} - \sum_{i=1}^{n-2} \lambda_{n-1}^{(n-1)} \frac{y_{i+1}}{y_i} \leq 1. \tag{9}$$

Then there exists a doubly Leslie matrix $\mathcal{L}$ *of the form (2), such that* $\lambda_j^{(j)}$ *is the maximal eigenvalue of* $\mathcal{L}_j$, $j = 1, \ldots, n$, $\left(\mathbf{y}, \lambda_{n-1}^{(n-1)}\right)$ *is an eigenpair of* $\mathcal{L}_{n-1}$, *and* $\left(\mathbf{x}, \lambda_n^{(n)}\right)$ *is an eigenpair of* $\mathcal{L}$.

The fourth algorithm shows as solve Theorem 4 by a construction of a specific order Double Leslie Matrix.

Algorithm 4: Double Leslie Matrix (Specific Order)

```
Clear["Global'*"]
k = 0;
Do[
 n = 5;
 \[Lambda]1[k + 1] =
{\[Lambda][1],
 \[Lambda][2],
 \[Lambda][3],
 \[Lambda][4],
 \[Lambda][5]} =
Sort[{RandomReal[{0, 1}],
RandomReal[{0, 1}],
RandomReal[{0, 1}],
```

```
RandomReal[{0, 1}],
RandomReal[{0, 1}]}];

x1[k + 1] = {x[1], x[2], x[3], x[4], x[5]} =
Sort[{RandomReal[{0, 1}],
RandomReal[{0, 1}],
RandomReal[{0, 1}],
RandomReal[{0, 1}],
RandomReal[{0, 1}]}];

y1[k + 1] = {y[1], y[2], y[3], y[4]} =
Sort[{RandomReal[{0, 1}],
RandomReal[{0, 1}],
RandomReal[{0, 1}],
RandomReal[{0, 1}]}];

p[0, \[Lambda]_] = 1;

a[1] = \[Lambda][1];

b[i_] := \[Lambda] [n - 1]*y[i + 1] /y[i];

a[n] = (\[Lambda] [n] - \[Lambda] [n - 1]) x[1] /x[n] - c[n];

a[j_] := (\[Lambda] [j]*p[j - 1, \[Lambda][j]])
/\!\(\*UnderoverscriptBox[\(\[Product]\), \(i = 1\), \(j - 1\)]\(b[i]\)\);

c[j_] := (\[Lambda] [n] - \[Lambda][n - 1])*x[n - j + 1] /x[n];

b[n - 1] = \[Lambda] [n]*x[n] /x[n - 1] - \!\(
\*UnderoverscriptBox[\(\[Sum]\), \(i = 1\), \(n - 2\)]\(\[Lambda]\ [n - 1]*
\*FractionBox[\(\(y[i + 1]\)\(\ \)\), \(y[i]\)]\)\);

p[j_, \[Lambda]_] := \[Lambda]*p[j - 1, \[Lambda]] - a[j]*\!\(
\*UnderoverscriptBox[\(\[Product]\), \(i = 1\), \(j - 1\)]\(b[i]\)\);

c[1] = \!\(\*UnderoverscriptBox[\(\[Sum]\), \(j = 1\), \(n - 2\)]
\(\[Lambda]\ [n - 1]*
\*FractionBox[\(\(y[j + 1]\)\(\ \)\), \(y[j]\)]*
\*FractionBox[\(\(x[n - 1]\)\(\ \)\), \(x[n]\)]\)\);

If[c[1] > 0 && c[2] > 0 && c[3] > 0 && c[4] > 0 && c[5] > 0 &&
0 < b[n - 1] \[VectorLessEqual] 1, k = k + 1];

If[c[1] > 0 && c[2] > 0 && c[3] > 0 && c[4] > 0 && c[5] > 0 &&
0 < b[n - 1] \[VectorLessEqual] 1,
A[k] = Sum[SparseArray[{{1, k} -> a[k]}, {n, n}], {k, 1, n - 1}] +
Sum[SparseArray[{{k + 1, k} -> b[k]}, {n, n}], {k, 1, n - 2}] +
Sum[SparseArray[{{j, n} -> c[n - j + 1]}, {n, n}], {j, 2, n - 1}]
+ SparseArray[{{n, n} -> c[1]}, {n, n}]
+ SparseArray[{{n, n - 1} -> b[n - 1]}, {n, n}]
+ SparseArray[{{1, n} -> a[n] + c[n]}, {n, n}]];

Clear[A1, A2, A3, A4];

If[c[1] > 0 && c[2] > 0 && c[3] > 0 && c[4] > 0 && c[5] > 00 &&
0 < b[n - 1] \[VectorLessEqual] 1, A1 = Drop[A[k], -1]];

If[c[1] > 0 && c[2] > 0 && c[3] > 0 && c[4] > 0 && c[5] > 00 &&
0 < b[n - 1] \[VectorLessEqual] 1,
A1 = Transpose[Drop[Transpose[A1], -1]]];

If[c[1] > 0 && c[2] > 0 && c[3] > 0 && c[4] > 0 && c[5] > 00 &&
```

```
0 < b[n - 1] \[VectorLessEqual] 1, A2 = Drop[A1, -1]];

If[c[1] > 0 && c[2] > 0 && c[3] > 0 && c[4] > 0 && c[5] > 00 &&
0 < b[n - 1] \[VectorLessEqual] 1,
A2 = Transpose[Drop[Transpose[A2], -1]]];

If[c[1] > 0 && c[2] > 0 && c[3] > 0 && c[4] > 0 && c[5] > 00 &&
0 < b[n - 1] \[VectorLessEqual] 1, A3 = Drop[A2, -1]];

If[c[1] > 0 && c[2] > 0 && c[3] > 0 && c[4] > 0 && c[5] > 00 &&
0 < b[n - 1] \[VectorLessEqual] 1,
A3 = Transpose[Drop[Transpose[A3], -1]]];

If[c[1] > 0 && c[2] > 0 && c[3] > 0 && c[4] > 0 && c[5] > 00 &&
0 < b[n - 1] \[VectorLessEqual] 1, A4 = Drop[A3, -1]];

If[c[1] > 0 && c[2] > 0 && c[3] > 0 && c[4] > 0 && c[5] > 00 &&
0 < b[n - 1] \[VectorLessEqual] 1,
A4 = Transpose[Drop[Transpose[A4], -1]]];

If[c[1] > 0 && c[2] > 0 && c[3] > 0 && c[4] > 0 && c[5] > 00 &&
0 < b[n - 1] \[VectorLessEqual] 1, \[Lambda]1[k]
= {\[Lambda][1], \[Lambda][2], \[Lambda][3], \[Lambda][4], \[Lambda][5]}];

If[c[1] > 0 && c[2] > 0 && c[3] > 0 && c[4] > 0 && c[5] > 00 &&
0 < b[n - 1] \[VectorLessEqual] 1, \[Lambda]2[k]
= {Re[Eigenvalues[A4][[1]]],
Re[Eigenvalues[A3][[1]]],
Re[Eigenvalues[A2][[1]]],
Re[Eigenvalues[A1][[1]]],
Re[Eigenvalues[A[k]][[1]]]}];

If[c[1] > 0 && c[2] > 0 && c[3] > 0 && c[4] > 0 && c[5] > 00 &&
0 < b[n - 1] \[VectorLessEqual] 1,x2[k]
= First[Re[Abs[Eigenvectors[A[k]]]]];, 20000]

In:= k
Out= 213
In:= A[k]//MatrixForm
```

In particular, in the Algorithm 4, here from 20.000 matrices, the algorithm found 213 doubly Leslie matrices; this number can change every time the algorithms run, to visualize the matrix only replacing k with a number, that is, for example, if $k = 6$ then A[6]// MatrixForm we obtain

$$\begin{pmatrix} 0.004797 & 0.09346 & 0.003333 & 0.09397 & 0.1727 \\ 0.2988 & 0 & 0 & 0 & 0.3452 \\ 0 & 0.3294 & 0 & 0 & 0.4827 \\ 0 & 0 & 0.3223 & 0 & 0.5254 \\ 0 & 0 & 0 & 0.3714 & 0.7034 \end{pmatrix}$$

Finally, we present the construction of a doubly companion matrix $\mathcal{C}$ of the form (2) with $b_i = 1,\ i = 1, 2, \ldots, n-1$, from the maximal eigenvalues of its all leading principal submatrices and one of its eigenpairs.

Theorem 5 ([11]) *Let the set of n real numbers* $\left\{\lambda_1^{(1)}, \lambda_2^{(2)}, \ldots, \lambda_{n-1}^{(n-1)}, \lambda_n^{(n)}\right\}$ *and the positive vectors* $\mathbf{y} = (y_1, y_2, \ldots, y_{n-1})^T$ *and* $\mathbf{x} = (x_1, x_2, \ldots, x_n)^T$ *be satisfying*

$$0 \le \lambda_1^{(1)} < \lambda_2^{(2)} < \cdots < \lambda_n^{(n)}, \tag{10}$$

$$0 < \lambda_{n-1}^{(n-1)} \frac{y_{i+1}}{y_i} \le 1,\ i = 1, 2, \ldots, n-2, \tag{11}$$

$$\frac{x_j}{x_{j+1}} = \frac{y_j}{y_{j+1}},\ j = 1, 2, \ldots, n-1, \tag{12}$$

and

$$0 < \lambda_n^{(n)} \frac{x_n}{x_{n-1}} - \sum_{i=1}^{n-2} \lambda_{n-1}^{(n-1)} \frac{y_{i+1}}{y_i} \le 1. \tag{13}$$

Then there exists a doubly Leslie matrix $\mathcal{L}$ *of the form (2), such that* $\lambda_j^{(j)}$ *is the maximal eigenvalue of* $\mathcal{L}_j,\ j = 1, \ldots, n$, $\left(\mathbf{y}, \lambda_{n-1}^{(n-1)}\right)$ *is an eigenpair of* $\mathcal{L}_{n-1}$, *and* $\left(\mathbf{x}, \lambda_n^{(n)}\right)$ *is an eigenpair of* $\mathcal{L}$.

The Algorithm 5 summarizes this discussion and proposes to solve Theorem 5.

Algorithm 5: Doubly companion matrix (Specific Order)

```
ClearAll["Global`*"]
k = 0;
Do[
 n = 5;
 \[Lambda]1[k + 1] =
{\[Lambda][1],
 \[Lambda][2],
 \[Lambda][3],
 \[Lambda][4],
 \[Lambda][5]} =
Sort[{RandomReal[{0, 1}],
RandomReal[{0, 1}],
RandomReal[{0, 1}],
RandomReal[{0, 1}],
RandomReal[{0, 1}]}];

x1[k + 1] = {x[1], x[2], x[3], x[4], x[5]} =
Sort[{RandomReal[{0, 1}],
RandomReal[{0, 1}],
RandomReal[{0, 1}],
```

```
RandomReal[{0, 1}],
RandomReal[{0, 1}]}];

p[0, \[Lambda]_] = 1;

a[j_] := \[Lambda] [j]*p[j - 1, \[Lambda][j]];

a[1] = \[Lambda][1];

b[i_] = 1;

a[n] = \[Lambda] [n]*x[1] /x[n] - \!\(
\*UnderoverscriptBox[\(\[Sum]\), \(j = 1\), \(n\  - \ 1\)]\(a[j]*
\*FractionBox[\(\(x[j]\)\(\ \)\), \(x[n]\)]\)\) - c[n];

c[j_] := \[Lambda] [n]*x[n - j + 1] /x[n] - x[n - j] /x[n];

p[j_, \[Lambda]_] := \[Lambda]*p[j - 1, \[Lambda]] - a[j];

If[c[1] > 0 && c[2] > 0 && c[3] > 0 && c[4] > 0 && a[5] + c[5] > 0, k = k + 1];

If[c[1] > 0 && c[2] > 0 && c[3] > 0 && c[4] > 0 && a[5] + c[5] > 0,
A[k] = Sum[
SparseArray[{{1, kk} -> a[kk]}, {n, n}], {kk, 1, n - 1}]
+ Sum[SparseArray[{{j, n} -> c[n - j + 1]}, {n, n}], {j, 1, n}]
+ Sum[SparseArray[{{kk + 1, kk} -> b[kk]}, {n, n}], {kk, 1, n - 1}]
+ SparseArray[{{1, n} -> a[n]}, {n, n}]];

Clear[A1, A2, A3, A4];

If[c[1] > 0 && c[2] > 0 && c[3] > 0 && c[4] > 0 && a[5] + c[5] > 0,
A1 = Drop[A[k], -1]];

If[c[1] > 0 && c[2] > 0 && c[3] > 0 && c[4] > 0 && a[5] + c[5] > 0,
A1 = Transpose[Drop[Transpose[A1], -1]]];

If[c[1] > 0 && c[2] > 0 && c[3] > 0 && c[4] > 0 && a[5] + c[5] > 0,
A2 = Drop[A1, -1]];

If[c[1] > 0 && c[2] > 0 && c[3] > 0 && c[4] > 0 && a[5] + c[5] > 0,
A2 = Transpose[Drop[Transpose[A2], -1]]];

If[c[1] > 0 && c[2] > 0 && c[3] > 0 && c[4] > 0 && a[5] + c[5] > 0,
A3 = Drop[A2, -1]];

If[c[1] > 0 && c[2] > 0 && c[3] > 0 && c[4] > 0 && a[5] + c[5] > 0,
A3 = Transpose[Drop[Transpose[A3], -1]]];

If[c[1] > 0 && c[2] > 0 && c[3] > 0 && c[4] > 0 && a[5] + c[5] > 0,
A4 = Drop[A3, -1]];

If[c[1] > 0 && c[2] > 0 && c[3] > 0 && c[4] > 0 && a[5] + c[5] > 0,
A4 = Transpose[Drop[Transpose[A4], -1]]];

If[c[1] > 0 && c[2] > 0 && c[3] > 0 && c[4] > 0 &&
a[5] + c[5] > 0, \[Lambda]1[k]
= {\[Lambda][1],
\[Lambda][2],
\[Lambda][3],
\[Lambda][4],
\[Lambda][5],
\[Lambda][6],
\[Lambda][7],
```

```
\[Lambda][8],
\[Lambda][9],
\[Lambda][10],
\[Lambda][11],
\[Lambda][12],
\[Lambda][13],
\[Lambda][14],
\[Lambda][15],
\[Lambda][16]}];

If[c[1] > 0 && c[2] > 0 && c[3] > 0 && c[4] > 0 &&
a[5] + c[5] > 0, \[Lambda]2[k] = {Re[Eigenvalues[A4][[1]]],
Re[Eigenvalues[A3][[1]]],
Re[Eigenvalues[A2][[1]]],
Re[Eigenvalues[A1][[1]]],
Re[Eigenvalues[A[k]][[1]]]}];

If[c[1] > 0 && c[2] > 0 && c[3] > 0 && c[4] > 0 && a[5] + c[5] > 0,
x2[k] = First[Re[Abs[Eigenvectors[A[k]]]]]];, 10000]

In: k
Out: 903
In:= A[k] // MatrixForm
```

Remark 7 Here from 10000 matrices, the Algorithm found 903 doubly companion matrices; this number can change every time the algorithms run, to visualize the matrix only replacing k with a number, that is, for $k = 705$, A[705]// MatrixForm we obtain

$$\begin{pmatrix} 0.1025 & 0.1912 & 0.06526 & 0.2604 & 0.08229 \\ 1 & 0 & 0 & 0 & 0.006633 \\ 0 & 1 & 0 & 0 & 0.194 \\ 0 & 0 & 1 & 0 & 0.004673 \\ 0 & 0 & 0 & 1 & 0.3954 \end{pmatrix}.$$

4 Examples

In this section, we present examples developed by Algorithms that realize Theorems 2, 3, 4, and 5, which construct Leslie, companion, doubly Leslie, and doubly Leslie matrices of order $n = 5$. Additionally, considering the error to compare, the standard error and the logarithmic error for these examples, we analyze the efficiency of the presented method.

Example 1 For Theorem 2, with $n = 5$

Let

$\lambda_1^{(1)}$	$\lambda_2^{(2)}$	$\lambda_3^{(3)}$	$\lambda_4^{(4)}$	$\lambda_5^{(5)}$
0.199718	0.468908	0.534272	0.644681	0.769578

and the vector

$$\mathbf{x} = (0.329241,\ 0.401052,\ 0.423291,\ 0.453416,\ 0.58821)^T$$

satisfying the conditions of Theorem 2, then, exist a Leslie matrix L given by

$$L = \begin{pmatrix} 0.1997 & 0.1437 & 0.03685 & 0.07754 & 0.1409 \\ 0.9374 & 0 & 0 & 0 & 0 \\ 0 & 0.8123 & 0 & 0 & 0 \\ 0 & 0 & 0.8243 & 0 & 0 \\ 0 & 0 & 0 & 0.9984 & 0 \end{pmatrix}$$

$$\begin{aligned} \sigma(L_1) &= \{\mathbf{0.1997}\}, \\ \sigma(L_2) &= \{-0.0.2692, \mathbf{0.4689}, \} \\ \sigma(L_3) &= \{-0.1673 - 0.1566, -0.1673 + 0.1566I, \mathbf{0.5343}\}, \\ \sigma(L_4) &= \{0.005459 - 0.4069, 0.005459 + 0.4069i, -0.4559, \mathbf{0.6447}\}, \\ \sigma(L_5) &= \{-0.4767 - 0.3122i,\ -0.4767 + 0.3122i,\ 0.1918 - 0.5626i, \\ &\quad 0.1918 + 0.5626i, \mathbf{0.76958}\} \end{aligned}$$

and $L\mathbf{x} = (\mathbf{0.769578})\mathbf{x}$.

Example 2 We use two types of errors to measure the efficiency of the algorithm, typical error and logarithmic error. For the test we consider the vector $\boldsymbol{\lambda}=\left(\lambda_1^{(1)}, \ldots, \lambda_n^{(n)}\right)$ and the positive vector $\mathbf{x} = (x_1, x_2, \ldots, x_n)^T$, whose components are real numbers chosen randomly according of Theorem 2, and we consider $\widetilde{L}$ the Leslie matrix $n \times n$ constructed and the vectors $\widetilde{\boldsymbol{\lambda}}$ and $\widetilde{\mathbf{x}}$, whose entries are the maximal eigenvalue of $\widetilde{L}_j$, $j = 1, 2, \ldots, n$, and the components of the eigenvector corresponding to the maximal eigenvalue of $\widetilde{L}$, respectively Figs. 2 and 3 shows the plots of both errors

$$e_\lambda = \log\left(\frac{\|\boldsymbol{\lambda} - \widetilde{\boldsymbol{\lambda}}\|}{\|\boldsymbol{\lambda}\|}\right), \quad e_x = \log\left(\frac{\|\mathbf{x} - \widetilde{\mathbf{x}}\|}{\|\mathbf{x}\|}\right)$$

and Figs. 4 and 5 shows the plot of

$$e_\lambda = \frac{\|\boldsymbol{\lambda} - \widetilde{\boldsymbol{\lambda}}\|}{\|\boldsymbol{\lambda}\|}, \quad e_x = \frac{\|\mathbf{x} - \widetilde{\mathbf{x}}|}{\|\mathbf{x}|}$$

for $n = 5$.

Example 3 Given

$\lambda_1^{(1)}$	$\lambda_2^{(2)}$	$\lambda_3^{(3)}$	$\lambda_4^{(4)}$	$\lambda_5^{(5)}$
0.3401	0.4786	0.5348	0.8111	0.9765

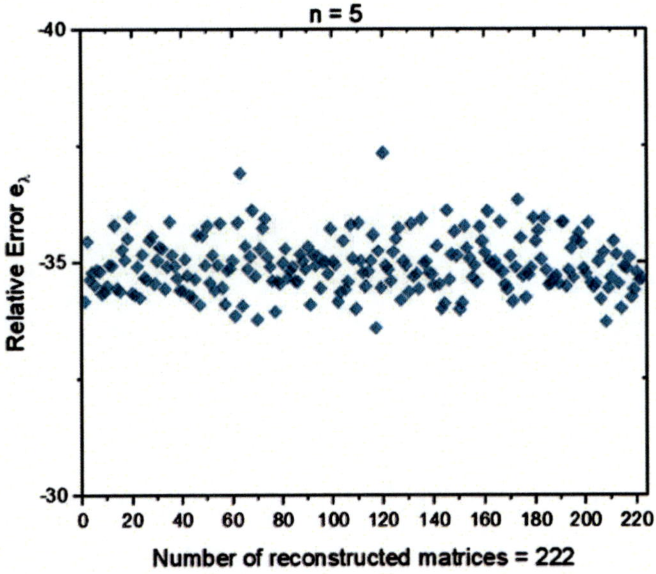

Fig. 2 The logarithmic error of 222 Leslie matrices of order 5 was constructed

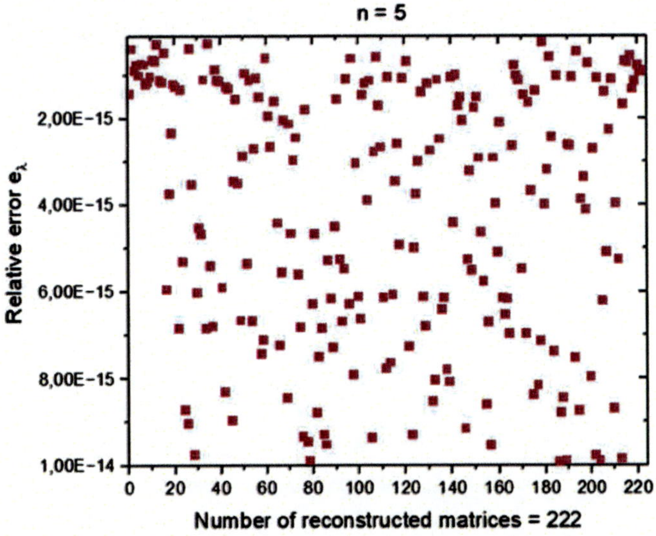

Fig. 3 Normal error of 222 constructed Leslie matrices of order 5 constructed

satisfying the condition of Theorem 3. We obtain the following companion matrix for $n = 5$.

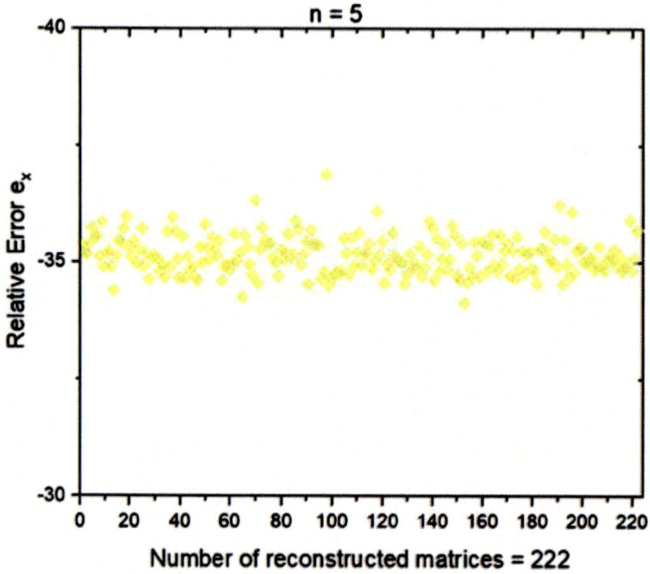

Fig. 4 Construction of 222 Leslie matrices of order 5

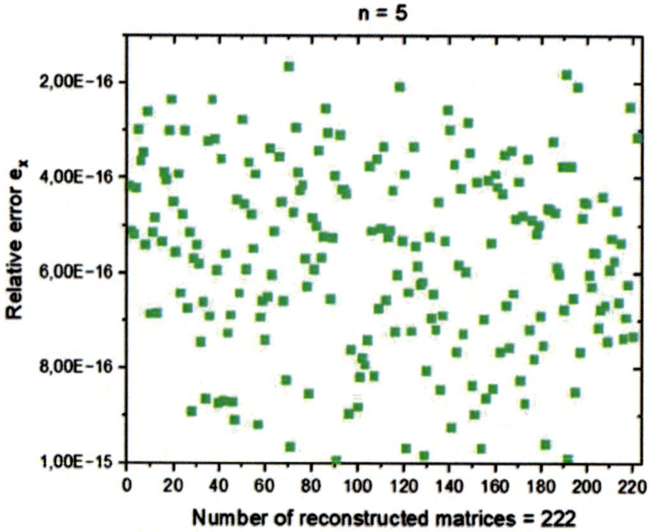

Fig. 5 Construction of 222 Leslie matrices of order 5

$$C = \begin{pmatrix} 0.3401 & 0.06629 & 0.02023 & 0.1913 & 0.3109 \\ 1.0000 & 0 & 0 & 0 & 0 \\ 0 & 1.0000 & 0 & 0 & 0 \\ 0 & 0 & 1.0000 & 0 & 0 \\ 0 & 0 & 0 & 1.0000 & 0 \\ 0 & 0 & 0 & 0 & 1.0000 \end{pmatrix}$$

where

$$\begin{aligned} \sigma(C_1) &= \{\mathbf{0.3401}\}, \\ \sigma(C_2) &= \{-0.1494,\ \mathbf{0.4786}\}, \\ \sigma(C_3) &= \{-0.2787 - 0.5038i,\ -0.2787 + 0.5038i,\ \mathbf{0.5348}\}, \\ \sigma(C_4) &= \{-0.5999,\ 0.0645 - 0.6236i,\ 0.0645 + 0.6236i,\ \mathbf{0.8111}\}, \\ \sigma(C_5) &= \{-0.5746 - 0.3836i,\ -0.5746 + 0.3836i,\ 0.2564 - 0.7755i, \\ &\quad 0.2564 + 0.7755i,\ \mathbf{0.9765}\}. \end{aligned}$$

Example 4 For $n = 5$, we construct a doubly Leslie matrix considering the following spectral information:

$$\begin{array}{ccccc} \lambda_1^{(1)} & \lambda_2^{(2)} & \lambda_3^{(3)} & \lambda_4^{(4)} & \lambda_5^{(5)} \\ 0.1543 & 0.4953 & 0.5112 & 0.6589 & 0.8966 \end{array}$$

and the vectors

$$\mathbf{x} = (0.1009,\ 0.2095,\ 0.292,\ 0.3056,\ 0.8929)^T$$

and

$$\mathbf{y} = (0.42997,\ 0.3779,\ 0.4813,\ 0.608)^T$$

satisfying Theorem 4.

$$\mathcal{L} = \begin{pmatrix} 0.1543 & 0.2033 & 0.009885 & 0.01145 & 0.02687 \\ 0.8308 & 0 & 0 & 0 & 0.05577 \\ 0 & 0.8391 & 0 & 0 & 0.07774 \\ 0 & 0 & 0.8324 & 0 & 0.08136 \\ 0 & 0 & 0 & 0.1174 & 0.8564 \end{pmatrix}$$

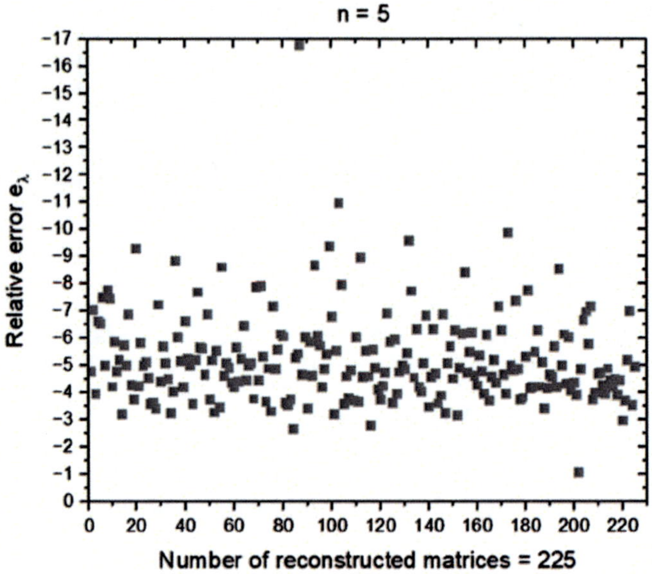

Fig. 6 Construction of 225 doubly companions matrices of order 5

with

$$\begin{aligned}
\sigma(\mathcal{L}_1) &= \{\mathbf{0.1543}\},\\
\sigma(\mathcal{L}_2) &= \{-0.341, \mathbf{0.4953}\},\\
\sigma(\mathcal{L}_3) &= \{-0.3139, \ -04294, \ \mathbf{0.5112}\},\\
\sigma(\mathcal{L}_4) &= \{-0.5444, \ 0.01994 - 0.4299i, \ 0.01994 + 0.4299i, \ \mathbf{0.6589}\},\\
\sigma(\mathcal{L}_5) &= \{-0.5455, -0.11885 - 0.433i, -0.11885 + 0.433i,\\
&\quad 0.6218, \mathbf{0.8966}\},
\end{aligned}$$

$\mathcal{L}\mathbf{x} = (0.8966)\mathbf{x}$ and $\mathcal{L}_4\mathbf{y} = (0.6589)\mathbf{y}$.

Example 5 Finally, we present an illustrative example of a doubly companion matrix, considering the vectors $\boldsymbol{\lambda}=\left(\lambda_1^{(1)}, \ldots, \lambda_n^{(n)}\right)$ and $\mathbf{x} = (x_1, x_2, \ldots, x_n)^T$, whose entries are real numbers chosen randomly. Let us $\widetilde{\mathcal{C}}$ the doubly companion matrix $n \times n$ satisfying conditions of Theorem 5, considering the vectors $\widetilde{\boldsymbol{\lambda}}$ and $\widetilde{\mathbf{x}}$, whose entries are the maximal eigenvalue of $\widetilde{\mathcal{C}}_j$, $j = 1, 2, \ldots, n$, and the components of the eigenvector corresponding to the maximal eigenvalue of $\widetilde{\mathcal{C}}$. The subsequent figures depict plots of e_λ and e_x in both the Companion and Doubly Companion matrices, as defined in Example 2, for the case when $n = 5$ (Figs. 6, 7, 8 and 9).

Funding: J. Pantáz was supported by Programa Regional MATHAMSUD MATH2020003 and by Fortalecimiento de la Investigación y Posgrado de la Universidad de Antofagasta, enfatizando la asociatividad con alianzas y programas de

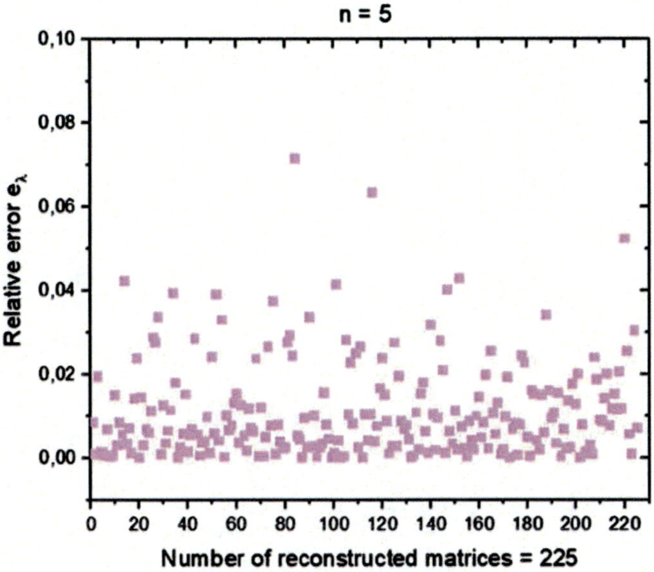

Fig. 7 Construction of 225 doubly companions matrices of order 5

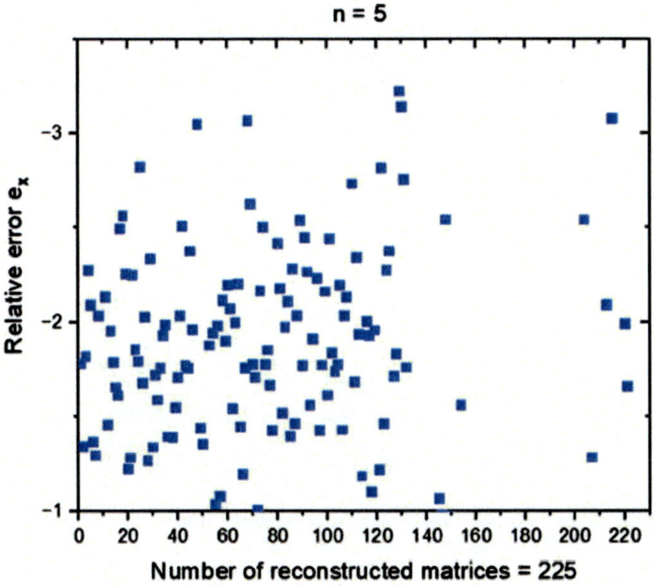

Fig. 8 Construction of 225 doubly companions matrices of order 5

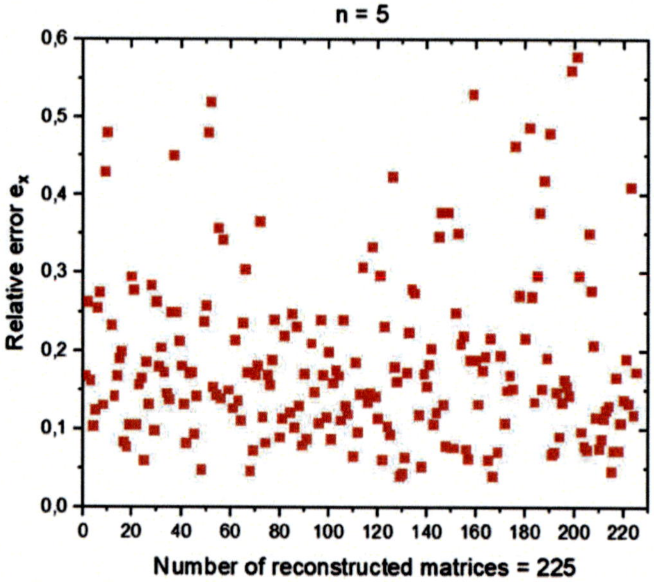

Fig. 9 Construction of 225 doubly companions matrices of order 5

posgrado que impacten en las necesidades regionales, ANT-20992. J. Rodríguez was supported by MINEDUC-UA project, code ANT-1899 and Funded by the Initiation Program in Research - Universidad de Antofagasta, INI-19-06 and Programa Regional MATHAMSUD MATH2020003. L. Medina was supported by Programa Regional MathAmSud código AMSUD220015.

References

1. Bacaër, N.: The Leslie matrix (1945). In: A Short History of Mathematical Population Dynamics. Springer (2011)
2. Benvenuti, L.: The inverse eigenvalue problem for Leslie matrices. Electron. J. Linear Algebra. **35**, 319–330 (2019)
3. Butcher, J., Wright, W.M.: Applications of doubly companion matrices. Appl. Numer. Math. **56**(3), 358–373 (2006)
4. Donovan, T.M., Welden, C.W.: Spreadsheet Exercises in Conservation Biology and Landscape Ecology. Oxford University Press, Incorporated (2002)
5. Minc, H.: non-negative Matrices, Wiley Series in Discrete Mathematics and Optimization (1988)
6. Medina, L., Nina, H., Valero, E.: A note on NIEP for Leslie and Doubly Leslie matrices. Mathematics **8**, 559 (2020)
7. Peng, J., Xi-Yan, Zhang, L.: Two inverse eigenvalue problems for a special kind of matrices **416**, 347–369 (2006)
8. Pickmann, H., Soto, R., Egaña, J., Salas, M.: An inverse eigenvalue problem for symmetrical tridiagonal matrices. Comput. Math. Appl. **54**, 699–708 (2007)

9. Pickmann, H., Egaña, J., Soto, R.: Extremal inverse eigenvalue problem for bordered diagonal matrices. Linear Algebra Appl. **427**, 256–271 (2007)
10. Pickmann, H., Egaña, J., Soto, R.: Extreme spectra realization by real symmetric tridiagonal and real symmetric arrow matrices. Electron. J. Linear Algebra **22**, 52 (2011)
11. Pickmann-Soto, H., Arela-Pérez, S., Nina, H., Valero, E.: Inverse maximal eigenvalues problems for Leslie and doubly Leslie matrices. Linear Algebra Appl. **592**, 93–112 (2020)
12. Rodin, E.Y.: A matrix model of population growth. Moth/Comput. Model. **10**(4), 299–306 (1988)
13. Wanicharpichat, W.: Explicit minimum polynomial, eigenvector and inverse formula of doubly Leslie matrix. J. Appl. Math. Inf. **33**, 247–260 (2015)

Statistical Approach to the Gompertz Growth Model and the Underlying Timescales

A. Samoletov and B. Vasiev

Abstract The Gompertz growth model is one of the most established paradigms in biology. It is widely used to describe the growth of bacteria and tumour cell populations, as well as plants and animals. Here, a statistical approach is used to model biological growth. This approach involves the probabilistic estimation of population–environment interactions. By applying this theory, we derive stochastic equations, of which the Gompertz growth model is a limiting case. This is confirmed by numerical simulations. We examine the dynamic modes in dependence on the timescale relation. By altering the invariant measure, we suggest modified forms of the stochastic Gompertz growth model.

Keywords Population dynamics · Statistical approach · the Gompertz growth model · Population- environment interaction

1 Introduction

Dynamics of a single species population is a prototype for mathematical modelling in ecology and is commonly described by the following differential equation,

$$\dot{x} = xf(x), \tag{1}$$

where x is the population density and $f(x)$ describes the per capita growth. The observed upper limit for population growth determines the level of saturation known as carrying capacity. The modelling of population dynamics using Eq. (1) is based on a special choice of mathematical expressions for the laws governing population growth, that is, the function $f(x)$, so that the environment is treated as a static reservoir. The simplest and most common form of the associated dynamics, the logistic

A. Samoletov · B. Vasiev
Department of Mathematical Sciences, University of Liverpool, Liverpool L69 7ZL, UK

A. Samoletov (✉)
Don Institute for Physics and Engineering, 283050 Donetsk, Ukraine
e-mail: A.Samoletov@liverpool.ac.uk

D. Vlachos (ed.), *Mathematical Modeling in Physical Sciences*, Springer Proceedings in Mathematics & Statistics 446, https://doi.org/10.1007/978-3-031-52965-8_5

equation [1, 2], is often and successfully used to model population growth. The logistic equation inherits the requested generic properties of the Eq. (1). To substantiate a choice of function $f(x)$, data of experimental observations and heuristic reasoning are usually used. Alternatively, we have recently suggested a statistical interpretation of a population growth curve in [3]. This statistical interpretation was thoroughly examined.

This article focuses on the statistical interpretation of the Gompertz growth model [2, 4, 5], which is the second most widely used growth model. The Gompertz model is often and successfully used to describe the growth of plants and animals, bacteria and cancer cells, and other processes (e.g., [6, 7]). We demonstrate that this model can be treated statistically in a similar way to the logistic model.

To mathematically formulate a statistical approach to population growth dynamics, it is important to recognise that natural populations do not exist and grow in isolation, but are in dynamic contact with the environment. Then, in order to gain a better grasp of population growth, dynamic models must account for population–environment interactions, environmental responses to the growing population, and internal population and environmental processes that may be only partially known, introducing a degree of uncertainty into population dynamics. Moreover, any population residing in the environment will perturb it to some extent, and will itself be subject to the feedback effects of that perturbation. Depending on the timescale of population growth, the environment can be divided into two parts: the part involved in joint dynamics with the population, and the unperturbed part, which determines the general statistical properties of population–environment coevolution and the long-term homeostasis of the population.

A precise overview of the mathematical assumptions and methods needed for the statistical interpretation of ecological growth models is outlined in [3], which combines the dynamic principle for ensemble control tools, the definition of ecological temperature, the corresponding invariant probability density, and the ergodic hypothesis [8–11]. We will not repeat the details here. Instead, we will utilise the definitions and notations present in [3] for the subsequent discussion.

2 Statistical Interpretation of the Gompertz Growth Model

2.1 *Preliminaries*

Consider a single species homogeneous population with density $x \in \mathbb{R}_{+}$, placed in an environment such that a state of equilibrium exists. This is not a static equilibrium. Population growth depends on available resources and other environmental conditions, as well as the size of the population itself, to ensure a dynamic statistical equilibrium. These processes, which involve the environment in coevolution, must have balanced dynamics where the fluctuation and relaxation processes must be appropriately specified. To formulate a statistical approach to population dynamics, we will assume that the state of equilibrium and the corresponding ecological temperature ϑ [3] are determined by the ecosystem as a whole, of which the population is a small part.

According to the theoretical scheme presented in [3], we must first establish the invariant probability density for the combined system of the population and the environment supposing that they are statistically independent in equilibrium.

The Environment Let us suppose that the environment involved in a joint motion with the population is characterised by a variable $y \in \mathbb{R}$ with a probability density $\sigma_\vartheta^*(y)$. By necessity, y is a collective variable that incorporates a number of environmental factors and processes. Thus, it is reasonable to guess the Gaussian statistics, that is,

$$\sigma_\vartheta^*(y) \propto \exp\left\{-\vartheta^{-1}\frac{1}{2}y^2\right\}, \tag{2}$$

where y is a dimensionless variable. The ecological temperature ϑ defines the intensity of the environmental fluctuations. The Gaussian statistics is completely characterised by the first two cumulants: $\mathbb{E}_\vartheta(y) = 0$, $\mathbb{E}_\vartheta\left\{[y - \mathbb{E}_\vartheta\{y\}]^2\right\} = \mathbb{E}_\vartheta\left\{y^2\right\} = \vartheta$. Therefore, a linear combination of y and $\left(y^2 - \vartheta\right)$ is a ϑ–expression as defined in [3].

Following the paper [3] we set the temperature expression associated with the environment in the form,

$$\Theta_G^*(y, \vartheta) = \lambda\vartheta y - \gamma\left(y^2 - \vartheta\right). \tag{3}$$

In the case of non-Gaussian statistics, we accept the probability density,

$$\sigma_\vartheta^*(y) \propto \exp\left\{-\vartheta^{-1}h^*(y)\right\},$$

where $h^*(y)$ is a suitable function, that is, $h^*(y)$ is a sufficiently smooth function, bounded from below and growing at infinity, $h^*(y) \geq a\,|y|^b$ for some $a > 0,\ b > 0$, that is, a coercive function. Then the ecological temperature expression takes the form (see paper [3] and references therein),

$$\Theta_{NG}^*(y, \vartheta) = \lambda\vartheta h^{*\prime}(y) - \gamma\left[\left(h^{*\prime}(y)\right)^2 - \vartheta h^{*\prime\prime}(y)\right].$$

If $h^*(y) = y^2/2$, then arrive at the Gaussian case (3).

The Population Let us accept the invariant probability density $\sigma_\vartheta(x)$ associated with the population in the form,

$$\sigma_\vartheta(x) \propto \exp\left\{-\vartheta^{-1}h(x)\right\},$$

where the coercive function $h(x)$ is a subject to define. The ϑ parameter is the same as in $\sigma_\vartheta^*(y)$. The simplest ϑ-expression associated with the population, that is, a first-order polynomial in ϑ with ϑ as the additive term [3], has the form,

$$\Theta(x, \vartheta) = xh'(x) - \vartheta, \tag{4}$$

where prime denotes derivative (Lagrange notation). The proof is straightforward,

$$\mathbb{E}_{\vartheta}\left\{xh'(x)\right\} \equiv \left(\int \exp\left\{-\vartheta^{-1}h(x)\right\}dx\right)^{-1} \int xh'(x)\exp\left\{-\vartheta^{-1}h(x)\right\}dx = \vartheta.$$

In this article, we will focus on this particular expression (4), leaving aside other possible θ–expressions (see [8, 9]).

We set the ecological temperature value by the θ–expression (3) associated with the environment, which seems quite reasonable. A θ–expression associated with the population is then generated during the solution of the dynamic principle functional equation [3]. In fact, in our case, it is expression (4).

2.2 *Gompertz Growth Model*

To proceed to the statistical interpretation of the Gompertz growth model, it is necessary to solve the functional equation of the dynamic principle as described in [3]. Note that if $\gamma \neq 0$, there are only stochastic equations among the solutions of the dynamic principle equation [3, 8]. The procedure to solve the dynamic principle equation has similarities to finding the first integrals for a given dynamical system. Discovering these integrals is a challenging task, but confirming whether a function meets the requirement for a first integral is easier. To derive a particular solution to the functional equation of the dynamic principle, we use the method previously discussed in [3, 8]. Here, we do not provide the corresponding calculations, but rather the justification that the particular equations of motion satisfy all the necessary conditions.

Let us consider the equations of motion:

$$\begin{aligned}\dot{x} &= \lambda xy,\\ \dot{y} &= -\lambda\left[xh'(x) - \vartheta\right] - \gamma y + \sqrt{2\gamma\vartheta}\,\xi(t),\end{aligned} \tag{5}$$

where $\xi(t)$ is the standard Gaussian white noise, $\mathbb{E}_{\omega}\{\xi(t)\} = 0$, $\mathbb{E}_{\omega}\left\{\xi(t)\xi(t')\right\} = \delta(t - t')$, $\lambda > 0$ and $\gamma > 0$ are parameters that actually define two time scales. Here $\mathbb{E}_{\omega}\{\cdots\}$ means the mathematical expectation over all realisation of the generalised random process $\xi(t)$. Or, in the case of a non-Gaussian statistics for the environmental variable y,

$$\begin{aligned}\dot{x} &= \lambda xh^{*\prime}(y),\\ \dot{y} &= -\lambda\left[xh'(x) - \vartheta\right] - \gamma h^{*\prime}(y) + \sqrt{2\gamma\vartheta}\,\xi(t).\end{aligned} \tag{6}$$

In this case the ecological temperature expression has the form $\Theta^{*}_{NG}(y, \vartheta) = \lambda\vartheta h^{*\prime}(y) - \gamma\left[\left(h^{*\prime}(y)\right)^{2} - \vartheta h^{*\prime\prime}(y)\right]$.

First we check the ϑ–expression for the environment, $\Theta^*_G(y,\vartheta)$, by direct calculation. That is, we calculate that

$$\begin{aligned}\Theta^*_G(y,\vartheta) &= \mathbb{E}_\omega\left\{h'(x)\,\lambda xy + y\left(-\lambda\left[xh'(x)-\vartheta\right]-\gamma y+\sqrt{2\gamma\vartheta}\xi(t)\right)\right\}\\ &= \lambda\vartheta y + \mathbb{E}_\omega\left\{-\gamma y^2+\sqrt{2\gamma\vartheta}y\xi(t)\right\}\\ &= \lambda\vartheta y - \gamma\left[y^2-\vartheta\right],\end{aligned}$$

where Novikov's formula [12] was used to obtain the final formula.

In the same way,

$$\begin{aligned}&\Theta^*_{NG}(y,\vartheta)\\ &= \mathbb{E}_\omega\left\{h'(x)\,\lambda xh^{*\prime}(y) + h^*(y)\left(-\lambda\left[xh'(x)-\vartheta\right]-\gamma h^{*\prime}(y)+\sqrt{2\gamma\vartheta}\xi(t)\right)\right\}\\ &= \lambda\vartheta h^{*\prime}(y) + \mathbb{E}_\omega\left\{-\gamma\left(h^{*\prime}(y)\right)^2+\sqrt{2\gamma\vartheta}h^{*\prime}(y)\xi(t)\right\}\\ &= \lambda\vartheta h^{*\prime}(y) - \gamma\left[\left(h^{*\prime}(y)\right)^2-\vartheta h^{*\prime\prime}(y)\right],\end{aligned}$$

where Novikov's formula [12] was once again used to obtain the final formula.

Then, we prove that the probability density,

$$\sigma^+_\vartheta(x,y) \propto \exp\left\{-\vartheta^{-1}h(x)\right\}\times\exp\left\{-\vartheta^{-1}\frac{1}{2}y^2\right\}, \tag{7}$$

is invariant for dynamics (5). Indeed, the Fokker-Planck equation corresponding to stochastic differential Eq. (5) (e.g., [12, 13]) has the form,

$$\frac{\partial}{\partial t}\sigma = \mathcal{F}^*\sigma, \tag{8}$$

where

$$\mathcal{F}^*\sigma = -\frac{\partial}{\partial x}\left[\lambda xy\sigma\right] - \frac{\partial}{\partial y}\left\{\left[-\lambda\left[xh'(x)-\vartheta\right]-\gamma y-\gamma\vartheta\frac{\partial}{\partial y}\right]\sigma\right\}, \tag{9}$$

is the Fokker-Planck operator. The practical way to write down the Fokker-Planck Eq. (8) and the corresponding Fokker-Planck operator (9) is to use the approach described in the book [12].

We prove the identity $\mathcal{F}^*\sigma^+_\vartheta(x,y)\equiv 0$ by direct calculation, implying that $\sigma^+_\vartheta(x,y)$ (7) is the invariant density for stochastic dynamics (5). Explicitly,

$$
\begin{aligned}
&-\frac{\partial}{\partial x}\left[\lambda xy\sigma_{\vartheta}^{+}(x,y)\right]-\frac{\partial}{\partial y}\left\{\left[-\lambda\left[xh'(x)-\vartheta\right]-\gamma y-\gamma\vartheta\frac{\partial}{\partial y}\right]\sigma_{\vartheta}^{+}(x,y)\right\}\\
&\qquad=-\frac{\partial}{\partial x}\left[\lambda xy\sigma_{\vartheta}^{+}(x,y)\right]-\frac{\partial}{\partial y}\left\{-\lambda\left[xh'(x)-\vartheta\right]\sigma_{\vartheta}^{+}(x,y)\right\}\\
&\qquad\qquad+\gamma\frac{\partial}{\partial y}\left\{y\sigma_{\vartheta}^{+}(x,y)+\vartheta\frac{\partial}{\partial y}\sigma_{\vartheta}^{+}(x,y)\right\}\\
&\qquad=\left\{-\lambda y+\lambda xyh'(x)\frac{1}{\vartheta}-\lambda\left[xh'(x)-\vartheta\right]y\frac{1}{\vartheta}\right\}\sigma_{\vartheta}^{+}(x,y)\\
&\qquad\qquad+\gamma\frac{\partial}{\partial y}\left\{\left[y-y\right]\sigma_{\vartheta}^{+}(x,y)\right\}\equiv 0.
\end{aligned}
$$

In the same way we prove that the probability density,

$$
\sigma_{\vartheta}^{+}(x,y)\propto\exp\left\{-\vartheta^{-1}h(x)\right\}\times\exp\left\{-\vartheta^{-1}h^{*}(y)\right\},
$$

is invariant for dynamics (6). One would expect that for the stochastic evolution Eqs. (5) and (6) the dynamics would be ergodic.

We have derived relatively simple equations of population–environment coevolution, that is, Eqs. (5) and (6). Now we can inquire how these stochastic dynamic equations relate to the conventional deterministic population growth Eq. (1), in particular the Gompertz growth model, that is, Eq. (14). To answer this question, consider the limiting case, $\gamma \gg 1$ (relaxation processes are extremely fast) and $\vartheta \to +0$ (the environment is in static equilibrium). In this procedure we follow an analogy with the Kramers problem [14, 15]. As a result, we arrive at deterministic dynamics involving only the population density variable. It is worth noting, however, that following the mathematical scheme of paper [15] we obtain not only the expressions given here, but also corrections to them of the next order of magnitude, which allow us to formulate mathematically the conditions of the limit transition, including the conditions on the smoothness of the function $h(x)$ [15]. In the context of our problem, there appears to be no need for special attention to be paid to such mathematical details [3]. For more mathematical details, please refer to [15] and [16], and the literature cited in them.

Thus, passing to the limit $\gamma \gg 1$, $\dot{y}=0$, and $\vartheta \to +0$, we get $y=-\gamma^{-1}\lambda xh'(x)$, and hence the correspondence

$$
xf(x)=-\frac{\lambda^{2}}{\gamma}x^{2}h'(x).
$$

For the Gompertz growth model we have to set the h–function in the form,

$$
h(x)=\frac{1}{2K}\left(\ln\frac{x}{K}\right)^{2}. \tag{10}
$$

Thus, the population growth rate takes the form

$$f(x) = -\frac{\lambda^2}{\gamma} \ln\left(\frac{x}{K}\right).$$

By setting

$$\lambda^2 = r\gamma,$$

we obtain the deterministic Gompertz growth model,

$$\dot{x} = -rx \ln\left(\frac{x}{K}\right). \tag{11}$$

Let us now substitute the Gompertz h–function (10) into the Eqs. (5) and (13). This provides a stochastic analogue of the Gompertz growth model, which to some extent describes the coevolution of population and environment. Explicitly, we obtain the following equations of motion:

$$\begin{aligned} \dot{x} &= \lambda xy, \\ \dot{y} &= -\lambda\left[\ln\left(\frac{x}{K}\right) - \vartheta\right] - \gamma y + \sqrt{2\gamma\vartheta}\xi(t), \end{aligned} \tag{12}$$

and

$$\begin{aligned} \dot{x} &= \lambda x h^{*\prime}(y), \\ \dot{y} &= -\lambda\left[\ln\left(\frac{x}{K}\right) - \vartheta\right] - \gamma h^{*\prime}(y) + \sqrt{2\gamma\vartheta}\xi(t), \end{aligned} \tag{13}$$

if the environment variable y does not follow a Gaussian distribution. Note that if we fix the parameter r in the deterministic Gompertz growth model (11), the two timescale parameters in the joint population–environment coevolution equations, λ and γ, become related, $\lambda^2 = r\gamma$.

To illustrate the distinction between the dynamics outlined by (11) and (12), we carried out a series of numerical simulations. These simulations also allowed us to test the validity of our basic assumptions.

We simulated the dynamic Eq. (12) using the Euler scheme with a time step $dt = 0.001$. In all the simulations shown in Figs. 1, 2, 3 and 4, we keep $K = 1000$, initial values: $x = 10$, $y = 0$, and the fixed relation between the parameters, $\lambda^2 = r\gamma$, to remain consistent with the Gompertz growth model (11), but vary the ecological temperature ϑ and the timescale parameters λ and γ.

The results of the simulations confirm our assumptions and expectations. Fig. 1 shows that the concept of a statistical approach to interpreting the Gompertz growth model is reasonable. With the parameter $\gamma = 50 \gg 1$, the agreement with the Gompertz growth curve, as described by Eq. (11), improves significantly as the ecological temperature decreases. Note that this good agreement corresponds to short relaxation times of the environmental variable, $\gamma^{-1} = 0.02 \ll 1$, which is in line with our theoretical assumptions.

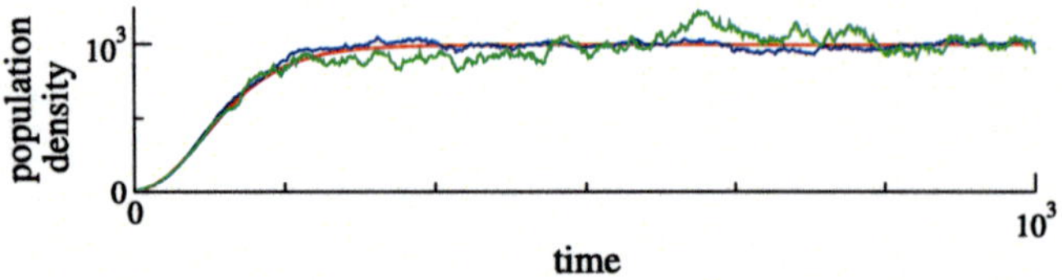

Fig. 1 (colour online) Population density versus time. The various curves correspond to the Gompertz growth model (11) (red curve) and a stochastic counterpart of the Gompertz growth model (12) at two values of the ϑ parameters: $\vartheta = 0.001$ (blue curve) and $\vartheta = 0.005$ (green curve). Other parameters: $r = 0.02$, $\lambda = 1$, $\gamma = 50$ so λ^2 equals $r\gamma$

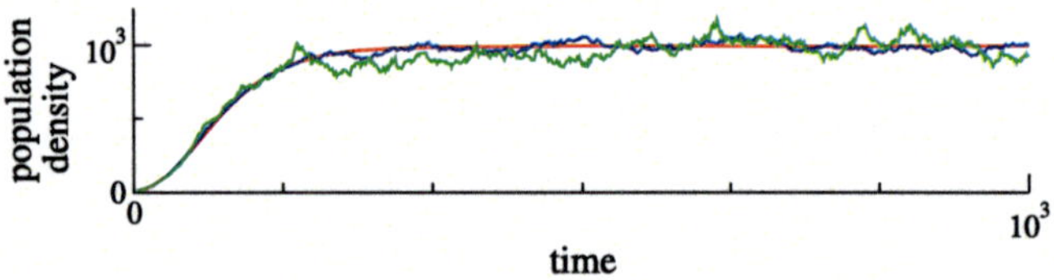

Fig. 2 (colour online) Population density versus time. The various curves correspond to the Gompertz growth model (11) (red curve) and a stochastic counterpart of the Gompertz growth model (12) at two values of the ϑ parameters: $\vartheta = 0.001$ (blue curve) and $\vartheta = 0.005$ (green curve). Other parameters: $r = 0.02$, $\lambda = 2$, $\gamma = 200$ so λ^2 equals $r\gamma$

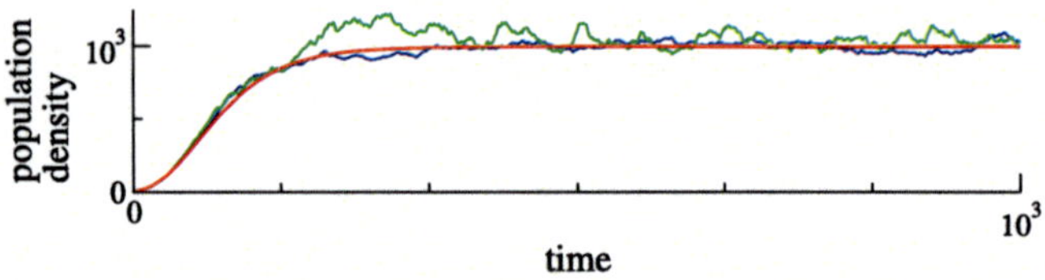

Fig. 3 (colour online) Population density versus time. The various curves correspond to the Gompertz growth model (11) (red curve) and a stochastic counterpart of the Gompertz growth model (12) at two values of the ϑ parameters: $\vartheta = 0.001$ (blue curve) and $\vartheta = 0.005$ (green curve). Other parameters: $r = 0.02$, $\lambda = \sqrt{0.02}$, $\gamma = 1$ so λ^2 equals $r\gamma$

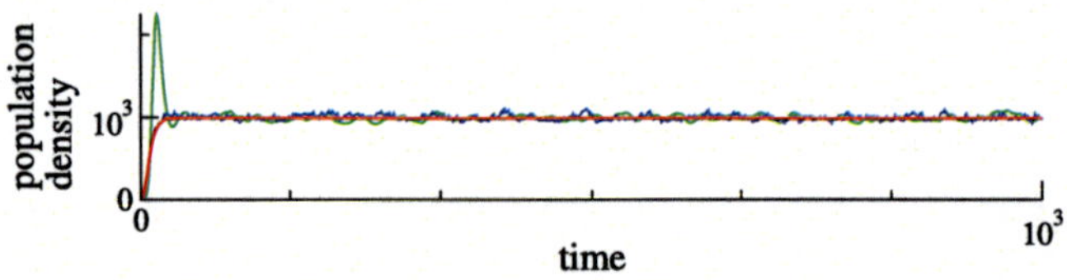

Fig. 4 (colour online) Population density versus time. The various curves correspond to the Gompertz growth model (11) (red curve) and a stochastic counterpart of the Gompertz growth model (12) at two values of the parameters: $\lambda = 1$, $\gamma = 5$ (blue curve) and $\lambda = 0.2$, $\gamma = 0.2$ (green curve). Other parameters: $r = 0.2$ so λ^2 equals $r\gamma$, $\vartheta = 0.001$

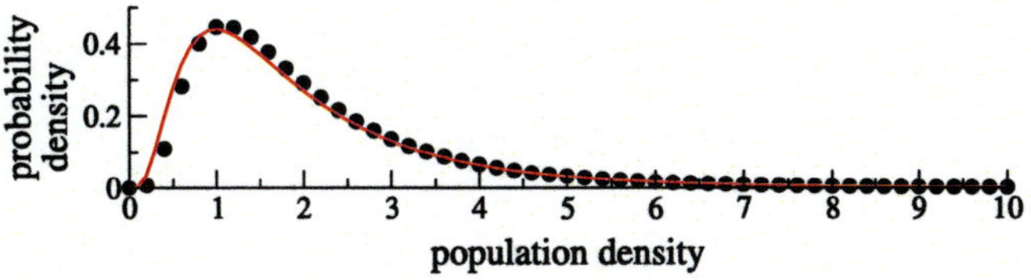

Fig. 5 (colour online) Population probability density versus population density. The (black) circles show the normalised histogram generated by the Gompertz growth model (12) at $\vartheta = 0.5$, $K = 1$, $\lambda = 1$ and $\gamma = 50$ running numerical simulation for quite a long time, up to $t_{max} = 10^6$. The (red) line shows the theoretical curve given by the Gompertz h–function at $\vartheta = 0.5$ and $K = 1$, that is, $\sigma = [\sqrt{\pi}\exp(0.25)]^{-1}\exp\{-(\ln x)^2\}$. Observe, the circles follow the theoretical curve

The changes in the growth curve where the time scales γ^{-1} and λ^{-1} were varied, keeping the parameter $r = 0.02$ and the relation $\lambda^2 = r\gamma$, are shown in Figs. 2 and 3. It is noticeable that the relaxation curve of the population density to its most probable value does not change significantly, despite significant changes in the parameters λ and γ. The reason for this is that the parameter r in the relation $\lambda^2 = r\gamma$ is kept constant at $r = 0.02$. In such circumstances, when transitioning to dimensionless time in equations of motion (12), the dimensionless value $\gamma/\lambda = \lambda/r$ remains quite large.

Therefore, the parameter r assigned to the Eq. (11) of the deterministic Gompertz growth model is of importance if the additional timescale parameters assigned to the stochastic Gompertz growth equations (12) satisfy the condition $\lambda^2 = r\gamma$. This is tested by using a higher value for the parameter r. In fact, let the parameter r be ten times larger, $r = 0.2$. Then we have simulated the Eq. (12) for two different sets of parameters λ and γ, i.e. $\lambda = 1; 0.2$ and $\gamma = 5; 0.2$, so that λ^2 is still equal to $r\gamma$. The results are shown in Fig. 4. It is clear to see: There are two very different evolution curves. The reason for this is that the dimensionless parameter $\gamma/\lambda = \lambda/r$ takes two qualitatively different values: 5 and 1. The latter value certainly does not satisfy the condition of consistency with the Gompertz growth model (11), i.e. $\gamma/\lambda = \lambda/r \gg 1$.

Finally, we assessed the validity of the ergodic hypothesis through conducting a simulation for an extended duration. In fact, the validity of the ergodic hypothesis is confirmed by Fig. 5. It is evident that the population density distribution obtained numerically conforms to the theoretical curve.

3 Concluding Comment

Finally, we should make a note about the random term in the dynamic systems (12) and (13). If we put $\gamma \equiv 0$, we obtain systems of ordinary differential equations,

$$\dot{x} = \lambda x y,$$
$$\dot{y} = -\lambda\left[\ln\left(\frac{x}{K}\right) - \vartheta\right], \tag{14}$$

and

$$\dot{x} = \lambda x h^{*\prime}(y),$$
$$\dot{y} = -\lambda \left[\ln \left(\frac{x}{K} \right) - \vartheta \right], \tag{15}$$

for which the appropriate probability density, $\sigma_{\vartheta}^{+}(x, y)$, is invariant. This can be verified by substituting the density $\sigma_{\vartheta}^{+}(x, y)$ into the corresponding Liouville equation, which gives the required result by direct calculation,

$$-\frac{\partial}{\partial x}\left[\lambda x y \exp\left\{-\frac{1}{\vartheta}\frac{1}{2K}\left(\ln\frac{x}{K}\right)^2 - \frac{1}{\vartheta}\frac{1}{2}y^2\right\}\right]$$
$$-\frac{\partial}{\partial y}\left[-\lambda\left[\ln\left(\frac{x}{K}\right)-\vartheta\right]\exp\left\{-\frac{1}{\vartheta}\frac{1}{2K}\left(\ln\frac{x}{K}\right)^2 - \frac{1}{\vartheta}\frac{1}{2}y^2\right\}\right] \equiv 0$$

However, the dynamics is not ergodic for the given probability density, since equations of motion (14) and (15) have the following integrals of motion, correspondingly,

$$I = \frac{1}{2K}\left(\ln\frac{x}{K}\right)^2 - \vartheta \ln x + \frac{1}{2}y^2 = const,$$

and

$$I = \frac{1}{2K}\left(\ln\frac{x}{K}\right)^2 - \vartheta \ln x + h^{*\prime}(y) = const.$$

The proof is obtained by direct calculation.

Let us calculate the ϑ–expressions corresponding to the equations of motion (14) and (15):

$$\Theta = \frac{1}{x}\ln\left(\frac{x}{K}\right)\lambda x y + y\left(-\lambda\left[\ln\left(\frac{x}{K}\right)-\vartheta\right]\right) = \lambda\vartheta y,$$

and

$$\Theta = \frac{1}{x}\ln\left(\frac{x}{K}\right)\lambda x \lambda x h^{*\prime}(y) + \lambda x h^{*\prime}(y)\left(-\lambda\left[\ln\left(\frac{x}{K}\right)-\vartheta\right]\right) = \lambda\vartheta\lambda x h^{*\prime}(y).$$

As previously stated, this form of ϑ–expression while acceptable, raises the question of ergodicity [3, 8]. It should be noted that these ϑ–expressions do not enable us to represent the ecological temperature ϑ in terms of average functions of dynamic variables. It seems crucial as it pertains to the operational meaning of ecological temperature.

In our specific case, the proof of the non-ergodicity of the dynamical equations (14) and (15) is obtained by identifying the corresponding integral of motion.

Acknowledgements This work has been supported by the EPSRC grant EP/S033211/1.

References

1. Kingsland, S.: The refractory model: The logistic curve and the history of population ecology. Q. Rev. Biol. **57**(1), 29–52 (1982)
2. Kot, M.: Elements of Mathematical Ecology. Cambridge University Press (2001)
3. Samoletov, A., Vasiev, B.: A statistical interpretation of biologically inspired growth models. arXiv:2302.08218 (2023)
4. Winsor, C.P.: The Gompertz curve as a growth curve. Proc. Natl. Acad. Sci. U.S.A. **18**(1), 1 (1932)
5. Tjørve, K.M., Tjørve, E.: The use of Gompertz models in growth analyses, and new Gompertz-model approach: An addition to the unified-richards family. PloS one **12**(6), e0178691 (2017)
6. Enderling, H., Chaplain, M.A.J.: Mathematical modeling of tumor growth and treatment. Curr. Pharm. Des. **20**(30), 4934–4940 (2014)
7. Avraam, D., Arnold, S., Vasieva, O., Vasiev, B.: On the heterogeneity of human populations as reflected by mortality dynamics. Aging (Albany NY) **8**(11), 3045 (2016)
8. Samoletov, A., Vasiev, B.: Dynamic principle for ensemble control tools. J. Chem. Phys. **147**, 204106 (2017)
9. Samoletov, A., Vasiev, B.: Advanced selection of ensemble control tools. J. Phys. Conf. Ser. **2090**, 012059 (2021)
10. Samoletov, A., Dettmann, C., Chaplain, M.: Thermostats for "slow" configurational modes. J. Stat. Phys. **128**(6), 1321–1336 (2007)
11. Samoletov, A., Dettmann, C., Chaplain, M.: Notes on configurational thermostat schemes. J. Chem. Phys. **132**, 246101 (2010)
12. Klyatskin, V.I.: Dynamics of Stochastic Systems. Elsevier (2005)
13. Gardiner, C.W.: Stochastic Methods: A Handbook for the Natural and Social Sciences. Springer, Berlin (2009)
14. Kramers, H.A.: Brownian motion in a field of force and the diffusion model of chemical reactions. Physica **7**(4), 284–304 (1940)
15. Samoletov, A.A.: A remark on the Kramers problem. J. Stat. Phys. **96**, 1351–1357 (1999)
16. Mel'nikov, V.I.: The Kramers problem: Fifty years of development. Phys. Rep. **209**(1–2), 1–71 (1991)

On 4-dimensional Hom-Lie Algebras

María Alejandra Alvarez and Sonia Vanesa Vera

Abstract In this paper we obtain the classification of complex multiplicative non-Lie Hom-Lie algebras of dimension 4 associated to a particular linear map. This provides the first isomorphism classes of this type of algebras in dimension 4.

Keywords Hom-Lie algebras · Classification · 4-dimensional

1 Introduction and Preliminaries

Hom-Lie algebras appeared in [4] in the study of deformations of the Witt and Virasoro algebras using σ-derivations. Hom-Lie algebras are a subclass of skew-symmetric algebras and a generalization of Lie algebras. Many authors have investigated different kinds of Hom-structures and applications.

A **Hom-Lie algebra** is a triple $\mathfrak{g} = (V, [\cdot,\cdot], \alpha)$ consisting of a vector space V over a field $\mathbb{K}$, a bilinear map $[\cdot,\cdot]\colon V \times V \to V$ and a linear map $\alpha\colon V \to V$ satisfying, for all $x, y, z \in V$:

$$\begin{aligned} [x, y] &= -[y, x] \qquad &\text{(skew-symmetry)} \\ \circlearrowleft_{x,y,z} [\alpha(x), [y, z]] &= 0 \qquad &\text{(Hom-Jacobi identity)} \end{aligned} \tag{1}$$

where $\circlearrowleft_{x,y,z}$ denotes the cyclic summation over x, y, z.

A Hom-Lie algebra is **multiplicative** if α is an algebraic map, i.e.,

$$\alpha([x, y]) = [\alpha(x), \alpha(y)]. \tag{2}$$

M. A. Alvarez (✉)
Departamento de Matemáticas, Facultad de Ciencias Básicas, Universidad de Antofagasta, Antofagasta, Chile
e-mail: maria.alvarez@uantof.cl

S. V. Vera
CIEM–FAMAF and FCEFyN, Universidad Nacional de Córdoba, Córdoba, Argentina

D. Vlachos (ed.), *Mathematical Modeling in Physical Sciences*, Springer Proceedings in Mathematics & Statistics 446, https://doi.org/10.1007/978-3-031-52965-8_6

Lie algebras are a special case of multiplicative Hom-Lie algebras where α is the identity map.

The obtaining of the algebraic classification of Hom-Lie algebras in low dimensions is still in early stages. The first classification of complex multiplicative non-Lie Hom-Lie algebras of dimension 3 appeared in [5]. Later, in [2, 3] the authors obtain the full classification of complex Hom-Lie algebras of dimension 3. The geometric classification of Hom-Lie algebras began in [8] and [9], where the authors proved, among other results, that one can associate a Hom-Lie structure to every 3-dimensional skew-symmetric algebra but this is not true in dimension 4. Next, in [1], rigid complex 3-dimensional Hom-Lie algebras are obtained.

The first step towards obtaining the algebraic classification of complex 4-dimensional Hom-Lie algebras is given in [7], where Hom-Lie structures on 4-dimensional Lie algebras are characterized. Later, in [6], the authors considered 3 and 4-dimensional algebras and describe the space of possible linear maps, with minimum dimension, which make these algebras Hom-Lie algebras. In particular, they characterize the products that ensure the obtaining of a 4-dimensional Hom-Lie algebra arising from the nilpotent map $\alpha = \begin{pmatrix} 0&1&0&0 \\ 0&0&0&0 \\ 0&0&0&0 \\ 0&0&0&0 \end{pmatrix}$. Since the obtaining of Hom-Lie algebras of dimension 4 involve difficult computations, explicit isomorphism classes have not been obtained so far.

In this work we contribute in this direction by presenting the classification of complex multiplicative non-Lie Hom-Lie algebras of dimension 4 associated to the linear map $\alpha = \begin{pmatrix} \lambda&1&0&0 \\ 0&\lambda&1&0 \\ 0&0&\lambda&1 \\ 0&0&0&\lambda \end{pmatrix}$. In particular, we obtain infinitely many isomorphism classes described in terms of 3 one-parameter families.

2 On the Classification of Hom-Lie Algebras

There have been two different approaches to obtain algebraic classifications. On the one hand, by first classifying the skew-symmetric products and then obtaining the linear maps; this has been done, for instance, in [2, 3, 7]. On the other hand, by first classifying the linear maps and then obtaining the skew-symmetric products (see, for instance, [1, 6]).

Two Hom-Lie algebras $(V, [\cdot,\cdot]_1, \alpha)$ and $(V, [\cdot,\cdot]_2, \beta)$ are said to be **isomorphic** if there exists a bijective algebra morphism $\phi : (V, [\cdot,\cdot]_1) \to (V, [\cdot,\cdot]_2)$ compatible with α and β, i.e.,

$$\phi([x, y]_1) = [\phi(x), \phi(y)]_2 \quad \text{and} \quad \phi \circ \alpha = \beta \circ \phi.$$

Remark 1 In particular, this implies that every Hom-Lie algebra $(V, [\cdot,\cdot], \alpha)$ is isomorphic to a Hom-Lie algebra $(V, \phi([\phi^{-1}(\cdot), \phi^{-1}(\cdot)]), \phi \circ \alpha \circ \phi^{-1})$, where $\phi \colon V \to V$ is a bijective map.

By Remark 1, it is enough to consider linear maps in their Jordan canonical form. In particular, we are interested in the linear map $\alpha = \begin{pmatrix} \lambda & 1 & 0 & 0 \\ 0 & \lambda & 1 & 0 \\ 0 & 0 & \lambda & 1 \\ 0 & 0 & 0 & \lambda \end{pmatrix}$. Then we obtain the following result:

Theorem 1 *Let* $\mathfrak{g} = (V, [\cdot,\cdot], \alpha)$ *be a multiplicative Hom-Lie algebra with* $\alpha = \begin{pmatrix} \lambda & 1 & 0 & 0 \\ 0 & \lambda & 1 & 0 \\ 0 & 0 & \lambda & 1 \\ 0 & 0 & 0 & \lambda \end{pmatrix}$. *If* $\lambda \neq 0$ *then* $(V, [\cdot,\cdot])$ *is a Lie algebra.*

Proof Consider a basis of V, $\{e_1, e_2, e_3, e_4\}$. Denote by

$$[\cdot,\cdot] = \begin{pmatrix} c^1_{1,2} & c^1_{1,3} & c^1_{1,4} & c^1_{2,3} & c^1_{2,4} & c^1_{3,4} \\ c^2_{1,2} & c^2_{1,3} & c^2_{1,4} & c^2_{2,3} & c^2_{2,4} & c^2_{3,4} \\ c^3_{1,2} & c^3_{1,3} & c^3_{1,4} & c^3_{2,3} & c^3_{2,4} & c^3_{3,4} \\ c^4_{1,2} & c^4_{1,3} & c^4_{1,4} & c^4_{2,3} & c^4_{2,4} & c^4_{3,4} \end{pmatrix}$$

a skew-symmetric product, i.e. $[e_i, e_j] = \sum_{k=1}^4 c^k_{i,j} e_k$.

Since $\mathfrak{g}$ is a multiplicative Hom-Lie algebra, Eq. (2) gives us:

$$0 = c^k_{i-1,j-1} + \lambda(c^k_{i-1,j} + c^k_{i,j-1}) + c^k_{i,j}(\lambda^2 - \lambda - 1), \qquad k = 1, 2, 3, \tag{3}$$

$$0 = c^4_{i-1,j-1} + \lambda(c^4_{i-1,j} + c^4_{i,j-1}) + c^4_{i,j}(\lambda^2 - \lambda). \tag{4}$$

For $k = 1, 2, 3$, Eq. (3) gives us

$$\begin{aligned} 0 &= c^k_{1,2}(\lambda^2 - \lambda - 1), \\ 0 &= \lambda c^k_{1,2} + c^k_{1,3}(\lambda^2 - \lambda - 1), \\ 0 &= \lambda c^k_{1,3} + c^k_{1,4}(\lambda^2 - \lambda - 1), \\ 0 &= c^k_{1,2} + \lambda c^k_{1,3} + c^k_{2,3}(\lambda^2 - \lambda - 1), \\ 0 &= c^k_{1,3} + \lambda(c^k_{1,4} + c^k_{2,3}) + c^k_{2,4}(\lambda^2 - \lambda - 1), \\ 0 &= c^k_{2,3} + \lambda c^k_{2,4} + c^k_{3,4}(\lambda^2 - \lambda - 1), \end{aligned} \tag{5}$$

Equation (4) gives us

$$\begin{aligned} 0 &= c^4_{1,2}\lambda(\lambda - 1), \\ 0 &= \lambda c^4_{1,2} + c^4_{1,3}\lambda(\lambda - 1), \\ 0 &= \lambda c^4_{1,3} + c^4_{1,4}\lambda(\lambda - 1), \\ 0 &= c^4_{1,2} + \lambda c^4_{1,3} + c^4_{2,3}\lambda(\lambda - 1), \\ 0 &= c^4_{1,3} + \lambda(c^4_{1,4} + c^4_{2,3}) + c^4_{2,4}\lambda(\lambda - 1), \\ 0 &= c^4_{2,3} + \lambda c^4_{2,4} + c^4_{3,4}\lambda(\lambda - 1), \end{aligned} \tag{6}$$

Assume first that $\lambda \neq 0, 1, \frac{1\pm\sqrt{5}}{2}$. Then (5) and (6) imply $c^k_{i,j} = 0$ for $1 \leq i, j, k \leq 4$. Therefore $[\cdot,\cdot] = 0$ and $(V, [\cdot,\cdot])$ is a Lie algebra.

Suppose now $\lambda = 1$. Then (5) and (6) give

$$[\cdot,\cdot] = \begin{pmatrix} 0 & 0 & 0 & 0 & 0 & 0 \\ 0 & 0 & 0 & 0 & 0 & 0 \\ 0 & 0 & 0 & 0 & 0 & 0 \\ 0 & 0 & 0 & 0 & 0 & 0 \\ 0 & 0 & c_{1,4}^4 & -c_{1,4}^4 & c_{1,4}^4 & c_{3,4}^4 \end{pmatrix}.$$

By the Hom-Jacobi identity (1), this is a Hom-Lie product only if $c_{1,4}^4 = 0$, which also implies that $[\cdot,\cdot]$ is a Lie product.

Finally, suposse $\lambda = \frac{1\pm\sqrt{5}}{2}$. Then, by (5), (6) and (1), we obtain a Hom-Lie product

$$[\cdot,\cdot] = \begin{pmatrix} 0 & 0 & c_{1,4}^1 & -c_{1,4}^1 & \frac{c_{1,4}^1}{\lambda} & c_{3,4}^1 \\ 0 & 0 & 0 & 0 & 0 & c_{3,4}^2 \\ 0 & 0 & 0 & 0 & 0 & c_{3,4}^3 \\ 0 & 0 & 0 & 0 & 0 & 0 \end{pmatrix} \text{ with } c_{1,4}^1(c_{1,4}^1 - c_{3,4}^3) = 0.$$

If $c_{1,4}^1 = 0$ or $c_{3,4}^3 = c_{1,4}^1$, $[\cdot,\cdot]$ is also a Lie product. □

By the previous result, the possible complex multiplicative non-Lie Hom-Lie algebras of dimension 4 associated with $\alpha = \left(\begin{smallmatrix} \lambda & 1 & 0 & 0 \\ 0 & \lambda & 1 & 0 \\ 0 & 0 & \lambda & 1 \\ 0 & 0 & 0 & \lambda \end{smallmatrix}\right)$ arise by considering $\lambda = 0$. Therefore we obtain the main result:

Theorem 2 *The classification of complex multiplicative non-Lie Hom-Lie algebras* $\mathfrak{g} = (V, [\cdot,\cdot], \alpha)$ *with* $\alpha = \left(\begin{smallmatrix} \lambda & 1 & 0 & 0 \\ 0 & \lambda & 1 & 0 \\ 0 & 0 & \lambda & 1 \\ 0 & 0 & 0 & \lambda \end{smallmatrix}\right)$ *is as follows:*

– $\mathfrak{g}_1(\gamma) = \left(V, \begin{array}{l} [e_1, e_4] = \gamma e_1, \\ [e_2, e_3] = e_1, \\ [e_3, e_4] = e_2. \end{array}, \left(\begin{smallmatrix} 0 & 1 & 0 & 0 \\ 0 & 0 & 1 & 0 \\ 0 & 0 & 0 & 1 \\ 0 & 0 & 0 & 0 \end{smallmatrix}\right) \right)$, $\gamma \neq 0, 1, -1$.

– $\mathfrak{g}_2(\gamma) = \left(V, \begin{array}{l} [e_1, e_4] = e_1, \\ [e_2, e_3] = e_1, \\ [e_2, e_4] = \gamma e_1, \\ [e_3, e_4] = e_2. \end{array}, \left(\begin{smallmatrix} 0 & 1 & 0 & 0 \\ 0 & 0 & 1 & 0 \\ 0 & 0 & 0 & 1 \\ 0 & 0 & 0 & 0 \end{smallmatrix}\right) \right)$,

– $\mathfrak{g}_3(\gamma) = \left(V, \begin{array}{l} [e_1, e_4] = e_1, \\ [e_2, e_3] = -e_1, \\ [e_2, e_4] = \gamma e_1, \\ [e_3, e_4] = -e_2. \end{array}, \left(\begin{smallmatrix} 0 & 1 & 0 & 0 \\ 0 & 0 & 1 & 0 \\ 0 & 0 & 0 & 1 \\ 0 & 0 & 0 & 0 \end{smallmatrix}\right) \right)$.

Moreover, $\mathfrak{g}_k(\gamma) \simeq \mathfrak{g}_k(\mu)$ *if and only if* $\gamma = \mu$ *for* $k = 1, 3$, *and* $\mathfrak{g}_2(\gamma) \simeq \mathfrak{g}_2(\mu)$ *if and only if* $\gamma = \mu$ *or* $\gamma + \mu = 2$.

Proof By Theorem 1, we only need to consider the nilpotent map $\alpha = \left(\begin{smallmatrix} 0 & 1 & 0 & 0 \\ 0 & 0 & 1 & 0 \\ 0 & 0 & 0 & 1 \\ 0 & 0 & 0 & 0 \end{smallmatrix}\right)$. Denote by

$$[\cdot,\cdot] = \begin{pmatrix} c_{1,2}^1 & c_{1,3}^1 & c_{1,4}^1 & c_{2,3}^1 & c_{2,4}^1 & c_{3,4}^1 \\ c_{1,2}^2 & c_{1,3}^2 & c_{1,4}^2 & c_{2,3}^2 & c_{2,4}^2 & c_{3,4}^2 \\ c_{1,2}^3 & c_{1,3}^3 & c_{1,4}^3 & c_{2,3}^3 & c_{2,4}^3 & c_{3,4}^3 \\ c_{1,2}^4 & c_{1,3}^4 & c_{1,4}^4 & c_{2,3}^4 & c_{2,4}^4 & c_{3,4}^4 \end{pmatrix}$$

a skew-symmetric product. We consider the set of unknowns

$$U = \{c_{1,2}^1, c_{1,3}^1, c_{1,4}^1, c_{2,3}^1, c_{2,4}^1, c_{3,4}^1, c_{1,2}^2, c_{1,3}^2, c_{1,4}^2, c_{2,3}^2, c_{2,4}^2, c_{3,4}^2, \\ c_{1,2}^3, c_{1,3}^3, c_{1,4}^3, c_{2,3}^3, c_{2,4}^3, c_{3,4}^3, c_{1,2}^4, c_{1,3}^4, c_{1,4}^4, c_{2,3}^4, c_{2,4}^4, c_{3,4}^4\}$$

and solve the system given by Eqs. (1) and (2). This gives us one solution:

$$[\cdot,\cdot] = \begin{pmatrix} 0 & 0 & c_{1,4}^1 & c_{3,4}^2 & c_{2,4}^1 & c_{3,4}^1 \\ 0 & 0 & 0 & 0 & 0 & c_{3,4}^2 \\ 0 & 0 & 0 & 0 & 0 & 0 \\ 0 & 0 & 0 & 0 & 0 & 0 \end{pmatrix}.$$

Moreover, it is easy to check that if $c_{3,4}^2 c_{1,4}^1 = 0$ then $(V, [\cdot,\cdot])$ is a Lie algebra.

Next, to obtain the isomorphism classes we will consider bijective maps ϕ satisfying $\phi \circ \alpha \circ \phi^{-1} = \alpha$. We have the following cases:

– If $c_{3,4}^2 c_{1,4}^1 c_{2,4}^1 (c_{3,4}^2 - c_{1,4}^1)(c_{3,4}^2 + c_{1,4}^1) \neq 0$, we take $\phi = \begin{pmatrix} c_{3,4}^2 & m_1 & m_2 & 0 \\ 0 & c_{3,4}^2 & m_1 & m_2 \\ 0 & 0 & c_{3,4}^2 & m_1 \\ 0 & 0 & 0 & c_{3,4}^2 \end{pmatrix}$, where

$$m_1 = \frac{c_{3,4}^2 c_{2,4}^1}{c_{1,4}^1 + c_{3,4}^2},$$

$$m_2 = \frac{c_{3,4}^2 \left(c_{1,4}^1 c_{3,4}^1 (c_{1,4}^1 + 2c_{3,4}^2) + c_{2,4}^1 c_{3,4}^2 (c_{1,4}^1 - c_{2,4}^1) + (c_{3,4}^2)^2 (c_{2,4}^1 + c_{3,4}^1)\right)}{(c_{1,4}^1 - c_{3,4}^2)(c_{1,4}^1 + c_{3,4}^2)^2}$$

to obtain the product

$$[e_1, e_4] = \gamma e_1, \quad [e_2, e_3] = e_1, \quad [e_3, e_4] = e_2, \quad \gamma = \frac{c_{1,4}^1}{c_{3,4}^2} \neq 0, 1, -1. \tag{7}$$

– If $c_{3,4}^2 c_{1,4}^1 (c_{3,4}^2 - c_{1,4}^1)(c_{3,4}^2 + c_{1,4}^1) \neq 0$ and $c_{2,4}^1 = 0$, consider $\phi = \begin{pmatrix} c_{3,4}^2 & 0 & m_1 & 0 \\ 0 & c_{3,4}^2 & 0 & m_1 \\ 0 & 0 & c_{3,4}^2 & 0 \\ 0 & 0 & 0 & c_{3,4}^2 \end{pmatrix}$, where $m_1 = \dfrac{c_{3,4}^2 c_{3,4}^1}{c_{1,4}^1 - c_{3,4}^2}$. Then we obtain again product (7).

– If $c_{3,4}^2 c_{1,4}^1 \neq 0$, $c_{3,4}^2 - c_{1,4}^1 = 0$ and $c_{2,4}^1 = 0$, consider $\phi = \begin{pmatrix} c_{1,4}^1 & m & 0 & 0 \\ 0 & c_{1,4}^1 & m & 0 \\ 0 & 0 & c_{1,4}^1 & m \\ 0 & 0 & 0 & c_{1,4}^1 \end{pmatrix}$, where $m^2 + c_{1,4}^1 m + c_{1,4}^1 c_{3,4}^1 = 0$. Then we obtain the product:

$$[e_1, e_4] = e_1, \quad [e_2, e_3] = e_1, \quad [e_2, e_4] = \gamma e_1, \quad [e_3, e_4] = e_2, \tag{8}$$

with $\gamma^2 - 2\gamma + \frac{4c^1_{3,4}}{c^1_{1,4}} = 0$.

– If $c^2_{3,4}c^1_{1,4} \neq 0$, $c^2_{3,4} + c^1_{1,4} = 0$ and $c^1_{2,4} = 0$, consider $\phi = \begin{pmatrix} c^1_{1,4} & 0 & -\frac{c^1_{3,4}}{2} & 0 \\ 0 & c^1_{1,4} & 0 & -\frac{c^1_{3,4}}{2} \\ 0 & 0 & c^1_{1,4} & 0 \\ 0 & 0 & 0 & c^1_{1,4} \end{pmatrix}$.

Then we obtain the product:

$$[e_1, e_4] = e_1, \quad [e_2, e_3] = -e_1, \quad [e_3, e_4] = -e_2. \tag{9}$$

– If $c^2_{3,4}c^1_{1,4}c^1_{2,4} \neq 0$ and $c^2_{3,4} - c^1_{1,4} = 0$, consider $\phi = \begin{pmatrix} c^1_{1,4} & m & 0 & 0 \\ 0 & c^1_{1,4} & m & 0 \\ 0 & 0 & c^1_{1,4} & m \\ 0 & 0 & 0 & c^1_{1,4} \end{pmatrix}$, where $m^2 + (c^1_{1,4} - c^1_{1,2})m + c^1_{1,4}c^1_{3,4} = 0$. Then we obtain product (8) with $\gamma = -\left(\frac{2m - c^1_{2,4}}{c^1_{1,4}}\right)$.

– If $c^2_{3,4}c^1_{1,4}c^1_{2,4} \neq 0$ and $(c^2_{3,4} + c^1_{1,4}) = 0$, consider $\phi = \begin{pmatrix} c^1_{1,4} & 0 & \frac{c^1_{3,4}}{2} & 0 \\ 0 & c^1_{1,4} & 0 & \frac{c^1_{3,4}}{2} \\ 0 & 0 & c^1_{1,4} & 0 \\ 0 & 0 & 0 & c^1_{1,4} \end{pmatrix}$. Then we obtain the following product:

$$[e_1, e_4] = e_1, \quad [e_2, e_3] = -e_1, \quad [e_2, e_4] = \gamma e_1, \quad [e_3, e_4] = -e_2, \tag{10}$$

with $\gamma = \frac{c^1_{2,4}}{c^1_{1,4}} \neq 0$.

Notice that product (9) is the missing case in product (10), i.e., when $\gamma = 0$. The same analysis shows that

- $\mathfrak{g}_1(\gamma) \simeq \mathfrak{g}_1(\mu)$ if and only if $\gamma = \mu$,
- $\mathfrak{g}_2(\gamma) \simeq \mathfrak{g}_2(\mu)$ if and only if $\gamma = \mu$ or $\gamma + \mu = 2$, and
- $\mathfrak{g}_3(\gamma) \simeq \mathfrak{g}_3(\mu)$ if and only if $\gamma = \mu$.

□

Acknowledgements M. A. Alvarez is supported by Vicerrectoría de Investigatión, Innovación y Postgrado of Universidad de Autofagasta, project "Sobre la estructura de álgebras" CR 4824.

References

1. Alvarez, M.A., Vera, S.: On rigid 3-dimensional Hom-Lie algebras. J. Algebra **588**, 166–188 (2021)
2. García-Delgado, R., Salgado, G., Sánchez-Valenzuela, O.A.: On 3-dimensional complex Hom-Lie algebras. J. Algebra **555**, 361–385 (2020)
3. García-Delgado, R., Salgado, G., Sánchez-Valenzuela, O. A.: Corrigendum to "On 3-dimensional complex Hom-Lie algebras". [J. Algebra **555**, 361–385 (2020)]. J. Algebra **562**, 286–289 (2020)
4. Hartwig, J., Larsson, D., Silvestrov, S.: Deformations of Lie algebras using σ-derivations. J. Algebra **295**, 314–361 (2006)
5. Li, X., Li, Y.: Classification of 3-dimensional multiplicative Hom-Lie algebras. J. Xinyang Norm. Univ. **25**(4), 427–430 (2012)
6. Ongon'a, E., Richter, J., Silvestrov, S.: Classification of Low-dimensional Hom-Lie algebras. Algebraic Struct. Appl. **317**, 223–256 (2020)
7. Ongon'a, E., Ongaro, J., Silvestrov, S.: Hom-Lie structures on complex 4-dimensional Lie algebras. Lie Theor. Appl. Phys. **335**, 373–381 (2020)
8. Remm, E., Goze, M.: On the algebraic variety of Hom-Lie algebras. arXiv:1706.02484 (2017)
9. Remm, E.: 3-Dimensional skew-symmetric algebras and the variety of Hom-Lie algebras. Algebra Colloquium **25**(4), 547–566 (2018)

Optimal PD-PD and State Feedback Control of Underactuated Ball and Beam System with Uncertainty and Disturbances

Yaman Sahu, Bibek Gupta, Mahmoud Wael, and Raafat Shalaby

Abstract This research paper investigates the optimal control of the ball and beam system, a challenging fourth-degree system that is commonly used in control studies. Four different control techniques, namely cascaded PD-P, cascaded PD-PD, optimized cascaded PD-PD control using a genetic algorithm, and state feedback control, are compared in terms of their performance. The study aims to achieve specific design criteria, including low rise time, less than three percent overshoot, and less than two seconds of settling time. In addition, the impact of disturbances, uncertainty, control action, and beam angle on the robustness of the system is also examined. MATLAB software is employed to implement and evaluate the performance of the different control methods. The findings contribute to the understanding of how to achieve optimal control of complex systems like the ball and beam system.

Keywords Optimal control · State feedback · Genetic algorithm · Ball and beam · Robust control

1 Background

The ball and beam system discussed in this paper is unstable [7]. Controlling unstable systems is increasingly becoming a priority in the control field, especially with the current control evolution and the evolution of technology. Ziegler and Nichols method [12] shows control applications widely utilize PID parameters calculated based on the ultimate gain and period of oscillation. The ball and beam system, whether implemented in hardware or software modeling, serves as a secure experi-

Y. Sahu (✉) · B. Gupta · M. Wael
MECT Program, School of Engineering and Applied Science, Nile University, Giza, Egypt

Erasmus Mundus Joint Master Mechatronic Engineering(EU4M), Oviedo, Spain
e-mail: yaman.sahu@eu4m.eu

R. Shalaby
Smart Engineering Systems Research Center (SESC), Nile University, Giza, Egypt

Faculty of Electronic Engineering, Menofia University, Menouf, Egypt

D. Vlachos (ed.), *Mathematical Modeling in Physical Sciences*, Springer Proceedings in Mathematics & Statistics 446, https://doi.org/10.1007/978-3-031-52965-8_7

ment with the necessary traits to simulate an unstable system. By controlling such an unstable system using various control theories, opportunities arise to determine the most effective control theory for such systems. This opens doors to successfully handle more intricate unstable systems in real-life scenarios, such as stabilizing an airplane during landing or managing turbulent airflow [3]. Numerous models were constructed for the ball and beam systems, considering diverse design parameters, as elaborated in the "Ball and Beam System" section. Consequently, a range of control methods, including Fractional Order PID (FOPID) [24], Optimal fractional-order PID [23], Fuzzy Control [17], and State Feedback- have been applied to such a system to control it according to different criteria [13] and uses a nonlinear state error feedback control law (NLSEF) [2]. By the synthesis of H_2 and H_{inf} controllers, authors have shown that H_{inf} controllers offer better performance [9]. Tayfun Abut [1] used Proportional-Derivative (PD) to control the ball's position. Tahrir et al. [27] study relies on PID controller and Pole placement(PP) method to stabilize and control the system. Kim et al. [15] developed a state observer to estimate a ball's and a motor's angular velocity. Ye et al. [29] showed new state and partial state feedback laws as global stabilizers of frictionless ball and beam systems. Guanghi et al. [14] used state control, a cascade structure used for the fuzzy controller, and considered disturbance compensation in the controller; hence they concluded that the algorithms have high precision and good stability. Gutierrez et al. [11], and Murugesan et al. [18] showed that a PID controller tuned using a genetic algorithm(GA) is more cost-efficient than a conventionally tuned PID controller. Murugesan et al. [18] compared the performance of Ziegler Nichols (ZN) PID and GA-PID in simulation and experimentation to show the competence of GA-PID with ZN-PID through performance error indices.

Mollabashi et al. [5] compared the Genetic algorithm(GA) with Decoupled sliding mode fuzzy neural network control (DSMFNNC) and found that with GA, the curves are smoother, and the response time is shorter. Ali et al. [4] showed that the GA-PI-PD controller is much more efficient for the set point tracking of ball and beam systems to SIMC-PID, PI-PD, and H_{inf} controllers. Grzegorz et al. [10] examined the significance of incorporating eccentric fixation into the models and also introduced a novel model that factors in friction resistance. Rosa et al. [21] The study demonstrated the application of a linear quadratic regulator (LQR) combined with a cascaded feedback linearization controller (FLC) to achieve optimal control of the BBS. The LQR controller was developed using the derived dynamics model of the BBS, while the FLC was utilized to achieve the desired tracking error dynamics. Additionally, deep reinforcement learning was employed as another approach to enhance the control system by [28] The approach incorporates a reward function, an intelligent positioning algorithm for the ball coordinate, and a reinforcement learning policy model to achieve linearization near the equilibrium point. Furthermore, experimental results validate the efficacy of the proposed methods in the control system of the BBS.

Implemention of hybrid optimization algorithms by [25] on PID controllers to improve the system performance with Two hybrid algorithms, PSO-GSA and PSO-GWO, controller parameters are adjusted using the tuning methods, resulting in enhanced system performance as demonstrated by the simulation outcomes.

The study examines the utilization of fuzzy logic controllers (FLCs) as intelligent control techniques, which exhibit PID-like behavior. It investigates three distinct rule sets for FLCs and explores various optimization techniques, including differential evolution, genetic algorithm, particle swarm optimization, and quantum-behaved particle swarm optimization [5]. The study focuses on the application of adaptive control, specifically the linear state-feedback model reference adaptive control (MRAC), to achieve synchronization between the states of the BBS and a given reference model. The research emphasizes the convergence response by defining initial feedback gains and setting limits to minimize oscillations [20]. Adaptive Neuro-Fuzzy Inference System (ANFIS) and PID control by [22] employs image processing techniques to achieve the goal of controlling the position of the ball on the beam and the angle of the beam in order to reach the desired position. Both simulation and experimental assessments indicated that the control system utilizing the Adaptive Neuro-Fuzzy Inference System (ANFIS) exhibited superior tracking performance and robustness to uncertainties compared to the PID-based control method.

2 Methods

2.1 Ball and Beam System Model

The research aims to achieve positional regulation of the ball on the beam by controlling the voltage supplied to the DC motor. This motor, upon rotating a certain angle, moves the lever and positions the ball at the desired reference point. The paper outlines a series of steps: initially, the mathematical description of the ball and beam system will be provided, considering specific parameters to derive a transfer function. Subsequently, the DC motor will be mathematically introduced in two components, mechanical and electrical, incorporating its parameters to obtain a transfer function. Finally, the two transfer functions will be combined in a cascaded manner to form the complete system. The research paper's objective is to assess various control methods using a standardized model. Specifically, it will explore the effectiveness of four techniques: genetic algorithm-optimized PD control, state feedback, cascaded PD-P, and cascaded PD-PD controller. The figure in the paper illustrates the ball and beam system as a Free-body diagram 1. The ball and beam model utilized in the paper employs Newtonian and Lagrangian techniques, in contrast to previous work by [16], and the force resolution based on [20] is utilized.

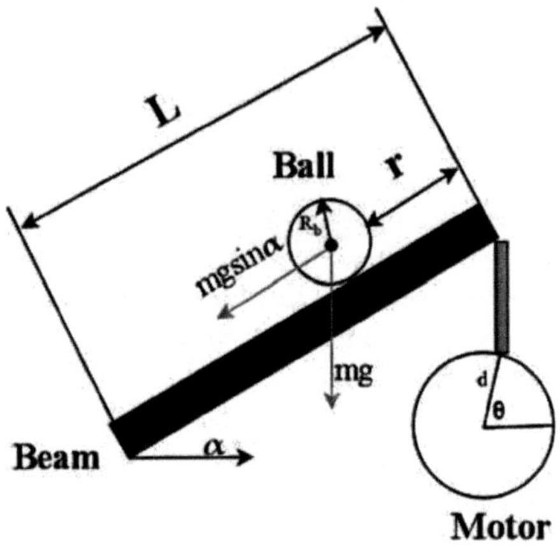

Fig. 1 Ball and beam system free body diagram

2.1.1 Ball and Beam Dynamics

If Newton's law is applied to the free body diagram the following equation can be deduced as:

$$mgsin\alpha - F_f = m\ddot{r}, \tag{1}$$

where,

$$F_f = \frac{J}{R^2}. \tag{2}$$

F_f is the inertial resistance force.
m is the mass of the ball.
r is the distance on the beam.
$\ddot{r}$ is the velocity.
R_b is the radius of the ball.
J is the measure of its rotational inertia.
α is the deflection of the beam.

In this case, α is assumed to be very small and hence the ball and beam system motion is linearized in the following equation:

$$mg\alpha + \frac{J}{R^2}\ddot{r} = m\ddot{r}(\frac{J}{R_b^2} + m)\ddot{r} = mg\alpha, \tag{3}$$

and

$$\alpha = \frac{d}{L}\theta, \tag{4}$$

so:

$$(\frac{J}{R_b^2} + m)\ddot{r} = mg\frac{d}{L}\theta. \tag{5}$$

and

$$\frac{R(s)}{\theta(s)} = \frac{mgd}{L(\frac{J}{R_b^2} + m)} * \frac{1}{s^2}. \tag{6}$$

The parameters in our system are as follows:
Mass of ball $(m) = 0.11$ kg.
Radius of the ball $(R_b) = 0.015$ m.
Lever Arm Offset $(d) = 0.03$ m.
Acceleration due to gravity $(g) = 9.81\text{ms}^{-2}$.
Length of the beam $(L) = 1.0$ m.
Ball's moment of inertia $(J) = 10^{-6}\text{kgm}^2$.

By substituting parameter values, we can derive the transfer function for the ball and beam system as shown:

$$\frac{R(s)}{\theta(s)} = \frac{0.21}{s^2}. \tag{7}$$

2.1.2 DC Motor

As mentioned previously, the positioning of the ball will be controlled using a DC motor. The objective is to achieve angular displacement as the desired output from the DC motor. In this section, we will derive the transfer function of the motor that will be utilized in the MATLAB model.
The parameters of the motor are as follows:

Armature Current $(i_a) = 2.0$ Amperes.
Motor Speed $(w_m) = 500\ \frac{\text{rad}}{\text{s}}$.
Static Friction Torque $(T_f) = 0.012$ Nm.
Torque Constant $(K_T) = 0.06\ \frac{\text{Nm}}{\text{A}}$.
Back EMF Constant $(K_E) = 0.06\ \frac{\text{Vs}}{\text{rad}}$.
Armature Resistance $(R_a) = 1.2$ ohms.
Armature Inductance $(L_a) = 0.02$ H.
Rotational inertia $(J_m) = 6.2*10^{-4}\frac{\text{Nms}^2}{\text{rad}}$.
Viscous Friction $(B_m) = 10^{-4}\frac{\text{Nms}}{\text{rad}}$.
Gear ratio $(G_r) = 10$.

The motor dynamics can be analyzed as follows:
Developed Torque:

$$T = K_T * i_a - T_f. \tag{8}$$

Developed Power in watts:

$$P = \omega_m * T. \tag{9}$$

Back EMF in volts:

$$E_b = K_e * \omega_m. \tag{10}$$

Armature Terminal Voltage:

$$E_a(t) = R_a * i_a + L * \frac{di_a(t)}{dt} + E_b(t). \tag{11}$$

Torque equation is obtained as:

$$T(t) = J_m * \frac{d\omega_m(t)}{dt} + B_m * \omega_m(t). \tag{12}$$

Applying laplace transform to Eqs. (11) and (12), we get:

$$E_a(s) = R_a * i_a(s) + Ls * i_a + E_b(s), i_a(s) = \frac{E_a(s) - E_b(s)}{(Ls + R_a)}. \tag{13}$$

and

$$T(s) = [J_m * s + B_m]\omega_m(s), \omega_m(s) = \frac{T(s)}{J_m * s + B_m}. \tag{14}$$

From Eqs. (13) and (16), transfer function is obtained as:

$$\frac{\omega_m(s)}{E_a(s)} = \frac{K_T}{(L_a s + R_a)(J_m s + B_m) + K_T * K_E}, \tag{15}$$

$$\frac{\omega_m(s)}{E_a(s)} = \frac{K_T}{s^2(L_a s * J_m) + s(R_a J_m + B_m L_a) + Y}, \tag{16}$$

and

$$\frac{\theta_m(s)}{E_a(s)} = \frac{K_T}{s^3(L_a s * J_m) + s^2(R_a J_m + B_m L_a) + sY}. \tag{17}$$

where, Y = $(K_T * K_E + R_a B_m)$.
The transfer function of the motor is obtained as:

$$\frac{\theta_m(s)}{E_a(s)} = \frac{K_s}{s(1 + s * \tau_s)}. \tag{18}$$

where,

$$K_s = \frac{K_T}{(K_T * K_E) + (R_a * B_m)}. \tag{19}$$

and

$$\tau_s = \frac{R_a J_m}{(K_T * K_E) + (R_a * B_m)}. \tag{20}$$

Next, the numerical value transfer function, motor gain constant, and motor time constant will be obtained by substituting the parameters of the DC motors in the final reduced form of the transfer function with a gear ratio of 1/10,

Motor gain constant = K_s = 16.13.

Motor time constant = τ_s = 0.2. Therefore, transfer function of the motor:

$$\frac{\theta_m(s)}{E_a(s)} = \frac{1.613}{s + 0.2s^2}. \tag{21}$$

2.2 *Ball and Beam's System Optimization Using Different Methods*

In this research paper, we put forth four distinct controllers aimed at enhancing the performance of the ball and beam system. These controllers include cascaded PD-P, cascaded PD-PD, cascaded PD-PD optimized by GA [15], and state feedback. We conduct a comprehensive comparative analysis of the results obtained from each technique and discuss the performance parameters associated with them.

2.3 *Ball and Beam System Control Using PD Controller*

The following parameters K_P and K_D in Table 1 are obtained from the PD tuner in Simulink, according to the design criteria [6]:

Overshoot $O < 3\%$, and Settling Time (T_s) <= $2s$.

Table 1 Values of K_p, K_i and K_d for the controllers

Values of gains		
	PD controller	P controller
Proportional K_p	0.1	2.5
Integral K_i	0	0
Derivative K_d	6	0

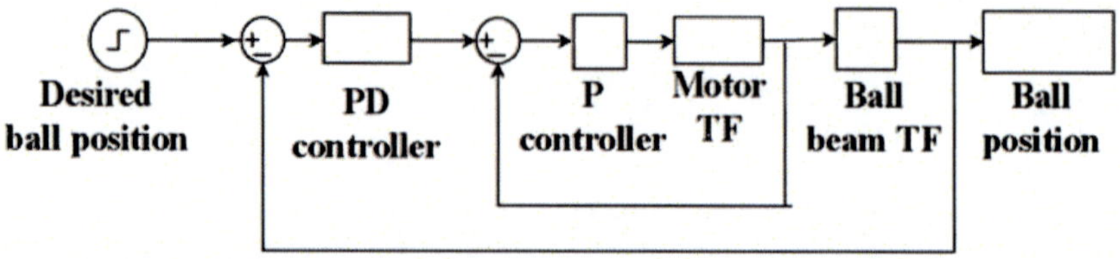

Fig. 2 Block diagram of PD-P system

Table 2 PD-P performance parameters

	Values of performance parameters
Rise time	0.778
Settling time	2.16
Overshoot	6.08%
Peak	1.06 %
RMS	0.7358

2.3.1 Cascaded PD-P Control

The block diagram for the PD-P controller is shown in Fig. 2.

The following values for K_p, K_i, and K_d are manually tuned for the outer and inner loops respectively to obtain the desired response as shown in Table 1.

The performance parameters for the cascaded PD-P system are shown in Table 2.

2.3.2 Cascaded PD-PD Control

The block diagram for the cascaded PD-PD controller connected with the block of motor and ball beam transfer function is shown in Fig. 3.

The following values for K_p, K_i, and K_d are obtained by manually tuning the inner loop of the PD controller for the motor as shown in Table 3.

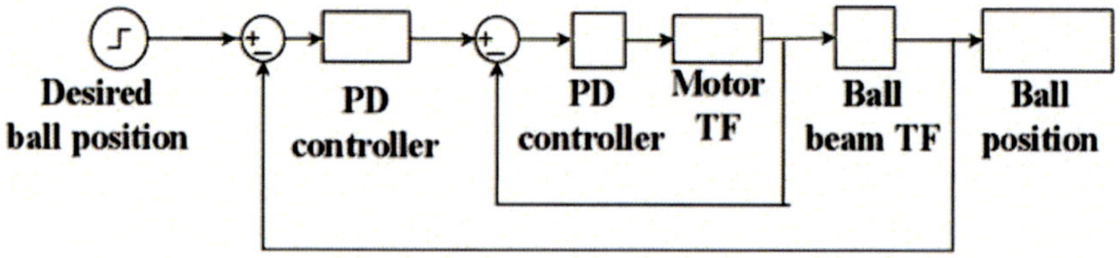

Fig. 3 Block diagram PD-PD control

Table 3 Values of K_p, K_i and K_d for the controllers

	Values of gains	
	PD controller	PD controller
K_p	0.1	1.5
K_i	0	0
K_d	15	15

Table 4 Performance Parameters for manually tuned PD

	Values of performance parameters
Rise time	0.689
Settling time	0.808
Overshoot	0.591
Peak	1.01
RMS	0.7903

Table 5 Results for the genetic algorithm(GA) optimization with performance parameters

G	P	K_p	K_d	T_r	T_s	O
100	30	0.85	37.7	0.69	1.067	0.257
100	50	1.64	52.82	0.559	0.726	0.115
200	30	2.741	67.145	0.414	0.51	0.2
200	50	3.055	80.839	0.387	0.46	0.232

G	P	RMS	e_{ss}
100	30	0.8323	≈ 0
100	50	0.8444	≈ 0
200	30	0.8562	≈ 0
200	50	0.8646	≈ 0

2.4 Optimized Cascaded PD-PD Through Genetic Algorithm

The PD-PD block diagram is shown in Fig. 3 involves the optimization of the outer loop PD controller using a Genetic Algorithm, which leverages the Integral of Time multiplied by Absolute Error (ITAE) function [19] (Table 4). The results in Table 5 have used generations as stopping criteria [26] with respective population sizes. Optimized tuned parameters for the PD response using a Genetic algorithm [8].

Now, the response for optimized PD is shown in Fig. 4.

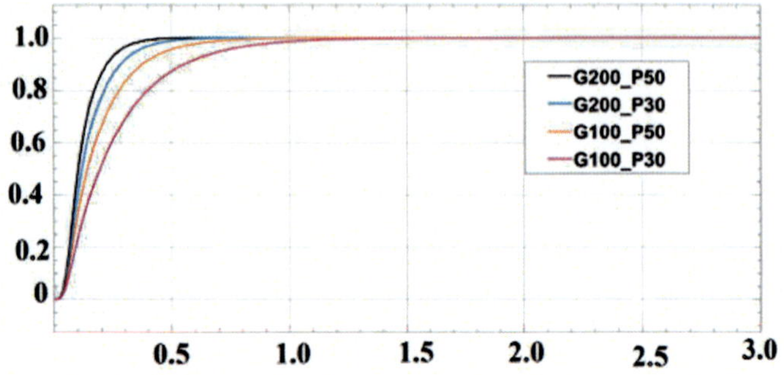

Fig. 4 Variation of response with population and generations where x-axis: time (s) and y-axis: Position

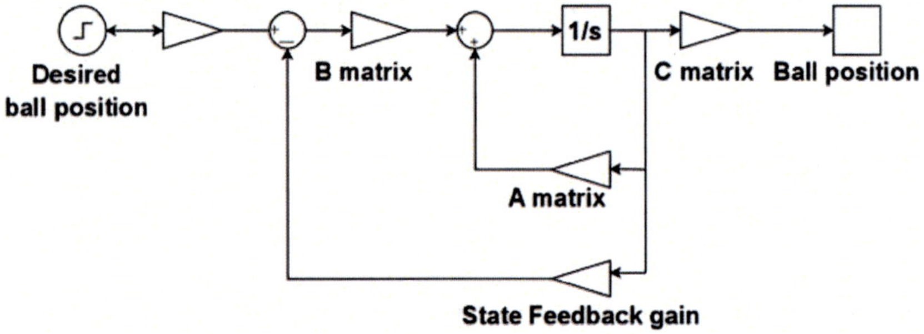

Fig. 5 Block diagram for state feedback

2.5 Ball and Beam System Control Using State Feedback

State feedback means that state measurements can be used to determine the control action [26]. Figure 5 shows the block diagram for state feedback.

As shown in the state feedback block diagram, we have utilized state-space representation for the matrices A, B, and C. The gain for the state feedback is obtained according to the required design criteria [6]. Since the transfer function is order four hence, as per root locus diagram, three poles are located at origin and one pole lies at left hand plane.
The transfer function:

$$\frac{Y(s)}{X(s)} = \frac{1.681}{s^4 + 4.962s^3}. \tag{22}$$

General form of state-space: $\dot{x} = Ax + Bu$, and
$y = Cx + Du$. where matrices A, B, and, C represent the system, input, and output matrices respectively. x, $\dot{x}$, y, and u The state vector, its derivative, the output vector, and the input vector are represented as part of the system. Matrix D is set to zero.

After incorporating the state vectors, the dynamic equation for the ball and beam system can be expressed in state-space form as follows:

$$\begin{bmatrix} \dot{r} \\ \ddot{r} \\ \dot{\theta} \\ \ddot{\theta} \end{bmatrix} = \begin{bmatrix} 0 & 1 & 0 & 0 \\ 0 & 0 & 0.21 & 0 \\ 0 & 0 & 0 & 1 \\ 0 & 0 & 0 & -4.96 \end{bmatrix} \begin{bmatrix} r \\ \dot{r} \\ \theta \\ \dot{\theta} \end{bmatrix} + \begin{bmatrix} 0 \\ 0 \\ 0 \\ 8 \end{bmatrix} E_a. \quad (23)$$

$$\mathbf{y} = \begin{bmatrix} 1 & 0 & 0 & 0 \end{bmatrix} \begin{bmatrix} r \\ \dot{r} \\ \theta \\ \dot{\theta} \end{bmatrix} \quad (24)$$

The eigenvalues serve as the poles of the system, including three poles located at the origin and a negative pole that influences the system to return to a stable equilibrium position. These poles align with our design criteria of [6] are $M_{OS} <=$ 3% overshoot, and $t_s <= 2s$, so according to the following expressions for the OS (25) and T_s (26), ξ and ω_n values are computed as 2.70 and 0.74, respectively.

$$M_{OS} = e^{-\frac{\xi\pi}{\sqrt{1-\xi^2}}} \quad (25)$$

$$t_s = 4\tau = \frac{4}{\xi\omega_n} \quad (26)$$

Based on the design criteria the poles are placed at the following locations: $\left[-8+0.1i \ -8-0.1i \ -20 \ -40\right]$.

Using Ackermann's formula the gains are calculated as: $\left[30466 \ 9900 \ 228 \ 9\right]$.

The reference gain for the input gain is calculated using the Matlab function `rscale` as 3.04661×10^4. Compared to other control techniques, the state feedback response exhibits minimal steady-state error. The state feedback response can be adjusted based on the design criteria, thereby allowing the manipulation of pole locations on the root locus diagram.

3 Results

3.1 Comparing the Performance of Various Control Techniques on Ball and Beam System

The comparison between the performance parameters of various control techniques is shown in Table 6, where Opt.PD-PD refers to optimized PD-PD controller by Genetic Algorithm and State Feedback.

Table 6 Detailed comparison between PD-P, PD-PD, Optimized PD-PD, and State Feedback

	T_r	T_s	O	RMS	e_{ss}
PD-P	0.778	2.16	6.08	0.7358	≈ 0
PD-PD	0.689	0.808	0.591	0.7903	≈ 0
Opt. PD-PD	0.387	0.46	0.232	0.8646	≈ 0
SF	0.438	0.9	0.505	0.676	0.000088

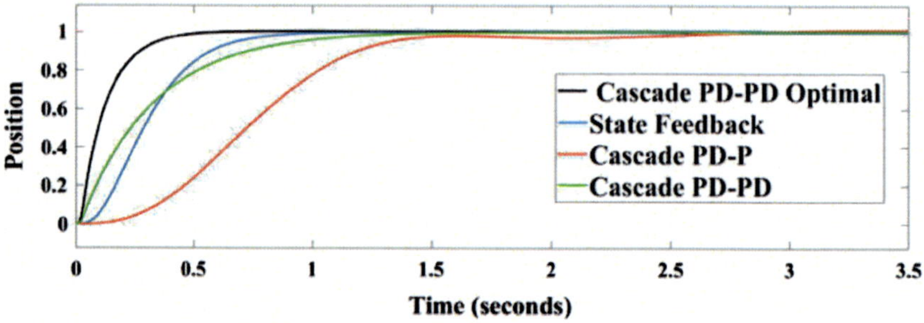

Fig. 6 Comparison of ball displacement response using PD-P, PD-PD, optimized PD, and State feedback control where x-axis: time (s) and y-axis: Ball displacement

In Fig. 6 Response of PD-P, PD-PD, optimized PD-PD by genetic algorithm (GA), and state feedback are compared by their respective performance where the x-axis and the y-axis represent time(s) and ball displacement, respectively.

3.2 Comparing the Performance of Various Control Techniques on Ball and Beam System After Applying the Sine Wave

The Fig. 7 displays the response observed when a sine wave with a frequency and amplitude of one is applied to the system.

3.3 Assessing the Impact of Disturbances on Performance Parameters in Ball and Beam Control System

The performance parameters of the ball and beam system, including stability, response time, and accuracy, can be impacted by external disturbances like sudden forces or vibrations. To counteract the effects of such disturbances and ensure stable and accurate performance, various control techniques can be implemented.

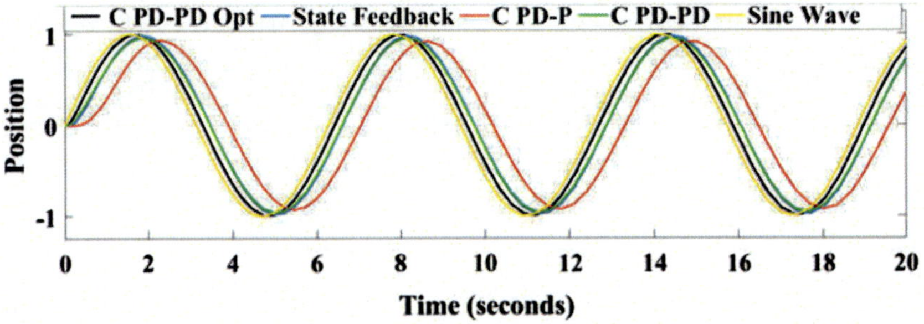

Fig. 7 Comparison between PD-P, PD-PD, optimized PD, and State feedback control by application of sine wave

Table 7 Performance parameters due to effect of disturbance on PD-P, PD-PD, Optimized PD-PD, and State Feedback

	T_r	T_s	$O\%$	RMS	e_{ss}
PD-P	0	2.5	22	0.145	≈ 0
PD-PD	0	1	20.5	0.275	≈ 0
Opt. PD-PD	0	0.6	20.5	0.431	≈ 0
SF	0	1	33	0.269	0.000088

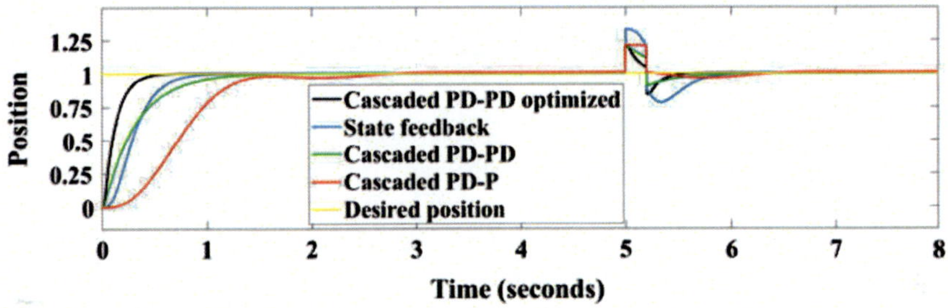

Fig. 8 Comparison between the performance of on PD-P, PD-PD, optimized PD, and State feedback control when subjecting the ball position to pulse disturbance

To simulate a similar disturbance effect, an impulse of 20% magnitude relative to the desired value is applied for a duration of 0.2 s. The system's response to this disturbance is presented in the Table 7 and Fig. 8.

Table 8 Performance parameters due to Uncertainty in gravity on PD-P, PD-PD, Optimized PD-PD, and State Feedback

	T_r	T_s	$O\%$	RMS	e_{ss}
PD-P	0	1	5.9	1.026	≈ 0
PD-PD	0	0.4	4.6	1.012	≈ 0
Opt. PD-PD	0	0.2	4.6	1.008	≈ 0
SF	0	0	0	1.037	0

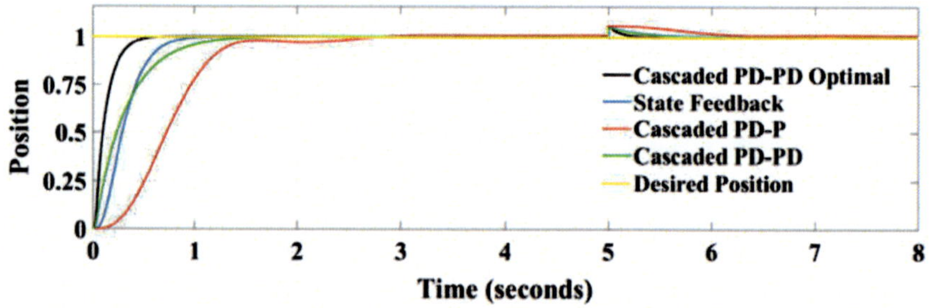

Fig. 9 Comparison between the performance of PD-P, PD-PD, optimized PD, and State feedback control when subjecting the system parameters to step change disturbance

3.4 *Analysis of Uncertainty Parameters in Ball and Beam Control System*

The robustness of a control system refers to its ability to maintain stability and performance under various conditions, such as changes in the environment, disturbances, and uncertainties in the system parameters. Uncertainties: The ball and beam system parameters, such as gravity, length, and friction coefficients, may need to be accurately known and can vary over time. To study the effect of acceleration due to gravity on the ball and beam system, we can perform an uncertainty parameter study by changing the acceleration due to gravity acting on a ball and beam system and observing the system response. For the simulation, acceleration due to gravity is increased by 20% of the initial value of 9.81 m/s^2 after 5 s, for which performance parameters and system response are shown in Table 8 and Fig. 9 respectively.

3.5 *Comparing Beam Angle for Control Techniques on a Ball and Beam System*

The ball and beam system has gained significant attention as a benchmark control problem due to its widespread use and ease of implementation in recent years. One of the significant obstacles in regulating the ball and beam system is selecting an

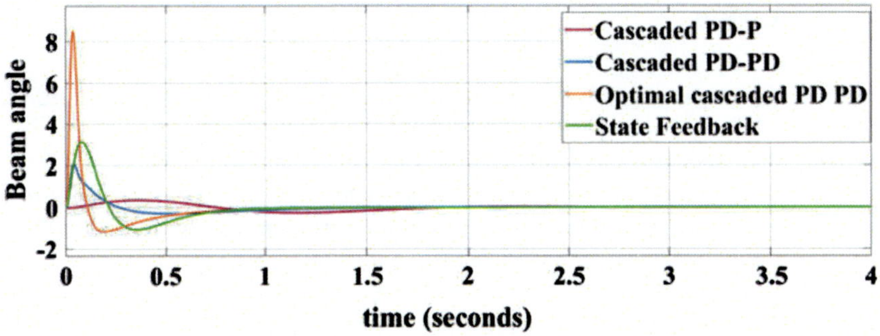

Fig. 10 Comparison of beam angle using PD-P, PD-PD, and optimized PD-PD

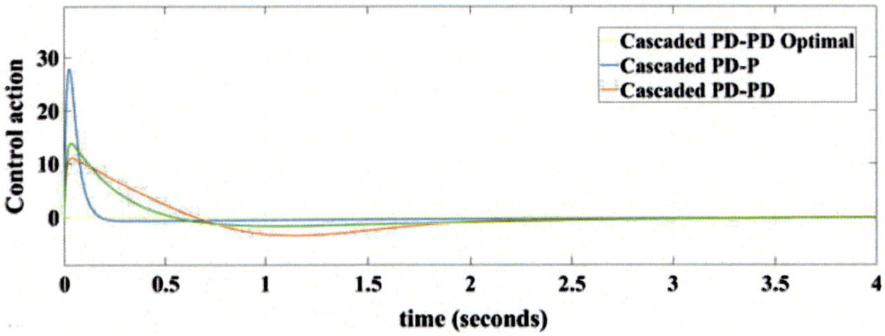

Fig. 11 Comparison of control action using PD-P, PD-PD, and optimal PD-PD

appropriate beam angle. Our research compares different beam angles for various ball and beam system control techniques. This is crucial for hardware implementation and hardware constraints. Figure 10 displays a comparison of the beam angle for the three techniques mentioned.

3.6 Comparing the Control Action for Different Techniques on a Ball and Beam System

The system is comprised of a horizontal axis rotating beam and a ball placed on it, with the objective of controlling the ball's position by regulating the beam's angle. The results of our study indicate that the output of the outer cascaded PD controller serves as the control action. Figure 11 shows the control actions obtained for various control techniques.

4 Discussion

The research findings indicate that the cascaded PD-P control systems exhibit a prolonged settling time and experience high overshoot. Conversely, the cascaded PD-PD control strategy tends to result in significant undershooting, primarily attributed to ball and beam system disturbances. Notably, the state feedback control output remains stable despite sudden changes in system parameters, such as mass.

According to the latest published papers on this topic, this paper has approached a more complex system (4th order), which contains both the ball and beam with the DC motor and applies more rigid criteria, which decreases the overshoot limit from 5% to 3%. The output/results displayed in this paper have proved to be the best (according to the criteria).

5 Conclusion

The research paper uses various techniques to optimize control strategies for the ball and beam system. The study concludes that an optimized cascaded PD-PD controller developed using a genetic algorithm outperforms other control techniques regarding rise time, settling time, and disturbance response. This suggests that the optimized PD-PD controller could be used in real-world scenarios where disturbances are likely to occur.

However, the study also highlights the importance of considering uncertainty in control system design. State feedback control produced the most favorable results in dealing with uncertainty parameters. Therefore, combining control strategies could enhance the system's performance and robustness in real-world scenarios. The study also indicates that further research is needed to explore the potential benefits of advanced control techniques such as Model Predictive Control and Adaptive Control.

Furthermore, the study suggests that alternative optimization algorithms such as ant colony optimization, particle swarm optimization, and simulated annealing should be explored to identify even more efficient control strategies for the ball and beam system. These optimization algorithms could enhance the system's performance and robustness in real-world scenarios.

6 List of Abbreviations

T_r Rise time
T_s Settling time
O Overshoot percentage
RMS Root mean square value
e_{ss} Steady state error

$PD-P$ Proportional derivative-proportional
$PD-PD$ Proportional derivative-proportional derivative
GA Genetic Algorithm
SF State Feedback
Opt Optimal control
G Number of generations
P Population size
C Cascaded
M_{OS} Maximum overshoot
LQR linear quadratic regulator
FLC Feedback linearization controller
BBS Ball and beam system
PSO Particle swarm optimization
GWA Grey Wolf Algorithm
GSA Gravitational Search Algorithm

References

1. Abut, T.: Position controller design and implementation of ball and beam system with smc and pd control methods. Energy Eng. Control Syst. **6**(2), 120–126 (2020)
2. Ali, E.T., Abdullah, S., Amir, M., Adeel, E.M.: Stability control of ball and beam system using heuristic computation based pi-d and pi-pd controller. Tech. J. **24**(01), 21–29 (2019)
3. Ali, S.S.: Position control of ball and beam system using robust h∞ loop shaping controller. Indonesian J. Electr. Eng. Comput. Sci. **19**(1), 91–98 (2020)
4. Ali, T., Malik, S.A., Adeel, M., Amir, M.: Set point tracking of ball and beam system using genetic algorithm based pi-pd controller. NUST J. Eng. Sci. **11**(1), 12–16 (2018)
5. Altinoz, O.T., Yilmaz, A.E.: Investigation of the optimal pid-like fuzzy logic controller for ball and beam system with improved quantum particle swarm optimization. Int. J. Comput. Intell. Appl. **21**(04), 2250025 (2022)
6. Abdul Aziz, N.S., Adnan, R., Tajjudin, M.: Design and Evaluation of Fuzzy Pid Controller for Ball and Beam System, pp. 28–32 (2017)
7. Banu Sundareswari, M., Then Mozhi, G., Dhanalakshmi, K.: Intelligent tuning of pid controller to balance the shape memory wire actuated ball and beam system. Phys. Mesomechanics **23**, 621–630 (2020)
8. Cao, F.: Pid controller optimized by genetic algorithm for direct-drive servo system. Neural Comput. Appl. **32**(1), 23–30 (2020)
9. Ding, M., Liu, B., Wang, L.: Position control for ball and beam system based on active disturbance rejection control. Syst. Sci. Control Eng. **7**(1), 97–108 (2019)
10. Grzegorz, G., Paweł, D., Kamil, L.: Modeling of underactuated ball and beam system-a comparative study. In: Actuators, vol. 12, p. 59. MDPI (2023)
11. Gutierrez, M.K., Choi, D.M., Jula, H.: Using Genetic Algorithms to Optimize Control of a Ball-and-beam System, p. 1–6 (2020)
12. Huba, M., Chamraz, S., Bistak, P., Vrancic, D.: Making the pi and pid controller tuning inspired by ziegler and nichols precise and reliable. Sensors **21**(18), 6157 (2021)
13. Kaan, C., Başçi, A.: Position control of a ball & beam experimental setup based on sliding mode controller. Int. J. Appl. Math. Electr. Comput. (Special Issue-1), 29–35 (2017)

14. Massao Kagami, R., Kovalski da Costa, G., Uhlmann, T.S., Antônio Mendes, L., Freire, R.Z.: A generic weblab control tuning experience using the ball and beam process and multiobjective optimization approach. Information **11**(3), 132 (2020)
15. Lee, K.-T., Choi, H.-L.: Switching control of ball and beam system using partial state feedback: Jacobian and two-step linearization methods. Trans. Korean Inst. Electr. Eng. **66**(5), 819–832 (2017)
16. Mahmoodabadi, M.J., Danesh, N.: Gravitational search algorithm-based fuzzy control for a nonlinear ball and beam system. J. Control Decis. **5**(3), 229–240 (2018)
17. Moezzi, R., Trieu Minh, V., Tamre, M.: Fuzzy logic control for a ball and beam system: fuzzy logic control for a ball and beam system. Int. J. Innovative Technol. Interdiscip. Sci. **1**(1), 39–48 (2018)
18. Murugesan, B.S., Mozhi Ganapathy, T., Kaliaperumal, D.: Design of genetic algorithm-based pid controller for a joule heated shape memory spring driven ball and beam system. Int. J. Mech. Autom. **7**(4), 182–191 (2020)
19. Rao, C.S., Santosh, S., et al.: Tuning optimal pid controllers for open loop unstable first order plus time delay systems by minimizing itae criterion. IFAC-PapersOnLine **53**(1), 123–128 (2020)
20. Romdlony, M.Z., Ridho Rosa, M., Muhammad Puji Syamsudin, E., Trilaksono, B.R., Wibowo, A.S.: Design and application of models reference adaptive control (mrac) on ball and beam. J. Mech. Electr. Power Veh. Technol. **13**(1), 15–23 (2022)
21. Rosa, M.R., Romdlony, M.Z., Riyanto Trilaksono, B.: The ball and beam system: Cascaded lqr-flc design and implementation. Int. J. Control Autom. Syst. **21**(1), 201–207 (2023)
22. Saleem, M.K., Laiq Ur Rahman Shahid, M., Nouman, A., Zaki, H., Atiq Ur Rehman Tariq, M.: Design and implementation of adaptive neuro-fuzzy inference system for the control of an uncertain ball and beam apparatus. Mehran Univ. Res. J. Eng. Technol. **41**(2), 178–184 (2022)
23. Shalaby, R., El-Hossainy, M., Abo-Zalam, B.: Fractional order modeling and control for under-actuated inverted pendulum. Commun. Nonlinear Sci. Numerical Simul. **74**, 97–121 (2019)
24. Shalaby, R., El-Hossainy, M., Abo-Zalam, B., Mahmoud, T.A.: Optimal fractional-order pid controller based on fractional-order actor-critic algorithm. Neural Comput. Appl. 1–34 (2022)
25. Srivastava, V., Srivastava, S.: Hybrid optimization based pid control of ball and beam system. J. Intell. Fuzzy Syst. **42**(2), 919–928 (2022)
26. Srivastava, V., Srivastava, S.: Hybrid optimization based pid control of ball and beam system. J. Intell. Fuzzy Syst. **42**(2), 919–928 (2022)
27. Tahir, N.M., Abolore Yakinni, L., Bature, U.I., Yahuza, I., Shuaibu, A.N., Sambo, I.U.: Control and stability studies of ball and beam system. **4**, 122–131 (2019)
28. Yao, S., Liu, X., Zhang, Y., Cui, Z.: Research on solving nonlinear problem of ball and beam system by introducing detail-reward function. Symmetry **14**(9), 1883 (2022)
29. Zaare, S., Reza Soltanpour, M.: The position control of the ball and beam system using state-disturbance observe-based adaptive fuzzy sliding mode control in presence of matched and mismatched uncertainties. Mech. Syst. Sig. Process. **150**, 107243 (2021)

Are Referees Unfair to Away Teams? Evidence from the Premier League

Gahyun Choi, Kwangwon Ahn, Hanwool Jang, and Daniel S. Kim

Abstract This study examines the contribution of the crowd effect to home advantage in the English Premier League. Using the coronavirus 2019 pandemic as a natural experiment, we test crowd effects based on the changes in the number of audiences. We find that the difference in the number of audiences has an impact on the players' performance, leading to the home advantage. Different from previous studies, which report a referee bias induced by crowd effect, our findings suggest that the crowd effect has no direct influence on referees. Specifically, more yellow cards against away teams results from more aggressive plays by these teams, a response to performance suppression by the home crowd. Findings suggest that social pressures caused by the crowd effect influence only players, not referees.

Keywords Home advantage · Referee bias · Football · Crowd effect

1 Introduction

Home advantage is a long-studied area in the field of sports statistics. Courneya and Carron [1] define home advantage as "the consistent finding that home teams in sports competitions win over 50% of the games played under a balanced home and away schedule". Such home team favoritism is shaped by crowd pressure [2, 3]. Endrich and Gesche [4] claim that the presence of spectators affects match results by increasing social pressure on referees and leading referees to decide in favor of home teams. Furthermore, Scoppa [5] finds that social pressure from spectators acts on both athletes and referees. In case of athletes, their performance measures

G. Choi · K. Ahn
Yonsei University, Seoul 03722, Republic of Korea

H. Jang
Glasgow Caledonian University, Glasgow G4 0BA, UK

D. S. Kim (✉)
Chung-Ang University, Seoul 06974, Republic of Korea
e-mail: danskim@cau.ac.kr

D. Vlachos (ed.), *Mathematical Modeling in Physical Sciences*, Springer Proceedings in Mathematics & Statistics 446, https://doi.org/10.1007/978-3-031-52965-8_8

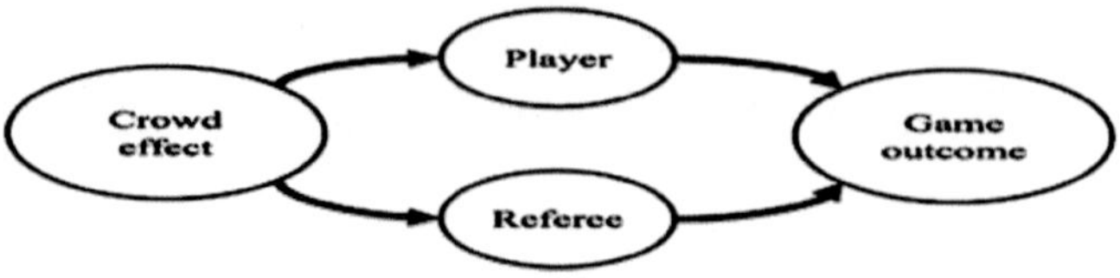

Fig. 1 Channels via which the crowd effect influences the match results

are improved as they are positively influenced by the cheering spectators. Overall, previous studies show that the spectators' influence on referees and players creates the home advantage, as shown in Fig. 1.

This study aims to further investigate the "home advantage" under quasi-natural experiment; the COVID-19 pandemic. Specifically, we first investigate whether the degree of home advantage changed before and after the pandemic using data from the English Premier League ("EPL"). During the pandemic, the number of spectators was restricted due to public health concerns, leading to dramatic drop in the number of spectators in EPL games. We further provide insight on which of the two channels, referees or players, plays more important role in materializing the "home advantage".

2 Data and Methodology

2.1 Data

Our sample period includes two EPL seasons; 2018/2019 and 2020/2021. The COVID-19 pandemic started in late 2019 affecting Europe in early 2020. The 2019/2020 season is excluded from our sample because (i) the pandemic only partially affected the season and (ii) the season serves as the event season.[1] Essentially, we compare pre-COVID 19 (2018/2019 season) with 776,369 weekly average audiences to post-COVID 19 (2020/2021 season) with 8,864 weekly average audiences. There are 15 football clubs included in our sample out of 25 clubs that played in EPL during 2018 to 2021. We require that all football clubs in our sample played all three seasons between 2018 and 2021. We utilize football club-season observations of the 15 football clubs that meet our filter in the analysis.

Our variables are calculated as the seasonal club average by home and away games. *Winning rate* is the percentage of matches that a football club won during a season, *Yellow cards* is the average yellow cards per game, *Shots on target* is the average number of shots that either i) scored a goal, or ii) were blocked by the goalkeeper/last-man per game, *Audience* is the average number of audience per game, and *Tackles* is the average number of tackles per game.

[1] The 2019/2020 season was abruptly ended early on March 13th, 2020, with the World Health Organization's declaration of pandemic.

Table 1 provides descriptive statistics on the main variables of our study. During the pre-COVID season in Panel A, Winning rate and Shots on target are significantly higher in home games than in away games. On the other hand, home teams receive significantly smaller number of yellow cards than away teams. The difference in Yellow cards stands out even more when we consider that the number of tackles in home and away games are not statistically different. On average, the number of audience of home games is larger than that of away games by more than 5,400 (about 14%). The larger audience in home games, although the difference is statistically insignificant, may generate the crowd effect resulting in home advantage.

In contrast, in Panel B, Winning rate is even lower in home games than in away games, albeit being statistically insignificant. The number of audience for home and away games are not statistically different again in the post-COVID season, which implies the home advantage is due to the absolute size of home audience rather than relative size of home and away audience. Also note that the difference in Yellow cards between home and away games is insignificant while the difference for Shots on target is still significant in post-COVID season. These results imply that the crowd effect leads to home advantage more via affecting players than referees. We explore these inferences more rigorously in the next subsection.

Table 1 Descriptive statistics

Variables	Home		Away		
	Mean	Std. Dev	Mean	Std. Dev	Difference
Panel A. 2018/2019 season					
Winning rate	52.98	20.91	38.95	17.77	14.03***
Yellow cards	27.53	7.79	32.13	5.40	−4.60**
Shots on target	5.03	1.31	4.07	0.91	0.96***
Audience ('000)	43.59	15.47	38.13	6.39	5.46
Tackles	17.31	2.14	17.05	1.71	0.26
Panel B. 2020/2021 season					
Winning rate	41.75	12.96	44.56	15.77	−2.81
Yellow cards	25.67	4.91	26.33	5.79	−0.66
Shots on target	4.72	0.95	4.22	0.81	0.50***
Audience ('000)	5.01	1.65	4.32	1.51	0.69
Tackles	15.27	2.26	15.30	1.96	−0.03

Note This table provides descriptive statistics of the main variables used in this study. Statistics are provided by pre- and post-COVID season, and by home and away status. Shots on target is the average number of shots that scored a goal, or were blocked by the goalkeeper or the last-man per game, Audience is the average number of audience per game, and Tackles are the average number of tackles per game. ***, **, and * indicate 1%, 5%, and 10% significance, respectively, of paired *t*-test.

2.2 Regression Analysis

In this section, we use ordinary least squares regression and random effects (RE) model to explore the effect of audience on home advantage, and the channel via which crowd effect leads to the home advantage. Specifically, we take the difference of variables between home and away games (*"home–away"*). $ATT_{h,i,t} - ATT_{a,i,t}$ is selected to extract the effect of crowd difference on dependent variables. We use the following RE model based on the result of Hausman test [6]:

$$y_{h,i,t} - y_{a,i,t} = \beta_1 \left(ATT_{h,i,t} - ATT_{a,i,t} \right) + \beta_2 P_{i,t} + \varepsilon_{i,t},$$

where subscripts h, a, i, and t refer to home, away, team, and season, respectively. For $y_{h,i,t} - y_{a,i,t}$, we use winning rate, number of yellow cards, and number of tackles. $ATT_{h,i,t} - ATT_{a,i,t}$ is the difference in the number of audiences.

To find out the channel of the crowd effect, we further examine following regression models:

$$ST_{i,t,k} = \beta_2 A_{i,t,k} + \varepsilon_{i,t,k},$$

$$TK_{i,t,k} = \beta_3 ST_{i,t,k} + \varepsilon_{i,t,k},$$

$$YC_{i,t,k} = \beta_4 TK_{i,t,k} + \varepsilon_{i,t,k},$$

where i, t, and k is the indicator of team, season, and home/away (*h*: home and *a*: away), respectively. $A_{i,t,k}$ is the away dummy variable, $ST_{i,t,k}$ is the number of shots on target, $TK_{i,t,k}$ is the number of tackles, and $YC_{i,t,k}$ is the number of yellow cards.

3 Results and Discussion

Table 2 presents the difference-in-difference estimation, that is, the gap across home and away match outcomes over pre- and post-COVID-19 seasons. The number of audiences, winning rate, the number of yellow cards, and shots on target are used as proxies for crowd effect, the magnitude of home advantage, referees' bias, and players' performance, respectively.

In Panel A of Table 2, it is evident that the difference in the number of audiences between home and away matches in pre-COVID 19 (2018–2019) and post-COVID 19 (2020–2021) seasons is statistically significant. The gap between home and away games is a proxy for relative home advantage, and the crowd effect should be stronger in matches where home audience dominates away audience in number. Thus, the result in Panel A indicates that home advantage, if any, should have been stronger in pre-COVID 19 season.

Table 2 Difference-in-differences model

	Pre-COVID 19	Post-COVID 19	Pre-Post
Panel A. Number of audiences			
Home	43.59	5.01	38.58***
Away	38.13	4.43	33.70***
Home-Away	5.46	0.58	4.88*
Panel B. Winning rate			
Home	52.98	41.75	11.23***
Away	38.95	44.56	−5.61*
Home-Away	14.03***	−2.81	16.84***
Panel C. Number of yellow cards			
Home	27.53	25.67	1.86
Away	32.13	26.33	5.8***
Home-Away	−4.60**	−0.66	−3.94
Panel D. Shots on target			
Home	5.03	4.72	0.31*
Away	4.07	4.22	−0.15
Home-Away	0.96	0.50***	0.46***

Note This table presents the differences in outcomes (number of audiences, winning rate, number of yellow cards, and shots on target) over time (before and after the outbreak of COVID 19 pandemic) across home and away matches. ***, **, and * represent significance at the 1%, 5%, and 10% levels, respectively.

Panels B, C, and D of Table 2 present the results of difference-in-difference analysis on various match outcomes. In Panel B, winning rates of home games are significantly higher in pre-COVID 19 season supporting claims that the home advantage exists in the English Premier League. As noted in earlier review on potential channels of home advantage, the higher winning rates of home teams before the pandemic may be due to either more pressure on referees to suppress aggressive actions of away teams or more support of home audience cheering players.

In Panel C, the gap between home and away games in the number of yellow cards is not statistically different before and after the pandemic. It implies that the pressure on referees to suppress away teams was not too different over the two seasons. However, in Panel D, relative shots on target of home team over away team is significantly higher in pre-COVID 19 season than post-COVID 19 season suggesting that relatively higher winning rate of home team before the pandemic may have been due to home players performing better with support of home audience. This finding contradicts those from [4] and [7], who find that the cheering of the audience can affect the referee leading to biased number of yellow cards to away teams.

Table 3 Random effects model

Variables	Dependent variables Winning rate (1)	Yellow cards (2)	Shots on target (3)
Audience	0.23*	−0.07	0.01**
Constant	4.43	−2.28	0.66***
Observations	30	30	30
R^2	0.10	0.03	0.18

Note This table presents the results from random effects model of winning rate, yellow cards, and shots on target. ***, **, and * indicate significance at the 1%, 5%, and 10% levels, respectively.

We next run random effects model to further examine the presence of home advantage and its transmission paths; players or referees.[2] In Table 3, columns (1), (2), and (3) use winning rate, number of yellow cards, and shots on target as dependent variables, respectively. Consistent with the results from Table 2, audience has significantly positive association with winning rates and shots on target, however, audience does not significantly influence the number of yellow cards. We can infer that the relative cheering of home audience affects the performance of players but not the referee's likelihood of giving yellow cards to away players. This result is the core of our analysis which is different from the findings of several earlier studies which report that referees issue more yellow cards on away teams due to home crowds [2–4, 7]. We next investigate the source of referee bias in away games.

Table 4 provides an explanation of the referee bias. The away games show fewer shots on target, fewer shots on target results in more tackles, and more tackles end up receiving more yellow cards. This result provides an additional interpretation of the cheering of home crowd on referee bias. Away players are depressed by the cheering of home fans resulting in poorer performance. As the game does not work out as planned, away players begin to play tough and tackles more to keep team morale buoyant. This increased number of tackles naturally leads to increase in the number of yellow cards. The so-called referee bias in away games, as claimed in prior literature, is in fact due to tougher plays of away teams rather than referees being influenced by the home crowd. The effect of home advantages on the referee is the result of the depressed performance of away players rather than the referee bias.

[2] We choose random effects model over the fixed effects model based on Hausman test.

Table 4 Disentangling referee bias

	Dependent variables		
Variables	Shots on target (1)	Tackle (2)	Yellow card (3)
Away	−0.73*	–	–
Shots on target	–	−0.51*	–
Tackle	–	–	0.05*
Constant	0.72**	4.87***	18.54***
Observations	60	60	60
R^2	0.09	0.12	0.06

Note ***, **, and * represent significance at the 1%, 5%, and 10% levels, respectively.

4 Conclusion

We show that cheering of home audience significantly affects on players' performance while referees are not necessarily biased against away teams. Home teams benefit from the support of home fans and enjoy higher shots on target and winning rates whereas away teams are depressed by home crowd resulting in fewer shots on target. The fewer shots on target leads to more tackles by away teams resulting in a greater number of yellow cards to away teams. Prior literature claimed dual channels of home advantage: players' performance and referee bias. We claim a single channel of home advantage; asymmetric effect on players' performance. Home crowd supports home players' performance while depressing away players' performance.

References

1. Courneya, K.S., Carron, A.V.: The home advantage in sport competitions: a literature review. J. Sport Exerc. Psychol. **14**(1), 13–27 (1992)
2. Pollard, R.: Home advantage in football: a current review of an unsolved puzzle. Open Sports Sci. J. **1**(1), 12–14 (2008)
3. Buraimo, B., Forrest, D., Simmons, R.: The 12th man? Refereeing bias in English and German soccer. J. R. Stat. Soc. Ser. A **173**(2), 431–449 (2010)
4. Endrich, M., Gesche, T.: Home-bias in referee decisions: Evidence from "Ghost Matches" during the Covid19-pandemic. Econ. Lett. **197**, 109621 (2020)
5. Scoppa, V.: Social pressure in the stadiums: do agents change behavior without crowd support? J. Econ. Psychol. **82**, 102344 (2021)
6. Cardellichio, P.A.: Estimation of production behavior using pooled microdata. Rev. Econ. Stat. **72**(1), 11–18 (1990)
7. Bryson, A., Dolton, P., Reade, J.J., Schreyer, D., Singleton, C.: Causal effects of an absent crowd on performances and refereeing decisions during Covid-19. Econ. Lett. **198**, 109664 (2021)

The Use of Twins in Isotopic Analysis

Tatiana Iavoruk and Alexander Bazhenov

Abstract Twin (from TW_ice IN_terval) is a set of two nested intervals. The first interval presents an internal estimation of the calculations result, the second one presents an external estimation. The most of chemical elements has two or more stable isotopes. A sample is characterized by the different ratio of stable isotopes, or the isotopic signature. This paper deals with an application of two types of twins for analysis of stable isotopic compositions. The first type of twins is the Nesterov twin, which semantically uses external estimates. With this type of twins we can say about the presence (existence) of the chemical element in the mixture. With another type of twins we can estimate the minimum and maximum contents of element that are possible in the sample. The suggested twin approach covers a very wide application set: the cosmology, global climatology, biology, Earth sciences, forensics sciences, etc.

Keywords Interval analysis · Twin · Isotopic signature · Isotopic landscape · Isotopic niche

1 Introduction

Isotope information are used in different areas of science and industry [1–7]. The variety of data inspires us to develop mathematical concepts, procedures, and software tools for describing data and carrying out computations. Data are frequently described with intervals. For instance, various isotopes have interval molecular weights [8, 9]. It is common for chemical elements to have multiple isotopes [10], so we would like to have a unified and clear way to describe them. One way to put it data as a twins. This approach can be used to analyze isotopic landscapes and niches as well as isotopic signatures.

T. Iavoruk (✉)
ITMO University, 197101 St. Petersburg, Russia
e-mail: tatianaiavoruk@gmail.com

A. Bazhenov
Ioffe Institute, 194064 St. Petersburg, Russia

Peter the Great St. Petersburg Polytechnic University, 195251 St. Petersburg, Russia

D. Vlachos (ed.), *Mathematical Modeling in Physical Sciences*, Springer Proceedings in Mathematics & Statistics 446, https://doi.org/10.1007/978-3-031-52965-8_9

1.1 Interval Arithmetic

We describe mathematical objects with uncertainty intervals. The standardized notation system for interval analysis is given in [11]. In interval analysis, the standard practice (also in this paper) is to present interval objects in bold italic type.

Classical interval arithmetic $\mathbb{IR}$ is an algebraic system formed by proper intervals $\boldsymbol{x} = [\underline{\boldsymbol{x}}, \overline{\boldsymbol{x}}] \subset \mathbb{R}$. For any arithmetic operation $\star \in \{+, -, \cdot, /\}$ the result of the operation between intervals is defined as $\boldsymbol{x} \star \boldsymbol{y} = \{x \star y \mid x \in \boldsymbol{x}, y \in \boldsymbol{y}\}$, where $x, y \in \mathbb{R}$.

More generally, the problems of interval analysis can be solved in complete interval arithmetic or Kaucher arithmetic $\mathbb{KR}$ [12]. This arithmetic is obtained from classical interval arithmetic $\mathbb{IR}$ by adding improper intervals $\boldsymbol{x} = [\underline{\boldsymbol{x}}, \overline{\boldsymbol{x}}]$ where $\underline{\boldsymbol{x}} \geq \overline{\boldsymbol{x}}$. It provides better algebraic properties, such as the possibility of taking the minimum by inclusion in the general case and the existence of inverse elements related to addition and multiplication. The use of improper intervals allows you to build original methods for solving problems with interval objects.

Let us introduce some operations with intervals $\boldsymbol{a} = [\underline{\boldsymbol{a}}, \overline{\boldsymbol{a}}]$, $\boldsymbol{b} = [\underline{\boldsymbol{b}}, \overline{\boldsymbol{b}}]$:

$$\texttt{dual element}: \ \mathrm{dual}(\boldsymbol{a}) = [\overline{\boldsymbol{a}}, \underline{\boldsymbol{a}}] \tag{1}$$

$$\texttt{interval width}: \ \mathrm{wid}(\boldsymbol{a}) = \overline{\boldsymbol{a}} - \underline{\boldsymbol{a}} \tag{2}$$

$$\texttt{minimum by inclusion}: \ \wedge(\boldsymbol{a}, \boldsymbol{b}) = \left[\max(\underline{\boldsymbol{a}}, \underline{\boldsymbol{b}}), \min(\overline{\boldsymbol{a}}, \overline{\boldsymbol{b}})\right] \tag{3}$$

$$\texttt{maximum by inclusion}: \ \vee(\boldsymbol{a}, \boldsymbol{b}) = \left[\min(\underline{\boldsymbol{a}}, \underline{\boldsymbol{b}}), \max(\overline{\boldsymbol{a}}, \overline{\boldsymbol{b}})\right] \tag{4}$$

1.2 Twins

In this section, we describe a twin formally and provide descriptions of Nesterov [13] and Sainz et al. [14] arithmetic operations.

Definition 1 Twin $\mathbb{T}$ by $\subseteq$-property is

$$\mathbb{T} = [\boldsymbol{t}_{in}, \boldsymbol{T}_{ex}] = \{\boldsymbol{x} \in \mathbb{KR} \mid \boldsymbol{t}_{in} \subseteq \boldsymbol{x} \subseteq \boldsymbol{T}_{ex}\},$$

where $\boldsymbol{t}_{in}$ and $\boldsymbol{T}_{ex}$ are internal and external estimations of $\boldsymbol{x}$.

Twin arithmetic operations using (1), (3), (4) are presented below. These representations can provide for more simple interpretations. We use the symbol $\star$ to denote the operations of addition, subtraction, multiplication, division.

Twin arithmetic by Nesterov for two objects $\mathbb{T}_k = \left[\left[\underline{t_k}, \overline{t_k}\right], \left[\underline{T_k}, \overline{T_k}\right]\right], k \in \{1, 2\}$:

$$\mathbb{T}_1 \star \mathbb{T}_2 = [\boldsymbol{t}_{in}, \boldsymbol{T}_{ex}] \tag{5}$$

$$\begin{aligned} \boldsymbol{t}_{in} = \vee(&\text{dual}(\wedge(\underline{t_1} \star \left[\underline{T_2}, \overline{T_2}\right], \overline{t_1} \star \left[\underline{T_2}, \overline{T_2}\right])), \\ &\text{dual}(\wedge(\underline{t_2} \star \left[\underline{T_1}, \overline{T_1}\right], \overline{t_2} \star \left[\underline{T_1}, \overline{T_1}\right]))) \end{aligned} \tag{6}$$

$$\boldsymbol{T}_{ex} := \left[\underline{T_1}, \overline{T_1}\right] \star \left[\underline{T_2}, \overline{T_2}\right]. \tag{7}$$

Twin arithmetic by Sainz et al. for two objects has the following form:

$$\begin{aligned} \boldsymbol{t}_{in} = \vee(&\vee(\underline{t_1} \star \left[\underline{t_2}, \overline{t_2}\right], \overline{t_1} \star \left[\underline{t_2}, \overline{t_2}\right]), \\ &\vee(\underline{t_2} \star \left[\underline{t_1}, \overline{t_1}\right], \overline{t_2} \star \left[\underline{t_1}, \overline{t_1}\right])) \end{aligned} \tag{8}$$

Additionally, we present multiplication by a scalar $\lambda \in \mathbb{R}$ for both arithmetic:

$$\lambda \mathbb{T} = [\lambda \boldsymbol{t}_{in}, \lambda \boldsymbol{T}_{ex}]. \tag{9}$$

To characterize the twin, we will use the Jaccard index [15], as it is determined by (10).

$$Ji(\mathbb{T}) = \frac{\text{wid}(\boldsymbol{t}_{in})}{\text{wid}(\boldsymbol{T}_{ex})} \tag{10}$$

2 Numerical Example

In this section, we consider numerical examples of using twins to calculate isotope signatures. This can also be applied in the study of isotopic niches and landscapes. The Jaccard index (10) will be used to characterize the traceability of the isotope signature.

All biological objects are composed from α amino acids. The pathways of the origin of amino acids are different. For example, there exist 3 types of the photosynthesis as C3, C4 and CAM. Each type of the photosynthesis has the specific isotopic signatures.

Let us consider aromatic alpha-amino acids: tyrosine $C_9H_{11}NO_3$, phenylalanine $C_9H_{11}NO_2$, tryptophan $C_{11}H_{12}N_2O_2$. We use IUPAC database [16] information to describe atomic weights. The data in the database [16] displays the atomic weight of an isotope as an interval.

Table 1 Twins for hydrogen, carbon, nitrogen and oxygen

Chemical element	Internal estimate	External estimate
Hydrogen	[1.0079656, 1.0079692]	[1.0078507, 1.0081074]
Carbon	[12.010764, 12.010778]	[12.0096623, 12.0115045]
Nitrogen	[14.0067371, 14.0067411]	[14.006435, 14.0072715]
Oxygen	[15.9993457, 15.9993636]	[15.9990295, 15.9997617]

Table 2 Inner estimation results

Acid	Internal estimate (Nesterov)	Internal estimate (Sainz et al.)
$C_9H_{11}NO_3$	[181.189917, 181.189581]	[181.189272, 181.189496]
$C_9H_{11}NO_2$	[165.190155, 165.190551]	[165.189926, 165.190132]
$C_{11}H_{12}N_2O_2$	[204.226833, 204.226622]	[204.226157, 204.226399]

Table 3 External estimation results

Acid	External estimate
$C_9H_{11}NO_3$	[181.176842, 181.199278]
$C_9H_{11}NO_2$	[165.177812, 165.199516]
$C_{11}H_{12}N_2O_2$	[204.211423, 204.237904]

To present the molecular mass as twin we use the interval mode [17] as an internal estimate. The interval mode is the interval that is the subset of maximal intervals of the interval sample. If the interval mode consists of a set of several intervals, then we use the mode with the widest interval. As an external estimate, we take an interval that includes all intervals and, at the same time, has a minimum width. The results are shown in Table 1 for each of the chemical elements used.

With the data of Table 1, we will compute twins of acids using two types of twins. Table 2 presents the inner estimates by Nesterov' and Sainz' twins. Table 3 gives external estimates both for Nesterov' and Sainz' twins.

Table 4 contains the Jaccard indexes by the formula (10). Negative values in Table 4 indicate that the internal estimate is an improper interval. For example, for tryptophan amino acid with Nesterov's twin we get

$$Ji(\mathbb{T}) = \frac{-204.226833 + 204.226622}{-204.211423 + 204.237904} = -0.0079. \tag{11}$$

The calculations have been provided by the Python language code using the library from https://github.com/Tatiana655/Twins.

Table 4 Jaccard indices of amino acids calculated using different twins

Acid	Type of twin	Jaccard index
Tyrosine	Nesterov	–0.01497
Fenilalanine	Nesterov	0.0183
Tryptophan	Nesterov	–0.0079
Tyrosine	Sainz et al.	0.00999
Fenilalanine	Sainz et al.	0.0095
Tryptophan	Sainz et al.	0.00913

3 Discussion

Let us discuss the results presented in Sect. 2. As one can see from Table 4, the calculations using the Nesterov' twins are more sensitive to origin of initial substances than with Sainz et al., arithmetics.

An interpretation of the arithmetic operations used for internal evaluation will help explain herewith. We denote the result of an arithmetic operation for two twins $\mathbb{T}_1, \mathbb{T}_2$ is described by the symbol $\mathbb{T} = [\boldsymbol{t}, \boldsymbol{T}]$.

When applying Sainz's arithmetic (8), only internal estimates are used. The Sainz twins only use proper intervals, so we obtain (12).

$$\forall z \in \boldsymbol{t} \quad (\forall x \in \boldsymbol{t}_1, \exists y \in \boldsymbol{t}_2) \quad or \quad (\forall x \in \boldsymbol{t}_2, \exists y \in \boldsymbol{t}_1) : x \star y = z \tag{12}$$

This type of twins has some good algebraic properties. One of them is that $\mathbb{T} + \mathbb{T} = 2\mathbb{T}$.

For Nesterov twin arithmetic (6) we have (13).

$$\forall z \in \boldsymbol{t} \quad (\forall x \in \boldsymbol{t}_1, \exists y \in \boldsymbol{T}_2) \quad or \quad (\forall x \in \boldsymbol{t}_2, \exists y \in \boldsymbol{T}_1) : x \star y = z \tag{13}$$

This type of twin (13) uses the external estimation in operations, so some good algebraic properties are lost: $\mathbb{T} + \mathbb{T} \neq 2\mathbb{T}$.

The external estimation can also be represented with maximum by inclusion (4). It is normally the proper interval. Using this assumption, we get (14):

$$\begin{aligned} \boldsymbol{T}_{ex} = \vee(&\vee(\underline{\boldsymbol{T}_1} \star [\underline{\boldsymbol{T}_2}, \overline{\boldsymbol{T}_2}], \overline{\boldsymbol{T}_1} \star [\underline{\boldsymbol{T}_2}, \overline{\boldsymbol{T}_2}]), \\ &\vee(\underline{\boldsymbol{T}_2} \star [\underline{\boldsymbol{T}_1}, \overline{\boldsymbol{T}_1}], \overline{\boldsymbol{T}_2} \star [\underline{\boldsymbol{T}_1}, \overline{\boldsymbol{T}_1}])) \end{aligned} \tag{14}$$

In more general context, one have to keep in mind that operations can have a different sense and algebraic properties depending on the semantics of the problem. It is also possible to mix arithmetic operations.

4 Conclusion

In this paper, we provide an interpretation of two types of twins: Sainz et.al. twin and Nesterov twin. These types differ in internal estimations obtained as a result of arithmetic operations: the first uses only internal estimates, the second uses internal and external estimations to obtain the result of arithmetic operations. Additionally, we have demonstrate how twins can be used to study the isotope analysis and highlighted the differences between them with Jaccard index.

For a more extended use of twins, it is necessary to develop numerical methods, which gives more information for researchers and the possibility to choose between different semantics of the calculations. Many questions still remain open.

The first set of the problems concerns the application features. How to initially choose an internal evaluation using different categories of source substrates? Are there more convenient internal and external estimates?

In the future, it is possible to develop a twin approach not only in the field of isotopic analysis, but also in other areas where two-sided estimations are used.

References

1. Ogbesejana, A.B., Liu, B., Ostadhassan, M.: Stable isotope geochemistry of the organic elements within shales and crude oils: A comprehensive review. Molecules **27**(1), 34 (2022). https://doi.org/10.3390/molecules27010034
2. Rowland, S.J., Sutton, P.A., Wolff, G.A.: Biosynthesis of ambrein in ambergris: evidence from isotopic data and identification of possible intermediates. Nat. Prod. Res. **35**, 1235–1241 (2019). https://doi.org/10.1080/14786419.2019.1644630
3. Madzunkov, S.M., Nikolić, D., Simcic, J., Belousov, A., Gonzalez, M., Darrach, M.R.: Data analysis and isotopic ratios measured onboard the spacecraft atmosphere monitor. Int. J. Mass Spectrom. (2022). https://doi.org/10.1016/j.ijms.2022.116847
4. Hallmann, N., Camoin, G., Webster, J.M., Humblet, M.: A standardized database of marine isotopic stage 5e sea-level proxies on tropical Pacific islands. Earth Syst. Sci. Data **13**, 2651–2699 (2021). https://doi.org/10.5194/essd-13-2651-2021
5. Nasr, E., Epova, E.N., Sebilo, M., Larivière, D., Hammami, M., Souissi, R., Abderrazak, H., Donard, O.F.: Olive oil traceability studies using inorganic and isotopic signatures: A review. Molecules **27** (2022). https://doi.org/10.3390/molecules27062014
6. Stantis, C.: Isotopic data for bioarchaeological samples from Polynesia (pre-Contact and modern periods). Data Brief **38** (2021). https://doi.org/10.1016/j.dib.2021.107419
7. Magozzi, S., Contina, A., Wunder, M., Vander Zanden, H., Bowen, G.: A global compilation of known-origin keratin hydrogen and oxygen isotope data for wildlife and forensic research. EGU General Assembly 2020, Online, 4–8 May 2020, EGU2020-9674 (2020). https://doi.org/10.5194/egusphere-egu2020-9674
8. Possolo, A., van der Veen, A., Meija, J., Hibbert, D.: Interpreting and propagating the uncertainty of the standard atomic weights (IUPAC Technical Report). Pure Appl. Chem. **2**(2018), **90**, 395–424 (2018). https://doi.org/10.1515/pac-2016-0402
9. Prohaska, T., Irrgeher, J., Benefield, J., Böhlke, J., Chesson, L., Coplen, T., Ding, T., Dunn, P., Gröning, M., Holden, N., Meijer, H., Moossen, H., Possolo, A., Takahashi, Y., Vogl, J., Walczyk, T., Wang, J., Wieser, M., Yoneda, S., Zhu, X., Meija, J.: Standard atomic weights of the elements 2021 (IUPAC Technical Report). Pure Appl. Chem. **94**(5), 573–600 (2022). https://doi.org/10.1515/pac-2019-0603

10. Holden, N., Coplen, T., Böhlke, J., Tarbox, L., Benefield, J., de Laeter, J., Mahaffy, P., O'Connor, G., Roth, E., Tepper, D., Walczyk, T., Wieser, M., Yoneda, S.: IUPAC periodic table of the elements and isotopes (IPTEI) for the education community (IUPAC Technical Report). Pure Appl. Chem. **90**(12), 1833–2092 (2018). https://doi.org/10.1515/pac-2015-0703
11. Kearfott, R.B., Nakao, M.T., Neumaier, A., Rump, S.M., Shary, S.P., van Hentenryck, P.: Standardized notation in interval analysis. Comput. Technol. **15**(1), 7–13 (2010). http://www.ict.nsc.ru/jct/content/t15n1/Shary_n.pdf
12. Shary, S.P.: A new technique in systems analysis under interval uncertainty and ambiguity. Reliab. Comput. **8**, 321–418 (2002). https://doi.org/10.1023/A:1020505620702
13. Nesterov, V.M.: Interval and twin arithmetics. Reliab. Comput. **3**, 369–380 (1997). https://doi.org/10.1023/A:1009945403631
14. Sainz, M., Armengol, J., Calm, R., Herrero, P., Jorba, L., Vehí, J.: Modal Interval Analysis: New Tools for Numerical Information, vol. 2091, 10th edn. (2013). https://doi.org/10.1007/978-3-319-01721-1
15. Bazhenov, A.N., Telnova, A.Y.: Generalization of jaccard index for interval data analysis. Meas. Tech. **65**, 882–890 (2023). https://doi.org/10.1007/s11018-023-02180-2
16. Coplen, T.B., Shrestha, Y.: Isotope-abundance variations and atomic weights of selected elements: 2016 (IUPAC Technical Report). Pure Appl. Chem. **88**, 1203–1224 (2016). https://doi.org/10.1515/pac-2016-0302
17. Hu, C., Hu, ZH.: On statistics, probability, and entropy of interval-valued datasets. Inf. Process. Manage. Uncertain. Knowl. Based Syst. **1239**, 407–421 (2020). https://doi.org/10.1007/978-3-030-50153-2_31

Metaheuristic Approaches to Tune PID Controller for Ball on Plate System

Mitterand Ekole, Omer Abdalla, Iman Shalabi, and Raafat Shalaby

Abstract This paper presents a comprehensive study on the optimization of PID controller parameters for a ball & plate system through the utilization of Genetic Algorithm (GA) and Particle Swarm Optimization (PSO). The objective of the study is to attain precise steady-state response while shortening settling time and minimizing overshoot. The assessment of controller performance is conducted using the Integral Absolute Error (IAE) cost function. The study highlights the limitations of conventional tuning techniques and the need for metaheuristic optimization algorithms, particularly when system models and variables are incomplete or imperfect, and demonstrates that both GA and PSO can improve the performance of the controller, but GA outperforms PSO in terms of fitness function value and optimal gains. The study also highlights the importance of accurate system models and variables in achieving optimal performance. The results of this study hold significant implications for the application of PID controllers and optimization algorithms based on metaheuristic principles in various applications, particularly in process industries. The study demonstrates the potential of metaheuristic algorithms for optimizing PID controllers and highlights the need for further research in this area.

Keywords PID · Metaheuristic · Optimization · Data-driven control · Ball on plate system

---

Omer Abdalla and Mitterand Ekole are equally contributed.

M. Ekole · O. Abdalla (✉)
Mechatronics Engineering (EU4M), Oviedo University, Oviedo, Asturias 33007, Spain
e-mail: omer.abdalla@eu4m.eu

M. Ekole
e-mail: mitterand.ekole@eu4m.eu

M. Ekole · O. Abdalla · I. Shalabi · R. Shalaby
Mechatronics Engineering, Nile University, Giza, Egypt
e-mail: ishalabi@nu.edu.eg

R. Shalaby
e-mail: rshalaby@nu.edu.eg

R. Shalaby
Faculty of Electronic Engineering, Menofia University, Shibin Al Kawm, Egypt

D. Vlachos (ed.), *Mathematical Modeling in Physical Sciences*, Springer Proceedings in Mathematics & Statistics 446, https://doi.org/10.1007/978-3-031-52965-8_10

1 Introduction

Throughout various industries such as manufacturing, robotics, aviation, and process control, the influence of feedback control mechanisms has been profound over time. Noteworthy among these is the proportional-integral-derivative (PID) controller, but other capable, resilient, and adaptable control methods have also been suggested [1–3]. Conversely, PID controllers have gained unparalleled recognition and standing within control systems. Depending on the precise demands of a given process, the application of a PID control system can take different forms, ranging from proportional-only (P-mode) to proportional-integral (PI mode), proportional-derivative (PD-mode), and even proportional-integral-derivative (PID mode) configurations [4]. Since the 1890s, PID controllers have found employment in industrial contexts, retaining their position as the dominant choice for process control applications [5]. Their enduring prevalence can be attributed to their robustness, simplicity, and the ease of adjusting control parameters [6]. Nonetheless, PID controllers do have drawbacks, including the potential for excessive speed overshooting and sluggish responses stemming from unforeseen load torque fluctuations and sensitivities to controller gains such as KI and KP [7]. The precision of system models and variables plays a pivotal role in determining the effectiveness of a PID controller's performance.

Over the course of time, various strategies for optimizing the performance of PID controllers have emerged. Notable among the traditional approaches are Ziegler Nichols' method of utilizing closed-loop oscillations and Cohen-Coon's approach involving process reaction curves. Ziegler-Nicholes' tuning technique is rooted in heuristics and aimed at refining the behavior of PID controllers. This method was originally formulated by the collaborative efforts of **John G. Ziegler** and **Nathaniel B. Nichols** [8]. Beyond these, there exist dynamic tuning techniques like those proposed in [9–11], methodologies based on sliding mode control [12–14], and approaches rooted in state-dependent Riccati equations (SDRE) [15–17]. These advanced techniques have shown promise in enhancing the efficiency of controllers. However, their practical application can pose challenges and even lead to unintended consequences, such as the occurrence of undesirable chattering in sliding PID control systems.

A metaheuristic algorithm is a search procedure designed to find an excellent solution to a complex and difficult-to-solve optimization problem. In this world of limited resources, it is critical to find a near-optimal solution based on imperfect or incomplete information (e.g., computational power and time) [18]. One of the most notable achievements in operations research over the last two decades has been the development of metaheuristics for solving such optimization problems. Metaheuristic algorithms mimic nature's way of solving complex problems. In Control and Process Engineering Research, metaheuristic-based algorithms have been used quite notably. Genetic Algorithms have been employed to optimize the parameter values of a PID controller designed for Brushed DC motors, the authors used a modified form of the Integral Time Absolute Error(ITAE) objective function to tune PID parameter gains [19]. Optimal PID parameters have been determined for a robotic arm manipulator using Ant Colony Optimization [20]. Particle Swarm Algorithms have also been employed to determine the optimal parameters of a PID controller for a DC-DC boost converter [21]. Particle Swarm Optimization, Ant Colony Optimization, and AntLion Optimizer are used for tuning the FO parameters of the incremental conductance-based MPPT [22].

This paper introduces a method for tuning a PID controller used in the a ball-on-plate system, employing metaheuristic-based algorithms. Two metaheuristic optimization algorithms are utilized to obtain the best possible PID controller parameters subject to cost function minimization. A full comparison is made between these algorithms to demonstrate their stability and efficiency. This paper is outlined thus: In Sect. 2 a detailed explanation of the used metaheuristic algorithms is presented. Section 3 presents the the ball on the plate system modeling using Lagrange equations. Section 4 shows the standard parameters which were used, parameters common to the two algorithms which were tested, alongside the standard Integral of Absolute of Error cost function. Results are discussed in Sect. 5. In Sect. 6 a conclusion on this work is presented.

2 Metaheuristic Algorithms

2.1 Particle Swarm Algorithm

Particle Swarm Optimization (PSO) was initially introduced by **James Kennedy** and **Russel C. Eberhart** [23]. It is a computational technique employed in the field of computational science, aiming to progressively enhance a potential solution based on a predefined quality measure. This method involves utilizing a population of potential solutions, referred to as particles, and guiding their traversal across the search space using a simple mathematical formula involving their position and velocity. Through this approach, complex problems can be tackled. Notably, the particles are not only guided towards the best-known positions within the search space, which are updated as other particles discover superior locations, but they are also influenced by their individual locally best-known positions. Hence, the swarm should migrate toward better answers as a result of this as demonstrated in Fig. 1. Mathematically and more intuitively, PSO works thus:

- Initialise a random population of agents(particles). Assuming we have P particles, each particle's position at iteration t is denoted by:

$$X^i(t) = (x^i(t), y^i(t)) \tag{1}$$

and its velocity is:

$$V^i(t) = (v_x^i(t), v_y^i(t)) \tag{2}$$

- Assess the position of each particle by evaluating the objective function. If a particle's current position proves superior to its previous one, proceed to update both its position and the associated velocity accordingly.

$$X^i(t+1) = X^i(t) + V^i(t+1) \tag{3}$$

- Update Each particle's velocity by the following rule

$$V^i(t+1) = wV^i(t) + c_1 r_1 (P_b^i - X^i(t)) + c_2 r_2 (G_b - X_i(t)) \tag{4}$$

where r_1 and r_2 represent random values within the range of 0 and 1 (they play a vital role in preventing premature convergence and enhancing the likelihood of identifying global optima). w, c_1, and c_2 are hyperparameters specific to the PSO algorithm. c_1 is a positive constant serving as an individual cognitive parameter, determining the significance of a particle's past experiences. It quantifies the importance of a particle's prior knowledge. c_2 functions as a social learning parameter, indicating the importance of collective learning within the entire swarm. The term $(G_b - X^i(t))$ operates as an attractive force, pulling a particle towards the optimal point until a predetermined stopping criterion is met. Here, G_b denotes the globally best position within the search space, representing the aggregated best position across all particles in the swarm. P_b^i signifies the finest position within the search space for *particle i*.

2.2 Genetic Algorithm

Genetic Algorithms (GAs) represent search and optimization techniques rooted in two biological principles: the process of natural selection and the mechanisms of natural genetics [24]. GAs function as stochastic global search methods, simulating the evolutionary process observed in nature [25]. This approach falls under the category of probabilistic optimization methods that leverage concepts of natural selection and genetic inheritance. GAs operate by initiating exploration from multiple independent points concurrently, thus sidestepping issues such as local minima and convergence to suboptimal solutions [26, 27]. The foundation of genetic algorithms is built upon an analogy drawn from the genetic composition and behavior of chromosomes. This analogy can be articulated as follows:

- Individuals within a randomized population engage in competition for resources and mating.
- The successful individuals(best fit) then mate to create more off-springs than others.
- Genes from the best-fit individuals propagate through the generations, Consequently, each subsequent generation becomes progressively better adapted to its environment than the previous one.

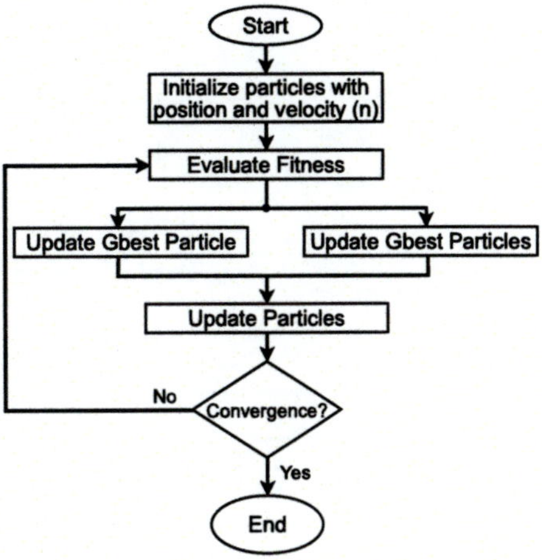

Fig. 1 Particle swarm optimization algorithm

The population of individuals is kept inside the search space. In the search space for a given problem, each individual represents a solution. Each person is coded as a finite-length vector (similar to a chromosome) of components. Genes are equivalent to these changeable components. As a result, a chromosome (individual) comprises multiple genes (variable components) (Fig. 2).

Each individual is assigned a Fitness Score, which indicates their capacity to "compete." Individuals with optimal (or near ideal) fitness scores are sought. The GAs keeps track of n individuals (chromosomes/solutions) and their corresponding fitness evaluations. Individuals with higher fitness ratings have a better probability of reproducing than others. Individuals with higher fitness ratings are chosen to mate and generate healthier offspring by merging their parents' chromosomes [28, 29].

3 Ball on Plate System

This system consists of two major components. The DC Motor model is an electromechanical system that receives input voltage signals and converts this into an angular inclination of the plate. The second component is the ball placed on the plate, it takes in as input the angular inclination of the plate and outputs a displacement of the ball (in a circular trajectory for the purpose of this paper).

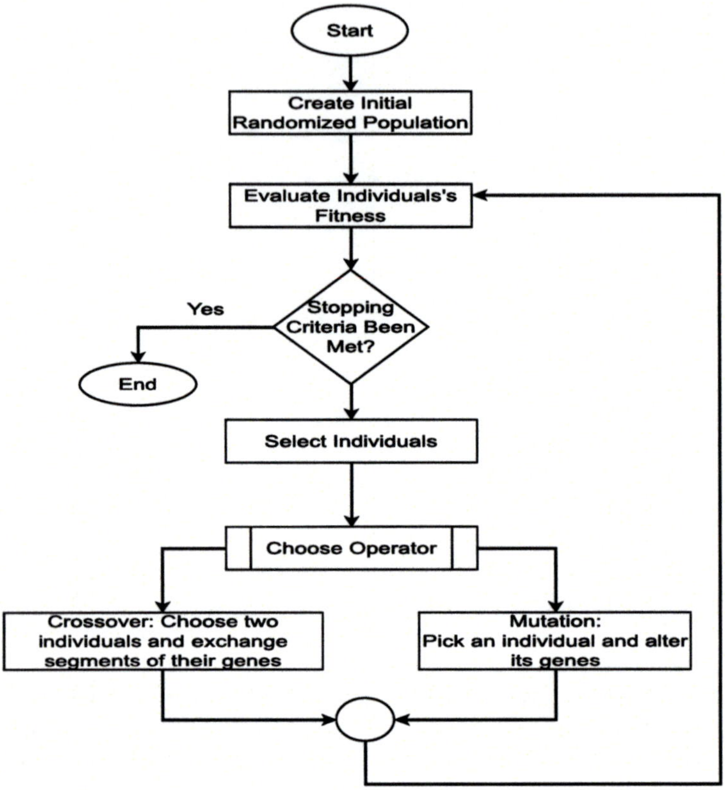

Fig. 2 GA optimization process

3.1 Modelling of the Ball & Plate System

In this section, we proceed by modeling the ball-plate system using the standard Lagrangian Equations [30–32]. Several assumptions are introduced to streamline the modeling procedure:

- The friction between the ball and the plate is disregarded
- The ball doesn't slip.
- Ball and Plate are in contact at every instant.

$$\frac{d}{dt}\frac{\partial K}{\partial \dot{q}_i} - \frac{\partial K}{\partial q_i} + \frac{\partial P}{\partial q_i} = Q_i \tag{5}$$

q_i represents the i-th directional coordinate, K denotes the kinetic energy, and P stands for the potential energy. Q represents the cumulative effect of forces acting upon the system. The system possesses four Degrees of Freedom (DOF): two related

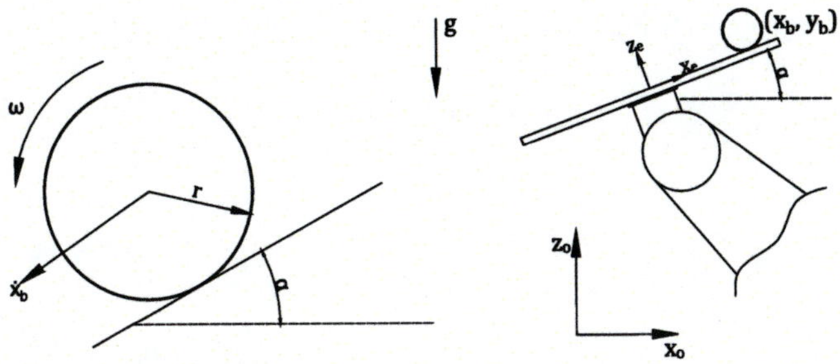

Fig. 3 Side view ball on plate system

to the ball's direction of motion and two associated with the plate's angle of inclination. x_b and y_b are assumed to signify the position of the ball on the plate concerning the x and y coordinate axes. α and β indicate the angular inclination of the plate (Fig. 3).

The sum of the transitional and rotational energies represent the combined kinetic energy of the ball (K_b):

$$K_b = \frac{1}{2}m_b(\dot{x}_b^2 + \dot{y}_b^2) + \frac{1}{2}I_b(w_x^2 + w_y^2) \tag{6}$$

where m_b is the mass of the ball, I_b is the moment of inertia of the ball, $\dot{x}_b^2$ and $\dot{y}_b^2$ represent the ball's translational velocities, w_x^2 and w_y^2 denote the ball's rotational velocities along the x and y axes respectively.

The connection between translational and rotational velocities is defined as follows:

$$\dot{x}_b = r_b w_x, \dot{y}_b = r_b w_y \tag{7}$$

By incorporating Eq. (10) into Eq. (9) and treating the ball as a point mass at (x_b, y_b), and accounting for its rotational energy relative to its center of mass, the plate's kinetic energy can be expressed as:

$$K_p = \frac{1}{2}(I_b + I_p)(\dot{\alpha}^2 + \dot{\beta}^2) + \frac{1}{2}m_b(x_b\dot{\alpha} + y_b\dot{\beta})^2 \tag{8}$$

Where α and β denote the plate's angles of inclination along the x and y axes, respectively. Its derivatives are the corresponding rotational velocities. The overall system's kinetic energy is given by:

$$K = K_p + K_b \tag{9}$$

The ball's potential energy is given by:

$$P_b = m_b g(x_b \sin\alpha + y_b \sin\beta) \tag{10}$$

Applying Lagrangian using Eqs. (9)–(12) we get the following system of equations:

$$L = K_b + K_p - P_b \tag{11}$$

$$\frac{\partial K}{\partial \dot{\alpha}} = (I_b + I_p)\dot{\alpha}_x^2 + m_b x_b(x_b\dot{\alpha} + y_b\dot{\beta}),\ \frac{\partial L}{\partial \alpha} = mg\sin\alpha \tag{12}$$

$$\frac{\partial K}{\partial \dot{\beta}} = (I_b + I_p)\dot{\beta}_x^2 + m_b y_b(x_b\dot{\alpha} + y_b\dot{\beta}),\ \frac{\partial L}{\partial \beta} = mg\cos\beta \tag{13}$$

$$\frac{\partial K}{\partial \dot{x_b}} = (m_b + \frac{I_b}{r_b^2})\dot{x}_b,\ \frac{\partial L}{\partial x_b} = m_b(x_b\dot{\alpha} + y_b\dot{\beta})\dot{\alpha} \tag{14}$$

$$\frac{\partial K}{\partial \dot{y_b}} = (m_b + \frac{I_b}{r_b^2})\dot{y}_b,\ \frac{\partial L}{\partial y_b} = m_b(x_b\dot{\alpha} + y_b\dot{\beta})\dot{\beta} \tag{15}$$

We assume the corresponding torques exerted on the plate as τ_x and τ_y. From Euler-Langrange equations we get:

$$\frac{d}{dt}\frac{\partial K}{\partial \dot{\alpha}} - \frac{\partial L}{\partial \alpha} = (I_b + I_p)\ddot{\alpha} + m_b x^2\ddot{\alpha} + 2m_b x_b\dot{\alpha} + m_b x_b y_b\ddot{\alpha} + m_b\dot{x}_b y_b\ddot{\beta} + m_b x_b\dot{y}\dot{\beta} - m_b g\cos\alpha \tag{16}$$

$$\frac{d}{dt}\frac{\partial K}{\partial \dot{\beta}} - \frac{\partial L}{\partial \beta} = (I_b + I_p)\ddot{\beta} + m_b x^2\ddot{\beta} + 2m_b x_b\dot{\beta} + m_b x_b y_b\ddot{\beta} + m_b\dot{x}_b y_b\ddot{\alpha} + m_b x_b\dot{y}\dot{\alpha} - m_b g\cos\beta \tag{17}$$

$$\frac{d}{dt}\frac{\partial K}{\partial \dot{x}_b} - \frac{\partial L}{\partial \dot{x}_b} = (m_b + \frac{I_b}{r_b^2})\ddot{x} - m_b(x_b\dot{\alpha} + y_b\dot{\beta})\dot{\alpha} + m_b g\sin\alpha = 0 \tag{18}$$

$$\frac{d}{dt}\frac{\partial K}{\partial \dot{y}_b} - \frac{\partial L}{\partial \dot{y}_b} = (m_b + \frac{I_b}{r_b^2})\ddot{y} - m_b(x_b\dot{\alpha} + y_b\dot{\beta})\dot{\alpha} + m_b g\sin\beta = 0 \tag{19}$$

We can represent the system's differential equation as follows:

$$(m_b + \frac{I_b}{r_b^2})\ddot{x}_b - m_b(x_b\dot{\alpha}^2 + y_b\dot{\alpha}\dot{\beta}) + m_b g\sin\alpha = 0 \tag{20}$$

$$(m_b + \frac{I_b}{r_b^2})\ddot{y}_b - m_b(y_b\dot{\beta}^2 + x_b\dot{\alpha}\dot{\beta}) + m_b g\sin\beta = 0 \tag{21}$$

$$\tau_x = (I_p + I_b + m_b x_b^2)\ddot{\alpha} + 2m_b x_b \dot{x}_b \dot{\alpha} + m_b x_b y_b \ddot{\beta} + m_b \dot{x}_b y_b \dot{\beta} + m_b g x_b \cos\alpha \tag{22}$$

$$\tau_y = (I_p + I_b + m_b y_b^2)\ddot{\beta} + 2m_b y_b \dot{y}_b \dot{\beta} + m_b y_b x_b \ddot{\alpha} + m_b \dot{x}_b y_b \dot{\alpha} + m_b g y_b \cos\beta \tag{23}$$

The two previous equations illustrate the impact of external torque on the ball and plate system. However, due to their complexity, in order to apply the Laplace transform to the system, it is necessary to linearize the model.

3.2 *Linearisation of Ball on Plate System*

We make use of an approximate value for the moment of inertia of a solid ball, given by:

$$I_b = \frac{2}{5} m_b r_b^2 \tag{24}$$

Therefore, Eqs. (23) and (24) can be rewritten as:

$$m_b[2\ddot{x}_b - (x_b\dot{\alpha}^2 + y_b\dot{\alpha}\dot{\beta}) + g\sin\alpha] = 0 \tag{25}$$

$$m_b[2\ddot{y}_b - (y_b\dot{\beta}^2 + x_b\dot{\alpha}\dot{\beta}) + g\sin\beta] = 0 \tag{26}$$

For small plate's inclinations angles, we approximate $\sin\alpha \approx \alpha$ & $\sin\beta \approx \beta$. We further assume that $\dot{\alpha} <<0$ & $\dot{\beta} <<0$ hence $\dot{\alpha}\dot{\beta} \approx 0$, $\dot{\alpha}^2 \approx 0$, $\dot{\beta}^2 \approx 0$. Therefore, Eqs. (28) and (29) become:

$$\frac{5}{7}\ddot{x}_b + g\alpha = 0 \tag{27}$$

$$\frac{5}{7}\ddot{y}_b + g\beta = 0 \tag{28}$$

Two separate differential equations for the x and y axes results from Linearizing the aforementioned equations. Considering alpha and beta as inputs to the system, we obtain the following transfer function:

$$P_x(s) = \frac{X_b(s)}{\alpha(s)} = \frac{g}{\frac{5}{7}s^2} \tag{29}$$

$$P_y(s) = \frac{Y_b(s)}{\beta(s)} = \frac{g}{\frac{5}{7}s^2} \tag{30}$$

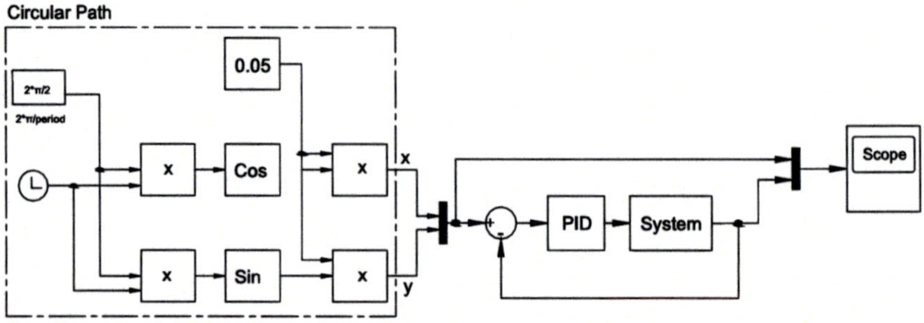

Fig. 4 Simulink block model for ball on plate system

3.3 *Simulink Model*

We consider a circular trajectory (of radius 0.05 m) for the ball in our system. This trajectory is generated as a result of the various angles of inclinations of the plate at time t. The Simulink model is illustrated below (Fig. 4):

4 PID Controller Tuning

Tuning a PID Controller using metaheuristic approaches to determine the best fitness function value (minimal) will result in optimal gain parameters for the controller. Conventionally used fitness functions are ITAE, Integral of Absolute Magnitude of Error (IAE), Mean Squared Error (MSE), and Integral Squared Error (ISE). For the purpose of this paper, we used the IAE function coupled with other parameters for the Genetic Algorithm, and Particle Swarm Optimization Algorithm. We assumed a common search space for the two algorithms (for each parameter of the search). Table 1 summarizes various parameters of our system.

Table 1 Metaheuristic algorithm parameters

Algorithm parameters		
Parameters	GA	PSO
Minimum limit	[0 0 0]	[0 0 0]
Maximum limit	[1000 1000 1000]	[1000 1000 1000]
Size	20	20
Cost	IAE	IAE
Stopping critetion	Max generation	Max iteration

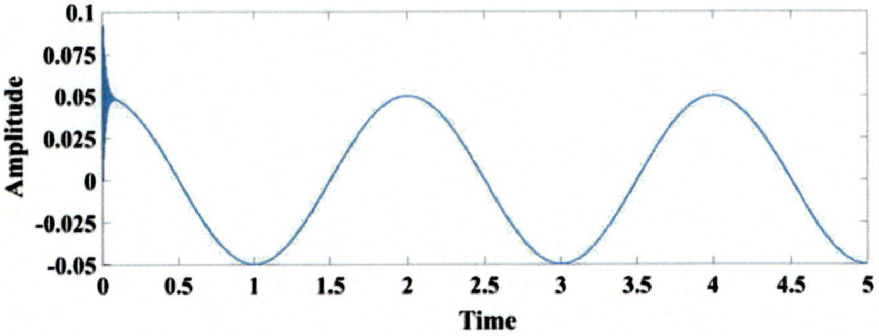

(a) Sustained Oscillations on X-axis due to GA

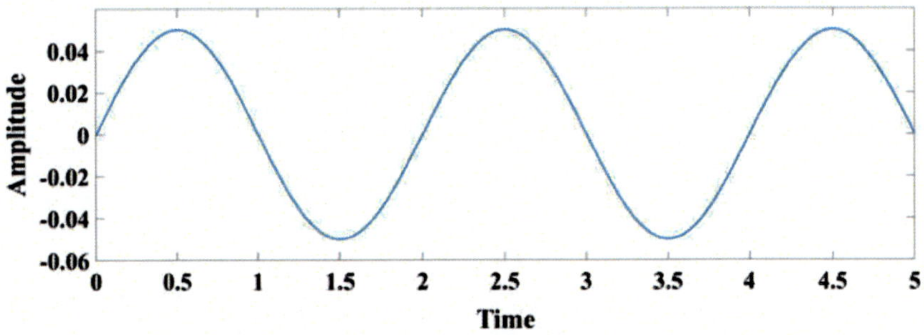

(b) Sustained Oscillations on Y-axis due to GA

Fig. 5 Oscillations due to GA

5 Results and Discussions

For the purpose of this paper, each algorithm with the general parameters mentioned above and some peculiar parameters to each was run for 10 different simulations. For each simulation, a fitness value and optimized gain parameters were found.

5.1 Genetic Algorithm

After 10 simulation runs, we found a series of values for the optimized PID Gains. These results are shown in Table 2. The associated sustained system oscillation for the x & y axis considering circular path input are illustrated in Fig. 5. However, it should be noted that for a circular path, the sustained oscillations are a series of cosine and sine waves in the x and y axis respectively.

Table 2 Optimized PID gains from GA

GA results				
Runs	Kp	Ki	Kd	Fitness function
1	0	0.0417	977.7021	0.0010
2	396.9864	0.0156	978.6324	0.0011
3	1.0477	59.6474	930.3387	0.0011
4	954.0604	42.1491	941.6162	0.0011
5	743.1393	63.6624	999.9196	0.0011
6	966.1418	344.5075	999.9447	0.0011
7	49.9890	0.2852	863.6238	0.0010
8	754.62	114.8941	945.20428	0.0011
9	719.8954	274.9312	999.7826	0.0011
10	963.3004	1649.7432	914.2055	0.0011

Table 3 Optimized PID gains from PSO

PSO results				
Runs	Kp	Ki	Kd	Fitness function
1	1000	1000	345.1874	0.0040
2	1000	1000	345.4524	0.0040
3	1000	1000	345.4631	0.0040
4	1000	1000	345.4640	0.0040
5	1000	1000	345.4632	0.0040
6	1000	1000	345.4584	0.0040
7	1000	1000	345.4635	0.0040
8	1000	1000	345.4622	0.0040
9	1000	1000	345.4615	0.0040
10	1000	1000	345.4688	0.0040

All optimized gains were tested on the system. The best optimized PID gains were found to be at Run 7, with a faster response time of 1.05e-3 s, and relative minimal overshoot as opposed to other runs.

5.2 Particle Swarm Optimization

The results obtained are illustrated in Table 3. The best of the optimized gains with a faster response time of 2.5e-3 s and 2% overshoot were found to be at run 9. Figure 6 illustrates the system's oscillation for the x & y axises after the optimized PSO gains were applied to the system.

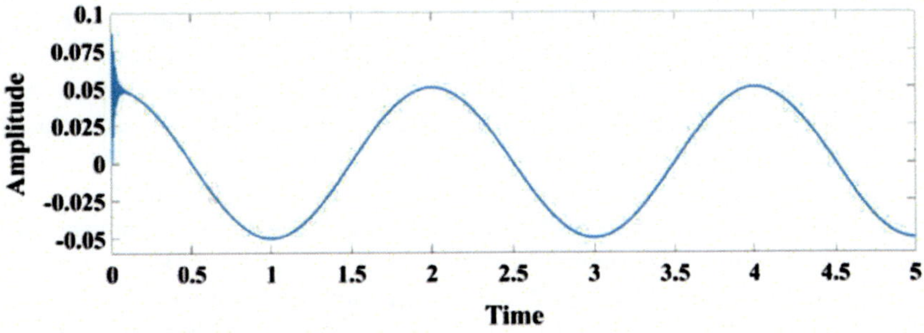

(a) Sustained Oscillations on X-axis due to PSO

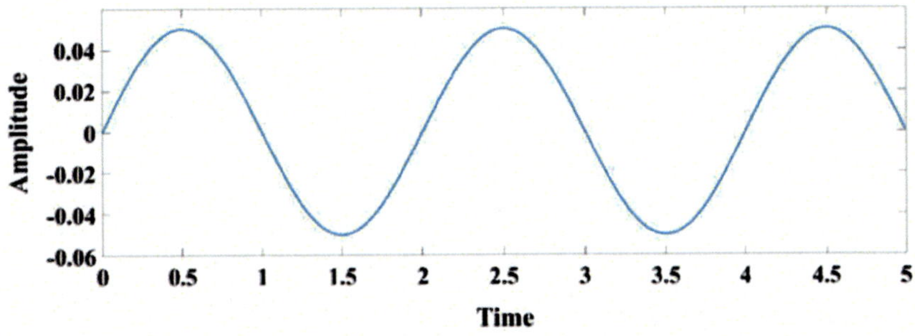

(b) Sustained Oscillations on Y-axis due to PSO

Fig. 6 Oscillations due to PSO

6 Conclusion

The study showcased in this paper highlights the effectiveness of metaheuristic optimization algorithms in optimizing PID controller parameters for a ball & plate system. The results show that GA and PSO can improve the performance of the controller, but GA outperforms PSO in terms of fitness function value and optimal gains.

The study also highlights the advantages of PID controllers in process industries, including their simplicity, transparency, reliability, and high efficiency. PID controllers have been a staple in the industrial domain for process control for more than a century, and they continue to be the predominant choice for controllers in various sectors of process and manufacturing industries at present.

However, the study suggests that the system model's accuracy is critical to the performance of a PID controller. Inaccurate or incomplete information can lead to poor performance, and conventional tuning techniques may not be sufficient to optimize the controller. This is where metaheuristic optimization algorithms, such as GA and PSO, can be particularly useful, as they can find near-optimal solutions based on imperfect or incomplete information.

The findings drawn from this research carry significant implications for the utilization of metaheuristic algorithms in control and process engineering research. The study demonstrates the potential for further optimization of PID controllers in various applications, and it highlights the need for accurate system models and variables to achieve optimal performance.

In conclusion, the study outlined in this paper showcases the effectiveness of metaheuristic optimization algorithms in optimizing PID controller parameters for a ball & plate system. The study also highlights the advantages of PID controllers in process industries and the potential for further optimization using metaheuristic algorithms. The outcomes of this study hold significant implications for the use of PID controllers in various applications and demonstrate the potential for further research in this area.

References

1. Joseph, S.B., Dada, E.G., Abidemi, A., Oyewola, D.O., Khammas, B.M.: Metaheuristic algorithms for PID controller parameters tuning: Review, approaches and open problems. Heliyon e09399 (2022)
2. Bharat, S., Ganguly, A., Chatterjee, R., Basak, B., Sheet, D.K., Ganguly, A.: A review on tuning methods for PID controller. In: Asian Journal For Convergence In Technology (AJCT) ISSN-2350-1146 (2019)
3. Zoppoli, R., Sanguineti, M., Gnecco, G., Parisini, T., Zoppoli, R., Sanguineti, M., Parisini, T.: Optimal control problems over an infinite horizon. Neural Approx. Optim. Control Decis. 471–511 (2020)
4. Bharat, S., Ganguly, A., Chatterjee, R., Basak, B., Sheet, D.K., Ganguly, A.: A review on tuning methods for PID controller. Asian J. Converg, Technol (2019)
5. Bansal, H.O., Sharma, R., Shreeraman, P.: PID controller tuning techniques: A review. J. Control Eng. Technol. **2**, 168–176 (2012)
6. Wang, L.: PID Control System Design and Automatic Tuning using MATLAB/Simulink. John Wiley &Sons (2020)
7. Purnama, H.S., Sutikno, T., Alavandar, S., Subrata, A.C.: Intelligent control strategies for tuning PID of speed control of DC motor-a review. In: Proceedings of the 2019 IEEE Conference on Energy Conversion (CENCON), pp. 24–30. IEEE (2019)
8. Ziegler, J.G., Nichols, N.B.: Optimum settings for automatic controllers. In: Transactions of the American Society of Mechanical Engineers (ASME), p. 64 (1942)
9. Hou, Z., Xiong, S.: On model-free adaptive control and its stability analysis. IEEE Trans. Autom. Control **64**(11), 4555–4569 (2019)
10. Asiain, E., Garrido, R.: Anti-chaos control of a servo system using nonlinear model reference adaptive control. Chaos Solitons Fract. **143**, 110581 (2021)
11. Borges, F.G., Guerreiro, M., Monteiro, P.E.S., Janzen, F.C., Corrêa, F.C., Stevan, S.L., Jr., Kaster, M.D.S.: Metaheuristics-based optimization of a robust GAPID adaptive control applied to a DC motor-driven rotating beam with variable load. Sensors **22**(16), 6094 (2022)
12. Yu, W., Rosen, J.: Neural PID control of robot manipulators with application to an upper limb exoskeleton. IEEE Trans. Cybern. **43**(2), 673–684 (2013)
13. Wei, L., Yanyu, Z., Yong, G., Xiuli, C.: Algorithm research and real-time simulation of neural network sliding mode position control. In: Proceedings of the 25th Chinese Control and Decision Conference (CCDC), pp. 1904–1907 (2013)

14. Sahu, P.K., Shaw, P., Maity, S.: Fixed-frequency sliding mode control for power quality improvement of a grid-connected inverter. In: Proceedings of the Annual IEEE India Conference (INDICON), pp. 1–6 (2015)
15. Nasiri, N., Fakharian, A., Menhaj, M.B.: A novel controller for nonlinear uncertain systems using a combination of SDRE and function approximation technique: Regulation and tracking of flexible-joint manipulators. J. Franklin Inst. **358**(10), 5185–5212 (2021)
16. Wang, M., Dong, X., Ren, X., Chen, Q.: SDRE based optimal finite-time tracking control of a multi-motor driving system. Int. J. Control **94**(9), 2551–2563 (2021)
17. Giernacki, W., Stepien, S., Chodnicki, M., Wroblewska, A.: Hybrid Quasi-Optimal PID-SDRE Quadrotor Control. Energies **15**(12), 4312 (2022)
18. Peres, F., Castelli, M.: Combinatorial optimization problems and metaheuristics: Review, challenges, design, and development. Appl. Sci. **11**(14), 6449 (2021)
19. Zahir, A.A.M., Hassan, M.A.H.: Genetic algorithm for PID controller for brushed DC motor. In: Hassan, M.A.H. (ed.) Intelligent Manufacturing and Mechatronics (2018)
20. Dhamane, S.A., et al.: Ant colony optimization algorithm based on optimal PID parameters for a robotic arm. In: International Journal of Control Systems and Robotics
21. Leon, J.A., Palacios, C.C., Vargas-Salgado, C., et al.: Particle swarm, genetic algorithm, grey wolf optimizer algorithms performance comparative for DC-DC boost converter. Adv. Sci. Technol. Eng. Syst. J. **6**(1), 619–625 (2021)
22. Ammar, H., Azar, A., Shalaby, R., Ibrahim, M.: Metaheuristic optimization of fractional order incremental conductance (FO-INC) maximum power point tracking (MPPT). Complexity 1–13 (2019) Hindawi Limited
23. Kennedy, J., Eberhart, R.C.: Particle swarm optimization. In: Proceedings of International Conference on Neural Networks (ICNN95) (1995)
24. Ahmmed, T., Akhter, I., Karim, S.R., Ahamed, F.S.: Genetic algorithm based PID parameter optimization. Am. J. Intell. Syst. **10**(1), 8–13 (2020)
25. Sreekanth, P., Hari, A.: Genetic algorithm based self-tuning regulator for ball and hoop system. In: 2016 Conference on Emerging Devices and Smart Systems (ICEDSS), pp. 147–152. IEEE (2016)
26. Barman, A., Dutta, S., Tiwari, K., Roy, S., Pain, S.: Genetic algorithm based adaptive PID tuning of time delay process. In: Artificial Intelligence: First International Symposium. ISAI 2022, Haldia, India, February 17–22, 2022, Revised Selected Papers, pp. 64–75. Springer Nature Switzerland, Cham (2023)
27. Skogestad, S.: Probably the best simple PID tuning rules in the world, vol. 77. AIChE Annual Meeting. Reno, Nevada (2001)
28. Hassanat, A., Almohammadi, K., Alkafaween, E.A., Abunawas, E., Hammouri, A., Prasath, V.S.: Choosing mutation and crossover ratios for genetic algorithms-a review with a new dynamic approach. Information **10**(12), 390 (2019)
29. Katoch, S., Chauhan, S.S., Kumar, V.: A review on genetic algorithm: past, present, and future. Mult. Tools Appl. **80**, 8091–8126 (2021)
30. Fan, X., Zhang, N., Teng, S.: Trajectory planning and tracking of ball and plate system using hierarchical fuzzy control scheme. Fuzzy Sets Syst. **144**(2), 297–312 (2004)
31. Lee, K.K., Btz, G., Wollherr, D.: Basketball robot: Ball-on-plate with pure haptic information. In: IEEE International Conference on Robotics and Automation (2008)
32. Yuan, D., Zhang, Z.: Modelling and control scheme of the ball-plate trajectory-tracking pneumatic system with a touch screen and a rotary cylinder. Control Theory Appl. IET **4**(4) (2010)

Frequency Criterion for the Existence of Sliding Processes in Control Systems with an Arbitrary Variable Structure

V. L. Kodkin and E. A. Kuznetsova

Abstract The article proposes a criterion for the existence of sliding processes according to the frequency characteristics of the control device and the control object. This approach is proposed for the first time, in contrast to the method of phase trajectories and state space equations used to analyze systems with a variable structure. This approach allows us to formulate several practical engineering provisions that are very important for the implementation of this control method in real electric drives. The main conclusions are confirmed by modeling.

Keywords Variable structure control systems (VSS) · Sliding modes (SM) · Slow and fast movements · Ideal slide · Phase trajectories · Control object

1 Introduction

Systems with a variable structure (VSS) with sliding modes (SM) are a vivid example of the effectiveness of the application of new mathematical methods in technical automation systems.

In these systems, the theory of which is exhaustively described in the books of Emelyanov [1, 2], Utkin [3], Emelyanov and Korovin [4], transient processes are formed by switching the feedback sign in the vicinity of the slip space (or trajectory).

The essence of this control is that in a certain region of the phase space around the switching surface, conditions of counter motions are created. The phase trajectory is "decomposed" into two movements: fast and slow. These movements have the necessary rigidity to external disturbances and changes in the system parameters. When the sign of the feedback changes, the control forms a switching line (surface), on which the phase trajectory of the automatic control system hits it and remains on it until the end of the transient process. In this case, the switching surface is a "slow" movement, and movements around the surface are "fast".

V. L. Kodkin (✉) · E. A. Kuznetsova
South Ural State University, Chelyabinsk 454080, Russia
e-mail: kodkinvl@susu.ru

D. Vlachos (ed.), *Mathematical Modeling in Physical Sciences*, Springer Proceedings in Mathematics & Statistics 446, https://doi.org/10.1007/978-3-031-52965-8_11

This principle of formation of the transient process seemed very promising for electromechanical systems, in which there are always fast and slow coordinates. Fast—armature current and motor armature voltage; slow—speed and angular displacement of the output shaft. However, studies of such systems very quickly confirmed that special criteria are required.

Moreover, some factors that were little taken into account in linear systems, for example, small delays of dynamic links, static external perturbations, limitations of mechanical moments, etc. in systems with a sliding mode can lead to qualitative changes in transient processes.

2 Formulation of the Problem. Conditions for Ideal Sliding VSS of the Second Order

Let us consider the sliding conditions in the simplest system of the second order, in which the object of control (CO) will be a double integrator, as the link most prone to instability, the controller (C) with a coefficient K, and the shaper of the switching trajectory (TG) is a forcing link with a constant time T1 (Fig. 1).

System of differential equations of VSS:

$$\begin{cases} T^2\ddot{x} + K|x|signS = 0 \\ S = T_1\dot{x} + x \end{cases} \quad (1)$$

The conditions for the existence of an ideal SM for a second-order VSS are as follows:

I. The condition for the trajectory to move to the switching surface $S = 0$ for $S < 0$ and $S > 0$;
II. The switching trajectory $S = 0$ is a stable monotone trajectory;
III. The condition of oncoming movement is in the vicinity of $S = 0$, at $S < 0$ the condition $S > 0$ must be satisfied.

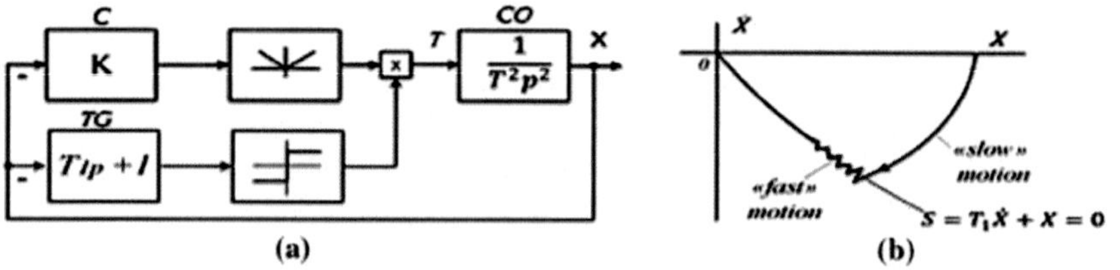

Fig. 1 Structural diagram of the VSS of the second order (**a**) and the phase trajectory of the process (**b**)

These conditions can be interpreted as stability conditions for the second Lyapunov method [5], for which the Lyapunov function will have a "variable structure".

The condition for the existence of sliding modes can be obtained using the equivalent control method from Kodkin [8].

The condition I is satisfied for any parameters T^2 and K. Equation (1) for $S > 0$, $X > 0$ takes the form:

$T^2X + KX = 0$ and $\dot{S} < 0$ always at $S > 0$ (Fig. 1b).

The condition II is always fulfilled for $T_1 > 0$;

The condition III looks like this: $\dot{S} > 0$ for $S \leq 0$;

$\dot{S}$ from the system of Eq. (1) is $\dot{S} = T_4\dot{X} + \ddot{X}$;

From condition $S = 0 - \dot{X} = -\frac{X}{T_1}$,

$$\dot{S} = \frac{T_1}{T^2}KX - \frac{X}{T_4} = \frac{X}{T^2}\left(T_1^2K - T^2\right) > 0;$$

Final condition:

$$T_1 > \sqrt{\frac{T^2}{K}}$$

Let's call $T_{1boun.} = \sqrt{\frac{T^2}{K}}$ the boundary speed of the initial system, the condition connects the sliding trajectory (time constant T_1), the parameters of the control object (T_2) and the parameters of the controller (K).

This condition makes it possible to see a number of advantages of VSS over conventional subordinate control systems, for example:

The choice of T_1 greater than $T_{1boun.}$ will ensure the fulfillment of the slip conditions when the equation of the electric drive system deviates from the initial one (1). It should also be noted that the speed of the system is determined only by the parameter T and is not related to the coefficient K.

That is, the dynamics of the VSS when the sliding conditions are met does not depend on possible variations of K.

Condition (2) determines the conditions for an ideal sliding process—"slow" movement occurs along a trajectory for all initial conditions, and "fast" movements are oscillations around the sliding trajectory with an infinitely high frequency and an infinitely small amplitude. Such a movement is ideal for any electric drive.

This condition is obtained by the phase plane method. This method can be very visual, including for non-linear automatic control systems, but is almost helpless in the presence of high-order links in automatic control systems. Since there are always such links, this is a very significant problem.

The phase space method is much more complicated than the phase plane method and is less clear. It can rarely be used in engineering calculations. Frequency methods are much more widely used, that is, methods that use the frequency characteristics of control objects and control elements. Let us consider the conditions for the existence of SM in the VSS by their logarithmic frequency characteristics (LFC).

3 Interpretation of the Sliding Conditions of VSS Links According to LFC

Slip condition for VSS according to formula (2):

$$T_1 \geq \sqrt{\frac{T^2}{K}} \tag{2}$$

Thus, in the SPS described by system (1) there will be an ideal sliding process around $S = 0$, provided that conditions (2) are met, which can be formulated as follows:

The time constant of the sliding trajectory must be greater than the time constant of the contour formed by the link of the control object and the link-regulator.

Note that the link with the cutoff frequency $\omega = \sqrt{\frac{K}{T^2}}$ is the control object CO covered by feedback with the controller C with the coefficient K.

Figure 2 shows the LFC and transient diagrams for various options for fulfilling condition (2).

The process in Fig. 2a corresponds to ideal slip. Condition (2) is satisfied. In this case, the equivalent phase shift of the TG links and the contour is not less than

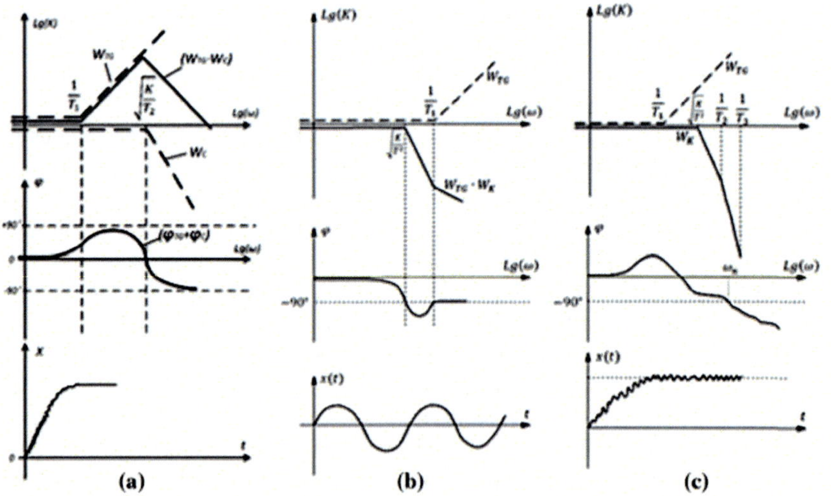

Fig. 2 LAFC, LPFC and transient processes in systems with ideal (a), critical (b) and real (c) slip

(−90°). This follows from the frequency characteristics of the SPS links shown in Fig. 2a.

If condition (2) is not met, the process is self-oscillatory (Fig. 2b). In this case, at a frequency less than the cutoff frequency of the sliding trajectory—$1/T_1$, the phase shift of the series connection of the TG links and the circuit reaches values less (−90°). There are no stable slow slips in VSS.

If condition (2) is satisfied for the values of the time constants of the control object, but there are delay or hysteresis links in the SPS that change the transfer functions of the control object (CO), then the real slip mode begins in the SPS—with finite values of frequency and amplitude. Due to the characteristics of the object, the phase shift reaches a value of (−90°) at a finite frequency and amplitude, which will determine the parameters of fast movements (Fig. 2c).

As mentioned above, the final condition connects the transfer functions of the control object—$\left(\frac{1}{T^2p^2}\right)$, the shaper of the switching trajectory—$(T_1p + 1)$ and the controller K. These transfer functions are strictly connected with the LFC of these links.

Note that the link with the cutoff frequency $\omega = \sqrt{\frac{K}{T^2}}$ is the object of regulation covered by feedback with the controller with coefficient K.

The sliding condition (2) can therefore be "transferred" to the frequency characteristics of VSS links as follows:

The ideal slip condition is met if two links—the slip shaper and the circuit including the controller and the controlled object form a series connection, the equivalent phase characteristic of which does not go over the value (−90°), and (−90°) is achieved at ω → ∞.

$$\varphi[W_K \cdot W_{TG}] > -90° \tag{3}$$

herein $W_K = \frac{W_{AM} \cdot W_{CO}}{1 + W_{AM} \cdot W_{CO}}$.

4 VSS Equivalent Diagram

The proposed frequency condition [6] will be satisfied if, when transferring the considered frequency characteristics to the complex space, the real part of their serial connection is positive:

$$\mathrm{Re}[W_K \cdot W_{TG}] > 0 \tag{4}$$

This condition is a special case of absolute stability [2] of an equivalent diagram containing the same links in a slightly different configuration, shown in Fig. 3.

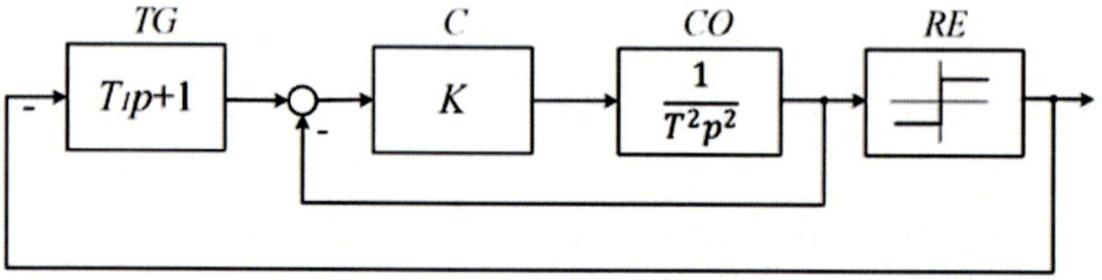

Fig. 3 Equivalent diagram of a second-order variable structure system

Figure 4 shows the timing diagrams of sliding transients in VSS for ideal (**a**), real (**c**), and broken (**b**) slips. Modeling of these regimes has been proposed repeatedly [8, 10].

Figure 4 shows the structure of the equivalent circuit model (4a) and the processes in this circuit—when the conditions of the ideal, real and frustrated sliding processes are met, this corresponds to the boundary stability of the equivalent circuit (4b), stability "with a margin" (4c), and the oscillatory process (4d). All modes fully comply with the provision on the stability of the VSS equivalent circuit proposed in this article.

For real electric drives, it is important to violate the conditions of ideal sliding, shown in Fig. 2c. Violation of the condition for the phase characteristic occurs in the high-frequency zone and can violate the conditions for the stability of the equivalent circuit at the same frequencies. Obviously, this is a condition of real slip, that is, with a finite frequency and a finite amplitude, and at the same time a condition of self-oscillations in the equivalent circuit. That is, this is an example of an VSS with a first order switching line and a control object of order higher than 2—the most important case for a real VSS. In the VSS theory, such cases are analyzed and the conditions are obtained, but it is quite difficult to apply them for engineering calculations. We offer an engineering approach.

Let us put forward a hypothesis that the stability condition (3), corresponding to the ideal slip conditions (2), will be preserved for any systems whose frequency characteristics satisfy condition (4). Then, applying the criterion of Popov [7], it is possible to obtain a very convenient interpretation of various variants of slip regimes, which is extremely important for the practical application of SM in VSS.

In a control system with a variable structure, with a general control object (Fig. 5), there will be a sliding mode if its equivalent diagram (Fig. 6) satisfies the condition of absolute stability. If self-oscillations with finite frequency and amplitude exist in the equivalent circuit, fast motions with the same parameters will exist in the VSS [8–10].

Such a sliding condition can be applied to electric drive systems with non-linear links and high-order links and be the basis of engineering criteria for sliding processes. By this criterion, it is possible to determine the parameters of real slip—their amplitude and frequency, which is very important for real electric drives.

The stability criterion in the equivalent circuit (Fig. 4) to assess the existence of slip in the original system (Fig. 1) can be used when analyzing SM in a system with

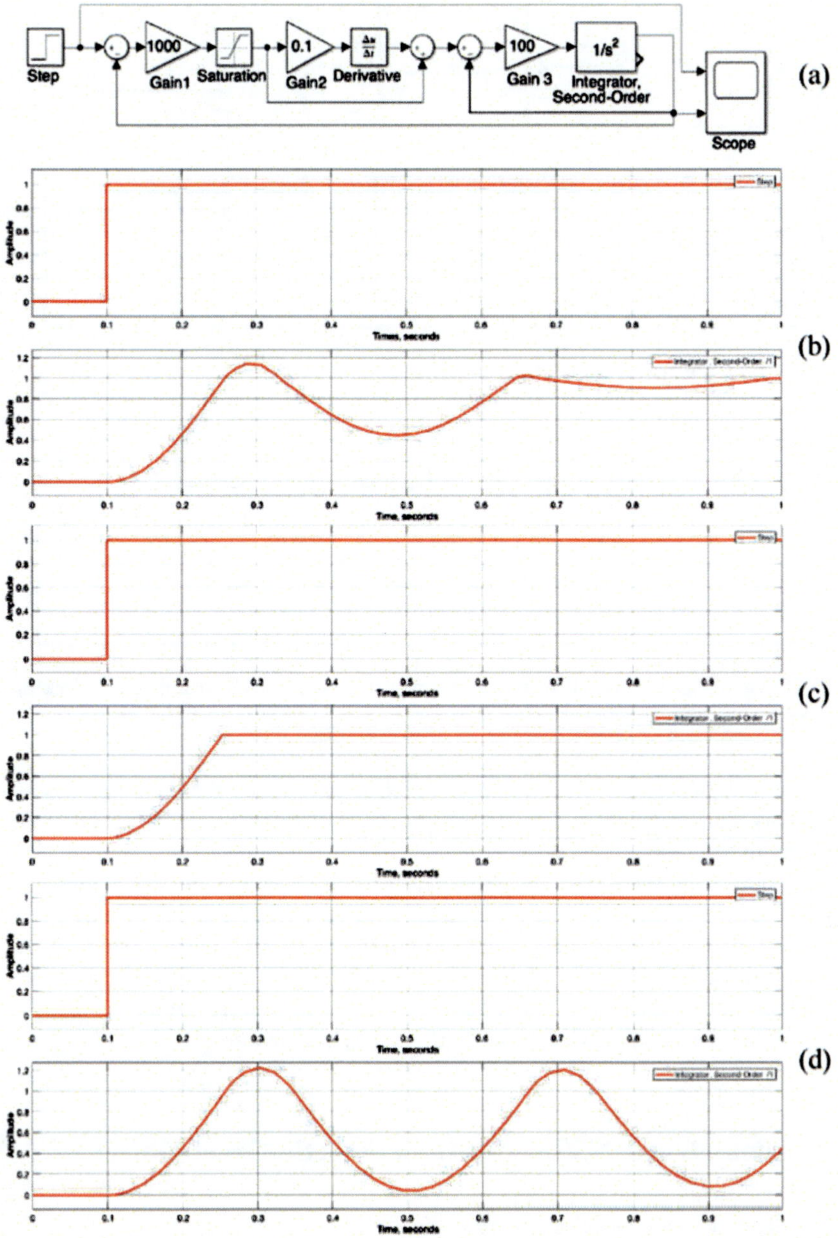

Fig. 4 Model of the VSS equivalent circuit with sliding processes and process diagrams in the model with a variable time constant of the differentiating link T: **b** T = 0,01s., **c** T = 0,1s., **d** T = 0,001s

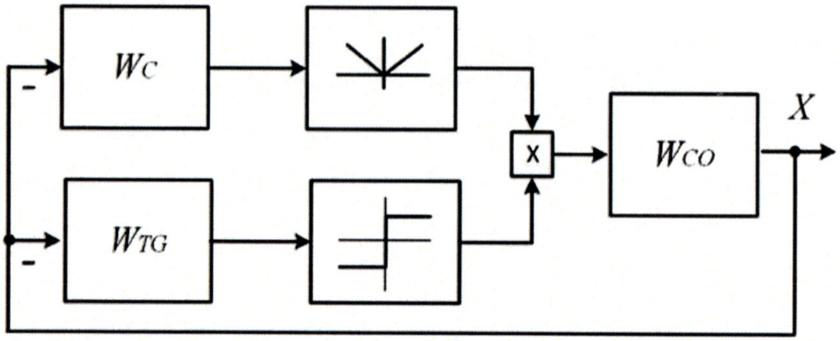

Fig. 5 VSS with dynamic links of an arbitrary order

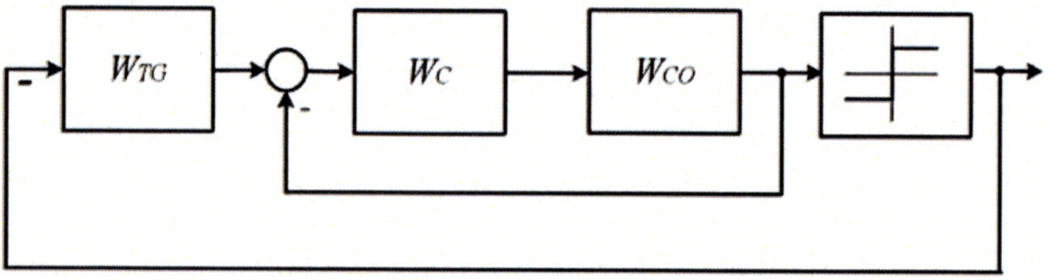

Fig. 6 Equivalent circuit of a control variable structure system and a control object of arbitrary order

almost any nonlinear object, i.e. any, which is covered by the theory of absolute stability, thus excluding a detailed analysis of each specific case, as is done in the theory of SM in the VSS [1–4].

It should be noted that the fulfillment of conditions (4) and (5) allows us to introduce the concept of "structural stability of the SPS" for the SPS, which is not violated with variations in the coefficients of the regulators. Since in real EMC it is always difficult to accurately determine a number of the most important parameters, the proposed concepts and conditions are of great practical importance.

Further, the main provisions are confirmed by simulation.

5 Simulation of VSS with a High (Fourth) Order Object

The diagram of the model is shown in Fig. 7.

Control object $T^2 = 1s^2$ and two inertial links with small time constants—0,001 s;

Regulator with K = 100;

Link forming slip—forcing link with T = 0,2s.

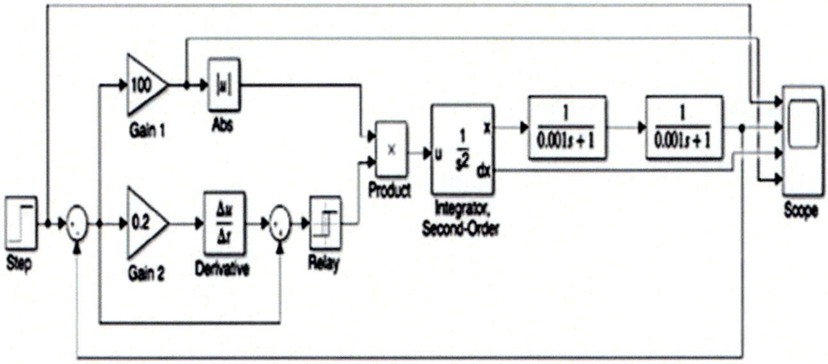

Fig. 7 Model diagram

Condition (2) is fulfilled:

$T_{1boun.} = 0{,}1\text{s}., T_1 = 0{,}2\text{s}.$

Figure 8 shows the processes in the VSS model.

Diagram 1—reference signal;

Diagram 2—adjustable coordinate;

Diagram 3—derivative of the adjustable coordinate;

Diagram 4—enhanced mismatch between the driving signal and the controlled coordinate.

In Fig. 8b, the processes in the same model with increased small time constants in the control object are up to 0,01s. The amplitude increased and the frequency of fluctuations of the controlled value derivative decreased (diagram 3). Controlled value diagram (Diagram 2) has not changed much.

The process in Fig. 9 is self-oscillations of all coordinates, there is no slip, the parameters of the slip shaper have been changed: the time constant has been reduced to 0,01s. Condition (2) is not satisfied.

6 Promising Offers

In control systems of real electric drives, most often the link of the slide path shaper is a forcing link of the first order, the time constant of which is determined by the requirements for the speed of the drive. The control object at the same time contains several inertial links with small time constants, except for the integrating link of the second order. It is quite difficult to accurately determine in this case the equation of the sliding trajectory using the equivalent control method [1]. Experiments have shown that for such a system, communication by the control signal of the electric drive—the control signal or the voltage at the motor armature—will be effective.

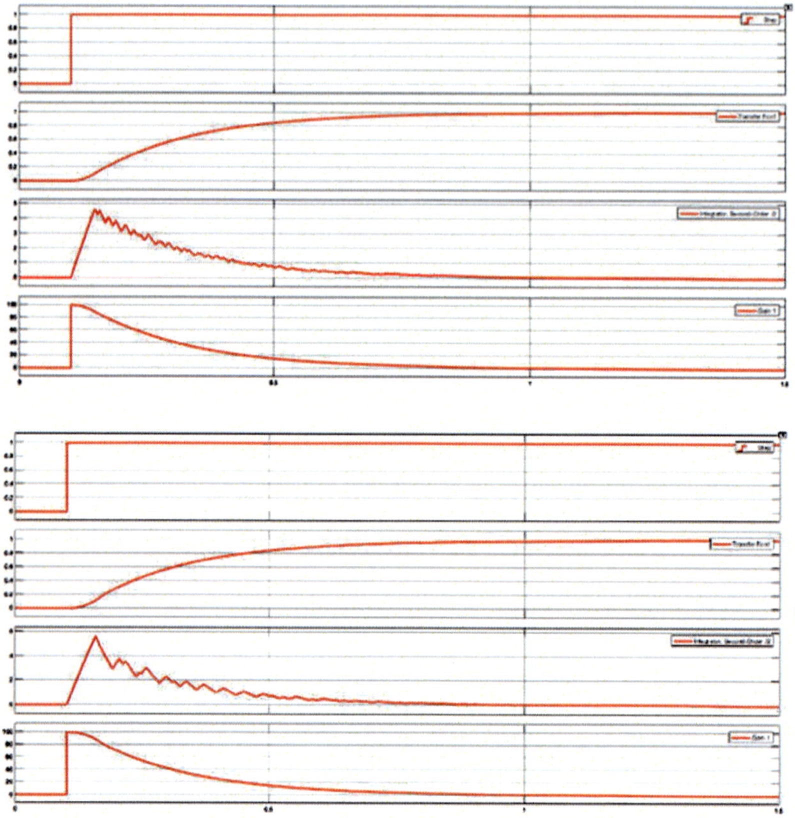

(a)

(b)

Fig. 8 Sliding processes in VSS with a high-order control object: **a** close to ideal slip (quasi-ideal slip); **b** real slip

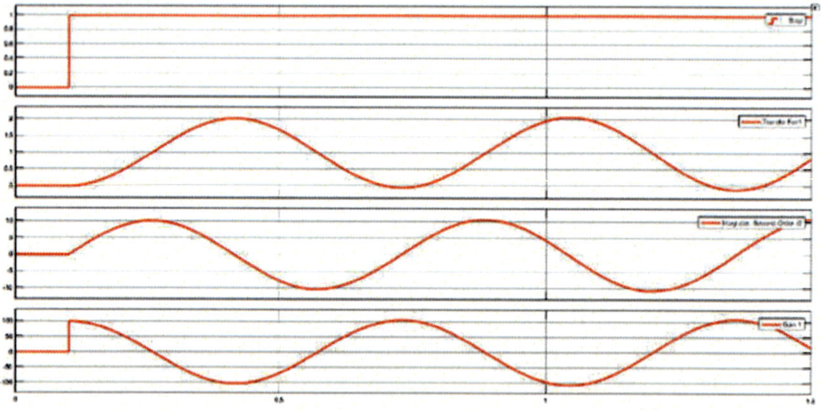

Fig. 9 Processes in VSS when condition (2) is violated

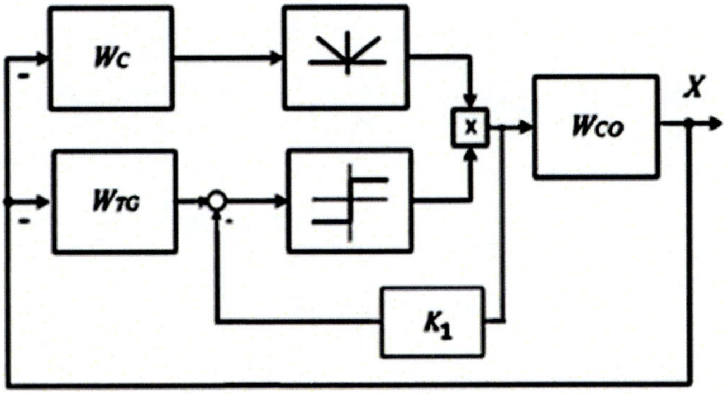

Fig. 10 VSS diagram

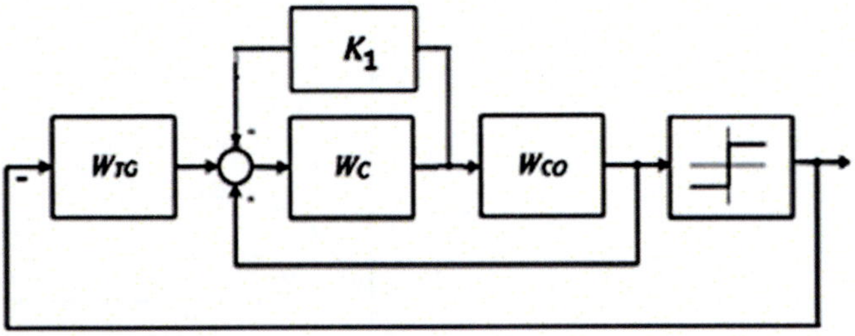

Fig. 11 Equivalent diagram

This relationship is shown in the VSS diagram (Fig. 10) and in the equivalent diagram (Fig. 11).

This signal in relation to the output will be a forced signal with the required parameters that will ensure the fulfillment of the sliding conditions (4) or the stability of the substitution circuit. Taking into account the connection, the condition will take the form (5).

$$\mathrm{Re}[W_k(W_{TG} + K_1 U)] > 0 \tag{5}$$

Since the conditions (2) will be met, the sliding will be very close to ideal with respect to the slow coordinate, the fast ones will have a minimum amplitude (Fig. 12).

Simulations were carried out confirming the effectiveness of cross-coupling. Figure 13a shows the VSS model with cross coupling and transient processes with a real sliding process with a coupling coefficient K1 equal to 0 (Fig. 13b).

Figure 13c shows processes with a coefficient K1 equal to 0.01, which makes the process quasi-ideal, i.e. without fluctuations around the slip line in the presence of high-order links in the control object. It was not possible to obtain such a result in VSS of an arbitrary order [8, 10].

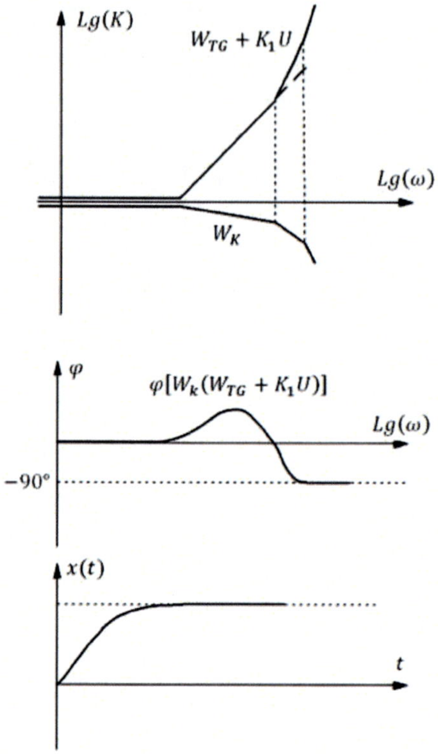

Fig. 12 LAFC, LPFC and transient process in the system

Figure 14 shows a model of the equivalent circuit of a system with slip and stable processes with changing parameters of self-oscillations in full accordance with the provisions of the article: self-oscillations decrease the amplitude, increase the frequency with the introduction of a cross-coupling, which reduces the slip amplitude to the parameters of quasi-ideal slip.

7 Conclusions

By the method of equivalent control, the conditions for ideal sliding in the SPS of the second order are obtained.

It is shown that these conditions can be interpreted by the conditions connecting the frequency characteristics of the control system elements: control objects, the controller and the link that forms the sliding trajectory.

It is shown that the VSS sliding condition is adequate to the stability conditions of the proposed nonlinear equivalent circuit.

It is suggested that this adequacy extends to VSS with objects of an arbitrary order.

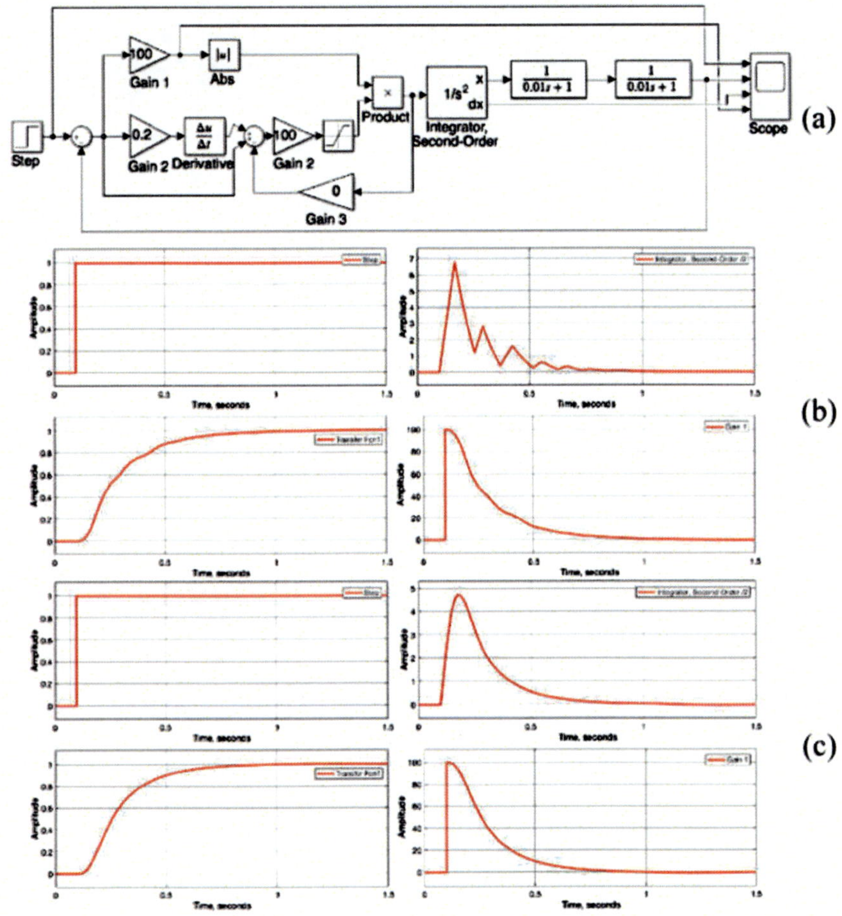

Fig. 13 VSS model (**a**), cross-coupling and transient processes with a real sliding process with a coupling coefficient K1 equal to 0 (**b**), processes with a coefficient K1 equal to 0.01 (**c**)

These provisions can become the basis for engineering criteria for the existence of sliding processes, quasi-ideal and real (with finite sliding parameters) in real electromechanical systems.

The results of modeling in VSS with the object of the 4th order are presented, confirming the main results and provisions of the article.

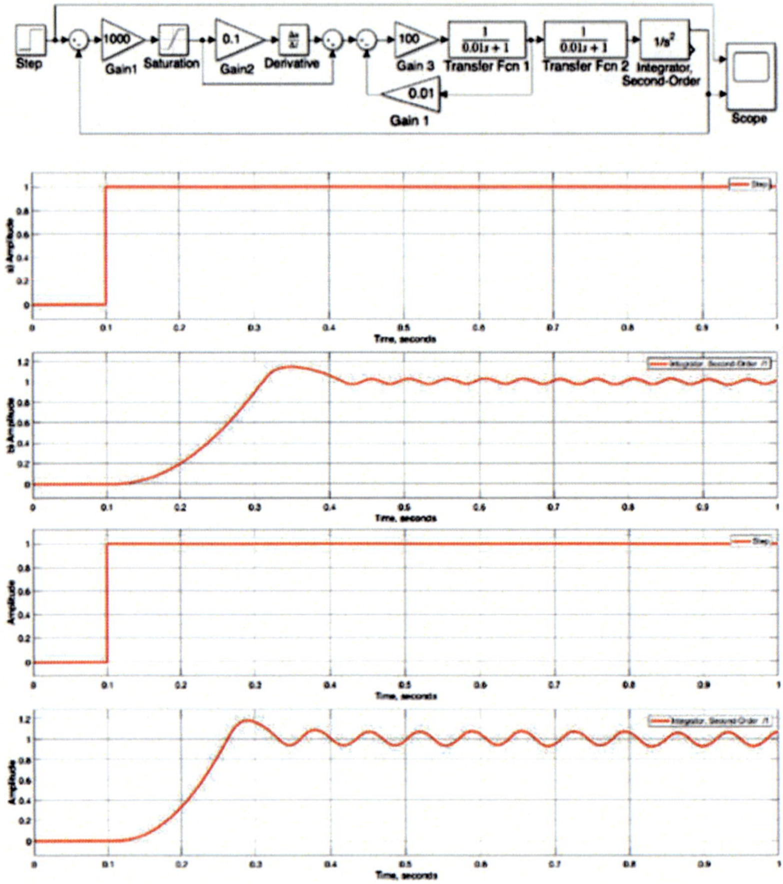

(a)

(b)

(c)

Fig. 14 Model of the equivalent circuit of a system with slip (**a**), transient processes with K1 equal 0 (**b**), with K1 equal 0,01 (**c**)

References

1. Emelyanov, S.V.: Automatic Control Systems with Variable Structure, p. 336. Moscow (1892)
2. Emelyanov, S.V. (ed.): Theory of Systems with Variable Structure, p. 368. Nauka, Moscow (1981)
3. Utkin, V.A.: Sliding modes in optimization and control problems, p. 368. Nauka, Moscow (1981)
4. Emelyanov, S.V., Korovin, S.K.: New Types of Feedback: Management in Conditions of Uncertainty, p. 352. Nauka, Fizmatlit, Moscow (1997)
5. Rush, N., Abets, P., Lalua, M.: Lyapunov's Direct Method in the Theory of Stability, p. 300. Mir, Moscow (1980)

6. Kodkin, V.L.: Methods for optimizing the speed and accuracy of complex guidance systems based on the equivalence of the field of attraction of automatic control systems and the unconditional stability of their equivalent circuits, Chelyabinsk, Bulletin of SUSU. Ser. Comput. Technol. Manag. Electron. **1**(1), 23–33 (2017)
7. Popov, V.M.: Hyperstability of Automatic Systems, p. 456. Nauka, Moscow (1970)
8. Kodkin, V.L.: Discreteness in time and evaluation of the effectiveness of automatic control systems: examples of the influence of discreteness on mathematical regularity. In: Control Theory in Engineering. IntechOpen. https://doi.org/10.5772/intechopen.91467
9. Kodkin, V.L., Anikin, A., Baldenkov, A.: Nonlinear Dynamics of Asynchronous Drive: Engineering Interpretation and Correction Techniques. IntechOpen, London (2020)
10. Kodkin, V.L., Anikin, A.: On the physical nature of frequency control problems of induction motor drives. Energies **14**, 4246 (2021)

Least Squares for Generalized Gauss–Laplace Distribution of the Error in Certain Nonlinear Regressions with Perpendicular Offsets

Carmen E. Stoenoiu and Lorentz Jäntschi

Abstract There are mainly two methods of calculating the parameters of the regression equations: the minimization of the squared errors (or least squares, LS) and the maximization of the likelihood. Regarding the distribution of the error of experimental observations, there are several theoretical distributions, but two of them are on the one hand better known and on the other easily generalizable into one (Gauss–Laplace, GL): the normal (or Gaussian) distribution and the double exponential (or Laplace) distribution. In the construction of the squared errors is possible to replace the classical vertical offsets (which are the sides of the squares of the errors) with perpendicular ones, more suited when all variables are equally subjected to experimental errors. In the present work, it is proposed to use an iterative algorithm for the calculation of the regression parameters using LS of perpendicular offsets under the assumption of GL distributed error. The method is exemplified on a non-linear regression model.

Keywords Parameters estimation · Least squares · Generalized normal distribution · Nonlinear regression · Perpendicular offsets

C. E. Stoenoiu (✉)
Department of Electric Machines and Drives, Technical University of Cluj-Napoca, 26 28 Barițiu Str., 400027 Cluj-Napoca, Cluj, Romania
e-mail: carmen.stoenoiu@emd.utcluj.ro
URL: http://carmen.academicdirect.ro

L. Jäntschi
Department of Physics and Chemistry, Technical University of Cluj-Napoca, 103-105 Muncii Blvd., 400641 Cluj-Napoca, Cluj, Romania
e-mail: lori@chimie.utcluj.ro
URL: http://lori.academicdirect.org

D. Vlachos (ed.), *Mathematical Modeling in Physical Sciences*, Springer Proceedings in Mathematics & Statistics 446, https://doi.org/10.1007/978-3-031-52965-8_12

1 Introduction

In physics, chemistry and biology and applied sciences symmetric distributions play an essential role [1]. In physics, if it remains unchanged by some space-time transformation, the thing is symmetric, so all conservation principles are examples of symmetry. In chemistry take for instance the water molecule, in biology the radial symmetry of stems, and in medical sciences the bilateral symmetry of body parts. Pillars of logics (true/false logical complements), probability theory (uniform distribution) and statistics (min. vs. max.) are build on symmetry.

Generalized Gauss-Laplace (GL) distribution is a symmetric tri-parametric distribution (parameters: location, scale, shape) for which the parameter estimation via maximum likelihood and the method of moments have been reported in [2]. Unfortunately the parameters estimates do not have a closed form and must be obtained numerically. Here is used a modified version of it [3] expressed as a natural extension from Gauss's [4] and Laplace's [5] symmetric distributions.

The least squares (LS) method was introduced in the context of identifying the parameters of elliptical planetary motion in [6] and reinvented under a different name in [7]. The classical approach of least squares (named ordinary LS) is to employ vertical offsets [8], which means that the side of the square of the sum of the squares of the errors is the distance between the observed and the estimated values measured on the axis (usually depicted vertically) representing the dependent variable. Unfortunately for the nonlinear case, the parameters estimates do not have a closed form and must be obtained numerically.

When the observations come from an exponential family with identity as its natural sufficient statistics and mild-conditions are satisfied (as in binomial, exponential, normal and Poisson distributions), least-squares and maximum-likelihood estimates are identical [9].

When perpendicular instead of vertical offsets are drawn, the residual is normal to the model [10]. Furthermore, by construction, perpendicular offsets makes the smallest squares and as such maximizes the likelihood.

One interesting fact is that the connection between the perpendicular offsets and the maximization of the likelihood has been discovered [11] even before the formulation of the maximum likelihood method [12, 13]. The solution employing perpendicular offsets for simple linear regression was communicated for the first time in [14] and revised in [15]. The parameters estimates from the perpendicular offsets of the nonlinear regression once again do not have a closed form and must be obtained numerically.

In this work, the least squares method (the correct formulation in this case is least hypercubes) is employed to minimize the residuals constructed from perpendicular offsets in the case of nonlinear regression, assuming that they (residuals) follow a Gauss–Laplace distribution. On a series of paired data coming from current and voltage simultaneous measurements involving a photovoltaic cell a nonlinear regression model was applied and the results are discussed.

2 Material and Method

Let's consider a general nonlinear model in two variables (x and y) in an explicit form of it (Eq. (1)).

$$y \sim f(x; c) \tag{1}$$

where c are known coefficients as an array of m values (c_0, ..., c_{m-1}) intended to be used to obtain a best fit with a series of n paired observed data ((x_0, y_0), ..., (x_{n-1}, y_{n-1})). In Eq. 1 semicolon express that the c values are fixed in an deterministic way ($\sim$) usually to provide a best fit with the observed data. In the context used in this paper x and y are one-dimensional (univariate) but the following treatise can be extended to apply for multi-dimensional (multivariate) case. In the classical vertical offsets the residuals are $y_0 - f(x_0; c)$, ..., $y_{n-1} - f(x_{n-1}; c)$.

Let be g such that ($g(x, y, z; c)$, Eq. 2):

$$g(x, y, z; c) = \sqrt{(x - z)^2 + (y - f(z; c))^2} \tag{2}$$

One approach to get the perpendicular offsets in each of the observation points ((x_k, y_k) for $k = 0, ..., n - 1$) is to find z_k for f, c and y_k given, for which $g(x_k, y_k, z_k)$ is minimum [10].

One should notice that in this instance $g(x_k, y_k, z_k)$ is the residual absolute error (ϵ_k, Eq. 3).

$$\epsilon_k = g(x_k, y_k, z_k; c) \tag{3}$$

Formally z_k values can be added to the (x_k, y_k) pair, but on the other hand, (x_k, y_k) is the pair of observed values while z_k values depends on the values of c coefficients (and by that in any estimation of them from experimental data on all paired values). Let's take a look to the case in which the perpendicular offset is minimum (Fig. 1).

The case illustrated in Fig. 1 doesn't apply for functions with sudden variations. However, there is an alternate route approximating the size of the perpendicular offset when the inverse of the function is available (Eq. 4).

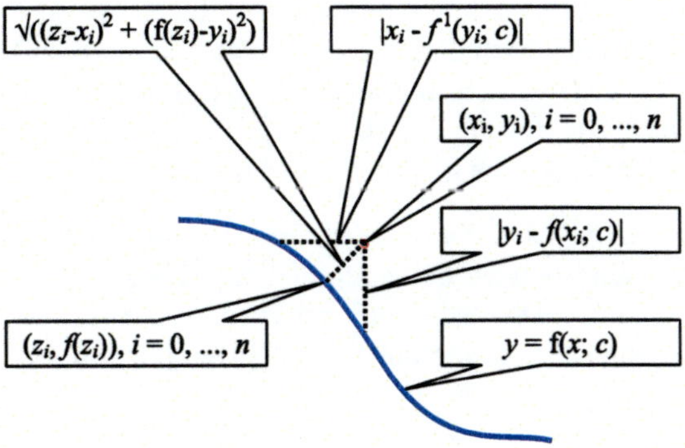

Fig. 1 Perpendicular offset is locally the shortest path from the observation point to the curve

$$\epsilon_k = \frac{s_h s_v}{\sqrt{s_h^2 + s_v^2}}, s_h = |x_k - f^{-1}(y_k; c)|, s_v = |y_k - f(x_k; c)| \tag{4}$$

GL distribution can be expressed [16] in terms of error standard deviation σ and power of the errors q by using the Gamma function (Γ) with Eq. 5.

$$GL(\epsilon, \sigma, q) = \frac{q}{2\sigma} \frac{(\Gamma(3/q))^{1/2}}{(\Gamma(1/q))^{3/2}} \exp\left(-\frac{|\epsilon|^q}{\sigma^q}\left(\frac{\Gamma(3/q)}{\Gamma(1/q)}\right)^{q/2}\right) \tag{5}$$

For convenience, a simple check reveals the two particular casses, Gaussian and Laplace distributions given in Eq. (6), where μ is the population mean.

$$\begin{aligned} GL(x - \mu, \sigma, 2) &= \frac{1}{\sqrt{2\pi}\sigma} \exp\left(-\frac{|x - \mu|^2}{2\sigma^2}\right), \\ GL(x - \mu, \sigma\sqrt{2}, 1) &= \frac{1}{2\sigma} \exp\left(-\frac{|x - \mu|}{\sigma}\right) \end{aligned} \tag{6}$$

With the data from [10] function f from Eq. (7) proven to provide a very good fit.

$$f(x; c) = c_0 - \exp(-c_1 + c_2 \ln(x)) \tag{7}$$

The same function is considered here. It's inverse is given in Eq. (8):

$$f^{-1}(x; c) = \exp\left(\frac{\ln(c_0 - y) + c_1}{c_2}\right) \tag{8}$$

The residual error (RSQ) from GL subjected to minimization can be considered the one from Eq. (9):

$$RSQ = \left(\sum_{k=0}^{n-1} \epsilon_k^q\right)^{q^{-1}} \tag{9}$$

With a powerfull nonlinear optimization library, from hereon, only the statement of the problem in the formalism of the library must be given. In the experience of the author, such libraries are available in Mathcad [17], Mathematica [18], Matlab [19] and open libraries (such as AlgLib [20]) usable under a programming environment (such as FreePascal in [10]). Here only the general procedure is given (for the algorithms to be slightly adapted see [10]):

- Do the initial estimates for the values of the parameters C from classical Gaussian vertical offsets of the selected model (Eq. (7)) of nonlinear regression;
- Add q to the list of parameters ($q \leftarrow 2$ initializes it);
- Run the minimization for RSQ (Eq. (9)) with ϵ_k from Eqs. (4) or (3).

Table 1 Initial estimates for the coefficients

n	Initial C	Statistics	PV cell parameters
3	$c_0 = 1.83$		$I_{sc} = 1.83000$ mA
	$c_1 = 22$	$RSS = 0.0019093$	$U_{oc} = 1576.51$ mV
	$c_2 = 3.07$	$r^2_{adj} = 0.9986$	$I_{xp} = 1.3804$ mA
	$q = 2.0$	$F = 9228$	$U_{xp} = 998.00$ mV
			$P_{xp} = 1.3776$ mW

$I_{sc} = f(0; c)$; $U_{oc} = f^{-1}(0; c)$; $U_{xp} = u \mid (u \cdot f(u; c))'_u = 0$; $I_{xp} = f(U_{xp}; c)$; $P_{xp} = U_{xp} I_{xp}$

3 Results and Discussion

The optimization should be feed with some initial data. The initial data can be taken from nonlinear regression with vertical offsets and Gaussian distribution of the error ($q = 2$). Using the experimental data given in [10], the values given in Table 1 were used for the initialization of the optimization.

In Table 1 also a series of solar cell parameters are derived: short circuit intensity (I_{sc}), open circuit voltage (U_{oc}), current at maximum power point (I_{xp}), voltage at maximum power point (U_{xp}), and power at maximum power point (P_{xp}).

Figure 2 provides the evolution of parameters values with the optimization advancement starting from Table 1 data when Eq. (4) is used.

Convergence is relatively fast when compared with the alternate algorithm [10] considering that here the double embedded optimization has been reduced to a single optimization (the inner optimization has been replaced by a simple analytic calculation). However, in terms of the number of steps the optimization has been increased in complexity–which is the expected outcome since a new parameter has been added–q–the power of the error (see Eq. (5)).

Optimization is smooth to the convergence (see Fig. 3).

The final optimized values of the parameters using Eq. (4) are given in Table 2.

If the optimization did not increased the adjusted determination coefficient, it significantly decreased the residual error (RSS, from 0.0019 in Table 1 to 0.0012 in Table 2). Also the total variance explained by the model has been increased (F value is 9228 in Table 1 and 9499 in Table 2). Some of the parameters suffered some change after the optimization (take for instance open circuit voltage, with more than 1 % change) a very little change (take for instance the maximum power, with less than 1 ‰ change).

When compared to the previously reported optimization (Table 3 in [10]), the statistics did not bring more explanatory power (take for instance F value, which is 9528 in [10] and 9499 reported here).

The alternate (to Eq. (4)) approach is to effectivelly find (by optimization) the normal to the courve (find z_k such that ϵ_k from Eq. (3)) to be minimum. The results regarding the evolution of the values of the parameters staring from the same initial values from Table 1 are provided in Fig. 4.

Optimization is smooth to the convergence (see Fig. 5).

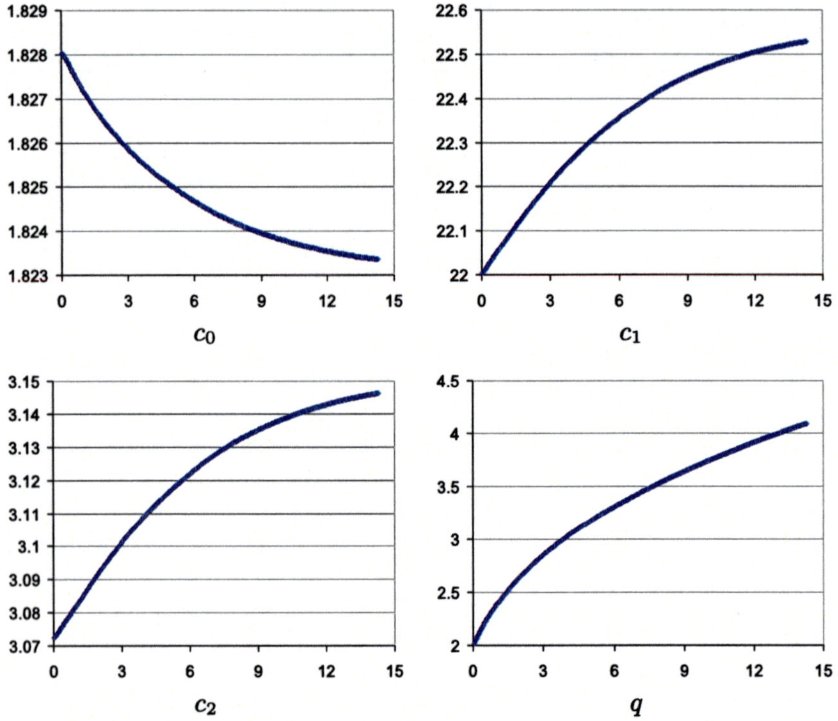

Fig. 2 Parameters of the model (c_0, c_1, c_2) and the power of the error (q) as functions of milions of iterations until convergence with Eq. (4)

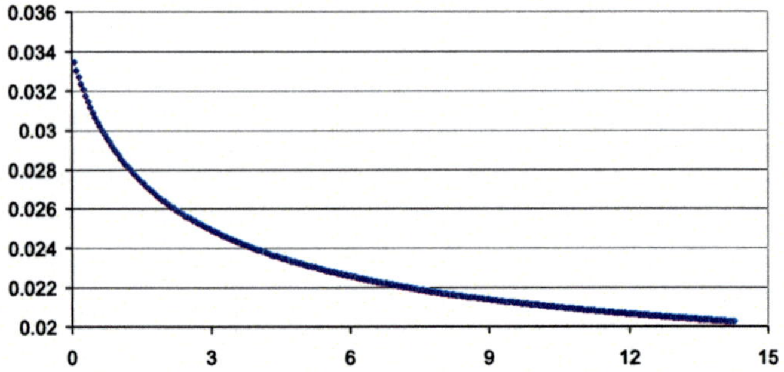

Fig. 3 Residual error RSQ as function of milions of iterations until convergence approximating ϵ_k with Eq. (4)

Table 2 Final (Eq. (4) optimum) estimates for the coefficients (after 14280311 iterations starting from Table 1 data)

n	Final C	Statistics	PV cell parameters
3	$c_0 = 1.82335$		$I_{sc} = 1.82335$ mA
	$c_1 = 22.5285$	$RSS = 0.001163$	$U_{oc} = 1558.56$ mV
	$c_2 = 3.14617$	$r^2_{adj} = 0.9986$	$I_{xp} = 1.3836$ mA
	$q = 4.09287$	$F = 9499$	$U_{xp} = 994.93$ mV
			$P_{xp} = 1.3766$ mW

I_{sc}, U_{oc}, U_{xp}, I_{xp}, P_{xp} derived in same manner as in Table 1

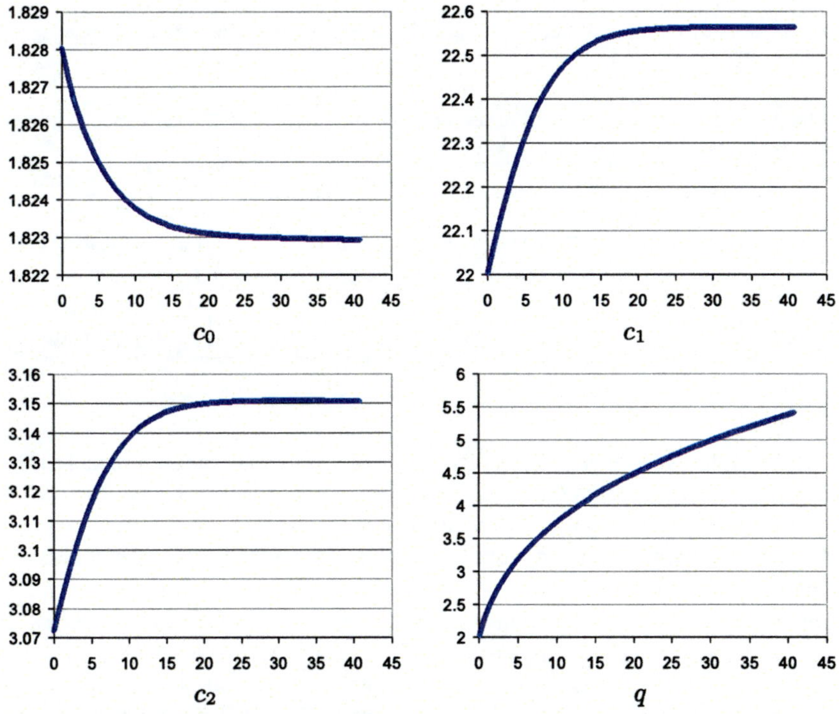

Fig. 4 Parameters of the model (c_0, c_1, c_2) and the power of the error (q) as functions of milions of iterations until convergence with Eq. (3)

A significantly higher number of iterations were necessary to get the exact solution (over 40 millions interations in Fig. 5, less than 15 millions iterations in Fig. 3). The exact solution is even more time consumming than the ratio of the two numbers shows, considering that to get the exact values of the perpendicular oofsets is another problem of optimization (for a detailed discussion please see [10]).

The final optimized values of the parameters using Eq. (3) are given in Table 3.

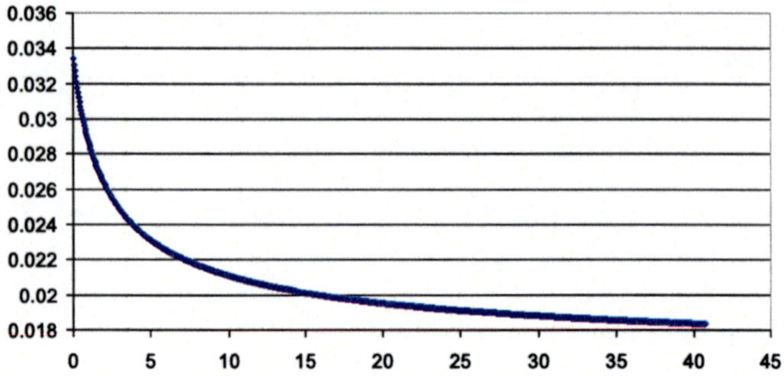

Fig. 5 Residual error as function of milions of iterations until convergence calculating ϵ_k with Eq. (3)

Table 3 Final (Eq. (3) optimum) estimates for the coefficients (after 40764550 iterations starting from Table 1 data)

n	Final C	Statistics	PV cell parameters
3	$c_0 = 1.82294$		$I_{sc} = 1.82294$ mA
	$c_1 = 22.5634$	$RSS = 0.001190$	$U_{oc} = 1558.48$ mV
	$c_2 = 3.15092$	$r^2_{adj} = 0.9986$	$I_{xp} = 1.3838$ mA
	$q = 5.14599$	$F = 9477$	$U_{xp} = 992.01$ mV
			$P_{xp} = 1.3727$ mW

Table 3 data reveals that the statistics for the regression optimum model in the supposition of generalized Gauss-Laplace distribution of the error have RSS and F values smaller than the values calculated under the supposition of Gaussian distribution of the error [10]. Even more, it can be mathematically proved that also r^2_{adj} from generalized Gauss-Laplace distribution supposition is less or equal with r^2_{adj} from Gaussian supposition, and this is simply because everywhere in these formulas (RSS, r^2_{adj}, and F) is about squared quantities, like in Gaussian distribution. This expected outcome does not imply that the obtained regression model is less performant, since the two models are derived in two different statistical suppositions. On the same argument, the residual sums of the Eq. (9) form with the optimum coefficients from Gaussian supposition are bigger (or equal when $q = 2$) than the residuals obtained with the optimum coefficients from generalized Gauss-Laplace supposition (for the sake of the argument, the residual sums of the Eq. (9) form, when q is 5.14599–with the coefficients from Table 3 in [10] is 0.019428 while the same with our Table 3 coefficients–also represented in Fig. 5–is 0.018369).

4 Conclusions

Generalized Gauss-Laplace distribution of the experimental error is always able to reduce further the residual error (by adding also one more parameter to be estimated, the power of the resuidual error), but as has been shon in the considered example case this addition does not always bring more explanatory power.

A negative result is also communicated here–about the simplification of the route to the perpendicular offets by the use of the approximation of them as the height of the triangle formed by the vertical and the horrizontal offsets (Eq. (4) in the paper)–even in a case with smooth variations and points close to the model, the approximation do not accuarely provide the optimum point.

The main result reported here is that a iterative optimization has been succesfully applied to the problem of generalized Gauss-Laplace distribution of the error in a case of paired data sets when both paired variables are affected by errors and perpendicular offsets were involved to identify the parameters for a nonlinear regression model.

Authors Statements

Author Contributions C.E.S.: formal analysis, investigation, resources, validation, writing—original draft preparation; L.J.: writing—review and editing, conceptualization, drawings, methodology, supervision.

Funding This research received no external funding.

Data Availability Data used in this paper is fully available online in Table 1 of [10].

Conflicts of Interest The authors declares no conflict of interest.

Acknowledgements The authors wish to express their gratitude to the IC-MSQUARE 2023 organizers for providing the oportunity to present this research under the frame of the event. Help from the reviewers in improving the work was highly appreciated.

References

1. Jäntschi, L.: Introducing structural symmetry and asymmetry implications in development of recent pharmacy and medicine. Symmetry **14**(8), 1674 (2022). https://doi.org/10.3390/sym14081674
2. Varanasi, M.K., Aazhang, B.: Parametric generalized Gaussian density estimation. J. Acoust. Soc. Am. **86**(4), 1404–1415 (1989). https://doi.org/10.1121/1.398700
3. Jäntschi, L., Bolboacă, S.D.: Observation vs. observable: maximum likelihood estimations according to the assumption of generalized gauss and laplace distributions. Leonardo Electron. J. Pract. Technol. **8**(15), 81–104 (2009). http://lejpt.academicdirect.org/A15/081_104.pdf
4. Gauss, C.O.: Theoria Motus Corporum Coelestium. Perthes et Besser, Hamburg (1809)
5. Laplace, P.: Thèorie analytique des probabilitès. Courcier, Paris (1812)

6. Legendre, A.M.: Sur la méthode des moindres quarrés, pp. 72–75. Firmin Didiot, Paris (1805). http://books.google.com/books?id=FRcOAAAAQAAJ
7. Griliches, Z., Ringstad, V.: Error-in-the-variables bias in nonlinear contexts. Econometrica **38**(2), 368–370 (1970). https://doi.org/10.2307/1913020
8. Jäntschi, L., Bolboacă, S.D.: Szeged matrix property indices as descriptors to characterize fullerenes. Ovidius Univ. Ann. Chem. **27**(2), 73–80 (2016). https://doi.org/10.1515/auoc-2016-0010
9. Charnes, A., Frome, E.L., Yu, P.L.: The equivalence of generalized least squares and maximum likelihood estimates in the exponential family. J. Am. Stat. Assoc. **71**(353), 169–171 (1976). https://doi.org/10.1080/01621459.1976.10481508
10. Jäntschi, L.: Symmetry in regression analysis: Perpendicular offsets-the case of a photovoltaic cell. Symmetry **15**(4), 948 (2023). https://doi.org/10.3390/sym15040948
11. Adcock, R.J.: Note on the method of least squares. Analyst **4**(6), 183–184 (1877). https://doi.org/10.2307/2635777
12. Fisher, R.A.: On an absolute criterion for fitting frequency curves. Messenger Math. **41**, 155–160 (1912). http://hdl.handle.net/2440/15165
13. Fisher, R.A.: On the mathematical foundations of theoretical statistics. Philos. Trans. Royal Soc. A **222**(594–604), 309–368 (1922). https://doi.org/10.1098/rsta.1922.0009
14. Adcock, R.J.: A problem in least squares. Analyst **5**(2), 53–54 (1878). https://doi.org/10.2307/2635758
15. Kummell, C.H.: Reduction of observation equations which contain more than one observed quantity. Analyst **6**(4), 97–105 (1879). https://doi.org/10.2307/2635646
16. Jäntschi, L., Pruteanu, L.L., Cozma, A.C., Bolboacă, S.D.: Inside of the linear relation between dependent and independent variables. Comput. Math. Methods Med. **2015**(360), 752 (2015). https://doi.org/10.1155/2015/360752
17. Jäntschi, L.: A test detecting the outliers for continuous distributions based on the cumulative distribution function of the data being tested. Symmetry **11**(6), 835 (2019)
18. Jäntschi, L.: Detecting extreme values with order statistics in samples from continuous distributions. Mathematics **8**(2), 216 (2020). https://doi.org/10.3390/math8020216
19. Joiţa, D.M., Tomescu, M.A., Bàlint, D., Jäntschi, L.: An application of the eigenproblem for biochemical similarity. Symmetry **13**(10), 1849 (2021). https://doi.org/10.3390/sym13101849
20. Bochkanov, S.: Alglib Project. 1994–2017. http://alglib.net, Last Accessed 19 July 2023 (2017)

On the Influence of Absorption on the Spatial Spectrum of Radio Waves Propagation in the Polar Ionosphere

George Jandieri, Nika Tugushi, and Irma Mindiashvili

Abstract Statistical characteristics of scattered radio waves propagating in the terrestrial high-latitude conductive ionosphere are investigated considering diffraction effects applying the modify smooth perturbation method. Polarization coefficients of scattered ordinary and extraordinary waves in the polar region of the terrestrial ionosphere are obtained for the first time. Correlation function and the variances of the phase fluctuations has been derived for arbitrary correlation function of electron density fluctuations. These statistical characteristics include the polarimetric parameters, Hall's, Pedersen and longitudinal conductivities, anisotropy parameters of elongated plasmonic structures and the velocity of a plasma flow. Numerical calculations are carried out for the anisotropic Gaussian correlation function containing anisotropy factor and the angle of inclination of these irregularities with respect to the lines of forces of a geomagnetic field using the satellite and remote sensing observation data for the polar ionosphere. Diffraction effects becomes important in proportion of distance propagating by a wave in the ionosphere. Spatial power spectrum has a maximum along the direction of an incident wave.

Keywords Conductivity · Irregularities · Statistical moments

1 Introduction

Peculiarities of radio waves propagation in different turbulent media are now full enough investigated. The features of the statistical moments of electromagnetic waves propagation and in random media are rather well studied [1]. Spatial power spectrum (SPS) in complex geometrical optics approximation was considered in detail [2]. Local minimum in the SPS has been revealed along the direction of elongated

G. Jandieri (✉) · I. Mindiashvili
Georgian Technical University, 77 Kostava Str., 0175 Tbilisi, Georgia
e-mail: george.jandieri@gtu.ge

N. Tugushi
Tbilisi State University, 1 Chavchavadze Av., 0128 Tbilisi, Georgia

D. Vlachos (ed.), *Mathematical Modeling in Physical Sciences*, Springer Proceedings in Mathematics & Statistics 446, https://doi.org/10.1007/978-3-031-52965-8_13

irregularities. Numerical calculations have been carried out by Monte-Carlo method. The peculiarities of waves propagation in random media with fluctuating parameters are solved, as a rule, using different approximate methods depending on a statement of the problem—weakly or strongly fluctuate medium parameters, relationship between the wavelength and linear scales of irregularities, geometry of the problem (path length, width of an incident wave beam), and so on. Nevertheless, much problems can be solved by means of developed approximate methods. If conditions of a single scattering (fluctuations in the medium are not enough weak, the scattered field is not small) are violate, two-, three, etc. multiple scattering of the field must be taken into account, i.e., it is necessary to build the theory of multiply scattered waves.

The features of the SPS of electromagnetic waves propagating in anisotropic magnetoplasma applying the ray-optics approximation were considered in [3–5]. The "Double-Humped Effect" in the collision nonconductive magnetoplasma was discovered in [6, 7].

In a high-latitude ionosphere geomagnetic lines of forces are oriented vertically upward and ionospheric plasma moves along the geomagnetic lines of forces generating elongated vertical plasmonic irregularities. Peculiarities of small-scale inhomogeneities are investigated by different statistical methods.

This paper is devoted to the analytical and numerical investigation of the "Double-Humped Effect" in the high-latitude conductive collision ionospheric magnetized (COCOIMA) plasma applying the modify smooth perturbation method taking into account the diffraction effects. For the first time we calculated analytically both refractive index and polarimetric parameters of scattered ordinary (O-) and extraordinary (E-) waves in a high-latitude COCOIMA plasma considering Hall's, Pedersen and longitudinal conductivities without any assumptions. Second order statistical moments of the phase fluctuations of electromagnetic wave are valid for an arbitrary cross-correlation function of electron concentration fluctuations. Numerical simulations are carried out applying the satellite and remote sensing observation data.

2 Introduction

Initial is the wave

$$\left(\frac{\partial^2}{\partial x_i\, \partial x_j} - \Delta\, \delta_{ij} - k_0^2\, \tilde{\varepsilon}_{ij}(\boldsymbol{r}) \right) E_j(\boldsymbol{r}) = 0 \tag{1}$$

where: $\tilde{\varepsilon}_{ij} = \varepsilon_{ij} - i\, 4\pi\, \sigma_{ij}/k_0\, c$, k_0 is the wave number of an original radio wave, Δ is the Laplacian, δ_{ij} is the Kronecker symbol. Geomagnetic lines of forces in a high-latitude ionosphere $\boldsymbol{H_0}$ are oriented vertically upwards along the Z-axis, wave vector of an original radio wave in the COCOIMA plasma is located in the main YOZ plane of the Cartesian coordinate system. Components of the permittivity tensor in a high-latitude COCOIMA plasma may be written as [8]:

$$\varepsilon_{xx} = \varepsilon_\perp - i\,(\sigma_\perp + s\,g),\ \varepsilon_{xy} = -\,\varepsilon_{yx} = \cos\theta\ [s\,æ\,\delta - i\,(æ + \sigma_H)],$$
$$\varepsilon_{yy} = (\varepsilon_\perp + p_0\,u\,\sin^2\theta) - i\left[(\sigma_\perp\cos^2\theta + \sigma_{||}\sin^2\theta) + s\,(g - p_0\,u\,g_1\sin^2\theta)\right]$$
$$\varepsilon_{zz} = (\varepsilon_\perp + p_0\,u\,\cos^2\theta) - i\left[(\tilde{\sigma}_{||}\cos^2\theta + \sigma_\perp\sin^2\theta) + s\,(g - p_0\,u\,g_1\cos^2\theta)\right] \quad (2)$$

here: $p_0 = \mathrm{v}/(1-u)$, $g = p_0(1+u)/(1-u)$, $\delta = 2/(1-u)$, $æ = p_0\sqrt{u}$, $g_1 = (3-u)/(1-u)$; θ is an angle between the wavevector of a refractive wave and an external magnetic field; $\mathrm{v}(\boldsymbol{r}) = \omega_p^2(\boldsymbol{r})/\omega^2$ and $u = (e\,H_0/m_e\,c\,\omega)^2$ are the magneto-ionic parameters, $\omega_p(\boldsymbol{r}) = \left[4\pi N_e(\boldsymbol{r})\,e^2/m_e\right]^{1/2}$—plasma frequency; $\sigma_\perp$ and σ_H are the Pedersen and Hall's conductivities, respectively, $\sigma_{||}$ is the longitudinal conductivity; $n_e(\boldsymbol{r}) = n_0 + n_1(\boldsymbol{r})$ is a random electron density, first is a constant term, $n_1(\boldsymbol{r})$ is a fluctuating function of the spatial coordinates; e and m_e are the charge and mass of an electron, $\nu_e = \nu_{e\,n} + \nu_{i\,n}$ collision frequency containing electrons and ions collision with the neutral plasma particles.

Wave field offer as $E(\boldsymbol{r}) = E_0\exp\{\Phi(\boldsymbol{r})\}$, where $\Phi(\boldsymbol{r})$ is the complex phase of the wave $\Phi(\boldsymbol{r}) = \varphi_0 + \varphi_1 + \varphi_2 + \cdots$, $\varphi_0 = i\,k_0 x + i\,k_\perp y$, which corresponds to propagation of an original wave along the x axis, $k_\perp << k_0$. φ_1 is of order of ε_1, φ_2 has order ε_1^2 and so on. First term describes the wave propagation of an original wave, the second term is associated with the diffraction.

2D Fourier transformation of the fluctuating phase is

$$\varphi_1(x, y, z) = \int_{-\infty}^{\infty} dk_x \int_{-\infty}^{\infty} dk_y\ \psi(k_x, k_y, z)\ \exp(i\,k_x\,x + i\,k_y\,y)$$

From Eq. (1) we obtain the stochastic differential equation

$$\frac{\partial\psi}{\partial z} + i\,\frac{k_x\,(k_y + k_\perp) + k_0\,k_y\,G_{1,2} - k_x^2\,P_{1,2}}{(k_y + k_\perp)\,G_{1,2} - 2k_0\,P_{1,2}}\,\psi = -\,i\,\frac{k_0^2}{(k_y + k_\perp)\,G_{1,2} - 2k_0\,P_{1,2}}$$
$$(\varepsilon_{yx}^{(1)} + \varepsilon_{yy}^{(1)}\,P_{1,2} + \varepsilon_{yz}^{(1)}\,G_{1,2}). \quad (3)$$

Right part of Eq. (3) contains terms of electron density fluctuations.
Polarization coefficient at $s << \varepsilon_{ij}$, $\tilde{\sigma}_{ij}$ are defined as

$$\mathrm{P}_{1,2} = \frac{< E_y >}{< E_x >} = \mathrm{P}'_{1,2} - i\,\mathrm{P}''_{1,2},\ G_{1,2} = \frac{< E_z >}{< E_x >} = -\,G'_{1,2} + i\,G''_{1,2}.$$

where: $P'_{1,2} = \frac{1}{T_3^2+T_4^2}(T_1T_3 - T_2T_4)$, $P''_{1,2} = \frac{1}{T_3^2+T_4^2}(T_1T_4 + T_2T_3)$, $G'_{1,2} = T_5 + T_7P'_{1,2} - T_8P''_{1,2}$, $G''_{1,2} = T_6 - T_7\,P''_{1,2} - T_8P'_{1,2}$, $T_1 = \left[(p_0 - \tilde{N}_1^2)p_0u - (\tilde{N}_2^2 - \tilde{\sigma}_\perp)(\sigma_{||} - \sigma_\perp) - (æ + \sigma_H)^2\right]\sin\theta\cos\theta$, $T_2 = \left[(p_0 - N_1^2)(\sigma_{||} - \sigma_\perp) + p_0u(N_2^2 - \sigma_\perp)\right]\sin\theta\cos\theta$, $T_3 = (æ + \tilde{\sigma}_H)(\tilde{N}_2^2 - \tilde{\sigma}_{||})\sin\theta$, $T_4 = (æ + \tilde{\sigma}_H)\left[p_0(1-u) - \tilde{N}_1^2)\right]\sin\theta$, $T_5 = (æ + \tilde{\sigma}_H)(\tilde{\sigma}_{||}\cos^2\theta +$

$\tilde{\sigma}_\perp \sin^2\theta)\sin\theta/\Delta$, $T_6=(æ+\tilde{\sigma}_H)(\varepsilon_\perp+p_0u\cos^2\theta)\cdot\sin\theta/\Delta$, $\Delta=(\varepsilon_\perp+p_0u\cos^2\theta)^2+(\tilde{\sigma}_{||}\cos^2\theta+\tilde{\sigma}_\perp\sin^2\theta)^2$, $T_7=\left[p_0u(\varepsilon_\perp+p_0u\cos^2\theta)+(\tilde{\sigma}_{||}\cos^2\theta+\tilde{\sigma}_\perp\sin^2\theta)(\tilde{\sigma}_{||}-\tilde{\sigma}_\perp)\right]\sin\theta\cos\theta/\Delta$, $T_8=\left[p_0u(\tilde{\sigma}_{||}\cos^2\theta+\tilde{\sigma}_\perp\sin^2\theta)-(\varepsilon_\perp+p_0\,u\,\cos^2\theta)\,(\tilde{\sigma}_{||}-\tilde{\sigma}_\perp)\right]\sin\theta\cos\theta/\Delta$.

We do not write out the expressions of complex the refractive index N^2 because of their bulkiness.

At $s << \varepsilon_{ij}$, $\tilde{\sigma}_{ij}$ from Eq. (3) we yield

$$\frac{\partial\psi}{\partial z}+\frac{k_0}{2\,\mathrm{P}^2}\,(\Lambda_1+i\,\Lambda_2)=\frac{k_0}{2\,\mathrm{P}^2}\,(\tilde{D}_3-i\,\tilde{D}_4)\,n_1 \tag{4}$$

here: $\Lambda_1=\tilde{D}_2\,\mathrm{P}'-\tilde{D}_1\,\mathrm{P}''$, $\Lambda_2=\tilde{D}_1\,\mathrm{P}'+\tilde{D}_2\,\mathrm{P}''$, $\mathrm{P}^2=\mathrm{P}'^2+\mathrm{P}''^2$, $\tilde{D}_2=\mathrm{P}''\,x^2+G''\,y$, $\tilde{D}_1=\mathrm{P}'\,x^2-(\mu+y)\,x+G'\,y$, $\tilde{D}_3=\mathrm{P}''\,e_0+\mathrm{P}'\,e_1$, $\tilde{D}_4=\mathrm{P}''\,e_1-\mathrm{P}'\,e_0$, $e_0=\mathrm{P}'\,p_0\,(1-u\,\sin^2\theta)$, $e_0=\mathrm{P}'\,p_0\,(1-u\,\sin^2\theta)+\mathrm{P}''\,(\tilde{\sigma}_{||}\sin^2\theta+\tilde{\sigma}_\perp\cos^2\theta)+\left[G'\,p_0\,u-G''\,(\tilde{\sigma}_{||}-\tilde{\sigma}_\perp)\right]\sin\theta\cos\theta$, $e_1=(p_0\sqrt{u}+\tilde{\sigma}_H)\cdot\cos\theta+\mathrm{P}''\,p_0\,(1-u\,\sin^2\theta)-\mathrm{P}'\,(\tilde{\sigma}_{||}\sin^2\theta+\tilde{\sigma}_\perp\cos^2\theta)+\left[G''\,p_0\,u+G'\,(\sigma_{||}-\sigma_\perp)\right]\sin\theta\cos\theta$.

For brevity further the 1 and 2 indexes in polarization coefficients are omitted. Ordinary and extraordinary waves in the COCOIMA plasma are elliptically polarized.

We are interested in the transversal correlation function of the wave field, which is determined as $W_{EE^*}(\boldsymbol{\rho})=<E(\boldsymbol{r})><E^*(\boldsymbol{r}+\boldsymbol{\rho})>+<E(\boldsymbol{r})\,E^*(\boldsymbol{r}+\boldsymbol{\rho})>$, $\boldsymbol{\rho}=\{\rho_y,\rho_z\}$. First term describes the mean fields, which can be ignored; and, hence, auto-correlation function of a scattered field we will calculate applying the second term $W_{EE^*}(\boldsymbol{\rho})=<E(\boldsymbol{r})\,E^*(\boldsymbol{r}+\boldsymbol{\rho})>$. Substituting $E=E_0\exp(\varphi_1+\varphi_2+i\,k_0x+i\,k_\perp y)$ in the last expression for a transversal auto-correlation functions of a scattered field we obtain [3, 9]

$$\begin{aligned}W_{EE^*}(\boldsymbol{\rho},k_\perp)&=E_0^2\exp(-i\rho_yk_\perp)\exp\left(2\mathrm{Re}<\varphi_2>\right)\\&\exp\left[\frac{1}{2}(<\varphi_1^2(\boldsymbol{r})>+<\varphi_1^{2*}(\boldsymbol{r}+\boldsymbol{\rho})>)\right]\cdot\\&\exp[<\varphi_1(\boldsymbol{r})\varphi_1^*(\boldsymbol{r}+\boldsymbol{\rho})>]\end{aligned} \tag{5}$$

where: E_0^2 is the intensity of an incident wave. SPS is easily calculated by Fourier transformation [1, 10].

Taking into account the boundary condition the solution of Eq. (4) can be written as

$$\begin{aligned}\psi(k_x,k_y,L)=\frac{k_0}{2\,\mathrm{P}^2}\,(\tilde{D}_3-i\,\tilde{D}_4)\,\int_0^L dz'\,n_1(k_x,k_y,z')\\\exp\left[-(\Lambda_1+i\,\Lambda_2)\,k_0\,(L-z')\right]\end{aligned} \tag{6}$$

The variance of a random phase for any auto-correlation function of electron density fluctuations $V_n(\boldsymbol{k})$ has the following form

$$< \varphi_1^2 > = \frac{\pi}{2} \frac{k_0^4 L}{\mathrm{P}^4} (\tilde{D}_3 - i \tilde{D}_4)^2 \int_{-\infty}^{\infty} dx \int_{-\infty}^{\infty} dy\, V_n[k_0 x,\ k_0 y,\ (i\, m_1 - m_2)] \tag{7}$$

where: $m_1 = (\mathrm{P}'' \mu x + a_1 y)/2\mathrm{P}^2, a_1 = \mathrm{P}' G'' - \mathrm{P}'' G', m_2 = (a_2 y - \mathrm{P}' \mu x)/2\mathrm{P}^2$, $a_2 = \mathrm{P}' G' + \mathrm{P}'' G''$, $\mu = k_\perp / k_0$ is the diffraction parameter. As usual, geometrical optics approximation neglects the diffraction effects at $\sqrt{\lambda L} << l_\varepsilon$. In a modify smooth perturbation method a distance L propagating by the wave in a COCOIMA plasma exceeds the linear scale of electron concentration irregularities, $L >> l$, and the diffraction effects must be taken into account.

Auto-correlation function of electron concentration can be expressed via 2D Fourier transform

$$< \varphi_1(\boldsymbol{r}) \varphi_1^*(\boldsymbol{r} + \boldsymbol{\rho}) > = \frac{\pi}{2} \frac{k_0^4 L}{\mathrm{P}^4} (D_3^2 + D_4^2) \int_{-\infty}^{\infty} dx \int_{-\infty}^{\infty} dy\, V_n(k_0 x,\ k_0 y,\ -\ k_0 \Lambda_2) \cdot$$

$$\exp(-i\ \eta_x x - i\ \eta_y y) \tag{8}$$

where: $\eta_x = k_0 \rho_x, \eta_y = k_0 \rho_y$; ρ_x and ρ_y are small distances between the observation points in the principle and perpendicular planes, respectively; $x = k_x / k_0, y = k_y / k_0$; the asterisk indicates the complex conjugate.

Second approximation for the phase fluctuations, can be written as

$$< \varphi_2(x, y, L) > = -\pi\, k_0^4 L\, (\tilde{D}_3 - i\, \tilde{D}_4)^2 \int_{-\infty}^{\infty} dx \int_{-\infty}^{\infty} dy$$

$$\left(\frac{\mathrm{P}'' - i\, \mathrm{P}'}{\mathrm{P}^2}\, x\, y + i\, x^2 \right) \frac{\alpha - i\, \beta}{\alpha^2 + \beta^2}$$

$$V_n[k_0 x,\ k_0 y,\ k_0(i\, m_1 - m_2)] \left\{ 1 - \frac{\alpha - i\, \beta}{k_0 L\, (\alpha^2 + \beta^2)} \left(1 - \exp[-\ (\alpha + i\, \beta)\, k_0 L]\right) \right\}. \tag{9}$$

here: $\alpha = \mathrm{P}'' x\, y/\mathrm{P}^2$, $\beta = (\mathrm{P}^2 x^2 - \mathrm{P}' x\, y)/\mathrm{P}^2$.

More exact is the statement is an energy conservation in a nonadsorbing medium. Average value of energy-flux density $< S >$ can be found using formula $< S > = < w \cdot \boldsymbol{U}_{gr} >$, where w is the energy density, $\boldsymbol{U}_{gr}$—group velocity. Longitudinal component of the group velocity decreases with distance due to scattering; therefore, in order to preserve energy stream in a nonadsorbing medium the value of w must be increased. It is known, that $w \propto W_{EE^*}(\boldsymbol{\rho} = 0)$. Therefore, the increase of wave intensity with distance not contradict energy conservation law.

SPS of a scattered field $W(k, k_\perp)$ is calculated by Fourier transform

$$W(k, k_\perp) = \int_{-\infty}^{\infty} d\rho_y\, W_{EE^*}(\rho_y, k_\perp) \exp(i\, k \rho_y) \tag{10}$$

where k is a transverse component of the wave vector of a scattered field.

On the other hand, if the SPS of an incident wave has a finite width and its maximum is directed along the Z-axis, SPS of a scattered radiation is given by the expression [2, 3, 6]

$$I(k) = \int_{-\infty}^{\infty} dk_{\perp}\, W(k, k_{\perp}) \exp(-\, k_{\perp}^2\, \beta^2). \tag{11}$$

where β characterizes disorder of an incident radiation of an incident radiation.

3 Numerical Calculations

Numerical simulation is carried out for 3 MHz incident radio wave ($k_0 = 6.28 \cdot 10^{-2} m^{-1}$), $v_0 = 0.28, u_0 = 0.22$. Fourier transformation of the anisotropic Gaussian correlation function of electron concentration fluctuations is:

$$V_n(k) = \frac{\sigma_n^2}{8\,\pi^{3/2}} \frac{l_{\|}^3}{\chi^2} \exp\left(-\frac{k_x^2\, l_{\perp}^2}{4} - p_1 \frac{k_y^2\, l_{\|}^2}{4} - p_2 \frac{k_z^2\, l_{\|}^2}{4} - p_3\, k_y\, k_z\, l_{\|}^2\right) \tag{12}$$

where: $p_1 = (\sin^2\gamma_0 + \chi^2 \cos^2\gamma_0)^{-1}\left[1 + (\chi^2 - 1)^2 \sin^2\gamma_0 \cos^2\gamma_0/\chi^2\right]$, $\xi = k_0\, l_{\|}$, $p_2 = (\sin^2\gamma_0 + \chi^2 \cos^2\gamma_0)/\chi^2$, $p_3 = (\chi^2 - 1)\, \sin\gamma_0 \cos\gamma_0/2\,\chi^2$, $\chi = l_{\|}/l_{\perp}$ is the anisotropy factor—the ratio of the longitudinal and transverse characteristic linear sizes of plasmonic irregularities, γ_0 is the inclination angle of elongated electron density fluctuations with respect to the geomagnetic field. Anisotropy shape of the ionospheric plasmonic structures is a result of the diffusion processes in the field aligned and field perpendicular directions.

Equation (5) can be rewritten as

$$\frac{W_{EE^*}(\xi, \eta, \mu)}{E_0^2} = \exp(-\, i\, \eta\, \mu) \exp\left[-\, 2\pi\, \sqrt{2}\, B_0\, (p_1 + p_2\, \mu^2 + 4\, p_3\, \mu)^{-1/2}\right] \cdot$$
$$\exp\left\{\exp\left(-\frac{\chi^2}{\xi^2}\, \eta_x^2 - i\, \eta_y^2\, t\right) \frac{\xi\, B_0}{2\sqrt{\pi}} \int_{-\infty}^{\infty} dt\, \exp\left(-\frac{\xi^2}{2}\left[\frac{1}{4}\, p_2\, t^4 + (p_2\, \mu + 2 p_3)\, t^3 + \right.\right.\right.$$
$$\left.\left.\left. + (p_1 + p_2\, \mu^2 + 4\, p_3\, \mu)\, t^2\right]\right)\right\} \tag{13}$$

where: $B_0 = \sigma_n^2\, \sqrt{\pi}\, \xi\, k_0\, L/4\, \chi$

The curves of a normalized intensity of a scattered field versus the direction of an incident wave at: $\mu = 200$, $s_{x0} = 0.07$, $\alpha_{\perp} = 5$ and different anisotropic factors are presented on Fig. 1. Maxima of an intensity are observed along the direction of an incident radio wave; small maximum corresponds to the reflected wave. Numerical simulation show, that a dip increases along the direction of elongated irregularities in proportion of the anisotropy factor χ. Figure 2 illustrates the normalized intensity of a scattered field for different anisotropy coefficients $\chi = 100$, $\mu = 200$ at various angles of an incident wave: $s_{x0} = 0, 07$ and $s_{x0} = 0, 1$. Increasing the incident angle on the anisotropic medium the amplitude of a reflected wave decreases, maximum of an intensity of a scattered wave does not vary and it is displaced to the right in a direction of propagation of an incident wave. The dip extends, but the direction of deepening does not vary.

Figure 3 represents the normalized SPS of a scattered radio waves as a function of nondimensional wavenumber parameter **k** at different anisotropy coefficient χ and at small angles γ_0; $\xi = 4000(l_{\|} \approx 60$ km), distance between the observation points $\rho_x \approx 160$ m. At $\chi = 40$ a double-humped shape in the SPS does not observed. Increasing anisotropy coefficient in the interval 50, 80, 100, 150, the double-humped shape broadens (the curves $2 \div 5$, respectively) and a gup becomes deeper.

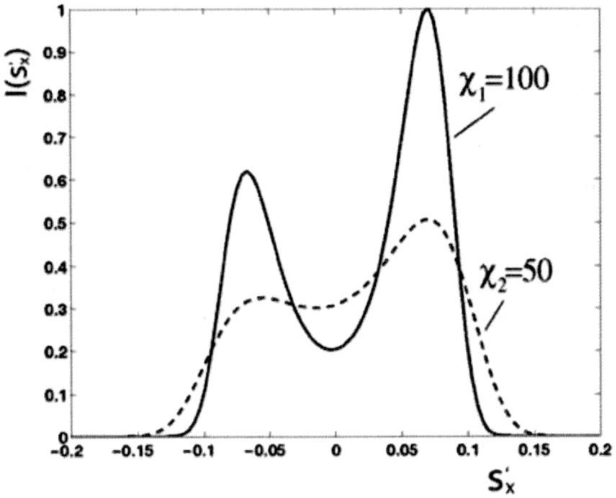

Fig. 1 Normalized intensity of a scattered radiation versus direction of a scattered radio wave at $\alpha_{\perp} = 5$, $s_{x0} = 0, 07$, $\mu = 200$. Solid curve corresponds to $\chi_1 = 100$, dotted line to $\chi_2 = 50$

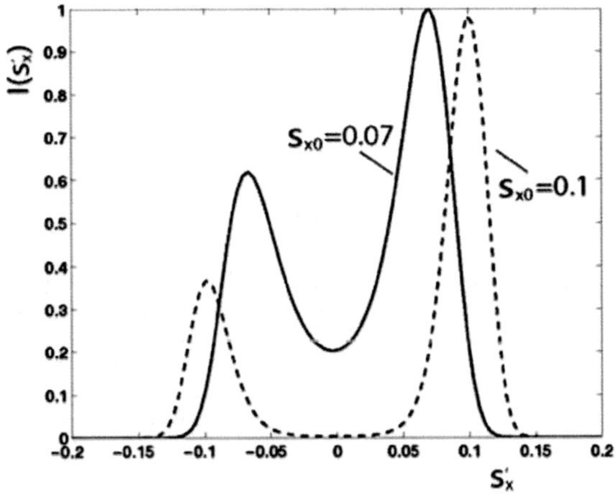

Fig. 2 Normalized intensity of scattered radiation versus the direction of a scattered wave at $\alpha_{\perp} = 5$, $\chi = 100$, $\mu = 200$. Solid curve corresponds to $s_{x0} = 0, 07$, dotted line to $s_{x0} = 0, 1$

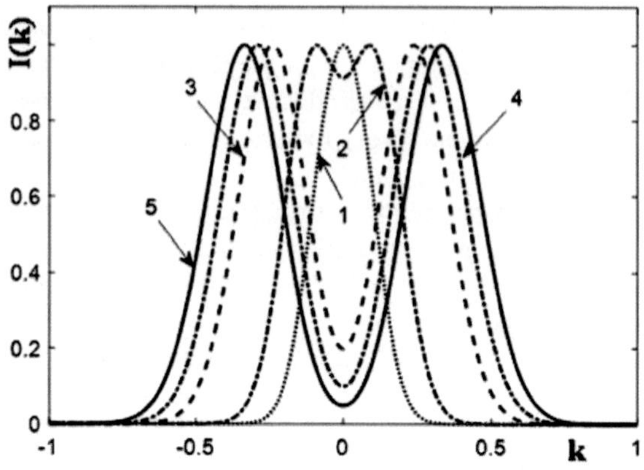

Fig. 3 The normalized SPS at different anisotropy factors

Increasing anisotropy coefficient, a gup decreases and at $\chi = 600$ the double-humped shape in the SPS is disappears. Figure 4 shows the evaluation of a double-humped shape at $\chi = 100$, $\gamma_0 = 2^0$, $\eta_x = 10$. Increasing characteristic spatial scale of electron density irregularities 16, 24 and 32 km (the curves 4–2, respectively), the depth of the SPS decreases and it narrows, at $l_{||} \approx 40$ km the double-humped shape is disappeared.

The "Double-Humped Effect" in a high-latitude COCOIMA plasma can be observed at small both inclination angles of elongated electron density irregularities with respect to the external magnetic field and linear scales of electron concentration inhomogeneities. Neglecting diffraction effects the double-humped effect is disappeared.

The obtained results will have applications at diagnostic of electron density irregularities in a high-latitude terrestrial ionosphere and remote sensing.

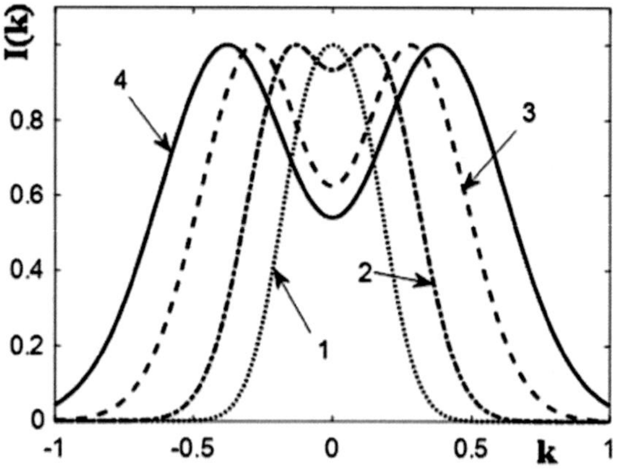

Fig. 4 The normalized SPS at different linear scales of electron density irregularities

4 Conclusions

Statistical characteristics of multiple-scattered radio waves of the "Double-Humped Effect" in a high-latitude COCOIMA plasma are investigated for the first-time considering the diffraction effects. Second order statistical moments of the phase fluctuations contain arbitrary correlation function of electron concentration fluctuations using the modify smooth perturbation method. Polarimetric parameters of the scattered ordinary and extraordinary waves have been derived for the first time in the high-latitude COCOIMA plasma of the terrestrial ionosphere. It was shown that strongly pronounced dip is exist in the SPS of scattered electromagnetic waves if the width of the SPS of an incident wave exceeds the anisotropy factor of elongated plasmonic structures. Phase function has a peak along the direction of an incident wave. The width of the SPS for the anisotropic Gaussian curve decreases inversely proportion of a disorder parameter. The double-humped shape substantially depends on the angle between the wave vector on an original wave and an axis of elongated electron density irregularities. Diffraction effects becomes essential in proportion the distance propagating by radio wave in the ionosphere. Formation of a dip in the SPS at different anisotropy parameters (inclination angle and anisotropy coefficient) of the elongated plasmonic structures in the polar COCOIMA plasma is considered for the first time. It was shown that location of spectrum maximum varies weakly and its width substantially increases in proportion of a distance passed by wave in the COCOIMA plasma.

Acknowledgements This work is supported by Shota Rustaveli National Science Foundation of Georgia (SRNSFG), grant NFR-21-316 "Investigation of the statistical characteristics of scattered electromagnetic waves in the terrestrial atmosphere and application".

References

1. Ishimaru, A.: Wave propagation and scattering in random media. In: Multiple Scattering, Turbulence, Rough Surfaces and Remote Sensing, vol. 2. IEEE Press, Piscataway, New Jersey (1997)
2. Gavrilenko, V., Aistov, A., Jandieri, G.: Some peculiarities of wave multiple scattering in a statistically anisotropic media. Waves Random Media **10**, 435–445 (2000)
3. Jandieri, G., Gavrilenko, V., Sarokin, A., Jandieri, V.: Some properties of the angular power distribution of electromagnetic waves multiply scattered in a collisional magnetized turbulent plasma. Plasma Phys. Rep. **31**, 604–615 (2005)
4. Jandieri, G., Ishimaru, A., Jandieri, V., Khantadze, A., Diasamidze, Zh.: Model computations of angular power spectra for anisotropic absorptive turbulent magnetized plasma. PIER **70**, 307–328
5. Jandieri, G., Ishimaru, A., Zhukova, N.: On the influence directional fluctuations of an external magnetic field on the phase and amplitude correlation variations of scattered radiation in a magnetized plasma slab. PIER B **22**, 121–143 (2010)
6. Jandieri, G.: PIER M.: "Double-humped effect" in the turbulent collision magnetized plasma. PIER M **48**, 95–102 (2016)

7. Jandieri, G., Ishimaru, A., Kharshiladze, O.: New features of the "double-humped effect" in the magnetized plasma. PIER M **62**, 1–9 (2017)
8. Ginzburg, V.: Propagation of electromagnetic waves in plasma. Gordon and Beach, New York, USA (1961)
9. Jandieri, G., Gavrilenko, V., Ishimaru, A., Jandieri, V.: Peculiarities of the spatial spectrum of scattered electromagnetic waves in anisotropic inhomogeneous medium. PIER B. **7**, 191–208 (2008)
10. Rytov, S., Kravtsov, Yu., Tatarskii, V.: Principles of statistical radiophysics. In: Waves Propagation Through Random Media, vol. 4. Springer, Berlin, New York (1989)

Computational Bar Size Optimization of Single Layer Dome Structures Considering Axial Stress and Shape Disturbance

Ahmed Manguri, Najmadeen Saeed, Farzin Kazemi, Neda Asgarkhani, Marcin Szczepanski, and Robert Jankowski

Abstract A computational method is proposed in this paper to minimize the material usage in the construction of modern spatial frame structures by prestressing a minimal number of members. The computational optimization is conducted in two steps. Firstly, a numerical model of a single-layer dome structure is used to minimize the cross-sectional area through several iterations. Different assumed ratios (r) ranging from 0.95 to 0.75 are multiplied by the designed cross-sectional area, and the optimal actual ratio (R) is determined through multiple steps using MATLAB. The selection of the optimum ratio is based on ensuring structural stability and considering various constraints. Secondly, a computational optimization is performed using the fmincon function in MATLAB, which employs an interior-point optimization algorithm to search for the minimum summation of the function. The algorithm is designed to exclude actuators with negligible actuation, thereby minimizing the number of actuators. Constraints are set on the stress of all members and the nodal displacements to maintain the desirable shape of the optimized structure. The obtained results demonstrate that the cross-sectional area of the numerical dome structure can be reduced by up to 18% by prestressing only nine members. The validity of the results is confirmed by comparing them with those obtained from MATLAB and SAP2000 software.

A. Manguri (✉) · F. Kazemi · N. Asgarkhani · M. Szczepanski · R. Jankowski
Faculty of Civil and Environmental Engineering, Gdansk University of Technology, Gdansk, Poland
e-mail: ahmed.manguri@pg.edu.pl

F. Kazemi
e-mail: farzin.kazemi@pg.edu.pl

N. Asgarkhani
e-mail: neda.asgarkhani@pg.edu.pl

M. Szczepanski
e-mail: marcin.szczepanski@pg.edu.pl

A. Manguri · N. Saeed
Civil Engineering Department, University of Raparin, Rania, Kurdistan Region, Iraq
e-mail: najmadeen_qasre@uor.edu.krd

D. Vlachos (ed.), *Mathematical Modeling in Physical Sciences*, Springer Proceedings in Mathematics & Statistics 446, https://doi.org/10.1007/978-3-031-52965-8_14

Keywords Computational optimization · Cross-sectional area · Prestressing · Dome structures

1 Introduction

In the modern world, designing symbolic architectural structures is popular. The aim of constructing such structures from an architectural point of view is to attract tourists [1–5]. From the structural engineering point of view, dome structures provide ample free column space. Meanwhile, the material weight ratio to the covering area is low compared to other forms [6–8].

Construction is one of the sectors that cause significant carbon emissions [9] and waste production [10]. Minimizing material consumption nowadays is challenging from environmental and resource points of view. Researchers have been studying for decades to find solutions to reduce material consumption for building structures [11]. Truss size and/or topology optimization have been investigated using different methods and algorithms [12–16]. Regarding dome structures, the natural frequency was considered a constraint for optimizing dome structures' sizes [17]. However, Structural stability mainly depends on the member sections in the structural design process; size and topology optimization of domes was performed, considering joint flexibility as a constraint [18].

Hitherto, dome structures were not prestressed to optimize the size of the bars. In this study, bar size optimization is performed, considering stress in members and the structure's overall shape as constraints. In recent years, the interior-point approaches have become highly effective strategies for addressing linear programming problems.

When tackling complex linear programming problems, the interior point approaches outperform simplex methods in terms of both apparent computing improvements and improved theoretical features. They are also found to be resistant to degeneracy [19].

The interior point is one of the algorithms used in fmincon function in MATLAB [20, 21]. The fmincon function was implemented to improve the design of box girder bridges with corrugated steel webs [22]. Heretofore, researchers have not considered the fmincon function to minimize actuation that minimizes the cross-sectional area of dome structures. Actuators are used to change the length of some active members.

The change in the length of some members may significantly affect the axial stress in members and the nodal coordinates [23]. Researchers changed the length of some active prestressable members for reshaping cable structures [24, 25], spherical shells [26], and single-layer egg-shaped frame structures [27]. Moreover, members of a space truss were prestressed to control their axial forces [28, 29]. Stress and shape control via prestressing was also examined on pin-jointed assemblies [30]. In this study, the bar size minimization is done via redistributing the stress in members by prestressing active members, which leads to reducing the stress in highly stressed members. At the same time, the shape disturbance of the dome is considered.

Researchers stated that the location of actuators substantially affects their efficiency of actuators. By selecting the optimum place of the actuators, the number of necessary actuators can be significantly reduced [27, 31]. Besides the bar size optimization, the number of used actuators was also minimized in this study. Optimization for the actuator numbers has been performed for large deployable mesh reflectors under thermal strain [32], cable net [33, 34], and truss [35, 36] structures. In this study, the fmincon function [37] that relies on interior point optimization is subjected to prestressing techniques to minimize the number of used actuators.

In this study, the cross-sectional area of a single-layer dome structure is minimized while the stress in all members is kept within the yield stress, and the nodal displacements remain within the desired level. The work is done by changing the length of the most active members using actuators. In line with the prestressing technique, the fmincon function which mainly relies on interior point algorithm is applied to identify and exclude passive actuators to minimize the necessary actuator numbers and total actuation.

2 Numerical Model

This section presents the structural geometry, physical properties of the used materials, formulations, and the study's procedure.

2.1 Structural Geometry

The numerical model is 1049 mm high and 2000 mm wide. It is formed by 145 joints and 384 members. For stability reasons, the height-to-width ratio is assumed to be within an acceptable range [6]. The structure is formed by six circles on top of each other, each circle shaped by 24 joints (see Fig. 1). The single top joint has an elevation of 1049 mm; members connect the joints of each level. Furthermore, the joints on top of each other are also connected, with members, as presented in Fig. 2. The bottom joints are restrained against translation, while they are free to get rotation in all directions.

2.2 Materials Properties and Loadings

The members are 6 mm in diameter, have a modulus of elasticity of 200GPa, and yield strength of 350 MPa. Furthermore, the allowable axial force moment and torsion are 9898N, 12,605 N.mm, and 2121.3 N.mm, respectively.

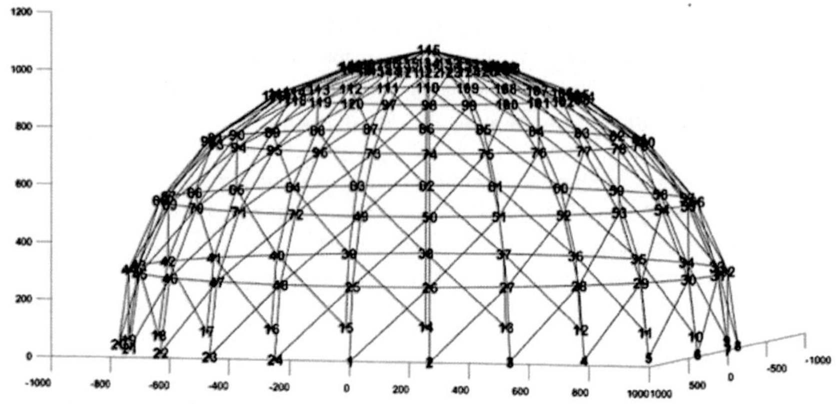

Fig. 1 Joint numbering of the dome

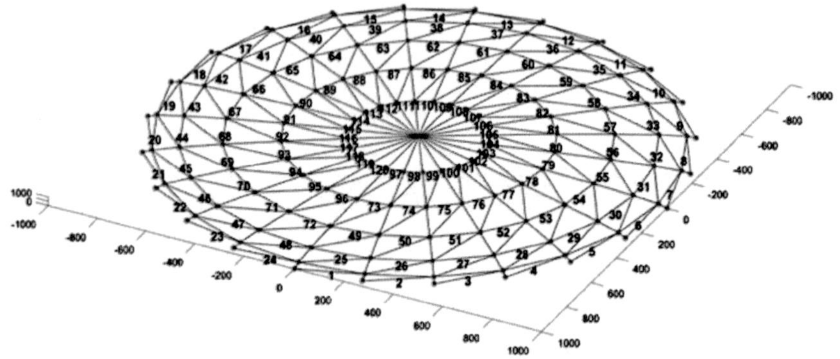

Fig. 2 Member numbering of the dome

Joints 121 to 145 are loaded in the Z direction with −1025N. In addition, Joints 25–36, 49–60, and 73–84 are loaded in the X direction with 1000 N, while joints 37–48, 61–72, and 85–96 are loaded in the Z direction with 1000 N.

2.3 *Prestressing and Computational Optimization*

In this subsection, the procedure of the work is detailed. First of all, the details about the structural geometry is inputted

$$
\begin{aligned}
-d_{all} \le Ye'd_p \le d_{all} \\
-t_{all} \le Ze't_p \le t_{all}
\end{aligned}
\tag{1}
$$

Y is the matrix formed to relate the nodal displacements and the actuators, while Z is designed to link the axial force in members and the actuators. Furthermore, e' is the actuation of members, d_t, and d_p are targeted and induced displacements, respectively. While t_p and t_t are induced and targeted axial forces in members, respectively. d_{all} and t_{all} are allowable nodal displacements and axial forces of the optimized structure.

$$\min f(x) = \sum_{i=1}^{n} e_i' \tag{2}$$

Equations (1) and (2) are coupled to find the optimum solution for the problem using a minimum actuation and the optimum number of actuators while (3) is preserved.

$$|L_b| \le |e_o| \le |U_b| \tag{3}$$

L_b and U_b are the lower and upper limits of the actuation of actuators. Their values depend on the length of the members. In this study, e' is assumed to be within the boundary of ±5 mm. Thus, the optimization algorithms, in most cases, interior point tries numbers between −5 mm and 5 mm for each actuator. In several iterations, the minimum of the sum of actuation is obtained. Meanwhile, the pseudocode shown in Fig. 3 excludes the actuators with less than 0.5 mm, reducing the number of actuators.

```
1. start
2. input Coordinates of the geometry (coor), Bar connectivity (bar), Materials property (E, A,fy),
   Loading (P), assumed ratio (r), k=[],i=1,nB= number of bars
3. Find Displacements (d) and Internal forces (t)
4. Find ttar=fy*A*r
5. for i=1:nB
6.       Find e'
7.       if e'>0.5, add an element to k
8. end
9. nB=length(k)
10. Find da=d+Y*e'; and ta=t-Z*e';
11.R=ta/tp
12.Ar=A*R
13.for i=1:nB
14.      Find e'
15.      if e'>0.5, add an element to k
16.end
17.strs2=ta/Ar, find abs(max(strs2))
18.If abs(max(strs2)) > fy, (R=R+0.001), and go to 12
19.optimum number of actuators = size (e')
20.optimum area = A*R
21.The optimization ratio = (1-R)*100
22.end
```

Fig. 3 The pseudocode of the computational procedure used in the current work

2.4 The Strategy of the Work

First, the structural geometry, material property, loadings, and an assumed ratio are inputted, then the absolute maximum of axial force (t_1) is observed. The maximum allowable axial force after optimization is obtained as follows

$$t_{all} = fy * A_1 * r \quad (4)$$

Secondly, the required prestressing is attained to achieve the targeted displacements and axial forces. Thirdly, after applying the set of actuation obtained from Eqs. (1) and (2) to the ordinary structure, the axial force after applying prestressing can be obtained as follow

$$t_a = t_p - Z * e' \quad (5)$$

Now, the actual optimization ratio can be gained by

$$R = \frac{\max(abs(t_a))}{t_p} \quad (6)$$

A new proposed cross-sectional area is found as follows

$$A_R = A * R \quad (7)$$

Then the absolute maximum stress (strs2) is also obtained. If the obtained stress is greater than fy, R should be reduced until strs2 <= fy. After that, for the sake of practice, the actuators with less than 0.5 mm actuation are excluded to remove inactive actuators. The algorithm itself identifies the passive actuators and dismisses them for the coming iteration. Finally, when both conditions (strs2 <= fy and e0 > 0.5 mm) are met, the optimum cross-sectional area is obtained, as presented in the pseudocode in Fig. 3. Furthermore, the optimization ratio is calculated by

$$\text{Optimization ratio} = (1 - R) * 100 \quad (8)$$

3 Results and Discussion

After applying the loads, the structure negligibly deformed, the maximum absolute displacements (max (abs (d))) are shown presented in Table 1. Furthermore, the maximum absolute (max (abs)) induced forces are shown in Table 2. One can see that the most critical parameter is axial forces.

Table 1 Maximum absolute induced displacements (max (abs (d)))

Directions	X	Y	Z
Max(abs(d)) (mm)	2.4383	1.6158	2.8097

Table 2 Maximum absolute induced forces and their corresponding allowable

Force	Axial	M1a	M1b	M2a	M2b	Torsion
Units	(N)			N.mm		
Max(abs)	9873.1	2299.2	2863.4	1911.3	2679.4	231.26
Allowable	9898	12,605	12,605	12,605	12,605	2121.3

The members are safe in terms of moment and torsion, while they are vulnerable to axial stress failure. In this work, attention is paid to the axial stress after minimizing the cross-sectional area of the members. In addition, the allowable nodal displacements were set to be 10 mm for the optimized structure.

The assumed minimization ratio (r) started at 0.95 and was reduced with 0.05 increments to see how the structure behaved. Meanwhile, the optimization of the number of actuators and the amount of actuation was also monitored.

3.1 *Bar Size Optimization*

Five different ratios were assumed (r) to minimize the cross-sectional area of the bars. Furthermore, several iterations were taken to find the optimum solution for each assumed ratio. The actual ratio (R) may not be the same as the assumed one. When the cross-sectional area changes, the stress distribution is also changed because the structure has multiple degrees of indeterminacy. Figure 4 shows the enhancement of R with decreasing the r. The figure illustrates that the optimum assumed ratio is 0.75, which gives R = 82%, which is 18% of bar size minimization.

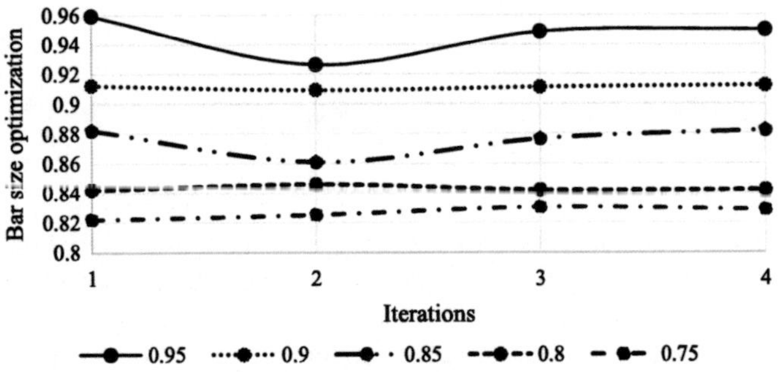

Fig. 4 Bars size optimization in four iterations

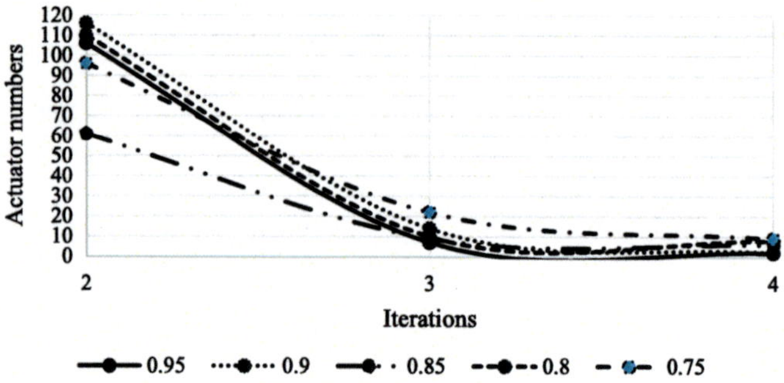

Fig. 5 Optimization of the actuator numbers in four iterations

There is a slight fluctuation in the optimization for the assumed ratios 0.95, 0.85, and 0.80 in four iterations, while the optimization remained unchanged for 0.9 and 0.75 ratios. One can see that the actual optimization ratios (R) are close to the assumed ratios (r) for the first two assumed ones. Nonetheless, the difference between the assumed ratio and the optimization increases by decreasing the assumed ratio.

3.2 Optimization of the Actuator Number

The optimization algorithms identify and exclude the inactive members in the subsequence iterations. Figure 5 illustrates the significant reduction of the actuator numbers in four iterations. In the first iteration, all 384 members participated as actuators. More than a third of the actuators participated with an actuation amount of less than 0.5 mm; thus, they were excluded for the second iteration.

Numerically speaking, for r = 0.85, only 60 active members remained for the second iteration. Moreover, the participated actuators in the second iteration were 106, 116, 110, and 96 for the assumed ratios of 0.95, 0.90, 0.80, and 0.75, respectively. In the last iteration, the active actuators were 2, 2, 6, 8, and 9 for the corresponding 0.95, 0.90, 0.85, 0.80, and 0.75 assumed ratios.

3.3 Optimization of the Actuation Amount

In line with the reduction in the actuator numbers, the amount of actuation also declined in four iterations. Figure 6 shows the substantial drop in the total actuation for prestressing active members. The total amount of actuation in the second iteration is four times less than that in the first iteration. There is a slight decrease in the actuation amount from the 2nd to the 3rd iteration, while the changes are almost null from the 3rd to the last iteration.

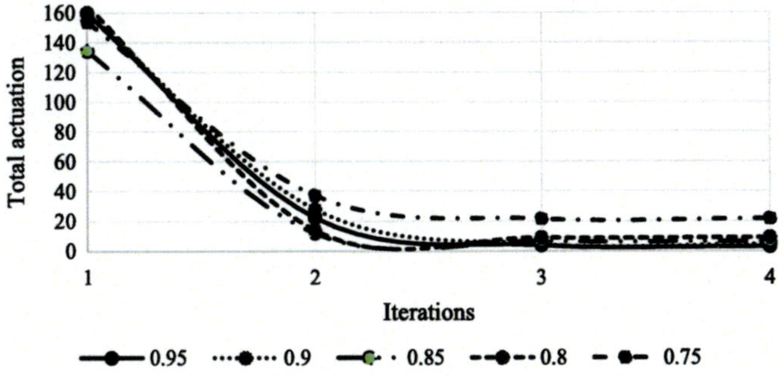

Fig. 6 Optimization in total actuation in four iterations

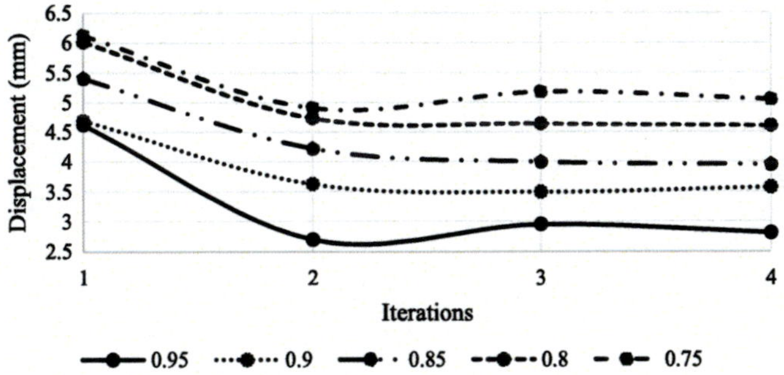

Fig. 7 Maximum absolute displacements in X-direction

3.4 Nodal Displacement Constraints

Besides the axial stress in the members, the joint displacements were set as constraints. The allowable displacement of each joint was set to be ±10 mm after prestressing. Figures 7, 8, and 9 show the maximum absolute nodal displacements in X, Y, and Z directions for each assumed ratio in four respective iterations. The figures show that the nodal displacements are within the limit after prestressing for all ratios and in the four iterations. Generally, the nodal displacements of the optimized structure are negligible.

3.5 Error

The errors were calculated by comparing the maximum absolute axial stress of the optimized structure to the yield stress as follows

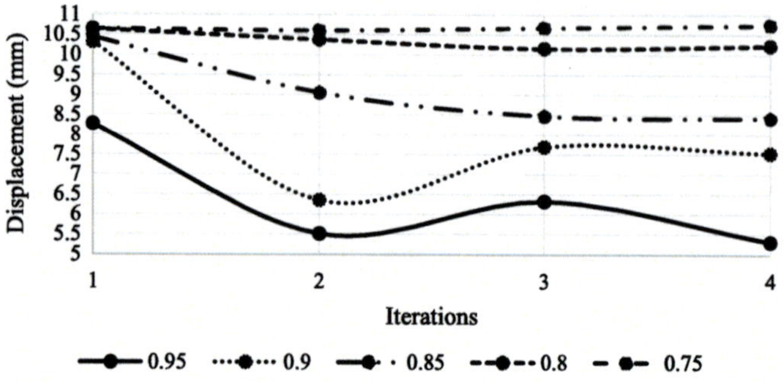

Fig. 8 Maximum absolute displacements in Y-direction

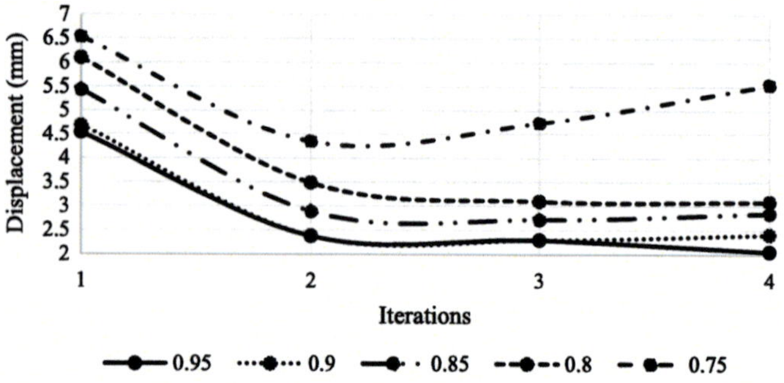

Fig. 9 Maximum absolute displacements in Z-direction

$$Error = \frac{\max(abs(strs2)) - fy}{fy} * 100 \quad (9)$$

Only positive numbers were considered since the opposing numbers indicate less stress than the allowable yield stress. Figure 10 shows the percentage of errors for each ratio in four iterations. One can see that the error for all ratios was negligible except for when r = 0.80. The percentage error for r = 0.75, 0.85, 0.9, and 0.95 were less than 1%, whereas it was just under 2.5% for r = 0.80.

3.6 *Optimal Scenarios*

The optimal case can be r = 0.75, which gives 18% optimization of the cross-sectional area using nine actuators with an actuation amount of 21 mm. It is followed by r = 0.8, which provides 16% bar size optimization using eight actuators with 8.9 mm actuation. For the cases r = 0.9 and r = 0.95, the active members were only two; however, their optimizations are 9% and 5%, respectively.

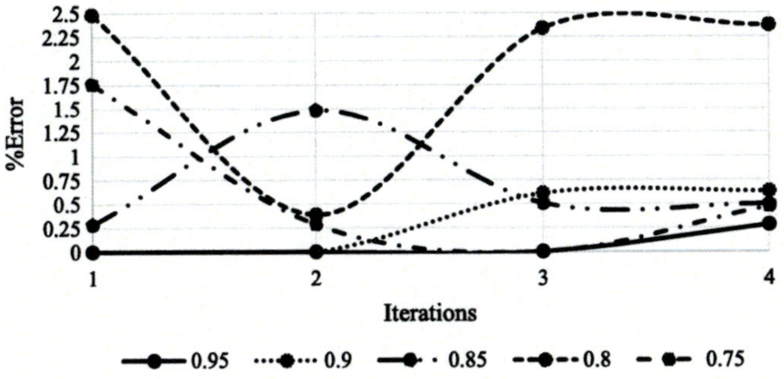

Fig. 10 The error percentage of the stress in four iterations

4 Conclusions

This paper uses a computational technique based on the force method to minimize the cross-sectional area of a numerical single-layer dome by redistributing the stress among all members and keeping the attractiveness of the structural shape. The prestressing is done by actuating the optimum number of actuators and the amount of actuation defined by the fmincon function in MATLAB that relies on the interior point optimization algorithm. The designed cross-sectional area was multiplied by several assumed ratios, and then the optimal ratio was determined. Using MATLAB, hundreds of steps were taken to find the minimum possible actual ratio (R) for each assumed ratio (r). The optimal case was selected based on the minimum possible cross-sectional area, the optimal number of actuators, and the total amount of actuation. It was found that the size of bars can be minimized up to 18% by actuating only nine members. Moreover, the stress in all members and the nodal displacements were kept within the limit.

References

1. Bradshaw, R., Campbell, D., Gargari, M., Mirmiran, A., Tripeny, P.: Special structures: past, present, and future. J. Struct. Eng. **128**(6), 691–709 (2002). https://doi.org/10.1061/ASCE0733-94452002128:6691
2. Winterstetter, T., Berger, R., Starz, F., Bechmann, R., Sobek, W.: Innovative Bautechnik im Herzen Asiens–die EXPO 2017 in Astana, Kasachstan. ce/papers **2**(1), 51–57 (2018). https://doi.org/10.1002/cepa.630
3. Knebel, K., Sanchez-Alvarez, J., Zimmermann, S.: The structural making of the Eden Domes. Space Struct. **5**(1), 245–254 (2002)
4. Imbert F et al. (2013) Concurrent geometric, structural and environmental design: Louvre Abu Dhabi. In: Advances in Architectural Geometry 2012, pp. 77–90 (2013), Springer. https://doi.org/10.1007/978-3-7091-1251-9_6
5. Chang, X., Haiyan, X.: Using sphere parameters to detect construction quality of spherical buildings. In 2nd International Conference on Advanced Computer Control (2010). Shenyang, China IEEE. https://doi.org/10.1109/ICACC.2010.5487073

6. Nie, G.-b., Zhi, X.-d., Fan, F., and Dai, J.-w.: Seismic performance evaluation of single-layer reticulated dome and its fragility analysis. J. Construct. Steel Res. 100, 176–182 (2014)
7. Manguri, A., Saeed, N., and Haydar, B.: Optimal shape refurbishment of distorted dome structure with safeguarding of member stress. In 7th international engineering conference "research & innovation amid global pandemic"(IEC), 90–95 (2021). Erbil, Iraq IEEE. https://doi.org/10.1109/IEC52205.2021.9476107
8. Saeed, N., Manguri, A., and Al-Zahawi, S.: Optimum geometry and stress control of deformed double layer dome for gravity and lateral loads. In 2021 7th International Engineering Conference "Research & Innovation amid Global Pandemic"(IEC), pp. 84–89 (2021). Erbil, Iraq IEEE. https://doi.org/10.1109/IEC52205.2021.9476094
9. Allwood, J. M., et al.: Sustainable materials: with both eyes open. Vol. 2012 (2012). UIT Cambridge Limited Cambridge, UK
10. Brütting, J., Desruelle, J., Senatore, G., Fivet, C.: Design of truss structures through reuse. Structures **18**, 128–137 (2019). https://doi.org/10.1016/j.istruc.2018.11.006
11. Manguri, A., Saeed, N., Szczepanski, M., and Jankowski, R.: Bending moment control and weight optimization in space structures by adding extra members in the optimal locations. In: Adv. Sci. Technol. Res. J. 17(4), 313–324 (2023). https://doi.org/10.12913/22998624/169573
12. Kaveh, A., Rahami, H.: Analysis, design and optimization of structures using force method and genetic algorithm. Int. J. Numer. Meth. Eng. **65**(10), 1570–1584 (2006)
13. Temür, R., Türkan, Y. S., and Toklu, Y. C., Geometrically nonlinear analysis of trusses using particle swarm optimization. In: Recent advances in swarm intelligence and evolutionary computation, pp. 283–300 (2015), Cham, Switzerland: Springer. https://doi.org/10.1007/978-3-319-13826-8_15
14. Cai, Q.: Ruoqiang Feng, and Zhijie Zhang: Topology optimization of trusses incorporating practical local buckling stability considerations. Structures **41**, 1710–1718 (2022). https://doi.org/10.1016/j.istruc.2022.05.109
15. Nguyen-Van, S., Nguyen, K.T., Luong, V.H., Lee, S., Lieu, Q.X.: A novel hybrid differential evolution and symbiotic organisms search algorithm for size and shape optimization of truss structures under multiple frequency constraints. Expert Syst. Appl. **184**, 115534 (2021). https://doi.org/10.1016/j.eswa.2021.115534
16. Manguri, A., Saeed, N., Kazemi, F., Szczepanski, M., Jankowski, R.: Optimum number of actuators to minimize the cross-sectional area of prestressable cable and truss structures. Structures **47**, 2501–2514 (2023). https://doi.org/10.1016/j.istruc.2022.12.031
17. Dede, T., Atmaca, B., Grzywinski, M., Rao, R.V.: Optimal design of dome structures with recently developed algorithm: Rao series. Structures **42**, 65–79 (2022). https://doi.org/10.1016/j.istruc.2022.06.010
18. Ye, J., Lu, M.: Design optimization of domes against instability considering joint stiffness. J. Constr. Steel Res. **169**, 105757 (2020). https://doi.org/10.1016/j.jcsr.2019.105757
19. Rao, S., Mulkay, E.: Engineering design optimization using interior-point algorithms. AIAA J. **38**(11), 2127–2132 (2000). https://doi.org/10.2514/2.875
20. Saeed, N., et al.: Using minimum actuators to control shape and stress of a double layer spherical model under gravity and lateral loadings. Adv. Sci. Technol. Res. J. 22(6) (2022). https://doi.org/10.12913/22998624/155214
21. Manguri, A., Saeed, N., Katebi, J., Mahmood, A., and Jankowski, R.: Optimal reshaping and stress control of double-layer spherical structures under vertical loadings. Arch. Civil Eng. 68(4) (2022). https://doi.org/10.24425/ace.2022.143056
22. Ji, W., Shao, T.: Finite element model updating for improved box girder bridges with corrugated steel webs using the response surface method and Fmincon algorithm. KSCE J. Civ. Eng. **25**(2), 586–602 (2021). https://doi.org/10.1007/s12205-020-0591-3
23. Irschik, H.: A review on static and dynamic shape control of structures by piezoelectric actuation. Eng. Struct. **24**(1), 5–11 (2002). https://doi.org/10.1016/S0141-0296(01)00081-5
24. You, Z.: Displacement control of prestressed structures. Comput. Meth. Appl. Mech. Eng. **144**(1), 51–59 (1997). https://doi.org/10.1016/S0045-7825(96)01164-4

25. Manguri, A.A., Kwan, A.S.K., Saeed, N.M.: Adjustment for shape restoration and force control of cable arch stayed bridges. Int. J. Comput. Meth. Exp. Meas. **5**(4), 514–521 (2017). https://doi.org/10.2495/CMEM-V5-N4-514-521
26. Wang, K. and Preumont, A.: Shape control of an adaptive spherical shell reflector under space environment. In Eighth Symposium on Novel Photoelectronic Detection Technology and Applications, 2462–2467 (2022). SPIE. https://doi.org/10.1117/12.2625829
27. Saeed, N., Manguri, A., Abdulkarim, S., and Shekha, A.: Shape restoration of deformed egg-shaped single layer space frames. In 2019 International Conference on Advanced Science and Engineering (ICOASE), pp. 220–225 (2019). Duhok, Kurdistan Region, Iraq IEEE. https://doi.org/10.1109/ICOASE.2019.8723714
28. Kwan, A., Pellegrino, S.: Prestressing a space structure. AIAA J. **31**(10), 1961–1963 (1993). https://doi.org/10.2514/3.11876
29. Saeed, N., et al.: Using minimum actuators to control shape and stress of a double layer spherical model under gravity and lateral loadings. Adv. Sci. Technol. Res. J. 16(6), 1–13 (2022). https://doi.org/10.12913/22998624/155214
30. Kawaguchi, K.-I., Hangai, Y., Pellegrino, S., Furuya, H.: Shape and stress control analysis of prestressed truss structures. J. Reinf. Plast. Compos. **15**(12), 1226–1236 (1996). https://doi.org/10.1177/073168449601501204
31. Joo, J. J., Sanders, B., Johnson, T., and Frecker, M. I.: Optimal actuator location within a morphing wing scissor mechanism configuration. In Smart Structures and Materials 2006: Modeling, Signal Processing, and Control, 24–35 (2006). SPIE. https://doi.org/10.1117/12.658830
32. Yuan, S., Yang, B.: Shape adjustment of large deployable mesh reflectors under thermal strain. In AIAA Scitech 2021 Forum, pp. 1148 (2021). https://doi.org/10.2514/6.2021-1148
33. Yuan, S., Jing, W.: Optimal shape adjustment of large high-precision cable network structures. AIAA J. **59**(4), 1441–1456 (2021). https://doi.org/10.2514/1.J059989
34. Saeed, N.M., Manguri, A.A.H., Adabar, A.M.: Shape and force control of cable structures with minimal actuators and actuation. Int. J. Space Struct. **36**(3), 241–248 (2021). https://doi.org/10.1177/09560599211045851
35. Manguri, A., Saeed, N., Szczepanski, M., Jankowski, R.: Buckling and shape control of prestressable trusses using optimum number of actuators. Sci. Rep. **13**(1), 3838 (2023). https://doi.org/10.1038/s41598-023-30274-y
36. Saeed, N.M., Manguri, A.A., Szczepanski, M., Jankowski, R., Haydar, B.A.: Static shape and stress control of trusses with optimum time, actuators and actuation. Int. J. Civil Eng. **21**(3), 379–390 (2023). https://doi.org/10.1007/s40999-022-00784-3
37. MATLAB. Find minimum of constrained nonlinear multivariable function. 2023 02–01–2023]; Available from: https://www.mathworks.com/help/optim/ug/fmincon.html

Algorithmic Feature Selection and Dimensionality Reduction in Signal Classification Tasks

Jan Zavadil, Václav Kůs, and Milan Chlada

Abstract This paper presents a research endeavour addressing the recognition of acoustic emission signals, aiming to enhance their utilisation in non-destructive defectoscopy and machining process control. The classification task can be accomplished through two approaches: representing signals using a suitable attribute set, or directly passing the signals in their entirety to the classification algorithm. Our primary focus was on the meticulous selection of methods and tools for automating the extraction of a comprehensive set of features from the signals, followed by dimensionality reduction techniques. Subsequently, we conducted a comprehensive performance evaluation by comparing various classifiers applied to the low-dimensional projections. Lastly, we put the feature based classification approach to the test with direct signal classification employing convolutional neural networks.

Keywords Time-series classification · Feature selection · Convolutional neural networks · Acoustic emission

1 Signal Classification and Experiments

Acoustic emission (AE) is a phenomenon of releasing elastic waves in materials due to irreversible internal changes, such as cracks or deformations caused by stress. AE finds applications in Structural Health Monitoring, offering non-destructive testing used for tasks like monitoring nuclear reactors, checking pressure vessel leaks, and even adapting manufacturing processes in real-time subject to variable operation conditions. AE is categorised as burst-like (distinct events) or continuous (noise-like) emission, recorded by means of piezoelectric sensors attached to material surfaces.

J. Zavadil (✉) · V. Kůs
Department of Mathematics, FNSPE, Czech Technical University in Prague, Prague, Czechia
e-mail: zavadja4@fjfi.cvut.cz

M. Chlada
Institute of Thermomechanics, Czech Academy of Sciences, Prague, Czech Republic

D. Vlachos (ed.), *Mathematical Modeling in Physical Sciences*, Springer Proceedings in Mathematics & Statistics 446, https://doi.org/10.1007/978-3-031-52965-8_15

In all these applications, signal recognition and classification is crucial for satisfactory material monitoring. Traditionally, some features (eg. dominant frequency) were extracted from signals in order to represent them in a low dimensional feature space [5]. The major downside of this approach lies in the necessity of manual specification of the set of features. This often requires deep insight into the specific classification problem at hand. Therefore, in the following text, we try to provide an automated feature extraction process working across various time-series classification setups.

Under recent developments in the fields of artificial intelligence and machine learning, we use the convolutional neural networks (CNNs) to perform the classification task directly to the original AE signals avoiding the need of feature space specification. This approach has already been successfully tested in [2]. The drawback here is the difficulty of interpreting the decision process due to the black-box nature of CNNs.

1.1 Two Defectoscopic Experiments

To obtain real sample data for our classification tasks, the following two experiments were carried out in cooperation with the Institute of Thermomechanics of the Czech Academy of Sciences.

In the first experiment, pressure vessel/tank equipped with four sensors was used. Acoustic emission was created by manually invoked impacts on the surface of the vessel. Five objects of different shape and material were used to produce burst-like AE, for each object 1000 impacts were recorded by the four sensors. The angles and forces of the impacts were not constant throughout the measurement, resulting in higher variability in observed signals for the same object. The goal of our work with this dataset is to correctly predict the used object based on recorded AE signals.

In the second experiment, drilling system was used to drill a hole into a metal block with four sensors attached to it. The system enabled drilling under constant pressure and angle, eliminating variability of these two factors and allowing us to focus on the sharpness of the drill itself. Five series of measurements were performed with the drill undergoing additional manual blunting using grinding stone after each session. Measured data is represented by a long continuous AE signal, thus, we randomly select several shorter parts of each signal to use them in classification task. Consequently, we strive for determining the level of bluntness of the drill based on these sections of emitted signals.

2 Conventional Algorithmic Feature Approach

Feature extraction. The traditional way of analysing time series data, like our acoustic signals, involves simplifying complex raw data into informative summaries called features. These elaborated functions of signal time or frequency domain enable easier interpretation of the classification problem. In order to avoid the demanding process of hand picking a set of features that perform well on the specific AE scanning configuration, we employed available software libraries that compute numerous features, benefiting from their versatility. This approach doesn't assume any signal structure and uses feature database from various applications. First, the MATLAB based package *hctsa* [6] was tested with relatively convincing achievements, but finally, we eventually decided to apply *python* based library TSFresh [1] which brings easier data flows.

TSFresh library provides an automated extraction of almost 800 features from time-series along with a feature selection algorithm. If labels supply the measured time-series, this selection algorithm estimates the distribution of every feature within each class and keeps only those features possessing different distributions. Each AE event in our datasets was represented by four separate time-series, after TSFresh processing we obtained 2921 and 1161 features for the first and second experiment, respectively.

Dimensionality reduction. The number of features extracted in the previous step remains too high for the application of statistical oriented classifiers. Furthermore, a lot of those features carry little discriminatory information to separate classes. Therefore, certain dimensionality reduction techniques are required. Five different methods were implemented for our purposes:

1. Sequential Feature Selection (SFS)
2. Principal Component Analysis (PCA)
3. Kernel Principal Component Analysis (KPCA)
4. Linear Discriminant Analysis (LDA)
5. Generalized Discriminant Analysis (GDA).

The algorithm SFS selects k best original features based on a defined criterion, in our case it was the accuracy of classification obtained by the method of Quadratic Discriminant Analysis (QDA). The other listed algorithms create k new features as linear combinations of original features which allows them to preserve more information. KPCA [7] is a non-linear generalisation of classical PCA method, both of them are unsupervised. GDA [8] is nonlinear generalisation of the well established supervised LDA technique. Both KPCA and GDA take advantage of the so-called kernel trick, rewriting PCA and LDA to the form of inner products computed using specific kernel function $k(x, y)$. Table 1 lists the kernels that were used for our dimensionality reduction.

Table 1 List of kernels used in dimensionality reduction

	Polynomial	Sigmoid	RBF	Cosine
$k(x, y)$	$(x^T y + 1)^\beta$	$\tanh(x^T y + 1)$	$\exp(-\|\|x - y\|\|^2)$	$\frac{x^T y}{\|\|x\|\| \|\|y\|\|}$

Classification methods. Now, when our AE events are represented as points x_i in feature space $\mathbb{R}^k$, we can employ number of classification methods available for this setup. For the reduced set of features we used statistical Bayes classifier QDA (Quadratic Discriminant Analysis) which assumes that the observations are normally distributed for all classes and assigns new observation through the posterior probability. Furthermore, we employed the CART Decision Tree (DT) and more stable ensemble learning Random Forest (RF) method. Finally we put to use fully-connected Deep Neural Networks (DNN).

Unfortunately, the dimension of unreduced feature space is too high for an efficient use of QDA and the decision tree turned out to be very unstable for this type of signal classification. Therefore we only classified our data on the full feature space using RF and DNN with corresponding augmentation of meta parameters (tree depth, number of layers, etc.).

3 Contemporary Deep Learning Approach

With the development of neural networks in the recent years, more straightforward approach of AE events classification is available through direct raw signal classification. The convolutional neural networks (CNN) were originally developed to work with 2D image data as they take into account the location of each data point (pixel in an image). The same principle is crucial for CNN one dimensional version which suits well to time-series data. In our setup, the CNN input is in the form of 4 signals from 4 sensors recording any AE event. These input signals are merged and transformed through convolutional and pooling layers, then flattened into a single vector and forwarded into a fully-connected neural network. Our specific architecture of CNN was inspired by LeNet [4]. It consists of two pairs of convolution and dropout layers, followed by flattening and three fully-connected layers, as can be seen in Fig. 1.

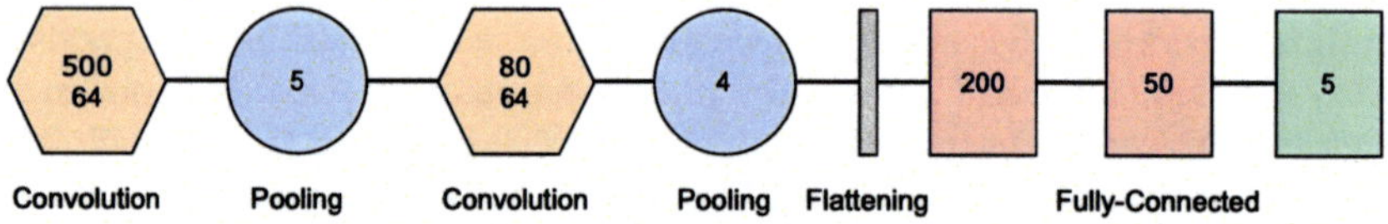

Fig. 1 Architecture of implemented convolutional neural network. The numbers inside the diagram indicate the length of filters and their amount for convolutional layers, length for pooling layers and number of neurons for dense layers

Table 2 Classification accuracy of data from the first AE impact experiment

		Dimension reduction technique				
		SFS	PCA	KPCA (cosine)	LDA	GDA (cosine)
Classifier	QDA	94.5%	76.3%	76.7%	99%	99.4%
	DT	89.8%	77.6%	77.9%	97.8%	98.4%
	RF	93.4%	84.5%	84.5%	98.9%	99.2%
	ANN	93.7%	85%	79.9%	99.4%	99.1%

4 Classification Results and Discussion

For data from our both experiments, we first reduced the dimensions of feature spaces to 4 using presented methods and then evaluated performance of our classifiers. Full feature set classification was performed by using random forest (RF) and deep neural networks (DNN) only.

Experiment 1. The results of classification on reduced feature spaces are summarised in Table 2. Accuracy of more than 99%, was obtained for supervised dimensionality reduction multiple classifiers, i.e. the recognition of objects based on their impact AE signatures is quite feasible. The DNN used had four layers with 22-22-12-5 neurons.

Full feature set contained 2921 features. RF with 1000 trees and maximum depth of each tree set to 20 attained accuracy of 99.3%. DNN consisting of five layers with 4000-2000-1000-200-5 neurons and early stopping regularisation was able to classify impacts with 99.7% accuracy, surpassing all the methods on reduced datasets.

For direct signal classification using CNN under architecture in Fig. 1, early stopping was implemented again. The network was able to classify with accuracy of 99.6% after 13 epochs of training. Performance of both the feature and direct approach was comparable in this case, probably thanks to relatively well separable data.

Experiment 2. The second drilling experiment focused on estimating the sharpness of a drill. Table 3 presents our classification results on reduced feature spaces. Unsupervised dimensionality reduction methods (SFS, PCA, KPCA) provide projections of data in which it is significantly more difficult to separate classes than for their supervised counterparts, therefore their use in this case is not viable. The best result of 75.7% accuracy is achieved combining GDA reduction with DNN classification. We consider this result to be very convincing and promising since the difference in sharpness of the drill between classes is not easily tracked.

For this second experiment, we obtained in total 1161 features. It tell us in itself that classification will be more difficult compared to the first experiment since TSFresh algorithm disqualified more features. Classification using RF with 1000 trees and unlimited depth of trees reached the accuracy of 65.3%, significantly lower than

Table 3 Classification accuracy of data from second AE drilling experiment

		Dimension reduction technique				
		SFS	PCA	KPCA (rbf)	LDA	GDA (rbf)
Classifier	QDA	38.5%	25%	28.2%	73.9%	75.5%
	DT	35%	25.3%	24.6%	64.8%	69.2%
	RF	41.9%	28%	29.7%	72.8%	74.2%
	ANN	45.3%	27.9%	29.9%	73.9%	75.7%

some classifiers on reduced feature spaces. We attribute this fact to potentially high number of features with low discriminant information. After extensive optimisation of DNN meta parameters, we make use of the network formed by 5 layers with 2200-2200-1400-600-5 neurons. This network classifies the sharpness of drilling with 76% accuracy, slightly better result than for classification based on the reduced sets.

Again, using the architecture of direct CNN, described in Fig. 1 with early stopping rule, we arrived at the accuracy of 78.9% after 17 epochs of training. Thus, the direct CNN classification algorithm significantly outperforms all the feature based approaches.

5 Conclusion

Classification on the complete feature space using random forests and neural networks achieved satisfying accuracy. Dimension reduction of feature spaces using supervised methods did not cause notable decrease in accuracy while providing better interpretability and possibility of visualisation. However, employing convolutional neural networks directly on signals yielded significant improvement in accuracy compared to the algorithms with extracted features, especially on data from the second drilling experiment. This direct method turned out to be the most effective approach for classifying acoustic emission signals, encouraging further research into more complex neural network architectures such as inception time networks or U-net.

References

1. Christ, M., Braun, N., Neuffer, J., Kempa-Liehr, A.W.: Time series featuRe extraction on basis of scalable hypothesis tests (tsfresh-a python package). Neurocomputing **307**, 72–77 (2018)
2. Kovanda, M.: Rozpoznávání ultrazvukových signálů pomocí konvolučních neuronových sítí. Bachelor thesis, DM FNSPE CTU, Prague (2020)
3. Baudat, G., Anouar, F.: Generalized discriminant analysis using a kernel approach. Neural Comput. **12**(10), 2385–2404 (2000)
4. Lecun, Y., Bottou, L., Bengio, Y., Haffner, P.: Gradient-based learning applied to document recognition. Proceed. IEEE **86**(11), 2278–2324 (1998)

5. Zavadil, J., Kůs, V.: Classification of AE signals based on statistics and machine learning. In: SPMS 2020/21–Proceedings of the international SPMS conference, Czech Technical University in Prague, pp. 175–182 (2021)
6. Fulcher, B.D., Jones, N.S.: Hctsa: a computational framework for automated time-series phenotyping using massive feature extraction. Cell Syst. **5**(5), 427–532 (2017)
7. Schölkopf, B., Smola, A., Müller, KR.: Kernel principal component analysis. In: Gerstner, W., Germond, A., Hasler, M., Nicoud, J.D. (eds.), Artificial Neural Networks–ICANN'97. Lecture Notes in Computer Science, vol. 1327. Springer, Berlin, Heidelberg (1997)
8. Baudat, G.: Anouar, F.: Generalized discriminant analysis using a kernel approach. Neural Comput. **12**(10), 2385–2404 (2000)

Two-Terminal Reliability of the K_4-Ladder—Revisited

Philippe Poulin, Simon R. Cowell, and Valeriu Beiu

Abstract The exact calculation of network reliability has been a practical, but difficult problem, solved for some periodic graphs—including ladders—using recursion and transfer matrices. We revisit the results for non-oriented K_4-ladders of arbitrary length, using Markov chains (transition matrices) instead of recursion. We consider three cases, where the edges and vertices of the ladder have independent reliabilities: (i) p and 1 respectively (3×3 transition matrix); (ii) p and ρ respectively (4×4 transition matrix); (iii) varying throughout the ladder (6×6 transition matrices, indexed). To use transition matrices seems rather new in the context of graph reliability. These matrices have a clearer interpretation than their analogs, the transfer matrices, obtained from recursion: their entries represent probabilities to transit from one state to another when eliminating a K_4 block. As a consequence, secondary results may be derived more easily—by modifying the initial and targeted states, or by invoking classical results in Markov chain theory.

Keywords Network reliability · Transition matrix · Finite Markov chain · Stochastic process

P. Poulin (✉)
Higher Colleges of Technology, Dubai, United Arab Emirates
e-mail: ppoulin@hct.ac.ae

S. R. Cowell · V. Beiu
"Aurel Vlaicu" University of Arad, 2-4 Elena Drăgoi Str., 310330 Arad, Romania
e-mail: simon.cowell@uav.ro

V. Beiu
e-mail: valeriu.beiu@uav.ro

D. Vlachos (ed.), *Mathematical Modeling in Physical Sciences*, Springer Proceedings in Mathematics & Statistics 446, https://doi.org/10.1007/978-3-031-52965-8_16

1 Introduction

Classical ladders (https://en.wikipedia.org/wiki/Ladder) are ancient tools known for over 10,000 years. It is about two centuries ago [1] that they have started lending their name to various electrical networks of resistors-inductors-capacitors (RLC, https://en.wikipedia.org/wiki/RLC_circuit).

And it is about a century later that they began to be used in relation to network reliability—a field established in the mid 1950's, when Moore and Shannon introduced hammocks/brick-wall networks as well as the reliability polynomial of a network in a seminal paper [2]. Reinforcing a classical ladder by diagonals was used for the ladders of fire trucks. As with the classical ladder, such ideas permeated first into electrical circuits [3–8] and later into networks/graphs, an example being the K_4-ladder [9] (K_4 being the complete graph on 4 vertices).

Lately, the constant progress of communication schemes/circuits (be they analog or digital), has pushed reliability aspects to center stage, with quite a few different networks being analyzed by Tanguy [9–14]. The steady progress in scaling transistors [15] (see Singer [16] for a 20-year vision), makes it that *locally connected regular networks/arrays* are matching advanced arrays of transistors (currently FinFET, moving towards MBCFET, GAAFET, CNT, etc.), as well as low level quantum architectures (e.g., the honeycomb/brick-wall lattice and its heavy hexagonal variation being both used and advocated by IBM [17–19], Google [20], and quite a few others). These have played a part in making network reliability a very lively research topic.

That is why we have considered revisiting the K_4-ladder for trying to gain additional insights which might show how they could fit advanced transistor structures, and which reliability benefits they might bring. For quantum computing the interest would be to see if and how the K_4-ladder might contribute to error correction, e.g., by enhancing reliability using surface codes.

While both of these are very practical applications, our approach is mathematically oriented following on Tanguy's results. As Tanguy [9], we shall use the contraction-deletion method to calculate the two-terminal reliability of K_4-ladders. However, we shall derive the general result from a Markov chain [21] instead of recursion. This last approach is rather new in the context of reliability of periodic graphs [22]. It has advantages as classical results about Markov chains may be invoked to deepen the study—as illustrated in Sect. 5.

The paper is organized as follows. After stating the problem (Sect. 2), we illustrate Tanguy's recursion method on the simplest example, where each edge of a K_4-ladder is reliable with probability p and each vertex is certainly reliable (Sect. 3). Then, we revisit the result using a Markov chain on three states (Sect. 4), and the two methods are compared (Sect. 5). We generalize the calculation to edges and vertices which are reliable with respective probabilities p and ρ—using a Markov chain on four states (Sect. 6). Finally, we generalize further by allowing the edges and vertices to have various reliabilities throughout the ladder—using a Markov chain on six states

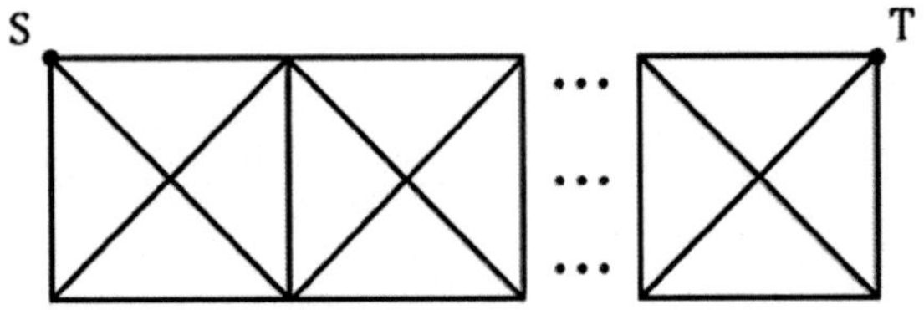

Fig. 1 The K_4-ladder, its source S (input) and terminus T (output)

with indexed components (Sect. 7). The elements of the transition matrix for this last, most general case are also presented.

2 Statement of the Problem

The *K_4-ladder of length n* consists of n adjacent undirected K_4 graphs joined to form a ladder, as illustrated in Fig. 1. We aim to study the probability (known as *two-terminal reliability*) that the source or input S and the terminus or output T are connected, given that the vertices and edges may fail independently from one another. This is more general than the original two-terminal reliability from Moore and Shannon [2], where only the edges could fail, while the vertices were considered perfectly reliable. Notice that by flipping the vertical edge incident to S—thus interchanging the horizontal edges with the diagonal edges in the first block—S may be positioned indifferently on the top or bottom row of the ladder without affecting the underlying unweighted graph.

Throughout the paper, we shall use the notation $x \circ y = x + y - xy$. Observe that, given a pair of edges which are reliable with respective probabilities x and y, $x \circ y$ is the probability that the endpoints are connected when the edges are in parallel, while $x \cdot y$ is the probability that they are connected when the edges are in series. We shall also use $\overline{x} = 1 - x$.

3 Contraction-Deletion Method

Let us illustrate the contraction-deletion method on the simplest example, where vertices never fail, while each edge has the same probability p to work. To find a recursive pattern, we eliminate the K_4 graph on the right-hand side of the ladder under the assumption that the last vertical edge has probability x to work—so $x = p$ initially.

We perform the elimination along the probability tree presented in Fig. 2. For instance, along the upper left branch, the edge of probability x is assumed to work, leading to a contraction. Along the upper right branch, it is assumed not to work, leading to a deletion.

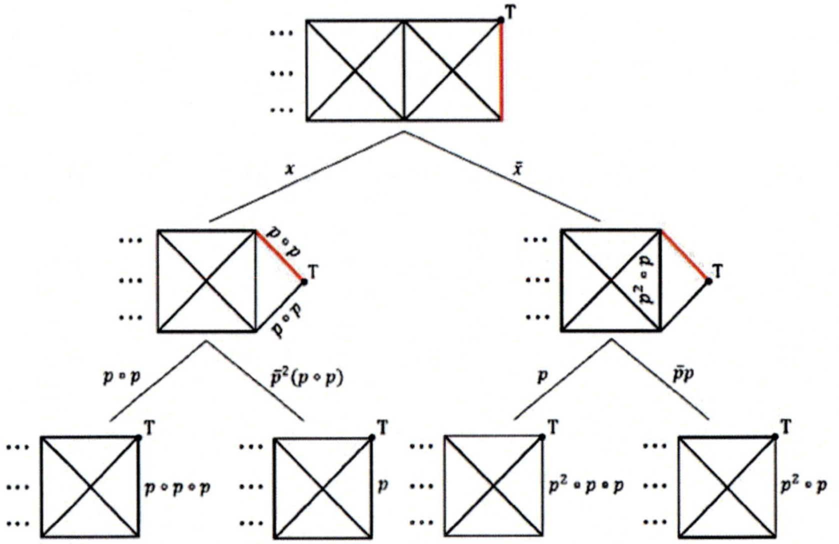

Fig. 2 Elimination by contraction-deletion of the right-hand K_4 block

Denote by $h_n(p)$ the probability that S and T are connected in the K_4-ladder of length n. The above probability tree yields the recurrence relation:

$$h_n(x) = x\big((p \circ p)h_{n-1}(p \circ p \circ p) + \overline{p}^2(p \circ p)h_{n-1}(p)\big) \\ + \overline{x}\big(ph_{n-1}\big(p^2 \circ p \circ p\big) + \overline{p}ph_{n-1}\big(p^2 \circ p\big)\big)$$

Substituting the ansatz $h_n(x) = \alpha_n x + \beta_n \overline{x}$ on both sides of the last relation yields $\begin{bmatrix} \alpha_n \\ \beta_n \end{bmatrix} = M \begin{bmatrix} \alpha_{n-1} \\ \beta_{n-1} \end{bmatrix}$, where the *transfer matrix* $M = [M_{ij}]$ as elements:

$$M_{11} = -2p^5 + 9p^4 - 14p^3 + 8p^2$$
$$M_{12} = 2p^5 - 10p^4 + 18p^3 - 14p^2 + 4p$$
$$M_{21} = 2p^5 - 4p^4 + 3p^2$$
$$M_{22} = -2p^5 + 4p^4 - 4p^2 + 2p$$

Thus, $\begin{bmatrix} \alpha_n \\ \beta_n \end{bmatrix} = M^{n-1} \begin{bmatrix} \alpha_1 \\ \beta_1 \end{bmatrix}$. Keeping in mind that $h_1(x) = \alpha_1 x + \beta_1 \overline{x}$, the suffix $\begin{bmatrix} \alpha_1 \\ \beta_1 \end{bmatrix}$ is obtained from the contraction and deletion presented in Fig. 3. It gives $\begin{bmatrix} \alpha_1 \\ \beta_1 \end{bmatrix} = \begin{bmatrix} -p^5 + 4p^4 - 5p^3 + p^2 + 2p \\ p^5 - 2p^4 + p^2 + p \end{bmatrix}$. Thus,

Fig. 3 Elimination of the last remaining block

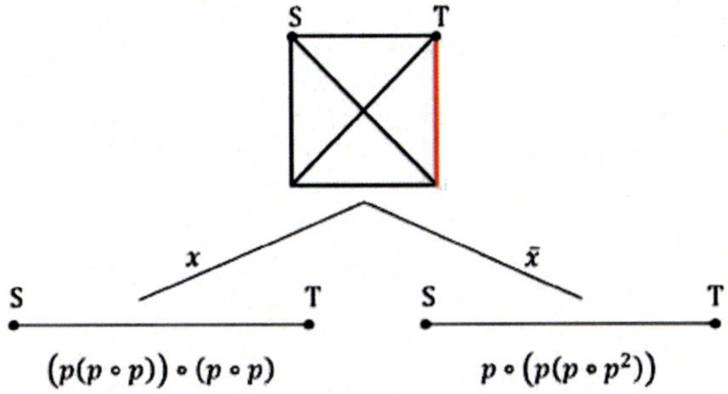

$$h_n(p) = \begin{bmatrix} p \ \overline{p} \end{bmatrix} M^{n-1} \begin{bmatrix} \alpha_1 \\ \beta_1 \end{bmatrix}. \tag{1}$$

4 Markov Chain Method

We now assume that the ladder is indefinitely long and eliminate the K_4 blocks from left to right, without touching their right-hand edge. After a block is eliminated, the ladder may be in any of the following states.

- **SS**: The source is known to access the two left-endpoints—then contracted.
- **S**: The source is known to access one of the two left-endpoints, but it did not access the other via the edges to the left of it—the infinite ladder is unchanged.
- **0**: The source does not access the ladder anymore.

It is worth mentioning that each state represents *what is known* immediately after eliminating a K_4 block. For instance, in the state **S**, it is known that—via edges which are already eliminated—the source accesses one of the two left endpoints. Since the graph is not oriented, the source *may or may not* access the other left endpoint—via the *remaining* edges, not to the left of it.

When any of the previous states is attained, transition to the next state does not depend on the past: the process is without memory. Therefore, it may be modeled by a Markov chain. Transition probabilities are calculated as before, using probability trees and contraction-deletion—as illustrated in Fig. 4 for the transitions from **SS**.

By ordering the states **SS** < **S** < **0**, we thus obtain the *transition matrix* $M = [M_{ij}]$ with the M_{ij} given below.

- Transitions from **SS**:

$$\text{SS} \mapsto \text{SS}: M_{11} = (p \circ p)^2 = p^4 - 4p^3 + 4p^2$$

$$\text{SS} \mapsto \text{S}: M_{12} = 2\overline{p}^2(p \circ p) = -2p^4 + 8p^3 - 10p^2 + 4p$$

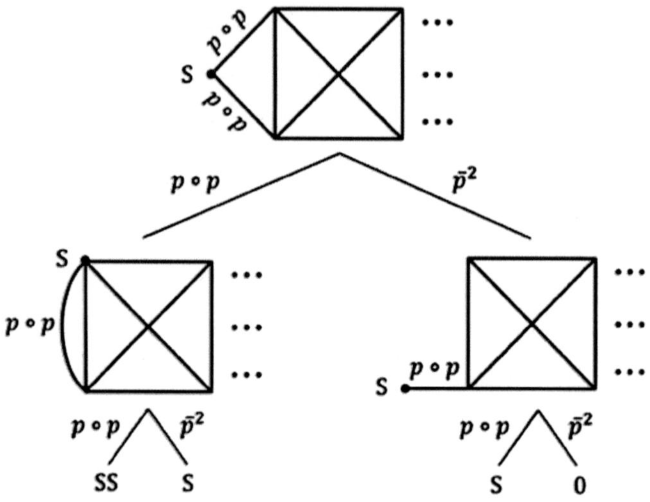

Fig. 4 Transitions from the state **SS**

$$\text{SS} \mapsto 0{:}M_{13} = \overline{p}^4 = p^4 - 4p^3 + 6p^2 - 4p + 1$$

- Transitions from **S**:

$$\text{S} \mapsto \text{SS}{:}M_{21} = 3p^5 - 8p^4 + 5p^3 + p^2$$

$$\text{S} \mapsto \text{S} : M_{22} = -4p^5 + 12p^4 - 10p^3 + 2p$$

$$\text{S} \mapsto 0{:}M_{23} = p^5 - 4p^4 + 5p^3 - p^2 - 2p + 1$$

- Transitions from **0**: $M_{31} = M_{32} = 0$ and $M_{33} = 1$.

To calculate the two-terminal reliability polynomial $h_n(p)$ for the K_4-ladder of length n, the prefix is clearly $\begin{bmatrix} 0 & 1 & 0 \end{bmatrix}$, since the initial state is **S**. Regarding the suffix, reliability is achieved if the process terminates in the state **SS**. It cannot be achieved if it terminates in **0**. Notice that in the original graph, the source may be drawn in a symmetric fashion along the horizontal, as shown in Fig. 5. By symmetry, if the process terminates in **S**, reliability is achieved half of the time. For the other half, the corresponding terminal state is a single edge from S to T, which has reliability p. Therefore, the suffix is $\begin{bmatrix} 1 & (1+p)/2 & 0 \end{bmatrix}$ transposed. In conclusion,

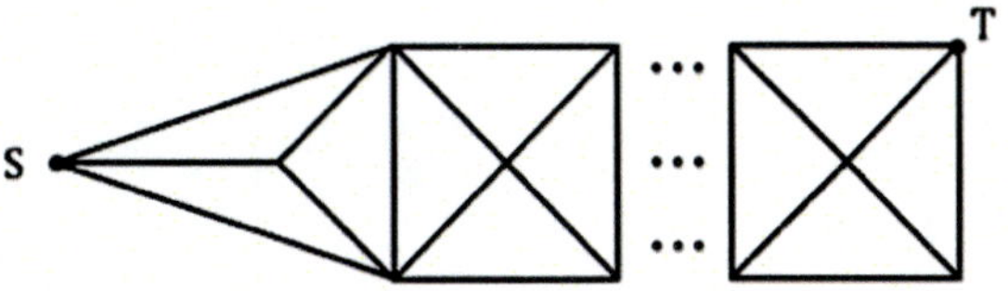

Fig. 5 Horizontal symmetry from the source S

$$h_n(p) = \begin{bmatrix} 0 & 1 & 0 \end{bmatrix} M^n \begin{bmatrix} 1 \\ (1+p)/2 \\ 0 \end{bmatrix}. \tag{2}$$

5 Comparison of the Two Methods

An advantage of the first method is the size of the transfer matrix: 2×2 only, while the Markov chain method requires a 3×3 matrix. However, the elements of the transfer matrix for the first method and even its size depend on choices made when developing the probability tree.

The Markov chain method is more systematic, namely, after states are identified, the transition matrix does not depend on choices made when eliminating a block. The elements of the transfer matrix have a clear interpretation as probabilities to transit from one state to another when eliminating a K_4 block. Hence, supplemental results are obtained by modifying the prefix or suffix. Indeed, all theory pertaining to finite Markov chains is available. If for instance one wonders after how many blocks on average an indefinitely long K_4-ladder loses reliability, it is given by the waiting time before absorption in the state **0** for a random walk on the graph shown in Fig. 6. A standard formula exists [21] (p. 25). With $M = \left[M_{ij}\right]$ as above, it gives:

$$\begin{bmatrix} 0 & 1 \end{bmatrix} \begin{bmatrix} 1 - M_{11} & -M_{12} \\ -M_{21} & 1 - M_{22} \end{bmatrix}^{-1} \begin{bmatrix} 1 \\ 1 \end{bmatrix} = \frac{3p^5 - 9p^4 + 9p^3 - 3p^2 + 1}{(p-1)^4 (2p^5 - 4p^4 + 2p^3 - 2p^2 + 2p + 1)}. \tag{3}$$

The expected length n at which the ladder loses reliability is illustrated in Fig. 7 for large values of p.

Finally, as shown by Tanguy [9], the first method still works when vertices may also fail independently. In fact, it still works when the edges and vertices have various

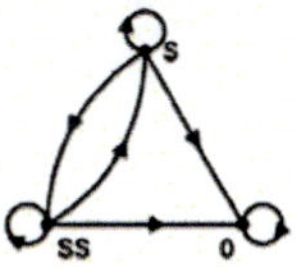

Fig. 6 Underlying random walk

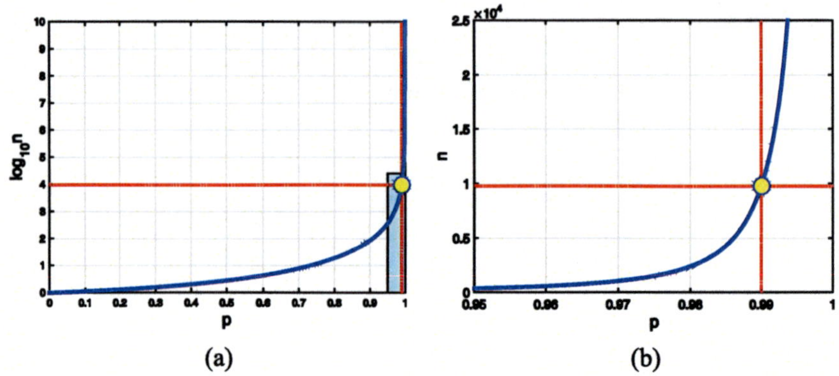

Fig. 7 Plot of Eq. (3): **a** logarithmic view; and **b** zoom on the region of interest (highlighted in cyan in Fig. 7a)

reliabilities in a K_4 block, or even throughout the ladder. The same is true for the Markov chain method, as we now show.

6 First Generalization: Unreliable Vertices

We first treat the case where edges are reliable with independent probability p and vertices are reliable with independent reliability ρ. Successive elimination of blocks may then yield four different states, which we label as follows: **SS** < **S** < **S0** < **0**.

In the state **SS**, the source is known to access the two left-endpoints, which are in addition known to be reliable. Observe that nothing is said about reliability of the source itself, which will be taken in charge by the prefix. Similarly, in the state **S**, the source is known to access one of the two left-endpoints, and nothing is known yet about the other—*which may or may not be reliable*, and which may or may not be accessed by the source (thanks to edges not to the left of it). The new state **S0** is reached when the source is known to access one left-endpoint, while the other endpoint is known not to be reliable. Finally, the state **0** is reached when the source does not connect to the ladder anymore.

Transition probabilities are computed as before by contraction-deletion. However, each time a reliable edge from S to a vertex P is contracted, the probability tree splits into two branches, since P may or may not be working, with respective probabilities ρ and $\overline{\rho}$.

- Transitions from **SS** to **SS**, **S**, **S0**, and **0** respectively (as illustrated in Fig. 8):

$$M_{11} = p^4\rho^2 - 4p^3\rho^2 + 4p^2\rho^2$$

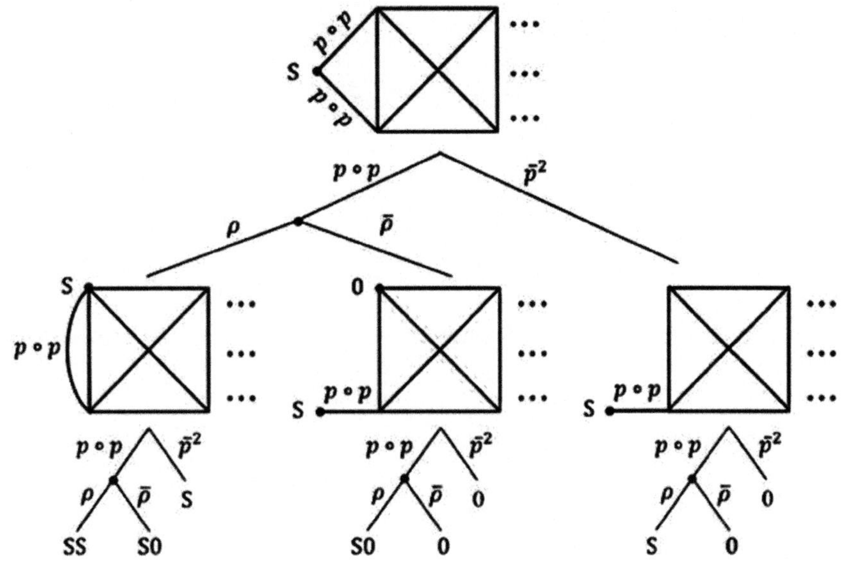

Fig. 8 Transitions from the state **SS** when vertices have reliability ρ

$$M_{12} = -2p^4\rho + 8p^3\rho - 10p^2\rho + 4p\rho$$

$$M_{13} = -2p^4\rho^2 + 2p^4\rho + 8p^3\rho^2 - 8p^3\rho - 8p^2\rho^2 + 8p^2\rho$$

$$M_{14} = p^4\rho^2 - 4p^3\rho^2 + 4p^2\rho^2 + 2p^2\rho - 4p\rho + 1$$

- Transitions from **S** to **SS**, **S**, **S0**, and **0** respectively:

$$M_{21} = 3p^5\rho^3 - 8p^4\rho^3 + 5p^3\rho^3 + p^2\rho^2$$

$$M_{22} = -4p^5\rho^2 + 12p^4\rho^2 - 10p^3\rho^2 + 2p^2\rho^2 - 2p^2\rho + 2p\rho$$

$$M_{23} = -4p^5\rho^3 + 4p^5\rho^2 + 12p^4\rho^3 - 12p^4\rho^2 - 8p^3\rho^3 + 8p^3\rho^2 - 2p^2\rho^2 + 2p^2\rho$$

$$M_{24} = p^5\rho^3 - 4p^4\rho^3 + 3p^3\rho^3 + 2p^3\rho^2 - p^2\rho^2 - 2p\rho + 1$$

- Transitions from **S0** to **SS**, **S**, **S0**, and **0** respectively:

$$M_{31} = p^2\rho^2$$

$$M_{32} = 2p\rho - 2p^2\rho$$

$$M_{33} = 2p^2\rho - 2p^2\rho^2$$

$$M_{34} = p^2\rho^2 - 2p\rho + 1$$

- Transitions from **0** to **SS**, **S**, **S0**, and **0** respectively:

$$M_{41} = M_{42} = M_{43} = 0$$

$$M_{44} = 1$$

Depending on whether or not the source node S works, the initial state is **S** or **0**, so the prefix is $\left[0\ \rho\ 0\ 1-\rho\right]$.

Regarding the suffix, in the final state **SS**, the source connects certainly to the vertex T, which is then known to work. In the final state **S**, half of the times the source connects to the vertex T and T is known to work. For the other half, the state **S** corresponds to a single edge from S to T, which works with probability $p\rho$. In the final state **S0**, half of the times the source connects to T and T is known to work, but reliability is impossible for the other half. Finally, reliability is impossible in the final state **0**. Therefore, the suffix is $\left[1\ (1+p\rho)/2\ 1/2\ 0\right]$ transposed.

In conclusion, with the 4×4 matrix $M = \left[M_{ij}\right]$ defined above the reliability polynomial of a K_4-ladder of length n is:

$$h_n(p,\rho) = \left[0\ \rho\ 0\ 1-\rho\right] M^n \begin{bmatrix} 1 \\ (1+p\rho)/2 \\ 1/2 \\ 0 \end{bmatrix}. \tag{4}$$

7 Second Generalization: Components of Various Reliabilities

This time, the vertices and the edges of a K_4 block may have different, independent reliabilities: a, b, c, d, e for the edges, A^*, B^*, B, A for the vertices, as in Fig. 9.

The Markov chain technique can still be applied, but with more states. As before, **SS** represents the state where the source is known to access A and B (which are

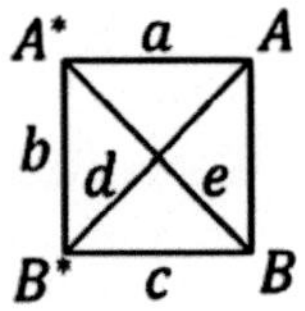

Fig. 9 Labeling of a general K$_4$ block

known to work), and **0** represents the dead state. When the source is known to access one vertex among A and B, each of the following four states may occur.

- **SB**: the source accesses A, known to work; it does not access B from the left.
- **AS**: the source accesses B, known to work; it does not access A from the left.
- **S0**: the source accesses A, known to work; B is not working.
- **0S**: the source accesses B, known to work; A is not working.

We order the states as **SS** < **SB** < **AS** < **S0** < **0S** < **0**. The transition matrix is then a 6×6 matrix $M = \left[M_{ij}\right]$ whose elements are computed by contraction-deletion. Their values are reported in Table 1 (on the next page).

It is straightforward to generalize further, with components whose reliability varies from block to block. For the first block, we index the reliabilities as follows: a_1, b_1, c_1, d_1, e_1 for the edges, and $A^* = A_0$, $B^* = B_0$, $B = B_1$, $A = A_1$ for the vertices. We then obtain a transition matrix $M_1 = \left[M_{ij}^{(1)}\right]$ which depends on these last component reliabilities only. We repeat the same for the second block, with component reliabilities a_2, b_2, c_2, d_2, e_2 for the edges, and $A^* = A_1$, $B^* = B_1$, $B = B_2$, $A = A_2$ for the vertices. We then get a transition matrix M_2, and so on and so forth.

The initial state is **SB** with probability A_0 (that the source works) or otherwise **0**, so the prefix is $\left[0 \; A_0 \; 0 \; 0 \; 0 \; 1 - A_0\right]$. The final states **SS**, **SB**, and **S0** yield reliability. The states **0S** and **0** do not yield reliability, the former, because $T = A_n$ is then known not to work. Finally, the state **AS** yields reliability with probability $b_{n+1}A_n$, where b_{n+1} is the probability that the vertical edge at the right-end of the ladder works. The suffix is thus $\left[1 \; 1 \; b_{n+1}A_n \; 1 \; 0 \; 0\right]$ transposed.

It follows that with the M_{ij} given in Table 1, the reliability polynomial is:

$$h_n = \left[0 \; A_0 \; 0 \; 0 \; 0 \; 1 - A_0\right] M_1 M_2 \ldots M_n \begin{bmatrix} 1 \\ 1 \\ b_{n+1}A_n \\ 1 \\ 0 \\ 0 \end{bmatrix}. \tag{5}$$

Table 1 Elements of the general transition matrix (for the kth block). (All variables are indexed by k except A^* and B^*, which are indexed by $k-1$)

Transitions from SS	
$M_{11} = ABac + ABae + ABcd + ABde - ABacd - ABace - ABade - ABcde + ABacde$ $M_{12} = Aa + Ad - Aac - Aad - Aae - Acd - Ade + Aacd + Aace + Aade + Acde - Aacde$ $M_{13} = Bc + Be - Bac - Bae - Bcd - Bce - Bde + Bacd + Bace + Bade + Bcde - Bacde$	
$M_{14} = Aac + Aae + Acd + Ade - ABac - ABae - ABcd - ABde - Aacd - Aace - Aade - Acde + ABacd + ABace + ABade + ABcde + Aacde - ABacde$	
$M_{15} = Bac + Bae + Bcd + Bde - ABac - ABae - ABcd - ABde - Bacd - Bace - Bade - Bcde + ABacd + ABace + ABade + ABcde + Bacde - ABacde$	
$M_{16} = Aad - Ad - Bc - Be - Aa + Bce + ABac + ABae + ABcd + ABde - ABacd - ABace - ABade - ABcde + ABacde + 1$	
Transitions from SB	
$M_{21} = ABae + ABB^*abc + ABB^*acd + ABB^*bcd + ABB^*bde + ABB^*cde - 2ABB^*abcd - ABB^*abce - ABB^*abde - 2ABB^*acde - 2ABB^*bcde + 3ABB^*abcde$	
$M_{22} = Aa - Aae + AB^*bd - AB^*abc - AB^*abd - AB^*acd - AB^*bcd - AB^*bde + 2AB^*abcd + AB^*abce + AB^*abde + AB^*acde + AB^*bcde - 2AB^*abcde$	
$M_{23} = Be - Bae + BB^*bc - BB^*abc - BB^*bcd - BB^*bce - BB^*bde - BB^*cde + BB^*abcd + BB^*abce + BB^*abde + BB^*acde + 2BB^*bcde - 2BB^*abcde$	
$M_{24} = Aae - ABae + AB^*abc + AB^*acd + AB^*bcd + AB^*bde - ABB^*abc - ABB^*acd - ABB^*bcd - ABB^*bde - 2AB^*abcd - AB^*abce - AB^*abde - AB^*acde - AB^*bcde + 2ABB^*abcd + ABB^*abce + ABB^*abde + ABB^*acde + ABB^*bcde + 2AB^*abcde - 2ABB^*abcde$	
$M_{25} = Bae - ABae + BB^*abc + BB^*bcd + BB^*bde + BB^*cde - ABB^*abc - ABB^*bcd - ABB^*bde - ABB^*cde - BB^*abcd - BB^*abce - BB^*abde - BB^*acde - 2BB^*bcde + ABB^*abcd + ABB^*abce + ABB^*abde + ABB^*acde + 2ABB^*bcde + 2BB^*abcde - 2ABB^*abcde$	
$M_{26} = ABae - Be - Aa - AB^*bd - BB^*bc + AB^*abd + BB^*bce + ABB^*abc + ABB^*bcd + ABB^*bde - ABB^*abcd - ABB^*abce - ABB^*abde - ABB^*bcde + ABB^*abcde + 1$	
Transitions from AS	
$M_{3j} = M_{2j}$ after interchanging $a \leftrightarrow d$, $c \leftrightarrow e$, and substituting A^* in B^*	
Transitions from S0	
$M_{41} = ABae$ $M_{42} = Aa - Aae$ $M_{43} = Be - Bae$	$M_{44} = Aae - ABae$ $M_{45} = Bae - ABae$ $M_{46} = ABae - Be - Aa + 1$
Transitions from 0S	
$M_{5j} = M_{4j}$ after substituting d in a, and c in e	
Transitions from 0	
$M_{61} = M_{62} = M_{63} = M_{64} = M_{65} = 0$	$M_{66} = 1$

8 Conclusion and Perspectives

We calculated the two-terminal reliability polynomials of the K_4-ladders at the same level of generality as in Tanguy [9], using Markov chains instead of recursion. Three cases were examined:

- the edges and vertices are reliable with respective probabilities p and 1;
- they are reliable with respective probabilities p and ρ;
- they are reliable with probabilities varying throughout the ladder.

All our results coincide with the ones in Tanguy [9], with a slight difference of notation in the last case—since we denote the reliabilities of vertical edges by $b_1, b_2, \ldots$ instead of $b_0, b_1, \ldots$.

In each case, our resulting transition matrices have one row and column more than the transfer matrices obtained in Tanguy [9]. However, our transition matrices are stochastic matrices: their entries have a clear interpretation as the probabilities to transit from one state to another when eliminating a K_4 block. Supplemental results may then be obtained. For instance, if one wants to know the probability that the source reaches both right endpoints in the most general K_4-ladder of length n, one simply calculates the probability to transit from **SB** to **SS** in n steps. This is done by replacing the suffix in Eq. (5) by $\begin{bmatrix} 1 & 0 & 0 & 0 & 0 & 0 \end{bmatrix}$ transposed. Indeed, all theory pertaining to finite Markov processes is available. It allows, for instance, to calculate the absorption time in the state **0**—that is, the expected number of blocks crossed when an indefinitely long K_4-ladder loses reliability.

The Markov chain method is also more systematic than the recursion method: transition matrices—contrary to their analogs, the transfer matrices—do not depend on which successive components are contracted or deleted when developing probability trees. It depends on the considered states only. This may help to find patterns for reliability polynomials of families of periodic graphs—for instance, hammocks.

The Markov chain method has been used in Jianu et al. [22] to approximate the reliability polynomials of hammocks of small widths. As we show in a work in progress [23], the method becomes exact when a different selection of states is used. In this way, we can extend the results from Cowell et al. [24] and provide the exact reliability polynomials of hammocks of size $w \times l$ for $w \leq 6$ and any l, or—by duality—for $l \leq 6$ and any w.

Lastly, it should be mentioned that hammocks as well as very similar networks have already been used for making quantum chips and, even more important, that both physical qubits as well as the ways they communicate are prone to failures—a fact which clearly makes an approach based on the Markov chain method (with both edges and vertices having independent reliabilities) even more attractive.

Acknowledgements This research was partly funded by a grant of the Romanian Ministry of Education and Research, CNCS-UEFISCDI, project no. PN-III-P4-ID-PCE-2020-2495, within PNCDI III (ThUNDER^2 = *Techniques for Unconventional Nano-Designing in the Energy-Reliability Realm*).

References

1. Kirchhoff, G.: Ueber die Auflösung der Gleichungen, auf welche man bei der Unter-suchung der linearen Vertheilung galvanischer Ströme geführt wird. Annalen der Physik und Chemie **148**(12), 497–508 (1847). (O'Toole, J.B.: On the solution of the equations obtained from the investigation of the linear distribution of galvanic currents. IRE Trans. Circ. Theory **5**(1), 4–7 (1958). https://doi.org/10.1109/TCT.1958.1086426)
2. Moore, E.F., Shannon, C.E.: Reliable circuits using less reliable relays–part I. J. Franklin Inst. **262**(3), 191–208 (1956). https://doi.org/10.1016/0016-0032(56)90559-2. (Preliminary version as: reliable circuits using crummy relays. Tech. Rep. Memo. 54-114-42, Bell Labs, Murray Hill, NJ, 29 Nov. 1954)
3. Dasher, D.J.: Synthesis of RC transfer function as unbalanced two-terminal pair networks. IRE Trans. Circ. Theory **PGCT-1**, 20–34 (1952). https://doi.org/10.1109/TCT.1952.1083572
4. Yarlagadda, R.: An application of tridiagonal matrices to network synthesis. SIAM J. Appl. Math. **16**(6), 1146–1162 (1968). https://doi.org/10.1137/0116096
5. Marshall, T.G.: Synthesis of RLC ladder networks by tridiagonalization. IEEE Trans. Circ. Theory **CT-16**(1), 39–46 (1969). https://doi.org/10.1109/TCT.1969.1082885
6. Hatayodom, T.: Ladder network transfer function characteristics. M.Sc. thesis, ECE Department, Missouri University Science & Technology, Rolla, MI (1971). https://scholarsmine.mst.edu/masters_theses/5455
7. Fowler, E.R., Yarlagadda, R.: A state-space approach to RLCT two-port transfer-function synthesis. IEEE Trans. Circ. Theory **CT-19**(1), 15–20 (1972). https://doi.org/10.1109/TCT.1972.1083388
8. Hua, T.: RC-gyrator ladder and RLC symmetrical lattice network synthesis: a state-space approach. PhD thesis, ECE Department, Missouri University Science & Technology, Rolla, MI (1974). https://scholarsmine.mst.edu/doctoral_dissertations/312
9. Tanguy, C.: What is the probability of connecting two points? J. Phys. A Math. Theor. **40**(47), 14099–14116 (2007). https://doi.org/10.1088/1751-8113/40/47/005
10. Tanguy, C.: Exact solutions for the two- and all-terminal reliabilities of a simple ladder network. CoRR abs/cs/0612143. https://doi.org/10.48550/arXiv.cs/0612143. Last accessed 28 Dec 2006
11. Tanguy, C.: Exact solutions for the two- and all-terminal reliabilities of the Brecht-Colbourn ladder and the generalized fan. CoRR abs/cs/0701005. https://doi.org/10.48550/arXiv.cs/0701005. Last accessed 30 Dec 2006
12. Tanguy, C.: Exact two-terminal reliability for the double fan. In: International Network Optimization Conference INOC, Spa, Belgium, 22–25 April 2007. https://www.euro-online.org/enog/inoc2007/Papers/author.29/paper/paper.29.pdf
13. Tanguy, C.: Exact two-terminal reliability of some directed networks. In: International Workshop Design & Reliable Communication Networks DRCN, La Rochelle, France, 7–10 October 2007. https://doi.org/10.1109/DRCN.2007.4762273
14. Tanguy, C.: Exact solutions for the two-terminal reliability of recursive structures: a few directions. In: International Conference of Mathematical Methods Reliable MMR, Moscow, Russia, 22–27 June 2009, pp. 220–224
15. IEEE: International Roadmap for Devices and Systems. IRDS™ (2023). https://irds.ieee.org/editions/2023
16. Singer, P.: 20-year semiconductor roadmap: tearing down the walls. Semicon. Digest 12 (2022). https://show-daily-day-one.semiconductordigest.com/?v=5. (for completeness see also https://www.imec-int.com/en/articles/20-year-roadmap-tearing-down-walls)
17. IBM: The IBM quantum heavy hex lattice. Technical note, 8 August 2021. https://research.ibm.com/blog/heavy-hex-lattice
18. Chamberland, C., Zhu, G., Yoder, T.J., Hertzberg, J.B., Cross, A.W.: Topological and subsystem codes on low-degree graphs with flag qubits. Phys. Rev. X **10**(1), 011022 (2020). https://doi.org/10.1103/PhysRevX.10.011022

19. Weidenfeller, J., Valor, L.C., Gacon, J., Tornow, C., Bello, L., Woerner, S., Egger, D.J.: Scaling of the quantum approximate optimization algorithm on superconducting qubit based hardware. Quantum **6**, 870–894 (2022). https://doi.org/10.22331/q-2022-12-07-870
20. Martinis, J., Ding, N., Babbush, R., Isakov, S.V., Neven, H., Smelyanskiy, V., Boixo, S.: Fidelity estimation for quantum computing systems. WO2017200536A1 (2016). https://patents.google.com/patent/WO2017200536A1/
21. Lawler, G.F.: Introduction to Stochastic Processes, 2nd edn. Chapman & Hall/CRC (2006). https://doi.org/10.1201/9781315273600
22. Jianu, M., Ciuiu, D., Dăuş, L., Jianu, M.: Markov chain method for computing the reliability of hammock networks. Probab. Eng. Info. Sci. **36**(2), 276–293 (2022). https://doi.org/10.1017/S0269964820000534
23. Poulin, P., et al.: On the exact reliability of thin hammock networks (Unpublished)
24. Cowell, S.R., Beiu, V., Dăuş, L., Poulin, P.: On the exact reliability enhancements of small hammock networks. IEEE Access **6**, 25411–25426 (2018). https://doi.org/10.1109/ACCESS.2018.2828036

Comparative Analysis of Authentication Using Formant Features of Vowels and Consonants

Yelena Belova

Abstract This article discusses 3 different biometric images of the user, which are based on the formant characteristics of the sounds of the Russian language. In the 1st biometric image, the formant characteristics of vowels (including diphthongs) and consonants are used, in the 2nd—vowels and diphthongs, and in the 3rd—consonants. A cluster of algorithms is presented, consisting of the Random Forest algorithm, K Nearest Neighbors, Logistic Regression, Multilayer Perceptron and Support Vector Machines. On the basis of a cluster of algorithms, models have been developed and trained. The mechanism of creation and training of each model is considered. To compare the effectiveness of user biometric images, an experiment was conducted in which 10 people participated. In total, users pronounced 46 sounds (6 vowels, 4 diphthongs and 36 consonants). Based on the provided data, samples were formed for each biometric image. Each sample, when it enters the input of a cluster of algorithms, is divided into training and test. A comparative analysis of the authentication procedure with each biometric image was carried out. For this, the following parameters were used: the average calculation time, the average accuracy value and the average accuracy value on the test sample.

Keywords Biometric authentication · Voice authentication · Biometric user image · Formant characteristics · Cluster of algorithms

1 Introduction

Today no enterprise can last without an authentication system. Historically social authentication would be used, because members of a certain circle knew one another by sight. However, today biometric authentication is getting increasingly popular. Principle-wise biometric authentication and social authentication have a lot in common. They imply a user's biometric image. The biometric authentication

Y. Belova (✉)
Ufa University of Science and Technology (UUST), Ufa, Russian Federation
e-mail: yelenabelova92@gmail.com

D. Vlachos (ed.), *Mathematical Modeling in Physical Sciences*, Springer Proceedings in Mathematics & Statistics 446, https://doi.org/10.1007/978-3-031-52965-8_17

system operates by the artificial neural network technology. It represents a machine intelligence imitating natural recognition human senses are usually responsible for.

This study is devoted to recognizing a user by voice. A user's biometric image is comprised of human speech formant features. The high efficacy of using formant features is proven in earlier studies [1, 2].

The aim of this research is to compare efficiency of an algorithms cluster using 3 different user's biometric images.

2 User's Biometric Image

The study applied comparative analysis to the user's biometric images based on the formant features below:

1. vowels and consonants (46 sounds);
2. vowels (10 sounds);
3. consonants (36 sounds).

The research considers both monophthongs and diphthongs.

The biometric image uses the following formant features:

1. First seven formants' frequencies—$F_1, F_2, F_3, F_4, F_5, F_6, F_7$.
2. Beginning and ending of the frequency range of each formant out of 7—$F_{1s}, F_{1f}, F_{2s}, F_{2f}, F_{3s}, F_{3f}, F_{4s}, F_{4f}, F_{5s}, F_{5f}, F_{6s}, F_{6f}, F_{7s}$ and F_{7f}.
3. Leading formant frequency (frequency with the maximum amplitude of outlier)—F_l.

Figure 1 shows the spectrogram of "A" sound. It clearly shows all 7 formants.

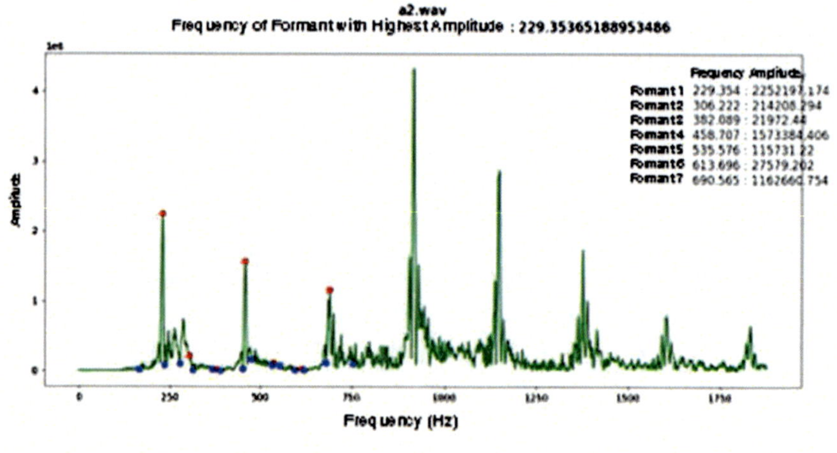

Fig. 1 "A"-sound spectrogram

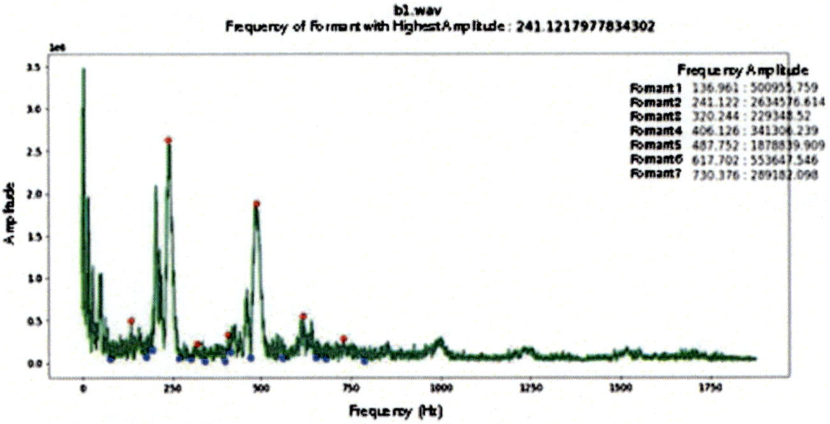

Fig. 2 "B"-sound spectrogram

Figure 2 shows the spectrogram of "B" sound.

As it is shown in Fig. 2 spectrograms of consonants are less informative. They are harder to derive high-amplitude frequencies from.

However, using consonants helps enrich the user's biometric image thanks to its growth.

3 Developing Sets

The research had 3 experiments. Each experiment had a set developed for it.

Authentication with vowel and consonant formant features uses vowel and consonant formant features derived from 46 Russian vowels and consonants pronounced by 10 users.

The set is recorded in csv. It consists of 461 lines (including the headline) and 31 columns.

28 columns are devoted to features of the first seven formants. 1 column contains users' numbers; 1 column contains sound numbers; 1 column contains the leading formant frequencies.

Authentication with vowel formant features uses formant features of vowels and diphthongs derived from 6 vowels and 4 diphthongs of the Russian language pronounced by 10 users.

The set of formant features and diphthongs is also recorded in csv. It consists of 101 lines (including the headline) and 31 columns. The columns have the same purpose.

Authentication with vowel and consonant formant features uses consonant formant features derived from 36 Russian consonants pronounced by 10 users.

The set consists of 361 lines (including the headline) and 31 columns.

All the sets are divided into training sets and testing sets with 80/20% ratio.

4 Developing Models Using Algorithms Cluster

The cluster of algorithms consists of:

1. Random Forest;
2. K-Nearest Neighbors;
3. Logistic Regression;
4. Multilayer Perceptron;
5. Support Vector Machine.

Figure 3 represents a structural graph of model training with Random Forest. The model is trained in the following manner:

1. The input receives the training set values (x_train—the entire training set).
2. Training parameters are set for training the model with Random Forest.
3. The training outcomes and attributes of the training set (y_train) undergo a grid search (GridSearchCV) to choose the most efficient model.
4. K-nearest neighbors help choose the best parameters. The model training is done.
5. The testing set (x_test) and its attributes (y_test) are received by the input of the trained model. The model's efficiency is determined.

Figure 4 illustrates a structural graph of the model training with k-nearest neighbors.

The model training includes the following steps:

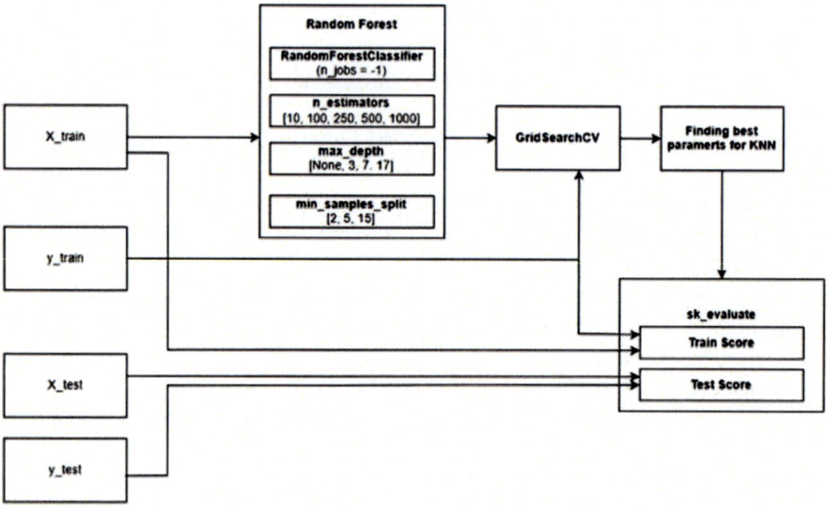

Fig. 3 Model training with random forest

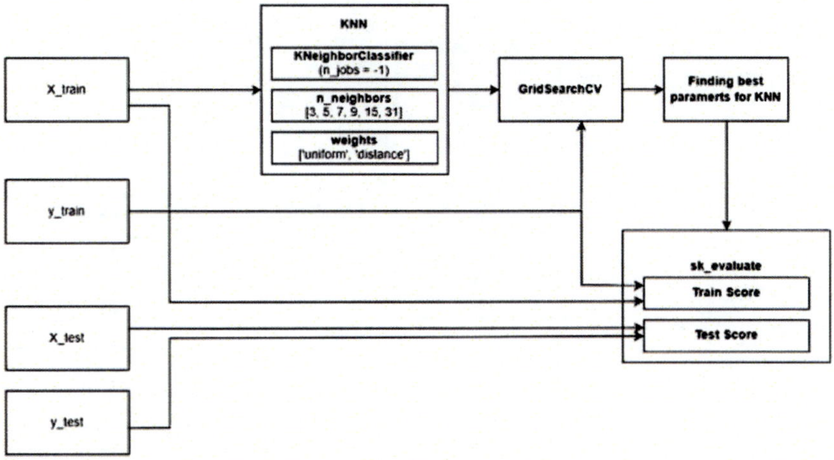

Fig. 4 Structural graph of model training with K-nearest neighbors

1. The input receives the training set values (x_train).
2. Training parameters are set for training the model with k-nearest neighbors.
3. The training outcomes and attributes of the training set (y_train) undergo a grid search (GridSearchCV) to choose the most efficient model.
4. K-nearest neighbors help choose the best parameters. The model training is done.
5. The testing set (x_test) and its attributes (y_test) are received by the input of the trained model to determine efficiency of the developed model.

A structural graph of the model training with logistic regression is given in Fig. 5. The model training consists of the following steps:

1. The input of the developed model receives the training set values (x_train).
2. The model training implies the stochastic gradient descent classifier (SGDClassifier) using all available processor cores. The tolerance for termination (tol) is entered; the log loss function is applied; the maximum number of iterations is selected for the model optimization; the dropout parameters are set.
3. The training outcomes and attributes of the training set (y_train) undergo grid search (GridSearchCV) resulting in determining the most efficient model.
4. K-nearest neighbors help choose the best parameters. So far, the model training is completed.
5. The testing set (x_test) and its attributes (y_test) are received by the input of the trained model to determine efficiency of the model.

A structural graph of the model training with multilayer perceptron is given in Fig. 6.

The model is trained in the following manner:

1. The input receives the training set values (x_train).

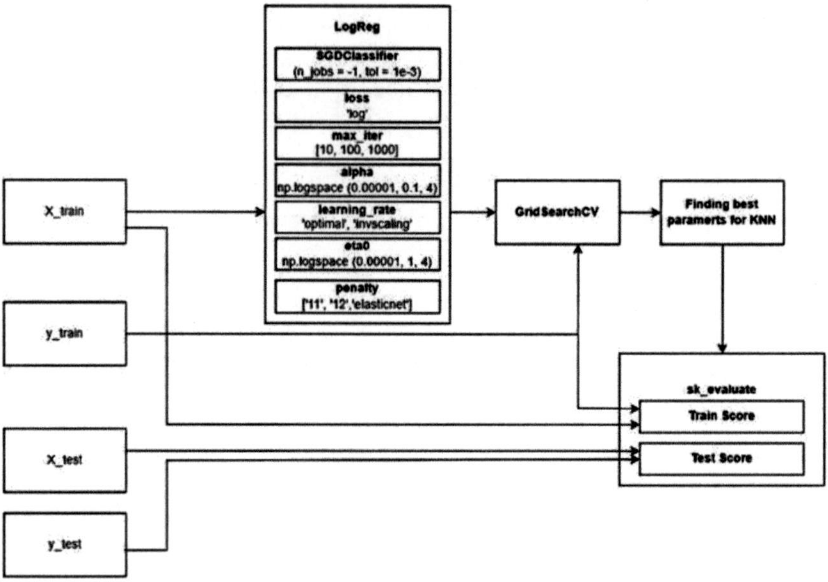

Fig. 5 Model training with logistic regression

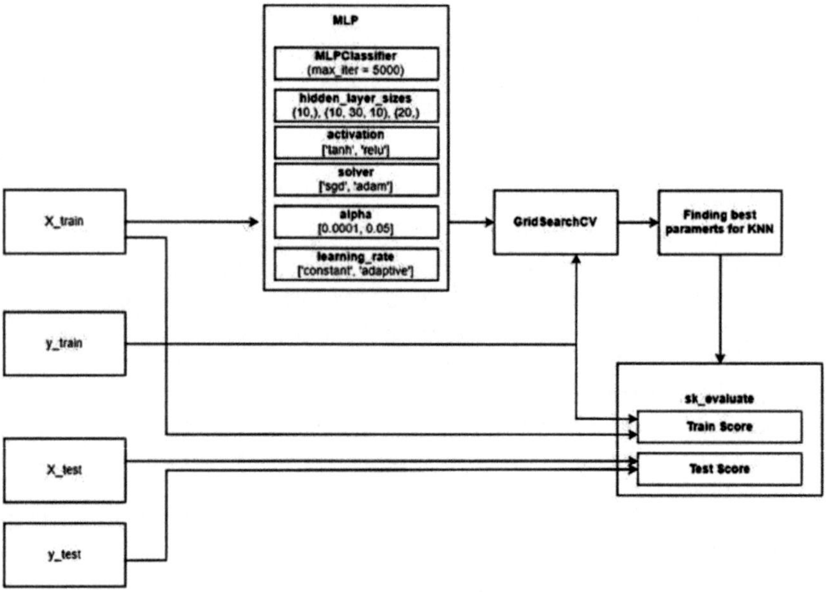

Fig. 6 Model training with multilayer perceptron

2. The model is trained with the following parameters: maximum number of iterations (max_iter) is 5,000; the input layer consists of 31 neurons (as many as columns in the set); the input layer is followed by the hidden one consisting of10 neurons; this layer is followed by 3 layers with 10, 30 and 10 neurons respectively; next hidden layer has 20 neurons and is followed by the output with 10 neurons (one for each user). The activation function of the hidden layer neurons is tanh (hyperbolic tangent) and relu (rectified linear activation function); the model is trained using the stochastic gradient descent (sgd) and the adaptive moment estimation (adam).
3. The training outcomes and attributes of the training set (y_train) undergo a grid search (GridSearchCV) to choose the most efficient model option.
4. The model training is completed with selecting the best parameters using k-nearest neighbors.
5. The testing set (x_test) and its attributes (y_test) are received by the input of the trained model to determine the model's efficiency.

Figure 7 represents a structural graph of the model training with support vector machine.

The model training includes the following steps:

1. The input receives the training set values (x_train).
2. The model is trained based on the selected parameters. The selected loss functions are hinge loss (hinge loss) and modified huber loss (modified_huber).

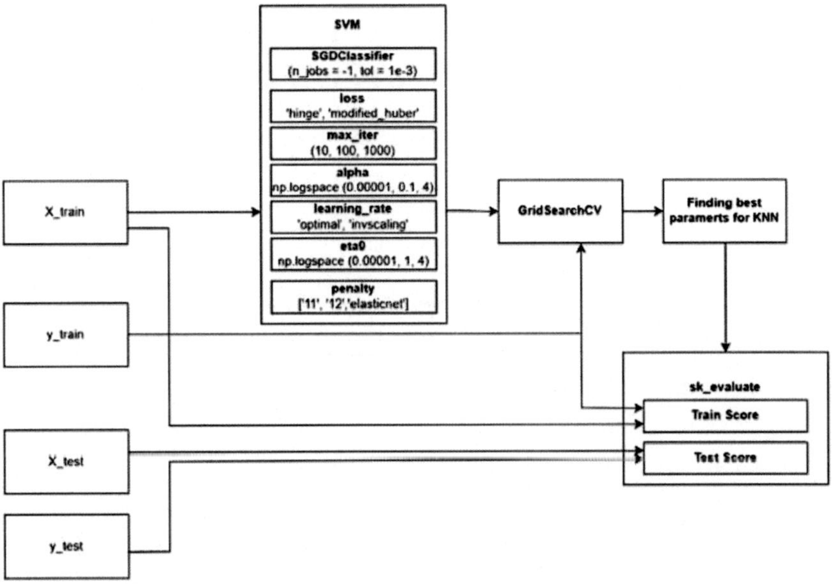

Fig. 7 Model training with support vector machine

3. The training outcomes and attributes of the training set (y_train) undergo a grid search (GridSearchCV) to choose the most efficient model.
4. The model training is completed with selecting the best parameters using k-nearest neighbors.
5. The testing set (x_test) and its attributes (y_test) are received by the input of the trained model to determine the model's efficiency.

Figure 8 represents a structural graph of building a model based on an algorithms cluster.

Each model is trained isolated from other models.

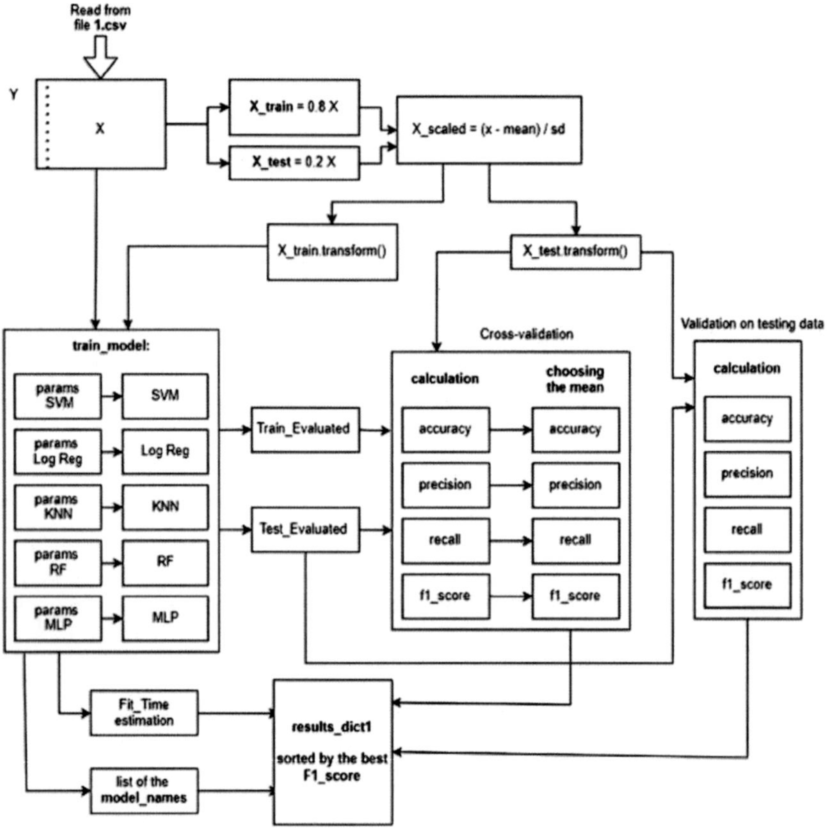

Fig. 8 Structural graph of building a model based on an algorithms cluster

5 Outcomes of Using 3 Different User's Biometric Images

3 different user's biometric images are compared based on the following parameters:

1. The average calculation time is the time a convolutional neural network takes to solve 1 problem. It is measured in seconds.
2. The average precision is precision of an artificial neural network's work on training and testing problems.
3. Average precision in the testing set.

Table 1 shows the results of using a user's biometric image based on formant features of vowels and consonants.

The multilayer perceptron takes the longest time to calculate the results. The time is 34.335736 s.

The maximum precision of 0.871467 and maximum average precision on the testing of 0.891304 set are achieved by the Random Forest.

Table 2 shows the results of using a user's biometric image based on formant features of vowels.

The maximum average calculation time of the multilayer perceptron is 7.003927 s.

The Random Forest achieves the maximum average precision of 0.91125 and the maximum average precision on the testing set of 0.915.

Table 3 contains the results of using a user's biometric image based on formant features of consonants.

Table 1 Results of using user's biometric image based on formant features of vowels and consonants

Model name	Average calculation time	Average precision	Average precision in testing set
Random forest	0.856373	0.871467	0.891304
K-nearest neighbors	0.003199	0.704348	0.729348
Logic regression	0.138321	0.533424	0.513043
Multilayer perceptron	34.335736	0.797554	0.816304
Support vector machine	0.149714	0.508152	0.548913

Table 2 Results of using user's biometric image based on formant features of vowels

Model name	Average calculation time	Average precision	Average precision in testing set
Random forest	0.248761	0.91125	0.915
K-nearest neighbors	0.001399	0.73125	0.760
Logic regression	0.030782	0.59125	0.625
Multilayer perceptron	7.003927	0.84375	0.870
Support vector machine	0.023986	0.57125	0.600

Table 3 Results of using user's biometric image based on formant features of consonants

Model name	Average calculation time	Average precision	Average precision in testing set
Random forest	0.719335	0.865972	0.873611
K-nearest neighbors	0.002599	0.693750	0.688889
Logic regression	0.093586	0.547569	0.541667
Multilayer perceptron	43.439978	0.797917	0.808333
Support vector machine	0.070960	0.535417	0.555556

The maximum average calculation time of 43.439978 s is achieved by the multilayer perceptron.

The maximum precision of 0.865972 and maximum average precision on the testing set of 0.873611 are achieved by the Random Forest.

The best performance is demonstrated in using a user's biometric image based on vowels.

6 Conclusion

The research compares users' biometric images based on vowels and consonants and used in biometric authentication by formant features with a cluster of algorithms. A user's biometric image based on vowels (diphthongs in particular) demonstrates the best performance. Therefore, it is advised to use it for authentication of all users of a local network.

Authentication by formant features of consonants is advised as an auxiliary authentication method. Further research plans to study all consonants and determine the most informative of those.

References

1. Belova, Ye.P., Mashkina I.V.: Architecture and work algorithm of biometric authentication system by voice. Autom Control Comput Sci **55**(8), 924–931 (2021). https://doi.org/10.3103/S014641162108006X
2. Mashkina, I.V., Belova E.P.: Developing neural network data base of biometric images for voice authentication system. Inf Secur Issues Comput. Syst **2**, 86–93 (2019). Saint-Petersburg
3. Kamalova, L.A.: Phoneme and phonological system of the Russian literary language in the Moscow and Leningrad (Saint-Petersburg) Phonological Schools. Lecture no. 3. http://kpfu.ru/docs/F1856424430/Kamalova.L.A.Lekciya_3..Fonetika.pdf. Accessed 18 Apr 2018

Hidden and Singular Attractors in Nonlinear Systems of Differential Equations

Nikolai Magnitskii

Abstract The concept of a "hidden" attractor is widely used in the modern literature on autonomous nonlinear chaotic systems of ordinary differential equations in the case when the system has an chaotic attractor and together with it has either no singular points or one or more stable singular points. It is shown in the paper that any hidden attractor of nonlinear system of differential equations is one of the singular attractors of the system in sense of the Feigenbaum-Sharkovsky-Magnitskii (FShM) universal bifurcation theory. That is any hidden irregular (chaotic) attractor is born during the implementation of FShM bifurcation scenario and is bounded non-periodic trajectory in phase space of finite or infinite dimension, which is the limit of the cycles of some cascade of Feigenbaum period-doubling bifurcations and contains in any of its neighborhood an infinite number of unstable periodic orbits.

1 Introduction

In modern literature on nonlinear and chaotic dynamics, most of examples of both regular and chaotic attractors are located in the neighborhoods of unstable singular points and can be found numerically starting from the basins of such singular points.

Other type of chaotic attractors called as hidden attractors were found in 3D autonomous chaotic systems with only one stable singular point [1, 2], without singular points [3, 4], and with infinite number of singular points [5–7].

It would seem impossible to determine the complexity of the attractor in this case, give its description and find its place in the family of other attractors of the system, and clearly say whether the solution is a chaotic attractor or a periodic trajectory of a large period.

In this regard, in modern scientific literature, the only method for analyzing solutions to systems with hidden attractors is used—the method of calculating the leading Lyapunov exponent. However, as follows from the results of the author and as

N. Magnitskii (✉)
Federal Research Center "Computer Science and Control", Russian Academy of Sciences, Moscow, Russian Federation
e-mail: nikmagn@gmail.com

D. Vlachos (ed.), *Mathematical Modeling in Physical Sciences*, Springer Proceedings in Mathematics & Statistics 446, https://doi.org/10.1007/978-3-031-52965-8_18

demonstrated by numerous examples in [8-11], there is one universal bifurcation scenario FShM (Feigenbaum-Sharkovsky-Magnitskii) for the transition to chaos in nonlinear systems of differential equations: dissipative and conservative, autonomous and non-autonomous, ordinary and with partial derivatives.

All irregular (chaotic) attractors generated by this scenario are exclusively singular attractors, that is, bounded non-periodic almost stable trajectories in phase space of finite or infinite dimension, in any neighborhood of which there is an infinite number of unstable periodic orbits. It is proved that any singular cyclic attractor generated by FShM—cascade of bifurcations of a singular cycle, belongs to the closure of its unstable invariant two-dimensional manifold, which is folded infinitely sheeted heteroclinic separatrix zigzag spanned on a separatrix tree (Möbius bands) of the Feigenbaum cascade and the following behind it cascades of bifurcations [8–11].

Therefore, any singular attractor has zero leading characteristic Lyapunov exponent. The numerical calculated positive leading Lyapunov exponent is exclusively a consequence of computational errors, since due to the existence of an everywhere dense set of non-periodic and unstable periodic orbits, numerical motion is possible along only the entire region in which the trajectory of a singular attractor is located, and not along its trajectory itself.

For the same reason, the numerically calculated leading Lyapunov exponent can also turn out to be positive for a stable periodic trajectory of some large period located in the neighborhood of a singular attractor, although theoretically it should be equal to zero. Thus, the found numerically positive Lyapunov exponent is not a characteristic of the chaotic motion.

The system with one stable singular point and an irregular attractor proposed in [12] was considered by the author in [10]. The system without singular points proposed in [13] was considered by the author in [11].

It was proved that the transition to chaos in both cases occurs in accordance with the FShM scenario. In this paper, we consider two new chaotic systems of equations: a system without singular points and a system with one stable singular point. It is proved that the so-called hidden attractors found in these two new systems are singular attractors generated by cascades of bifurcations of the FShM theory.

2 A System with no Singular Points

Consider the system proposed by S. Jafari at all in [14]

$$\dot{x} = y, \ \dot{y} = -x + z, \ \dot{z} = -0.8x^2 + z^2 + 2. \tag{1}$$

As system (1) has no singular points, the irregular (chaotic) attractor found starting from the initial point (0, 2.3, 0) and shown in Fig. 3a is hidden. To clearly demonstrate what actually occurs in the system (1) we consider this system with a bifurcation parameter a at the x^2 member

$$\dot{x} = y,\ \dot{y} = -x + z,\ \dot{z} = -ax^2 + z^2 + 2. \tag{2}$$

Chaotic attractor shown in Fig. 3a is the attractor of the system (2) at $a = 0.8$.

2.1 Analytical Study

Let us find an area of dissipativity of system (2) by calculating the divergence of its right part

$$divF = \left(\frac{\partial F_1}{\partial x} + \frac{\partial F_2}{\partial y} + \frac{\partial F_3}{\partial z}\right) = 2z < 0$$

for $z < 0$. In accordance with the bifurcation scenario of the universal FShM theory, bifurcation cascades begin with some stable cycle that exists in the system at some initial value of the bifurcation parameter.

This cycle can be born from a stable singular point as a consequence of Andronov-Hopf bifurcation, which can be investigated by analytical methods. But, in the absence of stable singular points in the system or the absence of any singular points, which takes place in system (2), the initial stable cycle can be born as a result of a saddle-node bifurcation, and it is not possible to carry out such an analysis by analytical methods.

Therefore, the study of cascades of cycle's bifurcations in the system (2) when the values of bifurcation parameter a increase will be carried out exclusively by numerical methods.

2.2 Numerical Analysis

Numerical analysis shows that the system (2) has stable (asymptotically orbital stable) cycle when the bifurcation parameter a is equal to 0.5. One can find this cycle solving the system with initial point (2.7; −2.6; −2). This cycle is shown in Fig. 1a. When the values of the bifurcation parameter a increase, the Feigenbaum cascade period-doubling bifurcations occurs, and then the complete cascade of subharmonic bifurcations occurs in accordance with the Sharkovsky order. Thus, at $a \approx 0.68$ a period two stable cycle is born in system (2), that is shown in Fig. 1b, at $a \approx 0.77$ a period four stable cycle is born in system (2), that is shown in Fig. 1c, at $a \approx 0.785$ a period eight stable cycle is born in system (2), that is shown in Fig. 2a.

The first simplest singular attractor—Feigenbaum attractor exists in the system (2) at $a = 0.789$ (Fig. 2b). When the values of the parameter a increase then complete subharmonic cascade of Sharkovsky bifurcations is realized in the system (2). Thus, a period six stable cycle exists in system (2) at $a = 0.7958$ which is shown in Fig. 2c,

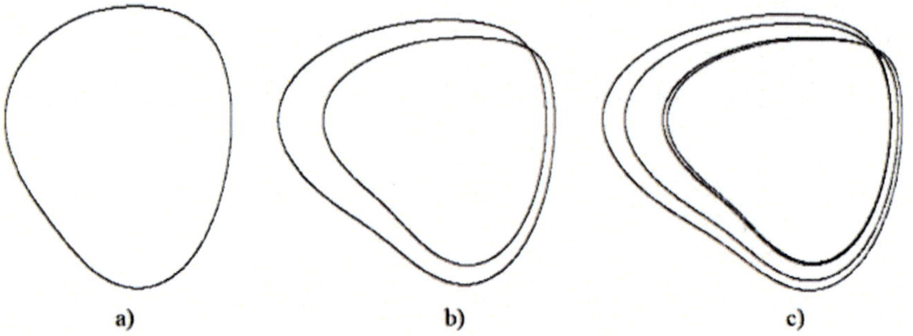

Fig. 1 Projections of periods one, two and four stable cycles of system (2) onto plane (x, y) at $a = 0.5$, $a = 0.76$ and $a = 0.78$

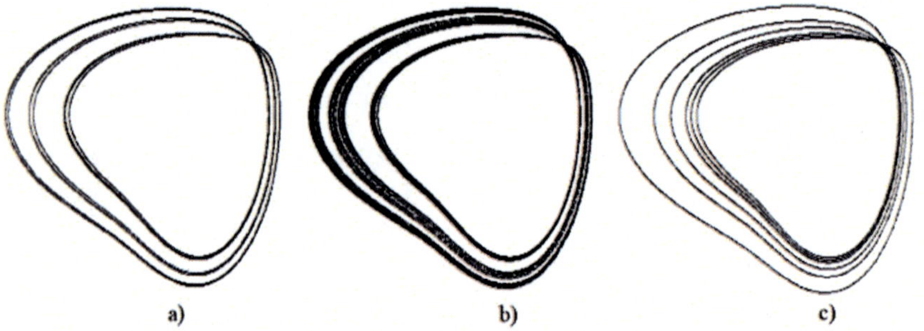

Fig. 2 Projections of period eight stable cycle, Feigenbaum attractor and period six stable cycle of system (2) onto plane (x, y) at $a = 0.785$, $a = 0.789$ and $a = 0.7958$

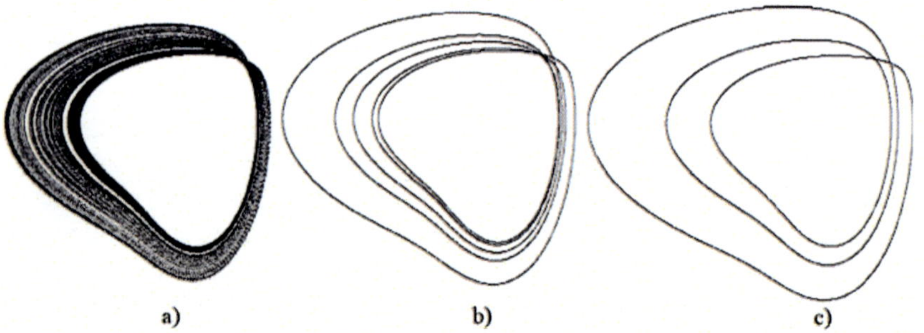

Fig. 3 Projections of hidden chaotic attractor and five and three periods stable cycles of Sharkovsky subharmonic cascade of bifurcations in system (2) onto plane (x, y) at $a = 0.8$, $a = 0.8095$ and $a = 0.8162$

a period five stable cycle exists at $a = 0.8095$, which is shown in Fig. 3b, and a period three stable cycle exists at $a = 0.8162$, which is shown in Fig. 3c.

Obviously, so called hidden attractor, found by authors of the paper [14] in system (2) at $a = 0.8$ and shown in Fig. 3a, is a simple singular attractor of the universal FShM theory lying between cycles of periods six and five in the Sharkovsky order. Maybe also it is not an irregular (singular) attractor, but it is stable periodic orbit of some large period.

3 A System with Only One Stable Singular Point

Consider another system analyzed by Dudkovskii at all in [15]

$$\dot{x} = y, \ \dot{y} = z, \ \dot{z} = -x - 0.6y - 2z + z^2 - 0.4xy \tag{3}$$

System (3) has one singular point $O = (0, 0, 0)$ and irregular (chaotic) attractor found starting from the initial point (4, -2, 0) and shown in Fig. 6c. To clearly demonstrate what actually occurs in the system (3) we consider this system with a bifurcation parameter b at the xy member

$$\dot{x} = y, \ \dot{y} = z, \ \dot{z} = -x - 0.6y - 2z + z^2 - bxy \tag{4}$$

Chaotic attractor shown in Fig. 6c is the attractor of the system (4) at $b = 0.4$.

3.1 Analytical Study

Let us find an area of dissipativity of system (4) by calculating the divergence of its right part

$$divF = \left(\frac{\partial F_1}{\partial x} + \frac{\partial F_2}{\partial y} + \frac{\partial F_3}{\partial z}\right) = -2 + 2z < 0$$

for $z < 1$. It is evident that point $O(0, 0, 0)$ is a singular point of the system (4). Let us show that it is asymptotically stable singular point of system (4) for all values of the parameter b.

Lemma. Singular point O is an asymptotically stable singular point (stable focus) of system (4) for all values of the parameter b.

Proof. Let us calculate the linearization matrix of system (4) at singular point O:

$$A = \begin{pmatrix} 010 \\ 001 \\ -1 - 0.6 - 2 \end{pmatrix}.$$

Calculating the characteristic polynomial at the singular point, we find that

$$\det(A - \lambda I) = -\lambda^3 - 2\lambda^2 - 0.6\lambda - 1.$$

Equating the characteristic polynomial to zero, we find that the linearization matrix at the singular point O has the following eigenvalues : $\lambda_1 \approx -1.955$, $\lambda_2 \approx -0.023+0.715i$, $\lambda_3 \approx -0.023-0.715i$. Therefore, for all b the linearization matrix A at the singular point O has one real negative eigenvalue and two complex conjugate eigenvalues with negative real parts. In accordance with Lyapunov theorem the singular point O is asymptotically stable. The lemma is proved.

Nevertheless, as is shown in paper [15], for $b = 0.4$ system (4) has a "hidden" attractor together with asymptotically stable singular point O. We shall show numerically that this "hidden" attractor presented in Fig. 6c is the result of cascade of bifurcations with respect to parameter b in accordance with the theory FShM.

3.2 *Numerical Analysis*

Further study of the dynamics of solutions to system (4) will be carried out by numerical methods. Numerical analysis shows that when parameter $b \approx 0.41437$ a stable cycle is generated in the system (4) as a result of saddle-node bifurcation. This cycle is shown in Fig. 4a for $b = 0.41$.

Thus, for every $b < 0.41437$ the system (4) has two attractors: stable singular point O and other regular or chaotic attractor. When the values of the bifurcation parameter b decrease, cascade of period-doubling bifurcations of the cycle occurs, and then complete subharmonic cascade of bifurcations occurs in accordance with the Sharkovsky order, and then incomplete homoclinic cascade of bifurcations occurs in accordance with the Magnitskii homoclinic order.

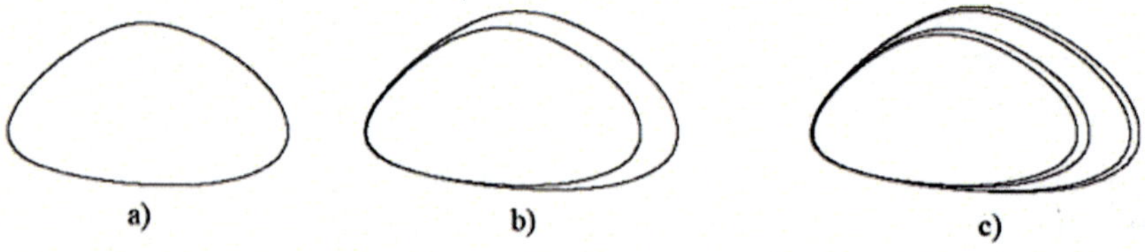

Fig. 4 Projections of periods one, two and four stable cycles of system (4) onto plane (x, y) at $b = 0.41$, $b = 0.406$ and $b = 0.4045$

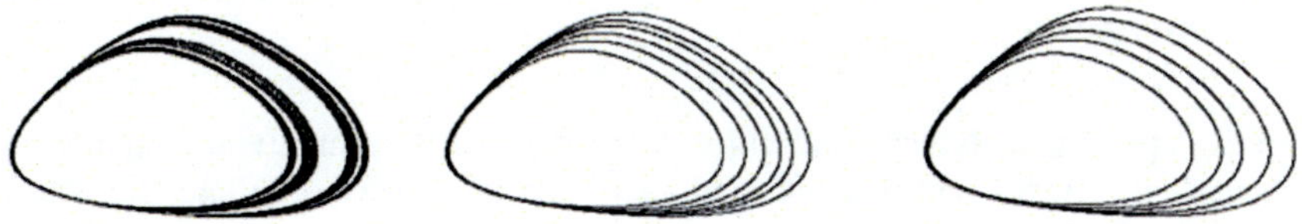

Fig. 5 Projections of Feigenbaum attractor and stable cycles of periods six and five of system (4) onto plane (x, y) at $b = 0.404$, $b = 0.4037$ and $b = 0.40305$

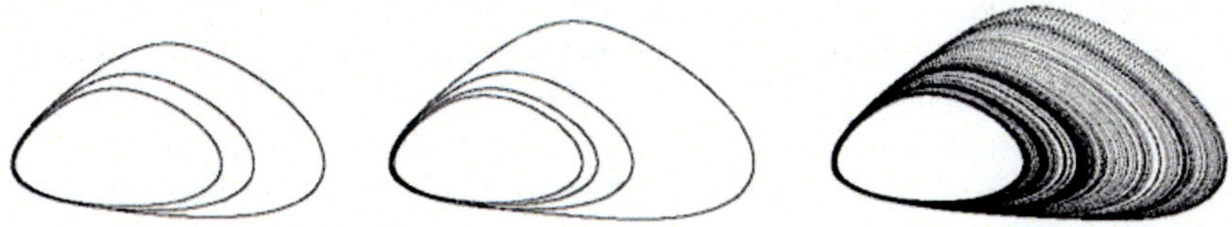

Fig. 6 Projections of period three and four stable cycles of cascade of homoclinic bifurcations and singular attractor of the system (4) onto plane (x, y) at $b = 0.4023$, $b = 0.4005$ and $b = 0.4$

Thus, at $b \approx 0.4075$ a period two stable cycle is born in system (4), that is shown in Fig. 4b for $b = 0.406$, at $b \approx 0.40475$ a period four stable cycle is born in system (4), that is shown in Fig. 4c for $b = 0.4045$.

At $b \approx 0.4042$ a period eight stable cycle is born in system (4). The first simplest singular Feigenbaum attractor exists in system (4) at $b \approx 0.404$ (Fig. 5a).

When the values of the parameter b decrease, complete subharmonic cascade of Sharkovsky bifurcations is realized in system (4). Thus, a period six stable cycle exists in the system (4) at $b = 0.4037$, which is shown in Fig. 5b, a period five stable cycle exists in the system (4) at $b = 0.40305$, which is shown in Fig. 5c, a period three stable cycle exists in the system (4) at $b = 0.4023$ which is shown in Fig. 6a.

For smaller values of bifurcation parameter b an incomplete cascade of homoclinic bifurcations of stable cycles is realized in system (4), converging to a homoclinic contour. So at $b = 0.4005$ insystem (4) there is a period four stable homoclinic cycle shown in Fig. 6b.

Obviously, so called hidden attractor, presented by authors of the paper [15] in system (4) at $b = 0.4$ and shown in Fig. 6c, is a complex singular attractor of the theory FShM lying between homoclinic cycles of periods four and five in the Magnitskii order. Maybe also it is not an irregular (singular) attractor, but it is stable periodic trajectory of very large period.

4 Conclusion

In this paper, an analytical and numerical research of the nature and structure of "hidden" attractors in autonomous nonlinear systems of ordinary differential equations done on the examples of two nonlinear systems one of them has no singular points at all and other system has stable singular point together with "hidden" chaotic attractor.

It is numerically shown that the generation of hidden attractors in these two nonlinear systems of equations occurs in accordance with the Feigenbaum-Sharkovsky-Magnitskii universal bifurcation scenario throught subgarmonic and then homoclinic cascades of stable cycles bifurcations, as in other chaotic systems of nonlinear differential equations.

In both cases, FShM—cascades of bifurcations begin with cycles generated by saddle-node bifurcations, and hidden attractors are simple or complex singular attractors of the FShM theory. That is any hidden irregular (chaotic) attractor is born during the implementation of FShM bifurcation scenario and is bounded non-periodic trajectory in phase space of finite or infinite dimension, which is the limit of the cycles of some cascade of Feigenbaum period-doubling bifurcations, and contains in any of its neighborhood an infinite number of unstable periodic orbits.

Acknowledgements The work is partially supported by Russian Science Foundation, project 23-21-00095.

References

1. Deng, Q. L., Wang, C. H., Yang, L. M.: 2020 Four-wing hidden attractors with one stable equilibrium point. Int. J. Bifurc. Chaos 30 (6) Article ID 2050086
2. Tian, H. G., Wang, Z., Zhang, P. J., Chen, M. S., Wang, Y.: 2021 Dynamic analysis and robust control of a chaotic system with hidden attractor. Complexity Article ID 8865522
3. Wei, Z.C.: Dynamical behaviors of a chaotic system with no equilibria. Phys. Lett. A **376**(2), 102–108 (2011)
4. Jafari, S., Sprott, J.C., Golpayegani, S.M.R.H.: Elementary quadratic chaotic flows with no equilibria. Phys. Lett. A **377**(9), 699–702 (2013)
5. Kingni, S.T., Pham, V.T., Jafari, S., Kol, G.R., Woafo, P.: Three-dimensional chaotic autonomous system with a circular equilibrium: analysis, circuit implementation and its fractional-order form *Circuits*. Circuit. Syst. Signal Process. **35**(6), 1933–1948 (2016)
6. Chen, H., Lei, T. F., Lu, S., Dai, W. P., Qiu, L. J., Zhong, L.: 2020 Dynamics and complexity analysis of fractional-order chaotic systems with line equilibrium based on adomian decomposition. Complexity, Article ID 5710765
7. Lai, Q., Wan, Z.Q., Kengne, L.K., Chen, P.D.K.C.Y.: Two-memristor-based chaotic system with infinite coexisting attractors. IEEE Trans. Circuits Syst. II Express Briefs **68**(6), 2197–2201 (2021)
8. Magnitskii, N.A.: On the topological structure of singular attractors of nonlinear systems of differential equations. Diff. Equat. **46**(11), 1552–1560 (2010)

9. Magnitskii, N. A.: Universality of transition to chaos in all kinds of nonlinear differential equations chapter in nonlinearity, bifurcation and chaos—theory and applications (Rijeca: InTech), P 133–174 (2012)
10. Magnitskii, N. A.: Bifurcation Theory of Dynamical Chaos Chapter in Chaos Theory (Rijeka: InTech) P 197–215 (2018)
11. Magnitskii, N.A.: Universal bifurcation chaos theory and its new applications. Mathemat. **11**, 2536 (2023). https://doi.org/10.3390/math11112536
12. Wang, X., Chen, G.R.: A chaotic system with only one stable equilibrium. Commun Nonlinear Sci. Numer. Simul. **17**, 1264–1272 (2012)
13. Sambas1 A, Mamat M, Vaidyanathan S, Mohamed M A, MadaSanjaya W S 2018 A New 4-D Chaotic System with Hidden Attractor and its Circuit Implementation. Int. J. Eng. & Tech. 7(3), 1245–1250
14. Jafari, S., Sprott, J., Golpayegani, S.: Elementary quadratic chaotic flows with no equilibria. Phys. Lett. A **377**, 699–702 (2013)
15. Dudkowski, D., Jafari, S., Kapitaniak, T., Kuznetsov, N.V., Leonov, G.A., Prasad, A.: Hidden attractors in dynamical systems. Phys. Rep. **637**(3), 1–50 (2016)

Complex Systems and Complex Networks

Nonlinear Features and Hybrid Optimization Algorithm for Automated Electroencephalogram Signal Analysis

Lyudmila Egorova, Lev Kazakovtsev, and Elena Vaitekunene

Abstract In this study a hybrid approach for EEG data classification for detecting epileptic seizures was proposed. The proposed hybrid model combines Extreme Gradient Boosting (XGB), Random Forest (RF) and Light Gradient Boosted Machine (LGBM) as base models and Support Vector Machines (SVM) as final model. It is shown that the hybrid approach provides higher recognition accuracy in binary classification of epilepsy seizure detection task then the base classification models. In order to provide optimal performance and time spent in models hyperparameters optimization the AutoML genetic programming-based system TPOT was used. The dynamic characteristics of the electroencephalogram (EEG) signal, the sample entropy, the Hurst exponent, and the senior Lyapunov exponent in combination with the spectral characteristics in delta, theta, alpha, beta and gamma bands were used as informative features for the classification algorithms.

Keywords EEG · Automatic seizure detection · Entropy · Hurst index · Lyapunov exponents · Hyperparameter optimization · Hybrid model

1 Introduction

The rapid development of information technology has led to the accumulation of a huge amount of biomedical data, which makes it possible to use them in artificial intelligence and machine learning systems to identify new, hidden dependencies in data and use this knowledge in medical automated systems.

The analysis of electroencephalogram (EEG) data is applied in such a promising area of research as the development of brain-computer interface (BCI) technology. This technology allows to evaluate the user's intention by analyzing brain signals, and then, in accordance with them, give commands to external hardware devices,

---

L. Egorova (✉) · L. Kazakovtsev · E. Vaitekunene
Siberian Federal University, 79 Svobodny Prospekt, 660041 Krasnoyarsk, Russia
e-mail: egorova_ld@rambler.ru

Reshetnev Siberian State University of Science and Technology, 31 Krasnoyarskiy Rabochiy Prospekt, 660037 Krasnoyarsk, Russia

D. Vlachos (ed.), *Mathematical Modeling in Physical Sciences*, Springer Proceedings in Mathematics & Statistics 446, https://doi.org/10.1007/978-3-031-52965-8_19

thus creating a channel between the human brain and the computer control system [1]. In addition, EEG is widely used in medical practice as a research method for diagnosing a wide range of diseases, such as various neurological disorders, epilepsy, strokes, etc.

An electroencephalogram is a recording of electrical potentials recorded using special electrodes attached to the surface of the patient's head. EEG data is a time series that reflects the total electrical activity of huge populations of neurons in the brain. Being a non-invasive, inexpensive and fairly simple method, EEG is an extremely informative way to gain knowledge about the information processes occurring in the brain.

Great efforts of researchers have recently been directed to the development of methods for automated analysis of EEG data in order to identify pathologies of the functioning of the brain. Manual analysis of electroencephalograms is a laborious process and requires a lot of time for a specialist with higher neurological training. The development of methods for automatic interpretation of EEG data with the accuracy necessary for use in clinical practice will significantly improve the quality of diagnostics, reduce the time of the study and its cost [2].

The brain is an integral system, which is a distributed environment consisting of active elements of the same type–neurons that interact with each other. Being a complex system, the brain exchanges matter and energy with the environment. Such distributed dissipative systems are characterized by the presence of self-oscillations (fluctuations). Self-oscillatory processes, the source of which are huge populations of neurons at the macro level, are expressed in rhythmic electrical activity, which is recorded using the EEG [3].

To consider complex systems, the chaos model is often used. Time series generated by chaotic systems are non-stationary and non-linear in nature, they are also often distorted by noise, which creates a significant difficulty for their analysis. The main parameters used to describe chaotic systems include the average rate of information generation by a random data source (entropy), Lyapunov exponents, and fractal characteristics [4].

To estimate the parameters of chaotic systems, optimization methods based on evolutionary algorithms and their hybrid modifications are used. Optimization methods are also used to optimize the architecture of neural networks, select the parameters of neural networks and weights to achieve the best performance in solving the problem of classifying epileptic events according to EEG data [5].

2 Problem Statement and Description of Methods

The EEG signal is a physical process that carries information about the functioning of the brain. It is believed that excessive the spontaneous synchronization of the spiking of neurons plays a fundamental role in the occurrence of epileptic seizures [6]. It can be assumed that the dynamic properties of the normal signal differ from the signal segments corresponding to an epileptic seizure. The chaotic properties of populations of neurons in the synchronization mode should differ from the normal mode of functioning.

Based on this assumption, the task of data analysis in this work was to search for such dynamic features of signal that make it possible to determine the presence of an epileptic seizure at a specific time interval. In machine learning problems, the problem of finding suitable informative features is often no less difficult than its solution [7].

The EEG signal registers the potential difference between two points on the subject's head. However, the signal amplitude itself is not a sufficient informative sign. The entropy, the Hurst exponent and the senior Lyapunov exponent calculated for the initial time series by the sliding window method were chosen as the features under study. The task was formulated as a supervised learning task for binary classification. The main positive class 1 was assigned an epileptic seizure (seiz), class 2–the norm (bckg).

2.1 Sample Entropy

Kolmogorov entropy (K-entropy). The Kolmogorov entropy is a fundamental quantity that characterizes chaotic motion in phase space. The Kolmogorov entropy is a measure of chaos and shows the average rate of information loss. The Kolmogorov entropy is defined as follows [8]:

Consider the trajectory of a dynamical system on a strange attractor in a d-dimensional phase space $X(t) = [x_1(t), x_2(t), ...x_d(t)]$. Let us partition the phase space into d-dimensional boxes of size l^d. We will determine the state of the system at equal time intervals τ. Let $P_{i_0} \dots P_{i_n}$ be the joint probability that $X(t = 0)$ is in cell i_0, $X(t = \tau)$ is in cell $i_1, \dots,$ and $X(t = n\tau)$ is in cell i_n. According to Shannon, the quantity

$$K_n = -\sum_{i_0 \dots i_n} P_{i_0 \dots i_n} \quad \log P_{i_0 \dots i_n} \tag{1}$$

proportional to the information needed to locate the system on a predetermined trajectory $i_1^* \dots i_n^*$ with accuracy l if only probabilities $P_{i_0 \dots i_n}$ are known a priori. Therefore $K_{n+1} - K_n$ is the additional information needed to predict which cell i_{n+1}^* the system will be in if it is known that it was previously in $i_1^* \dots i_n^*$. Thus, $K_{n+1} - K_n$ s the loss of information about the system in the time interval from n to $n + 1$. K-entropy is defined as the average rate of information loss over time:

$$K = \lim_{\tau \to 0} \lim_{l \to 0} \lim_{N \to \infty} \frac{1}{N\tau} \sum_{n=0}^{N-1} (K_{n+1} - K_n) =$$

$$= -\lim_{\tau \to 0} \lim_{l \to 0} \lim_{N \to \infty} \frac{1}{N\tau} \sum_{i_0 \dots i_{N-1}} P_{i_0 \dots i_{N-1}} \log P_{i_0 \dots i_{N-1}}. \tag{2}$$

Its value is 0 for regular motion, infinite for random systems, and constant for systems with deterministic chaos. From one time realization, one can determine the lower bound of the Kolmogorov entropy [9].

The Eq. (2) uses limit values for temporal and spatial partitions and an infinite length of the time series under study. This limits the use of the Kolmogorov entropy by analytical systems and makes it problematic to use it for short and noisy time series of real measurements. To calculate the entropy of such time series, special robust methods have been developed based, for example, on the concept of sample entropy (sampEn). The Sample entropy method can significantly reduce the dependence of the entropy estimation result on the sample length and can be applied to short and noisy time series [4, 10].

2.2 *Hurst Index*

Simple systems usually have exponential and Gaussian distributions. However, complex systems encountered in practice, which include the brain, have not Gaussian but power laws of distribution with "heavy" tails, or as they are also called 'paretians'. They often are mentioned as fractal distributions. Fractal forms are characterized by the presence of long-term correlations and trends [11].

One of the methods for evaluating the fractal properties of a time series is the rescaled range analysis (R/S analysis) proposed by the british hydrologist Harold Edwin Hurst [12]. This method consists in evaluating the Hurst exponent (H, Hurst index), which characterizes the fractal properties of the time series and is as follows [13]:

Let $\xi(t)$ be the considered random variable, the values of which are obtained in discrete time intervals t_i during some time period τ. The mean value of this random variable over the period τ is equal to

$$\langle \xi \rangle_\tau = \frac{1}{\tau} \sum_{t=1}^{\tau} \xi(t). \tag{3}$$

Let $X(t)$ be the accumulated deviation of $\xi(t)$ from the mean $\langle \xi \rangle_\tau$:

$$X(t, \tau) = \sum_{u=1}^{t} \{\xi(u) - \langle \xi \rangle_\tau\}. \tag{4}$$

The range R is the value between the maximum and minimum values of $X(t, \tau)$

$$R(\tau) = \max_{1 \le t \le \tau} X(t, \tau) - \min_{1 \le t \le \tau} X(t, \tau) \tag{5}$$

Since the range R depends on the considered period τ and grows with it, the range is normalized to the standard deviation (i.e., to the square root of the variance). The standard deviation is calculated by the following formula

$$S = \sqrt{\frac{1}{\tau}\sum_{t=1}^{\tau}\{\xi(t) - \langle\xi\rangle_\tau\}^2}. \tag{6}$$

The resulting normalized range R/S is a dimensionless quantity and for many time series is well described by the empirical relation

$$\frac{R}{S} = \left(\frac{\tau}{2}\right)^H. \tag{7}$$

where H is the Hurst index, which for many natural processes clusters around a value of 0.73 with a standard deviation of about 0.09. The values of the index $H > 0.5$, according to Hurst's assumption, indicate that the observations are not independent and there is a long-term memory in the system. The measure of this correlation can be calculated by the formula

$$C = 2^{2H-1} - 1. \tag{8}$$

From the formula (8) it follows that $H = 0.5$ corresponds to a random time series with uncorrelated events. The Hurst index $0 \leq H < 0.5$ corresponds to an antipersistent (ergodic) series, i.e. a series whose values return to their average, after the rise, there is a decline and vice versa. The values $0.5 < H < 1$ indicate the persistence of the series, i.e. about saving the trend by the system [14]. The fractal dimension D and the Hurst exponent H are related by the following relation

$$D = 2 - H. \tag{9}$$

2.3 Lyapunov Exponents

Lyapunov exponents serve as a measure of the average exponential rates of divergence in phase space of two nearby orbits of system in phase space over time. The Lyapunov exponents are among the most important dynamic characteristics of chaotic systems. A system for which at least one positive Lyapunov exponent is defined is chaotic. The magnitude of the exponent determines the time scale on which the dynamics of the system becomes unpredictable [9, 15, 16].

There are sufficiently reliable algorithms for calculating the Lyapunov exponents using numerical methods. To estimate the largest Lyapunov exponent from data time series in our work was used the algorithm of Rosenstein et al. [17].

3 Experimental Design

The data mining process organization is presented at the Fig. 1. In this work, the Temple University Hospital (TUH) EEG Seizure Corpus (TUSZ) dataset was used. This dataset is the world's largest set of annotated epilepsy-related EEG recordings [18, 19]. This set is a collection of clinical EEG recordings manually annotated by medical experts. It was developed specifically for the purpose of application in the field of machine learning. The set files contain an annotation for the seizure/background binary classification. For the study a reduced size sample was created from the files of original dataset. The subset used for the experiments contained 75 files among which 9 contained epileptic seizures and 66 were normal.

3.1 Data Understanding

The initial data are a time series of a multichannel EEG signal (22 derivations in a monopolar scheme). Since the EEG is a multichannel signal, significant events, such as epileptiform activity, may not occur on all channels simultaneously, but only on one channel or on a subset of channels. Figures 2 and 3 show the sample entropy distribution for the dataset, grouped by montages. Figures 2 and 3 shows that in the distribution of sample entropy values for windows labeled as pathology (seiz) and windows labeled as normal (bckg) there are differences. The values of the sample entropy for pathological sections of the record have a greater range, and also the minimum value is close to zero, in contrast to the normal sections of the record. A decrease in the minimum value in the entropy distribution for pathological areas may indicate a decrease in the dynamic complexity of the signal during an epileptic attack due to excessive synchronization of neurons.

The average statistical indicators of the calculated informative features without splitting by montages are presented in the Table 1. The Table 1 also shows differences in the distribution of sample entropy for pathology and norm. For windows marked as pathology ('seiz'), the range of values is greater than for normal ('bckg'). Also, you can see that the kurtosis for 'seiz' and 'bckg' is noticeably different.

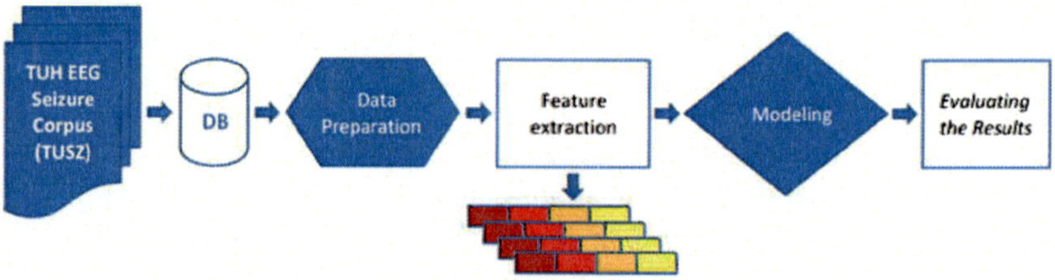

Fig. 1 Data mining process organization

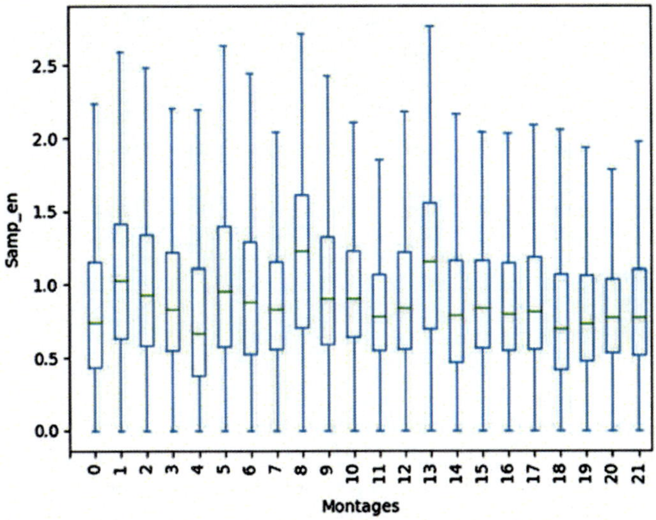

Fig. 2 Distribution of sample entropy values splitting by montages for windows marked as pathology (seiz)

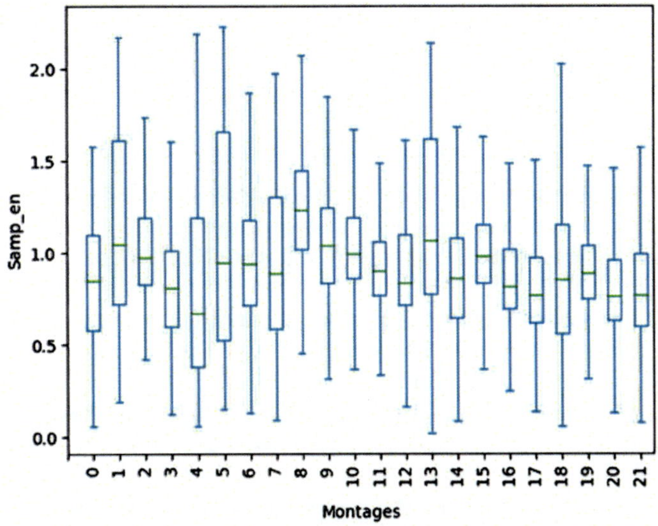

Fig. 3 Distribution of sample entropy values splitting by montages for windows marked as normal (bckg)

Table 1 Statistical indicators of distributions of calculated informative features

Label	Min	25%	Mean	Median	75%	Std	Max	Var	Kurtosis	Skew
Sample entropy										
Bckg	**1.24**	0.54	0.92	0.84	1.24	0.49	**3.08**	0.54	**–0.15**	0.59
Seiz	**0.02**	0.68	0.97	0.91	1.19	0.44	**2.23**	0.46	**0.17**	0.64
Hurst index										
Bckg	–0.05	0.70	0.79	0.82	0.90	0.15	1.31	0.19	0.38	–0.90
Seiz	0.25	0.72	0.79	0.81	0.88	0.12	1.08	0.15	0.77	–0.82
Lyapunov exponent										
Bckg	–0.059	0.001	0.009	0.005	0.012	0.013	0.282	1.450	6.401	1.791
Seiz	–0.048	0.001	0.008	0.006	0.012	0.012	0.087	1.463	6.341	1.700

The number of windows corresponding to the main class 'seiz' is 13024 (0.82% of the total), the number of windows corresponding to the second class 'bckg' is 1571020 (99.18% of the total). A low percentage of occurrence of objects of one of the classes indicates an imbalance in the data set. Since the objects of the positive class (seiz) are less than one percent, this task can be considered as an anomaly detection task.

The imbalance in the set comes from the nature of the task, since epileptic seizures are a much more rarely diagnosed event than the norm.

3.2 Data Preparation

In order to facilitate data manipulation, origin dataset files have been converted from .edf format and loaded into a Postgres database. The data were then converted from a monopolar scheme to a bipolar scheme and was cleared by deleting portions of records with null values.

As noted earlier, the data set is imbalanced, since the proportions of objects of different classes in the sample differ significantly. To make the classes balanced, the data was rebalanced. It consisted in replacing the large class bckg (normal) with a subsample equal in power to the small class seiz (pathology). This way of rebalancing is called undersampling the majority class. For the rebalancing the data was used the RandomUnderSampler method.

After calculating the features, they were normalized and centered.

3.3 Feature Extraction

Informative features were calculated using the sliding window method with a window width of 4 s. Lee et al. [2] and a shift of 1 s. The label was set by the last value. The sample entropy, the Hurst exponent, and the senior Lyapunov exponent were used as informative features. The Python library 'nolds' [20] was used to calculate this features. In addition, the spectral power was calculated in different frequency ranges as a measure of the ratio of the frequencies of the signal components. For the calculation, the fast Fourier transform (FFT) algorithm was used. Spectral analysis was made for the following frequency ranges: delta, theta, alpha, beta and gamma bands.

3.4 Selecting Modeling Techniques

The choice of an appropriate classifier is a complex process that depends on many factors, such as the size of the dataset, the nature of the data contained in it, the quality of data preparation, the number of informative features, and many other factors.

An efficient solution of the optimization problem for both model selection and selection of hyperparameters of algorithms is necessary to obtain satisfactory results for most practical problems. Common approaches for this task are grid search, random search, Bayesian optimization and evolutionary algorithm [21]. Typical supervised learning algorithms include: Logistic Regression (LR), Naive Bayes (NB), Support Vector Machines (SVM), K-Nearest Neighbor (KNN), Decision Tree (DT), Extreme Gradient Boosting (XGB), Random Forest (RF), and Artificial Neural Networks (ANN).

For model selection and hyperparameters optimization we used grid search and the open source Automated Machine Learning tool TPOT [21]. TPOT was used to achieve optimal performance and reduce the time spent on a preliminary assessment of the capabilities of various models for solving a problem on a given dataset. This AutoML package uses genetic programming to select optimal classification models and hyperparameter optimization. As a result of the preliminary experiments, the models with the best performance indicators were selected. The selected algorithms performance evaluation results are presented in Table 2. To improve the performance of selected classification algorithms, a hybrid approach was used in obtaining a resulting model. With this approach, the results of several models at once are involved in the formation of the final result. The output data of models included in the hybrid model are placed in an array of the training set in the form of columns of the target massif. After that, the model selected as the final learns on this new array of training data. This approach is called stack of estimators with a final classifier and allows to reduce the systematic error and retraining of the model. As the base estimators which are stacked together in hybrid model was chosen Extreme Gradient Boosting (XGB), Random Forest (RF) and Light Gradient Boosted Machine (LGBM). Support Vector Machines (SVM) was chosen as a final classifier which was used to combine the base estimators. During training, we used 5-fold cross-validation. As can be seen from the Table 2 the hybrid model showed a higher classification result (accuracy = 0.77, recall bckg = 0.77) than the starting models that are part of it. However, Recall Seiz turned out to be slightly lower (0.68) than the basic models.

Table 2 Testing performance for different ML techniques using the same dataset for the binary classification on the test dataset of normal and pathology windows

ML technology	Accuracy	Recall	
		Bckg	**Seiz**
XGB	0.70	0.70	0.72
RF	0.71	0.71	**0.74**
LGBM	0.71	0.71	0.71
Hybrid (SVM+XGB+RF+LGBM)	**0.77**	**0.77**	0.68

4 Conclusion

The algorithm was tested on a subset of Open Dataset TUH EEG Seizure Corpus (TUSZ) files. Because the study used imbalanced data set, the data was rebalanced by undersampling the majority class. For the rebalancing the data was used the RandomUnderSampler method. After this the dynamic characteristics of the EEG signal was calculated (sample entropy, Hurst exponent, senior Lyapunov exponent). In addition, the spectral power in delta, theta, alpha, beta and gamma bands was calculated. These dynamic and spectral characteristics were used as informative features. For model selection and hyperparameters optimization the Automated Machine Learning tool (TPOT) was used with the genetic optimization algorithm. Finally, for further experiments the following machine learning algorithms was chosen: Extreme Gradient Boosting (XGB), Random Forest (RF) and Light Gradient Boosted Machine (LGBM). A hybrid model (stack of estimators with a final classifier) has been proposed based on XGB, RF and LGBM as base models and Support Vector Machines (SVM) as final model. It was determined that the hybrid model gives the best accuracy results than the base models.

Acknowledgements The work was supported by the Ministry of Science and Higher Education of the Russian Federation (Grant No. 075-15-2022-1121).

References

1. Zafar, A., Kallu, K.D., Yaqub, M.A., Ali, M.U., Byun, J.H., Yoon, M., Kim, K.S.: A hybrid GCN and filter-based framework for channel and feature selection: an fNIRS-BCI study. Int. J. Intell. Syst. **2023**, 14 (2023). https://doi.org/10.1155/2023/8812844
2. Lee, K., Jeong, H., Kim, S., Yang, D., Kang, H., Choi, E.: Real-time seizure detection using EEG: a comprehensive comparison of recent approaches under a realistic setting. In: Proc. of Machine Learning Research. Conf. on Health, Inference, and Learning (CHIL), vol. 174, pp. 311–337 (2022)
3. Loskutov, A.Y., Mikhailov, A.S.: Fundamentals of the Theory of Complex Systems. Institute of Computer Sciences, ICR, Moscow-Izhevsk (In Russian) (2007)

4. Pincus, S.M.: Approximate entropy as a measure of system complexity. Proc. Natl. Acad. Sci. USA **88**, 2297–2301 (1991)
5. Loshchilov, I., Hutter, F.: SGDR: Stochastic gradient descent with warm restarts. In: 5th International Conference on Learning Representations, ICLR, Toulon, France. Conference Track Proceedings (2017)
6. Haken, H.: Brain Dynamics: Synchronization and Activity Patterns in Pulse-Coupled Neural Nets with Delays and Noise. Springer (2002)
7. Goodfellow, I.J., Bengio, Y., Courville, A.: Deep Learning. Cambridge, MIT Press, MA, USA (2016)
8. Schuster, H.G., Just, W.: Deterministic Chaos: An Introduction. Wiley-VCH (2005)
9. Potapov, A.A.: Fractals in Radiophysics and Radiolocation. Logos, Moscow (2002)
10. Richman, J.S., Moorman, J.R.: Physiological time-series analysis using approximate entropy and sample entropy. Am. J. Physiol. Heart Circul. Physiol. **278**(6), H2039–H2049 (2000)
11. Potapov, A.A.: Fractals, scaling, textures, fractional operators and deterministic chaos as the physical and mathematical components of the new conceptions and methods in radar and radio physics. J. Eng. **2019**, 7204–7209 (2019). https://doi.org/10.1049/joe.2019.0520
12. Hurst, H.E.: Long-term storage capacity of reservoirs. Trans. Am. Soc. Civil Eng. **116**, 770–799 (1951). https://doi.org/10.1061/TACEAT.0006518
13. Feder, J.: Fractals. Plenum Press, New York, London (1988)
14. Peters, E.E.: Chaos and Order in the Capital Markets: a New View of Cycles, Prices, and Market Volatility. John Wiley & Sons (1996)
15. Wolf, A., Swift, J., Swinney, H., Vastano, J.: Determining Lyapunov exponents from a time series. Phys. D: Nonlinear Phenom. **16**, 285–317 (1985). https://doi.org/10.1016/0167-2789(85)90011-9
16. Eckmann, J.P., Kamphorst, S.O., Ruelle, D., Ciliberto, S.: Lyapunov exponents from time series. Phys. Rev. A **34**, 4971–4979 (1986)
17. Rosenstein, M.T., Collins, J.J., De Luca, C.J.: A practical method for calculating largest Lyapunov exponents from small data sets. Phys. D **65**, 117–134 (1993)
18. Obeid, I., Picone, J.: The temple university hospital EEG data corpus. Front. Neurosci. **10**, 196 (2016)
19. Harati, A., López, S., Obeid, I., Picone, J.: The TUH EEG corpus: a big data resource for automated EEG interpretation. In: Signal Processing in Medicine and Biology Symposium (SPMB), pp. 1–5. IEEE (2014)
20. Nonlinear measures for dynamical systems (based on one-dimensional time series) https://pypi.org/project/nolds/
21. Le, T.T., Fu, W., Moore, J.H.: Scaling tree-based automated machine learning to biomedical big data with a feature set selector. Bioinformatics, Oxford, England **36**, 250–256 (2020)

Properties of Model Networks Generated by Modified Small-World Scale-Free Fractal Tree Algorithm

Nobutoshi Ikeda

Abstract We modified the small-world scale-free fractal tree algorithm by introducing the length distribution of branches that grow from the tree structure. The new algorithm provides a convenient tool for studying the self-organisation of small-world fractal networks, taking advantage of the simplicity of the model, where only the parameter γ, the power-law exponent describing the degree distribution, can produce various small-world graphs with different values of the fractal dimension. The numerical result suggests the γ-dependency of the fractal box dimension, $d_b \simeq 1 + (\gamma - 2)^{-0.5}$, and demonstrates that, when d_b becomes less than 2 in response to γ being greater than 3, it becomes difficult to maintain fractal properties while maintaining small-world properties. These results are in contrast to those for non-small-world fractal networks, where fractality can be maintained regardless of the values of γ and d_b. Understanding the relationship between γ and d_b in various types of fractal networks is important for interpreting empirical data from real networks.

1 Introduction

Recent studies on complex networks have revealed the ubiquity of fractal structures in networked systems [1–3]. Thus, the self-organisation of fractal structures in networks has become a crucial factor in the study of model networks. In our previous study, we showed that random processes occurring in networks, such as random walks, can generate the fractal structure of networks [4–6]. We developed a small-world scale-free fractal (SSF) tree algorithm by imitating the stratified structure produced by the growth models of fractal networks [7]. However, the fractal property produced by this algorithm tends to become increasingly obscured, as the value of the power-law exponent γ describing the degree distribution, $P(k) \sim k^{-\gamma}$ where k denotes the vertex degree, increases from 2 to 3. In this study, we modify the original algorithm by introducing the length distribution of branches that grow from the tree structure

N. Ikeda (✉)
Tohoku Seikatsu Bunka Junior College, 1-18-2 Niji-no-Oka, Izumi-ku, Sendai 981-8585, Japan
e-mail: ikeda.nobu@nifty.com

D. Vlachos (ed.), *Mathematical Modeling in Physical Sciences*, Springer Proceedings in Mathematics & Statistics 446, https://doi.org/10.1007/978-3-031-52965-8_20

into the model, and show that this modification makes the model exhibit clearer fractality than the previous model.

Several scale-free network models exhibit fractal properties. For some of these networks, the relation between the fractal box dimension d_b and the power-law exponent γ has been obtained [8]. (The definition of d_b is described in Sect. 3). However, most of these networks are non-small-world networks, where the relation $V \sim \bar{d}^{d_d}$ holds, V denotes the number of vertices in the graph, and $\bar{d}$ is the mean distance between the vertices in the graph. In contrast, the growth model of the networks proposed in this study explains the coexistence between the small-world and fractal properties, where the relation $V \sim \exp(\bar{d}/\beta)$ ($\bar{d} = \beta \log V$) holds. The extent to which the small-world property coexists with fractality is an important problem because the small-world property is more prevalent in real-world networks than fractality. Furthermore, the proposed model requires only the tuneable parameter γ, which regulates all the other properties of the model network. The simplicity of the model makes it easier to study the emergence of fractals than when using our previous network models, such as random walk models, because several parameters are assumed to describe the random phenomena occurring in the network.

In this study, taking advantage of the simplicity of the model mentioned above, we consider the dependence of the fractal box dimension d_b on γ and the restrictions on the coexistence of the small-world and fractal properties. In addition, to highlight the properties specific to the proposed model, we examined a typical non-small-world fractal tree model, the critical branching tree [9], and compared the results with those of the proposed model. Our approach is important for studying real fractal networks with or without small-world networks because our study will reveal what type of model network adequately approximates real networks.

The remainder of this paper is organised as follows. Section 2 describes the modified SSF (M-SSF) tree algorithm and compares the numerical results for the stratified structure and mean distance with those for the critical branching tree. Sections 3 and 4 present the numerical results of fractal analysis using box counting [1] and the cluster-growing method [2], respectively. In these sections, the dependence of the fractal box dimension d_b on γ is given, and the limitation of the coexistence of small-world and fractal properties is examined. Section 5 summarises the results.

2 M-SSF Tree Algorithm and Stratified Structure

The growth model of networks usually assumes that the network evolves initially from connected core vertices or, particularly for growth tree structures, from one root vertex. In the next step ($t = 2$), a new vertex is attached to the root vertex through an edge. At each time step $t (= 3, 4, \cdots)$, a new vertex is attached via one edge to one of the vertices that already exists in the current tree structure. In growing tree graphs, the vertices can be classified into several layers by measuring the path length from the root vertex to each vertex. In typical cases, the number of vertices in the lth layer M_l is governed by the following equation:

$$\frac{dM_{l+1}}{dt} = \frac{B_l}{t} M_l, \tag{1}$$

where B_l denotes the mean ability of vertices in the l-th layer to gain a new edge per unit of time. Note that M_l/t is the ratio of the number of vertices in the lth layer, because $V = t$.

In the SSF tree algorithm [7], B_l is assumed to have the functional form $\sim l^{-\alpha}$. (For simplicity, only $\alpha = 0$ is assumed in this study.) This functional form was deduced from a detailed observation of the local network structures generated by local rules that provide fractal properties. Note that (1) is only available for determining the sequence of numbers M_l. For the emergence of the scale-free property with the intended value of γ, the new vertex should preferentially be attached to a vertex with a probability proportional to the degree of the vertex, and the ratio of the numbers of terminal vertices V_1 and other vertices V_2, V_1/V_2, must be fixed to $8/(\gamma+3)(\gamma-2)$ as closely as possible [7]. That is,

$$\frac{V_1}{V_2} \simeq \frac{8}{(\gamma+3)(\gamma-2)}. \tag{2}$$

In summary, the SSF tree algorithm is described as follows: (The following excerpts are from Ref. [7].)

SSF tree algorithm

1. Consider a root vertex with a degree k_{root} as an initial graph $G_{k_{\text{root}}+1}$, where $t = k_{\text{root}} + 1$. At this stage, $M_0 = 1$, $M_1 = k_{\text{root}}$, $M_l = 0$ for $l \geq 2$, and the maximum integer l that satisfies $M_l > 0$, $l_{\max}$, is 1.
2. At each time step t, one of the following two operations is performed on G_t:
   (a) If G_t satisfies the condition $V_1/V_2 \leq 8/(\gamma+3)(\gamma-2)$, an integer $l = 0$ is selected with the probability $k_{\text{root}}/\sum_{i \in G_t} s_i$, where s_i is the number of offspring of the vertex i. Otherwise (this event has a probability of $1 - k_{\text{root}}/\sum_{i \in G_t} s_i$), an integer is selected from $l = 1, 2, \cdots, l_{\max}$ with a probability proportional to $M_l B_l/t$. The selected l is referred to as l_s. Then, a new vertex is linked to one of the vertices at the depth l_s with a probability proportional to s_i, unless all the vertices at the depth l_s have no offspring. If all the vertices at the depth l_s have no offspring, a new vertex is randomly linked to one of the vertices at the depth l_s.
   (b) If G_t satisfies the other condition $V_1/V_2 > 8/(\gamma+3)(\gamma-2)$, an integer is selected from $l = 1, 2, \cdots, l_{\max}$ with a probability proportional to $M_l B_l/t$ (the probability that the number $l = 0$ is selected as 0). The selected l is referred to as l_s. Then, a new vertex is linked randomly to one of the vertices without offspring at the depth l_s, unless all the vertices at the depth l_s have offspring. If all the vertices at the depth l_s have offspring, a new vertex is linked to a vertex i at the depth l_s with a probability proportional to s_i.
3. An integer 1 is added to M_{l_s+1} ($M_{l_s+1} \to M_{l_s+1} + 1$) and to t ($t \to t+1$). If $l_s = l_{\max}$, then an integer 1 is added to $l_{\max}$ ($l_{\max} \to l_{\max} + 1$).

4. Repeat, starting with the second step.

In this study, we modified the SSF algorithm by replacing the edge attached at each time step with a branch with a length distribution. Here, we assume that the length n of the branch is subject to a geometric distribution,

$$P(n) = q(1-q)^{n-1}. \tag{3}$$

Distribution (3) describes an irregular extension of the layered structure of the graph and determines the mean length of the branches $\bar{n}$ as $1/q$. Note that (1) is valid only when $\bar{n} = 1$, because the factor B_l affects not only M_{l+1} but also $M_{l+2}, M_{l+3}, \cdots$, if $\bar{n} > 1$.

When using (3), note that the ratio V_1/V_2 has a maximum value for a fixed value of $\bar{n}$ because, as shown in Fig. 1, the ratio V_1/V_2 is never greater than $1/(\bar{n}-1)$ even if the process shown in Fig. 1c only occurs. According to (2), corresponding to the condition $V_1/V_2 < 1/(\bar{n}-1)$, the tuneable range of γ is also limited to $\gamma_{\min} < \gamma$, where $\gamma_{\min}$ is given by

$$\gamma_{\min} = \frac{-1+\sqrt{32\bar{n}-7}}{2}. \tag{4}$$

The value $\bar{n}$ satisfying (4) is the maximum mean length of the branches required to realise the power-law exponent $\gamma(=\gamma_{\min})$. In the numerical calculation, we use $\bar{n}$, which satisfies (4), to make $\bar{n}$ as large as possible for the intended value of γ.

Figure 2 shows numerical results for M_l and the dependence of the mean shortest path length $\bar{d}$ on V. As shown in Fig. 2b, the mean path length $\bar{d}$ is described by the small-world property $\bar{d} = \beta \log V$, for any γ. The stratified structures M_l appear to show a particular functional form that is different from the Poisson distribution $M_l = V^{1/2}\{\log(V)/2\}^{l-1}/(l-1)!$, which describes the M_l of the trees generated by the preferential attachment rule (BA model) [10] (Fig. 2c). The rate of increase in

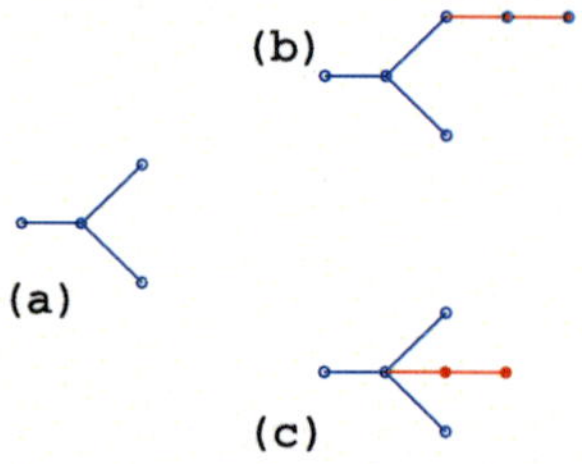

Fig. 1 Growth process of the tree by attaching a branch with the length $n = 2$ (filled circles). **a** Tree structure before the branch is attached. **b** If the branch is attached to a terminal vertex, then the increment of the number of terminal vertices, ΔV_1, is equal to $1-1=0$, and the increment of the number of other vertices, ΔV_2, is equal to $n=2$. On average, this case leads to $\Delta V_1 = 1-1 = 0$ and $\Delta V_2 = \bar{n}$. **c** If the branch is attached to a non-terminal vertex, then $\Delta V_1 = 1$ and $\Delta V_2 = 2-1 = 1$. On average, this case leads to $\Delta V_1 = 1$ and $\Delta V_2 = \bar{n}-1$. To maximise the ratio V_1/V_2, the process **c** should be selectively performed. Therefore, the maximum value of V_1/V_2 is $1/(\bar{n}-1)$

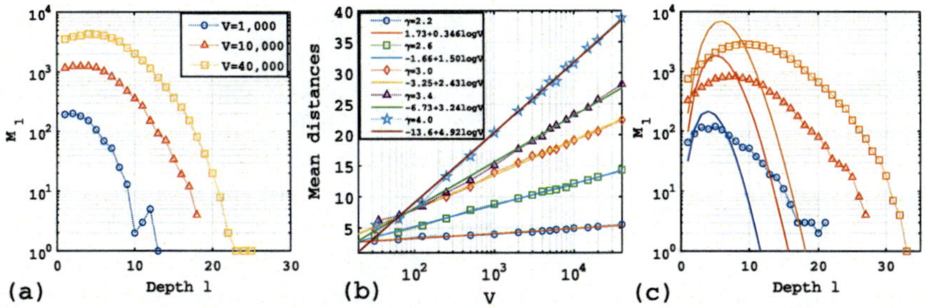

Fig. 2 Shell structures M_l at $V = 1,000$ (circles), $10,000$ (triangles), and $40,000$ (squares), and the small-world property of M-SSF trees. **a** M_l when $\gamma = 2.4$. **b** Dependence of the mean distances $\bar{d}$ on V and the fitted lines $\sim \beta \log V$ for various values of γ (= 2.4, 2.6, 3.0, 3.4, and 4.0). **c** M_l when $\gamma = 2.8$ compared with the analytical forms $M_l = V^{1/2} \frac{\{\log(V)/2\}^{l-1}}{(l-1)!}$ for the BA model [10] (solid lines)

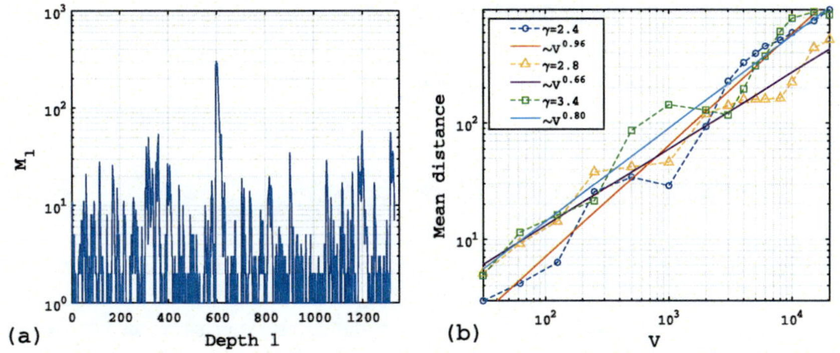

Fig. 3 Typical shell structure M_l of the critical branching tree **a**, and the log-log plots of $\bar{d}$ versus V **b**. In **a**, $f(s) \sim s^{-2.6}$ and $V \simeq 10,000$. In **b**, $\gamma = 2.4$, 2.8, and 3.4 ($f(s) \sim s^{-\gamma}$) are used. The solid lines are the non-small-world properties $\bar{d} \sim V^{1/d_{\mathrm{d}}}$ fitted to the data

the number of layers was enhanced by the M-SSF tree (SSF-tree) algorithm. This enhancement was expected because the fractal property requires a faster increase in the diameter than the non-fractal network such as the BA model.

To highlight the structure specific to the M-SSF tree, the numerical results for the critical branching tree are presented as a typical example of a non-small-world fractal network in Fig. 3. The critical branching tree is defined by a distribution $f(s)$, which is used to describe the probability distribution $P(s)$ that regulates the number of offspring s for each vertex.

$$P(s) = \begin{cases} c_1 & (s = 0) \\ c_2 f(s) & (s = 1, 2, \cdots,) \end{cases} \tag{5}$$

where the constants c_1 and c_2 are determined by the normalised condition $c_1 + c_2 \sum f(s) = 1$ and the mean number of offspring $\bar{s} = c_2 \sum s f(s)$. The tree graph emerges from a root vertex, and the vertices in the lth layer originate from each vertex in the $l-1$th layer according to the distribution (5). When the critical condition $\bar{s} \simeq 1$ holds, the resulting tree graph exhibits fractal behaviour. Figure 3 shows a typical result for M_l and $\bar{d}$ when $f(s) = s^{-\gamma}$. (Note that the stratified structure of the critical branching tree may be interrupted with the probability $c_1^{M_l}$ at the l-th layer. In numerical calculations, when the layer was interrupted, we executed repeated calculations to avoid interruption. Therefore, the critical branching trees under consideration are special samples that have not been accidentally interrupted in any layer.)

The critical branching tree is regarded as a static model, because once the degree of the vertex is determined, that is, once M_l is determined, the degree of the vertex and M_l will not change over time. The value of M_l for the critical branching tree changes drastically and irregularly with respect to l (Fig. 3a), in contrast to M_l observed in the M-SSF tree (Fig. 2a and c). This is because of the large fluctuation in the s of each vertex in the critical branching tree. In addition, the critical branching tree produces only a non-small-world fractal structure $\bar{d} \sim V^{1/d_{\mathrm{d}}}$ (Fig. 3b).

3 Results for the Box-Counting Method

The box-counting method explores the fractal property of networks by checking the power-law relation $N_{\mathrm{b}}(l_{\mathrm{b}}) = l_{\mathrm{b}}^{-d_{\mathrm{b}}}$, where $N_{\mathrm{b}}(l_{\mathrm{b}})$ is the number of boxes required to tile the graph, l_{b} is the smallest integer longer than all the shortest paths between all pairs of vertices in each box, and the constant d_{b} is the fractal box dimension. (We employed the compact-box-burning algorithm [11] at least five times.).

The M-SSF tree algorithm exhibited fractal properties more clearly in the context of the box-counting method compared with the SSF tree. As shown in Fig. 4a, although the original SSF tree shows the fractal property when γ is sufficiently close to 2, as the value of γ moves away from 2, the range of l_b supporting the fractal power-law relationship becomes relatively narrow. Figure 4a shows that the M-SSF tree widens the range of l_b supporting the fractal power-law relation. However, even the M-SSF tree cannot support an apparent fractal relation for the entire range of l_{b} as γ approaches and exceeds 3 (Fig. 4b).

The power-law forms of $N_{\mathrm{b}}(l_{\mathrm{b}})$ performed by the M-SSF tree allow us to calculate the dependence of d_{b} on γ. Figure 4c shows a numerical result, which indicates that the value of d_{b} decreases from greater than 5 to less than 2 as γ increases from 2. In Fig. 4c, the values of d_{b} were estimated by fitting $N_{\mathrm{b}}(l_{\mathrm{b}}) \sim l_{\mathrm{b}}^{-d_{\mathrm{b}}}$ to the data in two different ranges, a range of small l_{b} ($2 \le l_{\mathrm{b}} \le 11$) and the whole range of l_{b}, which satisfies $N_{\mathrm{b}}(l_{\mathrm{b}}) \ge 2$. The difference in d_{b} between these two extreme estimations indicates the incompleteness of the power-law form. Considering the decrease in d_{b} with respect to γ ($d_{\mathrm{b}} \to 1 (\gamma \to \infty)$) and the increase in d_{b} with respect to V at $\gamma \simeq 2$,

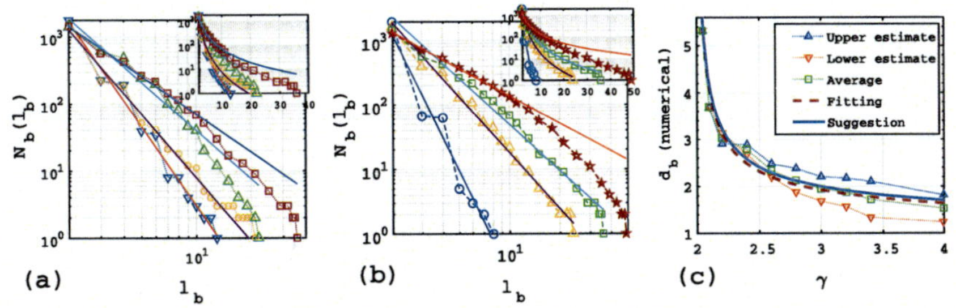

Fig. 4 Results for the box-counting method. **a** Comparison between the results for the SSF and M-SSF trees at $\gamma = 2.2$: SSF, downward triangles; and M-SSF, circles, and at $\gamma = 2.8$: SSF, upward triangles; and M-SSF, squares. **b** Results for the M-SSF trees with different values of γ (2.05, 2.4, 2.8, and 3.4). The fitted values of d_b are 5.32, 2.89, 2.38, and 1.42, respectively. Subpanels are semi-log plots for the same data. **c** Numerical results for γ versus the box-dimension d_b: downward triangles, obtained by fitting to the data in $2 \leq l_b \leq 11$; upward triangles, obtained by fitting to the data, which satisfy $N_b(l_b) \geq 2$. The mean values between these two extremes are plotted using squares. The fitted line $d_b = 1 + \frac{a}{(\gamma-2)^\theta}$ using $a = 0.92$ and $\theta = 0.52$ is illustrated by the dashed line, which suggests $a \simeq 1, \theta = 0.5$ (solid line). All the calculations were performed on the networks at $V = 2,000$

as reported in previous studies [7], a possible functional form $d_b = 1 + a(2-\gamma)^{-\theta}$ is suggested. Figure 4c indicates that $a \simeq 1$ and $\theta \simeq 0.5$ provide a good explanation for the data. However, the fractal property is obscured when γ exceeds 3 ($d_b < 2$). When γ exceeds 3, the functional form $d_b = 1 + (\gamma - 2)^{-0.5}$ describes only the behaviour between the two extreme estimations of d_b.

4 Results for the Cluster-Growing Method and the Small-World Property

The cluster-growing method explores the fractal property of networks by checking the power-law relation $\langle M_c(l_c)\rangle = \langle k+1\rangle\, l_c^{d_c}$, where $\langle M_c(l_c)\rangle$ is the mean size of several clusters (each cluster consists of vertices reachable from a randomly chosen seed vertex by l_c steps), and the constant d_c denotes the fractal cluster dimension. The cluster-growing method allows us to understand how a fractal structure can coexist with small-world properties. Figure 5 shows the results for the cluster-growing method applied to M-SSF trees, with superpositions of the mean distance $\bar{d}$ versus V and the diameter D versus V.

Figure 5a, b show that the range of l_c supporting the power-law relation $\langle M_c(l_c)\rangle \sim l_c^{d_c}$ is widened by increasing the value of γ from 2 to 3. In this situation, although the small-world property $V = \exp(\bar{d}/\beta)$ appears to contradict the fractality $\langle M_c(l_c)\rangle \sim l_c^{d_c}$, this apparent contradiction is solved by considering d_c as a V-dependent value. As reported in Refs. [4, 5], the change in d_c in response to an increase in V should have the form $d_c \simeq \log(V/\langle k+1\rangle)/\log \bar{d}$.

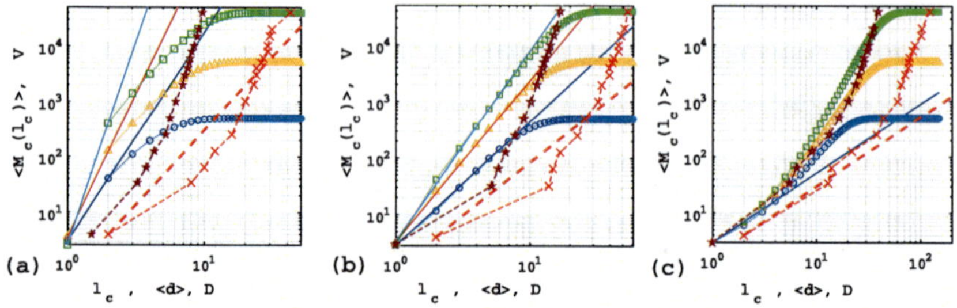

Fig. 5 Results for the cluster-growing method at $V = 500$ (circles), $V = 5,000$ (triangles), and $V = 40,000$ (squares). The solid lines denote the lines fitted to them. The mean distance $\bar{d}$ versus the number of vertices V (stars) and the diameter D versus V (crosses) are superimposed. The ideal behaviours of D, $V \sim D^{d_b}$ are illustrated by dashed lines. **a** When $\gamma = 2.4$ (the estimated box dimension $d_b = 2.69$), the estimated values of d_c fitted to the data in $1 \le l_c \le 2$ are 3.76, 5.15, and 7.29, respectively. **b** When $\gamma = 2.8$ (the estimated box dimension $d_b = 1.87$), the estimated values of d_c fitted to the data in $1 \le l_c \le 7$ are 2.19, 2.84, and 3.35, respectively. **c** When $\gamma = 4.0$ (the estimated box dimension $d_b = 1.24$), $\langle M_c(l_c)\rangle$ loses the power-law behaviour. The estimated values of d_c fitted to the data in $1 \le l_c \le 2$ are $1.0 \sim 1.2$ (solid line)

The diameter D shown in Fig. 5a, b does not show complete consistency with the ideal behaviour, $V \sim D^{d_b}$ [12], which is suggested by the fractal property in the context of the box-counting method. The gap between the actual and ideal behaviours of D gradually increases with increases in γ and V. The introduction of the length distribution of branches can clarify the emergence of the fractal property. However, the effect diminishes as V increases. There may still be room for improvement in the adequate functional form of the length distribution.

Figure 5c shows that the fractal property in the context of the cluster-growing method is also difficult to be maintained while maintaining the small-world property $V \sim \exp(\bar{d}/\beta)$ when γ is greater than 3 (d_b is less than 2). The slope $d_c = \log\langle M_c(l_c)\rangle/\log l_c$ does not show a size dependency, as shown in Fig. 5a, b. Instead, the l_c-dependence of $\langle M_c(l_c)\rangle$ begins to exhibit a small-world behaviour similar to $V \sim \exp(\bar{d}/\beta)$. The difficulty in maintaining the relationship $\langle M_c(l_c)\rangle \sim l_c^{d_b}$ is due to the disappearance of the fractal property in the context of the box-counting method. In fact, the actual behaviour of D shown in Fig. 5c is far from the ideal behaviour $V \sim D^{d_b}$. In the condition $\gamma > 3$, instead of the disappearance of fractality in the context of the cluster-growing method, the asymptotic behaviour of $\langle M_c(l_c)\rangle$ for a small l_c is consistent with the behaviour of D and its ideal behaviour $V \sim D^{d_b}$ at a small V. These results are different from those obtained under the condition $\gamma < 3$ ($d_b > 2$). In this situation, the restoration of fractality is hardly expected, even if the length distribution of the branches is adequately modified.

To show that the disappearance of fractal properties in response to a decrease in d_b is specific to small-world fractal networks, we consider the fractal behaviour of the critical branching tree. The critical branching tree is a typical example of non-small-world fractal networks where the non-small-world behaviour $V \sim \bar{d}^{d_d}$ is consistent with other quantities [12], such as the mean cluster size $\langle M_c(l_c)\rangle \sim l_c^{d_c}$,

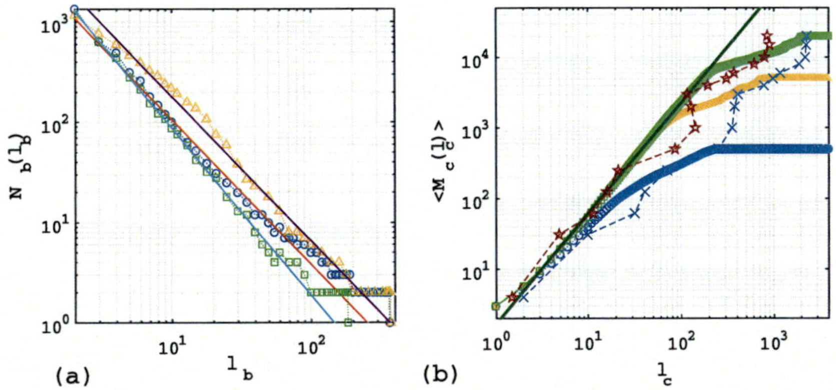

Fig. 6 Fractality of the critical branching tree. **a** Results for the box-counting method applied to the cases $\gamma = 2.4$ (circles, $d_b \simeq 1.44$), $\gamma = 3.4$ (triangles, $d_b \simeq 1.44$), and an exponential functional form $f(s) \sim \exp(-2s)$ (squares, $d_b \simeq 1.67$). **b** Result for the cluster-growing method applied to the case $\gamma = 3.4$ at $V = 500$ (circles), $V = 5,000$ (triangles), and $V = 20,000$ (squares). The cluster dimension is estimated as $d_c \simeq 1.57$ at $V = 20,000$. The behaviours, $\bar{d}$ versus V (stars) and D versus V (crosses), are also explained by the similar power-law form

the number of boxes to tile the graph $N_b(l_b) \sim l_b^{-d_b}$, and the diameter $D \sim V^{1/d_D}$. Figure 6 shows that the power-law behaviour of these quantities can be described by a similar power-law exponent, that is, $d_b \simeq d_c \simeq d_d \simeq d_D$. In contrast to the M-SSF tree, the consistency between the fractal dimensions did not restrict the values of the fractal dimensions. Consequently, the condition $d_b \simeq d_c < 2$ can be realised without difficulty even if this condition requires a very rapid increase in $\bar{d}$ and D with respect to V. For the M-SSF tree, it becomes difficult for the small-world property $\bar{d} = \beta \log V$ to coexist with the fractal property $N_b(l_b) \sim l_b^{-d_b}$ as d_b becomes smaller, because the gap between the behaviours of $\bar{d} = \beta \log V$ and $D \sim V^{1/d_D} \simeq V^{1/d_b}$ must be widened to support the small-world property, as d_b becomes smaller.

Additionally, critical branching trees do not always require a scale-free property to support the fractal property. As shown in Fig. 6a, the fractal property can be realised regardless of the value of γ, or even when $f(s)$ has an exponentially decaying form. This is because the fractality of the critical branching tree originates in the criticality of the offspring $\bar{s} = 1$, which mimics the property of non-small-world fractal giant components at the percolation transition points. The critical behaviour $\bar{s} = 1$ can be realised regardless of the functional form of $f(s)$.

5 Conclusion

We proposed a modified version of the SSF tree algorithm, in which the length distribution of branches that grow from the tree structure is introduced. This modification makes the fractal properties of the tree structure more apparent than those of the original algorithm, and allows a detailed investigation of the emergence of the fractal properties of small-world networks. The numerical results suggest a relationship between the fractal box dimension d_b and the power-law exponent γ describing the degree distribution $d_b \simeq 1 + (\gamma - 2)^{-0.5}$. However, as γ increases from 2 to values larger than 3, the fractal behaviour tends to be obscured in the larger region of l_b. In such a situation, the relationship only describes the mean value between the two extreme estimates of d_b. It may be possible to improve the diameter behaviour by adequately assuming the length distribution of the branches. However, the disappearance of the fractal property in the context of the cluster-growing method in response to γ being greater than 3 should be considered an essential change in the graph property. This result suggests that the coexistence of the fractal and small-world properties is limited under the condition $\gamma > 3$ ($d_b < 2$). By contrast, the non-small-world fractal tree has no restrictions on the values of fractal box dimensions d_b and γ. This is because there was no frustration between the behaviours of the mean distance and diameter, as observed in the M-SSF trees. A comparison of M-SSF trees with other non-small-world fractal network models will help us understand small-world and non-small-world fractal networks in the real world. We plan to investigate real networks using the knowledge of the difference in the relationship among d_b, d_c, and γ between small-world and non-small-world fractal networks.

References

1. Song, C., Havlin, S., Makse, H.A.: Nature **433**, 392 (2005)
2. Csányi, G., Szendröi, B.: Phys. Rev. E **70**, 016122 (2004)
3. Rosenberg, E.: Fractal Dimensions of Networks. Springer International Publishing, Cham, Switzerland (2020)
4. Ikeda, N.: Phy. A **521**, 424 (2019)
5. Ikeda, N.: Phys. A **537**, 122743 (2020)
6. Ikeda, N.: J. Phys. Conf. **2090**, 012085 (2021)
7. Ikeda, N.: Phys. A **583**, 126299 (2021)
8. Rozenfeld, H.D., Havlin, S., Ben-Avraham, D.: New J. Phys. **9**, 175 (2007)
9. Kim, J.S., Goh, K.I., Salvi, G., Oh, E., Kahng, B., Kim, D.: Phys. Rev. E **75**, 016110 (2007)
10. Ben-Naim, E., Krapivsky, P.L.: J. Phys. A **42**, 475001 (2009)
11. Song, C., Gallos, L.K., Havlin, S., Makse, H.A.: J. Stat. Mech. **03**, P03006 (2007)
12. Ikeda, N.: Chaos Solitons Fract. **137**, 109837 (2020)

Slow-Fast Dynamical Systems with a Load Variation

Elena Savenkova, Sergey Vakulenko, and Ivan Sudakow

Abstract We analyze a category of slow-fast dynamical systems that exhibit complex bifurcations as the dynamic parameter, representing the load on the system, undergoes variation. We establish that within such systems featuring distinct structurally stable dynamical states, a transition from the first regime to the second ensues as a consequence of load variation. We extend these concepts to investigate bifurcations in conceptual climate models, specifically those resulting from restrictions on greenhouse gas emissions.

Keywords Quadratic systems · Dynamical systems · Vector field realization · Bifurcation · Climate · Carbon · Emission

1 Introduction

In this paper, we explore a class of slow–fast dynamical systems with complex long-term behaviors and bifurcations, which arise from variations in specific dynamical parameters. The conceptual climate models discussed below serve as prime examples of this particular category of dynamical systems. Many conceptual climate models involve the decomposition of physical variables into distinct slow and fast modes. As time progresses significantly, the rapid dynamics of the fast mode become effectively governed by the more stable slow dynamics within a slow-fast system. The behavior of the slow variables plays a crucial role in shaping the long-term climate evolution,

E. Savenkova (✉)
The Center for Research and Invention, Veliky Novgorod 173008, Russia
e-mail: savenkova.en@mail.ru
URL: http://www.tsnir.ru/

S. Vakulenko
Institute of Problems in Mechanical Engineering, Russian Academy of Sciences, St. Petersburg 199178, Russia

I. Sudakow
School of Mathematics and Statistics, The Open University, Milton Keynes MK7 6AA, United Kingdom

D. Vlachos (ed.), *Mathematical Modeling in Physical Sciences*, Springer Proceedings in Mathematics & Statistics 446, https://doi.org/10.1007/978-3-031-52965-8_21

influenced by external (astronomical) factors, for example, insolation. Conversely, the fast modes are linked to rapid influences such as daily shifts in atmospheric dynamics and inherent alterations in the biosphere. To illustrate this concept, we can introduce a simplified instance using a two-layer model of the ocean [6]:

$$C_s \frac{dT_s}{dt} = -\beta T_s - \eta\gamma(T_s - T_o) + h(t), \tag{1}$$

$$C_o \frac{dT_o}{dt} = \gamma(T_s - T_o) \tag{2}$$

Equation (1) outlines the transient variations in the global mean near-surface temperature denoted as T_s, which corresponds to the fast mode of the system. Meanwhile, Eq. (2) delineates the extended alterations in the deep-ocean temperature T_o, governing the behavior of the slow component. The respective heat capacities are represented as C_s and C_o. The parameters γ and β serve as climate feedback coefficients, with η signifying the impact of elevated ocean temperatures on climate feedback. The term $h(t)$ represents a radiative forcing that is directly proportional to the system's temperature, regulated by the feedback parameter β.

The model defined by Eqs. (1) and (2) can be expanded by incorporating additional layers representing different subsystems of the climate system, as explored in [4]. Furthermore, it is possible to establish connections between the climate system and Earth's biosphere through the parameter h or other relevant parameters. For instance, the albedo of plants can exert an influence on the climate system, while in a reciprocal manner, the populations of plants are influenced by the temperature T_s. In desert regions, an elevation in T_s might lead to the extinction of plant life, while in temperate zones, the same temperature increase could foster the growth of plant populations. Consequently, we can express h as a combination of two components: an industrial contribution denoted as h_{ind}, and a biotic contribution labeled as h_{bio}:

$$h = h_{ind}(t) + h_{bio}, \quad h_{bio} = \sum_{i=1}^{m} b_k x_k, \tag{3}$$

where b_k are some coefficients and x_k is the population of the k-th plant species. We can formulate dynamical equations for the variable x_k, which incorporate T_s as a parameter

$$\frac{dx_k}{dt} = x_k X_k(x, T_s, v), \quad x = (x_1, ..., x_m) \tag{4}$$

and $v = (v_1, , , , v_n)$ are resources. The resource dynamics is defined by equations

$$\frac{dv_j}{dt} = V_j(v, x, T_s). \tag{5}$$

Then we obtain a coupling with Eqs. (1) and (2). It's important to note that systems featuring multiple temperature compartments can encompass polynomial nonlinearities extending up to the fourth order, owing to the influence of the Stefan-Boltzmann law, which incorporates terms involving T^4 as highlighted in [17].

Another significant slow-fast climate model is the low-order model developed by K.A. Maasch and B. Saltzman [8]

$$\frac{dx}{dt} = -x - y, \tag{6}$$

$$\frac{dy}{dt} = ry - pz + sz^2 - yz^2, \tag{7}$$

$$\frac{dz}{dt} = -qx - qz, \tag{8}$$

where the variables x, y, and z represent anomalies, signifying deviations from long-term averages, for the total global ice mass, atmospheric carbon dioxide concentration, and North Atlantic Deep Water (NADW) volume, respectively. The coefficients q, r, p, and s are positive parameters. This particular model, which has been extensively examined recently in works such as [5], effectively elucidates climate transitions in relation to insolation forcing and exhibits distinct bifurcation behaviors. It is categorized as a slow-fast model when $q >> 1$, as noted in [5].

All the examples of climate models presented in this introduction can be analyzed in the context of quadratic systems of a large dimension,

$$u_t = Au + B(u) + h, \tag{9}$$

where $u \in H = \mathbb{R}^N$ is a phase space, A is a linear operator, B is a quadratic operator, and h is an external forcing (we will also call this parameter a *load* following the terminology from the ecological literature).

It was demonstrated in [14] that the dynamics of this gradual system exhibit a remarkable diversity. Specifically, by selecting an appropriate value for parameter h, this dynamics can be transformed into another quadratic system featuring only n variables, where $n << N$. The resulting reduced systems can take on a practically arbitrary form and showcase a range of dynamic behaviors, including chaos, Lorenz dynamics (for $n = 3$), saddle-node and Andronov-Hopf bifurcations (for $n = 2$), along with potential bifurcations between these regimes.

Moreover, recent findings indicate that these systems have the capacity to generate all types of structurally stable dynamics, such as hyperbolic behavior, as outlined in [14]. This insight implies the emergence of phenomena like Ruelle-Takens strange attractors, Smale horseshoes, and other forms of stable chaotic behavior.

The novel finding can be succinctly summarized as follows. We keep all parameters fixed, except for the variable h, which we can vary as the load. Consequently, the nature of the slow system becomes contingent on the value of h. For any pair of distinct structurally stable dynamic states-let's designate the first as a periodic cycle and the second as a strange attractor-a corresponding system (referred to as Eq. (9)) exists. Through variation of the parameter h, this system undergoes a bifurcation from the first dynamic state to the second. For a more rigorous articulation of this concept, refer to Theorem 1.

These mathematical findings can be interpreted in the following manner. We postulate that the dynamics of the climate in a conceptual model can undergo unpredictable variations contingent upon the load parameter h. This load can be directly associated with the magnitude of carbon emissions, as outlined in [11, 12].

Our primary technical approach involves the utilization of the method of vector field realizations (RVF), introduced by P. Pol'ačik [9]. The main idea of the RVF method is to consider, instead of a single semiflow, a family of local semiflows $S^t(P)$ depending on some parameters P involved in the problem (see Sect. 2 for more detail).

The structure of the paper is as follows. In Sects. 2 and 3, we delve into the RVF method and demonstrate that adjustments to the load parameter h can give rise to potential bifurcations, leading from one form of structurally stable behavior to another. Section 4 outlines several straightforward scenarios of bifurcations in climate models that may arise due to variations in h.

2 Method of Realization of Vector Fields

Allow us to elucidate the method known as the realization of vector fields (RVF), as introduced by P. Poláčik [9]. We consider a family of local semiflows $S^t(P)$ in a Banach space $\mathbf{H}$. Suppose these semiflows depend on a parameter $P \in \mathbf{E}$, where $\mathbf{E}$ is another Banach space. Let $B^n(R)$ be the ball $B^n(R) = \{q : |q| \leq R\}$ in $\mathbf{R}^n$ of radius $R > 0$ centered at 0, where $q = (q_1, q_2, ..., q_n)$ and $|q|^2 = q_1^2 + ... + q_n^2$. We write simply B^n for $R = 1$. Let's examine a system of differential equations on the ball $B^n(R)$:

$$\frac{dq}{dt} = Q(q), \tag{10}$$

where

$$Q \in C^1(B^n), \quad \sup_{q \in B^n(R)} |\nabla Q(q)| < 1. \tag{11}$$

Suppose the vector field Q is directed strictly inward to the ball $B^n(R)$ at its boundary $\partial B^n(R) = \{q : |q| = R\}$:

$$Q(q) \cdot q < 0 \quad for\ q \in \partial B^n(R). \tag{12}$$

Then Eq. (10) defines a finite-dimensional global semiflow on $B^n(R)$.

Definition 1 Let ϵ be a positive number. We say that the family of the local semiflows $S^t(P)$ in $\mathbf{H}$ realizes the field Q with accuracy ϵ (briefly, ϵ-realizes) on B^n, if there exists a parameter $P = P(Q, \epsilon, n)$ such that

(i) the semiflow $S^t(P)$ has a locally invariant manifold $\mathcal{M}_n(P)$. This manifold is defined by a map $Z : B^n(R) \to \mathbf{H}$

$$z = Z(q), \quad q \in B^n(R), \quad z \in \mathbf{H}, \quad Z \in C^{1+r}(B^n), \tag{13}$$

where $r \geq 0$;

(ii) the restriction of the semiflow $S^t(P)|_{\mathcal{M}_n(P)}$ on $\mathcal{M}_n(P)$ is defined by the system of differential equations

$$\frac{dq}{dt} = Q(q) + \tilde{Q}(q, P), \quad \tilde{Q} \in C^1(B^n(R)), \tag{14}$$

where

$$|\tilde{Q}(\cdot, P)|_{C^1(B^n(R))} < \epsilon. \tag{15}$$

We assume here that all the locally invariant manifolds $\mathcal{M}_n(P)$ are locally attracting.

Remark These manifolds $\mathcal{M}_{\mathcal{P}}$ can be globally attracting, i.e., inertial, positively invariant and locally attracting, or invariant. The theory concerning invariant and inertial manifolds has been extensively developed, as documented in the relevant literature [2, 3, 7, 18].

We refer to the families $S^t_{\mathcal{P}}$ of the semiflows, which are capable to ϵ-realize all finite-dimensional fields Q with any accuracy ϵ, as *maximally dynamically complex families of semiflows* or, for brevity, simply maximally complex semiflows (MCFs). According to the definition of structural stability, all finite-dimensional dynamics that are stable under small perturbations can be generated by an MCF (up to orbital topological equivalencies).

Let's revisit the fundamental concept of structural stability [10] We consider a smooth vector field F on a compact domain $\mathbb{D}^n$ of $\mathbb{R}^n$ with a smooth boundary (or on a compact smooth manifold M of dimension n). Assume that $F \in C^1(\mathbb{D}^n)$ and consider all ϵ-small perturbations $\tilde{F}$ such that

$$|\tilde{F}|_{C^1(\mathbb{D}^n)} < \epsilon. \tag{16}$$

Next, we consider systems of differential equations $dx/dt = F(x)$ and $dx/dt = F(x) + \tilde{F}(x)$ and suppose that they define global semiflows S^t_F and $S^t_{F+\tilde{F}}$ on $\mathbb{D}^n$. The system $dx/dt = F(x)$ is called structurally stable if there exists an ϵ_0 such that for all positive $\epsilon < \epsilon_0$ trajectories of semiflows S^t_F and $S^t_{F+\tilde{F}}$ are orbitally topologically conjugated (there exists a homeomorphism, which map trajectories of the first system into trajectories of the second one). Roughly speaking, the original system is structurally stable if any sufficiently small C^1 perturbations of that system conserve the topological structure of its trajectories, for example, the equilibrium point stays an equilibrium (maybe, slightly shifted with respect to the equilibrium of a non-perturbed system), or the perturbed cycle is again a cycle (maybe slightly deformed and shifted).

It's important to recognize that structurally stable dynamics can possess a form of "chaotic" behavior. Notably, there exists a considerable divergence in the definitions of "chaos." In principle, any notion of chaos can be applicable in this context. If this chaos is persistent under small C^1 -perturbations, this kind of chaos occurs in the dynamics of an MCF. Here, following the classical mathematical tradition [10], we focus our attention on compact invariant sets with hyperbolic chaotic dynamics. We use only the following basic property of hyperbolic sets, the so-called persistence [10]. This means that the hyperbolic sets are, in a sense, robust: if (10) generates the hyperbolic set Γ and ϵ is sufficiently small, then the dynamics of Eq. (14) also generates another hyperbolic set $\tilde{\Gamma}$. The dynamics of Eqs. (10) and (14) restricted to Γ and $\tilde{\Gamma}$ respectively, are topologically orbitally equivalent.

Thus, all hyperbolic sets can appear in the dynamics of an MCF, for example, hyperbolic rest points, cycles and also chaotic hyperbolic sets: the Smale horse-shoes, Anosov flows, the Ruelle-Takens-Newhouse chaos, see [10]. Examples of MCF families can be given by some reaction-diffusion systems [15]. Neural net-works provide instances of MCFs that hold significant practical applications [13]. The RVF method can be applied to systems whose dynamics permit a decomposition into slow and fast components.

3 Quadratic Systems

Let us consider a class of quadratic systems with slow and fast variables, which exhibit complicated bifurcations when the load h changes. First, we examine the system

$$\frac{dy_i}{dt} = (\mathbf{K}_i y, y) + \mathbf{B}_i y + h_i, \tag{17}$$

where $y = (y_1(t), ..., y_N(t))$ is a unknown vector function and we, for brevity, use the notation

$$(\mathbf{K}_i x, y) = \sum_{m=1}^{N} K_{ilm} x_l y_m, \quad \mathbf{B}_i y = \sum_{k=1}^{N} B_{ik} y_k,$$

where $i = 1, ..., N$.

We will use the following result obtained in [14, 16]. System (17) exhibits maximally complex dynamics, in the sense of the previous section. Parameters are coefficients K_{ijl}, B_{ik}, h_i and N.

More precisely, let us consider a dynamical system defined by (10) satisfying (12) and $\epsilon > 0$. Let A_Q be an attractor of that system. To ϵ-realize the vector field Q (the system (10)) we can take a sufficiently large N, fix coefficients K_{ijl} satisfying some conditions, and then adjust the matrix $\mathbf{B}$ and the load vector h.

Let $\epsilon > 0$ and $A_{\bar{Q}}$ be an attractor of the system (18). To ϵ-realize the vector field $\bar{Q}$ we can take a sufficiently large N, take the same coefficients K_{ijl} as that above and adjust another matrix $\bar{\mathbf{B}}$ and another load vector $\bar{h}$.

This construction can be used to show that a quadratic system can exhibit complicated bifurcations between structurally stable dynamics under changes of the load h and for fixed K_{ijl}, N, $\mathbf{B}$. Let us consider a second dynamical system

$$\frac{dq}{dt} = \bar{Q}(q), \tag{18}$$

where $\bar{Q}$ satisfies the same conditions (11) and (12) that Q satisfies.

By results [14] one can obtain the following:

Theorem 1 *Let (10) and (18) be two systems of differential equations satisfying conditions (11) and (12) and generating global semiflows with local attractors A and $\bar{A}$, respectively. Then for each $\epsilon > 0$ there exists a quadratic system (17), which ϵ-realizes both systems (10) and (18) by a variation of h for fixed N, $\mathbf{K}$ and $\mathbf{B}$. If the systems generate structurally stable semiflows, then there exists a quadratic system (17), which has a local attractor topologically equivalent to A for some choice of the load h and another local attractor $\bar{A}$ for another choice of h.*

Proof Let us take a small $\epsilon > 0$ and by [14] construct ϵ-realization of those systems by a quadratic systems

$$\frac{dy_i}{dt} = (\mathbf{K}_i y, y) + \mathbf{B}_i^{(l)} y + h_i^{(l)}, \tag{19}$$

where for $l = 1$ that system $\epsilon/2$-realizes 18 and for $l = 2$ it $\epsilon/2$- realizes Eq. (10). Let us consider the system

$$\frac{dy_i}{dt} = (\mathbf{K}_i^{(2)} y, x) + (\mathbf{K}_i y, y) + \mathbf{C}_i^{(1)} x + \bar{h}_i, \tag{20}$$

$$\frac{dx_j}{dt} = \mathbf{A}_j x + \tilde{h}_j, \tag{21}$$

where $i, j = 1, \ldots, N$, $\mathbf{A}, \mathbf{C}$ are square matrices and we use notation introduced above,

$$(\mathbf{K}_i^{(p)} x, y) = \sum_{m=1}^{N} K_{ilm}^{(p)} x_l y_m, \quad p = 1, 2, 3.$$

$$\mathbf{A}_i x = \sum_{k=1}^{N} A_{ik} x_k.$$

Suppose that the spectrum *Spec* $\mathbf{A}$ of the matrix $\mathbf{A}$ lies in a negative half-plane:

$$\sigma(\mathbf{A}) < -\delta_0 < 0.$$

Then

$$|x - x^*| < c \exp(-\delta_0 t), \quad x^* = \mathbf{A}^{-1}\tilde{h}.$$

Therefore, for large times we can replace the system (20) and (21) by the system (17), where

$$B_{ij}(x^*) = \sum_{l=1}^{N} K_{ilj} x_l^* + C_{ij}^{(1)},$$

and

$$h_i = \bar{h}_i + \mathbf{C}_i x^*.$$

For each i consider the linear maps $g_i : x^* \to z$ defined on $\mathbb{R}^N$ by relations

$$z_j = \sum_{l=1}^{N} K_{ilj}^{(2)} x_l^*.$$

Suppose that the ranges of all these maps have dimension N. It is clear that for generic coefficients K_{ilj} this property is fulfilled. Then we see that for an appropriate choice $x^* = x^{(1)}$ and $x^* = x^{(2)}$ we obtain systems (19) with $l = 1$ and $l = 2$, respectively; this proves Assertion 1. The second one follows from the first and the definition of structural stability.

4 Simple Bifurcations Scenarios

In this section, we analyze the potential impact of constraints on climate feedback (such as limitations on greenhouse gas emissions) on climate bifurcations. We will illustrate various scenarios, including:

A. Restrictions on greenhouse gas emissions delay an inevitable catastrophe; a bifurcation occurs, but its timing is shifted to a later period.
B. Restrictions on greenhouse gas emissions enable us to avert a climate bifurcation.
C. Restrictions on greenhouse gas emissions result in a climate bifurcation.

To show the possibility of these effects, we use a simple dynamic model involving a fast variable $x \in \mathbf{R}^m$ (for example, a near-surface temperature anomaly, as in [11]) and slow variables $q \in \mathbf{R}^n$, respectively:

$$\frac{dx}{dt} = -B(q)x + f(x, q), \tag{22}$$

$$\frac{dq}{dt} = \epsilon(g(x, q) + eh), \tag{23}$$

where h, ϵ are parameters, where $0 < \epsilon << 1$. The parameter h defines the intensity of different greenhouse gases emissions and $e \in \mathbf{R}^m$ is a non-zero vector with non-negative components that defines the direction of that emission (for example, e_1 can correspond to the emission of carbon dioxide, e_2 can correspond to the emission of black carbon, and e_3 to the emission of methane). We assume that f, g are smooth functions such that $f = O(x^2)$ as $x \to 0$. The bifurcations in the x dynamics can arise at q such that $Det\ B(q) = 0$. Suppose that the set of the root of the Eq. (22) $Det\ B(q) = 0$ consists of two connected components, M_1 and M_2. Outside bifurcation points, Eq. (22) implies that for large times t the fast variables are defined by $x = 0$ and for slow variables we then have

$$\frac{dq}{dt} = \epsilon(g(0, q) + eh). \tag{24}$$

For $|g| << h$ scenarios **A-C** depend on the locations of M_1 and M_2 and the direction vector e. These different scenarios **A**, **B**, **C** are illustrated in Fig. 1.

To understand the mechanisms behind scenarios **A**, **B**, and **C**, let's consider the simplest case with $g = 0$. In scenario **A**, the possibility arises when the emission

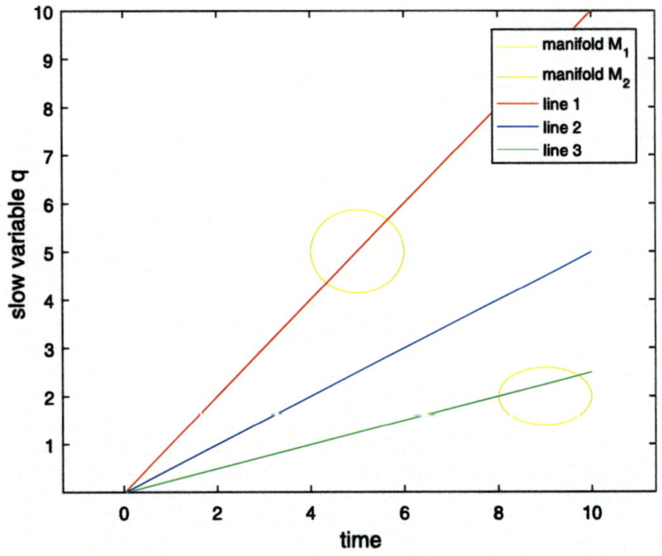

Fig. 1 Scenario of various bifurcations contingent upon the magnitude and direction of the load parameter h

coefficient h decreases (e.g., the industrial carbon dioxide emission is limited), yet the emission direction vector **e** remains unchanged (for instance, all greenhouse gas emissions decrease proportionally). As a result, the trajectory represented by the line 1 intersects the manifold M_1 at a later point (as depicted in Fig. 1).

Scenario **B** emerges when h remains constant, but the direction vector **e** changes (for instance, carbon dioxide emissions are restricted while methane emissions increase). In this case, the trajectory shifts from the line 1 to the line 2. Similarly, scenario **C** follows the same underlying mechanism but involves transitions from line 2 to line 3, and from line 1 to line 3, respectively.

5 Concluding Remarks

In this paper, we consider slow-fast dynamical systems exhibiting complex behavior, which emerge due to changes in specific dynamical parameters. Our investigation reveals that when we maintain all parameters constant except for the variable, which serves as a load that can be varied, the nature of the slow system becomes dependent on the value of the load. We prove the theorem where we show that for any pair of distinct structurally stable dynamic states, some corresponding system should exist. then by changing the load parameter, this system undergoes a bifurcation from the initial dynamic state to another state.

We introduce a conceptual climate model in which the load parameter, corresponding to the intensity of various greenhouse gas emissions, can influence the onset of climate bifurcation. In certain scenarios, the occurrence of bifurcation can be postponed or entirely prevented. Realizing these outcomes requires the careful selection of a load parameter, which translates to the implementation of a meticulous strategy for limiting greenhouse gas emissions [1].

Acknowledgements This research was funded by the Russian Science Foundation (RSF), project No. 21-71-10052. Part of this research was performed while I.S. was visiting the Institute for Mathematical and Statistical Innovation (IMSI), which is supported by the National Science Foundation (Grant No. DMS-1929348) and the Kavli Institute for Theoretical Physics (KITP), which is supported by the National Science Foundation (Grant No. PHY-1748958).

References

1. Cael, B.B., Bloch-Johnson, J., Ceppi, P., Fredriksen, H.-B., Goodwin, P., Gregory, J.M., Smith, C.J., Williams, R.G.: Energy budget diagnosis of changing climate feedback. Sci. Adv. **9**(16), eadf9302 (2023)
2. Chow, S.-N., Kening, L.: Invariant manifolds for flows in Banach spaces. J. Diff. Equ. **74**(2), 285–317 (1988)
3. Debussche, A., Temam, R.: Some new generalizations of inertial manifolds. In: Discrete & Continuous Dynamical Systems–A, pp. 543 (1996)

4. Kerry, E.: A simple model of multiple climate regimes. J. Geophys. Res. Atmos. **107**(D9), ACL 4–1–ACL 4–10 (2002)
5. Engler, H., Kaper, H.G., Kaper, T.J., Vo, T.: Dynamical systems analysis of the Maasch-Saltzman model for glacial cycles. Phys. D Nonlinear Phenom. **359**, 1–20 (2017)
6. Gregory, J.M.: Vertical heat transports in the ocean and their effect on time-dependent climate change. Clim. Dyn. **16**(7), 501–515 (2000)
7. Henry, D.: Geometric Theory of Semilinear Parabolic Equations. Lecture Notes in Mathematics. Springer-Verlag (1981)
8. Maasch, K.A., Saltzman, B.: A low-order dynamical model of global climatic variability over the full pleistocene. J. Geophys. Res. Atmos. **95**(D2), 1955–1963 (1990)
9. Poláčik, P.: Complicated dynamics in scalar semilinear parabolic equations in higher space dimension. J. Diff. Equ. **89**(2), 244–271 (1991)
10. David, R.: Elements of Differentiable Dynamics and Bifurcation Theory. Elsevier (2014)
11. Seshadri, A.K.: Fast-slow climate dynamics and peak global warming. Clim. Dyn. **48**(7), 2235–2253 (2017)
12. Sudakov, I., Vakulenko, S.A.: A mathematical model for a positive permafrost carbon–climate feedback. IMA J. Appl. Math. **80**(3, 05), 811–824 (2014)
13. Sudakow, I., Vakulenko, S.A., Grigoriev, D.: Excitable media store and transfer complicated information via topological defect motion. Commun. Nonlinear Sci. Numer. Simul. **116**, 106844 (2023)
14. Vakulenko, S., Grigoriev, D., Weber, A.: Reduction methods and chaos for quadratic system of differential equations. Stud. Appl. Math. **135**, 225–247 (2015)
15. Vakulenko, S.: Dissipative systems generating any structurally stable chaos. Adv. Diff. Equ. **5**, 01 (2000)
16. Vakulenko, S., Sudakov, I.: Complex bifurcations in Bénard-Marangoni convection. J. Phys. A Math. Theoret. **49**(42), 424001 (2016)
17. Vakulenko, S.A., Sudakov, I., Petrovskii, S.V., Lukichev, D.: Stability of a planetary climate system with the biosphere species competing for resources. Phys. Rev. E **103**, 022202 (2021)
18. Wegner, B., Guekenheimer, J., Holmes, P.: Nonlinear Oscillations, Dynamical Systems, and Bifurcations of Vector Fields. Springer-Verlag (1986)

The Problem of Premature Convergence in Engineering Optimization Problems

K. A. Ponomareva, I. P. Rozhnov, and L. A. Kazakovtsev

Abstract Optimization is a complex problem in mathematics and engineering applications, which has so far been solved using a variety of approaches based on heuristic methods, methods inspired by nature, methods based on the principles of fuzzy logic, and methods based on rigorous mathematical tools. The most important task is to combine various methods of modeling, optimization, design and management of complex systems within the framework of an integrated approach for a holistic description of the phenomena under study. The objective function of optimization problems combines continuous and discrete variables and various constraints, which indicates its complexity, as well as difficulties in solving such problems. Algorithms that are used in solving optimization problems may suffer from premature convergence when they stop at the optimal solution earlier than required. Based on the problem of premature convergence, a hybrid approach combining a sine–cosine algorithm (SCA) and an artificial bee colony algorithm (ABC) was proposed in the paper. In the proposed algorithm, called the hybrid sine–cosine algorithm (HSCA), both algorithms are executed alternately until the convergence criterion is satisfied.

Keywords Optimization · Premature convergence · Hybrid approach · Sine–cosine algorithm · Artificial bee colony algorithm

1 Introduction

Practical problems of modeling and optimization often have a multilateral character and, in some parts, allow classical approaches, while in others they require universal methods. At the same time, it is not known which approach at what point and to what extent it is necessary to apply.

K. A. Ponomareva (✉) · I. P. Rozhnov
Siberian Federal University, Krasnoyarsk, Russia
e-mail: ponomareva_katerina@mail.ru

L. A. Kazakovtsev
Reshetnev Siberian State University of Science and Technology, Krasnoyarsk, Russia

D. Vlachos (ed.), *Mathematical Modeling in Physical Sciences*, Springer Proceedings in Mathematics & Statistics 446, https://doi.org/10.1007/978-3-031-52965-8_22

More and more tasks for making optimal decisions are taken over by the computer. There are two types of methods for this: mathematically sound (usually very accurate and effective, but «sharpened» for a narrow range of tasks) and heuristic (for solving a wide range of tasks, but not so accurate).

All this determines the relevance of the development of hybrid approaches that allow timely and efficient transition from some methods of solving modeling and optimization problems to others and vice versa. The most obvious approach is hybrid optimization methods combining precise and mathematically sound methods and heuristic/metaheuristic approaches to optimization and modeling problems.

One of the main reasons for getting an incorrect answer as a result of the algorithm is the possible premature convergence of the algorithm. This happens when all the individuals forming the next population are grouped in the «neighborhood» of the local extremum of the objective function, and at the same time, no efforts of the crossing, mutation and selection operators lead to the «displacement» of the fittest individuals in the «direction» of the desired optimal solution that delivers the global extremum of the objective function. An increase in the number of populations also does not give the expected effect, and a significant increase in their number will differ little from the method of a complete search of options [1].

In this paper, an attempt is made to avoid premature convergence of algorithms by using a combination of a sine–cosine algorithm and an artificial bee colony algorithm.

2 Optimization Algorithms

2.1 *Sine–Cosine Algorithm*

In the sine–cosine algorithm, the optimization process begins with randomly generated solutions. The method updates other solutions from the aggregate using Eq. (1).

$$x_{i,j}^{t+1} = \begin{cases} x_{i,j}^{t} + r_1 \sin(r_2)\left|r_3 y_j^t - x_{i,j}^t\right|, r_4 < 0.5 \\ x_{i,j}^{t} + r_1 \cos(r_2)\left|r_3 y_j^t - x_{i,j}^t\right|, r_4 \geq 0.5 \end{cases} \quad (1)$$

where $i = 1, \ldots, SP$, $j = 1, \ldots, D$, so that SP—the population size, D—the number of design variables; $x_{i,j}$—jth component of the solution x_i at iteration t; y_j—the jth component of the best achieved solution at iteration t. Specific parameters the controls, r_1, r_2, r_3 and r_4, are randomly selected.

The sine–cosine algorithm has good search capabilities, a simple implementation, a small number of control parameters and a good execution time, but it lacks the accuracy of the final solution when solving specific real optimization problems. In addition, it may suffer from premature convergence, so it must be combined with some other metaheuristic algorithm to solve specific real optimization problems [2].

2.2 Artificial Bee Colony Algorithm

The artificial bee colony algorithm is a swarm intelligence algorithm based on imitation of the behavior of a colony of bees, can be used in optimization problems. A necessary condition for its application is the presence of some topological distance or its analogue on the solution domain [3].

The main idea of the bee colony paradigm is to use a two-level search strategy. At the first level, with the help of scout bees, many promising positions (sources) are formed, at the second level, with the help of foraging bees, studies of the surroundings of these sources are carried out. The goal of a bee colony is to find a source containing the maximum amount of nectar, which corresponds to the objective function in the optimization problem. In the algorithms of the problems under consideration, each solution is represented as a point (position) in the search space. The amount of nectar found represents the value of the objective function at this point [4].

Both the «working» stage and the «observer» stage use the same search strategy to create a suitable solution vi from the current solution xi. This strategy is described by the equation

$$v_{ij} = x_{ij} + \varphi(x_{ij} - x_{kj}) \quad (2)$$

where j—a randomly selected index; $x_{i,j}$—the jth component of the current solution x_i; x_k—a randomly selected solution other than x_i; φ—a random number between (−1, 1).

3 Proposed Approach

When developing a new combined approach, a new mutation operator was proposed, based on Eq. (1) and designed to «knock out» the population from the local extremum and protect against premature convergence. The new mutation operator provides a more productive search by maintaining a balance between exploitation and exploration.

$$\vartheta_{i,j}^{t+1} = \begin{cases} x_{n1,j}^{t} + r_1 \sin(r_2) \cdot \left| y_{i,j}^{t} - x_{i,j}^{t} \right| + rand_i \cdot (y_j^t - x_{n2,j}^{t}),\ R_{i,j} < 0.5 \\ x_{n1,j}^{t} + r_1 \cos(r_2) \cdot \left| y_{i,j}^{t} - x_{i,j}^{t} \right| + rand_i \cdot (y_j^t - x_{n2,j}^{t}),\ R_{i,j} \geq 0.5 \end{cases} \quad (3)$$

where x_{n1} and x_{n2}—two neighborhood solutions randomly selected from the aggregate; y—the best solution achieved so far; $rand_i$—a random number from (0, 1); j = 1, …, D.

The proposed mutation strategy is constructed as follows:

1. Two vectors x_{n1} and x_{n2} are randomly selected from the population.
2. Between these two vectors, the base vector is also randomly selected.

3. Two difference vectors are added to the base vector.
4. A randomly chosen solution and the best solution form the third term of Eq. (3), which leads to a better perturbation.

When a new v_i-candidate solution is created using Eq. (3), the feasibility rules developed by Deb are applied to determine whether the new solution will be used in combination [5]:

1. If both solutions are feasible, the one that has the best value of the objective function is chosen.
2. Any feasible solution is given preference over any unworkable solution.
3. Between two impracticable solutions, the one with the smallest amount of constraint violations (CV) is chosen.

In the new hybrid algorithm, the artificial bee colony algorithm does not include three separate stages, i.e., only the stage of using the bee is performed.

In that way, the hybrid algorithm is a combination of a sine–cosine algorithm and two search equations for an artificial bee colony algorithm using three feasibility rules.

4 Conclusion

The performance of the proposed hybrid sine–cosine algorithm was evaluated using several complex engineering optimization problems, including a pressure vessel, a side impact of a car, a hollow shaft, etc. The results achieved using the new hybrid approach were compared with the results achieved by other optimizers when solving the same complex optimization problems.

The conducted experiment showed that the hybrid approach converges faster than the two considered metaheuristics. Also, the new algorithm showed high stability of results, effective convergence rate, satisfying various complex design constraints.

The analysis of the results also showed that the combination of the sine–cosine algorithm and the artificial bee colony algorithm effectively maintains a balance between exploitation and exploration, while not increasing the number of control parameters.

Acknowledgements The work was supported by the Ministry of Science and Higher Education of the Russian Federation (Grant No. 075-15-2022-1121).

References

1. Permyakova, T., Morozenko, V.: Combined method for solving the backpack problem. In: International Conference Knowledge-Dialogue-Solutions, p. 7 (2017)
2. Mirjalili, S.: SCA: A sine cosine algorithm for solving optimization problems. Knowl. Based Syst. 120–133 (2016)
3. Vodolazsky, I.A.: Swarm Intelligence and Its Most Common Methods of Implementation, vol. 4. Young Scientist (2017)
4. Kureychik, V.V., Zhilenkov M.A.: Bee algorithm for solving problems with an explicit objective function. Inform. Comput. Eng. Eng. Educ. **1**(21) (2019)
5. Deb, K.: An efficient constraint-handling method for genetic algorithms. Comput. Methods Appl. Mech. Eng. 311–338 (2016)

Multiple Fuzzy Soft Graphs Based on Maps and Consider Their Applications in Decision-Making

Shadia Majeed Noori, Shuker Khalil, and Abd Ghafur Ahmad

Abstract One of the most important hyper models in mathematics is the measure of fuzzy soft sets, which can handle more uncertainty associated with real-life problems. Therefore, it has a wider range of applications. The purpose of this work is to investigate and study a new model called fuzzy measure soft graphs (FMSG). Based on this model, we defined several notions like the totally regular fuzzy measure soft graph, uniform edge fuzzy measure soft graph, complement of fuzzy measure soft graph, and uniform vertex fuzzy measure soft graph. Further, using these ideas, we solved one decision-making problem for selecting the perfect parameter for universal V. In addition, several kinds of fuzzy measure topological spaces of mappings with strongly closed graphs, with a focus on $FM\mho - \Phi$ (resp. $\Omega - \Psi\Phi$) $-$ closed graphs, employing the thoughts for $FM\mho - \Phi$ (resp. $\Omega - \Psi\Phi$) $-$ open sets and $FM\mho - \Phi$ (resp. $\Omega - \Psi\Phi$) $-T_2$. Finally, some properties of $FM\mho - \Phi$ (resp. $\Omega - \Psi\Phi$) $-$ closed graphs and almost $\varphi-$ minimized spaces have also been. investigated.

Keywords Soft sets theory · Fuzzy sets · Measure of fuzzy sets · Soft graphs

S. M. Noori
Department of Mathematics, College of Education for Pure Science, University of Tikrit, Salah Al Deen, Tikrit 34001, Iraq
e-mail: aljabary88@tu.edu.iq

S. Khalil (✉)
Department of Mathematics, College of Science, University Basrah, Basrah 61004, Iraq
e-mail: shuker.khalil@uobasrah.edu.iq

S. M. Noori · A. G. Ahmad
Department of Mathematical Sciences, Faculty of Science and Technology, Universiti Kebangsaan Malaysia, UKM, 43600 Bangi, Selangor DE, Malaysia
e-mail: ghafur@ukm.edu.my

D. Vlachos (ed.), *Mathematical Modeling in Physical Sciences*, Springer Proceedings in Mathematics & Statistics 446, https://doi.org/10.1007/978-3-031-52965-8_23

1 Introduction

Decision-making (DM) in particular, with the application of multi-criteria decision-making (MCDM) approaches, provides a gateway to coping with intricate challenges relating to ordinary life circumstances. Inside, the most satisfying option is chosen from a range of options after viewing competing criteria with a high level of enjoyment. Several options were interpreted using the views of decision makers (DMs). The number of criteria that can be employed in MCDM is not limited; the crucial issue is how the various alternatives are integrated and evaluated. Regardless, numerous scholars have made substantial progress in this field while employing MCDM. Decision-makers use evaluated information in crisp numbers in (DM) procedures. Since the complexities of real life, all data cannot be represented by precise numbers. To improve (DM) methodologies for coping with fuzziness and uncertainty in evaluated information, Zadeh [1] developed the fuzzy set (FS) theory. This development, utilized to cope with a variety of situations and MCDM problems, is an exciting topic these days by numerous researchers [2–7].

For many qualities, the value of attribute weights in the (DM) process cannot be emphasized [8]. One of the most critical issues in the multi-attribute (DM) process is determining how to calculate attribute weights [9]. Ding et al. [10] developed a multi-objective approximation system based on coefficients of closeness to collect attribute weights in a probabilistic, uncertain, and fuzzy situation. There is already some material evaluating attribute weights using the Pythagorean fuzzy (PF) structure. Such approaches, however, do not fully exploit data in (DM). The methods for obtaining attribute weights are currently insufficient. To supplement existing methodologies, it is necessary to offer some unique methods for determining attribute weights in (PF) multi-attribute decision-making problems. We employed a new mechanism for determining attribute weights under PF circumstances, inspired by Chen and Li's [11] entropy weight method.

In additionally, a novel mathematical concept called the soft set [12] theory was developed by Molodtsov (1999) for solving uncertainties. Many researchers have further applied soft sets in different domains. For example, Maji et al. [13] discussed fuzzy soft sets and their application in MCDM. Al-Sharqi et al. [14] employed a modern algorithm for solving MCDM based on the extension of fuzzy soft sets. Al-Masarwa and Abu Qamar [15] introduced some new notions of fuzzy soft graphs like complement and $\mu-$ complement fuzzy soft graphs. Chellamani et al. [16] proposed an approach to decision-making by fuzzy soft graphs.

In other side, the significance of open sets in topologies and their applications is now generally acknowledged. These are the areas where many topologists have focused their study, yielding a slew of significant and useful results. Many researchers discussed topological spaces in soft sets [17, 18], permutation sets [19–21], nano sets [22], and nuetrosophich sets [23–25]. In this regard, the various types of coherence that have been modified, and so on. Over time, many hypotheses about the concept of topological spaces have been considered and broadly examined. Ekici [26] and Hatir and Noiri [27] among the majority notable ideas the thoughts of $e-$ open sets and $\delta - \beta_\theta-$ open sets were presented [28, 29] separately as a speculation of shut sets, $e-$ shut sets, $\delta - \beta_\theta-$ shut sets, and the connected sets were presented and contemplate. Alaa et al. [30] using the idea of $e-$ (resp. $\delta - \beta_\theta)-$ open sets and e $-$ (resp. $\delta - \beta_\theta-$) closed sets by conclusion and individually of this set to provide the ideas of $e - \theta$(resp.$\delta - \beta_\theta)-$ open sets using the notion of $\mho - \Phi$ (resp. $\Omega - \Psi\Phi$) $-$ closed sets and $\mho - \Phi$ (resp. $\Omega - \Psi\Phi$) $-$ closed sets administrator in (TS).

The goal of this work is to investigate and study fuzzy measure graph and fuzzy measure soft graph. Some properties of them have also been investigated. The totally regular fuzzy measure soft graph, uniform edge fuzzy measure soft graph, and uniform vertex fuzzy soft graph are investigated and their application are explained in decision making using examples, for example the decision making to select perfect parameter for set V under given fuzzy measure soft set using the characters of uniform vertex fuzzy measure soft graph ($UVFMSG$), uniform edge fuzzy measure soft graph ($UEFMSG$), and complement of fuzzy measure soft graph ($FMSG$) are investigated. We found and discussed another class of mappings with strongly closed charts in this study. These mappings are acquired by acquiring some properties and investigating properties. We research and study different types of mappings with strongly closed graphs in fuzzy measure topological spaces, with a focus on $FM\mho - \Phi$ (resp. $\Omega - \Psi\Phi$) $-$ closed graphs, using the ideas for $FM\mho - \Phi$ (resp. $\Omega - \Psi\Phi$) $-$ open sets and $FM\mho - \Phi$ (resp. $\Omega - \Psi\Phi$) $-T_2$. Also, some characteristics of $FM\mho - \Phi$ (resp. $\Omega - \Psi\Phi$) $-$ closed graphs and almost $\varphi-$ minimized spaces were also studied.

To do this, the remainder of this paper has been structured as follows. The first section introduces some fundamental topics. Section 2 introduced innovative measure of fuzzy soft set, and Sect. 3 introduced novel aggregation based on these notions. The recommended applications were discussed in Sect. 4. Finally, in Sect. 5, final thoughts are discussed.

2 Preliminaries

In this section, we will recall some definitions of some concepts that we will depend on in our work.

Definition 2.1 [27] Assume that (X, τ) is a topological space (TS) and $A \subseteq X$. We say A is δ-open set ($\delta - OZ$), if there is a subset S of X satisfies $x \in S \subset A, \forall x \in A$, where S is a regular open set (ROS).

Definition 2.2 [30] The union of all regular open sets (ROSs) which are contained in A is termed $\delta-$ interior of A and is referred by $\text{Int}_\delta(A)$ and A is ($\delta - OZ$), if $A = \text{Int}_\delta(A)$, also its complement $A^c = B$ is supposed to be $\delta-$ closed set ($\delta - CZ$) and $BCl_\delta(B)$.

Remark 2.3 [27] A point $x \in X$ is known as a $\delta-$ group points of A, if $A\cap \text{Int}(Cl(S)) \neq \varnothing$ for any open set S with $x \in S$. The arrangement of all $\delta-$ group points of A is known as the ($\delta - CZ$) of A and is meant by $Cl_\delta(A)$.

Definition 2.4 Liu et al. [9] Let (X, τ) be a (TS) and $A \subseteq X$. We say A is $e-$ open set $(e - OZ)$ (resp., $\delta - \beta_\theta-$ open set $(\delta - \beta_\theta - OZ)$) if $A \subset Cl(Int_\delta(A)) \cup Int(Cl_\delta(A))$(resp., $A \subset Cl(Int(\delta - Cl(A)))$, and A^c is termed e-closed $(e - CZ)$ (resp., $\delta - \beta_\theta-$ closed $(\delta - \beta_\theta - CZ)$). The intersection of all e$-$closed sets $(e - CZz)$ (resp., $\delta - \beta_\theta-$ open sets $(\delta - \beta_\theta - CZz)$) of X containing A is termed the e$-$closure (resp., $\delta - \beta_\theta-$ closure) of A and is referred by $e - CL(A)$(resp., $\delta - \beta_\theta - Cl(A)$). Moreover, the union of all $(e - OZz)$ (resp., $(\delta - \beta_\theta - OZz)$ of X contained in A is termed the e $-$ interior (resp., $\delta - \beta_\theta-$ interior) of A and is referred by $e - int(A)$(resp., $\delta - \beta_\theta - int(A)$).

Definition 2.5 [26] Let (X, τ) be a (TS) and $A \subseteq X$. We say A is e-regular set $(e - RZ)$ (resp., $\delta - \beta_\theta$-regular set $(\delta - \beta_\theta - RZ)$) if it is $(e - OZ)$ and $(e - CZ)$ (resp., $(\delta - \beta_\theta - OZ)$ and $(\delta - \beta_\theta - CZ)$).

Remark 2.6 [9] The family of all $(e - OZz)$ (resp., $(e - CZz)$, $(e - RZz)$, $(\delta - \beta - OZz)$, $(\delta - \beta_\theta - CZz)$, $(\delta - \beta_\theta - RZz)$) of X with $x \in X$ is referred by $E\Sigma(X, x)$ (resp., $EC(X, x)$, $ER(X, x)$, $\delta - \beta\Sigma(X, x)$, $\delta - \beta_\theta C(X, x)$, $\delta - \beta_\theta R(X, x)$). Also, they are referred in a space X by $E\Sigma(X, T)$(resp., $EC(X, T)$, $ER(X, T)$, $\delta - \beta\Sigma(X, T)$, $\delta - \beta_\theta C(X, T)$, $\delta - \beta_\theta R(X, T)$).

Definition 2.7 Al-Jumaili et al. [31] Let $x \in X$, we say x is an $e - \backslash\theta$-closure (resp. $\delta - \beta_\theta$-closure) point of A if $eCl(K) \cap A \neq \varnothing$(resp.$\delta - \beta_\theta - Cl(K) \cap A \neq \varnothing$), for every $K \in E\sum(X, x)$(resp.$K \in \delta - \beta\sum(X, x)$). The set of all e $-\theta$-closure (resp. $\delta - \beta_\theta-$closure [21]) of A is termed e-\\$\theta$-closure (resp. $\delta - \beta_\theta$-closure) of A and is referred by $e - Cl_0(A)$ (resp. $\delta - \beta_\theta - Cl_0(A)$). A subset A is termed $e - \theta-$ closed set $(e - \theta - CZ)$ (resp. $\delta - \beta_\theta-$ closed set $(e - \beta_\theta - CZ)$]) if $A = e - Cl_0(A)$ (resp. $A = \delta - \beta_\theta - Cl_0(A)$). The complement of an $(e - \theta - CZ)$ (resp. $(e - \beta_\theta - CZ)$) set is termed $e - \theta-$ open set $(e - \theta - OZ)$(resp. $\delta - \beta_\theta - OZ$)).

Definition 2.8 Caldas et al. [32] Let $f : (X, \tau) \to (Y, \tau^*)$ be any mapping. We say $G(f) = \{(x, f(x)) : x \in X\}$ is graph of f, where $G(f)$ is a subset of the space $(X \times Y, \tau \times \tau^*)$.

Definition 2.9 Molodtsov [12] Let U be a discourse universe and A be a parameter set. Let a power set of U be P(U). A soft set over U is called an ordered pair (F, A), where F is a mapping given by $F : A \to P(U)$.

3 Certain Types of Fuzzy Soft Graphs

First, we give the following definitions and some theorems related our work:

Definition 3.1 Assume that $V \neq \varnothing$ is a finite set and $\sigma: V \to [0; 1]$ and $\mu: V \times V \to [0; 1]$ satisfies: $\mu(x, y) \leq 1 - (\sigma(x) \wedge \sigma(y)) \forall (x, y) \in V \times V$. We say $G = (\sigma, \mu)$ is a fuzz y graph over the set V. Also, σ and μ are respectively said to be fuzzy vertex and fuzzy edge of the fuzzy graph $G = (\sigma, \mu)$.

Definition 3.2 We say $G = (M, \sigma, u)$ is a fuzzy measure graph (FMG), if it satisfy the following conditions: $u : V \times V \to [0, 1]$ such that $u(x, y) \leq \sigma(x) \wedge \sigma(y) \forall (x, y) \in V \times V$, where V is a nonempty finite set.

Definition 3.3 We say $(F, M, G, *)$ is a fuzzy measure topological group if it such that:

(1) F is satisfy conditions of fuzzy.
(2) M is satisfy condition of measure.
(3) G is satisfy condition of a group, $(G, *)$ is a group.
(4) G is satisfy condition of a (TS).
(5) The mappings $(x, y) \to x * y and x \to x^{-1}$ are continous for all $x, y \in G$.

Definition 3.4 If $(F, M, G, *)$ is a fuzzy measure topological group, $N \Delta\ G$, the quotient fuzzy measure topological group G by $N is \frac{G}{N} such that \frac{G}{N}$ such that $\frac{G}{N}$ is $\frac{G}{N}$ such that $\frac{G}{N}$ is a fuzzy measure group.

Definition 3.5 The underlying crisp graph of a (FMG) $G = (m, \sigma, \mu)$ is denoted by $G = (m^*, \sigma^*, \mu^*)$, where $\sigma^* = \{\mu \in V : \sigma(u) > 0\}$ and $\mu^* = \{(u, v) \in V \times V : \mu(u, v) > 0\}$.

Definition 3.6 A fuzzy measure soft graph $(FMSG)$ $\tilde{G} = (M, G^*, \tilde{F}, \tilde{K}, A)$ is a $5-$tuple if satisfies the following:

1. $G^* = (M, V, E)$ is a simpl e graph,
2. $A \neq \varnothing$ is a set of parameters,
3. $(M, \tilde{F}, A)$ is a fuzzy measure soft set $(FMSS)$ over V,
4. $(M, \tilde{K}, A)$ is a $(FMSS)$ over E,
5. $(\tilde{F}(a), \tilde{K}(a))$ is a fuzzy measure (sub)graph of G^*, $\forall a \in A$. That is,

$$\tilde{K}(a)(x, y) \leq 1 - min\{\tilde{F}(a)(x), \tilde{F}(a)(y)\}$$

Definition 3.7 Assume that $G^* = (M, V, E)$ is a crisp measure graph (CMG) and $\tilde{G}$ be a $(FMSG)$ of G. We say $\tilde{G}$ is a uniform vertex fuzzy soft graph $(UVFSG)$ if $\tilde{H}(e)$ is a uniform vertex fuzzy measure graph $(UVFMG)$, $\forall e \in A$.

Example 3.8 Let $G^* = (M, V, E)$ be a (CMG) with $V = \{x_1, x_2, x_3\}$ and $E = \{x_1x_2, x_2x_3, x_3x_1\}$. Let $A = \{e_1, e_2, e_3\}$ be a parameter set and let $(M, \tilde{F}, A)$ be a $(FMSS)$ ove r V with its approximate function $\tilde{F} : A \to P(V)$ defined by:

$$\tilde{F}(e_1) = \{x_1|0.8, x_2|0.8, x_3|0.8\},$$

$$\tilde{F}(e_2) = \{x_1|0.5, x_2|0.5, x_3|0.5\},$$

$$\tilde{F}(e_3) = \{x_1|0.2, x_2|0.2, x_3|0.2\},$$

Assume that $(M, \tilde{K}, A)$ is a $(FMSS)$ over E with its approximate function $\tilde{K}$: $A \to P(E)$ defined as

$$\tilde{K}(e_1) = \{x_1x_2|0.1, x_2x_3|0.4, x_3x_1|0.5\},$$

$$\tilde{K}(e_2) = \{x_1x_2|0.3, x_2x_3|0.4, x_3x_1|0.3\},$$

$$\tilde{K}(e_3) = \{x_1x_2|0.4, x_2x_3|0.5, x_3x_4|0.3\},$$

Regular calculations make it simple to see that fuzzy measure graphs $(FMGs)$.

$$\tilde{H}(e_1) = (\tilde{F}(e_1), \tilde{F}(e_1))$$

$$\tilde{H}(e_2) = (\tilde{F}(e_2), \tilde{F}(e_2))$$

$$\tilde{H}(e_3) = (\tilde{F}(e_3), \tilde{F}(e_3))$$

$(UVFMGs)$. Hence, $\tilde{G} = (\tilde{H}(e_1), \tilde{H}(e_2), \tilde{H}(e_3))$ is a uniform vertex fuzzy measure soft graph $(UVFMSG)$ of G^* (Fig. 1).

Hence, the decision making in this example to select perfect parameter for set V under given $(FMSS)$ is found by using the character of $(UVFMSG)$.

Definition 3.9 Let $G^* = (M, V, E)$ be a (CMG) and $\tilde{G}$ be a $(FMSG)$ of G.We say $\tilde{G}$ is a uniform edge fuzzy measure soft graph $(UEFMSG)$ if $\tilde{H}(e)$ is a uniform edge fuzzy measure graph, $\forall e \in A$.

Example 3.10 Consider a (CMG) $G^* = (M, V, E)$ such that $V = \{x_1, x_2, x_3\}$ and $E = \{x_1x_2, x_2x_3, x_3x_1\}$. Let $A = \{e_1, e_2, e_3\}$ be a parameter set and let $(M, \tilde{F}, A)$ be a $(FMSS)$ over V with its approximate function $\tilde{F} : A \to P(V)$ defined as:

$$\tilde{F}(e_1) = \{x_1|0.2, x_2|0.5, x_3|0.6\},$$

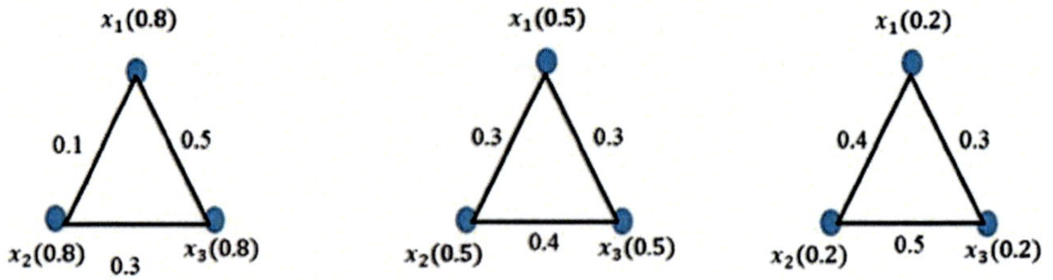

Fig. 1 Uniform vertex fuzzy soft graph

$$\tilde{F}(e_2) = \{x_1|0.3, x_2|0.5, x_3|0.7\},$$

$$\tilde{F}(e_3) = \{x_1|0.4, x_2|0.7, x_3|0.9\},$$

Let $(\tilde{K}, A)$ be a $(FMSS)$ over E with its approximate function $\tilde{K} : A \to P(E)$ defined as

$$\tilde{K}(e_1) = \{x_1x_2|0.4, x_2x_3|0.4, x_3x_4|0.4, x_4x_1|0.4\},$$

$$\tilde{K}(e_2) = \{x_1x_2|0.5, x_2x_3|0.5, x_3x_4|0.5, x_4x_1|0.5\},$$

$$\tilde{K}(e_3) = \{x_1x_2|0.6, x_2x_3|0.6, x_3x_4|0.6, x_4x_1|0.6\},$$

Regular calculations make it simple to see that (FMG).

$$\tilde{H}(e_1) = (\tilde{F}(e_1), \tilde{F}(e_1))$$

$$\tilde{H}(e_2) = (\tilde{F}(e_2), \tilde{F}(e_2))$$

$$\tilde{H}(e_3) = (\tilde{F}(e_3), \tilde{F}(e_3))$$

are $(UEFMGs)$.

Then, $\tilde{G} = (\tilde{H}(e_1), \tilde{H}(e_2), \tilde{H}(e_3))$ is a $(UEFMSG)$ of G^* (Fig. 2).

Hence, the decision making in this example to select perfect parameter for set V under given $(FMSS)$ is found by using the character of $(UEFMSG)$.

Definition 3.11 Assume that $\tilde{G}$ is a $(FMSG)$ of G^*. The degree of a vertex u is

$$deg_{\tilde{G}}(u) = \sum_{e_i \in A}\left(\sum_{v \in V, v \neq u} \tilde{K}(e_i)(u, v)\right).$$

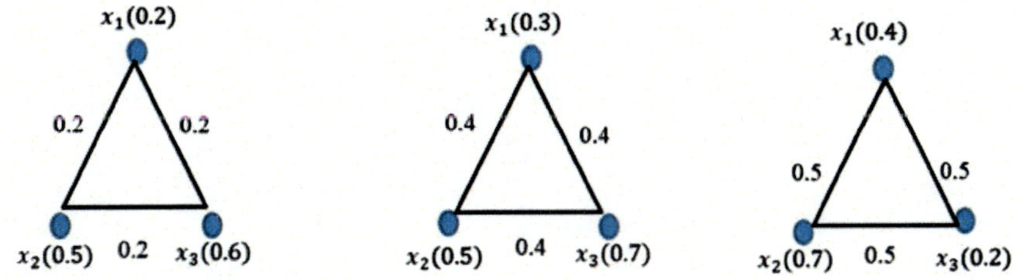

Fig. 2 Uniform edge fuzzy soft graph

Definition 3.12 Assume that $\tilde{G}$ is a $(FMSG)$ of G^*. The total degree of a vertex u is

$$deg_{\tilde{G}}(u) = \sum_{e_i \in A}\left(\sum_{v \in V, v \neq u} \tilde{K}(e_i)(u, v)\right) + \sum_{e_i \in A} \tilde{F}(e_i)(u).$$

$$deg_{\tilde{G}}(u) = deg_{\tilde{G}}(u) + \sum_{e_i \in A} \tilde{F}(e_i)(u).$$

Definition 3.13 Assume that $\tilde{G}$ is a $(FMSG)$ of G^*.The complement of $\tilde{G}$ is a $(FMSG)$ $\overline{\tilde{G}} = (M, \tilde{G}, \overline{\tilde{F}}, \overline{\tilde{K}}, A)$ where $\overline{\tilde{F}}(e_i)(u) = \tilde{F}(e_i)(u)$ and $\overline{\tilde{K}}(e_i)(u, v) = min\{\tilde{F}(e_i)(u), \tilde{F}(e_i)(v)\} - \overline{\tilde{K}}(e_i)(u, v) \forall u, v \in V, \forall e_i \in A$, for $i = 1, 2, 3, \ldots n$.

Example 3.14 Consider a (CMG) $G^* = (M, V, E)$ such that $V = \{x_1, x_2, x_3\}$ and $E = \{x_1x_2, x_2x_3, x_3x_1\}$. Let $A = \{e_1, e_2\}$ be a parameter set and let $(M, \tilde{F}, A)$ be a $(FMSS)$ over V with its approximate function $\tilde{F} : A \to P(V)$ defined as

$$\tilde{F}(e_1) = \{x_1|0.4, x_2\,|0.7, x_3|0.5\,\}, \tilde{F}(e_2) = \{x_1|0.5, x_2\,|0.3, x_3|0.7\,\}$$

Let $(M, \tilde{K}, A)$ be a $(FMSS)$ over E with its approximate function $\tilde{K} : A \to P(E)$:

$$\tilde{K}(e_1) = \{x_1x_2|0.3, x_2x_3|0.2, x_3x_1|0.1\},$$

$$\tilde{K}(e_2) = \{x_1x_2|0.4, x_2x_3\,|0.5, x_3x_1|0.6\},$$

It easy to see that $(FMGs)$

$$\tilde{H}(e_1) = \left(\tilde{F}(e_1), \tilde{F}(e_1)\right),$$

$$\tilde{H}(e_2) = (\tilde{F}(e_2), \tilde{F}(e_2))$$

Are $(FMGs)$. Hence, $\tilde{G} = (\tilde{H}(e_1), \tilde{H}(e_2))$ is a $(FMSG)$ (Fig. 3).
By means of standard calculations, a vertex's degree, x_1, is

$$deg_{\tilde{G}}(x_1) = \sum_{e_i \in A}\left(\sum_{v \in V, x_1 \neq v} \tilde{K}(e_i)(x_1, v)\right)$$

$$= \sum_{e_i \in A} \tilde{K}(e_i)(x_1, x_2) + \tilde{K}(e_i)(x_1, x_3)$$

$$= (0.3 + 0.1) + (0.4 + 0.6)$$

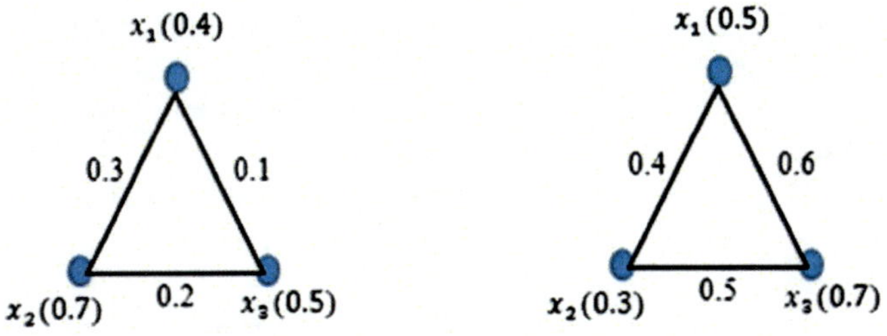

Fig. 3 Fuzzy measure soft graph

$$= 0.4 + 1.10 = 1.$$

The total degree of a vertex x_1 is

$$tdeg_{\tilde{G}}(x_1) = deg_{\tilde{G}}(x_1) + \sum_{e_i \in A} \tilde{F}(e_i)(x_1)$$

$$= 1 + (0.4 + 0.6) = 1 + 1 = 2.$$

Also, the complement of a $(FMSG)$ $\tilde{G}$ is shown in Fig. 4.

Hence, the decision making in this example to select perfect parameter for set V under given $(FMSS)$ is found by using the character of complement of $(FMSG)$.

Theorem 3.15 *Assume that* $G^* = (M, V, E)$ *is a simple measure graph* (SMG). *If* $\tilde{G}$ *is a regular fuzzy measure soft graph* $(RFMSG)$ *and* $(UVFMSG)$ *on* G, *then the complement of* $(FMSG)$ $\tilde{G}$ *is also regular*.

Proof Suppose that V is a finite set of vertices. Since V is finite, we have $\tilde{F}^* = \{x \in V : \tilde{F}(e_i)(x) > 0 for some e_i \in A\}$ is also finite. Let $|\tilde{F}^*| = n$. Given that

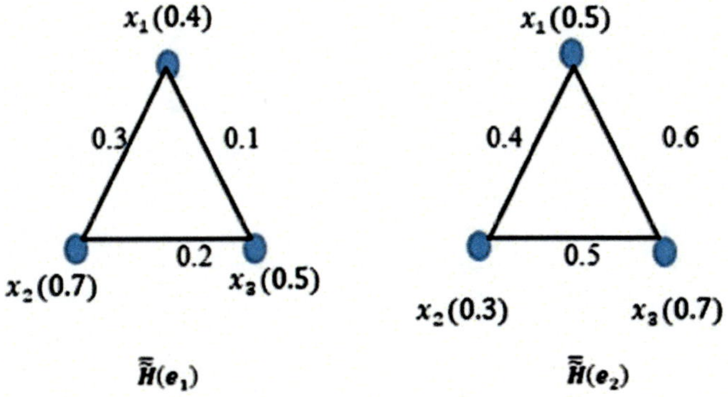

Fig. 4 Complement fuzzy measure soft graph

$\tilde{G}$ is a $(RFMSG)$. Then $deg_{\tilde{G}}(a) = r_i$ in fuzzy graphs $\tilde{H}(e_i)\forall a \in \tilde{F}^*$ for some real number $0 < r_i$ and $\forall e_i \in A$ for $i = 1, 2, 3, \ldots n$. Also, given that $\tilde{G}$ is a $(UVFMSG)$. Then $\tilde{F}(e_i)(x) = k_i$ in fuzzy measure graphs $\tilde{H}(e_i)\forall a \in \tilde{F}^*$ for some real number k_i where $0 < k_i \leq 1$ and $\forall e_i \in A$ for $i = 1, 2, 3, \ldots n$.

Let $x \in \tilde{F}^*$. $deg_{\overline{\tilde{G}}}(x) = \sum_{x \neq y} \overline{\tilde{K}}(e_i)(u, y)$ in $(FMGs)$ $\overline{\tilde{H}}(e_i)\forall e_i \in A$ for $i = 1, 2, 3, \ldots n$. Now,

$$deg_{\overline{\tilde{G}}}(x) = \sum_{x \neq y}(min\tilde{F}(e_i)(x), \tilde{F}(e_i)(y) - \tilde{K}(e_i)(x, y))$$

$$= \sum_{x \neq y} min\{\tilde{F}(e_i)(x), \tilde{F}(e_i)(y)\} - \sum_{x \neq y} \tilde{K}(e_i)(x, y)$$

$$= \sum_{x \neq y} k_i - deg_G(x) = k_i(n-1) - r_i$$

In $(FMGs)$ $\tilde{H}(e_i)\forall x \in \tilde{F}^*$ and $\forall e_i \in A$ for $i = 1, 2, 3, \ldots n$. Thus, $deg_{\overline{\tilde{G}}}(x) = k_i(n-1) - r_i, \forall x \in \tilde{F}^*$ and $\forall e_i \in A$ for $i = 1, 2, 3, \ldots n$. Therefore, $\tilde{G}$ is also$(RFMSG)$.

Theorem 3.16 *Let $G^* = (M, V, E)$ be a (SMG). If $\tilde{G}$ is a totally regular fuzzy measure soft graph $(TRFMSG)$ and $(UVFMSG)$ on G, then the complement of $(FMSG)$ $\tilde{G}$ is also totally regular.*

Theorem 3.17 *Assume that $G^* = (M, V, E)$ is a (SMG). If $\tilde{G}$ is both $(UEFMSG)$ and $(UVFMSG)$ on G, then $\tilde{G}$ is a $(TRFMSG)$.*

Proof Let V be a finite fuzzy measure set of vertices. Since V is finite, we have $\tilde{F}^* = \{x \in V : \tilde{F}(e_i)(x) > 0 forsome e_i \in A\}$ is also finite. Let $|\tilde{F}^*| = n$.

Given that $\tilde{G}$ is a $(UEFMSG)$. Then $\tilde{K}(e_i)(x, y) = k_i$ in $(FMGs)$ $\tilde{H}(e_i)\forall(x, y) \in \tilde{F}^* \times \tilde{F}^*$ with $x \neq y$ for some real number k_i where $0 < k_i \leq 1$ and $\forall e_i \in A$ for $i = 1, 2, 3, \ldots n$.

Also, given that $\tilde{G}$ is a $(UVFMSG)$. Then $\tilde{F}(e_i)(x) = s_i$ in $(FMGs)$ $\tilde{H}(e_i)\forall x \in \tilde{F}^*$ for some real number s_i where $0 < s_i \leq 1$ and $\forall e_i \in A$ for $i = 1, 2, 3, \ldots n$.

Let $u \in \tilde{F}^*$. Now, $dge_{\tilde{G}}(u) = dge_{\tilde{G}}(u) + \tilde{F}(e_i)(u) = \sum_{u \neq v} \tilde{K}(u, v) + \tilde{F}(e_i)(u) = \sum_{x \neq y} k_i + s_i$ in $(FMGs)$ $\tilde{H}(e_i)\forall u \in \tilde{F}^*$ and $\forall e_i \in A$ for $i = 1, 2, 3, \ldots n$. Thus, $dge_{\tilde{G}}(u) = (n+1)k_i + s_i \forall u \in \tilde{F}^*$ and $\forall e_i \in A$ for $i = 1, 2, 3, \ldots n$. Therefore, $\tilde{G}$ is $(RFMSG)$.

Theorem 3.18 *Let $G^* = (V, E)$ be a (SMG). If $\tilde{G}$ is a complete of $(FMSG)$ and $(UVFMSG)$ on G, then $\tilde{G}$ is $(RFMSG)$.*

Proof Let V be a finite fuzzy measure set of vertices. Since V is finite, we have $\tilde{F}^* = \{x \in V : \tilde{F}(e_i)(x) > 0 \text{ for some } e_i \in A\}$ is also finite. Let $|\tilde{F}^*| = n$.

Given that $\tilde{G}$ is a complete of $(FMSG)$. Then $\tilde{K}(e_i)(u, v) = min\{\tilde{F}(e_i)(u), \tilde{F}(e_i)(v)\}$ in $(FMGs)$ $\tilde{H}(e_i)\forall u, v \in \tilde{F}^*$ and $\forall e_i \in A$ for $i=1, 2, 3, \ldots n$.

Also, given that $\tilde{G}$ is a $(UVFMSG)$. Then $\tilde{F}(e_i)(x) = k_i$ in $(FMGs)\tilde{H}(e_i)\forall x \in \tilde{F}^*$ for some real number k_i where $0 < k_i \leq 1$ and $\forall e_i \in A$ for $i = 1, 2, 3, \ldots n$.

Let $u \in \tilde{F}^*$. Now.

$$dge_{\tilde{G}}(u) = \sum_{u \neq v} \tilde{K}(e_i)(u, v)$$

$$= \sum_{u \neq v} min\{\tilde{F}(e_i)(u), \tilde{F}(e_i)(v)\}$$

$$= \sum_{x \neq y} k_i = (n-1)k_i$$

in $(FMGs)$ $\tilde{H}(e_i)\forall u \in \tilde{F}^*$ and $\forall e_i \in A$ for $i = 1, 2, 3, \ldots n$. Thus, $tdeg_{\overline{\tilde{G}}}(u) = (n+1)k_i - s_i, \forall u \in \tilde{F}^*$ and $\forall e_i \in A$ for $i = 1, 2, 3, \ldots n$. Therefore, $\tilde{G}$ is also $(TRFMSG)$.

Corollary 3.19 *Let $G^* = (M, V, E)$ be a (SMG). If $\tilde{G}$ is a complete of $(FMSG)$ and $(UVFMSG)$ on G^*, then $\tilde{G}$ is $(RFMSG)$.*

Proof Let V be a finite fuzzy measure set of vertices. Since V is finite, we have $\tilde{F}^* = \{x \in V : \tilde{F}(e_i)(x) > 0 forsomee_i \in A\}$ is also finite. Let $|\tilde{F}^*| = n$. Given that $\tilde{G}$ is a complete of $(FMSG)$. Then $\tilde{K}(e_i)(u, v) = min\{\tilde{F}(e_i)(u), \tilde{F}(e_i)(v)\}$ in $(FMGs)$ $\tilde{H}(e_i)\forall u, v \in \tilde{F}^*$ and $\forall e_i \in A$ for $i = 1, 2, 3, \ldots n$. Also, given that $\tilde{G}$ is a $(UVFMSG)$. Then $\tilde{F}(e_i)(x) = k_i$ in $(FMGs)$ $\tilde{H}(e_i)\forall x \in \tilde{F}^*$ for some real number k_i where $0 < k_i \leq 1$ and $\forall e_i \in A$ for $i = 1, 2, 3, \ldots n$. Let $u \in \tilde{F}^*$. $dge_{\tilde{G}}(u) = dge_{\tilde{G}}(u) + \tilde{F}(e_i)(u) = \sum_{u \neq v} \tilde{K}(e_i)(u, v) + \tilde{F}(e_i)(u)$ in $(FMGs)$ $\tilde{H}(e_i)\forall u \in \tilde{F}^*$ and $\forall e_i \in A$ for $i = 1, 2, 3, \ldots n$.

$$dge_{\tilde{G}}(u) = \sum_{u \neq v} min\{\tilde{F}(e_i)(u), \tilde{F}(e_i)(v)\} + \tilde{F}(e_i)(u)$$

$$= \sum_{u \neq v} k_i + k_i$$

$$= (n-1)k_i + k_i$$

$$= nk_{i.}$$

In $(FMGs)$ $\tilde{H}(e_i)\forall u \in \tilde{F}^*$ and $\forall e_i \in A$ for $i = 1, 2, 3, \ldots n$. Thus, $dge_{\tilde{G}}(u) = nk_i\forall u \in \tilde{F}^*$ for $i = 1, 2, 3, \ldots n$. Therefore, $\tilde{G}$ is also $(TRFMSG)$.

Corollary 3.20 *Let $G^* = (M, V, E)$ be a (SMG). If $\tilde{G}$ is a $(RFMSG)$ and $(UVFMSG)$ on G^*, then $\tilde{G}$ is $(TRFMSG)$.*

Proof As same proof of Corollary 3.19.

4 Properties of e $-\theta$ and $\delta - \beta_\theta - \theta$ Open Sets

The other novel kinds of mappings in measure topological spaces with strongly closed graphs via $e - \theta$ and $\delta - \beta_\theta - \theta$ open sets

Definition 4.1 A fuzzy measure topological space $(FMTS)(F, M, T)$ is called:

(a) $FMe - \theta - T_2$(resp. $FM\delta - \beta_\theta - T_2$) if $\forall x \neq y \in F$, there exist two $FMe - \theta -$open (resp. $\delta - \beta_\theta -$open) sets U and V with $x \in U, y \in V$ and $U \cap V = \emptyset$.
(b) $FMe - T_2$ $(resp.FM\delta - \beta_\theta - T_2)$ if $\forall x \neq y \in F$, there exist $U \in FME\sum(F, x)$ (resp. $U \in FM\delta - \beta\sum(F, x)$) and $V \in FME\sum(F, x)$) (resp. $V \in FM\delta - \beta\sum(F, x)$) with $U \cap V = \emptyset$.

Theorem 4.2 *The following terms are equivalent for a* $(FMTS)$ (F, M, T) :

(a) *If* $x \neq y \in F$, *then there exist* $U \in FME\theta\sum(F, x)$ *(resp.* $U \in FM\delta - \beta - \theta\sum(F, x)$) *and* $V \in FME\theta\sum(F, y)$*(resp.* $V \in FM\delta - \beta - \theta\sum(F, y)$) *with* $FMe - Cl_\theta(U)$ $(resp. FM\delta - \beta - Cl\theta(U)) \cap FMe - Cl_\theta(V)$ $(resp. FM\delta - \beta - Cl_\theta(V)) = \emptyset$.
(b) (F, M, T) *is* $FMe - \theta - T_2(resp. FM\delta - \beta_\theta - T_2)$;
(c) (F, M, T) *is* $FMe - T_2(resp. FM\delta - \beta - T_2)$;
(d) *If* $x \neq y \in F$, *then there exist* $U, V \in FME\sum(X, T)$ *(resp.* $U, V \in FM\delta - \beta\sum(X, T)$) *with* $x \in U, y \in V$ *and* $FMe - Cl(U)$ *(resp.* $FM\delta - \beta - Cl(U)) \cap FMe - Cl(V)(resp. FM\delta - \beta - Cl(V)) = \emptyset$;
(e) *If* $x \neq y \in F$, *then there exist* $U, V \in FMER(F, M, T)(resp. U, V \in FM\delta - \beta R(F, M, T))$ *such that* $x \in U, y \in V$ *and* $U \cap V = \emptyset$.

Proof $(a) \Rightarrow (b)$. This apparent proof was left away.

$(b) \Rightarrow (c)$. Since $FME\theta\sum(F, M, T)$(resp.$FM\delta - \beta - \theta\sum(F, M, T) \subset FME\sum(F, T)$
(resp. $FM\delta - \beta\sum(F, M, T)$, Thus, this evidence is clear.

$(c) \Rightarrow (d)$. Is obvious respectively.

$(d) \Rightarrow (e)$. for every $U \in FME\sum(F, M, T)$(resp.$U \in FM\delta - \beta\sum(F, M, T)$) we have $FMe - Cl(U)$ (resp.$FM\delta - \beta - Cl(U)) \in FMER(F, M, T)$(resp.$FM\delta - \beta R(F, M, T)$) therefore the proof follows immediately.

$(e) \Rightarrow (a)$. Since any FMe-regular (resp.$FM\delta - \beta -$ regular) set is a $FMe - \theta$(resp.$FM\delta - \beta_\theta$)$-$ open set and $FMe - \theta$(resp.$FM\delta - \beta_\theta$) $-$ closed set. Thus, this evidence is clear.

Definition 4.3 A mapping $f : (F_1, M_1, T_1) \rightarrow (F_2, M_2, T_2)$ is said to be:

(a) $FMPre - e - \theta -$open if $f(A) \in FME\theta\sum(F_2, M_2, T_2)$ for all $A \in FME\theta\Sigma$ (F_1, M_1, T_1).
(b) $FMPre - \delta - \beta_\theta -$ open if $f(A) \in FM\delta - \beta - \theta\Sigma(F_2, M_2, T_2)$ for all $A \in FM\delta - \beta - \theta\sum(F_1, M_1, T_1)$.

Remarks 4.4

(1) If a mapping $f:(F_1, M_1, T_1) \to (F_2, M_2, T_2)$ is a bijective $FMpree - \theta -$ (resp.FMpre$\delta - \beta\theta$)−open. Then $f(B) \in FME\theta C(F_2, M_2, T_2)$ (resp.$f(B) \in FM\delta - \beta - \theta C(F_2, M_2, T_2)$) for any$B \in FME\theta C(F_1, M_1, T_1)$(resp.$B \in FM\delta - \beta - \theta C(F_1, M_1, T_1)$).
(2) The following theorem shows that $FMe - \theta - T_2 -$ (resp.$FM\delta - \beta_\theta - T_2$) spaces remain invariant under bijective $FMpre\ e - \theta -$ (resp.$FMpre\delta - \beta_\theta$)− open mappings.

Theorem 4.5 *Let $f : (F_1, M_1, T_1) \to (F_2, M_2, T_2)$ be a bijective $FMpree - \theta -$ (resp. $FMpre\delta - \beta_\theta$)-open mapping and F_1 be a $FMe - \theta - T_2$(resp. $FM\delta - \beta_\theta - T_2$) then F_2 be a $FMe - \theta - T_2$ (resp. $FM\delta - \beta_\theta - T_2$).*

Proof Suppose that $y_1 \neq y_2 \in F_2$. Thus, we get $f^{-1}(y_1), f^{-1}(y_2) \in F_1$ and $f^{-1}(y_1) \neq f^{-1}(y_2)$ [since f is bijective]. Also, since F_1 is a $FMe - \theta - T_2$(resp.$FM\delta - \beta_\theta - T_2$), there exists a sets $U \in FME\theta \sum(F_1, M_1, f^{-1}(y_1))$(resp.$U \in FM\delta - \beta - \theta \sum(F_1, M_1, f^{-1}(y_1))$) and $V \in FME\theta \sum(F_1, M_1, f^{-1}(y_2))$(resp.$V \in FM\delta - \beta - \theta \sum(F_1, M_1, f^{-1}(y_2))$) such that $U \cap V = \emptyset$. Therefore FM pree − θ(resp. FM pre$\delta - \beta\theta$) − openness of f demonstrates the presence of two sets. $f(U) \in FME\theta \sum(F_2, M_2, y_1)$(resp.$f(U) \in FM\delta - \beta - \theta \sum(F_2, M_2, y_1)$) and $f(V) \in FME\theta \sum(F_2, M_2, y_2)$(resp.$f(V) \in FM\delta - \beta - \theta \sum(F_2, M_2, y_2)$) such that $f(U) \cap f(V) = \emptyset$. This shows that F_2 is $FMe - \theta - T_2 -$ (resp.$FM\delta - \beta_\theta - T_2$).

Definition 4.6 The grap h (F, M, G) of a mapping $f : (F_1, M_1, T_1) \to (F_2, M_2, T_2)$ is called:

(a) Strongl y $FMe - \theta$(resp.$FM\delta - \beta_\theta$) − closed if for each$(x, y) \in (F_1 \times F_2)(F, M, G)$, there exist $U \in FME\theta \sum(F_1, M_1, x)$(resp.$U \in FM\delta - \beta - \theta \sum(F_1, M_1, x)$) and $V \in FME\theta \sum(F_2, M_2, y)$(resp.$V \in FM\delta - \beta - \theta \sum(F_2, M_2, y)$) such that $f(U) \cap FMe - Cl_\theta(V)$ (resp.$FM\delta - \beta - Cl_\theta V$)) = $\emptyset''$
(b) Strongly e − closed [23] if $\forall(x, y) \in (F_1 \times F_2)(F, M, G), \exists U \in FME \sum(F_1, M_1, x)$ and an open set V in F_2 containing y with $(FMe - Cl(U) \times V) \cap G(f) = \emptyset$.
(c) Strongly $FM\delta - \beta-$ closed [24] if for each $(x, y) \in (F_1 \times F_2)(F, M, G)$, there exist $U \in FM\delta - \beta \sum(F_1, M_1, x)$ and a fuzzy measure open set V in F_2 containing y with $(FM\delta - \beta - Cl(U) \times V) \cap (F, M, G) = \emptyset$.

Theorem 4.7 *Let (F, M, G) be the graph of a mapping $f:(F_1, M_1, T_1) \to (F_2, M_2, T_2)$. Then following terms are equivalent:*

(a) *A (FMG) (F, M, G) of f is strongly $e - \theta -$ (resp. $FM\delta - \beta_\theta$)−closed;*
(b) *For any $(x, y) \in (F_1 \times F_2)(F, M, G)$, $\exists U \in FME\theta \sum(F_1, M_1, x)$(resp. $U \in FM\delta - \beta - \theta \sum(F_1, M_1, x)$)and $V \in FME\theta \sum(F_2, M_2, y)$(resp. $V \in FM\delta - \beta - \theta \sum(F_2, M_2, y)$) with $f(U) \cap V = \emptyset$;*

(c) *For any* $(x, y) \in (F_1 \times F_2)(F, M, G)$, $\exists U \in FME\sum(F, x)(resp. U \in FM\delta - \beta\sum(F_1, M_1, x))$ *and* $V \in FME\sum(F_2, M_2, y)(resp. V \in FM\delta - \beta\sum(F_2, M_2, y))$ *with* $f(e - Cl(U)(resp. FM\delta - \beta - Cl(U))) \cap V = \emptyset$;

(d) *For each* $(x, y) \in (F_1 \times F_2)(F, M)$, *there exist* $U \in FMER(F_1, M_1, x)(resp. U \in FM\delta - \beta R(F_1, M_1, x))$ *and* $V \in FMER(Y, y)(resp. V \in FM\delta - \beta R(Y, y))$ *with* $f(U) \cap V = \emptyset$;

(e) *A* (FMG) (F, M, G) *of* f *is fuzzy measure strongly* $FMe(resp. FM\delta - \beta) -$ *closed.*

Proof (a) $\Rightarrow$ (b).This apparent proof was left away.

(b) $\Rightarrow$ (c). $\forall U \in FME\theta\sum(F_1, M_1, x)\big(\text{resp.} U \in FM\delta - \beta - \theta\sum(F_1, M_1, x)\big)$, $\exists Uo \in FME\Sigma(F_1, M_1, x)\big(\text{resp.} Uo \in FM\delta - \beta\sum(F_1, M_1, x)\big)$ Satisfies $x \in Uo \subset e - Cl(Uo)(\text{resp.} FM\delta - \beta - Cl(Uo)) \subset U$.

(c) $\Rightarrow$ (d). For each$(x, y) \in (F_1 \times F_2)(F, M, G)$, by using part (3) there exist $U \in FME\sum(F_1, M_1, x)(\text{resp.} U \in FM\delta\beta\sum(F_1, M_1, x))$ and $V \in FME\theta\sum(F_2, M_2, y)(\text{resp.} V \in FM\delta - \beta - \theta\sum(F_2, M_2, y))$ such that$f(FMe - Cl(U)(\text{resp.} FM\delta - \beta - Cl(U)))IV = \emptyset$. we have$FMe - Cl(U)(\text{resp.} FM\delta - \beta - Cl(U)) \in FMER(F_1, M_1, T_1)(\text{resp.} FM\delta - \beta R(F_1 M_1, T_1))$ and there exist $Vo \in FMER(F_2, M_2, T_2)(\text{resp.} Vo \in FM\delta - \beta R(F_2, M_2, T_2))$ such that$y \in Vo \subset V$; thus$f(FMe - Cl(U)(\text{resp.} FM\delta - \beta - Cl(U)))IVo = \emptyset$.

(d)$\Rightarrow$(e). This apparent proof was left away.

(e)$\Rightarrow$(a). For each $(x, y) \notin (F, M, G)$, there exist $U \in FME\sum(F_1, M_1, x)(\text{resp.} U \in FM\delta - \beta\sum(F_1, M_1, x))$ and $V \in FME\sum(F_2, M_2, y)(\text{resp.} V \in FM\delta - \beta\sum(F_2, M_2, y))$ Such that $f(FMe - Cl(U)(\text{resp.} FM\delta - \beta - Cl(U))) \cap FMe - Cl(V)(\text{resp.} FM\delta - \beta - Cl(V)) = \emptyset$. Since, $x \in U \subset FMe - Cl(U)(\text{resp.} FM\delta - \beta - Cl(U) \in FMER(F_1, M_1, x)(\text{resp.} FM\delta\beta R(F_1, M_1, x)) \subset E\theta\sum(F_1, M_1, x)(\text{resp.} FM\delta - \beta - \theta\sum(F_1, M_1, x))$ and, $y \in V \subset FMeCl(V)(\text{resp.} FM\delta - \beta - Cl(V)) \in FMER(F_2 M_2, y)(\text{resp.} FM\delta - \beta R(F_2 M_2, y)) \subset FME\theta\Sigma(F_2 M_2, y)(\text{resp.} FM\delta - \beta - \theta\sum(F_2 M_2, y)) \cap FME\theta C(F_2 M_2, y)(\text{resp.} FM\delta - \beta - \theta C(F_2 M_2, y))$, $f(FMe - Cl(U)(\text{resp.} FM\delta - \beta - Cl(U))) \cap FMe - Cl\theta(FMe - Cl(V))(\text{resp.} FM\delta - \beta - Cl\theta(FM\delta - \beta - Cl(V)) = f(FMe - Cl(U)(\text{resp.} FM\delta - \beta - Cl(U))) \cap FMe - Cl(V))(\text{resp.} FM\delta - \beta - Cl(V)) = \emptyset$.

Definition 4.8 A mapping f is called fuzzy measure weakly e$-$irresolute $(FMW - e - Ir)$ [resp. fuzzy measure weakly $\delta - \beta -$ irresolute $(FMW - \delta - \beta - Ir)$] if $\forall x \in F_1$ and $\forall V \in FME\sum(F_2, M_2, f(x))$ (resp. $V \in FM\delta - \beta\sum(F_2, M_2, f(x))$), $\exists U \in FME\sum(F_1, M_1, x)(\text{resp.} U \in FM\delta - \beta\sum(F_1, M_1, x))$ satisfies $f(U) \subset FMe - Cl(V)(\text{resp.} f(U) \subset FM\delta - \beta - Cl(V))$.

Theorem 4.9 *Let* $f: (F_1, M_1, T_1) \rightarrow (F_2, M_2, T_2)$ *be* $(FMW - e - Ir)$ *(resp.* $(FMW - \delta - \beta - Ir)$*) mappings and* F_2 *be a* $FMe - \theta - T_2$ $(resp. FM\delta - \beta_\theta - T_2)$ *then* (F, M, G) *is strongly* $FMe - \theta(resp. FM\delta - \beta_\theta)-$ *closed.*

Proof Let$(x, y) \in (F_1 \times F_2)(F, M, G)$. The $FMe - \theta - T_2$(resp.$FM\delta - \beta_\theta - T_2$) of Y gives the existence of a set $V \in FME\theta \sum(F_2 M_2, y)$(resp.$V \in FM\delta - \beta - \theta \sum(F_2, M_2, y)$) such that$f(x) \notin FMe - Cl_\theta(V)$(resp.$FM\delta - \beta - Cl_\theta(V)$). Now,$F_2 \backslash FMe - Cl_\theta(V) \in FME\theta \sum(F_2, M_2, f(x))$(resp.$F_2 \backslash FM\delta - \beta - Cl_\theta(V) \in FM\delta - \beta - \theta \sum(F_2, M_2, f(x))$. Then, from the $(FMW - e - Ir)$ [resp. $(FMW - \delta - \beta - Ir)$]of f, we get $\exists U \in FME\theta \sum(F_1, M_1, x)(resp. U \in FM\delta - \beta - \theta \sum(F_1, M_1, x))$ satisfies $f(U) \subset F_2 \backslash FMe - Cl_\theta(V)(resp. f(U) \subset Y \backslash FM\delta - \beta - Cl_\theta(V)$. Consequently , $f(U) \cap FMe - Cl_\theta(V) = \emptyset$(resp.$f(U) \cap FM\delta - \beta - Cl_\theta(V) = \emptyset$) and therefore (F, M, G) is Strongly $FMe-\theta$(resp.$FM\delta-\beta_\theta$)− closed.

Definition 4.10 A $(FMTS)$ (F_1, M_1, T_1) is called a $FMe-\theta-$(resp.$FM\delta-\beta_\theta$)− space if the union of any two $FMe - \theta -$ (resp.$FM\delta - \beta_\theta$) − closed measure sets is a $FMe - \theta - (resp. FM\delta - \beta_\theta)$ − closed measure set.

Proposition 4.11 *Let $f, g: (F_1, M_1, T_1) \to (F_2, M_2, T_2)$ be mappings, F_2 a $FMe-\theta - T_2$ (resp.$FM\delta - \beta_\theta - T_2$) and Let F_1 be a $FMe-\theta-$(resp.$FM\delta-\beta_\theta$)− space. If $f \& g$ $(FMW - e - Ir)$ (resp. fuzzy measure weakly $FM\delta - \beta-$ irresolute), then $\underset{\circ}{A} = \{x \in F_1 : f(x) = g(x)\}$ is $FMe - \theta -$ (resp.$FM\delta - \beta_\theta$)− closed in F_1.*

Proof If $x \in F_1 \backslash \underset{\circ}{A}$, then $f(x) \neq g(x)$. Since F_2 is $FMe - \theta - T_2$ (resp.$FM\delta - \beta_\theta - T_2$), there exist $V_1 \in FME\theta\Sigma(F_2, M_2, f(x))$(resp.$V_1 \in FM\delta - \beta - \theta\Sigma(F_2, M_2, f(x))$ and $V_2 \in FME\theta \sum(F_2, M_2, g(x))$(resp.$V_2 \in FM\delta - \beta - \theta\Sigma \sum(F_2, M_2, g(x))$ such that $V_1 \cap V_2 = \emptyset$. By the fact that $f\&g$ are measure weakly e−irresolute (resp. measure weakly$\delta-\beta-$irresolute), $f^{-1}(V_1)$ and $g^{-1}(V_2)$ are $FMe-\theta-$(resp.$FM\delta-\beta_\theta$)− open. Since F_1 is a $FMe-\theta-$(resp.$FM\delta-\beta\theta$)− space, so $x \in f^{-1}(V_1) \cap g^{-1}(V2) \in FME\theta\Sigma(F_1, M_1, T_1)$(resp.$x \in f^{-1}(V_1) \cap g^{-1}(V2) \in FM\delta - \beta - \theta\Sigma(F_1, M_1, T_1)$). Put $U = f^{-1}(V_1) \cap g^{-1}(V2)$, therefore $U \in FME\theta\Sigma(F_1, M_1, T_1)$(resp.$U \in FM\delta-\beta-\theta\Sigma(F_1, M_1, T_1)$) and $U \cap \underset{\circ}{A}, = \emptyset$. This explains that $\underset{\circ}{A}$ is $FMe - \theta -$ (resp.$FM\delta - \beta_\theta$)− closed in F_1.

Definition 4.12 A subset $\underset{\circ}{A}$ of a $(FMTS)$ (F_1, M_1, T_1) is called fuzzy measure nearly Φ − compact relative $(FMN - \Phi - CR)$ to F_1 if any cover $\{V_\lambda : \lambda \in \Delta\}$ of $\underset{\circ}{A}$ via $FMe - \theta -$ (resp.$FM\delta - \beta_\theta$)− open sets of F_1, there exists a finite subset Δ_0 of Δ satisfies.

$\underset{\circ}{A} \subset U\{FMe - Cl(V_\lambda)$(resp.$FM\delta - \beta - Cl(V_\lambda)) : \lambda \in \Delta_0\}$. If $\underset{\circ}{A} = F_1$, the space F_1 is called fuzzy measure nearly Φ− compact $(FMN - \Phi - C)$.

Theorem 4.13 *For a $(FMTS)$ (F_1, M_1, T_1), the following terms are equivalent:*

(a) *F_1 is $(FMN - \Phi - C)$;*
(b) *Any cover of F_1 via $FMe -$ $(resp. FM\delta - \beta)$− Regular sets has a finite sub-cover;*
(c) *For every family $\{V_\lambda : \lambda \in \Delta\}$ of $FMe -$ $(resp. FM\delta - \beta)$− Regular sets of (F_1, M_1, T_1) such that $\cap\{V_\lambda : \lambda \in \Delta\} = \emptyset$, there exists a finite subset Δ_0 of Δ satisfies $\cap\{V_\lambda : \lambda \in \Delta\} = \emptyset$.*

Proof (a) $\Rightarrow$ (b) $\Rightarrow$ (c) : This proof was omitted because it is clear.

(c) $\Rightarrow$ (a) : Let $\{V_\lambda : \lambda \in \Delta\}$ be a cover of F via $e-\theta-$ (resp.$\delta-\beta_\theta$)$-$ open sets of F. Thus, $\{FMe - Cl_\theta(V_\lambda)(\text{resp.}FM\delta - \beta - Cl_\theta(V_\lambda)) : \lambda \in \Delta\}$ is $FMe-(resp.\ FM\delta-\beta)-$ Regular cover of F_1.

Therefore $\{F_1\backslash FMe - Cl_\theta(V_\lambda)(\text{resp.}F_1\backslash FM\delta - \beta - Cl_\theta(V_\lambda)) : \lambda \in \Delta\}$" is a family of $FMe-$(resp.$FM\delta-\beta$)$-$ Regular sets of F_1 having the empty intersection.

Via (c), there exists a finite subset Δ_0 of Δ with $\cap\{F_1\backslash FMe - Cl_\theta(V_\lambda)(\text{resp.}F_1\backslash FM\delta - \beta - Cl_\theta(V_\lambda)) : \lambda \in \Delta_0\} = \emptyset$; thus $F_1 = U\{FMe - Cl_\theta(V_\lambda)(\text{resp.}FM\delta - \beta - Cl_\theta(V_\lambda)) : \lambda \in \Delta_0\}$. This shows that F_1 is fuzzy measure nearly $\Phi-$ compact.

Theorem 4.14 *Let $f : (F_1M_1, T_1) \to (F_2M_2, T_2)$ be a mapping has a fuzzy measure strongly $FMe-\theta-(resp.\ FM\delta-\beta_\theta)-$ closed measure graph and Let (F_1M_1, T_1) be a $FMe-\theta-(resp.\ FM\delta-\beta_\theta)-$ space, then it has the subsequent characteristic:*

$(P^*)-$ If M is $(FMN - \Phi - CR)$ to (F_2M_2, T_2), then $f^{-1}(M)$ is $eFM-\theta-(resp.\ FM\delta-\beta_\theta)-closed$ subset of (F_1M_1, T_1).

Proof Assume that $f^{-1}(M)$ is not $FMe-\theta-(resp.\ FM\delta-\beta_\theta)-$ closed in F_1. Then there exists $x \in FMe - Cl_\theta(f^{-1}(M))\backslash f^{-1}(M)(\text{resp.}x \in FM\delta - \beta - Cl_\theta(f^{-1}(M))\backslash f^{-1}(M))$.

Let $y \in M$. then $(x, y) \in (F_1 \times F_2)(F, M, G)$. Strong $FMe-\theta-$ (resp.$FM\delta-\beta_\theta$)$-$ closedness of (F, M, G) gives the existence of $U_y(x) \in FME\theta\sum(F_1, x)(\text{resp.}U_y(x) \in FM\delta - \beta - \theta\sum(F_1, x))$ and $V_y \in FME\theta\Sigma(F_2, M_2, y)(\text{resp.}V_y \in FM\delta - \beta - \theta\Sigma(F_2, M_2, y))(s.t)f(U_y(x)) \cap FMe - Cl_\theta(V_y)(\text{resp.}f(U_y(x)) \cap FM\delta - \beta - Cl_\theta(V_y)) = \emptyset''$.

Clearly $\{V_y : y \in M\}$ is a cover of M by $FMe-\theta-$ (resp.$FM\delta-\beta_\theta$) $-$ open sets inF_2. The fuzzy measure near $\Phi-$ compactness of M relative to F_2 assures that exists $FMe-\theta-$ (resp.$FM\delta-\beta_\theta$)$-$ open sets $V_{y_1}, V_{y_2}, V_{y_3}, \ldots, V_{y_n}$ in F_2 (s. t)

$$M \subset \bigcup_{i=1}^{n} FMe - Cl_\theta\left(V_{y_1}\right)(\text{resp.}M \subset \bigcup_{i=1}^{n} FM\delta - \beta - Cl_\theta\left(V_{y_1}\right))$$

$$LetU = \bigcap_{i=1}^{n} U_{y_1}(x)$$

where $U_{y_i}(x)$ are the $FMe-\theta-$ (resp.$FM\delta-\beta_\theta$)$-$ open sets in F_1 satisfying (a), moreover $U \in FME\theta\sum(F_1, M_1, x)(\text{resp.}U \in FM\delta - \beta - \theta\Sigma(F_1, M_1, x))$. Now we have

$$f(U) \cap M \subset f\left[\bigcap_{i=1}^{n} U_{y_1}(x)\right] \cap \left(\bigcap_{i=1}^{n} FMe - Cl_\theta\left(V_{y_1}\right)\right)$$

$$\subset \bigcap_{i=1}^{n} (f\left[U_{y_1}(x)\right] \cap FMe - Cl_\theta\left(V_{y_1}\right) = \varphi.$$

Respectively:

$$f(U)\cap M\subset f\left[\bigcap_{i=1}^{n}U_{y_1}(x)\right]\cap\left(\bigcap_{i=1}^{n}FM\delta-\beta-Cl_\theta\left(V_{y_1}\right)\right)$$
$$\subset\bigcap_{i=1}^{n}(f\left[U_{y_1}(x)\right]\cap FM\delta-\beta-Cl_\theta\left(V_{y_1}\right)=\varphi.$$

But $x\in FMe-Cl_\theta(f^{-1}(M))$(resp.$x\in FM\delta-\beta-Cl_\theta(f^{-1}(M))$, so $U\cap f^{-1}(M)\neq\emptyset$. which defies the conclusion made above. The result is thus valid.

Proposition 4.15 *Let* (F_1,M_1,T_1) *be a* $FMe-\theta-$ (resp.$FM\delta-\beta_\theta$)$-$ *space and Let* (F_2,M_2,T_2) *be fuzzy measure nearly* $\Phi-$ *compact, then a mapping* $f:(F_1,M_1,T_1)\to(F_2,M_2,T_2)$ *with a fuzzy measure strongly* $FMe-\theta-$(resp.$FM\delta-\beta_\theta$)$-$ *closed measure graph is* $(FMW-e-Ir)$ *(resp. fuzzy measure weakly* $\delta-\beta-$*irresolute).*

Proof Suppose that Q is $FMe-\theta-$ (resp.$FM\delta-\beta_\theta$) $-$ closed set of F_2. Since F_2 is nearly $\Phi-$ compact, then Q is $(FMN-\Phi-CR)$ to F_2 and it considers by Proposition 4.15 that $f^{-1}(Q)$ is $FMe-\theta-$ (resp.$FM\delta-\beta_\theta$) $-$ closed set in F_1. then we get f is $(FMW-e-Ir)$ (resp. $(FMN-\delta-\beta-Ir)$) mapping.

Definition 4.16 If (F,M,F) is fuzzy measure set and $C=\{Ci:i\in I\}$ c lasses of a set F, Then Ci can be represented by the same graph G. The graph number of a grap h G e quals $min\{|\bigcup Ci|:i\in I\}$. Then (F,M,F,C) is a fuzz y measure graph numbe r $(FMGN)$ of a (FMG) G.

Definition 4.17 If (F,M,F,C) is $(FMGN)$ of a (FMG) G. Let G be a graph and $vi,vj\in V(G)$. If the vertice s vi is represented by a set A and vj is represented by a set B in a set graph G, then $N(vi,vj)$ is defined by$|A\cap B|$. It is clear that $N(vi,U)=|A|$ and $N(vj,U)=|B|$, where U is the universe set of vertices.

Theorem 4.18 *If* (F,M,F,C) *is* $(FMGN)$ *of a* (FMG) *G. If* $G1$ *and* $G2$ *are two graphs which represented by two parallel classe s* $C1$ *and* $C2$*, respectively, then* $G1$ *and* $G2$ *are isomorphic.*

Proof Let (F,M,F,C) be $(FMGN)$ of a (FMG) G. If $C1$ and $C2$ are two parale l classes, then there exists a bijective a function F from $C1$ to $C2$ i.e., $F:C1\to C2$, such that $F(C1)=C2$, where $C1=\{Ai:i\in I\}$ and$C2=\{F(Ai):i\in I\}$. Then every vertex vAi in$G1$, the (FMG) generated by$C1$, have a corresponding v ertex $v_F(Ai)$ in$G2$, the (FMG) generated by $C2$.

So $N(vAi,vAj)=|Ai\cap Aj|=|F(Ai\cap Aj)|=|F(Ai)\cap F(Aj)|=N(vF(Ai),vF(Aj))$.

The converse of Example 4.19 may not be true.

Example 4.19 If (F,M,F,C) is $(FMGN)$ of a (FMG) G, the two graphs of classes $C1$ and $C2$ are isomorphic. However, the set graphs cannot be represented by two parallel classes.

Definition 4.20 If (F, M, F, C) is $(FMGN)$ of a (FMG) G, any isolated point can be represented by an empty set φ in a new d efinite class C.

Definition 4.21 If (F, M, F, C) is $(FMGN)$ of a (FMG) G. A fuzzy measure pseudograph $(FMP)P$, with degree$n$, can be represented by a fuzzy measure topology for a set $X = \{1, 2, 3, \ldots, n\}$.

Example 4.22 The fuzzy measure topological graph $(FMTG)$ on a discrete graph D will be constructed if we delete only the oops in a (FMP) P. In simple graphs, Every adjacent vertex has a single edge drawn between them in discrete fuzzy measure graph $(DFMG)$ D.

Theorem 4.23 *If* (F, M, F, C) *is* $(FMGN)$ *of a* (FMG) *G. Then the following satisfying:*

(i) *The number of edges* $|E_P(G)|$ *of a* (FMP) *G of the set* $X = \{1, 2, \ldots, n\}$ *equals* $n2n - 2(2n - 1 - 1) + n2n - 1$.
(ii) *The number of edges* $|Ed(G)|$ *of a* $(DFMG)$ *arising from a* (FMP) *i.e., by deleting loops in P equals* $n2n - 2(2n - 1 - 1)$.
(iii) *The number of edges* $|Es(G)|$ *of a simple fuzzy measure graph arising from the* $(DFMG)$ *i.e., by drawing only one edge joint two adjacent vertices in D equals* $\frac{1}{2}(22n - 2n - 3n + 1)$.

Proof In Fig. 5, the relationship between the groups is depicted.

(i) Let $A \subseteq X$, $|A| = m$ and $\left|A^C\right| = n - m$

$$\begin{aligned} deg(V_A) &= \sum R(V_A, V_B) + \sum R(V_A, V_E) + \sum R(V_A, V_L) + \sum R(V_A, V_K) \\ &= \sum |A \cap B| + \sum |A \cap E| + \sum |A \cap L| + \sum |A \cap K| \\ &= |B| + \sum |A| + \sum |A \cap L|. \end{aligned}$$

$$\sum |B| = \sum_{i=1}^{m-1} i^m C_i = \sum_{i=1}^{m-1} i \frac{m!}{i!(m-1)} = m \sum_{i=1}^{m-1} \frac{(m-1)!}{(i-1)!(m-1)!} = m \sum_{i=1}^{m-1m-1} C_{i-1}$$

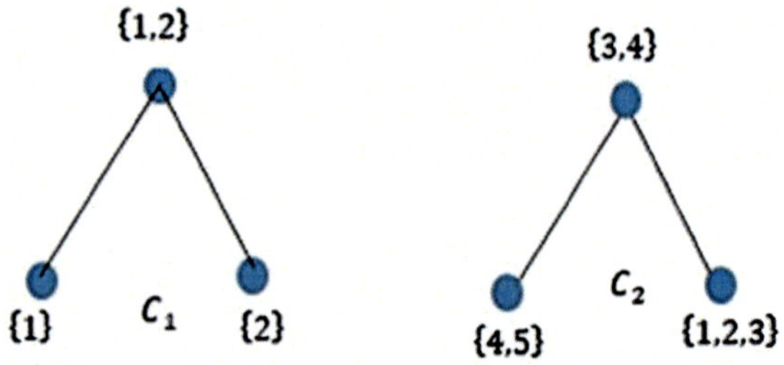

Fig. 5 $C1$ and $C2$ are isomorphic

$$= m\sum_{j=0}^{m-2}\left({}^{m-1}C_j - 1\right) = m\left(2^{m-1} - 1\right).$$

Since $B \subseteq A$, then $\sum A = m\left(2^{m-1} - 1\right)$.

But $A \subseteq E$, then by complement $\sum A = m\left(2^{n-m} - 1\right)$ and $\sum A \cap L = \left(2^{n-m} - 1\right)\sum_{i=1}^{m-1} {}^{i m}C_i = m\left(2^{n-m} - 1\right)\left(2^{m-1} - 1\right)$.

So $deg(V_A) = \left(2^{n-m} - 1\right) + 2m$.

It follows that the number of edges of $(FMPs)$ G equals

$$E_p(G) = \frac{1}{2}\sum_{m-1}^{n} {}^{n}C_m deg(V_A) = \frac{1}{2}\sum_{m-1}^{n} {}^{n}C_m (m(\left(2^{n-1} - 1\right) + 2m)$$

$$= \frac{1}{2}\sum_{m-1}^{n} {}^{n}C_m m\left(2^{n-1} - 1\right) + \frac{1}{2}\sum_{m-1}^{n} {}^{n}C_m 2m$$

$$= n2^{n-2}\left(2^{n-1} - 1\right) + n2^{n-1}.$$

(ii) $deg(V_A) = m\left(2^{n-1} - 1\right)$.

It follows that the number of edges of discrete (FMGs) equals

$$|E_d(G)| = \frac{1}{2}\sum_{A} deg(V_A) = \frac{1}{2}\sum_{A} {}^{n}C_m m * \left(2^{n-1} - 1\right)$$

$$= \frac{1}{2}\left(2^{n-1} - 1\right)\sum_{m-1}^{n} m^n C_m$$

$$= \frac{1}{2}\left(2^{n-1} - 1\right)\sum_{m-1}^{n} m^n C_m$$

$$= \frac{1}{2}\left(2^{n-1} - 1\right)\sum_{m-1}^{n} m\frac{n!}{m(n-m)!}$$

$$= \frac{n}{2}\left(2^{n-1} - 1\right)\sum_{m-1}^{n} \frac{n!}{((m-1)!(n-m)!)}$$

$$= \frac{n}{2}\left(2^{n-1} - 1\right)\sum_{m-1}^{n} {}^{n-1}C_{m-1}$$

$$= \frac{n}{2}\left(2^{n-1} - 1\right)\sum_{j=0}^{n-1} {}^{n-1}C_j$$

$$= \frac{n}{2}\left(2^{n-1} - 1\right)2^{n-1}$$

$$= n2^{n-2}\left(2^{n-1} - 1\right).$$

(iii) $deg(V_A) = 2^n - 2^{n-m} - 1$. *It follows that the number of edges of simple fuzzy measure graphs G equals* $|E_s(G)| = n/2\Sigma_{m=1}^{n} C_m deg(V_A) = 1/2\Sigma_{m=1}^{n} C_m(2^n - 2^{n-m} - 1) = 1/2(2^n - 1)\Sigma_{m=1}^{n} C_m - 1/2(2)^n \Sigma_{m=1}^{n} C_m 2^{-m} = 1/2(2^n - 1)(2^{n-1} - 1) - 2^{n-1}\Sigma_{m=1}^{n}(1/2)^{mn} C_m = 1/2(2^{2n} - 2^{n+1} + 1) - 2^{n-1}((1 + 1/2)^n - 1) = 1/2((2^{2n} - 2 * 2^n + 1) - (3^n - 2^n)) = 1/2(2^{2n} - 2^n - 3^n + 1)$.

Theorem 4.24 *If* (F, M, F, C) *is* $(FMGN)$ *of a* (FMG) *G. Let G be a* (FMG) *such that:*

(i) *G contains only one isolated vertex* v_0.
(ii) *G contains a vertex v adjacent with each vertex in* $G\backslash\{v_0\}$ *and* $N(vi, v) \leq N(vi, U) \leq N(v, U), \forall vi \in G\backslash\{v_0\}$.
(iii) *If* $vi \neq vj \in V(G)$, *then* $vi \cup vj \in V(G)$ *and* $vi \cap vj \in V(G)$. *Then the class* τ *of vertices is a* $(FMTG)$.

Proof Assume that G is a (FMG). We need only to prove three conditions:

(i) From Definition 2.4, we can represent G by different classes. Assume that the $(FMGN)$ of G is$m$. Then there exists aclass, sayτ, with $\tau \subseteq P(\{1, 2, 3, \ldots, m\}) = 2\{1, 2, 3, \ldots, m\}$ and represents$G$. Each $Ai \in \tau$ represents vi and $F \in \tau$ represents the set ofverticesV(G). Then, from Definition 2.6,$N(vi, V(G)) = N(vi, U)$. If $N = \{1, 2, 3, \ldots, m\}$, then $|A \cap F| = |A \cap N| = |A|$. This means that each $A \in \tau$ satisfies that$A \subseteq F$. Since every vertex in G is a graph subset of$V(G)$, i.e., for every $vi \in V(G)$ the singleton $\{vi\} \subseteq V(G)$ and each element in τ is a subset from$F \in \tau$, then$F = \{1, 2, \ldots, m\}$. Also, the isolated vertex $v0$ can be represented by $\varphi \in \tau$.
(ii) Let $v_1, v_2, \ldots$ be an arbitrary different vertex in $V(G)$ represented by $Av1, Av2, \ldots$. Since $v_1 \cup v_2 \cup \cdots = v(Av_1 \cup Av_2 \cup \cdots)$, then $Av_1 \cup Av_2 \cup \cdots \in \tau$.
(iii) If vi and vj are two different vertices in $V(G)$ and represented by Ai and Aj, respectively. Since $vi \cap vj = vAi \cap Aj$, then$Ai \cap Aj \in \tau$. Thus τ is a fuzzy measuretopology.

Here, we use the $(FMTG)\tau$ to calculate the number of edges in G using.

Theorem 4.25 *The number of edges of a topological graph represented by a fuzzy measure topology*$\tau = \{\varphi, \{1\}, \{1, 2\}, \ldots, \{1, 2, \ldots, n\}\}$ *is defined as:* $16n(n2 - 1)$.

Proof Let $Am = \{1, 2, 3, \ldots, m\}, 1 < m < n$,

$$degVAm = m(n - m) + \sum_{r=1}^{m-1} r$$

$$= m(n - m) + \frac{1}{2}m(m - 1)$$

$= mn - m^2 + \frac{1}{2}m^2 - \frac{1}{2}m = mn - \frac{1}{2}m^2 - \frac{1}{2}m$. The number of edges is:

$$|Edges| = \frac{1}{2}\sum_{m=1}^{n} degVAm$$

$$= \frac{1}{2}\sum_{m=1}^{n}(mn - \frac{1}{2}m^2 - \frac{1}{2}m)$$

$$= \frac{1}{2}n\sum_{m=1}^{n} m - \frac{1}{4}\sum_{m=1}^{n} m^2 - \frac{1}{4}\sum_{m=1}^{n} m$$

$$= n^2(n+1) - \frac{1}{24}n(n+1)(2n+1) - \frac{1}{8}n(n+1)$$

$$= \frac{1}{24}n(n+1)(6n-3-2n-1)$$

$$= \frac{1}{24}n(n+1)(4n-4) = \frac{1}{6}n\left(n^2-1\right) = {}^{n+1}C_3.$$

We show Examples 4.26 and 4.27, which explain the results of Theorem 4.24 and Example 4.26.

Example 4.26 Let $F = \{1, 2, 3\}$ with a $(FMTG)\tau = \{\varphi, \{2\}, \{2, 3\}, F\}$. The number of edges equals 4. See Fig. 6.

Example 4.27 The isomorphism by $(FMTG)$ is given in Fig. 7. By Table 1, we get a class $C\{\{1, 2, 3\}, \{1, 4, 5\}, \{4, 6, 7\}, \{6, 8, 9\}, \{3, 9, 10\}, \{2, 12, 13\}, \{5, 14, 15\}, \{7, 11, 12\}, \{8, 13, 14\}, \{10, 11, 15\}\}$, which represents a (FMG) $G1$. The same class is obtained by Table 2 in a similar fashion, which represents a graph $G2$. Table 3 proves the homeomorphic between $(FMTGs)$ of $G1$ and $G2$.

A. M. Kozae et al./New types of graphs induced by topological spaces. And the theory of (FMGs) became the most crucial areas of mathematics as a result.

Example 4.28 Let G1 and G2 be two $(FMGs)$ which s shown in Fig. 7. From Tables 4 and 5, we have the same $(FMTG)$. $\tau = \{\varphi, \{1\}, \{2\}, \{3\}, \{1, 2\}, \{1, 3\}, \{2, 3\}, \{1, 2, 3\}\}$ for graphs G1 and G2. We show that G1 is an isomorphic with $G2$, in Table 6.

Given that topological graphs in Table 6 are homomorphic, G1 and G2 in Fig. 7 are isomorphic.

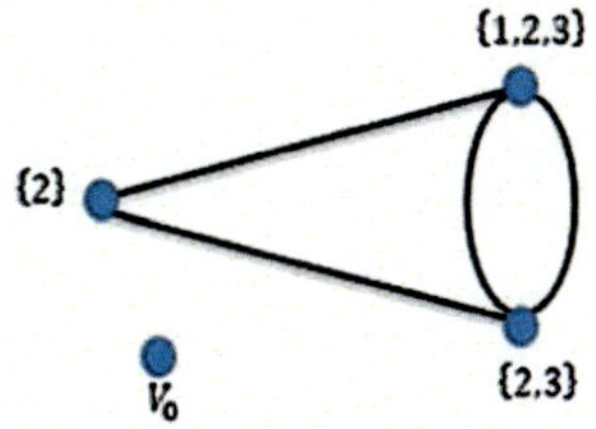

Fig. 6 A topological graph for 3 points

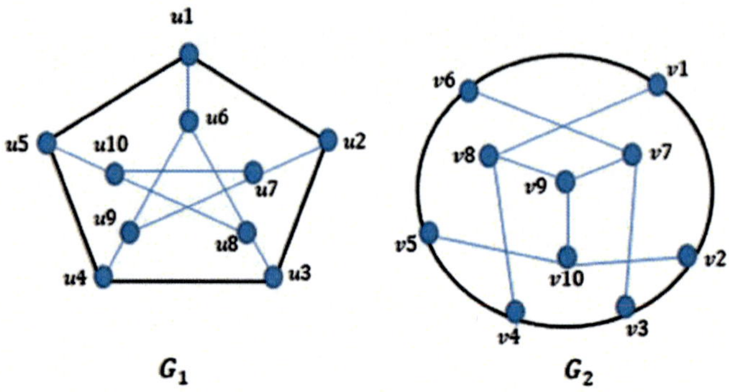

Fig. 7 Two isomorphic graphs G_1 and G_2

Table 1 A topological graph of G_1

Step/ Vertex	1	2	3	4	5	6	7	8
u1	{1, 2, 3}							
u2	{1,	{1, 4, 5}						
u3		{1,	{4, 6, 7}					
u4			{6,	{6, 8, 9}				
u5	{3,			{3, 9,	{3, 9, 10}			
u6	{2,					{2, 12, 13}		
u7		{5,					{4, 14, 15}	
u8			{7,			{7, 12,		{7, 11, 12}
u9				{8,		{8, 13,	{8, 13, 11}	
u10					{10,		{10, 15,	{10, 11, 15}

Table 2 A topological graph of G_2

Step/ Vertex	1	2	3	4	5	6	7	8
v1	{1, 2, 3}							
v2	{1,	{1, 4, 5}						
v3		{4,	{4, 6, 7}					
v4			{7,	{7, 11, 12}				
v5	{3,			{11,	{10, 11, 15}			
v6					{3, 10,	{3, 9, 10}		
v7						{6, 9,	{6, 8, 9}	
v8			{6,				{8,	{8, 13, 14}
v9	{2,			{8,	{2, 12	{2, 12		{2, 12, 13}
v10		{5,			{5, 15	{5, 15		{5, 14, 15}

Table 3 Homeomorphic between topological graph of G_1 and G_2

Set	G_1	G_2
{1, 2, 3}	u1	v1
{1, 4, 5}	u2	v2
{4, 6, 7}	u3	v3
{6, 8, 9}	u4	v4
{3, 9, 10}	u5	v5
{2, 12, 13}	u6	v6
{5, 14, 15}	u7	v7
{7, 11, 12}	u8	v8
{8, 13, 11}	u9	v9
{10, 11, 15}	u10	v10

Table 4 A topological graph of G_1

Step/Vertex	1	2	3	4
u0	ϕ			
u1		{1}		
u2			{2}	
u3				{3}
u4		{1,	{1, 2}	
u5		{1,		{1, 3}
u6			{2,	{2, 3}
u7		{1,	{1, 2,	{1, 2, 3}

Table 5 A topological graph of. G_2

Step/Vertex	1	2	3	4
u0	ϕ			
u1		{1}		
u2			{2}	
u3				{3}
u4		{1,	{1, 2}	{1, 3}
u5		{1,		{2, 3}
u6			{2,	
u7		{1,	{1, 2,	{1, 2, 3}

Table 6 Homeomorphic between topological graph of G_1 and G_2

Set	G_1	G_2
ϕ	$u0$	v0
$\{1\}$	$u1$	v1
$\{2\}$	$u2$	v2
$\{3\}$	$u3$	v3
$\{1, 2\}$	$u4$	v6
$\{1, 3\}$	$u5$	v5
$\{2, 3\}$	$u6$	v4
$\{1, 2, 3\}$	$u7$	v3

A. M. Kozae et al./New types of graphs induced by topological spaces questions regarding (FMTGs), either explicitly or indirectly. Topological spaces and the theory of (FMGs) became the most crucial areas of mathematics as a result. On the other hand, quantum physics, high energy physics, and superstring theory [31] all heavily rely on geometry. In order to calculate the degree of vertices and the number of edges of graphs, we examine the fuzzy measure topological structure of graphs and investigate its potential applications in superstring theory and quantum physics. Additionally, the ideas put forth in this work can be expanded to fuzzy topological structures [29], allowing for a more conclusive resolution of decision-making issues [32–34] in practical applications.

5 Conclusion

This research studied a technique for fuzzy measure soft graph by developing some novel fuzzy soft rules. Well-defined operational laws are critical in (DM) problems (DMPs). On the other hand, the goal of this project is to research and explore fuzzy measure graphs and fuzzy measure soft graphs. Some of their properties have also been investigated. The totally regular fuzzy measure soft graph, uniform edge fuzzy measure soft graph, and uniform vertex fuzzy soft graph are investigated and their application in (DM) is explained using examples. By selecting the perfect parameter for set V under a given fuzzy measure soft set using the characters of uniform vertex fuzzy measure soft graph (UVFMSG), uniform edge fuzzy measure soft graph (UEFMSG), and complement of fuzzy measure soft graph (FMSG) are investigated.

We presented and discussed new mapping kinds of strongly closed graphs in (TS), which is called absolutely $\mho - \Phi -$ $(\Omega - \Psi\Phi) -$ Graphs that are closed, in this work. These mappings have some properties that can be determined. Moreover, the numerical hypothesis of fuzzy sets has advanced significantly and is frequently used. A fuzzy measure can also be described as a fluffy perspective, which is how K_n, which is used in the hypothesis for a variety of applications, is provided. As a result, the fuzzy topological variation of the ideas and conclusions presented in this research is critical.

Future research will extend the approach suggested in this study to additional ambiguous fields, like linguistic term sets, probabilistic linguistictermsets, hesitantfuzzysets, and so on. The proposed approach can also be applied to other sectors, such as medical nutrition diagnostics, sustainable supplier selection, and so on.

Compliance with Ethical Standards

Funding Not applicable.

Availability of data and materials Not applicable.

Competing interests The authors declare that they have no competing interests.

Authors' contributions All authors read and approved the final manuscript.

Data Availability Statement My manuscript has no associated data.

References

1. Zadeh, A.: Fuzzy sets. Inf. Control. **8**, 338–353 (1965)
2. Ahmed, D., Dal, B., Mostafa, A.: Possibility Fermatean fuzzy soft set and its application in decision-making. J. Intell. & Fuzzy Syst. **44**(2), 1565–1574 (2023)
3. Khalil, S.: Decision making using algebraic operations on soft effect matrix as new category of similarity measures and study their application in medical diagnosis problems. J. Intell. & Fuzzy Syst. **37**, 1865–1877 (2019). https://doi.org/10.3233/JIFS-179249
4. Madad, K., Saima, A., Sarfraz, A., Muhammad, Z.: Computational bipolar fuzzy soft matrices with applications in decision making problems. J. Intell. & Fuzzy Syst. **44**(6), 10241–10253 (2023)
5. Khalil, M.S., Sharqi, S.: Some applications in decision-making using cosine maps and the relevance of the Pythagorean fuzzy. Eng. Appl. Artif. Intell. **122**, 106089 (2023)
6. Yu, W., Zhang, Z., Zhong, Q., Sun, L.: Extended TODIM for multicriteria group decision making based on unbalanced hesitant fuzzy linguistic term sets. Comput. Ind. Eng. **114**, 316–328 (2017)
7. Yu, W., Zhang, Z., Zhong, Q.: Consensus reaching for MAGDM with multi-granular hesitant fuzzy linguistic term sets: a minimum adjustment-based approach. Ann. Oper. Res. **1**, 1–24 (2019). https://doi.org/10.1007/s10479-019-03432-7
8. Pena, J., Nápoles, G., Salgueiro, Y.: Explicit methods for attribute weighting in multi-attribute decision-making: a review study. Artif. Intell. Rev. **1**, 1–26 (2019)
9. Liu, X., Wang, Z., Zhang, S., Hetzler, A.: CRM-based dynamic decision-making with hesitant fuzzy information for the evaluation of rangelands. Technol. Econ. Dev. Econ. **24**(5), 1979–2002 (2018)
10. Ding, J., Xu, Z., Zhao, N.: An interactive approach to probabilistic hesitant fuzzy multi-attribute group decision making with incomplete weight information. J. Intell. & Fuzzy Syst. **32**(3), 2523–2536 (2017)
11. Chen, T.Y., Li, C.H.: Determining objective weights with intuitionistic fuzzy entropy measures: a comparative analysis. Inf. Sci. **180**, 4207–4222 (2010)

12. Molodtsov, D.: Soft set theory-first results. Comput. Math. Appl. **37**, 19–31 (1999)
13. Maji, P.K., Roy, A.R., Biswas, R.: Fuzzy soft sets. J. Fuzzy Math. **9**, 589–602 (2001)
14. Al-Sharqi, F., Ahmad, A.G., Al-Quran, A.: Fuzzy parameterized-interval complex neutrosophic soft sets and their applications under uncertainty. J. Intell. Fuzzy Syst. **44**, 1453–1477 (2023)
15. Al-Masarwah, A., Qamar, M.A.: Some new concepts of fuzzy soft graphs. Fuzzy Inf. Eng. **8**(4), 427–438 (2016)
16. Chellamani, P., Ajay, D., Broumi, S., Ligori, T.A.A.: An approach to decision-making via picture fuzzy soft graphs. Granul Comput. 1–22 (2021)
17. Hasan, M.A., Abbas, N.M., Khalil, S.M.: On Soft α*-open sets and soft contra α*-continuous mappings in soft topological spaces. J. Interdiscip. Math. **24**, 729–734 (2021). https://doi.org/10.1080/09720502.2020.1861786
18. Hasan, M.A., Mahmood, S., Abbas, N.M.A.: Characteristics of the soft-(1 2) - GPRW-closed sets in soft bi-topological spaces. In: Conference IT-ELA 2020, vol. 9253110, pp. 103–108 (2020). https://doi.org/10.1109/IT-ELA50150.2020.9253110
19. Mahmood, S., Suleiman, E., Modhar, M.T.: Generated new classes of permutation I/B-algebras. J. Discr. Math. Sci. Cryptogr. **25**(1), 31–40 (2022). https://doi.org/10.1080/09720529.2021.1968110
20. Khalil, S., Suleiman, E., Abbas, N.M.: New technical to generate permutation measurable spaces. In: 2021 1st Babylon International Conference on Information Technology and Science (BICITS), pp. 160–163 (2021). https://doi.org/10.1109/BICITS51482.2021.9509892
21. Mahmood, S., Abbas, N.M.: Applications on new category of the symmetric groups. AIP Conf. Proc. **2290**, 040004 (2020). https://doi.org/10.1063/5.0027380
22. Khalil, S., Abbas, N.M.A.: On nano with their applications in medical field. AIP Conf. Proc. **2290**, 040002 (2020). https://doi.org/10.1063/5.0027374
23. Nivetha, A.R., Vigneshwaran, M., Abbas, N.M., Khalil, S.: On N g*α-continuous in topological spaces of neutrosophy. J. Interdiscip. Math. **24**(3), 677–685 (2021). https://doi.org/10.1080/09720502.2020.1860288
24. Abbas, N.M., Mahmood, S., Vigneshwaran, M.: The neutrosophic strongly open maps in neutrosophic bi-topological spaces. J. Interdiscip. Math. **24**(3), 667–675 (2021). https://doi.org/10.1080/09720502.2020.1860287
25. Khalil, S.: On neurosophic delta generated per-continuous functions in neutrosophic topological spaces. Neutrosophic Sets Syst. **48**, 122–141 (2022)
26. Ekici, E.: On e-open sets, DP*-sets and DPE* -sets and decompositions of continuity. Arab. J. Sci. Eng. **33**(2A), 269–282 (2008)
27. Hatir, E., Noiri, T.: Decompositions of continuity and complete continuity. Acta Math. Hungar. **113**(4), 281–287 (2006)
28. Ekici, E.: New forms of contra-continuity. Carpathian J. Math. **24**(1), 37–45 (2008)
29. Hatir, E., Noiri, T.: On δ-β-continuous functions. Chaos Solitons Fractals **42**, 205–211 (2009)
30. Alaa, M.F., Jumaili, A.L.: On some weak and strong forms of irresolute mappings in topological spaces via E-open and δ-β-open sets. In: International Conference on Mathematics, Engineering and Industrial Applications 2018 (ICoMEIA 2018), AIP Conference Proceedings (2013)
31. Al-Jumaili, A.F., Abed, M.M., Al-Sharqi, F.: Other new types of mappings with strongly closed graphs in topological spaces via e-θ and $\delta-\beta-\theta$-open sets. J. Phys. Conf. Ser. **1234**(1), 1–11 (2019)
32. Caldas, M., Jafari, S., Noiri, T., Simoes, M.: A new generalization of contra-continuity via Levine's g-closed sets. Chaos Solitons & Fractals **32**, 1597–1603 (2007); The 1st International Scientific Conference on Pure Science IOP Conference Series: Journal of Physics: Conference Series, vol. 1234, p. 012101 (2019). IOP Publishing. https://doi.org/10.1088/1742-6596/1234/1/012101
33. Zhan, J., Sun, B., Alcantud, J.R.: Covering based multi-granulation (I, T)-fuzzy rough set models and applications in multi-granulation group decision-making. Inf. Sci. **476**, 290–318 (2019)
34. Zhang, I., Zhan, J., Xu, Z.: Covering-based generalized IF rough sets with applications to multi-attribute decision-making. Inf. Sci. **478**, 275–302 (2019)

Cross-Border Blockchain Transactions as a Tool for Adapting the National Economy of the Russian Federation to International Sanctions

M. R. Safiullin, A. A. Dinmukhametova, and R. T. Burganov

Abstract In this study, we focused on the mechanisms for ensuring the sustainable development of the Russian economy in the context of correction of external financial flow owing to the adjustment of foreign economic supply chains. By constructing a system of recursive equations, it was established that a 1% reduction in volume of transnational financial flow creates the potential for a 0.45% decrease in Russia's gross domestic product. The localization of these risks can be achieved by implementing blockchain technologies in the Russian system of cross-border payments to provide a new foundation for financial flow under emerging systemic transformations in the external environment.

Keywords Blockchain · Transnational payment systems · Import substitution · Economic externalities · Sanction pressure · International supply chains · Economic sustainability

1 Introduction

The current conditions for the development of the national economy in Russia, characterized by the observed transformations in established value-added chains owing to disruptions in export and import flow under sanction confrontation, necessitate the search for new mechanisms and directions for intensifying economic growth and mitigating threats in export–import operations. In this regard, we agree with the

M. R. Safiullin (✉) · A. A. Dinmukhametova
Kazan Federal University, Kazan, Russia
e-mail: c.p@tatar.ru

A. A. Dinmukhametova
Center for Advanced Economic Research, Academy of Sciences of the Republic of Tatarstan, Kazan, Russia

Kazan National Research Technological University, Kazan, Russia

R. T. Burganov
Volga Region State University of Physical Culture, Sports and Tourism, Kazan, Russia

D. Vlachos (ed.), *Mathematical Modeling in Physical Sciences*, Springer Proceedings in Mathematics & Statistics 446, https://doi.org/10.1007/978-3-031-52965-8_24

position of several Russian scientists and experts [1–3]. that issues related to mitigating sanction threats become extremely relevant. Importantly, among the listed directions of strategic development, particular significance is given to the search for mechanisms to ensure the sustainability of external (relative to the Russian economy) supply chains. This problem is multifaceted and has a wide range of solutions ranging from mechanisms for import substitution to the creation of alternative international payment systems that are not controlled by Western countries and that do not pose threats to countries participating in cross-border payments with Russia.

The issue of creating alternative payment systems at the supranational level becomes especially relevant under sanctions imposed by individual countries on national economic systems. For example, intensified international sanction pressure on the Russian economy in 2022 led to restrictions on Russia's access to international payment systems. For instance, major Russian banks (Sberbank, VTB, Bank "Otkritie," VEB.RF, Moscow Credit Bank, Rosselkhozbank, etc.) and corporations were disconnected from SWIFT on June 3, 2022, significantly complicating international financial transactions to and from Russia. The severity of the problem is underscored by the fact that as of 2021, approximately 300 Russian financial organizations were using the SWIFT system. In terms of this indicator, Russia ranked second in the world after the United States. Therefore, it is not difficult to imagine the scale of the consequences for the Russian economy resulting from the disconnection from this data processing and financial messaging service. The disconnection of financial institutions from SWIFT means that their ability to carry out payments and currency operations is significantly limited, which affects the foreign trade turnover and disrupts the sustainability of supply chains.

Therefore, the question of alternative mechanisms for organizing international supply chains and the accompanying cross-border transactions is currently highly relevant for the Russian economy.

2 Methods

Before proceeding to the construction of models aimed at finding solutions to the abovementioned problems, it is important to consider the main trends that reveal the characteristics of foreign exchange flow in the Russian economy (Fig. 1).

Of note, a significant share of the total volume of cross-border cash flow is attributed to individuals. Their share accounts for approximately 40%. Moreover, the majority of transnational money transfers (approximately 85%) in the overall volume of financial transactions are directed to countries in the Western bloc (Fig. 2).

Overall, the results of the descriptive express analysis indicate the presence of significant volumes of cross-border cash flow in the Russian economy. This, on the one hand, reflects a high level of integration of Russia's economy into the global economic environment, and on the other hand, indicates a high level of dependence of the national economic system on potential disruptions in accessing international clearing services and cross-border supply chains.

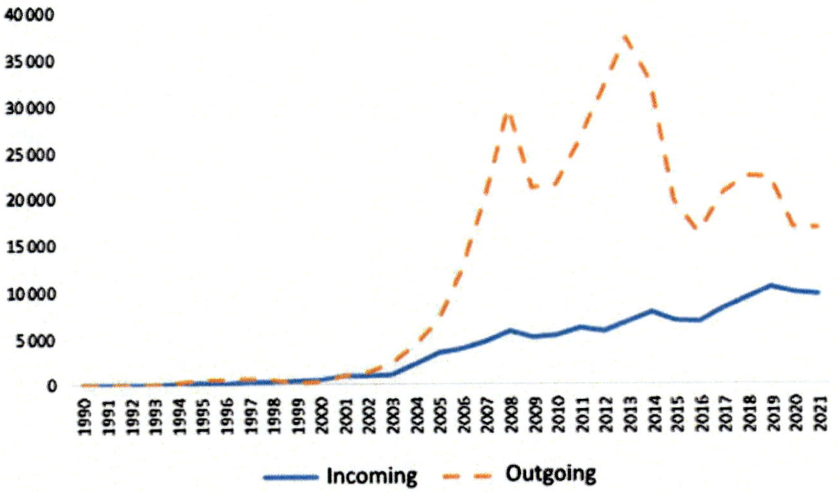

Fig. 1 Dynamics of incoming and outgoing cash flow, in millions of US dollars. *Data source* World Bank / URL: https://www.worldbank.org/en/topic/labormarkets/brief/migration-and-remittances (Accessed on 15.05.2023)

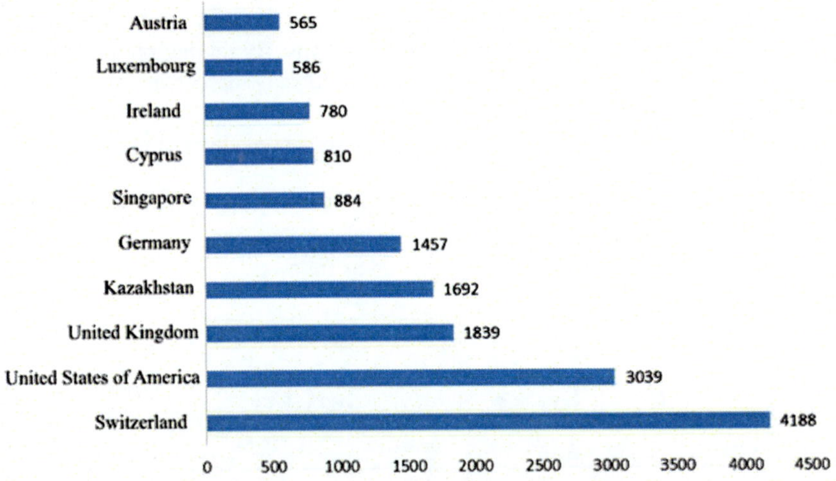

Fig. 2 Cross-border transfers from Russia in favor of individuals from the top ten countries in 2021, in millions of US dollars. *Source* Central Bank statistics URL: https://cbr.ru/statistics/macro_itm/svs/TGO_obz/crossborder_ind_2021. (Accessed on 15.05.2023)

Table 1 Share of imports in the Russian economy, % of GDP, 2010–2022 (calculated based on the available data)

Year	2010	2015	2021
Share of imports, % of GDP	21.1	20.6	21.3

Source Unified Interdepartmental Information and Statistical System (EMISS).URL: https://www.fedstat.ru/indicator/31057 (Accessed on: 15.05.2023)

Any disruptions to established external connections may have negative consequences, which is particularly relevant in the context of the substantial dependence of the Russian economy on imports, accounting for approximately 21% of its gross domestic product (GDP) (Table 1). Therefore, changes in the volumes of cross-border cash flow, including those caused by sanction restrictions, have negative consequences for other countries and can significantly disrupt the structure and dynamics of economic growth. To mitigate such risks and maintain stable supply channels of imported components, raw materials, and finished products, it is necessary to assess the possible consequences and propose mechanisms that ensure localization of identified threats.

The abovementioned problem can be solved by answering two key questions. The first question aims to identify patterns that reveal the desired relationship between Russia's GDP and the dynamics of transnational money flow. The second question focuses on developing possible directions to mitigate threats to sustainable development and disruptions in international supply chains. By relying on this methodological research concept, the following methodological approaches are presented to achieve these goals.

Methodological toolkit and assessment of the impact of changes in transnational money flow on Russia's GDP.

To study the dependence of GDP on incoming transnational money flow, a system of recursive equations is proposed.

In structural form, the model is represented as follows:

$$\begin{cases} y_1 = a_{01} + a_{11}x_1 + b_{12}y_2 + \varepsilon_1 \\ y_2 = a_{02} + a_{22}x_2 + a_{23}x_3 + \varepsilon_2 \\ y_3 = a_{03} + b_{31}y_1 + a_{32}x_3 + \varepsilon_3 \end{cases}$$

y_1—Incoming money flow

y_2—Consumer price index
y_3—Gross domestic product
x_1—Expenditure on digital technologies
x_2—Net exports
x_3—Dollar exchange rate
a_{01}, a_{02}, a_{03}—Constants

Thus, in the model, the incoming money flow, consumer price index, and gross domestic product indicators act as endogenous (dependent) variables. The expenditure on digital technologies, consumer price index, net exports, and dollar exchange rate indicators act as predetermined (independent) variables.

3 Result

Using the proposed methodological toolkit, the following system of equations is obtained, revealing the characteristics and regularities between the investigated factors:

$$\begin{cases} y_1 = 24497,3 + 0,0013x_1 - 176,76y_2 \\ y_2 = 112,24 - 0,0009x_2 + 0,034x_3 \\ y_3 = 5031662 + 5826,77y_1 + 2,68x_3 \end{cases}$$

To solve the system, a two-stage least squares (2SLS) method is applied because the instrumental variables estimation (IV) approach does not provide unique estimates for the over-identified system.

Based on the calculations, the following system of equations is obtained:

$$\begin{cases} y_1 = 0,04x_1 - 145,79y_2 \\ y_2 = -0,0125x_2 + 0,084x_3 \\ y_3 = 6025,8y_1 + 3,15x_3 \end{cases}$$

Based on the calculations, following the principles and approaches of the proposed methodological toolkit, and relying on modeling methods of the studied processes, Table 2 presents the forecasted estimates of the growth of Russia's GDP owing to the intensification of business activity and the corresponding changes in transnational money flow.

Development of possible directions to mitigate threats to the disruption of stable trends in the international supply chain organization.

Table 2 Modeling of GDP growth resulting from the intensification of business activity and the corresponding changes in transnational money flow (based on 2021 data)

№	Growth rates of the cumulative values of cross-border money flow	GDP growth, in %
1	Current value	–
2	+1.0%	+0.45
3	+3.0%	+1.34
4	+5.0%	+2.24

Source Calculated by the authors.

Owing to the sanctions imposed on the Russian economy, it is necessary to acknowledge the need to search for mechanisms to mitigate the threats of slowing down foreign trade turnover, primarily within the framework of sanction pressure that restricts international transactions. Undoubtedly, this task is complex. Currently, experts and scholars propose various mechanisms to address this issue, ranging from redirecting supply chains from the West to the East to intensifying scientific and technological development to stimulate import substitution processes. However, certain approaches with a significant potential for solving the problem often fall outside the scope of research focus. These approaches include the creation of alternative channels and gateways that facilitate cross-border payments, providing the ability to bypass sanction pressure. Specifically, blockchain platforms that provide functionality for international payments should be considered.

The market for blockchain technologies and the business operations based on them, including financial transactions, are rapidly developing worldwide. Economic relationships built using digital currencies (DCs) are becoming increasingly integrated into the system of economic relations [4–9]. This is occurring because blockchain is a highly effective tool for addressing the existing shortcomings in the international practice of interbank payments and the accompanying processes for global trade. It can be asserted with a certain degree of confidence that blockchain is an alternative mechanism to traditional banking methods for facilitating cross-border and domestic transactions. This technology is a cost-effective and secure alternative for international payment systems. It allows optimization of cross-border supply chains while maintaining the highest level of security for international financial transfers using a secure distributed ledger that restricts unauthorized access to financial operations [10–14].

Currently, the following issues are relevant for Russia: the prospects of transitioning the national payment system to blockchain, the effects associated with this process, and the intensifying international situation, which leads to increasing sanction pressure and restrictions on access to international clearing companies, payment systems, and the SWIFT system. In this new reality, the task of intensifying the search for alternative channels for conducting international and national transactions becomes strategically significant. The possibility and potential of replacing foreign technological solutions with domestic solutions determine not only Russia's global competitiveness but also lay the foundation for ensuring the sustainability of its national financial system and intensifying socio-economic development processes, taking into account existing constraints.

The necessity of building new channels for cross-border payments is currently becoming more relevant because economic relationships based on blockchain and DCs are becoming increasingly integrated into the system of economic relations worldwide. The survey conducted by the Bank for International Settlements [15] indicates that the majority of central banks are conducting their own research and are at various stages of evaluating the prospects for the introduction of DCs into circulation [16]. In particular, 86% of central banks are studying the possibilities of introducing DCs into circulation, 60% are experimenting with technologies, and 14% are at the stage of applied development or piloting [16]. Active positions on this issue have been taken by the European Central Bank, the Bank of Sweden, the Bank of Canada, the Central Bank of Russia, the Bank of England, the Reserve Bank of Australia, and the People's Bank of China. These regulators view DC as «a new form of fiat money issued in digital form by a central bank and serving as legal tender» [17].

According to estimates by the Bank for International Settlements [16], by 2026, more than 20% of central banks worldwide will launch DCs into circulation. According to the available data,2 as of the beginning of 2023, 87 countries are actively working on and discussing the launch of DCs into economic circulation (Fig. 3). However, the number of countries already using DCs in economic relations is currently considerably smaller than the number of countries in the discussion stage and pilot launch of DCs into economic circulation (Fig. 4).

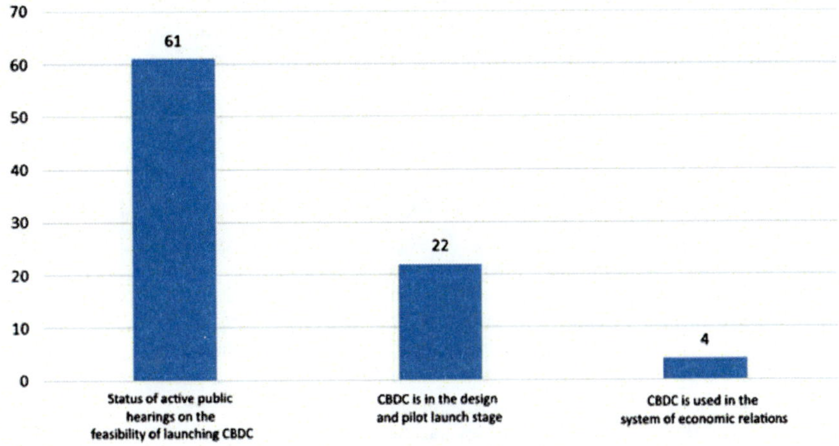

Fig. 3 Current status of CBDC launch. *Source* Developed based on (Boar, 2021, p.12)

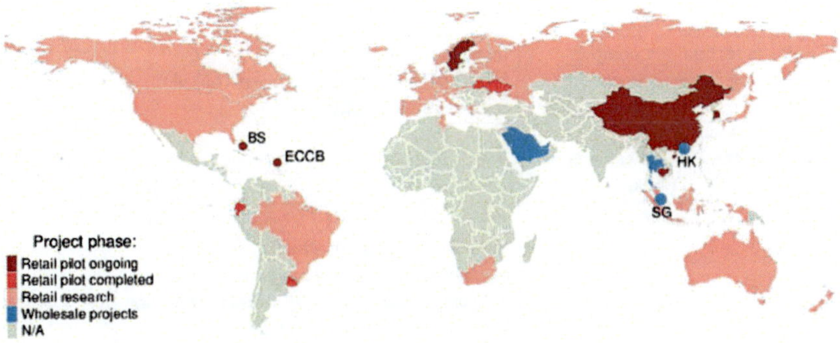

Fig. 4 Current status of CBDC launch. *Source* Developed based on [17]

Owing to the significance and role of blockchain-based payment systems, including at the cross-border level, it is necessary to note that despite the conflicting approaches and positions of experts regarding the feasibility and potential of using such payment systems in economic activities, some countries and economic entities are already actively pursuing and developing "blockchainization" in international supply chains (Table 3).

The intensification of blockchain's spread and integration into the business environment, including international trade, has occurred because this technology significantly facilitates the process of establishing economic relationships between parties involved in international transactions. The application of blockchain in cross-border payments to ensure the stability of international supply chains becomes even more relevant owing to the imposed sanctions that limit the potential of foreign economic operations.

Of note, the Russian economy has effectively neutralized threats from Western sanctions in terms of pressure on the national payment system. The creation of its own financial messaging systems (SPFS), which allow the autonomous processing

Table 3 Examples of application of blockchain ecosystems in the supply chains in international trade

Partner organizations	Functionality and purpose of the system	Project launch date
Banks of Canada, Singapore, and JP Morgan (an American bank, multinational financial holding)	Implementation of financial cross-border operations using DCs to accelerate transaction processes, reduce costs, and enhance security within the framework of blockchain utilization	2019
KlickEx, in collaboration with Stellar.org and IBM blockchain: KlickEx is a monetary payment system of the Central Bank of Polynesia that utilizes the interbank retail system Smart Market	Utilization of blockchain platforms for money transfers	2017
Circle—a technology company specializing in peer-to-peer payments, currently managing the popular stablecoin USDC, a cryptocurrency whose value is pegged to the US dollar	Instant cross-border transactions in the format of fee-free P2P payments based on the Ethereum blockchain	2017
ReiseBank (Germany) and (ABT) Canada	These banks utilize the Ripple blockchain for organizing and conducting international payments. The blockchain platform enables financial institutions to process cross-border transactions instantly and without fees	2016

Source Compiled from Ref. [18].

of financial transactions within the country, the establishment of sovereign mechanisms for interbank interaction, and the sustainable functioning of national card products have served as key instruments for adapting payment operations in the Russian economy.

Unfortunately, the payment system for facilitating cross-border payments has not been as successfully adapted to Western sanctions. The attempt to transfer these types of financial operations to the SPFS system, bypassing SWIFT, did not fully resolve the problem.

In this regard, the question of alternative mechanisms for organizing international supply chains and the accompanying international transactions remains highly relevant for the Russian economy [2]. This conclusion is particularly significant owing to the abovementioned assessments, which demonstrate the impact of disruptions in Russia's external trade relations on the prospects for GDP growth.

4 Conclusion

It is necessary to reiterate that the level of influence of changes in foreign economic money flow on macroeconomic dynamics is significant for Russia. This study revealed that a 1% increase in the volume of transnational money flow corresponds to a 0.45% increase in GDP. Accordingly, a decrease in this indicator will lead to a corresponding slowdown in economic growth.

On the other hand, if Russia integrates blockchain technology on a larger scale into global transactional chains, the mechanisms for mitigating sanction threats will receive active development and contribute to ensuring the independence of the Russian economy from any restrictions imposed by global regulators.

Overall, it is also important to emphasize that the proposed solutions, based on the integration of blockchain technologies into transnational transactions, for mitigating the risks of disruptions in international supply chains owing to sanction limitations will significantly contribute to the development of conceptual approaches in traditional schools of economic theory that focus on researching factors of economic growth. The intensification of transformation processes of productive factors in the era of the sixth technological paradigm dictate the need to explore new development approaches (including in the field of economic relationships) that are based on the application of blockchain and the construction of new international trade models.

Acknowledgements This research was funded by a grant from the Russian Science Foundation, Grant No. 23-28-00587, Kazan (Volga Region) Federal University, Kazan, Russia.

References

1. Masalskiy, M. G.: Economic sanctions. The impact of economic sanctions on the Russian Federation. Ekonomika i Socium 4–3(95), 95–97 (2022)
2. Dubrovskaya, E.S.: Russian economy under sanctions: current situation and prospects for innovative development. Kreativnaya Ekonomika **16**(11), 4553–4564 (2022). https://doi.org/10.18334/ce.16.11.116637
3. Safiuillin, M. R., Yelshin, L. A.: Sanction pressure on the Russian economy: ways to overcome costs and benefits of confrontation within import substitution. Finansy: Teoriya i Praktika 27(1), 150–161 (2023). DOI: https://doi.org/10.26794/2587-5671-2023-27-1-150-161
4. Kochergin, D.: Central bank digital currencies: global experience. Mirovaya Ekonomika i Mezhdunarodnye Otnosheniya **65**(5), 68–77 (2021). https://doi.org/10.20542/0131-2227-2021-65-5-68-77
5. Sakharov, D. M.: Central bank digital currencies: key characteristics and impact on the financial system. Finansy: Teoriya i Praktika 25(5), 133–149 (2021). DOI: https://doi.org/10.26794/2587-5671-2021-25-5-133-149
6. Safiuillin, M.R., Savelychev, M.V., Yelshin, L.A.: The place and evolution of blockchain technologies in the system of economic development. Nauchnoe Obozrenie: Teoriya i Praktika **9**(8), 1191–1200 (2019). https://doi.org/10.35679/2226-0226-2019-9-8-1191-1200
7. Sidorenko, E.: Digital currency of central banks: economic scenarios and forecasts. Mezhdunarodnye Processy **19**(2), 151–165 (2021). https://doi.org/10.17994/IT.2021.19.2.65.8
8. Engert W., Fung B.S.C.: Central bank digital currency: Motivations and implications. Bank of Canada Staff Discussion Paper (16)(2017). https://www.bankofcanada.ca/wp-content/uploads/2017/11/sdp2017-16.pdf
9. Alonso, N., Luis, S., Javier, J.-V., Forradellas, R., Francisco, R.: Central banks digital currency: Detection of optimal countries for the implementation of a CBDC and the implication for payment industry open innovation. J. Open Innov. Technol. Market Complex **7**(1), 1–21 (2021). https://doi.org/10.3390/joitmc7010072
10. Allen, D.W.E., Berg, C., Davidson, S., Novak, M., Potts, J.: International policy coordination for blockchain supply chains. Asia Pacific Policy Studies **6**(3), 367–380 (2019)
11. Zhang, T., Huang, Z.: Blockchain and central bank digital currency. ICT Express (2021). https://doi.org/10.1016/j.icte.2021.09.014
12. Ganne, E., Patel, D.: Blockchain & DLT in trade: a reality check, world trade organization (WTO) and trade finance global (TFG) (2019)
13. Tapscott, D., Tapscott, A.: How blockchain will change organizations. MIT Sloan Manag. Rev. **58**(2), 10–13 (2017)
14. Safiuillin, M. R., Yelshin, L. A., Prygunova, M. I.: Volatility and competitiveness of regions in the face of foreign policy and conjunctural changes: a case study of the regions of the Volga Federal District. Vestnik Rossiyskogo Ekonomicheskogo Universiteta imeni G.V. Plekhanova 12(78), 115–127 (2014)
15. Boar, C., Holden, H., Wadsworth, A.: Impending arrival—a sequel to the survey on central bank digital currency. Bank for International Settlements. BIS Papers (107) (2020). https://www.bis.org/publ/bppdf/bispap107.pdf
16. Boar, C., Wehrli, A.: Ready, steady, go? Results of the third BIS survey on central bank digital currency. BIS Papers (114) (2021). https://www.bis.org/publ/bppdf/bispap114.pdf
17. Allessie, D., Sobolewski, M., Vaccari, L., Pignatelli, F.: Blockchain for digital government, JRC (2019)
18. Glaser, F., Hawlitschek, F., Notheisen, B. (2019) Blockchain as a Platform. In: Treiblmaier H., Beck R. (eds) Business Transformation through Blockchain. Palgrave Macmillan, Cham, pp.121–143

Sociophysics and Econophysics—Mathematical Models of Human Behaviour and Economics

Remote Control of Mobile Robotic System Over Web-Based Interface for Supporting People with Physical Disabilities

Petko Stoev, Denis Chikurtev, and Rosen Ficherov

Abstract The incapacity of people with physical disabilities to lead fulfilling lives is a global problem. Physical limitations lead to stress and discomfort, as may require the need for a caregiver in everyday activities. Mobile robotic systems can address and give solution to these problems, thus improving the quality of life. This paper presents an innovative method for human–robot interaction and collaboration: a concept for a mobile robotic system that aims to assist the people with physical disabilities. A unique feature is having two modes within the platform, which have separate tele-operation techniques, methods and functions. The proposed system provides direct remote control for the robot, using selectable predefined actions. The system's feedback provides information, about the state of the robot, the camera and a visual representation of the data coming from the sensors.

Keywords Robot · Robot collaboration · Tele-operation · Human–robot interaction · Mobile robotic system · Web- based interface

1 Introduction

According to the Council of the European Union, by 2023 there will be over 87 million people with varying degrees of disability in the EU. Some of them are permanently confined to a wheelchair. The World Health Organization predicts that the number of people requiring wheelchairs will increase in the coming years. The impossibility of

P. Stoev · D. Chikurtev (✉)
Institute of Information and Communication Technologies, Bulgarian Academy of Science, Sofia, Bulgaria
e-mail: denis.chikurtev@iict.bas.bg

P. Stoev
e-mail: petko_ivanov_stoev@abv.bg

R. Ficherov
Faculty of German Engineering Education and Industrial Management, Technical University of Sofia, Sofia, Bulgaria

D. Vlachos (ed.), *Mathematical Modeling in Physical Sciences*, Springer Proceedings in Mathematics & Statistics 446, https://doi.org/10.1007/978-3-031-52965-8_25

leading a fulfilling lifestyle caused by physical limitations leads to stress, discomfort and a feeling of inferiority, and it requires companions in everyday activities. One of the ways to master and solve these problems is by creating a robotic mobile system that enables a more fulfilling and independent lifestyle.

Such an innovative solution has the potential to restore the satisfaction of fully independent living. By implementing the robotic system, people with paralysis of the lower limbs will be able to move in an upright state, interact with high-placed objects and communicate comfortably with people in different environments, without depending on assistants in their daily tasks. The interaction between the user and the robotic system is solved by applying the remote-control method.

In the context of mobile robots, teleoperation represents various methods for remote control and robot guidance. This holds immense significance in the realm of robotic mobility for individuals with mobility impairments. Through the implementation of teleoperation in mobile robots for supporting people with physical disabilities, individuals who may be confined to wheelchairs or have lower limb paralysis can gain the ability to navigate their environment, interact with objects, and engage with others in ways that were previously inaccessible.

This empowers them to regain a sense of autonomy, independence, and empowerment by remotely controlling robots that can move, interact, and perform tasks on their behalf [1]. This concept fundamentally shifts the traditional boundaries of mobility, making exploration, interaction, and participation in various activities feasible for individuals with mobility challenges.

Through the implementation of these methods, humans are allowed to extend their reach and capabilities beyond their physical limitations. There is a wide range of methods for implementing teleoperation in mobile robots. Each method has its own specific advantages and challenges, contributing to the diverse range of teleoperation possibilities.

There are several standard diverse methods for teleoperation that are well explored and developed and can aid in the field of healthcare for individuals with mobility impairments:

- Joystick-Based Control: Utilizing traditional physical or virtual joysticks to control robotic movements, especially beneficial for ground-based robots in controlled environments [2].
- Gesture-Based Control: Enabling users to guide robots through gestures, facilitating intuitive interaction and communication [3].
- Haptic Feedback Control: Integrating haptic feedback mechanisms that allow users to sense and control robotic movements, ensuring precision in tasks like surgical procedures [4].
- Augmented Reality (AR) and Virtual Reality (VR): Immersing users in a virtual environment where they can control robots using intuitive gestures or commands [5].
- Voice Command Control: Allowing users to navigate robots and perform tasks using voice commands, suitable for scenarios where hands-free control is essential [6].

Apart from the mentioned above there are many more methods for the implementation of teleoperation in mobile robots like those based on brain-computer interface and artificial intelligence [7]. However, these methods are very complex, costly and need further development in order to be implemented in a mobile robot for supporting people with physical disabilities in everyday tasks in different environments [8, 9].

2 Description of the Robotic Mobile System

The system will have two modes of operation: wheelchair and support assistant. In wheelchair mode, the robotic system attaches to the wheelchair, allowing the wheelchair user to move and perform daily activities in an upright position. After attaching the system to the wheelchair, the user will be able to benefit from better and stable mobility when moving outdoors and at home, shown in Fig. 1.

In the assistant mode, the robotic system automatically transforms into a mobile robot with direct tele-control, with which the user will be able to get what he needs without needing an assistant or performing traumatic and effortful movements (Fig. 2).

Through the robotic mobile system, people with paralysis and different degrees of physical disabilities in the lower limbs will be given the opportunity to move and interact freely and fully in different living environments and situations.

The construction of the robot is designed to ensure full operation in the two mentioned modes and the automatic transformation between them. The mechanical

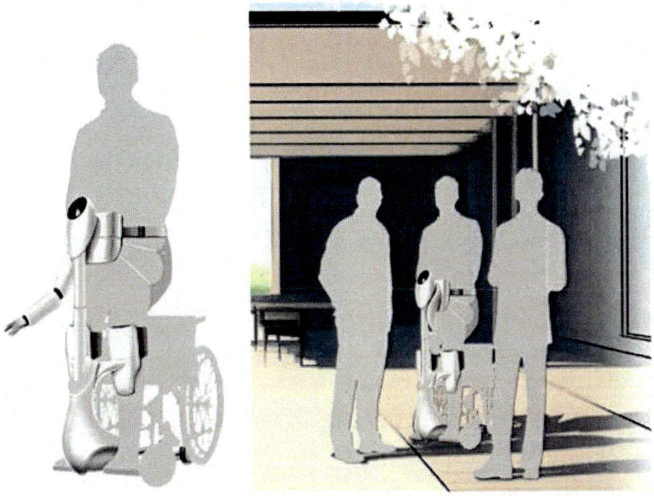

Fig. 1 Overview of the wheelchair mode of the robot and the user position on it

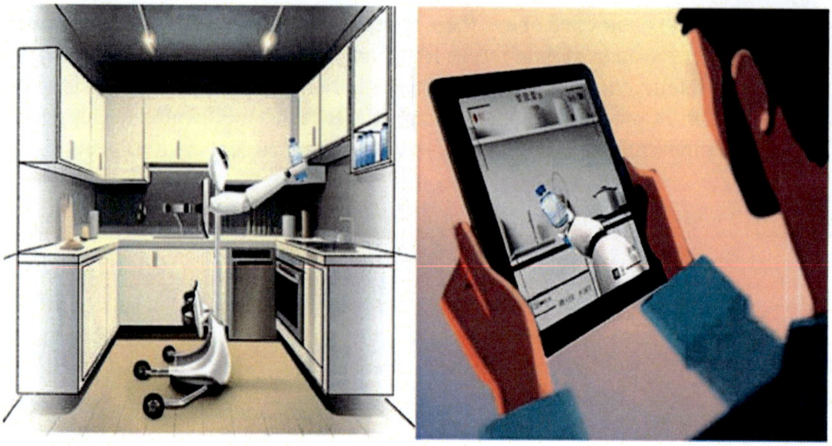

Fig. 2 Overview of the assistant mode of the robot and the user remote-control options

system will allow the user to benefit from better and stable mobility in outdoor and domestic movements after attaching the robot to the wheelchair.

Through the stabilization and lifting mechanisms implemented in the robot, the user will be allowed to stand upright and move safely and reliably, thus fully interacting with the environment. The user fixation and safety systems will be implemented according to the different ergonomic requirements, in order to guarantee maximum adaptability to the individual user and convenience when performing various activities.

3 Robot Remote-Control System

When examining the remote-control system, one must take into account the fact that we are looking at a robotic system that is not a standard robo-assistant. In our case, we have a "transformer" robot that transforms and has two working modes. These two modes are the main criteria in defining human interaction requirements and functions.

Interaction method—graphical interface, with integrated buttons and software joystick, visualization of robot video and robot status.

Interface functions

- automatic menu switching for the different operating modes, receiving data on the robot's status
- autonomous navigation
- button control
- additional interactive capabilities: notification by voice messages, control by voice commands

Remote control is control of the mobile platform using a speed controller. Interface commands are generated by a software joystick or buttons. According to the submitted commands, a regulator takes care of the execution of the task. The regulator has an acceleration limit and a maximum speed limit.

In assistant mode, the robot is controlled remotely via a tablet or phone. In wheelchair mode, control is provided by a built-in touch display or again by the user's device. The graphical interface visualizes the buttons and software joystick, video from the camera, as well as a map that the robot has made, shown in Fig. 3. So, the user has different methods of interaction and can be guided by both the video data and the map made.

In autonomous navigation mode, available in both forms of the robot, the user can very easily set a desired destination, and the robot will reach it autonomously.

An architecture of the overall control and telecommunication system has been developed. This architecture is divided into three levels—frontend, backend and robot software. The general appearance of the architecture is presented in Fig. 4.

The frontend part represents the visualization of the graphical interface for remote control, and functions for sending commands to the robot are also implemented in it. The backend part is a web server and a MQTT broker and database, this layer serves as a connection between the robot and the user. The database stores information such as a map of the room, saved locations from the map, settings for the robot and personal settings for the interface. All information from the interface to the robot and vice versa passes through them. An Internet connection to the robot can be used through this layer. The last layer is the software of the robot, which is packages for controlling the individual elements of the robot such as actuators and sensors. All data information is communicated through messages to the other layers.

Fig. 3 Overview of the remote-control graphical interface

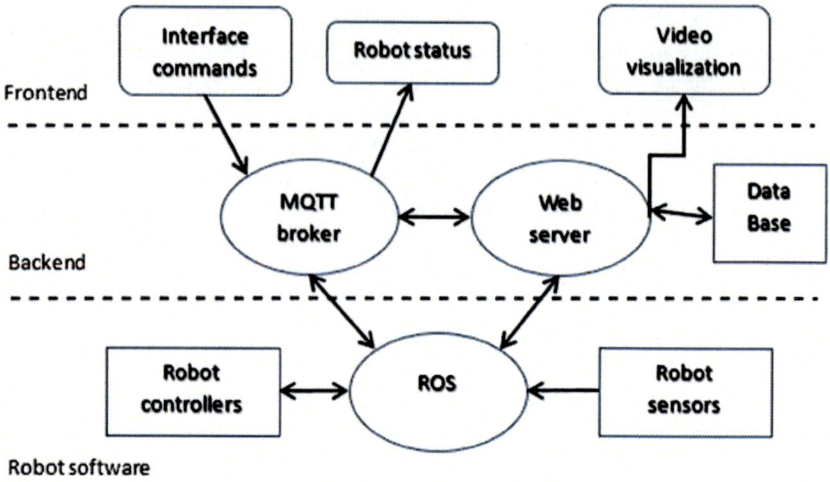

Fig. 4 Overview of the entire system architecture

3.1 Forming Messages from the Frontend to the Robot

In this study, we have already mentioned that we use two remote control methods: by direct movement commands from a software joystick and by using a map by sending information about a desired destination. The different remote-control methods are necessary because of the different operating modes of the robotic system.

In the first method, we send messages to the robot that contain movement commands—movement direction and movement speed for the entire robot. These velocities from the joystick are transformed by a law of motion of a differential mobile robot to each wheel—Eq. (1).

$$\begin{bmatrix} v \\ \omega \end{bmatrix} = 2\pi r \begin{bmatrix} 1/2 & 1/2 \\ -1/d & 1/d \end{bmatrix} \begin{bmatrix} \dot{\theta}_L \\ \dot{\theta}_R \end{bmatrix}, \tag{1}$$

where:

v is the vehicle's linear speed;

ω is the vehicle's rotational speed;

$\theta_{L,R}$ are the individual wheel speeds in revolutions per second;

r is the wheel radius;

d is the distance between the two wheels.

The inverse kinematics is derived from the previous formula, solving for the individual wheel speeds. It tells us the required wheel speeds for a desired vehicle motion (linear and rotational speed). We can find the inverse kinematics by inverting the 2 × 2 matrix of the forward kinematicso (2):

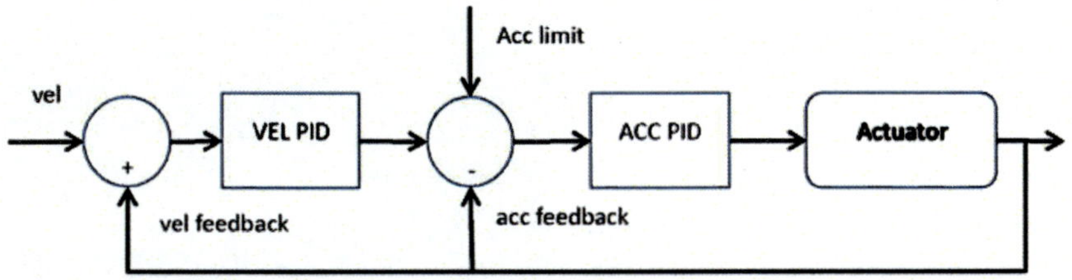

Fig. 5 Velocity and acceleration controller for each robot wheel

$$\begin{bmatrix} \dot{\theta}_L \\ \dot{\theta}_R \end{bmatrix} = \frac{1}{2\pi r}\begin{bmatrix} 1 & -{}^{d}/_{2} \\ 1 & {}^{d}/_{d} \end{bmatrix}\begin{bmatrix} v \\ \omega \end{bmatrix}, \quad (2)$$

Control commands processed by PID controllers are then applied to each driven wheel. We use a speed control controller to provide smooth stopping and starting by limiting the power supplied to the actuators. As an input parameter of the system, we use desired speed and acceleration limit, and for feedback, the data received from the encoders of the motors are used. Additionally, an internal power limiter has been added to the controller, which uses data from an acceleration sensor for feedback.

Movement speed is sent in messages that contain two vectors: for speed and for direction. Each vector consists of three variables x, y, z. Given the type of mobile platform, valid parameters use variable x from the velocity vector and variable z from the direction vector. All other variables must be set to 0.the format of the messages is shown in Fig. 5.

3.2 Controlling the Robot in Autonomous Navigation Mode

In this control mode, the means of interaction between the robot and the user is the map made and the menu for selecting a waypoint. This method provides very convenient options for working with the robot, choosing a desired destination, the user only needs to send a start command (Fig. 6). The robot then autonomously goes to the set location and reports that it has arrived. Another important option is saving destinations. At any position of the robot, as long as it is not moving, new entries can be added.

The map visualization also includes marking the location of the robot on the map, marking different objects from the environment, as well as visualization of the already saved destinations. The map itself is also interactive and allows, in addition to selecting new destinations to save, choosing a direct destination to reach, marking already detected objects and visualizing new objects in real time.

With this type of control, a message is sent to the robot, which contains information about the coordinates of the selected destination from the map made, as well as information about the desired orientation of the robot, shown in Fig. 7. The message

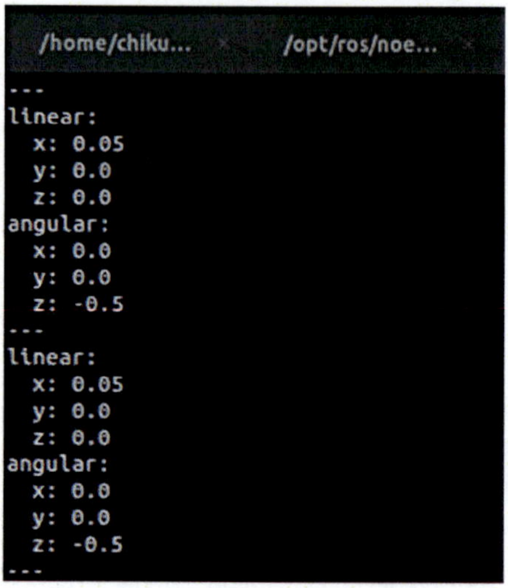

Fig. 6 Velocity commands send to the robot vie the web interface

about the position of the robot is in the form of a vector with three components—x, y, z. The orientation message is a quaternion with elements—x, y, z, w. All saved destinations are saved in this format and when selecting a destination, the corresponding parameters are sent (Fig. 8).

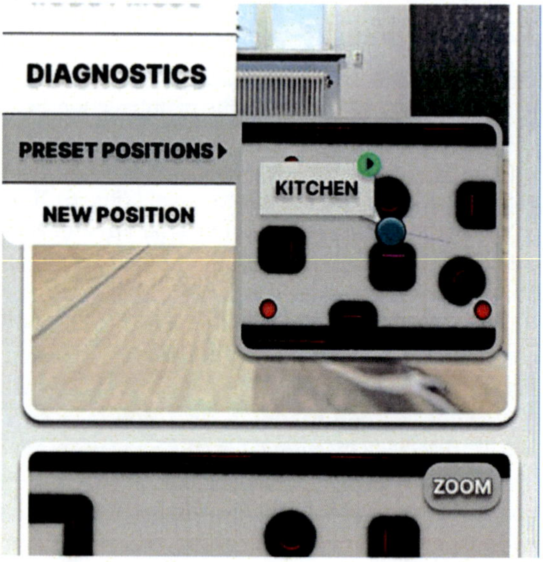

Fig. 7 Working with saved destinations over the map

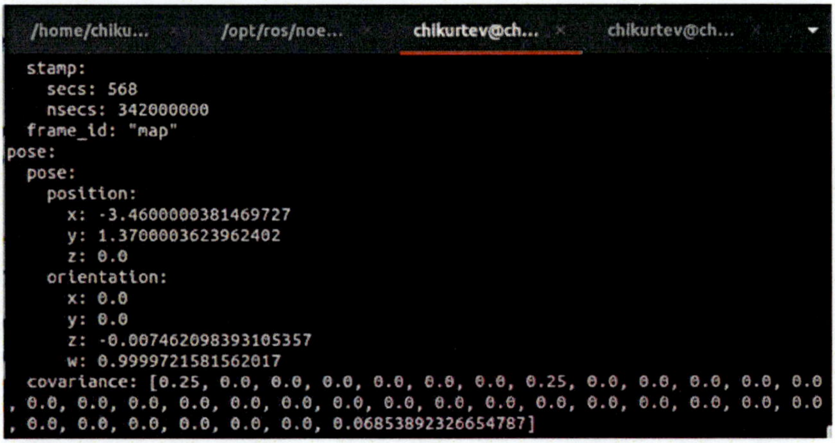

Fig. 8 The position and orientation message send to the robot navigation system

4 Conclusion

The development will contribute to the significant improvement of the quality of life of people with physical disabilities in the lower limbs. The developed robotic system will allow users to live more independently and interact with their living environment fully and freely without the need for a companion. Through the established work modes, users will be able to perform daily tasks much more comfortably, easily and efficiently without the need for stressful and health-straining movements. Their mobility will be optimized, more comfortable and reliable through the implemented robotic system.

The presented remote-control system allows the user to control the robot precisely and easily, while at the same time he can directly observe the events or use the information provided to him in a graphical interface. Development of voice interactive features and automation of delivery processes is envisaged for future research.

Acknowledgements This research is carried out as part of the project funded from the Bulgarian Ministry of Education and Science under the National science program "INTELLIGENT ANIMAL HUSBANDRY", grant agreement # D01-62/18.03.2021.

References

1. Olatunji, S., Potenza, A., Kiselev, A., Oron-Gilad, T., Loutfi, A., Edan, Y.: Levels of Automation for a Mobile Robot Teleoperated by a Caregiver (2021)
2. Cho, S.K., Jin, H.Z., Lee, J.M., Yao, B.: Teleoperation of a mobile robot using a force-reflection joystick with sensing mechanism of rotating magnetic field. IEEE/ASME Trans. Mechatron. **15**(1), 17–26 (2010). https://doi.org/10.1109/tmech.2009.2013848

3. Che, Y., Culbertson, H., Tang, C.W., Aich, S., Okamura, A.M.: Facilitating human-mobile robot communication via haptic feedback and gesture teleoperation. ACM Trans. Hum.-Robot. Interact. **7**(3), 1–23 (2018). https://doi.org/10.1145/3243503
4. Soyguder, S., Abut, T.: Haptic industrial robot control with variable time delayed bilateral teleoperation. Ind. Robot. Int. J. **43**(4), 390–402 (2016). https://doi.org/10.1108/ir-12-2015-0213
5. Jankowski, J., Grabowski, A.: Usability evaluation of VR interface for mobile robot teleoperation. Int. J. Hum.-Comput. Interact. **31**(12), 882–889 (2015). https://doi.org/10.1080/10447318.2015.1039909
6. Poncela, A., Gallardo-Star, L.: Command-based voice teleoperation of a mobile robot via a human-robot interface. Robotics **33**(1), 1–18 (2014). https://doi.org/10.1017/s0263574714000010
7. Nandikolla, V., Ghoslin, B., Matsuno, K., Medina Portilla, D.A.: A brain-computer interface for teleoperation of a semiautonomous mobile robotic assistive system using SLAM. J. Robot. **2022**, 1–12 (2022). https://doi.org/10.1155/2022/6178917
8. Labonte, D., Michaud, F., Boissy, P., Corriveau, H., Cloutier, R., Roux, M.A.: A pilot study on teleoperated mobile robots in home environments. In: 2006 IEEE/RSJ International Conference on Intelligent Robots and Systems, Beijing, China, 2006, pp. 4466–4471. https://doi.org/10.1109/IROS.2006.282082
9. Luo, J., Lin, Z., Li, Y., Yang, C.: A teleoperation framework for mobile robots based on shared control. IEEE Robot. Autom. Lett. **5**(2), 377–384 (2020). https://doi.org/10.1109/LRA.2019.2959442
10. Herve, S.A.: Computer interface for remote teleoperated mobile robot based on LabVIEW and open source hardware. Des. Constr. Maint. **2**, 115–124 (2022). https://doi.org/10.37394/232022.2022.2.17

Methodology for Development of Virtual Reality Based Resources for Education

Denis Chikurtev, Ava Chikurteva, and Nina Bogdanova

Abstract The modern development of education is directly related to the development of information and communication technologies and their application in the learning process. Virtual reality is completely changing the way knowledge is transmitted and acquired. The application of technology in learning provides an opportunity for students to learn through experience. The use of virtual reality as an educational resource aims to visualize objects that in the real world are dangerous or difficult to monitor. Technology uses computer-generated objects with which the user can interact in real time. When applying the technology in the learning process, we can differentiate the following problems: need for simulated environment, need for simulated object models and need for interactive functions. This study explores systems and tools for modeling and developing educational resources designed for virtual reality. Software tools and applications for creating and processing three-dimensional models are presented. Interaction methods have been studied providing human-virtual objects interaction in a simulated environment. A methodology for optimizing the process of creating educational resources with virtual reality has been developed. The methodology presents all stages of the process of creating these resources—development of a simulated environment, development of objects and development of interactive functions for the user. The proposed methodology has the potential to improve the process of creating digital resources for virtual reality and increase their integration in learning.

Keywords Virtual reality · 3D modeling · Educational resources · Intelligent animal husbandry

D. Chikurtev (✉) · A. Chikurteva
Institute of Information and Communication Technologies, Bulgarian Academy of Science, Sofia, Bulgaria
e-mail: denis.chikurtev@iict.bas.bg

A. Chikurteva
e-mail: ava.chikurtev@iict.bas.bg

N. Bogdanova
Department of Microelectronics, Faculty of Electronic Engineering and Technologies, Technical University of Sofia, Sofia, Bulgaria
e-mail: spasova@ecad.tu-sofia.bg

D. Vlachos (ed.), *Mathematical Modeling in Physical Sciences*, Springer Proceedings in Mathematics & Statistics 446, https://doi.org/10.1007/978-3-031-52965-8_26

1 Introduction

Education and technology have developed rapidly in the last two decades. Education produces prepared personnel for the IT industry, who constantly create and improve new technologies that are integrated into education. This becomes a cycle of knowledge that supports the development of education, and from there all other branches of science, industry, medicine, etc. Many studies point to the advancement of science and technology as the main transformer of education, and future education—full of imagination and innovative technologies [1, 2].

One of the technologies that managed to change the way of visualization and perception of objects and processes is virtual reality. Integrated into an educational environment, this technology has provided 'access' not only to remote, inaccessible or dangerous places, objects and processes, but also to virtual worlds, a figment of the imagination.

An important feature of the technology is interaction, the ability to interact and change the virtual environment. Integrating virtual and augmented reality in education changes the organization and methodology of the educational process, also changes the role of the teacher, who becomes an innovator, makes the connection between formal and informal learning, connects theory with practice, and activates creativity in his students [3, 4].

Virtual reality is a new way of human–computer interaction. Technology can greatly improve the accuracy of educational knowledge by making full use of the Internet platform to quickly and conveniently update content.

According to various studies, VR is an excellent tool for engineering education, in the form of a blended learning approach compared to the traditional learning approach [5].

Because of its pedagogical and cognitive benefits, it improves student understanding of subjects, performance and grades, as well as the educational experience [6].

In higher education, the largest application of VR technology is observed in education in the fields of medicine and engineering [7–10]. An intersection of these two major fields of knowledge is teaching students to use, retrieve information and apply various innovative technologies during their work.

In the field of animal husbandry, the training of veterinarians, animal engineers and even farm support staff can be aided by technology. VR will enable the formation of knowledge and skills for working with animals—rearing, feeding, observing behavior, etc. On the other hand, the technology will allow conducting studies and experiments without using animals [11].

In this research, we present the concept of applying virtual reality in animal husbandry as a means of training students, farmers and specialists. The idea for this concept is embedded in the National Scientific Program Intelligent Animal Husbandry. The aim is to prove that virtual reality can support and improve educational processes.

1.1 Benefits and Limitations of Using VR in Education

Recent years have seen increased interest in the use of VR in education, with most research focusing on the benefits and less on the limitations and problems associated with the use of the technology.

The main benefits are the compelling, immersive experience that motivates participants to use the technology again. The possibility to "experience" the observed processes, interaction with the observed objects and environment, "access" to hard-to-reach or dangerous objects and environments for observation, developing the imagination, etc. [12].

Limitations and problems in using VR for education have also been found in quite a few studies [13, 14]. If a few years ago the main obstacle to the implementation and use of VR in education was the price, today the main concerns are related to the lack of a clear methodology to link the technology to the goals of education, the lack of educational resources that meet personal needs of the specific teacher and group of learners, the evaluation—as in most studies it is stated that the result is reported through Multiple choice questions (MCQ), which many researchers consider to be an inappropriate and insufficient tool for evaluating results.

Therefore, the use of technology requires researchers and educators to carefully consider their assessment tools, perhaps incorporating mixed methods to provide a more holistic view of learning achieved [15]. It is also noted that learning theories are often not considered during VR application design [13] and the need for research to evaluate VR technology with emerging teaching and learning approaches.

And last but not least—there is no conclusive evidence that technology leads to an increase and improvement of students' knowledge. All these points should be investigated further to prove the effectiveness of VR as an educational technology. Until then, we suggest that it should be considered as a motivational technology, arousing students' interest in the learning process, and an illustrative technology [14].

2 Modeling and Developing Educational Resources Designed for Virtual Reality

VR is a 3D environment created by a combination of software and compatible hardware. This enables the user to interact with the virtual world in a 'real' way. When using VR for training, we can distinguish two main categories: desktop VR (D VR) and immersive VR (I-VR). In D-VR, the user interacts with the virtual environment on a computer screen with traditional keyboard and mouse hardware. I-VR provides a sense of immersion in the environment through 360° visualizations using a head-mounted display (HMD) and auditory stimulation through the use of headphones.

Typically, an HMD features an array of embedded liquid crystal displays (LCDs) that present each eye with an image from a slightly different angle. This mimics natural optical function, allowing the user to view a stereoscopic image full of depth perception and a wide field of view. Mobile VR headsets can achieve the same effect by using a single display, splitting the screen down the middle and presenting each half to the corresponding eye [13].

Depending on the learning needs, we can also distinguish three specific subcategories of VR: games, simulations and virtual worlds. Games provide free movement in the virtual world; they are suitable for testing hypotheses and increase motivation to achieve goals. Simulations are designed to recreate a real-world environment that allows testing of variables and resulting outcomes [14].

All subcategories have their place in the educational process, and the choice of a game, simulation or virtual world depends entirely on the educational goals and preferences of the teacher.

To achieve the expected results of training involving VR, several important aspects must be taken into account, which we can tentatively distinguish between technical and pedagogical. They are mutually dependent and reflect the peculiarities of educational systems and the individuality of teachers.

Pedagogical aspects:

- Type of lesson—lecture, exercise, seminar, etc. It depends on the type of the lesson, what are its goals, which determine the choice of the type of VR;
- Teaching methods—To achieve results in learning with VR, the technology needs to be connected to the didactic framework. VR can be the only method in the lesson, but it can also be combined with other methods. This requires good organization of study time;
- Age groups—The age of the students matters in the right selection of the type of VR and the length of the "experience".
- Familiarity with the technology—if students are using the technology for the first time, especially if it is I-VR with an HMD, they need to be prepared in advance to ensure their safety in the space during and after using the HMD.
- Time management—Depending on the main objective of the lesson, the VR experience may require repetition. In such cases, the goal is to acquire a specific knowledge or skill, and the experience in a virtual environment is part of the training. According to a number of studies [8, 9, 13, 16], virtual learning can be a successful precursor to real-world implementation [18]. VR technology can be useful in learning about hazardous occupational subjects without risk to yourself or others.

Technical aspects:

- The level of technical knowledge and skills of the teacher is important when working with VR. If this is a new technology for the teacher, then he needs instructions on how to work with it;
- Selecting the appropriate VR category and subcategory;

- Correct selection of appropriate educational resources, meeting the specific goals and objectives of the lesson and the teacher. This is often a difficult task because educators have individual resource requirements. Creating other types of digital resources does not require deep technical knowledge, but creating VR resources does. Often, educators use ready-made resources from various educational platforms without being able to edit or upgrade them.
- Duration of the VR experience—set the duration limit of an experience (educational resource).

3 Systems and Software for Creating Virtual Reality

Virtual reality systems are complex software products that integrate multiple libraries, hardware support, models, and more. They are usually made up of a number of additional software modules and subsystems, allowing remote access or connection via the Internet. Some of the main subsystems are presented in Table 1. Popular systems for creating virtual reality are Unity, Nvidia Omniverse, RoT Studio and others. Here we should clarify that virtual reality creation systems provide the full set of tools for creating, implementing, building and distributing ready-made virtual worlds.

Another important element is programs for modeling and creating three-dimensional objects. They are usually used by designers and allow them to create models of objects and scenes by using the methods of engineering and graphic design.

Table 1 Types of VR subsystems

Systems	Description
VR visualization	This type of software allows users to experience aggregated data in a virtual environment
VR Content Management Systems	Tools to collect, store and analyze VR content in a centralized location
VR SDK	Software Development Kits for VR (SDK)—provide the necessary base for designing, building and testing the technology
VR game engine	This software provides developers with the essentials for creating VR video game experiences
VR social platforms	Tools that allow users to collaborate in the VR environment, regardless of their location
VR training simulator	Employee training tools in a fully immersive environment

3.1 Virtual Reality Software Features

VR software is equipped with numerous features so that developers can create a complete VR experience. Several categories of functions can be defined according to the way of working with the software and the types of operations.

- **Content management**—Many tools allow users to upload either raw 3D content that will later be edited into a VR experience, or existing VR content directly into the platform. Users can manage and store content on these platforms and sometimes publish directly from them.
- **Editing content**—Most VR solutions have editing capabilities. Users can edit raw 3D content or existing VR experiences. Some editing features have drag-and-drop capabilities, meaning users can edit their VR content with little or no coding experience.
- **Hardware integration**—Any VR solution must integrate with hardware that supports VR experiences. These devices are usually headphones, but can also include mobile phones.
- **Cooperation**—VR tools allow multiple users to remotely access the solution at once so they can collaborate in real time. While collaborating, users should be able to interact on the same things at the same time, as well as watch games or events together.
- **Analysis**—Some VR solutions will provide users with analytics capabilities. This will allow businesses to better understand the behavior of audiences accessing VR content.

4 Methodology for Optimizing the Process of Creating Educational Resources with Virtual Reality

As mentioned above, the selection of an appropriate educational resource is an important element of the preparation of the learning process, which guarantees the achievement of the lesson's objectives. Creating VR educational resources has several stages. Software developers create the virtual world, then it is the visualization that allows interaction with the objects created by the developers. These 3D objects react to changes in the user's movement and the interactions mimic those in the real world.

There are additional hardware components, accessories that can simulate additional senses. Integrated into education, VR can take hands-on learning to the next level.

Instead of watching a teacher do a task, a student or learner can do it virtually. Some VR tools allow users to design, sculpt, model, paint, and create tangible objects in a VR environment. Users can interact with 3D models from any angle with tools like these.

The purpose of the methodology is to assist teachers in preparing lessons that include virtual reality. As here, it should be noted that to work with virtual reality software, you need to go through specialized training. Working with similar applications and software is specific.

In this research, we propose a simplified methodology that includes five main steps (Fig. 1): modeling a 3D object, modeling a 3D environment, integrating the object into the environment, modeling interactive functions between a participant and the 3D object and the environment, determining interrelationships between the interactive functions and the 3D objects. The order of execution of the steps may not be strictly defined, even in most practices the steps are executed in parallel.

Object modeling is the process of creating three-dimensional objects, usually through 3D modeling software. In this stage, an object must be created that corresponds to a real one, with specific dimensions, strokes, colors and other characteristics. An already created object can be integrated into any virtual world.

A 3D environment is usually modeled in virtual reality platforms. These platforms provide all the necessary features and capabilities for modeling and building 3D worlds. After creating the virtual world and the object we want to use for a given purpose, the next step is to integrate the object into the 3D environment. These three steps are common knowledge and apply to any content creation.

It then moves on to modeling interactive features. This means defining the methods that will be used for interaction. Such methods are: direct contact with an object or environment, buttons, sliders, drop-down menus and others. With the help of the various interactive methods, we determine what the user's possibilities will be to change and observe the virtual environment.

When we have defined interactive methods, it follows creating relationships between the interactive functions and the 3D objects. Here we define how the objects will react when using the various interactive methods. Limits, automated changes and other features are set.

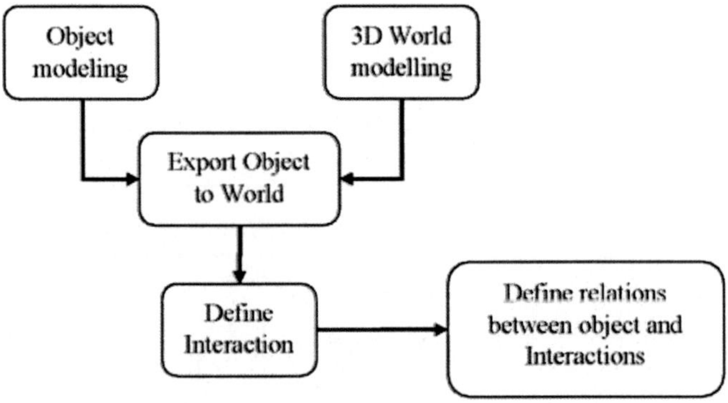

Fig. 1 General overview of the proposed methodology

5 Development of Virtual Reality According to an Example Lesson

The presented example is a continuation of previous research related to learning using modern methods and digital technologies [17]. This research is directly related to the Intelligent Animal Husbandry project, dealing with research in the direction of developing technologies to improve animal husbandry processes, improve the qualifications of veterinarians, and improve the educational process for students. The project develops new approaches to automating and robotizing processes, tracking the condition and vital signs of animals, and educational resources for training. The models presented in this study can be used for animal visualization, simulation of different impacts and training. Specifically, this article looks at models and techniques for visualizing data and interacting with the simulated models.

For interaction, we use a virtual window that has various options—a drop-down menu, a slider, a button, a switch. All these options give a different method of interaction depending on what we want to change. In order for the user to operate these interactive tools, an HMD must be used. In order for a human to interact with the simulated model, links must be added between the interactive window and the model. The general appearance of the user-virtual world interaction system is presented in Fig. 2. The user interacts with the environment and the object by clicking on buttons, selecting menus, or directly touching the object. Changes made to the environment and the object are perceived through visual contact, and specific parameters are written on an information panel. In order to make contact, the user must use a virtual reality hardware set.

Scenario of the developed resource: the user is an observer who interacts with the object, the object of observation is an animal—a cow. The animal can go through different phases of its development and the observer can track the changes in size, shape, color and more. The user can set the age or gender of the cow from an interactive menu. When a specific parameter is selected, the 3D model of the animal

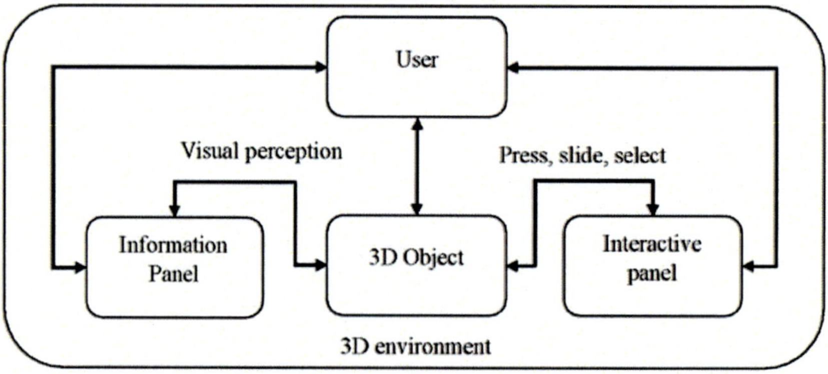

Fig. 2 Relations between user and virtual environment objects

changes, along with the information in the information panel. In addition, the user can directly modify the cow itself, thereby changing its age and weight. At this stage of the interactive panel, age, gender, weight, pregnancy and stage of pregnancy can be set. An example of the interactive window is shown in Fig. 3a. The left column of the window shows the different types of buttons and attributes that can be used for interaction. On the left are the elements themselves and supporting text for them. At the bottom of the figure, you can see the joysticks that are in the hands of the user. By pressing buttons on them and pointing to the various items, he can make changes.

Changes in the parameters of the animal are implemented through special interactive menus. For the correctness of the data, information from verified sources—textbooks on veterinary medicine—was used. Some of the data used are presented in Tables 2 and 3.

Information about the specific animal is visualized in an information panel: sex, age, age group, pregnancy, kilograms, recommended amount of food and others. It is presented in Fig. 3b. The values and units of measurement for each parameter are

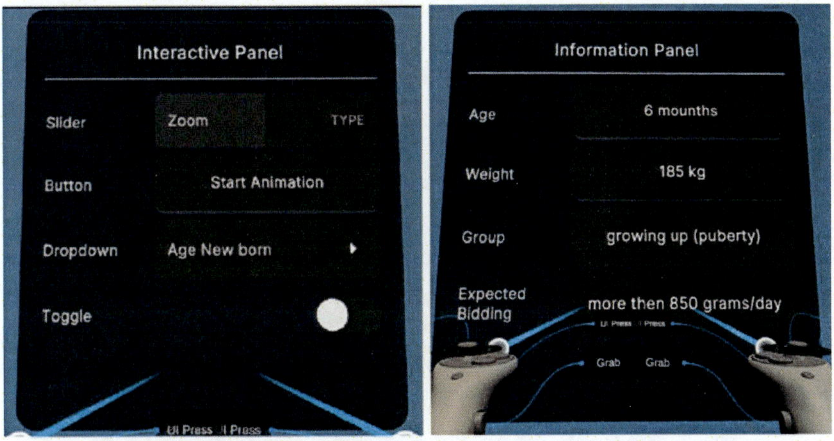

(a) (b)

Fig. 3 **a** Interactive panel. **b** Interactive panel

Table 2 Parameters for cow age, weight and bidding

Addiction		Bidding
Age (months)	Weight	
3	95–110	>850 gr
6	180–200	>850 gr
12	330–350	800 (<900)
15	380–400	600–700 (<800)
24	600–620	750–850

Table 3 Cow parameters for educations

Age groups	Age (months)	Physical assessment	Coefficient
Growing up	4–6	Calving	3.5
Growing up	6–12 (puberty)	Fertilization	2.75–3
Pregnant	12–22	Late lactation	3.25
Close-up	22–24	Dry period	3.5

clearly visible. This panel intentionally has no interactive features, only visualizing information.

The 3D model of the cow is shown in Fig. 4. The user can walk around it, observe it from different directions and make changes by again using the joysticks. Two interactive features have been added to the cow model itself to change its age and weight. Actions can also be added to show the cow moving, grazing, giving birth and more. Additional 3D models are required for these functions, which will be included in a later stage.

In order to achieve simultaneous interaction with the 3D object and the interactive panel, an architecture was developed that connects these two objects and they actually share the same parameters, shown in Fig. 5. The state of the parameters can be changed by any object, with the criterion for change being the last action performed.

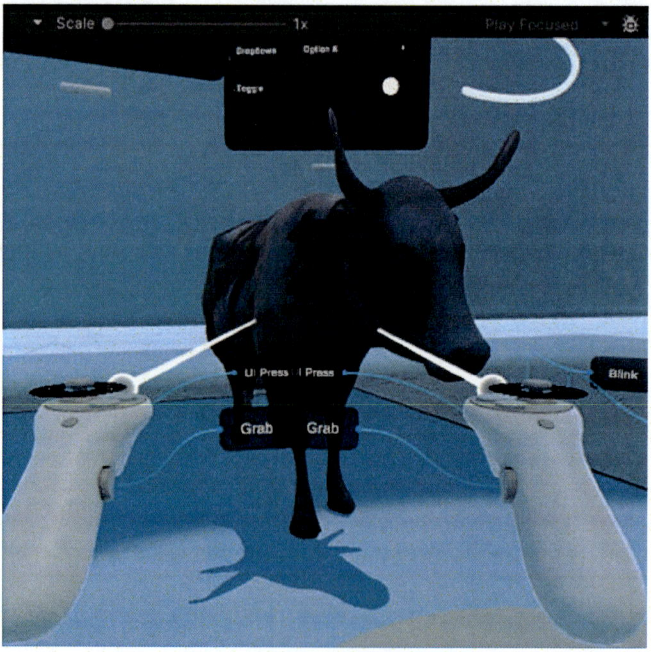

Fig. 4 Integrated 3D cow model in the virtual environment and interaction options

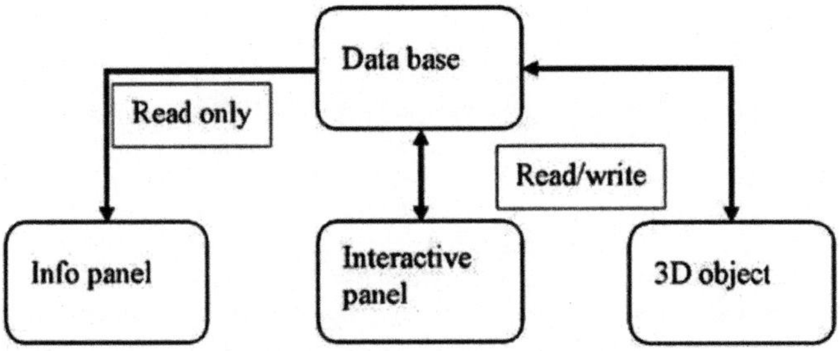

Fig. 5 Architecture design of the virtual reality objects and database

Objects have rights to read and write to the database. And at this time, the information panel shows the current status of the parameters, it only reads the information. This functionality is of great importance for the dynamics of processes and the correct visualization of all objects in the project.

The system thus developed can be upgraded with multiple objects and various interactive functions. The database can be expanded and cover absolutely all necessary parameters, according to the type of lesson or exercise. Interrelationships between objects can be made by sharing a common database.

6 Conclusion

The research in this paper presented a methodology for building interactive lessons based on virtual reality. The methodology provides clarity on the necessary steps and actions of the development process. In support of the methodology, an educational resource developed in an environment for creating virtual reality, where all the described steps are performed, is presented. A 3D model of a cow, an interactive panel and an information panel are integrated into the developed one. For the functioning of the simulation, an algorithm for data exchange between individual objects is implemented.

Such a development can significantly support the training processes of students and specialists in the field of animal husbandry. They will be able to easily familiarize themselves with the anatomy of the animal, the correct nutritional norms and standards. At this time, it is not necessary to visit stables and make long and tiring observations.

Another important aspect is the variety of possible worlds and models. With the help of virtual reality, ready-made simulations can be upgraded and in-depth, interesting and effective training can be achieved.

Acknowledgements This research is carried out as part of the project funded from the Bulgarian Ministry of Education and Science under the National science program "INTELLIGENT ANIMAL HUSBANDRY", grant agreement # D01-62/18.03.2021.

References

1. Zhang, Q., Wang, K., Zhou, S.: Application and Practice of VR virtual education platform in improving the quality and ability of college students. IEEE Access **8**, 162830–162837 (2020)
2. Stoeva, M.: Educational design for building the competence "dealing with conflicts" in the learning process through projects in technical subjects. In: 2022 20th International Conference on Emerging eLearning Technologies and Applications (ICETA), Stary Smokovec, Slovakia, pp. 607–612 (2022). https://doi.org/10.1109/ICETA57911.2022.9974893
3. Shopova, V.P., Dimitrov, I.M.: The place of augmented reality in the learning process
4. Stoeva, M., Stoev, P.: Educational technology integrating the project and problem-based approach in technical disciplines for the formation of creativity and creative thinking in engineering students. In: 2022 International Conference Automatics and Informatics (ICAI), Varna, Bulgaria, pp. 378–382 (2022).https://doi.org/10.1109/ICAI55857.2022.9960093
5. Serin, H.: Virtual reality in education from the perspective of teachers. Amazon Investig **9**(26), 291–303 (2020). https://doi.org/10.34069/AI/2020.26.02.33
6. Soliman, M., Pesyridis, A., Dalaymani-Zad, D., Gronfula, M., Kourmpetis, M.: The application of virtual reality in engineering education. Appl. Sci. **11**(6), 2879 (2021)
7. Oubahssi, L., Mahdi, O.: VEA: a virtual environment for animal experimentation. In: 2021 International Conference on Advanced Learning Technologies (ICALT), July 2021. IEEE, pp. 422–424 (2021)
8. Plotzky, C., Lindwedel, U., Sorber, M., Loessl, B., König, P., Kunze, C., Kugler, C., Meng, M. (2021). Virtual reality simulations in nurse education: a systematic mapping review. Nurse Educ. Today **101**, 104868
9. Baniasadi, T., Ayyoubzadeh, S.M., Mohammadzadeh, N.: Challenges and practical considerations in applying virtual reality in medical education and treatment. Oman Med. J. **35**(3), e125 (2020)
10. Stoeva, M., Stoev, P.: Educational concept of integrated project-oriented learning for the formation of "problem solving" competence in electrical engineering disciplines. IFAC-PapersOnLine **55**(39), 187–191 (2022). https://doi.org/10.1016/j.ifacol.2022.12.053, ISSN 2405-8963
11. Sekiguchi, T., Makino, M.: A virtual reality system for dissecting vertebrates with an observation function. In: 2021 International Conference on Electronics, Information, and Communication (ICEIC), January 2021. IEEE, pp. 1–4 (2021
12. Kiryakova, G.: To read or interact with textbooks-what is better for learners? Math. Inform. **62**(5), 517–526 (2019)
13. Hamilton, D., McKechnie, J., Edgerton, E., Wilson, C.: Immersive virtual reality as a pedagogical tool in education: a systematic literature review of quantitative learning outcomes and experimental design. J. Comput. Educ. **8**(1), 1–32 (2021)
14. Jensen, L., Konradsen, F.: A review of the use of virtual reality head-mounted displays in education and training. Educ. Inf. Technol. **23**, 1515–1529 (2018)
15. Raja, M., Priya, G.G.: Conceptual origins, technological advancements, and impacts of using virtual reality technology in education. Webology **18**(2) (2021)
16. Kiryakova, G., Angelova, N., Yordanova, L.: The potential of augmented reality to transform education into smart education. TEM J. **7**(3), 556 (2018)
17. Chikurteva, A., Atanasova, T.: Application of project-based learning in animal husbandry using flipped classroom and virtual reality. In: ICERI2022 Proceedings. IATED, pp. 3786–3795 (2022)
18. Dimitrova, K.: Project-based education in primary school with the use of virtual and augmented reality. In: ICERI2022 Proceedings. IATED, pp. 2975–2983 (2022)

Non-euclidean Geometries and Arts in the Beginning of the Twentieth Century: An Interdisciplinary Approach for High School Students

Adamantia Mavrogianni and Georgios Kokorelis

Abstract The action we are going to describe was carried out in a class of 27 students in the second grade of the High School (16–17 years old) IONIDIOS MODEL LYCEUM OF PIRAEUS, in the school year 2018–2019. The cooperative action of the mathematician and the philologist of the class, resulted from the common finding of a deficiency in the curricula which resulted in the students' inability to approach holistically the changes imposed on society by the inseparable relation between art, science and technology. With this in mind, the action was designed and focused primarily on presenting trends in art and painting from Impressionism to Van Gogh in parallel with the progress and changes in science and society during that time. It was then attempted to link science, art and technology in the early twentieth century through an approach of artistic manifestations, principles and theories about mathematics and non-Euclidean geometries as they influenced architecture, sculpture, painting, the Bauhaus movement. The action was accomplished by achieving its objectives, since the artworks of high aesthetics produced by students, effectively incorporated the theory of modern mathematics and artistic avant-garde and highlighted the link between the scientific and artistic achievements of human creation.

1 Introduction

Playing with the words, students of the second grade of **Ionidios Model Lyceum of Piraeus**, named our project Schoo-π-(**Skou-pi-dia,** meaning **rubbish** in Greek language) trying to embed the main goal of this experimental teaching action. The interdisciplinary/cross-curricular course was designed in 4 modules of 2 or 3 teaching hours each (totally at about 12 h):

A. Mavrogianni (✉)
Sivitanidios Public School of Arts and Professions, Kallithea, Greece
e-mail: amanda.mavrogianni@gmail.com

G. Kokorelis
Ampeteios High School of Cairo, Cairo, Egypt

D. Vlachos (ed.), *Mathematical Modeling in Physical Sciences*, Springer Proceedings in Mathematics & Statistics 446, https://doi.org/10.1007/978-3-031-52965-8_27

Introductory section (Philologist)

Part A: The reality up to Van Gogh.

Part B: Social and Political Reality on the beginning of the Twentieth Century.

Connecting science, art and technology at the beginning of the 20th century/ manifestos, avant-garde, architecture, mathematics, natural sciences, sculpture, cubism, Bauhaus, Duchamp (Mathematician and Philologist co-teaching).

Case Study:

(A) Non-Euclidean Geometries and Art. Theory of Basic Mathematical Concepts and Demonstration of Specific Artworks identifying Influence and Connection (Mathematician).
(B) Manifestos: Analysis, Related Texts (Philologist).

Applications

Student groups were designed to create texts, videos, visual or sculptural works or constructions, using knowledge and influence from what was learned during the action (all students and teachers).

2 The Project

From the outset and for a smooth and meaningful transition to the action, was noted, preceded by a relevant introductory skills assessment questionnaire for students, the need for a concise, but at the same time dense and holistically given, consideration of the basic parameters that conditioning the development and shaping of European culture, as it was consolidated with the Industrial Revolution.

2.1 Part A

Already in the late nineteenth century, Europe had transformed itself. Similar institutions had been created almost across the continent. The company, the bank, the stock exchange, the insurance companies, the securities and the shares, the market had defined their position in the capitalist system imposed by the gradual change of economic and social structures that the industrial Revolution had imposed. Businessmen, traders, engineers, bankers could communicate with each other after talking about the same things that little—much worked the same way. The common place is profit, prosperity, progress.

And it is this notion of progress that had triumphed over the century as a result of the rapid evolution of science and technology. Chemistry, electricity does not only cause top changes in the industry, but it establishes a new way of life, gradually transforming the everyday life of the European resident-citizen and making change and evolution a permanent goal, proof of social and professional development.

It was, of course, preceded by the middle of the century Darwin's theory of the evolution of species, whereby the best gifted species survive, those that manage to adapt to change. Darwin's biological evolution theory has imprinted the new social reality. The laboratory of the biologist, the chemist, the physicist, every scientist becomes a pole of attraction like everything innovating. As the novels of Jules Verne testify, the professor became a very important, for the industry, form, while his laboratory became an integral element of industrial development. Science penetrates the industry and the educational system is becoming more and more vital to industrial development [6].

European citizens are trained, learn foreign languages, move, travel, produce and consume. In the period 1875–1914, with the increase of population, of urbanization and real incomes, the mass market, which until then was mainly in the basic survival of the species, began to impose its sovereignty on the consumer industries goods. This development was more important than the notable increase in the consumption of the rich. Revolution in the automotive industry did not bring the Rolls Royce but the Ford Model T [7].

On the other hand, students should realize that historical events and long durations are governed by reactions to regularities that are the result of processes that go beyond social constants and intersect with philosophical pursuits that frame ideological revisions to interpret and disrupt the harmony of the social fabric.

It was therefore necessary to refer to socialistic ideas that appeared at the beginning of the nineteenth century and they were approached philosophically with the Communist Manifesto, in the middle of the same century, which practically called to a revolution all those who their toil contributed, but without any share, to the wealth of the bourgeoisie. The overthrow of the social and political regime established by the consolidation of the industrial revolution was the pursuit of the communist movement founded by the Communist Manifesto (Marx and Engels 1848).

Because this aspect of Europe—which was the broad context and the canvas of our action—of the majority of the people of toil, who, by reflection or only as a dream, saw the rise of the less others, the workers who watched themselves as observers of a world that changes rapidly, employed our approach that briefly turned to literature to illustrate the depiction of everyday life within the pages of the emblematic texts of the time. Starting from the pages of Stendhal and Flaubert, we passed to Dickens and Zola to make transparent the spacious mosaic of the social and political upheavals of the nineteenth century.

The art of speech, serving realism with the nuances of morality and romance and the excesses of naturalism, helped students to understand the European culture that was shaping. The social dimension of the changes that occurred in the nineteenth century Europe was also approached through the leading literary texts. So they were able to feel, as long as possible, the momentum of developments at the scientific, technological and economic level and the consequences that this momentum had on the level of life of different social groups.

This approach to aspects of Europe's history and culture has, so far, been based on positivism and determinism, the visible consequences caused by the reaction to external stimuli. On the basis of the positivism, all the sciences were initiated and the changes that were inflicted during the nineteenth century were provoked. The improvement of the conditions of life, that is, was clearly due to observations of the visible world, the generalization of which as well as the proof that came from their repetition led to certainties that determined the how and the why of everyday life.

The same certainties, as it was natural, had influenced the art of the time and, in terms of painting, limited in the representation of the visible world. In the last quarter of the nineteenth century, the Impressionists, initially, targeted a painting of the objects of the outer world. The light that is visible imposes a vision of the world tailored to what with the senses is perceived. And how else could it be since the deification of the accusative had contained the artistic expression in a singular representation of the universe, based on the proven assumptions of positivism.

These same assumptions led to the European bourgeoisie, which in the meantime had, marginally, exploited every challenge and led European markets to growth saturation, in search of new, enlarged fields of action. The expansion outside Europe's boundaries was an one-way street. The goal is to conquer the rest of the world. The most striking consequence was the sovereignty over the land of some western regimes that has no precedent in history. This enlargement had the character of a colonial spread. In front of the merchants, locomotives, ships and cannons of the West, the ancient civilizations and empires of the world retreated and collapsed [8].

In the early twentieth century, Europe was torn apart by political confrontation and upheaval. Nationalism has taken over all the people of Europe. The period of optimism of European capitalism is retreating. The economic dominance of European forces is transformed into political and cultural. The widening imposition of European forces in Africa and Asia with their technology, science and ideas has led to competition which has charged relations with each other. The causal rules were probably unable to act as soothing and everywhere seemed to be the impasse.

At the same time the progress of the natural sciences shook the sense that positivism had for life. In new scientific theories the large and unified universe crumbles. The principles of randomness and uncertainty replace certainty and prediction. The subject, the scientist he observes, is considered a part of the system he observes. Therefore, the results of scientific observations in time and space cannot exist without the observer. The world-machine that exists beyond the scientist collapses.

At this point of action, as the evolution of social and scientific upheavals has been articulated and the holistic approach of reality has been assimilated, the order of art, which is the aim, is to highlight the issues that they arose from and with the friction of people with their inner world since the objective reality slowly gives way to the subjectivity of the moment and the particular personal view of things.

As early as the middle of the nineteenth century, the ingenious French poet Charles Baudelaire declares that the visible reality may be hiding another, a world of free imagination that is not linked to matter, which is less about the eye that it is catalogued and the hand that pals and more with the morphing force, with the autonomous creative disposition of the human spirit. Thus, Baudelaire formed the prerequisites for that subjective aesthetic that was to be fully developed in the twentieth century.

At the turning point of the twentieth century, the progress of natural science shook the sense that positivism had for life. The artists, influenced by the climate of the time and the introversion, who came after the saturation from the progress, highlight the interior of the objects and claim the personal look in a changing world that cannot be, anymore, delimited. These multi-layered changes are difficult to detect and even harder to interpret. The man, lever and catalyst of any creative approach, becomes the subject of observation and especially self-observation since Freud proclaims the psychoanalytic vision that explores the unconscious and tries to decode the riddle of human desires and choices.

The choices of an artist, the Impressionist Monet, over the course of thirty years, highlight the transformations that are performed in the European scene of art. The projection and examination of his tables, "Breakfast in the Grass", 1866, "The Madame Monet and her son", 1875, "The Bridge of Charing Cross", 1903, showed students not only the change in the artist's technique, but the way the changes of his environment altered his attitude towards things. While the first target table has the imprint of the outer world, the reality gradually degrades to highlight the object, as shown on the Bridge of 1903, as a vector of fleeting light in its alternating gradients [5].

The most important of the Impressionists is Cezanne since it is believed that he has established, with his work, the modern art. His work does not mimic nature, but he stands as a worthy testimony beside her. This proves the triumph of the human spirit in the seemingly disorganized multiplicity of the natural world that surrounds us. He built his compositions according to the geometric order and tried to make the orphic richness of nature in the most simple geometric symbols, sphere, cone, cylinder and prism.

Then followed the projection and examination of the work of artists who refused to merely portray their observations. They considered that the artist's job is to shape realities and to format his own individual visions. As very characteristically wrote Paul Klee "not to attach the visible but to make visible what is not". This phrase could be the first definition of twentieth century art.

To enable students to understand the unsettling changes observed in art during the first three decades of the twentieth century, we dealt, apart from the paintings, with texts by the artists themselves clarifying not only their ideas, but the principles and the course that followed in order to touch their artistic goal. Leger refers in 1911: "To be free, but without losing touch with reality, this is the drama of the epic man we usually call inventor, artist or poet. Days and nights, dark or bright, sitting in a blinding illuminated bar. Renewed things of shapes and objects bathed in artificial light. Trees are no longer trees, a shadow falls on the hand that touches the counter, an eye distorted by light, the changing figures of passers-by. The life of the extracts: a red fingernail, an eye, a mouth. The elastic game of complementary colours that transform the object into another reality. He inflates, gets drunk with every moment that gets through from all his sides. It is a sponge, a feeling of sponge, transparency, acidity, new realism" (Reed 1978).

In the text of Leger, where it is described very well its own personal imprint in art, are visible its origins, which touch on Cubism, but mainly are projected the way of search and personal introspection, the kind of questions concerning the artist, the absence of certainties, the changing perspective, the tendency to overturn, a modernist view on things that encapsulates the surrounding atmosphere and shows how best the artist acts as catalyst in the endless melting pot of the multi-level life.

Leger, apart from a painter, as shown in the above quote, is also a writer, struggling to detect his archetypal origins, to observe the way he sees reality, doubts, flames from questions concerning not only his art, but his world, his deepest being, in order to explicitly imprint, what at that time and in the phase of that search, his own reality is explained by his preconscious or unconscious. So the pupils -at the next step of the action- were confronted with the manifestos of Futurism, Dadaism and Surrealism- and captured, as far as possible, holistically the artistic production, to feel the agony of an artist for his existence itself, to understand the need to publish the manifesto with which the artists proclaimed the basic principles of their art.

Artists who chatted through their works, created groups to discourse and create many times in selected places. There was the need for common expression and drafting of principles and ideas that had a kinematics perception based on shared experiences and perceptions. The need to propagandize their collective action led them to the joint declaration of principles and the formation of an ideological framework that was reflected in the manifestos of artistic movements.

It was then emphasized to the students that, what characterizes all the avant-garde artistic movements of the early twentieth century, was a series of poets and writers played a very important role in their shaping. Thus, a series of manifestos were approached such as the Futurism Manifesto, a movement that was theoretically founded and organized by the Italian poet Philip Tommaso Marinetti. A group of painters was gathered around him and in their manifesto proclaimed that "we must get rid of the moldy and worn issues, so that the vertigo of modern life can be expressed, which is a life of steel, fever, pride and crazy speedu" (Reed 1978).

Among many texts we approached, we mention "in my Pegasus" by Marinetti, "the Mr. AA, anti-philosopher" by Tzara, "Cloud in Trousers" by Mayakovski, "the wounded dove and the fountain" by Apollinaire, pieces of Pessoa's "Book of Concern". Our attempt was to make students to understand that in a dramatically changing world, the art is unified and all the differences that may have a futuristic poem by Mayakovski with a poem by Apollinaire, what are proving is the common agony to express the unspeakable, to break down forms, verses and measures in order to illustrate the tendency for constant searching and the desire to give the new that depicts a piece of the new reality.

The dilation of the verse and the breaking of the form in the way it was presented in action when we played by placing the words in a naughty and insensible flow, following the instructions of Tzara "How to Make a Dada poem", showed the impetuous and dynamically modifying power of the tongue and the lust and demand against everything traditional and expressed, the breakdown of the logical sequence, a sequence, which at least until then, was one and exclusive against anything it was escaping the rule.

The shattered Europe of World War I, as a result of the most absurd and abominable positivism, was the canvas for the frantic creativity of the artists. The Dada artists wanted to get rid of every shipment of artistic and social tradition. They wanted to trigger the bourgeoisie. Duchamp put a mustache on the Mona Lisa, Picabia painted irrational machines that their only mission was the spoof of Science (Reed 1978).

With the elaboration of the manifestos and the projection of works made of their great representatives, the action paid off. The students understood that the approaches of natural sciences, such as relativity and quantum mechanics, changed the whole way of understanding the natural world while art absorbed, in a world that was changing at a frantic pace, every change, by radiating such its boundaries as any given cosmic certainty and assumption.

2.2 *Part B*

Euclidean Geometry is the Geometry that is mainly taught in Secondary education in Greece, but also in Europe. The teaching hours are not sufficient to deepen this lesson, and no reference is made to geometries other than Euclidean. In this didactic-pedagogical action great importance was given in acquainting students with non-euclidean geometries in a way that would excite them in an educational way but also to produce results if possible. This is how art, and especially the art of the early twentieth century, was chosen as the intermediary for this acquaintance.

In Greek language the two terms, art and technique (and today technology) have a common root, the verb tikto (= create, birth). These are two areas in which man strives to transcend nature and become a creator himself. While art is an exclusive and individual creation, it is implemented by means of scientific or technological knowledge, research and application. Also, the presence of multi-talented people moving at the same time in all areas (science, technology and art) reveals the close relationship and connection of science, technology and art. Science and art have been decaying for many years and only in the twentieth century they reconnect again.

Reconnecting science and art can be done in two ways:

Initially, artists use the ideas of science and technology and then their tools. At the beginning of the twentieth century, an art school, the Deutsche Werkbund, was founded in Germany, aimed at the co-operation of fine arts and crafts with industry. Thus was born the industrial designer whose role is to improve the product in terms of functionality and also to enhance its appearance to provide aesthetic enjoyment to the user.

Subsequently, the Bauhaus School in Berlin from 1919 resulted from the merger of the School of Arts and Crafts with the Graduate School of Fine Arts. The school's products were architectural works (eg Sommorfent villa in Berlin, experimental house in Weimar), furniture (steel pipe seats, chairs), lighting (ceiling and table), ceramics, fabrics, graphics (logos).At the same time, in the 1920s, in Russia, we have the beginning of a new art based on the principles of painting, sculpture and architecture resulting in the creation of monumental constructions that could be in every-day use by the people.

The artist had to have more knowledge of architecture and engineering than knowledge of the rules of photography. Art unites with life, monitors the structure and organization of society and aims to find solutions to practical and social problems. At first, fabrics, porcelain, furniture, clothes and even building installations are designed and later on utopian architectural constructions called "Architectures" are created. The highlight of the relationship between art and technology is Vladimir Tatlin's "Letatlin" art plan, a plan for a flying machine designed to release man from gravity.

In this project, we wanted to see how the artists' efforts to understand and capture difficult mathematical concepts became apparent through the artworks, thereby causing students to be impressed and to assist in their own understanding. Typical works of art have been chosen to perceive the influence of Non-Euclidean geometries on them, as well as to emphasize the inseparable relation between all creatures of human activity.

Initially there was a brief introduction to how the non-Euclidean geometries began, with reference to Euclid's fifth request. Here students hear for the first time in their student life how the challenge of the fifth claim of Euclid has led to new mathematical approaches. **The aim was not to refer to mathematical formalism in detail, but to scientific need.** Thus an initial approach was made through the observation: *New developments in Mathematics lead to the so-called 'foundation crisis'.*

That is, there are paradoxes that must be resolved at all costs, since mathematical practice—as a strict practical or, more properly, as the standard of rigor—obviously cannot tolerate contradiction, logical inconsistency and therefore paradox.

The need was thus identified and the request was made for the re-foundation of Mathematics. On this basis and over a relatively short period of time, three major efforts were undertaken: Logicism (Frege and Russell), formalism (Hilbert) and Intuitionism (Brouwer).Starting in the mid-nineteenth century, because there have been rapid and unexpected developments in Mathematics since then we refer to:

- the birth of non-euclidean geometries, that is, the possibility of having geometries that cannot be understood on the basis of space surveillance,
- Cador's set theory,
- the mathematical ideas underlying differential and integral calculus, that is, mathematical analysis, not through Geometry and space surveillance, but through Arithmetic, and here we imply a whole set of developments associated with the great names in the history of Mathematics.

All these developments, both in the part of Mathematics which we would formally call Geometry and in the part of Mathematics which we would formally call Arithmetic, no longer seem to fit the existing philosophical notions of the kind of knowledge Mathematics provide us. The twentieth century began with Hilbert setting 23 unsolved problems as the basis for further development of Mathematics.

He and his successors sought to standardize and rigorously base the whole mathematical construct through the "formal axiom method", inaugurating the philosophical school of Formalism. At the same time Poincaré founded General Topology and laid the foundations for the later philosophical school of Intuitionism. Russell tries to support Mathematics in Logic (the school of Logicism) and paradoxically shakes the Theory of Sets. Gödel will be the one to create a crisis similar to and maybe greater than that of asymmetries in ancient Greece, by Godel's incompleteness theorem.

From now on, mathematics and their philosophy will never be the same again. Paradoxes and incompleteness will give rise to new mathematical theories such as Place Theory, Category Theory, Proof Theory, and Model Theory. Following the brief overview above, the areas of positive and negative curvature in the Lobachevsky-Bolyai and Riemann geometries were then described to the students and specific references and examples provided:

First by Gauss and by the mathematicians Lobachevsky and Bolyai, who, working independently, formed the first non-euclidean system of geometry. The basic principle was explained to the students that given a straight line and a point outside a straight line, there could be more than one parallel to the straight line. These parallels will tend to the straight line, but will never intersect, with the sum of the angles of a triangle less than 180 degrees (Fig. 1).

Correspondingly, in a Riemann space, the lines are always intersected, while the sum of the angles of a triangle is greater than 180 degrees (Fig. 2).

The introduction to the idea of dimensions over 3, which was really quite difficult to approach, was chosen to do—**always with the aim of introducing it into the modern art of the early twentieth century**—initially with the demonstration of the development of the hypercube. In this case, the students saw how the three-dimensional cube could be expanded to four-dimensional. It was a big surprise for them. We have always to think that we are targeting 16–17 years old students (Fig. 3).

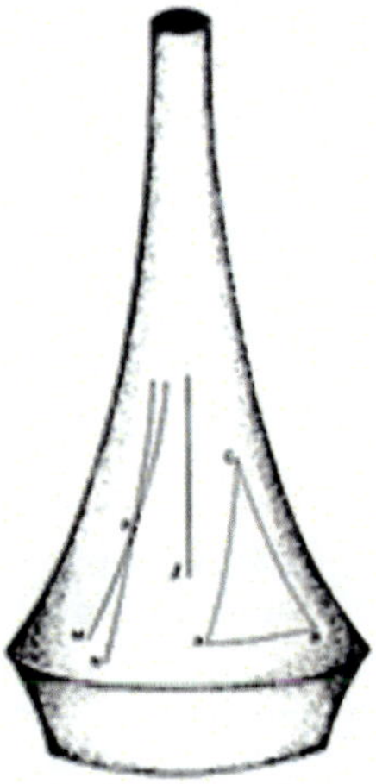

Fig. 1 Pseudosphere of Beltrami

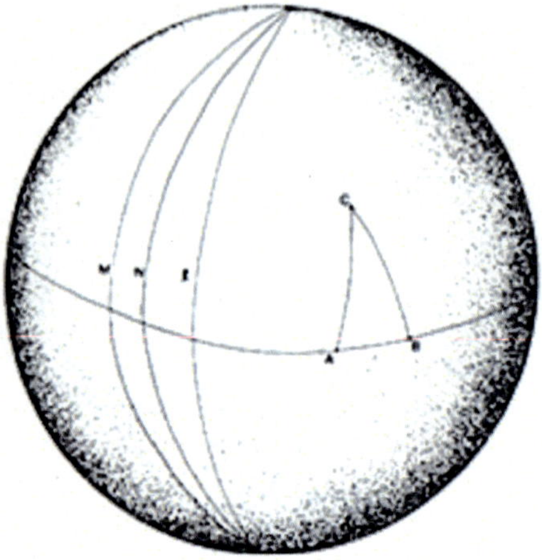

Fig. 2 Sphere of constant positive curvature

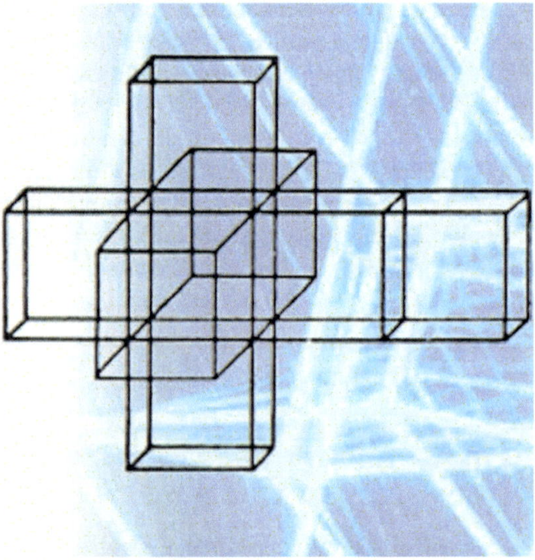

Fig. 3 Super cube spread

Also for further understanding of the dimensions above 3, reference and presentation was made in E. Jouffet's 1903 work, *Traite elementaire de la geometrie a quatre dimensions* (Fig. 4) and W.I. Stringham, 1881, *Regular Figures in N-dimensional Space* (Fig. 5).

A very important part of this teaching project was the selection of targeted artworks. Some examples in Figs. 6, 7 and 8.

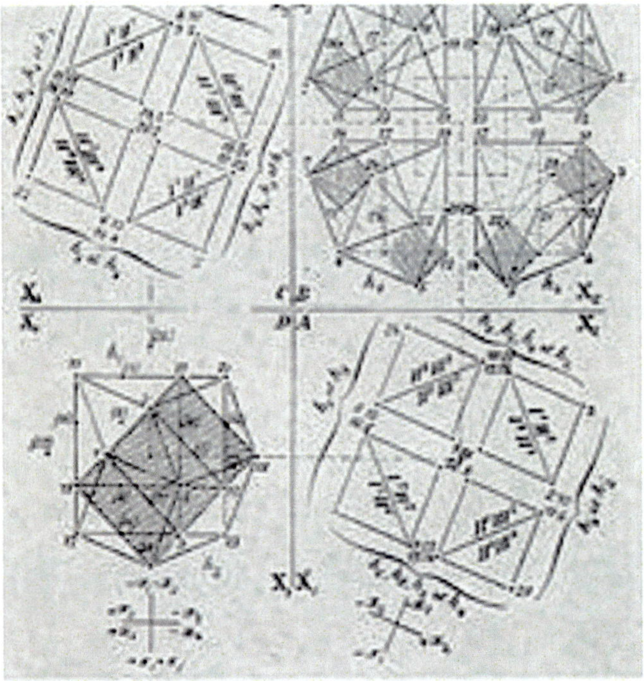

Fig. 4 E.Jouffet,1903, Traite elementaire de la geometrie a quatre dimensions

3 Results

Below we can see some selected artworks created by the students of the second class of the Lyceum, after the ending of this cross-sectional project (Figs. 9, 10 and 11).

Fig. 5 W.I. Stringham,1881, Regular Figures in N-dimensional Space

Fig. 6 Albert Gleizes, 1910, woman with flowers

Fig. 7 Le Fauconnier, 1910 abundance

Fig. 8 Juan Gris, 1912 guitar and flowers

Fig. 9 Conversational artworks through mirrors, by Nefeli Grigiratou, 17 and Marianni Dionisidou, 17 May 2019, Piraeus

Fig. 10 Realtime artwork product by Stelios Vagianos, 17, Dimitris Giannetos, 17 and Alexandra Didagelou, 17 May 2019, Piraeus

Fig. 11 Artwork using pc software by Ioanna Zaharia, 17 (She describes that it is influenced by Bauhaus movement after our project) May 2019, Piraeus

4 Conclusion

The aim of the action was the cross-curricular and interdisciplinary approach to art, language and mathematical subjects, through a targeted sample of carefully selected material from the teachers. The purpose was for the students to be awaken and recognize the link that exists between the scientific, technological and artistic achievements of human creation. The students created impressive works of high aesthetic value and they, effectively, incorporated modern mathematical and artistic theory.

We, the teachers, have often been overwhelmed by the response of high school students to particularly difficult subjects in understanding and applying, for which ones they had to combine multiple skills. In our opinion, the students' deep need to express themselves and to acquire new knowledge through non-trivial ways has proved to be very clear.

References

1. Stringham, W. I.: Regular Figures in N-dimensional Space Regular Figures in n-Dimensional Space. Am. J. Math. V 3, 16–7 (1881)
2. Jouffet, E.: Traite elementaire de la geometrie a quatre dimensions. Gauthier-Villars, Paris (1837)
3. Mpaltas, A.: (2004) https://helios-eie.ekt.gr/EIE/bitstream/10442/554/1/M01.033.07.pdf
4. Ritou, M.: (2015) http://ikee.lib.auth.gr/record/279293/files/GRI-2015-15430.pdf

5. Jaffe, H., Roters, E.: 20th century painting. Nefeli, Athens (1984)
6. Hobsbawm, E.J.: The age of capital. MIET, Athens (2003)
7. Hobsbawm, E.J.: The age of empire. MIET, Athens (2002)
8. Hobsbawm, E.J.: The age of revolution. MIET, Athens (1990)
9. Read, H.: Modern art history. Ypodomi, Athens (1977)
10. Kandinsky, W.: For the spiritual in art. Nefeli, Athens (1981)

The Impact of Sanctions Pressure on the Stability and Prospects for the Development of the National Economy of the Russian Federation

L. A. Elshin, M. R. Gafarov, and A. A. Dinmukhametova

Abstract The main purpose of the study is to systematize and analyze the key parameters of the economic growth of the Russian Federation under the sanctions pressure of the 2022 model and substantiate the policy of intensification of import substitution as a key mechanism for ensuring sustainable development in the medium and long term in the new reality. The subject of the study is the sanctions restrictions imposed by a number of Western countries in relation to the Russian economy, the costs they generate and the opportunities for building a new model of economic growth. As the main results of the study, it is necessary to highlight the systematization of sanctions imposed on Russia in 2022; identified trends in the formation of key macroeconomic parameters of the Russian economy in the conditions of foreign economic perturbations, including the projection of the results of empirical analysis on theoretical models of economic dynamics (IS-LM, AD-AS).

1 Introduction

Given the increased demand for research aimed at finding and substantiating the mechanisms of technological sovereignty and the implementation of import substitution policies both at the macro and meso levels, studies on this issue have begun to appear more and more often in the domestic scientific literature.

In addition, it is necessary to emphasize the fact that work on the problems of sustainable development of the national economy in the context of systemic transformations (including sanctions pressure on the national economy) is, in general,

L. A. Elshin (✉) · A. A. Dinmukhametova
Kazan Federal University, Kazan, Russia
e-mail: leon316@yandex.ru

L. A. Elshin · M. R. Gafarov · A. A. Dinmukhametova
Center for Advanced Economic Research, Academy of Sciences of the Republic of Tatarstan, Kazan, Russia

A. A. Dinmukhametova
Kazan National Research Technological University, Kazan, Russia

D. Vlachos (ed.), *Mathematical Modeling in Physical Sciences*, Springer Proceedings in Mathematics & Statistics 446, https://doi.org/10.1007/978-3-031-52965-8_28

fragmentary. This is largely due to the position on the opportunistic component of this kind of research, which does not predetermine their fundamental novelty. However, the growing restrictions on the part of the collective West imposed on the Russian economy and the resulting structural transformations in the socio-economic environment, the transformation of models of economic dynamics and the violation of traditional approaches to the study of economic growth as a result of artificial external interference in the formation of a market economy, in many ways neutralize this approach and elevate the fundamental and applied significance of scientific work in the area under consideration.

2 Literature Review

One of the first questions in the Russian scientific literature about sustainable economic development under conditions of external pressure on the economic system, which gives rise to a high level of restrictions on the import of technologies and goods, was raised by Academician L. I. Abalkin. In accordance with the views of the scientist, "the state, especially in a force majeure crisis situation, is forced even at a loss to produce what it cannot do without" [1]. Thus, actualizing the priority of the country's economic security and focus on the need to stimulate import substitution policy, especially in the "acute" phases of the cyclical development of the economy.

Osipov V. S. in his study "Import substitution in the economic policy of Russia" substantiates the position according to which "as a result of the adoption of the recipes of the Washington Consensus, the countries of the "third world" lost their industry, as they opened up to world markets too quickly. The influx of imported goods into the national market, first of all, ruined the industry, which led to a decrease in the income of the population (after all, increasing returns were lost, while diminishing returns remained)" [2].

The problem of significant dependence on imports in the manufacturing sector is also substantiated in the study by Tsypin [3]. In his work "Method of statistical assessment of the most risky subjects of the Russian Federation in terms of the development of economic potential depending on the change in the structure of commodity flows" [3], the author proposed a conceptual scheme for identifying (monitoring) risky subjects that are highly likely to face (if not already faced) with the problem of import dependence in the field of science-intensive and technologically complex machines and equipment, which, according to the author, characterizes the high vulnerability of this sector of the national economy of the Russian Federation.

It is also important to note that the issues of theoretical and applied understanding and research of the import substitution policy and the formation of stable cooperation ties, little affected by the global situation, have received considerable attention in recent years among foreign researchers. The most notable of these include: studies by Susan Stone, Dorothee Flaig and Frank Van Tongeren "Modeling Local Content Requirements: Quantitative Restrictions in a CGE Model" [4], Stone, S., Messent, J. and Flaig, D. "Emerging Policy Issues: Localization Barriers to Trade" [5], Hanna

Deringer, Fredrik Erixon, Philipp Lamprecht, Erik van der Marel "The Economic Impact of Local Content Requirements: A Case Study of Heavy Vehicles" [6], Bas M., Strauss-Kahn V. Does Importing More Inputs Raise Exports? Firm-Level Evidence from France» [7] and others.

In general, it must be stated that, despite individual works devoted to the problems of technological and product sovereignty of national economic systems, at the current time, the issues of both theoretical and empirical substantiation of import substitution models and the formation of stable cooperation ties in the context of systemic restrictions that are little affected by the global situation, have not received due development both at the national and regional levels. This causes and actualizes the need to increase efforts in the field of research in this area to fill the existing, to a certain extent, vacuum. These issues are especially acute on the current agenda in the Russian Federation, which is faced with external barriers in the supply of imports and the need to find directions and mechanisms for modeling new cooperation chains in the context of sanctions restrictions that have exacerbated the issues of intensifying import substitution processes.

3 Materials and Methods

The relevance of the identified problem of ensuring import substitution processes is determined, along with the above arguments and the fact that as of October 2022. 8 packages of sanctions were adopted against Russia, which made it the "leader" in the world in this indicator. In total, more than 10,000 sanctions measures have been imposed on the country that have a direct impact on the development of certain industries and types of economic activity.

It is important to note that the attitude of the countries of the world to these processes is far from unambiguous (Fig. 1).

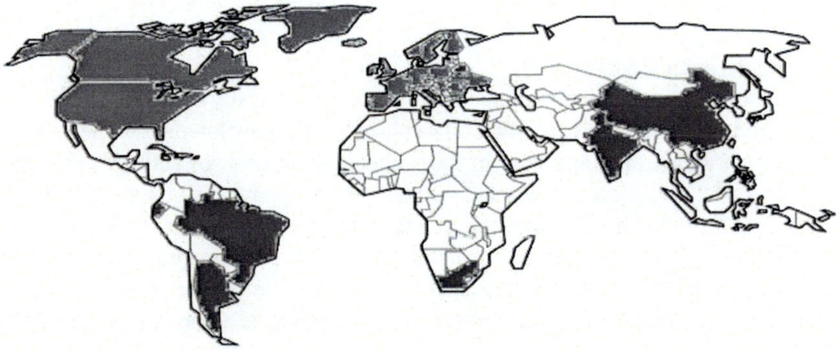

Fig. 1 The attitude of the countries of the world to the sanctions pressure on the Russian economy. *Source* developed according to [8–10]

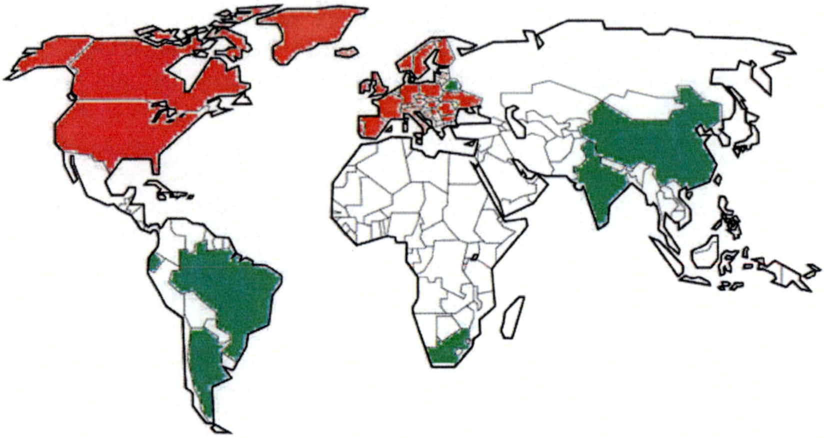

Fig. 2 Change in the key rate of the Central Bank

The sanctions pressure on the national economy of the Russian Federation and subsequent adjustments to key macroeconomic indicators demonstrate two stages of adaptation to systemic transformations. The first of them is characterized by increased volatility and a reaction to a foreign policy "shock" (large-scale sanctions in February–March 2022). The second is characterized by systemic adaptation to the perturbations that have occurred. The dichotomous basis of the adaptation processes to the sanctions pressure on the Russian economy was most clearly seen in the financial markets (Fig. 2).

The Central Bank of Russia raised the key interest rate by 2.5 times in the 1st quarter of 2022 in order to combat the emerging turbulence in the financial markets, which led to a rapid increase in loan prices (from 8.5% (February 2022) to 20% (March 2022)). Subsequently, the rate decreased and reached the level of 7.5% (October 2022).

The most important consequence of the rise in the discount rate to 20% was the reduction in GDP and investment activity, which fits perfectly into the IS-LM economic growth model (Fig. 3).

The most important feature of this crisis is the diametrically opposed adjustments of underlying assets relative to other crisis phases of the development of the Russian economy. If within the framework of the cyclical global crisis of 2008 and the sanctions year of 2014, energy prices and the ruble exchange rate fell by more than two times, in 2022 diametrically opposite trends were observed (Table 1).

It is important to note that the adaptation of the national economy of the Russian Federation in 2022 was more stable and effective compared to the previous crises of 2014 and 2020. This largely testifies to the development of some kind of immunity on the part of business, the population, and public authorities to overcome crisis processes, including within the framework of the coronavirus pandemic and sanctions pressure on the country's sovereignty by the collective West. These mechanisms of

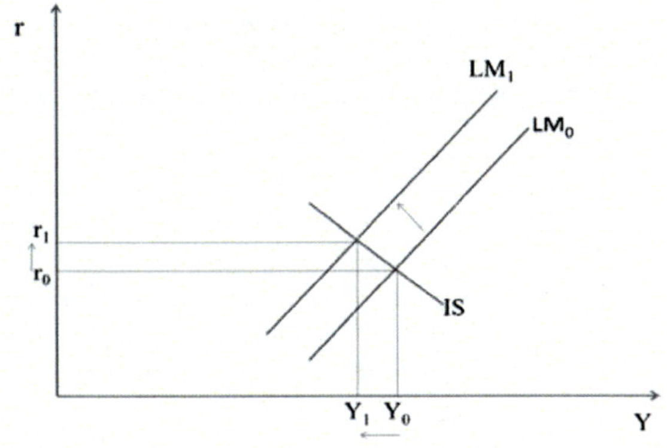

Fig. 3 Graphical interpretation of interest rate growth in 2022 (IS-LM model)

Table 1 Comparative analysis of the fall in oil prices and the growth of the dollar in 2008 and 2022

Price	2008		2014		2022	
	$	%	$	%	$	%
Futures for Brent oil	From 139 to 45	−68 ↓	From 112 to 53	−53 ↓	From 80 to 104	+30 ↑
US $	From 23.4 to 35.9	+53 ↓	From 34 to 64.3	+89 ↓	From 76 to 60	−21 ↑

adaptation to the new reality were especially noticeable in the initial phase of the crisis in 2022, when sanctions were most clearly manifested in pressure on the monetary and financial systems (freezing of gold and foreign exchange reserves) of the Russian economy, as well as in limiting mechanisms for the development of the real sector within the framework of the perturbation of cooperation, logistics, production ties and value chains.

Thus, it can be argued that the Russian economy has very successfully adapted to the structural and systemic changes that have taken place.

4 Results

The sustainability of economic growth that emerged by the end of 2021, characterized by positive signals of an exit from the post-COVID state of the global economy and an increase in the trajectory of the progressive development of the national economy of the Russian Federation, ensured, along with state regulation measures to adapt to sanctions pressure, the stability of the Russian economy to systemic transformations, including and sanctions.

At the same time, a number of macroeconomic indicators based on the results of the three quarters of 2022 still show very noticeable signals of emerging crisis manifestations. In general, when implementing a descriptive analysis of the main trends in the system of socio-economic development, it is necessary to single out a number of emerging key risks, judging by the development trajectories of some macro indicators. The most significant should include:

- reduction of technological sovereignty as a result of the ban on the import of a number of technologies and final/intermediate consumption goods;
- a decrease in export volumes in the trade balance (as a result of the introduction of marginal prices for Russian energy resources in 2023 and the systematic abandonment of some of them), which predetermines a reduction in budget revenues and forms the risks of a budget deficit;
- a decrease in net exports as a result of a reduction in export earnings predetermines a decline in GDP;
- the decisions of the G7 on the marginal prices for Russian oil, adopted in November, predetermine the risks of a decrease in export earnings, sequestration of net exports and, accordingly, an increase in the ruble exchange rate adjustment, including as part of budget balancing;
- violation of the financial stability of the banking sector as a result of a decrease in real incomes of the population and a decrease in the solvency of economic agents;
- a decrease in the overall level of competition in the economy as a result of the withdrawal of foreign companies from the national market predetermines the deterioration of the competitive environment and the decrease in the quality of organizational and managerial decisions; and etc.

In addition, the most important, from the point of view of generating risks, the consequences of sanctions pressure on the Russian economy were:

1. Many Russian companies have lost the opportunity to do business at the international level.
2. Quotes of shares of a number of Russian enterprises collapsed.
3. Air transportation to the EU countries, the USA, Great Britain and a number of other states is closed.
4. Foreign companies leave the country. At the same time, thousands of employees are at risk of being left without work, and Russian companies are losing contracts and suppliers.
5. The consequences of sanctions within the country are expressed in the shortage of many goods and rising prices.
6. The value of the trade balance has multiplied, as a result of which the ruble exchange rate against the dollar has significantly strengthened.

Turning to the specific values of the indicators that form the above list of risks, it is necessary to dwell separately on the most important of them. Thus, for example, the most sensitive macro indicator to ongoing upheavals was the Consumer Price Index, the dynamics of which showed a very tangible growth—20%.

Rising prices have hit consumers on at least two fronts:

1. Loans have become more expensive, as a result of which, for example, retail trade has fallen significantly in 2022.

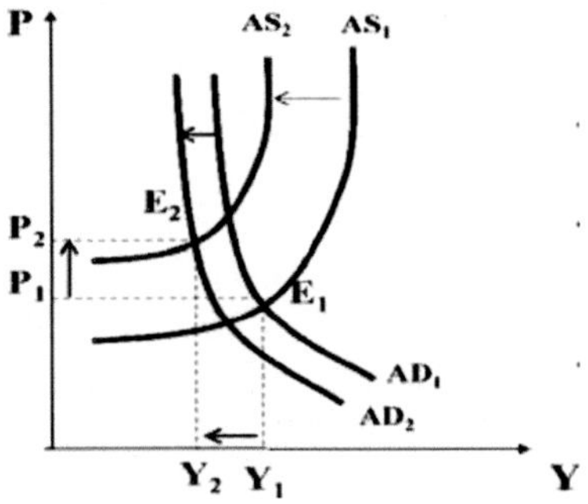

Fig. 4 AD-AS model as a tool to justify the shifts that have occurred in the Russian economy (sample for 2–3 quarters of 2022)

2. The depreciation (Q2 2022) of the ruble led to an increase in import prices.

The significant inflation that unfolded in 2022, in accordance with the law of the Keynesian model of economic growth, could not but affect the level of GDP (Fig. 4).

(A) The rise in prices provided a decrease in aggregate demand (shift of the AD curve from position 1 to position 2)
(B) Decrease in demand led to a reduction in aggregate supply (shift of the AS curve from position 1 to position 2)
(C) Shift of the macroeconomic equilibrium from point E1 to point E2, which indicates that the rise in prices provoked a decrease in GDP (shift of the equilibrium point that determines the volume of GDP at a fixed price level from Y1 to Y2)

5 Discussion

Despite the identified risks of sustainable development of the national economy of the Russian Federation under sanctions pressure, the emerging upheavals form a number of prospects for its development, which contributes to the opening of the so-called "windows of opportunity". First of all, they should include:

- an active import substitution policy will ensure the growth of technological sovereignty and the mobilization of internal resources of the national economy, which will largely contribute to strengthening economic sovereignty, economic growth and competitiveness in the presence of effective mechanisms to stimulate the import substitution policy;
- an increase in demand for goods produced in Russia as a result of a reduction in imports;
- restrictions on travel to Europe and the United States will contribute to the development of domestic tourism;
- scientific and technological development and more active integration of the Russian economy into the agenda of the sixth technological order as part of the implementation of the policy of endogenous economic growth;

- new opportunities for business development as a result of emerging niches; and etc.

6 Conclusion

In general, it should be noted that the current situation, characterized by the need to stimulate economic dynamics in the face of sanctions pressure and the need to replace foreign technologies and goods, involves focusing on the implementation of key mechanisms of the import substitution policy, based not only on the search for new cooperative ties in creating reproduction chains, but and strategic orientation towards scientific and technological development.

Meanwhile, it is important to note the need to differentiate and prioritize approaches to the implementation of import substitution mechanisms in the current conditions based on five major factors:

- the degree of vulnerability and dependence of sectors of the economy on the supply of imports of technologies and final/intermediate consumption goods (share of imports from unfriendly countries);
- the speed of reconfiguring new logistics channels for the supply of imported raw materials and intermediate consumption goods;
- the possibility and prospects of import substitution as part of the intensification of scientific and technological development and the formation of new cooperative value chains;
- research potential in certain areas of life as a tool for operational and strategic import substitution;
- Orientation of the industry to foreign and domestic markets.

These factors should largely determine the stages of the implementation of the import substitution policy in accordance with the concept of transition from solving extra-urgent tasks to the tasks of reaching export-oriented import substitution [11].

Acknowledgements The research was carried out with a grant from the Russian Science Foundation and with financial support from the Cabinet of Ministers of the Republic of Tatarstan within the framework of scientific project No. 23-28-10123, https://rscf.ru/project/23-28-10123/».

References

1. Абалкин Л.И.: Избранные труды: в 4-х т. Т. IV: В поисках новой стратегии. ВЭО России. — М.: Экономика. 799 с (2000)
2. Осипов В.С.: Отраслевые и операционные конкурентные преимущества в международном разделении труда: матрицы структуризации промышленной политики // Экономика и предпринимательство. № 6. — С. 26–33 (2014)

3. Цыпин А.П.: Влияние состояния промышленности на продовольственную безопасность региона / А.П. Цыпин // Азимут научных исследований: экономика и управление. Т. 5. – № 3 (16). – С. 209–211 (2016)
4. OECD: Trade and agriculture directorate. https://www.gtap.agecon.purdue.edu/resources/download/7203.pdf. Accessed 07 Mar 2021
5. OECD Trade Policy Papers, no. 180. OECD Publishing, Paris. https://doi.org/10.1787/5js1m6v5qd5j-en. Accessed 07 Mar 2021
6. https://ecipe.org/publications/the-economic-impact-of-local-content-requirements/. Accessed 5 Mar 2021
7. Bas, M., Strauss-Kahn, V.: Does importing more inputs raise exports? Firm-level evidence from France. Rev. World Econ. **150**(2), 241–275 (2014)
8. Санкции против России 2022 и их последствия для бизнеса. https://www.gd.ru/articles/12177-sanktsii-dlya-biznesa?ysclid=la9jj5fqle388143546 (Дата обращения: 05.11.2022г.)
9. Изменения в экспорте и импорте РФ в 2022 году: влияние санкций на международную торговлю. https://delprof.ru/press-center/open-analytics/izmeneniya-v-eksporte-i-importe-rf-v-2022-godu-vliyanie-sanktsiy-na-mezhdunarodnuyu-torgovlyu/?ysclid=lab0lhemj5811160288. (Дата обращения: 05.11.2022г.)
10. Экономические санкции Запада против России и российские антисанкции: издержки и выгоды конфронтации. https://www.rea.ru/ru/org/cathedries/Kafedra-otraslevojj-i-biznes-statistiki/Documents/Нуреев%20P.M._MP-2017.pdf
11. Сафиуллин, М.Р.: Санкционное давление на экономику России: пути преодоления издержек и выгоды конфронтации в рамках импортозамещения / М. Р. Сафиуллин, Л. А. Ельшин. Финансы: теория и практика, Т. 27, № 1. – С. 150–161 (2023). https://doi.org/10.26794/2587-5671-2023-27-1-150-161

Methodology for Assessing the Impact of Changes in Transnational Cash Flow on GDP Growth (on the Example of the Russian Federation)

M. R. Safiullin, A. Dinmukhametova, and A. R. Sharapov

Abstract The restrictions imposed by Western countries on the access of the Russian financial system to international clearing systems, if they do not localize the potential of foreign economic activity, then significantly complicate the processes of international trade, especially in terms of the growth of transaction costs. At the same time, an extremely important task is to assess the possible consequences caused by the correction of the movement of transnational cash flows in order to understand the possible risks of the sustainable development of national economic systems. This study is devoted to an overview of the emerging effects generated as a result of changes in the volumes of foreign economic supply chains. In it, within the framework of economic and mathematical modeling of perturbations in the field of correction of cash flows of the external contour, empirical estimates of a possible change in the growth rate of Russia's GDP are implemented.

1 Introduction

The formed conditions for the development of the national economy of the Russian Federation, characterized in many respects by the observable processes of transformation of the existing value chains as a result of the disruption of export and import flows within the framework of the sanctions confrontation, predetermine the search for new mechanisms and directions for intensifying economic growth and leveling threats in the sphere of restrictions on export–import operations.

M. R. Safiullin · A. Dinmukhametova (✉)
Kazan Federal University, Kazan, Russia
e-mail: aliyaabdukaeva@mail.ru

A. Dinmukhametova
Center for Advanced Economic Research of the Academy of Sciences of the Republic of Tatarstan, Kazan, Russia

Kazan National Research Technological University, Kazan, Russia

A. R. Sharapov
Volga Region State University of Physical Culture, Sports and Tourism, Kazan, Russia

D. Vlachos (ed.), *Mathematical Modeling in Physical Sciences*, Springer Proceedings in Mathematics & Statistics 446, https://doi.org/10.1007/978-3-031-52965-8_29

In this regard, in solidarity with the position of a number of Russian scientists and experts [1–4], the issues of import substitution, scientific and technological development, leveling sanctions threats are becoming extremely relevant on the agenda.

At the same time, it is important to emphasize that a special role in the list of designated areas of strategic development is played by the search for mechanisms to ensure the sustainability of supply chains, within the framework of an external contour in relation to the Russian economy.

Undoubtedly, this problem is multifaceted, and includes a wide range of tasks and directions for their solution: from mechanisms for the development of import substitution and ending with the creation of alternative international payment systems that are not controlled by the countries of the Western bloc and do not pose a threat of pressure on third countries involved in the implementation of cross-border payments from Russia.

The issues of creating alternative payment systems at the supranational level are of particular relevance in the context of sanctions restrictions imposed by individual countries in relation to national economic systems.

Thus, the increased international sanctions pressure on the national economy of the Russian Federation in 2022 led to the restriction of the access of the Russian financial system to international payment systems.

For example, the largest banks of the Russian Federation (Sberbank, VTB, Otkritie Bank, VEB.RF, Moscow Credit Bank, Rosselkhozbank, etc.) and corporations were disconnected from SWIFT on June 3, 2022, which significantly complicated the implementation of international financial transactions from Russia and to Russia.

The problem is exacerbated by the fact that, according to data for the pre-sanctions year 2021, about 300 Russian financial institutions used the SWIFT system. According to this indicator, the Russian Federation ranked second after the United States. In this regard, it is not difficult to assess the scale of the consequences for the Russian economy of disabling data processing and transmission of financial messages from this service.

The disconnection of financial institutions from SWIFT implies that the potential for their implementation of payment transactions and foreign exchange transactions is significantly limited, which accordingly affects the movement of foreign trade turnover and disruption of the stability of supply chains.

In this regard, the issue of alternative mechanisms for organizing international supply chains and accompanying cross-border transactions is quite acute on the current agenda for the Russian economy [5, 6].

Before proceeding to the processes of constructing models aimed at solving the tasks set here, it is very important to consider the main trends that reveal the features of the formation of foreign economic cash flows in relation to the Russian economy and identify key trends on this basis (Fig. 1).

It is important to note that a very significant share in the total volume of cross-border cash flows falls on individuals. Their share is about 40% (Table 1). In the total volume of money transactions, the countries of the Western bloc account for the bulk of transnational money transfers (about 85%) (Fig. 2).

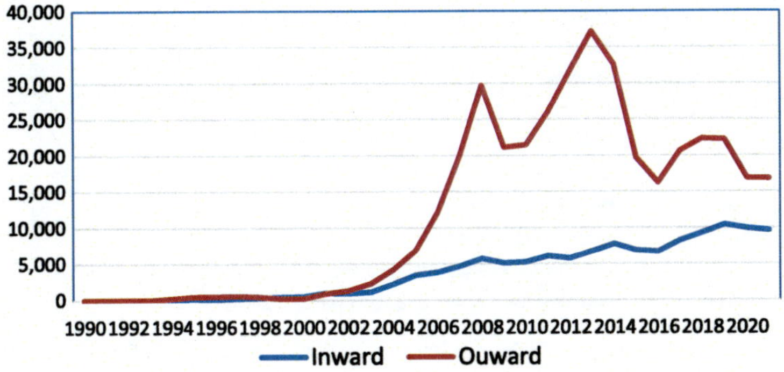

Fig. 1 Dynamics of incoming and outgoing cash flows, million USD (developed according to [7])

Table 1 Cross-border transfers made through payment systems by individuals, mln USD

	Transfers by individuals from the Russian Federation	To foreign countries	To CIS countries	Receipts in favor of individuals in the Russian Federation
2010	12 835	1 755	11 080	1 975
2011	17 533	2 390	15 143	2 767
2012	48 588	29 384	19 205	14 056
2013	23 589	2 808	20 780	3 972
2014	20 856	2 565	18 292	4 174
2015	9 719	1 740	7 978	3 319
2016	7 335	1 739	5 596	2 222
2017	9 186	2 183	7 004	2 609
2018	9 485	2 022	7 463	2 666
2019	7 541	1 445	6 096	3 329
2020	7 400	1 207	6 193	3 119
2021	8 662	1 085	7 577	3 237

Source [8]

2 Methods

In general, the results of a descriptive express analysis indicate the presence of very significant volumes of cross-border cash flows in the Russian economy, which, on the one hand, means a high level of integration of the Russian economy into the global economic environment, and on the other hand, a high level of dependence of the national economic systems from possible perturbations in the field of access to international clearing services.

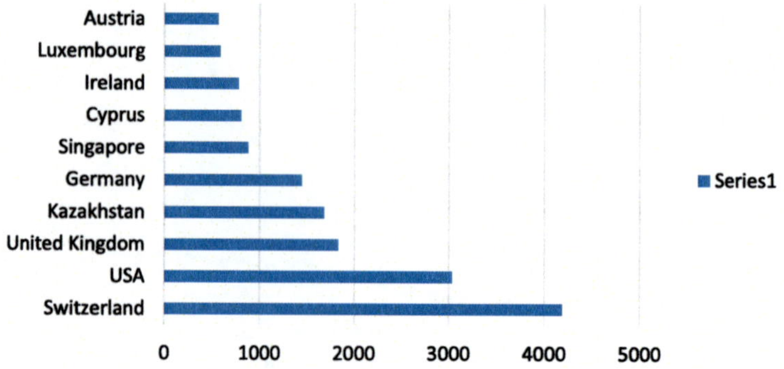

Fig. 2 Cross-border transfers in favor of individuals in the ten main counterparty countries of the Russian Federation in 2021, USD million, USD million (developed based on data [9])

Any disruption of well-established external relations can lead to very negative consequences, which is especially important in the context of the significant dependence of the Russian economy on imports and related supply chains. For example, the share of imports in the national economy of Russia is about 21% of GDP [8].

In this regard, the ongoing changes in the volume of foreign economic cash flows can significantly disrupt the pattern and dynamics of the formation of economic growth.

In order to mitigate such risks, it seems appropriate to assess the possible consequences and form a set of solutions that ensure the localization of the threats under consideration in order to maintain stable channels for the supply of imported components, raw materials, and finished products.

In this regard, further methodological approaches are presented for solving the problem of finding relationships between changes in gross foreign economic cash flows and the growth rate of Russia's GDP.

To study the dependence of GDP on incoming cash flows, a system of recursive equations is proposed. Structurally, the model looks like this:

$$\begin{cases} y_1 = a_{01} + a_{11}x_1 + b_{12}y_2 + \varepsilon_1 \\ y_2 = a_{02} + a_{22}x_2 + a_{23}x_3 + \varepsilon_2 \\ y_3 = a_{03} + b_{31}y_1 + a_{32}x_3 + \varepsilon_3 \end{cases} \quad (1)$$

где

y_1—Incoming cash flow.

y_2—Consumer price index.

y_3—Gross domestic product.

x_1—Digital spending.

x_2—Net export.

x_3—dollar exchange rate.

a_{01}, a_{02}, a_{03}—free members.

Thus, in the mathematical model, the following indicators act as endogenous (dependent) indicators: incoming cash flow, consumer price index, gross domestic product. As predetermined (independent): digital spending, consumer price index, net exports, dollar exchange rate.

As a rule, the use of Least Squares Method (LSM) for estimating the parameters of a structural model gives inconsistent and biased estimates, so the system of equations must be reduced to the reduced form. In the framework of this study, the given form will take the following form:

$$\begin{cases} y_1 =_{11} x_1 +_{12} y_2 +_{13} x_2 +_{14} x_3 +_{15} y_1 \\ y_2 =_{21} x_1 +_{22} y_2 +_{23} x_2 +_{24} x_3 +_{25} y_1 \\ y_3 =_{31} x_1 +_{32} y_2 +_{33} x_2 +_{34} x_3 +_{35} y_1 \end{cases} \quad (2)$$

When passing from the reduced form of the model to the structural one, it is important to check the identification condition, i.e., the uniqueness of the correspondence between the reduced and structural forms of the model.

The following situations are possible.

(1) Structural coefficients are uniquely expressed through the coefficients of the reduced form of the model. In this case, the structural model is said to be exactly identifiable.
(2) Some of the structural coefficients are not expressed in terms of the coefficients of the reduced form of the model. Such a structural model is called unidentifiable.
(3) Structural coefficients are ambiguously expressed through the coefficients of the reduced form of the model. Then the structural model is called overidentified.

Let us present the necessary and sufficient conditions for identifiability (over-identifiability). These conditions apply to each equation of the structural model, i.e., each equation of the system is checked for identifiability. If all equations of a system are exactly identifiable, then the system will be exactly identifiable.

If one or more equations of the system are over-identifiable, and the rest are exactly identifiable, then the system will be over-identifiable. If at least one of the equations is unidentifiable, then the system is unidentifiable.

Necessary condition for identification. Let K be the number of exogenous variables not included in the equation, but present in the system, and M be the number of endogenous variables included in the equation. If the condition is fulfilled: K ≥ M-1, then the equation in the structural model can be identified.

As part of the use of the proposed methodological tools, the following system of equations was obtained, revealing the features and patterns between the studied factors:

Table 2 Simulation modeling of GDP growth due to the intensification of business activity and the corresponding change in transnational money flows (according to 2021 data)

№	Growth rates of the aggregate value of cross-border cash flows	GDP growth, in %
1	Present value	–
2	+1.0%	+0,45
3	+3.0%	+1,34
4	+5.0%	+2,24

$$\begin{cases} y_1 = 24497,3 + 0,0013x_1 - 176,76y_2 \\ y_2 = 112,24 - 0,0009x_2 + 0,034x_3 \\ y_3 = 5031662 + 5826,77y_1 + 2,68x_3 \end{cases} \quad (3)$$

Each of the equations of the system is over-identifiable, therefore the system is over-identified. To solve the system, a two-step LSM was used, since the use of QLS (indirect LLS) will not give unambiguous estimates for the over-identified system. Based on the calculation results, the following system of equations was obtained:

$$\begin{cases} y_1 = 0,04x_1 - 145,79y_2 \\ y_2 = -0,0125x_2 + 0,084x_3 \\ y_3 = 6025,8y_1 + 3,15x_3 \end{cases} \quad (4)$$

Following the principles and approaches of the proposed methodological tools, as well as relying on the methods of simulation of the processes under study.

Table 2 presents predictive estimates of the growth in the gross domestic product of the Russian Federation as part of the intensification of business activity and the corresponding change in transnational cash flows.

3 Result

The estimates obtained, which characterize the growth potential of the gross domestic product of the Russian Federation due to the disruption of foreign economic cash flows and, accordingly, trade turnover, indicate a very high level of dependence of the national economy on changes in their volumes.

Following the data presented in Table 2, a 1% change in the volume of cash flows creates a GDP growth potential of 0.45% and vice versa.

In this regard, any adjustments in the international supply chains in the Russian Federation will have a very sensitive impact on the prospects for the sustainable development of the national economic system.

The issues of creating alternative payment systems at the transnational level are of particular relevance in the context of sanctions restrictions imposed by individual countries in relation to national economic systems. Thus, the increased international sanctions pressure on the national economy of the Russian Federation in 2022 led to the restriction of the access of the Russian financial system to international payment systems.

For example, the largest banks of the Russian Federation (Sberbank, VTB, Otkritie Bank, VEB.RF, Moscow Credit Bank, Rosselkhozbank, etc.) and corporations were disconnected from SWIFT on June 3, 2022, which significantly complicated the implementation of international financial transactions from Russia and to Russia.

The problem is exacerbated by the fact that, according to data for the pre-sanctions year 2021, about 300 Russian financial institutions used the SWIFT system. According to this indicator, the Russian Federation ranked second after the United States. In this regard, it is not difficult to assess the scale of the consequences for the Russian economy of disabling data processing and transmission of financial messages from this service.

Given the above, the issue of alternative mechanisms for organizing international supply chains and the accompanying cross-border transactions is quite acute on the current agenda for the Russian economy [10, 11].

Acknowledgements The study was supported by a grant from the Russian Science Foundation (project no. 23-28-00587). Kazan (Volga Region) Federal University, Kazan, Russia. The work is performed according to the KFU development program "Priority-2030".

References

1. Safiullin M.R., Elshin, L.A., Prygunova, M.I.: Volatility and competitiveness of regions in the context of foreign policy and market changes (on the example of the regions of the Volga Federal District). Bull. Russ. Univ. Econ. (Named after G.V. Plekhanov) **12**(78), 115–127 (2014)
2. Elshin L.A.: Comparative analysis of cyclic fluctuations of regional economic systems: modeling, identification, forecasting. Bull. Inst. Econ. Russ. Acad. Sci. **4**, 138–156 (2017)
3. Osipov, V.S., Zeldner, A.G., Pankova, S.V., et al.: Economics of import substitution: assessment of the impact of the structure of foreign trade flows on the development of economic potential and import substitution in the Russian Federation. Intellect Innov. Invest. **7**, 31–44 (2017)
4. Abalkin L.I.: Selected works: in 4 volumes. T. IV: In search of a new strategy. VEO of Russia. - M.: Economics, 799p. (2000)
5. Masalsky, M.G.: Economic sanctions. The impact of economic sanctions on the Russian Federation. Econ Soc **4–3**(95), 95–97 (2022)
6. Dubrovskaya, E.S.: Russian economy under sanctions: current situation and prospects for innovative development. Creat Econ **16**(11), 4553–4564 (2022). https://doi.org/10.18334/ce.16.11.116637
7. World Bank. https://www.worldbank.org/en/topic/labormarkets/brief/migration-and-remittances. Accessed 15 May 2023
8. Kochergin, D.: Digital currencies of central banks: world experience. World Econ Int Relat **65**(5), 68–77 (2021). https://doi.org/10.20542/0131-2227-2021-65-5-68-77
9. Cross-border transfers of individuals in 2021. https://cbr.ru/statistics/macro_itm/svs/TGO_obz/crossborder_ind_2021/. Accessed 15 May 2023
10. Safiullin, M.R., Abdukaeva, A.A., Elshin, L.A.: Integral multi-component assessment of the development of the blockchain technology market in the national economy of Russia. Innovations **7**(249), 41–49 (2019)

11. Sakharov D.M.: Central bank digital currencies: key features and impact on the financial system. Finance: Theory Pract **25**(5), 133–149 (2021). https://doi.org/10.26794/2587-5671-2021-25-5-133-149

The Impact of Trading Environments on Commodity Futures: Evidence from Biofuel Feedstocks' Network

Minhyuk Jeong, Kyohun Joo, Jinu Kim, Juyub Kim, Joohyung Kim, and Kwangwon Ahn

Abstract This study analyzes two types of networks constructed through correlation and Granger causality, consisting of eight commodity futures that can be used as biofuel feedstock. In the correlation network, futures in emerging markets are isolated from other futures, for example, crude palm oil (CPO) and sunflower futures, whereas those in mature markets are strongly correlated. In the Granger causality network, however, CPO and sunflower futures markets are influential in the network; CPO and sunflower futures provide and receive information with high degree centrality inside the network. In particular, CPO futures is an important bridge node connecting other markets with high betweenness centrality. Findings suggest that futures markets of biofuel feedstock commodities are significantly integrated, and the role of each market cannot be ignored regardless of the trading environment.

Keywords Biofuel · Commodity futures · Network · Trading environments

1 Introduction

With the growing trend of carbon neutrality, the demand for renewable energy is increasing [1]. Bioenergy, one of the renewables, is generated from natural organisms, that is, plants and animals. We can directly burn a mass of organisms to generate energy, or we can use them as biofuel feedstock.

In the U.S. and Europe, a significantly large portion of biofuel is produced from well-known food commodities, for example, canola, palm oil, and soybeans [2–4], which have been actively traded in the markets for decades. With global efforts toward sustainability, their price changes would be commonly influenced by the demand for biofuel, resulting in their stronger market integration. Accordingly, their

---

M. Jeong · J. Kim · J. Kim · K. Ahn (✉)
Yonsei University, Seoul 03722, Republic of Korea
e-mail: k.ahn@yonsei.ac.kr

K. Joo · J. Kim
KAIST, Daejeon 34141, Republic of Korea

D. Vlachos (ed.), *Mathematical Modeling in Physical Sciences*, Springer Proceedings in Mathematics & Statistics 446, https://doi.org/10.1007/978-3-031-52965-8_30

recent correlation and causal relationship would help us understand the role of each market in the price formation of biofuel.

In this study, we aim to analyze the relationship among different commodity futures that are used as biofuel feedstock through the aspect of two different networks: (1) correlation network [5] and (2) Granger causality network [6, 7]. Correlation is a widely used measure of market integration [8, 9], and it can capture how similarly or differently the returns of any two markets move. However, correlation cannot explain the direction of information flow. Moreover, studies, such as Pukthuanthong and Roll [10], criticized correlation as a measure of market integration because a weak correlation can sometimes be observed among completely integrated markets. Thus, we would explain the causal relationship among markets using the Granger causality network, which can make up for the weakness of the correlation network through directed edges. This study uses centrality metrics [5, 11, 12] to explain the role of each node (market) in the network of biofuel feedstock commodity futures in correlation and Granger causality cases.

2 Data and Methodology

2.1 Data

We select eight biofuel feedstock commodities that occupy a significant part of the total global biofuel production and have financialized futures markets [2–4]. As the front month futures are usually the most liquid contract, we use their prices to represent the futures prices of the commodities. The data extend from Jan 2016 to Aug 2022 with daily frequency.

We replace the returns of the rollover days with zeros because non-continuous price changes appear on rollover days and most rollover days of the eight markets do not coincide. Table 1 presents a briefly summary of the selected biofuel feedstock with their exchanges, trading currencies, and fuel types after refining.

2.2 Granger Causality

Let us assume that the vector autoregression model with lag p, namely, VAR(p), holds between two time-series variables X_t and Y_t:

$$X_t = \alpha + \sum_{i=1}^{p} \beta_i Y_{t-i} + \sum_{j=1}^{p} \gamma_j X_{t-j} + \epsilon_t,$$

Table 1 Selected biofuel feedstock

Commodity	Exchange	Currency	Fuel type
Canola	ICE/CA	CAD	Biodiesel
Corn	CBOT	USD	Bioethanol
Crude palm oil (CPO)	MDEX	MYR	Biodiesel
Rapeseed	Euronext	EUR	Biodiesel
Soybean oil (SBO)	CBOT	USD	Biodiesel
Soybeans	CBOT	USD	Biodiesel
Sunflower	SAFEX	ZAR	Biodiesel
Wheat	CBOT	USD	Bioethanol

Note Intercontinental Exchange Canada (ICE/CA), Chicago Board of Trade (CBOT), Malaysia Derivatives Exchange Euronext (MDEX), South African Futures Exchange (SAFEX), Canadian Dollar (CAD), U.S. Dollar (USD), Malaysian Ringgit (MYR), Euro (EUR), and South African Rand (ZAR)

where α is a constant; β_i denotes a regression coefficient quantifying the extent to which Y_{t-i} explains X_t; γ_j is an autoregressive coefficient quantifying the extent to which X_{t-j} explains X_t; ϵ_t represents error term; and p indicates the largest lag order obtained from the Akaike information criterion (AIC) [13].

Then, the null hypothesis of the Granger causality test is that "Y_t does not Granger cause X_t" and can be specified as follows [6, 7]:

$$H_0 : \beta_1 = \beta_2 = \cdots = \beta_p = 0.$$

2.3 Centrality Metrics

The centrality metrics quantify the influence or focality of a node in the network [11, 12]. Among various metrics, we use degree, betweenness [5, 11, 12], and eigenvector centrality [14, 15].

Degree centrality. Suppose a graph $G(V, E)$ is given, where V and E are sets of nodes and edges, respectively. Then, the number of edges in each node is defined by the degree centrality [11]. For a given node i, $D(i)$ denotes the degree centrality as:

$$D(i) = \sum_j w_{ij} m(i, j),$$

where w_{ij} is the weight of the edge between nodes i and j. $m(i, j)$ is defined by 1 if an edge exists between nodes i and j, and 0 if otherwise. Specifically, a node has a high degree centrality if it is connected to many neighbors; the higher the degree centrality, the more central the node.

However, the edges of directed graphs have directions from one node to another. In this case, a given node i has two types of degree centrality: out- and in-degree centralities. The out-degree centrality $D_{out}(i)$ is a proxy of an activity and can be expressed as follows:

$$D_{out}(i) = \sum_j w_{ij} m_{out}(i, j),$$

where $m_{out}(i, j)$ takes 1 if an edge is directed from nodes i to j and 0 if otherwise. From another aspect, the in-degree centrality $D_{in}(i)$ is a proxy of popularity and can be assessed as follows:

$$D_{in}(i) = \sum_j w_{ij} m_{in}(i, j),$$

$m_{in}(i, j)$ takes 1 if an edge is directed from nodes j to i and 0 if otherwise.

Betweenness centrality. The proportion of the shortest paths passing through a given node can be measured by the betweenness centrality of each node [11] and is defined as follows:

$$B(i) = \sum_{a,b} g_{aib} / g_{ab},$$

where, for any two nodes a and b, the number of the shortest paths passing through a given node i is defined by g_{aib}, whereas the total number of shortest paths between nodes a and b is represented by g_{ab}. Betweenness centrality is the metric that identifies a node acting as a bridge or broker in the network. In particular, information flow or network resources are controlled by a node with high betweenness centrality by mediating others with minimum paths.

Eigenvector centrality. The eigenvector centrality of a given node i represents the i th entry in the normalized eigenvector x, corresponding to the largest eigenvalue of the adjacent matrix for a given graph [14, 15]. The eigenvector centrality increases with nodes that have high correlation coefficients with many other nodes [14]. Unlike the degree centrality, nodes have high eigenvector centrality if they are connected to other nodes that are central in the network.

3 Results and Discussion

We investigate two different network structures of eight commodity futures returns: (1) correlation network, which is weighted but undirected; and (2) Granger causality network, which is directed but unweighted. The results are presented with (i) prices denominated in their trading currency, and (ii) prices denominated in USD. We use daily exchange rates to make the prices denominated in USD.

3.1 Correlation Network

Figure 1 shows the correlation network among eight futures markets of biofuel feedstock, where the weights of edges are Spearman rank correlation coefficients between the two markets. The red line presents the link between each asset, and the thicker line indicates that the two markets are highly correlated. Our results confirm that the correlation among the six markets traded in North America and Europe, which are considered mature markets, is very closely linked. However, Malaysia's CPO and the Republic of South Africa's sunflower market, which are considered emerging markets, appear to have a low correlation with the rest of the markets. The findings are consistent and robust regardless of considering the currency risk.

Furthermore, the link structure of the biofuel feedstock futures markets can be explored through the centrality metrics of the correlation network, as shown in Table 2. Biofuel feedstock commodity futures, traded in mature markets, indicate high degree and eigenvector centrality, for example, soybeans, SBO, canola, corn, rapeseed, and wheat, in the order of values in Panel A. Both centrality metrics show that the results considering the currency in Panel B are not different. On the other hand, the CPO and sunflower futures markets are somewhat isolated from the international biofuel feedstock markets.

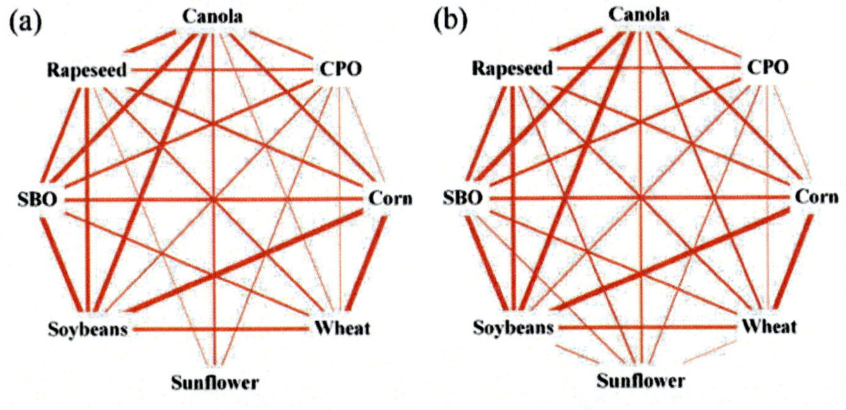

Fig. 1 Correlation network between futures markets of eight biofuel feedstock commodity. We use Spearman rank correlation coefficients to represent correlations. Panel (a) plots the correlation network of the futures prices denominated in their own trading currency, and Panel (b) plots that denominated in USD. We display links whose p-values are less than 0.05. The greater the correlation coefficient, the thicker the link

Table 2 Centrality metric of the correlation network

Centrality	Canola	Corn	CPO	Rapeseed	SBO	Soybeans	Sunflower	Wheat
Panel A. All prices are denominated in their own trading currency								
Degree	2.288	2.133	1.163	2.083	2.315	2.533	0.250	1.667
Eigenvector	0.155	0.152	0.079	0.142	0.159	0.175	0.014	0.123
Panel B. All prices are denominated in USD								
Degree	2.559	2.146	1.161	2.253	2.506	2.666	0.778	1.784
Eigenvector	0.157	0.142	0.074	0.139	0.154	0.166	0.049	0.119

3.2 Granger Causality Network

Figure 2 shows the causal relationship among the eight commodity futures markets, where the arrow from A to B is plotted if A Granger causes B. The CPO and sunflower futures markets interact with other markets but are mainly influenced by others.

Specifically, Panel A of Table 3 shows that the returns of CPO and sunflower futures are Granger caused by those of seven and six biofuel markets, respectively. Interestingly, in addition, Panel B, which shows the currency-considered results, confirms that the CPO and sunflower futures even influence other five and four futures markets, respectively, implying that they interact with other mature markets. Moreover, the CPO futures has relatively high betweenness centrality; that is, it plays an important role in spreading information among markets in the network regardless of the currency denomination. These results indicate that the CPO and

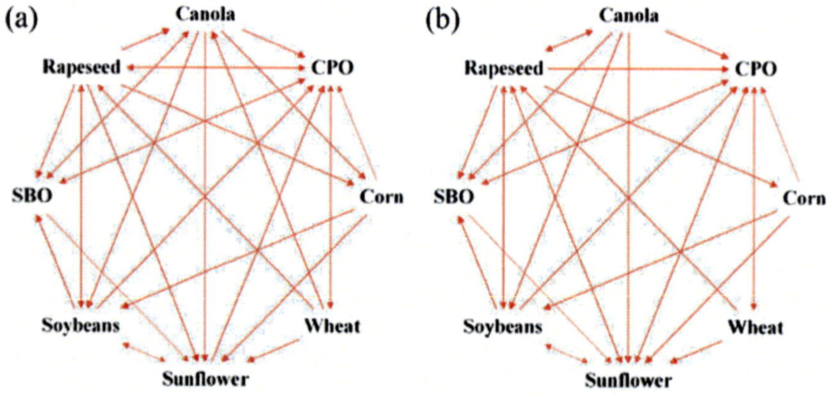

Fig. 2 Granger causality network between futures markets of eight biofuel feedstock commodity. The optimal lag lengths are chosen according to AIC. Panel **a** plots the Granger causality network of the futures prices denominated in their own trading currency, and Panel **b** plots that denominated in USD. Each arrow $A \rightarrow B$ means that A Granger causes B. A two headed arrow means that there exists bidirectional Granger causality between nodes. We only display arrows whose p-values are less than 0.05

Table 3 Centrality metric of the Granger causality network

Centrality	Canola	Corn	CPO	Rapeseed	SBO	Soybeans	Sunflower	Wheat
Panel A. All prices are denominated in their own trading currency								
Out-degree	5.000	3.000	3.000	6.000	3.000	4.000	2.000	4.000
In-degree	3.000	2.000	7.000	3.000	4.000	4.000	6.000	1.000
Betweenness	2.667	0.000	11.533	7.133	1.967	3.800	0.833	1.067
Panel B. All prices are denominated in USD								
Out-degree	5.000	5.000	5.000	6.000	2.000	4.000	4.000	3.000
In-degree	2.000	2.000	7.000	5.000	6.000	4.000	7.000	1.000
Betweenness	0.250	1.167	10.750	7.333	0.000	0.250	3.250	0.000

sunflower futures markets are well linked as members of the international network of the biofuel feedstock markets, and their importance cannot be overlooked.

4 Conclusion

This study provides the results of the network analysis of biofuel feedstock commodity futures. In the correlation network, returns in the emerging markets (i.e., CPO and sunflower futures) are not correlated with those in the mature markets. This finding implies a weak integration between the two markets, particularly in different regions. However, in the Granger causality network, we confirm the significant role of CPO and sunflower futures markets; their returns lead to price changes in several other markets in the network, and vice versa. Moreover, the CPO futures market is an essential bridge node connecting other markets with high betweenness centrality. Accordingly, we find that futures markets of biofuel feedstock commodities are strongly integrated regardless of their trading environments.

References

1. Lee, S., Yi, E., Cho, Y., Ahn, K.: The path to a sustainable palm oil futures market. Energy Rep. **8**, 6543–6550 (2022)
2. Kim, D.S., Hanifzadeh, M., Kumar, A.: Trend of biodiesel feedstock and its impact on biodiesel emission characteristics. Environ. Prog. Sustainable Energy **37**, 7–19 (2017)
3. Cadillo-Benalcazar, J.J., Bukkens, S.G.F., Ripa, M., Giampietro, M.: Why does the European Union produce biofuels? Examining consistency and plausibility in prevailing narratives with quantitative storytelling. Energy Res. Soc. Sci. **71**, 101810 (2021)
4. U.S. Energy Information Administration: Monthly Biofuels Capacity and Feedstock Update, https://www.eia.gov/biofuels/update/ (2022). Last accessed 2022/10/26
5. Kim, H., Ha, C.Y., Ahn, K.: Preference heterogeneity in Bitcoin and its forks' network Chaos. Solitons Fractals **164**, 112719 (2022)

6. Granger, C.W.: Investigating causal relations by econometric models and cross-spectral methods. Econometrica **37**, 424–438 (1969)
7. Jang, S.M., Yi, E., Kim, W.C., Ahn, K.: Information flow between Bitcoin and other investment assets. Entropy **21**, 1116 (2019)
8. Goetzmann, W.N., Li, L., Rouwenhorst, K.G.: Long-term global market correlations. J. Bus. **78**, 1–38 (2005)
9. Quinn, D.P., Voth, H.J.: A century of global equity market correlations. Am. Econ. Rev. **98**, 535–540 (2008)
10. Pukthuanthong, K., Roll, R.: Global market integration: An alternative measure and its application. J. Financ. Econ. **94**, 214–232 (2009)
11. Freeman, L.C.: Centrality in social networks conceptual clarification. Soc. Netw. **1**, 215–239 (1978)
12. Opsahl, T., Agneessens, F., Skvoretz, J.: Node centrality in weighted networks: Generalizing degree and shortest paths. Soc. Netw. **32**, 245–251 (2010)
13. Akaike, H.: Fitting autoregressive models for prediction. Ann. Inst. Stat. Math. **21**, 243–247 (1969)
14. Bonacich, P.: Factoring and weighting approaches to status scores and clique identification. J. Math. Sociol. **2**, 113–120 (1972)
15. Maharani, W., Adiwijaya, Gozali, A. A.: Degree centrality and eigenvector centrality in Twitter. In: 8th International Conference on Telecommunication Systems Services and Applications (TSSA), pp. 1–5. IEEE (2014)

Finite Element and Meshfree Methods

Natural Convection Flow in an Inclined Wavy Porous Medium in the Presence of an Inclined Periodic Magnetic Field

Shafqat Hussain and Bengisen Pekmen Geridonmez

Abstract In this study, unsteady natural convection flow in an inclined porous wavy cavity under the effect of an inclined periodic magnetic field (MF) is investigated. Mathematical model for the porous medium is chosen as Brinkman-extended Darcy model. Galerkin weighted residual finite element method is implemented to simulate the governing unsteady dimensionless equations. For discretization of time derivatives, Crank-Nicolson method is utilized. The obtained discrete nonlinear algebraic systems are iteratively solved by using the adaptive Newton's method. The behavior of fluid flow and heat transfer is examined in distinct parameters as inclination angle of the cavity ($\theta_c = 0° - 90°$), amplitude ($A = 0.025 - 0.2$) of wavy wall, the number of undulations ($N = 1 - 4$), Hartmann number ($Ha = 0 - 100$), angle ($\theta_b = 0° - 90°$) and period (λ=0.1-1) of the periodic MF. The rise in Hartmann number has a dampening effect on both heat transfer and fluid flow. Period $\lambda = 1$ and amplitude $A = 0.2$ has the most reducing influence on both fluid velocity and convective heat transfer. Oblique angle of periodic MF is efficient on fluid flow and heat transfer at $\theta_b = 45°$. Average Nusselt number is the largest at $\theta_c = 45°$ when the periodic MF affects the system at $\theta_b = 0°$.

Keywords Inclined cavity · Wavy cavity · Inclined periodic magnetic field · Porous medium

1 Introduction

Natural convection (NC) flow in a porous media is a subject taken great interest for many years due to diverse engineering applications such as insulation of buildings, solar collectors, food grain processing units and biological applications, etc. The

S. Hussain
Institut für Angewandte Mathematik (LS III), Fakultät für Mathematik, Technische Universität Dortmund, Dortmund, Germany
e-mail: shafqat.hussain@math.tu-dortmund.de

B. P. Geridonmez (✉)
Department of Mathematics, TED University, Ankara, Turkey
e-mail: bengisenpekmen@gmail.com

D. Vlachos (ed.), *Mathematical Modeling in Physical Sciences*, Springer Proceedings in Mathematics & Statistics 446, https://doi.org/10.1007/978-3-031-52965-8_31

presence of MF is also concerned in order to be able to control flow of fluid and transfer of heat. In view of geometry of the concerned structure, complex geometries are investigated in the last years in order to find better systems providing better heat transfer. In particular, numerous numerical studies involving wavy cavity may be mentioned.

Khanafer et al. [11] conducted NC flow in a square cavity having a heated wavy wall. They report that the wavy wall has small influence on the average Nu number comparing to a smooth wall when the amplitude is 0.05. The dependence of heat transfer to a geometry parameter as a result of concerned wavy geometry filled with a nanofluid is shown in [2]. In a complex wavy geometry, NC flow of a nanofluid is concerned by Mansour et al. [13]. They used smoothed particle hydrodynamics for simulation of the problem, and indicated that mean Nusselt (Nu) number increases with the augmentation in wave amplitude while Rayleigh number less than 10^3. Control volume based FEM (CVFEM) is performed in [20] for investigation of NC flow of a nanofluid in an inclined cavity with a wavy wall under the effect of a uniform MF. Their results show that the heat transfer enhancement between the presence and the absence of nanoparticles is maximum at inclination angles zero and sixty degrees. Sheremet et al. [21] applied finite difference method (FDM) in a vertical wavy walled cavity to simulate NC flow considering radiation, too. They found that average Nu number decreases with the rise in undulation number of waves. Ogut et al. [15] also used a commercial software to address the solution of NC flow in an inclined wavy cavity, and found that the rise in undulation number increases the rate of heat transfer. Chamkha et al [7] investigated NC flow in a cavity having triangle wave on the left vertical wall under the effect of a vertical MF. They reported that the average Nu number is a decreasing function of corrugation frequency in their problem. Pekmen Geridonmez [18] also numerically examined free convection flow in a cavity, of which vertical walls are differentially heated inward or outward wavy walls, under the influence of a magnetic source by using radial basis function based collocation method. The results report that the reduction in average Nu number with the rise in undulation number. Dogonchi et al. [8] studied on NC flow of Al2O3-water nanofluid in a porous wavy cavity in crown shape in the presence of an inclined uniform MF and a hot cylinder utilizing FEM. Their results declare that average Nu is an increasing function of radiation parameter as well as the left positioned cylinder. Aldulkadhim et al. [1] analysed NC flow of Cu-water nanofluid in a cavity with vertical cold wavy walls and a hot cylinder considering heat generation and absorption denoted by heat coefficient $q \in [-10, 10]$. FEM is preferred for numerical solution, and the enclosure is subjected to a horizontal uniform MF. They showed that heat absorption for $q < 10$, average Nu number boosts. Alsabery et al. [3] also applied FEM to solve 3D NC flow of a nanofluid in a porous cavity having wavy bottom wall and including an inner conducting solid block. Their results present that the entropy generation significantly occurs at higher undulation numbers. Mourad et al. [14] inquired the MHD NC flow in a two-layered cavity including hybrid nanofluid, Fe3O4-MWCNT/water, and a heated hole in ellipse shape applying Galerkin FEM. Their results state that the circulation in fluid flow increases with the rise in the amplitude and undulation number of waves. Parveen et al. [16] explored the significance of convex upper wavy wall on average Nu number in a cavity heated from the bottom straight wall and having cold vertical walls. The presence of wavy-shaped inner cylinder in a curve-shaped cavity is taken into account in Parvin et al. [17] considering the subjection to a uniform MF. Their results show that inner cylinder without waviness gives better heat transmission rate than with waviness. Shaik et al. [19] progressed FVM to simulate NC flow of hybrid nanofluid, Cu-TiO2/water, under the influence of an inclined uniform MF, radiation and a heat source in an oblique cavity having a wavy wall. In their results, the larger heat generation parameter ascends the heat transfer rate. Biswas et al. [4] performed FVM for simulation of NC flow of a Cu-Al2O3/water nanofluid in a cavity in various inclination angles. With the presence of hot left vertical wavy wall and imposition of horizontal uniform MF, their results reveal that the rise in undulation number (≤ 4) results in maximum fluid velocity as well as average Nu number. Partial MF is also taken into account by Biswas et al. [5] in an inclined wavy cavity involving a hybrid nanofluid.

In aforementioned studies, uniform MF is frequently used, and rare studies with periodic MF are found in literature. To the best of Authors' knowledge, in the current work, numerical investigation on free convection flow in an inclined wavy porous cavity is firstly done in the presence of an inclined periodic magnetic field.

2 Model of the Problem

The current problem is considered as a two-dimensional, time dependent natural convection flow in an inclined porous wavy cavity under the effect of an inclined periodic MF. The fluid is laminar and incompressible. The porous medium is isotropic and homogeneous. Problem configuration may be presented as in Fig. 1. Walls with T_h and T_c refer to the hot and cold walls, respectively. Wavy walls are insulated walls where there is zero temperature gradient. The angle θ_c denotes the inclination angle of the cavity. When $\theta_c = 0°$, the upper wavy wall has an equation $y = 1 + A\sin(N\pi x)$ and the lower wavy wall has an equation $y = -A\sin(N\pi x)$ in which A and N denotes the amplitude and the number of undulations, respectively. The angle θ_b stands for the inclination angle of the periodic MF.

The non-dimensional equations are derived as

$$\frac{\partial u}{\partial x} + \frac{\partial v}{\partial y} = 0 \tag{1a}$$

$$\begin{aligned} Pr\nabla^2 u = {} & \frac{\partial u}{\partial t} + u\frac{\partial u}{\partial x} + v\frac{\partial u}{\partial y} + \frac{\partial P}{\partial x} + \frac{Pr}{Da}u - RaPrT\sin(\theta_c) \\ & - Ha^2 Pr\sin^2\left(\frac{\pi}{\lambda}\tilde{A}\right)[v\sin(\theta_b)\cos(\theta_b) - u\sin^2(\theta_b)] \end{aligned} \tag{1b}$$

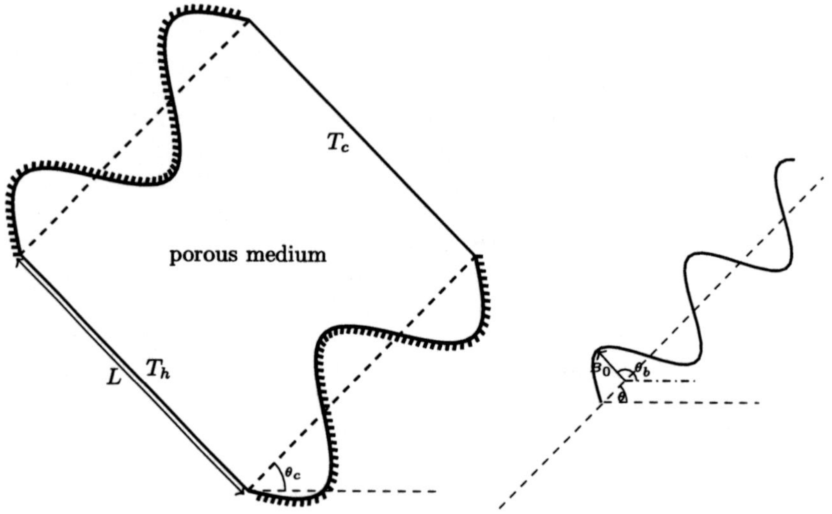

Fig. 1 Problem configuration

$$
\begin{aligned}
Pr\nabla^2 v = \frac{\partial v}{\partial t} + u\frac{\partial v}{\partial x} + v\frac{\partial v}{\partial y} + \frac{\partial P}{\partial y} + \frac{Pr}{Da}v - Ra\,Pr\,T\cos(\theta_c) \\
- Ha^2 Pr\sin^2\left(\frac{\pi}{\lambda}\tilde{A}\right)[u\sin(\theta_b)\cos(\theta_b) - v\cos^2(\theta_b)] \qquad (1c)
\end{aligned}
$$

$$
\nabla^2 T = \frac{\partial T}{\partial t} + u\frac{\partial T}{\partial x} + v\frac{\partial T}{\partial y}, \qquad (1d)
$$

where $\tilde{A} = x\cos(\theta) + y\sin(\theta)$, and Prandtl, Rayleigh, Hartmann and Darcy numbers have the definitions as

$$
Pr = \frac{\nu}{\alpha},\; Ra = \frac{g\beta\Delta T L^3}{\alpha\nu},\; Ha = B_0 L\sqrt{\frac{\sigma}{\mu}},\; Da = \frac{\kappa}{L^2}. \qquad (2)
$$

No-slip velocity boundary conditions, $u = v = 0$, are set on each walls. Insulated temperature boundary conditions are inserted on wavy walls while hot and cold walls are $T_h = 1$ and $T_c = 0$.

The average Nu number on the hot wall is calculated by

$$
\overline{Nu} = \int_0^1 \frac{\partial T}{\partial n} ds, \qquad (3)
$$

where $\partial T/\partial n$ means the normal gradient of temperature T, and ds is either dx or dy according to the tendency or location in the hot wall due to the inclination in cavity.

3 Numerical Progress

3.1 Validation and Grid Independency

The proposed problem is governed by the dimensionless system of time dependent nonlinear partial differential equations. The well-known second order Crank-Nicolson scheme has been employed for the time discretization and higher order Galerkin FEM has been implemented for the spatial discretization. In particular, an LBB-stable [6, 10] pair of finite elements is utilized to obtain the corresponding approximations. For further details regarding the designed solver, see [12, 22] for details.

The validation is done comparing average Nu number with a reference study [9] in the absence of nanofluid ($\phi = 0$) along the heated wall in a square enclosure in the presence of a uniform MF. Well agreement appears in Table 1.

Grid Independency Table 2 checks grid independency and in eighth row, 18666 number of elements are used in all computations.

Table 1 Comparison between $\overline{Nu}$ between Present and Ref. [9] when $\phi = 0$

	$\overline{Nu}$		$\lvert\psi\rvert_{\max}$	
Ra	Present	Ref. [9]	Present	Ref. [9]
10^3	1.002	1.002	0.127	0.128
10^4	1.182	1.183	1.191	1.193
10^5	3.137	3.150	5.696	5.710
10^6	7.811	7.907	13.946	14.088
10^7	16.325	16.929	29.909	30.813

Table 2 The grid test for $\theta_c = 0°$, $A = 0.1$, $N = 3$, $Ra = 10^5$, $Pr = 1$, $Da = 0.01$, $Ha = 25$, $\theta_b = 90°$, $\lambda = 1$

ℓ	#EL	#DOFs	Nu_{avg}
1	268	3100	3.49450493984
2	424	4770	3.50206018091
3	698	7443	3.51342851095
4	1176	12237	3.52521473701
5	1658	16807	3.53275555304
6	2684	26216	3.53931143726
7	7158	68115	3.55441153052
8	18666	172813	3.56308640495
9	28080	252832	3.56300548559

4 Numerical Outcomes and Analysis

In this part, numerical visualization results are presented as streamlines and isotherms in different parameter variations. Pr=1, Ra=1e+5, Da=1e-2 are kept fixed. Varying pertinent parameters are in the ranges θ_c=0°-90°, A=0.025-0.2, N=1-4, Ha=10-100, θ_b=0°-90°, λ=0.1-1. Average Nu number variations are also tabulated.

Figure 2 shows the behavior of fluid flow and heat transfer in the change of inclination angle of the cavity. Also, $\theta_b = 0°$ refers to the horizontally exposed MF. Counter clockwise circulation in streamlines at $\theta_c = 0°$ is turned into clockwise circulation at $\theta_c = 90°$. Velocity of flow also is declined as θ_c is changed from 0° to 90° as is noted from $|\psi|_{\max}$ values. The fastest fluid flow is noted at $\theta_c = 45°$. Further, secondary cells emerge at $\theta_c = 90°$. In isotherms, NC behavior at $\theta_c = 0°$ is changing into Rayleigh-Benard type convection at $\theta_c = 90°$ where temporal gradient occurs close to the hot bottom wall.

The increment in amplitude of wavy wall is analyzed in Fig. 3. Core vortex in primary cell in streamlines is squeezed inside cavity as A is ascended. The most convective behavior with faster fluid flow is obtained at $A = 0.1$ comparing to the other A values.

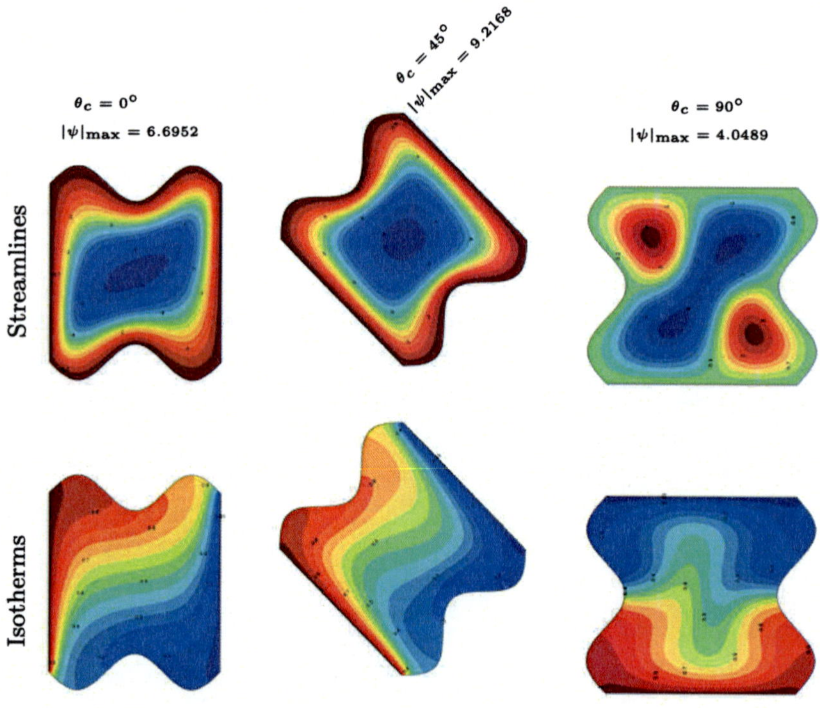

Fig. 2 Variation in θ_c values with $A = 0.1$, $N = 3$, $Ha = 25$, $\theta_b = 0°$, $\lambda = 1$

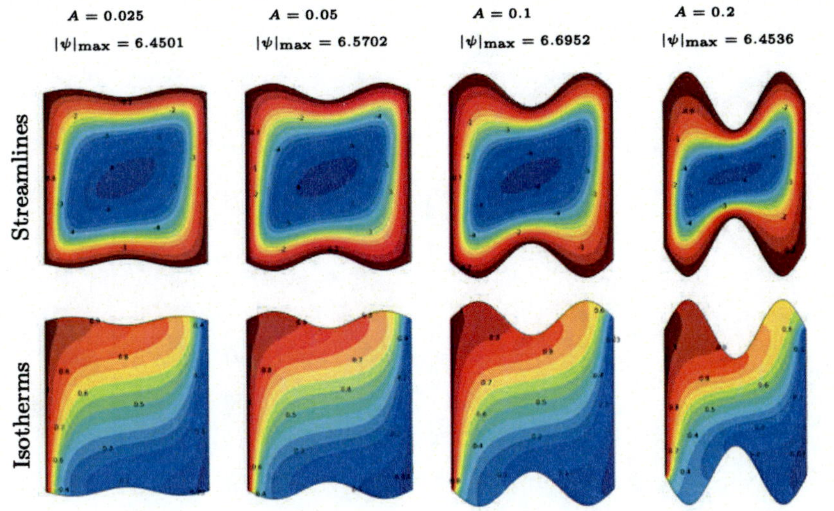

Fig. 3 Variation in A values with $\theta_c = 0°$, $N = 3$, $Ha = 25$, $\theta_b = 0°$, $\lambda = 1$

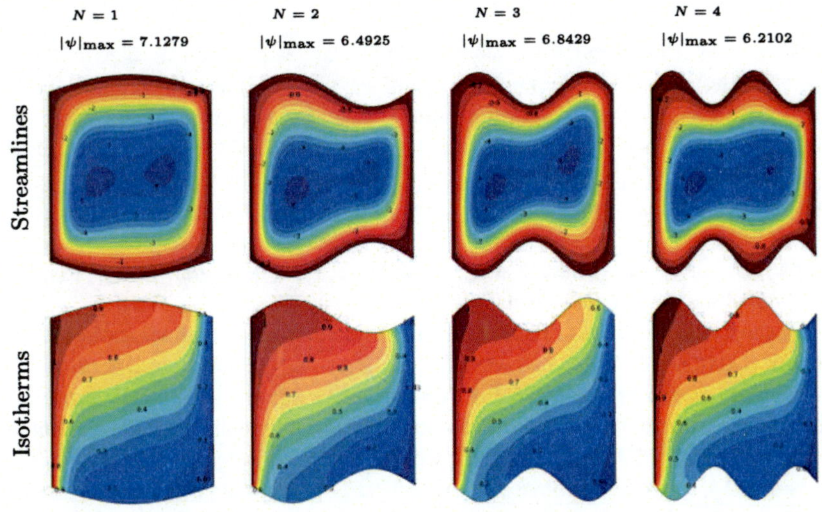

Fig. 4 Variation in N values with $\theta_c = 0°$, $A = 0.1$, $Ha = 25$, $\theta_b = 90°$, $\lambda = 1$

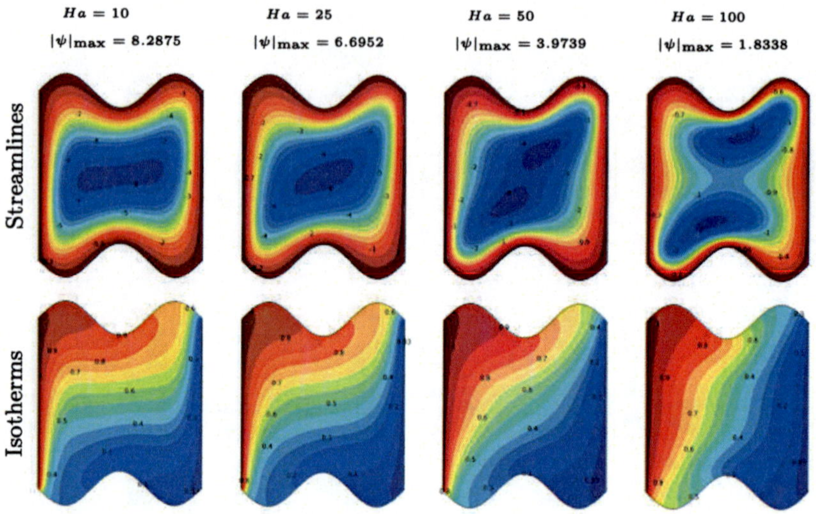

Fig. 5 Variation in Ha numbers with $\theta_c = 0°$, $A = 0.1$, $N = 3$, $\theta_b = 0°$, $\lambda = 1$

The variance in the number of undulations in wavy wall is also checked in Fig. 4. As can be noted in streamlines, inner wave in wavy wall suppresses fluid flow. From $N = 1$ to $N = 2$, symmetric central cells become asymmetric at $N = 2$ while the right central cell is more inhibited than the other left cell. At $N = 3$, again central symmetric two cells are realized. But, comparing to $N = 1$, inner one wave in wavy wall at $N = 3$ causes symmetricity as well as weakened fluid velocity. Since inner waves rise at $N = 4$, the reduction in fluid velocity becomes more pronouncing. In isotherms, convective behavior at $N = 1$ drops at $N = 4$. In view of NC, thermal gradient along the left wall seems most promising at $N = 1$.

The influence of various values of Hartmann numbers on streamlines and isotherms is indicated in Fig. 5. Notice that the MF is coming along the left hot wall periodically with period 1 and amplitude $A = 0.1$. In streamlines, this one wave tends to separate the core vortex into two parts as Ha number rises. Owing to the restraining effect of Lorentz force, fluid flow is dampened with this rise as is noted from $|\psi|_{\max}$ values. In isotherms, convective behavior at $Ha = 10$ is converted into conductive behavior at $Ha = 100$. That is, large Lorentz force inhibits the convective heat transfer.

Figure 6 displays the flow of fluid and transfer of heat in distinct values of inclination angle of periodic MF. The most increasing effect of this angle on fluid velocity is obtained at $\theta_b = 45°$. A little bit more convective behavior occurs also at $\theta_b = 45°$ in isotherms comparing to the other angles.

For various values of period of periodic MF, Fig. 7 is depicted. Period $\lambda = 1$ has the most weakening effect on fluid velocity while the largest values of $|\psi|_{\max}$ is obtained at $\lambda = 0.5$. Not much change in isotherms is noted.

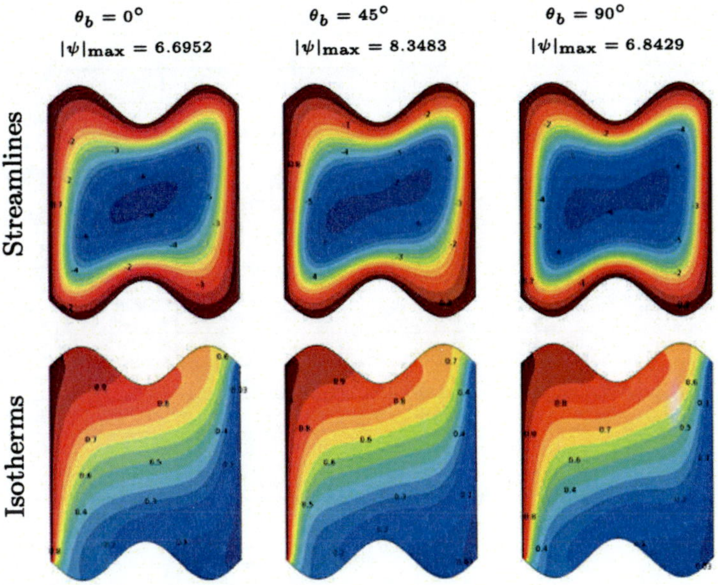

Fig. 6 Variation in θ_b values with $\theta_c = 0°,\ A = 0.1,\ N = 3,\ Ha = 25,\ \lambda = 1$

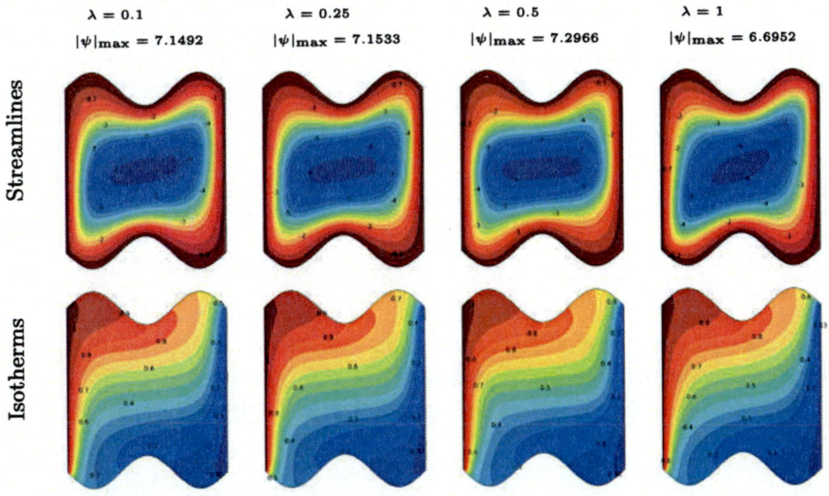

Fig. 7 Variation in λ values with $\theta_c = 0°,\ A = 0.1,\ N = 3,\ Ha = 25,\ \theta_b = 0°$

Table 3 Behavior of $\overline{Nu}$ in different parameter variations

θ_c	A	N	Ha	θ_b	$\lambda = 1$	$\overline{Nu}$
0°	0.1	3	25	0°	1	3.5631
45°						4.0554
90°						2.1301
0°	0.025	3	25	0°	1	3.5506
	0.05					3.5886
	0.1					3.5631
	0.2					3.2450
0°	0.1	1	25	0°	1	3.7874
		2				3.2856
		3				3.5631
		4				3.1189
0°	0.1	3	10	0°	1	4.1886
			25			3.5631
			50			2.4979
			100			1.4355
0°	0.1	3	25	0°	1	3.5631
				45°		4.0137
				90°		3.9519
0°	0.1	3	25	0°	0.1	3.6740
					0.25	3.6660
					0.5	3.6681
					1	3.5631

Table 3 presents $\overline{Nu}$ results along the heated wall in cases observed in aforementioned figures. The larger $\overline{Nu}$ is, the more convective heat transfer (CHT) occurs. In variation of

- θ_c, the most CHT occurs at $\theta_c = 45°$.
- A, CHT is the smallest one when $A > 0.1$.
- N, CHT rises from $N = 1$ to $N = 3$, but then CHT decreases from $N = 3$ to $N = 4$.
- Ha, CHT is the smallest at $Ha = 100$ in which the Lorentz force is the largest.
- θ_b, periodic MF coming horizontally along the heated wall ($\theta_b = 0°$) has the most suppressing effect on CHT.
- λ, period $\lambda = 1$ has the most inhibitive influence on CHT.

5 Conclusions

In this study, the influence of the inclination in a wavy porous cavity and in periodic MF on flow and temperature behavior is investigated. The enclosure has wavy insulated walls while the straight walls are differentially heated. Porous medium is modeled by Brinkman-extended Darcy model. Galerkin FEM is applied discretizing time derivatives by the second order Crank-Nicholson method.

The inclination angle of the cavity at $\theta_c = 45°$ has the most accelerating effect on fluid flow and CHT is also the largest value at $\theta_c = 45°$. The increment in amplitude of wavy wall (when $A \geq 0.05$) results in diminish in CHT. The presence of more inner waviness in wavy wall causes fluid to flow slowly. CHT increases till $N = 3$ but then it decreases. The larger the Lorentz force is, the more reduction in CHT and $|\psi|_{\max}$ is developed. The influence of inclination angle of periodic MF on CHT and on the fluid velocity is the most improving at $\theta_b = 45°$. Period $\lambda = 1$ of periodic MF has the most reducing effect on CHT and fluid velocity.

References

1. Abdulkadhim, A., Hamzah, H., Ali, F.H., Yildiz, C., Abed, A.M., M., A.E., Arici, M.: Effect of heat generation and heat absorption on natural convection of cu-water nanofluid in a wavy enclosure under magnetic field. Int. Commun. Heat Mass Transfer **120**(024), 105 (2021). https://doi.org/j.icheatmasstransfer.2020.105024
2. Abu-Nada, E., Oztop, H.: Numerical analysis of al2o3/water nanofluids natural convection in a wavy walled cavity. Numer. Heat Transfer Part A: Appl. **59**, 403–419 (2011). https://doi.org/10.1080/10407782.2011.552363
3. Alsabery, A.I., Ismael, M.A., Chamkha, A.J., Hashim, I., Abulkhair, H.: Unsteady flow and entropy analysis of nanofluids inside cubic porous container holding inserted body and wavy bottom wall. Int. J. Mech. Sci. **193**(106), 161 (2021). https://doi.org/10.1016/j.ijmecsci.2020.106161
4. Biswas, N., Mandal, D.K., Manna, N., Gorla, R., Chamkha, A.J.: Magnetohydrodynamic thermal characteristics of water-based hybrid nanofluid-filled non-darcian porous wavy enclosure: effect of undulation. Int. J. Numer. Methods Heat Fluid Flow **32**, 1742–1777 (2022). https://doi.org/10.1108/HFF-03-2021-0190
5. Biswas, N., Mondal, M.K., Mandal, D.K., Manna, N., Gorla, R., Chamkha, A.: A narrative loom of hybrid nanofluid-filled wavy walled tilted porous enclosure imposing a partially active magnetic field. Int. J. Mech. Sci. **217**(028), 107 (2022). https://doi.org/10.1016/j.ijmecsci.2021.107028
6. Boffi, D., Brezzi, F., Fortin, M.: Finite Elements for the Stokes Problem, pp. 45–100. Springer, Berlin (2008). https://doi.org/10.1007/978-3-540-78319-0_2
7. Chamkha, A., Selimefendigil, F.: Mhd free convection and entropy generation in a corrugated cavity filled with a porous medium saturated with nanofluids. Entropy **20**, 846 (2018). https://doi.org/10.3390/e20110846
8. Dogonchi, A., Sadeghi, M.S., Ghodrat, M., Chamkha, A.J., Elmasry, Y., Alsulami, R.: Natural convection and entropy generation of a nanoliquid in a crown wavy cavity: Effect of thermo-physical parameters and cavity shape. Case Stud. Thermal Eng. **27**(101), 208 (2021). https://doi.org/10.1016/j.csite.2021.101208
9. Ghasemi, B., Aminossadati, S., Raisi, A.: Magnetic field effect on natural convection in a nanofluid-filled square enclosure. Int. J. Thermal Sci. **50**, 1748–1756 (2011)

10. Hussain, S., Schieweck, F., Turek, S.: Efficient Newton-multigrid solution techniques for higher order space-time Galerkin discretizations of incompressible flow. Appl. Numer. Math. **83**, 51–71 (2014). https://doi.org/10.1016/j.apnum.2014.04.011, http://www.sciencedirect.com/science/article/pii/S0168927414000713
11. Khanafer, K., Al-Azmi, B., Marafie, A., Pop, I.: Non-darcian effects on natural convection heat transfer in a wavy porous enclosure. Int. J. Heat Mass Trans. **52**, 1887–1896 (2009). https://doi.org/10.1016/j.ijheatmasstransfer.2008.08.040
12. Logg, A., Mardal, K.A., Wells, G.: Automated Solution of Differential Equations by the Finite Element Method: The FEniCS Book. Lecture Notes in Computational Science and Engineering, p. 84. Springer (2012). http://gen.lib.rus.ec/book/index.php?md5=94ace312a0b9e30a2f712c871d88d4db
13. Mansour, M., Bakier, M.: Free convection heat transfer in complex-wavy-wall enclosed cavity filled with nanofluid. Int. Commun. Heat Mass Transfer **44**, 108–115 (2013). https://doi.org/10.1016/j.icheatmasstransfer.2013.02.015
14. Mourad, A., Aissa, A., Mebarek-Oudina, F., Jamshed, W., Ahmed, W., Ali, H.M., Rashad, A.M.: Galerkin finite element analysis of thermal aspects of fe3o4-mwcnt/ water hybrid nanofluid filled in wavy enclosure with uniform magnetic field effect. Int. Commun. Heat Mass Transfer **126**(105), 461 (2021). https://doi.org/10.1016/j.icheatmasstransfer.2021.105461
15. Ogut, E.B., Akyol, M., Arici, M.: Natural convection of nanofluids in an inclined square cavity with wavy side walls. J. Thermal Sci. Technol. **37**, 139–150 (2017). https://doi.org/10.1016/j.ijheatmasstransfer.2017.07.057
16. Parveen, R., Mahapatra, T.R., Saha, B.C.: Study of entropy generation and magnetohydrodynamic (mhd) natural convection in a curved enclosure having various amplitude and filled with cu-tio2/water hybrid nanofluid. J. Nanofluids **10**, 339–354 (2021). https://doi.org/10.1166/jon.2021.1794
17. Parvin, S., Roy, N.C., Saha, L.K.: Magnetohydrodynamic natural convection of a hybrid nanofluid from a sinusoidal wavy cylinder placed in a curve-shaped cavity. AIP Adv. **11**(029), 085 (2021). https://doi.org/10.1063/5.0061503
18. Pekmen Geridonmez, B.: Free convection in a wavy walled cavity with a magnetic source using radial basis functions. J. Heat Transfer **141**(042), 501–1 (2019). https://doi.org/10.1115/1.4042782
19. Shaik, J., Polu, R.A.B., Ahmed, M., Mohamed, R.A.: Characteristics of moving hot block and non-fourier heat flux model on sinusoidal wavy cavity filled with hybrid nanofluid. European Phys. J. Plus **137**, 131 (2022). https://doi.org/10.1140/epjp/s13360-022-02361-y
20. Sheikholeslami, M., Gorji-Bandpy, M., Ganji, D., Soleimani, S.: Mhd natural convection in a nanofluid filled inclined enclosure with sinusoidal wall using cvfem. Neural Comput. Appl. **24**, 873–882 (2014). https://doi.org/10.1007/s00521-012-1316-4
21. Sheremet, M.A., Miroshnichenko, I.V.: Effect of surface radiation on transient natural convection in a wavy-walled cavity. Numer. Heat Transfer Part A: Appl. **69**, 369–382 (2016). https://doi.org/10.1080/10407782.2015.1080577
22. Zienkiewicz, O.C., Taylor, R.L., Zhu, J.: The Finite Element Method: its Basis and Fundamentals, 7 edn. Butterworth-Heinemann (2013). http://gen.lib.rus.ec/book/index.php?md5=51571575047a0b5ae432cc374dba5c86

Topics in Mathematical Physics

Pseudogeometric Version of the Traveling Salesman Problem, Its Application in Quantum Physics Models and Some Heuristic Algorithms for Its Solution

Boris Melnikov and Dmitrii Chaikovskii

Abstract The geometric version of the traveling salesman problem (TSP) has been extensively studied, leading to the development of various approaches for solving its special cases. However, these algorithms often fall short when applied to problems beyond the geometric TSP. In this paper, we explore the pseudo-geometric TSP version, a generalization of the geometric TSP, and propose an adapted geometric algorithm for solving its specific instances. We leverage the knowledge of error bounds to estimate the reconstruction error of the TSP solution even when using geometric approaches for the pseudo-geometric TSP. This allows us to achieve reliable results despite uncertainties or noise in the data. We provide a concise description of our algorithmic adaptation and present the results of computational experiments to demonstrate its effectiveness.

Keywords Mathematical model · Hamilton cycles · Traveling salesman problem · Pseudo-geometric version · Geometric approach · Heuristic algorithms

1 Introduction and Motivation. Hamilton Cycles in Quantum Physics Models

The Traveling salesman problem (TSP) is a well-known optimization problem that seeks to find the shortest possible route that visits a set of points (or cities) and returns to the starting point. The TSP can be formulated in different ways, depending on the problem instance [1–5]. According to the classification given in [3], the traveling salesman problem is an example of an optimization problem that belongs to the most complex class $NPO(V)$. This class contains all optimization tasks that, when given an additional assumption such as $P \neq NP$, have a time complexity that cannot be

B. Melnikov (✉) · D. Chaikovskii
Shenzhen MSU–BIT University,Shenzhen, China
e-mail: bormel@mail.ru
URL: https://szmsubit.ru/en/

D. Chaikovskii
e-mail: mitichya@yandex.ru

D. Vlachos (ed.), *Mathematical Modeling in Physical Sciences*, Springer Proceedings in Mathematics & Statistics 446, https://doi.org/10.1007/978-3-031-52965-8_32

limited to any polylogarithmic function for all possible polynomial algorithms. Let us give a description of TSP according to [2, 3]; in this paper, we shall not repeat the detailed description of the problem statement.

One of the most common variants of the TSP is the geometric TSP [3, 5, 6], which is characterized by a set of points and their distances, where the distances are defined by the Euclidean distance between the points. In the case of the geometric TSP, we can use a number of algorithms to find an approximate solution to the problem. One of the most popular algorithms is the nearest neighbor algorithm, which starts at a random point and repeatedly selects the closest unvisited point as the next point to visit. Another popular algorithm is the convex hull method, which uses the convex hull of the set of points to reduce the number of points that need to be considered. These algorithms can be used to find approximate solutions to the problem with a high degree of accuracy.

The TSP can be defined also as a random problem, if the points are not defined by their coordinates, but by a set of random numbers. In this case, it is not possible to use geometric algorithms to find solutions, and instead, more general optimization techniques must be used. For example, a common approach is to use the branch and bound method, when we systematically enumerate all possible solutions to a problem and determine which is the best based on some objective criteria. This is done by dividing the problem into smaller sub-problems (or branches) and solving each sub-problem separately. The solutions to these sub-problems are then compared, and the best solution is chosen as the overall best solution.

However, in some cases, the problem instance may not be defined by the Euclidean distance between the points, but by some other metric that has some degree of randomness or noise built into it. This variant of the TSP is called the pseudo-geometric TSP [5–7], where we form initial data in the following way: for a given geometric TSP, we added a vector $R = (r_1, \ldots, r_m)$ where $m = |E|$; all the r_i are independent identically distributed random variables with a normal distribution with $\mu = 1$ and a predefined "acceptable" value of the square deviation σ. Each element of the cost matrix $(c(\{u, v\})$ (let it be c_i for some $i \in \{1, \ldots, m\}$) is replaced by $\max(c_i \cdot r_i, 0)$. It is important to note that the geometric TSP-version can be regarded as a special case of pseudo-geometric one with the value $\sigma = 0$.

The "ecological niche" of the geometric version of the TSP is well-studied, with many approaches to solving its particular cases developed over the years, [4] etc. Some examples of algorithms belonging to this group are the "onion-peeling" algorithms [8] and the "elastic network" [9], which have a relatively low computational complexity and can provide good solutions for special cases with millions of points that are practically optimal. However, these algorithms are usually not applicable outside of the geometric TSP-version, which is why this paper aims to explore the possibility of applying similar algorithms to other TSP-versions. Therefore, our goal is to create a reliable heuristic approach to solve pseudo-geometric TSP utilizing techniques from geometric TSP. By doing so, we can better understand the complexities of this challenging optimization problem and gain insight into more efficient ways of solving it. Additionally, this research can provide a platform for developing more efficient algorithms in the future.

Mathematical models and algorithms originally developed to solve the traveling salesman problem were designed to construct optimal routes for traffic and the choice of the optimal trajectory. However, these models and algorithms are now also used to solve other applied problems, such as sequencing nucleotide sequences of biopolymers [10], calculating the similarity of strings [11, 12] constructing practical research algorithms of specially defined infinite grammatical structures [13, 14], and constructing evolutionary trees [15]. Furthermore, the TSP has been applied to the analysis of DNA; after sequencing, a set of small-length strings is obtained, and an oriented graph is created with weights that represent the degree of similarity between them. In this setting, the longest common superstring has the maximum cost. This problem is close to the pseudo-geometric TSP-version, as the length of overlapping sections can be arbitrary (in certain boundaries). Thus, there is potential for further work in this area.

The pseudo-geometric approach can be used to solve other tasks, such as finding the optimal route for manipulating the state of one or more photons in a *quantum system*. In this approach, the *quantum states* are represented by points in a geometric space. However, the distance between two states is not determined by the distance between their corresponding points in the space. Instead, the distance is defined using a metric that captures the similarity or dissimilarity between the states. For example, the TSP can be used with points corresponding to different states of photons and finding the optimal route for manipulating the state of the photons. In addition, it possible to use *quantum computers* for solutions of such problems, with the usage of technologies such as quantum graph walking and *quantum annealing*. The *quantum graph walking* (QWG) is a quantum algorithm that uses quantum states to find the shortest path in a graph, while quantum annealing is used to control quantum systems through various influences such as electric and magnetic fields. By combining quantum graph walking and quantum annealing, we can solve the traveling salesman problem in a quantum system. This approach can be used to find the optimal path of a *quantum particle* in quantum mechanics, and then optimize and control the trajectory using quantum annealing. These techniques can be used to manipulate the state of photons in a quantum system and implement various quantum operations such as encryption and decryption in quantum cryptography and measurements in quantum metrology.

The organization of this paper is outlined as follows. In Sect. 2, we provide a theoretical foundation for our approach. In Sect. 3, we detail the method we developed to solve the pseudo-geometric TSP, including the techniques for classifying input data and placing points pseudo-optimally. Finally, in Sect. 4, we present the results of numerical experiments using our method and summarize the results.

2 The Theoretical Substantiation for the Application of a Heuristic Method

Thus, we use the geometric approaches for the pseudo-geometric TSP.

In the current paper, we propose to use some of the geometric approaches for solution of the pseudo-geometric TSP, in which the distance between two points is defined by a function that takes into account both the Euclidean distance between the points and a random variable, which is referred to as the "dispersion". The dispersion represents the degree of randomness or noise in the problem, and it is used to introduce a level of uncertainty into the problem, which makes it more difficult to solve.

For the TSP in 2D, we need to reconstruct the noisy path and estimate the error bound for the total distance. However, this is a it is an ill-posed problem and even a small error in the initial data can result in a significant error in the calculated derivatives [12]. Let's consider a set of points $P = (x_i, y_i)$ with $i = 1, \dots, N$ in the unit square, where the noisy coordinates are (x_i^δ, y_i^δ). Moreover, we assume the following:

$$\max\left(|y(i) - y_i^\delta|, |x(i) - x_i^\delta|\right) \le \delta,$$

where δ is a known level of noise or dispersion in the data. As we know, the TSP requires finding the shortest closed path that visits each point in P exactly once.

We can represent the true path as a sequence of indices $s = (s_1, s_2, \dots, s_N)$, where the index s_i refers to the i-th point in the true path. This sequence of indices can be found by techniques such as *nearest neighbors* or *greedy algorithms*. The noisy path will be given by a sequence of noisy coordinates $(x_{(s_i)}^\delta, y_{(s_i)}^\delta)$. We shall consider the problem in a piece-wise manner using smoothing cubic splines.

In some applications, considering a smooth path using splines over nearest neighbor or linear interpolation can be beneficial for several reasons. Smooth paths are important when the path is meant to be followed by a vehicle or a robot, as they ensure feasible and efficient movement without challenging or impossible-to-navigate sharp turns or discontinuities. Additionally, spline interpolation can provide a more accurate approximation of the true underlying path by reducing the impact of noise on the final result, whereas linear interpolation may be more sensitive to noise and nearest neighbor interpolation may not even create a continuous path. Smooth paths also tend to be more visually appealing and easier to understand when visualizing TSP results. Lastly, if the true underlying path is smooth, using splines can yield a better approximation than linear interpolation or nearest neighbor methods.

To construct a piece-wise cubic spline for the 2D TSP-path, we can consider a parametric representation of the path. Let t be the parameter that varies from 0 to 1, with t_i corresponding to the point (x_{s_i}, y_{s_i}).

For each consecutive pair of points (x_{s_i}, y_{s_i}) and $(x_{s_{i+1}}, y_{s_{i+1}})$, we can define two cubic functions $X_i(t)$ and $Y_i(t)$ for $t \in [t_i, t_{i+1}] \equiv \Omega_i$ in the following way.

$$X_i(t) = a_{x,i} + b_{x,i}(t - t_i) + c_{x,i}(t - t_i)^2 + d_{x,i}(t - t_i)^3,$$
$$Y_i(t) = a_{y,i} + b_{y,i}(t - t_i) + c_{y,i}(t - t_i)^2 + d_{y,i}(t - t_i)^3.$$

The coefficients $a_{x,i}$, $b_{x,i}$, $c_{x,i}$, $d_{x,i}$, $a_{y,i}$, $b_{y,i}$, $c_{y,i}$, and $d_{y,i}$ can be determined by imposing the following conditions.

1. The cubic functions pass through the points:

$$X_i(t_i) = x_{s_i}, \quad X_i(t_{i+1}) = x_{s_{i+1}},$$
$$Y_i(t_i) = y_{s_i}, \quad Y_i(t_{i+1}) = y_{s_{i+1}}.$$

2. The first and second derivatives are continuous at the junctions:

$$X_i'(t_{i+1}) = X_{i+1}'(t_{i+1}), \quad X_i''(t_{i+1}) = X_{i+1}''(t_{i+1}),$$
$$Y_i'(t_{i+1}) = Y_{i+1}'(t_{i+1}), \quad Y_i''(t_{i+1}) = Y_{i+1}''(t_{i+1}).$$

We denote the sets of piece-wise functions as:

$$X(t) = \begin{cases} X_1(t), & \text{for } t \in [t_1, t_2], \\ X_2(t), & \text{for } t \in [t_2, t_3], \\ \ldots, & \\ X_N(t), & \text{for } t \in [t_N, t_{N+1}]. \end{cases}$$

and

$$Y(t) = \begin{cases} Y_1(t), & \text{for } t \in [t_1, t_2], \\ Y_2(t), & \text{for } t \in [t_2, t_3], \\ \ldots, & \\ Y_N(t), & \text{for } t \in [t_N, t_{N+1}]. \end{cases}$$

Then we define the functional $\Phi[X, Y]$ as follows:

$$\Phi[X, Y] \equiv \frac{1}{N} \cdot \sum_{i=1}^{N} \left(\left(x_{s_i}^{\delta} - X(t_i) \right)^2 + \left(y_{s_i}^{\delta} - Y(t_i) \right)^2 \right) + \alpha \cdot \left(\left| X''(t) \right|_{L^2(\Omega)}^2 + \left| Y''(t) \right|_{L^2(\Omega)}^2, \right), \tag{1}$$

here, the domain Ω is the union of all the segments' domains, i.e., $\Omega = \bigcup_{i=1}^{N} \Omega_i$.

Let α be such that the minimizing elements $X_\alpha(t)$ and $Y_\alpha(t)$ of $\Phi[X, Y]$ satisfy:

$$\frac{1}{N} \cdot \sum_{i=1}^{N} \left(\left(x_{s_i}^{\delta} - X_\alpha(t_i) \right)^2 + \left(y_{s_i}^{\delta} - Y_\alpha(t_i) \right)^2 \right) = \delta^2.$$

We find the minimizing elements $X_\alpha(t)$ and $Y_\alpha(t)$ of $\Phi[X, Y]$:

$$(X_\alpha(t), Y_\alpha(t)) = \arg\min_{X,Y} \Phi[X, Y]. \tag{2}$$

These minimizing cubic spline functions $X_\alpha(t)$ and $Y_\alpha(t)$ will provide the best approximation of the true path, given the noisy data and the chosen smoothness parameter α.

The total distance can be approximated as follows:

$$D_{TSP} \approx \int_{t_1}^{t_N} \sqrt{(X'_\alpha(t))^2 + (Y'_\alpha(t))^2}\, dt\,.$$

With the piece-wise cubic spline, it is possible obtain a smooth approximation of the TSP path that is less sensitive to the noise in the data points, leading to a more accurate estimation of the total distance.

Then according to [16], the total error bound for the TSP path can be expressed in the following way:

$$\begin{aligned} &\left\|(X_\alpha(t), Y_\alpha(t)) - (X_{s_i}(t), Y_{s_i}(t))\right\|_{H^1(\Omega)} \leq \\ &\quad \sqrt{8}\Big(h\big(\|X''_\alpha(t)\|_{L^2(\Omega)} + Y''_\alpha(t)\|_{L^2(\Omega)}\big) \\ &\quad + \sqrt{\delta}\big(\|X''_\alpha(t)\|^{1/2}_{L^2(\Omega)} + Y''_\alpha(t)\|^{1/2}_{L^2(\Omega)}\big)\Big). \end{aligned} \tag{3}$$

This error bound depends on the smoothness of the true path segments and the noisy level δ. If the and δ are sufficiently small, the reconstruction error will be minimal, resulting in an accurate approximation of the total distance for TSP.

It is important to note that the error bound provided here depends on the assumption that we know the correct order of points in the TSP path. In practice, solving the TSP involves finding the optimal order of points, which is a computationally challenging problem. The error bound given here can be used as an estimate of the error for the TSP when the correct order of points is known or assumed.

If we have knowledge of the error bounds, it enables us to estimate the reconstruction error of the TSP solution, even when applying geometric approaches to a pseudo-geometric version of the TSP problem. This means that, by understanding the limits of our approximation, we can still achieve reliable results when addressing TSP instances that possess inherent geometric properties, despite any uncertainties or noise in the data.

3 Geometric Approach: Pseudo-Optimum Placement of the Points

The problem of pseudo-recovery the original coordinates of a set of points is more significant than the problem of classifying them (i.e. determining their class based on the known cost matrix in the case of the TSP).

Our algorithm for pseudo-recovery involves a simpler problem: restoring the location of points in a geometric TSP-version using the distance matrix defined with the function

$$c : E \to \mathbb{N}_0. \tag{4}$$

This problem can be easily solved using the following method.

- Select two arbitrary points v_1 and v_2 from the set of vertices of the graph V.
- Place v_1 at the origin, and v_2 at the point $(0, c(v_1, v_2))$.
- To find the coordinates of each of the following points (let it be u with the coordinates (x, y)), choose two points (let v and w) that have already been placed and solve the system of equations:

$$\begin{cases} (x - x_v)^2 + (y - y_v)^2 = (c(u, v))^2 \\ (x - x_w)^2 + (y - y_w)^2 = (c(u, w))^2. \end{cases} \tag{5}$$

However, when this algorithm is applied to the distance matrix (4) in the pseudo-geometric TSP-version, it generally does not result in the original location of the points in the geometric TSP-version. Additionally, due to the possible violation of the triangle inequality in the given distance matrix, the algorithm may not be applicable.

As we already said the main focus of this paper is a modified algorithm proposed by the authors for solving a particular case of TSP which involves the use of the above algorithm for restoring the location of points in the case of a geometric TSP-version. Additionally, it is worth noting that this algorithm can be applied to any particular case of TSP, but its use is hardly advisable in most "random" or different from pseudo-geometric special cases of TSP.

Thus, despite the fact that the application of algorithms similar to the above is, generally speaking, impossible for the pseudo-geometric TSP-version, we are trying the same algorithms to solve the problem of the location of cities, actually solving the minimization problem for the following specially computed discrepancy ("badness")

$$\sqrt{\frac{2}{n \cdot (n-1)} \cdot \sum_{i=1}^{n-1} \sum_{j=i+1}^{n} \left(c(u_i, u_j) - \tilde{c}(u_i, u_j)\right)^2}, \tag{6}$$

where:

- u_i $(i = 1, \ldots, n)$ are the points; we shall denote coordinates of the i-th point by (x_i, y_i);
- $c(u_i, u_j)$ is the (i, j)-th element of the *given* cost matrix;
- $\tilde{c}(u_i, u_j)$ is the (i, j)-th element of the *obtained* cost matrix.

As the heuristic algorithm for this problem, the authors propose the following one. *In fact, we are solving the same minimization problem, not paying attention to the fact that the points are not located according to a geometric law.*

Algorithm for pseudo-optimal placement of points.
Input. Matrix $c : E \to \mathbb{N}_0$(the matrix of weights of edges of a complete weighted graph with vertices $V = u_1, \ldots, u_n$); the value $N \in \mathbb{N}$.
N is the number of selectable pairs from the already allocated points. These pairs are chosen to accommodate each new point, starting with the 4-th one.
Step 1. $x_1 := 0$; $y_1 := 0$; $x_2 := 0$; $y_2 := c(u_1, u_2)$.
Step 2. For each point $k = 3, \ldots, n$ (we shall also denote the current k-th point under consideration simply by u), compute the coordinates (x_k, y_k) by the way of steps 3-7 below.
Step 3. $M := \min\left(\frac{(k-1)\cdot(k-2)}{2}, N\right)$.
Step 4. For each value $l = 1, \ldots, M$, implement steps 5 and 6 below.
Step 5. Randomly select uniformly distributed values $i, j \in \{1, \ldots, k-1\}$, where $i \neq j$. We shall also denote the i-th point by v, and the j-th point by w, and, additionally, denote $c_1 = c(u, v)$, $c_2 = c(u, w)$.
Step 6. If $c_1 + c_2 \leq c(v, w)$ (i.e., for the points u, v and w, the triangle inequality is violated), then we select

$$x_u = \frac{c_1 x_w + c_2 x_v}{c_1 + c_2}, \qquad y_u = \frac{c_1 y_w + c_2 y_v}{c_1 + c_2}.$$

In the event of a degenerate situation, where both c_1 and c_2 are equal to 0, the coordinates of the point are chosen to be in the middle of the segment between v and w. The coordinates are selected proportionally to the elements of the cost matrix.

However, if $c_1 + c_2 > c(v, w)$ (this option can be called a natural one), then two variants of pairs x_k and y_k are calculated according to Eq. (5). If $k = 3$, an arbitrary pair of coordinates is selected, otherwise, a pair is chosen according to one randomly chosen point among the numbers $1, \ldots, k-1$, that are not i and j. The selected pair of coordinates is added to the collection of pairs.
Step 7. Calculate the final values x_k and y_k as the arithmetic mean of the values l of the corresponding coordinates of the generated collection.
Output. Coordinates (x_1, y_1), ..., (x_n, y_n).
End of the algorithm description.

The goal of the above-mentioned algorithm is to apply a simple method of searching through all possible solutions, by sequentially placing the points $u_1, \ldots, u_n$ while minimizing the value

$$\sum_{i=1}^{k-2} \sum_{j=i+1}^{k-1} \left(c(u_i, u_j) - \tilde{c}(u_i, u_j)\right)^2,$$

included in Eq. (6). Once all the points have been placed, standard methods for solving the geometric TSP version are used (such as the "onion-peeling" algorithm), as explained in more detail in [7].

4 Conclusion. Some Results of Computing Experiments

For a given dimension (i.e. a given number of points) we considered only such a variant in which each point was randomly thrown into a unit square; more precisely, both coordinates of the point were independent equally distributed random variables with a uniform distribution law on the segment [0, 1].

Next, we applied what was called the pseudogeometric version of special cases of problems above—but at the same time the standard deviation was taken very small, which made the problem close to the geometric version.

A distance matrix was fed to the input of the algorithm—and after the location of the points according to the algorithms described in the article, we received a new distance matrix. The averaged discrepancy was calculated for all pairs of corresponding elements of these matrices.

In the table below, only decimal degrees of the obtained residual values are given in the cells: for example, the number 14 corresponds to the residual values from 10^{-14} to 10^{-13}, and so on. The selected dimension is in the rows of the table, the values σ are in the columns (Table 1).

We can say that the results are quite expected.

After that, all calculation results were compared with the results of the genetic algorithm used by us (which, as a rule, gives the least good results). Namely, the result of the work of the genetic algorithm was recorded (i.e., the value of the found route for a given variant of the TSP); the result of each of the other algorithms was considered in relation to the result of the genetic algorithm.

We used some variants of the following heuristic algorithms: genetic algorithm (GA); algorithm of simulated annealing (SA); algorithm of local search (LS); truncated branch and bound method (BBM); a hybrid algorithm, i.e., a special combination of three algorithms considered before: simulated annealing, local search and truncated branch and bound method (Hyb); algorithm of ant colony optimization (ACO); geometric algorithm (Geom), which is briefly described before.

The results of computational experiments are briefly described by following Table 2. As we can see, the results depend little on the dimensions of the problem under consideration.

Table 1 Discrepancy depending on dimension and standard deviation

	0	0.001	0.002	0.003	0.004	0.005
49	16	12	11	10	9	8
99	16	12	11	10	8	8
299	16	12	11	10	8	8
499	16	13	11	10	8	7

Table 2 Comparative results of various heuristic algorithms

dim	GA	SA	LS	BBM	Hyb	ACO	Geom
20	1	0.94	0.82	0.87	0.80	0.47	0.53
40	1	0.91	0.80	0.87	0.77	0.46	0.57
60	1	0.92	0.79	0.85	0.77	0.47	0.56
80	1	0.90	0.81	0.84	0.75	0.46	0.57
100	1	0.90	0.79	0.83	0.74	0.49	0.58

Acknowledgements This work was partially supported by a grant from the scientific program of Chinese universities "Higher Education Stability Support Program" (chapter "Shenzhen 2022—Science, Technology and Innovation Commission of Shenzhen Municipality").

References

1. Garey, M., Johnson, D.: Computers and Intractability. W. H. Freeman and Company, USA (1979), 388 p
2. Hromkovič, J.: Theoretical Computer Science. An Introduction to Automata, Computability, Complexity, Algorithmics, Randomization, Communication, and Cryptography. Springer (2003), 321 p
3. Hromkovič, J.: Algorithmics for Hard Problems. Introduction to Combinatorial Optimization, Randomization, Approximation, and Heuristics. Springer, Berlin (2003), 538 p
4. Gutin G., Punnen A. (eds.): The Traveling Salesman Problem. Kluwer Academic Publishers, Boston (1997), 856 p
5. Melnikov, B.: Multiheuristic approach to discrete optimization problems. Cybern. Syst. Anal. **42**(3), 335–341 (2006)
6. Melnikov B., Radionov A., Gumayunov V.: Some special heuristics for discrete optimization problems. In: Proceedings of 8th International Conference on Enterprise Information Systems, ICEIS-2006, pp. 360–364
7. Melnikov, B., Romanov, N.: Once again about heuristics for the traveling salesman problem. Theor. Probl. Inform. Expansions **4**, 81–92 (2001). (in Russian)
8. Liew, S.: Introducing Convex Layers to the Traveling Salesman Problem. Preprint: arXiv:1204.2348 (2012) (Internet resource)
9. Somhom, S., Modares, A., Enkawa, T.: Competition-based neural network for the multiple travelling salesmen problem with minimax objective. Comput. Oper. Res. **26**(4), 395–407 (1999)
10. Korostensky, C., Gonnet, G.: Near optimal multiple sequence alignments using a travelling salesman problem approach. In: String Processing and Information Retrieval Symposium & International Workshop on Groupware (1999), pp. 105–114
11. Melnikov, B., Panin, A.: On a parallel implementation of the multi-heuristic approach in the problem of comparison of genetic sequences Vektor Nauki of Togliatti State University (2012), vol. 4, no. 22, pp. 83–86 (in Russian)
12. Makarkin, S., Melnikov, B., Panin, A.: On the metaheuristics approach to the problem of genetic sequence comparison and its parallel implementation. Appl. Math. **4**(10A), 35–39 (2013)
13. Melnikov, B.: An algorithm for proving the equivalence of the infinite iteration of finite languages. Vestnik of Moscow University. Series 15: Computing Mathematics and Cybernetics (1996), no. 4, pp. 49–53 (in Russian)

14. Melnikov, B., Kashlakova, E.: Some grammatical structures of programming languages as simple bracketed languages. Informatica (Lithuania) **11**(4), 441–453 (2000)
15. Korostensky, C., Gonnet, G.: Using traveling salesman problem algorithms for evolutionary tree construction. Bioinformatics **16**, 619–627 (2000)
16. Hanke, M., Scherzer, O.: Inverse problems light: numerical differentiation. Am. Math. Mon. **108**, 512–521 (2001)

Solution to Infinity Problem of Scattering Matrix Using Time-Evolution Operators Without Needing Renormalization

Chol Jong

Abstract The current situation of research challenging the demanding tasks of renormalization implies that the present framework of quantum scattering theory does not offer good prospect and therefore in parallel with the development of renormalization, it is desirable to attempt to formulate a new theory able to solve the infinity problem fundamentally in a general way. Our purpose is to construct an alternative mathematical formulation capable of ensuring the convergence of the scattering matrix without relying on renormalization theory, thus preventing overlapping divergences of the scattering matrix in principle. We present alternative representations of the scattering matrix in terms of the local and global time-evolution operators which replace the Dyson series and do not need the Feynman diagram. Importantly, the obtained results clarify that substantially, there does not exist the infinity problem of the scattering matrix within the framework of our formulation. Ultimately, we draw the successful conclusion that it is possible to conceive of an alternative to the conventional scattering theory and our formalism as a new proposal can contribute to formulating a consistent theory without infinity and renormalization.

Keywords Scattering matrix · Dyson series · Infinity problem · Renormalization · Time-evolution operator

1 Introduction

In our view, one of the key questions of quantum field theory is whether renormalization theory is able to reach the ultimate goal to resolve the divergence problem of scattering matrix in a general way within the present theoretical framework. Such an opinion seems paradoxical and challenging but the present situation of research showing Odyssei in renormalization naturally makes it burgeon.

The Feynman diagrams as the graphical representations of Dyson's formula for the scattering matrix (S-matrix) which the Wick theorem produces had promoted

C. Jong (✉)
Kim Chaek University of Technology, Pyongyang, Democratic People's Republic of Korea
e-mail: jch59611@star-co.net.kp

D. Vlachos (ed.), *Mathematical Modeling in Physical Sciences*, Springer Proceedings in Mathematics & Statistics 446, https://doi.org/10.1007/978-3-031-52965-8_33

quantum scattering theory to a physically elaborated theory [1]. However, as a rule, we encounter formidable divergence problems on calculating the scattering matrix based on the Feynman diagram. Renormalization acknowledged as an astounding mathematical trick enables one to overcome some overlapping divergences in the Feynman diagrams.

It is possible to understand this situation by considering the historical milestones of the development of renormalization. Freeman Dyson's research which had begot the Dyson series had contributed significantly to the development of quantum scattering theory [2, 3]. As a result, the foundations of the future research on the scattering problem had been laid out, but at the same time, the problem of overlapping divergences arose. The conventional method for solving the problem of overlapping divergence of the scattering matrix is renormalization, which provides an approach to calibrating fundamental physical quantities so that computational results of the scattering matrix could coincide with experiments.

The researches on renormalization have spawned a diversity of approaches. John Ward's approach to renormalization featured in the advantage to be considerably simple [4] and the Yang-Mills method for finding a finite renormalized amplitude contributed to the early development of renormalization theory. An innovative attempt to resolve the problem of overlapping divergences was made by Salam [5], which attracted a lots of attention because the connection with significant work by Steven Weinberg [6] could provide a mathematically coherent version of renormalization. Another foundation for renormalization was presented by Stueckelberg and Green [7]. The mathematically important questions on the overlapping divergence had been proposed in detail by Bogoliubov and Parasiuk [8]. Klaus Hepp's review of an important theorem of Bogoliubov and Parasiuk's work was significant for the research in that direction [9]. On the other hand, Wolfhart Zimmermann discovered an explicit formula for Bogoliubov's iterative method [10], which was finalized as Bogoliubov, Parasiuk, Hepp and Zimmermann's (BPHZ) method. The proposition of Gelfand and Yaglom and Cameron's comment were an important success in making the Feynman history integral rigorous [11, 12].

The recent researches show the ramifications of renormalization technique which covers a wide range of the studies including scattering amplitudes and renormalization group [13–23], renormalization group function and equation [24–30], renormalized perturbation theory [31], renormalization theory on the perturbative Feynman graph expansion [32], and on the other side, connections between these different techniques of renormalization [33]. Recently, the development of the BPHZ renormalization is remarkable because it introduces massless fields instead of an auxiliary mass term [34]. The coverage of renormalization continues to extend inasmuch as it should satisfy the Lorentz and gauge symmetry and the requirement of cosmological spacetime as well [35–38].

Renormalization theory which has its long and challenging history from perturbative quantum electrodynamics developed by Dyson, Gell-Mann, Low and others to Wilson's formulation and Polchinski's functional equation has enjoyed so much successes in addressing overlapping divergences of the scattering matrix [39]. On the other hand, nonperturbative theory of quantum electrodynamics has turned into an interesting area for solving the infinity problem in a different way [40]. At present, renormalization maintains its dominant status in resolving the infinity problem of the scattering matrix. Nevertheless, the heart of renormalization theory still has not been finished and the centre of research continues to shift [41, 42]. The facts, in a sense, imply that it is necessary to explore for a new sphere of scattering theory in parallel with the development of renormalization [39, 43–46].

The present situation leads to a practical view that it is desirable to find out a certain general method available to all the case of calculations of the scattering matrix without dealing with the infinity problem. In this regard, it is remarkable that there are attempts to construct a new formulations of scattering theory without infinity [39, 45].

With such an understanding, we aim to construct a new mathematical formulation of quantum scattering theory free of infinity. Our formulation is based on two new time-evolution operator called the local time-evolution operator and the global time-evolution operator. The local time-evolution operator show that there is no infinity problem within the framework of our formulation. The global time-evolution operator, in essence, is the improved Heisenberg picture defining a correct scattering operator. These operators is characterized by being independent of the overlapping divergence, thus not needing renormalization. Conclussively, we show that it is possible to conceive of scattering matrices which are always convergent.

2 Infinity Problem of Scattering Matrix and Analysis

To help to understand the motivation of our work, let us examine the Dyson series from the mathematical aspect. The Schrödinger equation in the interaction picture is

$$i\hbar\frac{\partial}{\partial t}\Phi(t) = H_{int}(t)\Phi(\mathrm{t}). \tag{1}$$

For convenience, hereafter H_{int} is denoted briefly by H. For given initial state $\Phi(t_0)$, the unique solution of the above equation is represented as

$$\Phi(t) = S(t, t_0)\Phi(t_0).$$

Intuitively, the functions $\Phi(t_0)$ and $\Phi(t)$ correspond to an incoming free particle state Φ_{in} and an outgoing free particle state Φ_{out}, respectively.

The real number

$$W(t, t_0) = |\langle\Phi_{out}|\Phi_{out}\rangle|^2 = |\langle\Phi_{out}|S(t, t_0)\Phi_{in}\rangle|^2 .$$

represents the transition probability from an incoming free state to an outgoing free state. Here, the complex number $\langle\Phi_{\text{out}}|S(t, t_0)\Phi_{\text{in}}\rangle$ is called the matrix amplitude. As a rule, solving the Schrödinger Eq. (1) by means of the method of successive approximation, one can obtain a representation of the scattering operator. Integrating both sides of Eq. (1) with respect to time yields

$$|\Phi(t)\rangle = |\Phi(t_0)\rangle - \frac{\mathrm{i}}{\hbar}\int_{t_0}^{t} \hat{H}(t')\Phi(t')dt'. \tag{2}$$

The zeroth approximation becomes

$$|\Phi^{(0)}(t)\rangle = |\Phi(t_0)\rangle, \tag{3}$$

which is the initial state function. It is necessary to improve approximation, taking this as the zeroth approximation.

By means of the method of successive approximation, one can obtain the first approximation as

$$|\Phi^{(1)}(t)\rangle \approx |\Phi(t_0)\rangle - \frac{\mathrm{i}}{\hbar}\int_{t_0}^{t} \hat{H}(t')dt'|\Phi(t_0)\rangle \tag{4}$$

and then by successive substitution of the first approximation into the integrand can determine the second approximation. The iteration done in this way improves the approximation of the solution. The sufficient iteration may lead to an exact solution.

Finally, the nth order approximation is written as

$$|\Phi^{(n)}(t)\rangle \approx \Bigg[1 - \frac{\mathrm{i}}{\hbar}\int_{t_0}^{t} \hat{H}(t_1)dt_1 + \left(-\frac{\mathrm{i}}{\hbar}\right)^2 \int_{t_0}^{t} \hat{H}(t_1)dt_1 \int_{t_0}^{t_1} \hat{H}(t_2)dt_2 + \cdots + \left(-\frac{\mathrm{i}}{\hbar}\right)^n \cdot \int_{t_0}^{t} \hat{H}(t_1)dt_1 \int_{t_0}^{t_1} \hat{H}(t_2)dt_2 \cdots \int_{t_0}^{t_{n-1}} \hat{H}(t_n)dt_n\Bigg]|\Phi(t_0)\rangle. \tag{5}$$

The complete solution is represented with the aid of an infinite series as

$$|\Phi(t)\rangle = \sum_{n=0}^{\infty}\left(-\frac{\mathrm{i}}{\hbar}\right)^n \int_{t_0}^{t} dt_1 \int_{t_0}^{t_1} dt_2 \cdots \int_{t_0}^{t_{n-1}} dt_n \cdot \hat{H}(t_1)\hat{H}(t_2)\cdots\hat{H}(t_n)|\Phi(t_0)\rangle, \tag{6}$$

where using the Hamiltonian density $\mathcal{H}$, we write $\hat{H} = \int \mathcal{H}(x)\mathrm{d}^3x$.

Taking into consideration the domain of integration and the permutability of integration variables, Eq. (6) is modified as a multidimensional integral with like limits of integration, $[t_0, t]$:

$$|\Phi(t)\rangle = \sum_{n=0}^{\infty} \frac{1}{n!} \left(-\frac{\mathrm{i}}{\hbar}\right)^n \int_{t_0}^{t} dt_1 \int_{t_0}^{t} dt_2 \cdots \int_{t_0}^{t} dt_n \cdot \hat{H}(t_1)\hat{H}(t_2)\cdots\hat{H}(t_n)|\Phi(t_0)\rangle. \tag{7}$$

Considering that the Hamilton operators $\{\hat{H}(t_k)|k = 1, 2, \ldots, n\}$ may be noncommutative, i.e., $\left[\hat{H}(t_i), \hat{H}(t_j)\right]_- \neq 0$, one writes the final result with the help of the time-ordering operator $\mathcal{T}$ as

$$|\Phi(t)\rangle = \sum_{n=0}^{\infty} \frac{1}{n!} \left(-\frac{\mathrm{i}}{\hbar}\right)^n \int_{t_0}^{t} dt_1 \int_{t_0}^{t} dt_2 \cdots \int_{t_0}^{t} dt_n \cdot \mathcal{T}\hat{H}(t_1)\hat{H}(t_2)\cdots\hat{H}(t_n)|\Phi(t_0)\rangle. \tag{8}$$

Equation (8) can be written in a simple form as

$$|\Phi(t)\rangle = S(t, t_0)|\Phi(t_0)\rangle. \tag{9}$$

Here, the operator called the Dyson series:

$$S(t, t_0) = \sum_{n=0}^{\infty} \frac{1}{n!} \left(-\frac{\mathrm{i}}{\hbar}\right) \int_{t_0}^{t} dt_1 \int_{t_0}^{t} dt_2 \cdots \int_{t_0}^{t} dt_n \cdot \mathcal{T}[\hat{H}(t_1)\hat{H}(t_2)\cdots\hat{H}(t_n)] \tag{10}$$

is defined as the scattering operator. With the help of this series, it is possible to calculate elements of the scattering matrix in terms of the perturbation expansion.

It is usual to calculate higher-order approximation of the scattering matrix to consider concrete effects. We then encounter such difficulties that higher-order terms of the $S-$matrix diverge. If terms of higher-order approximation diverge, the perturbation expansion method obviously is not of significance. In this connection, renormalization theory is regarded as the best way to solve the problem of the overlapping divergence of scattering matrix. This method is characterized by the manner that based on analyzing causes of divergence, a series of physical quantities such as mass and charge and the like is redefined, and then infinity is removed by applying a definite procedure.

Renormalization theory claims the following principles. Since measured physical quantities really are finite, infinity which occurs on calculating scattering matrices should be necessarily eliminated by applying a proper method, even if it may be factitious. After renormalization, within the limits of the approximation of one-particle state, i.e., a free state of particle, all physical quantities should approximate those of a free particle. The long history of the investigation of renormalization method showed examples of successful solution of a series of problems.

However, this method cannot be assessed as providing the final solution to the problem of overlapping divergence for reasons given by physics and mathematics.

It can be reviewed from the mathematical and physical point of view whether or not renormalization theory possesses generality.

First, let us consider the problem from the physical point of view. If one should renormalize a certain series of physical quantities universally defined, it implies that a physical formula in which these quantities are involved is not perfect, since renormalization virtually indicates a modification of the physical formula. In this sense, renormalization should be considered a formal method for modifying a given scheme of successive approximation in such a way that change invariable physical quantities entering the formula so as to provide a proper result. This is, in essence, the modification of the Hamiltonian in consideration of unknown interactions not contained in $\hat{H}$ defined.

From the physical aspect, it is unreasonable to make a particular physical process correspond to every higher-order approximation of the scattering operator by using the Feynman diagram. If an arbitrary higher-order approximation corresponds to a definite real physical process, then we must give answer to the question about how to explain immense space and time necessary for internal processes. From the mathematical point of view, we in principle may imagine infinite-order approximation of the scattering operator. In this case, we should consider that a higher-order physical process is required to proceed locally and instantaneously without the constraints of space-time. However, it is impossible to think of such a physical process. Therefore, the Feynman diagram cannot be assessed to be legitimate in the context of the Dyson series. On that account, renormalization theory may give an ad hoc solution to some problems but cannot arrive at the final solution.

A particular higher-order scattering process that the Feynman diagram represents may be involved in a given scattering. In this case, the whole scattering outcomes to a superposition of possible higher-order scattering processes. In this sense, the Feynman diagram is of significance. However, it is not realistic to consider that all higher-order scattering processes that the Feynman rule dictates are possible. In fact, the mathematical concept of the higher-order of the Dyson series and the physical picture of the Feynman diagram are different matters.

Next, let us consider the problem from the mathematical point of view. In our view, in order to solve the infinity problem, it is necessary to correct the mathematical formulations causing the infinity problem rather than to resort to renormalization which treats invariable physical quantities as adjustable parameters.

To understand the truth of the infinity problem, let us start from Eq. (2). By the mean value theorem of integral calculus:

$$\int_a^b f(x)g(x)dx = f(c)g(c)(b-a) \qquad (c \in [a, b]),$$

it is possible to take

$$|\Phi(t)\rangle = |\Phi(t_0)\rangle - \frac{i}{\hbar}(t - t_0)\hat{H}(\bar{t})|\Phi(\bar{t})\rangle, \tag{11}$$

where $\bar{t} \in [0, t]$.

For $\hat{H}(\bar{t})|\Phi(\bar{t})$, we can imagine an eigenvalue equation with parameter $\bar{t}$:

$$\hat{H}(\bar{t})|\Phi(\bar{t})\rangle = E(\bar{t})|\Phi(\bar{t})\rangle. \tag{12}$$

Then, we have

$$|\Phi(t)\rangle = |\Phi(t_0)\rangle - \frac{\mathrm{i}}{\hbar}(t - t_0)E(\bar{t})|\Phi(\bar{t})\rangle. \tag{13}$$

Since the Schrödinger equation presupposes finiteness of solution, Eq. (12) is finite and thus Eq. (13) is finite as well.

In this case, in a formal manner, the scattering operator should be taken as

$$S(t, t_0) = 1 - \frac{\mathrm{i}}{\hbar}(t - t_0)E(\bar{t})\frac{|\Phi(\bar{t})\rangle}{|\Phi(t_0)\rangle}. \tag{14}$$

Of course, since $|\Phi(\bar{t})\rangle$ is unknown, the solution is formal, but Eq. (14) is enough to verify that the scattering operator should be finite. In fact, according to the definition of the Schrödinger equation, $|\Phi(\bar{t})\rangle$, $|\Phi(t_0)\rangle$ and $E(\bar{t})$ should be finite and non-zero valued. The convergence of the scattering matrix can be confirmed by

$$|\langle\Phi(t)|S(t, t_0)\Phi(t_0)\rangle| \le \int |S(t, t_0)\Phi(t_0)|^2 dx^3 =$$
$$\int \left|1 - \frac{\mathrm{i}}{\hbar}(t - t_0)E(\bar{t})\frac{\Phi(\bar{t})}{\Phi(t_0)}\right|^2 |\Phi(t_0)|^2 dx^3 \le$$
$$\int \left[1 + \left|\frac{1}{\hbar}(t - t_0)E(\bar{t})\frac{\Phi(\bar{t})}{\Phi(t_0)}\right|^2\right] |\Phi(t_0)|^2 dx^3 \le$$
$$\left(1 + A^2\right) \int |\Phi(t_0)|^2 dx^3 = (1 + A)^2,$$

where A is the maximum of $\left|\frac{1}{\hbar}(t - t_0)E(\bar{t})\frac{\Phi(\bar{t})}{\Phi(t_0)}\right|$ as an finite value. Consequently, the scattering matrix is finite. Considering in this way, we can draw the conclusion that there is no infinity problem so far as $\hat{H}$ ensures finiteness of solution to the wave equation.

It is important to review renormalization theory from the aspect of the convergence of successive approximation. Purely from the mathematical point of view, we cannot ensure that solution (8) obtained with the help of the method of successive approximation is always convergent. For example, let us consider the solution of a simultaneous linear equation in the case of applying the method of successive approximation. Let a simultaneous linear equation $\mathcal{A}\mathbf{X} = 0$ be able to be transformed into the form of $\mathbf{X} = \mathcal{B}\mathbf{X}$. Here, $\mathcal{A}$ and $\mathcal{B}$ denote matrices of coefficients and n, the order of successive approximation. Then the formula of the method of successive approximation for obtaining the solution of the simultaneous linear equation is written as

$$\mathbf{X}^{(n+1)} = \mathcal{B}\mathbf{X}^{(n)}. \tag{15}$$

For the solution of this equation by means of the method of successive approximation to converge, it is necessary to choose a proper initial point $\mathbf{X}^{(0)}$ and to apply a proper method of iteration, $\mathcal{B}$. If not so, the solution of the equation may diverge. In this case, mathematicians do not alter $\mathcal{B}$ in Eq. (15) in a factitious way without mathematical justification for the purpose of obtaining a convergent solution. In other words, they do not modify the equation itself, i.e., $\mathcal{A}\mathbf{X} = 0$. On the contrary, according to renormalization theory, physicists vary invariable physical quantities entering the equation, e.g., mass and charge etc., thus changing the formula itself describing physics. From the mathematical point of view, we cannot guarantee that choosing an arbitrary initial wave function always leads to an exact convergent solution for a problem of scattering on applying the method of successive approximation. Therefore, it is important to make the exact choice of initial wave function and the correct determination of iterative operation for the successive approximation so as to ensure the convergence of solution.

It is important to examine whether the time-ordering operator is needed or not. For brevity, we start with

$$\int_{t_0}^{t} dt_1 \int_{t_0}^{t_1} dt_2 \hat{H}(t_1)\hat{H}(t_2).$$

Since the value of a definite integral is independent of the notation of integration variable, we identify

$$\int_{t_0}^{t} dt_1 \int_{t_0}^{t_1} dt_2 \hat{H}(t_1)\hat{H}(t_2) = \int_{t_0}^{t} dt_2 \int_{t_0}^{t_2} dt_1 \hat{H}(t_2)\hat{H}(t_1).$$

With this, we get

$$\int_{t_0}^{t} dt_1 \int_{t_0}^{t_1} dt_2 \hat{H}(t_1)\hat{H}(t_2) =$$
$$\frac{1}{2}\left(\int_{t_0}^{t} dt_1 \int_{t_0}^{t_1} dt_2 \hat{H}(t_1)\hat{H}(t_2) + \int_{t_0}^{t} dt_2 \int_{t_0}^{t_2} dt_1 \hat{H}(t_2)\hat{H}(t_1)\right) =$$
$$\frac{1}{2}\left(\int_{t_0}^{t} dt_1 \int_{t_0}^{t} dt_2 \Theta(t_1 - t_2)\hat{H}(t_1)\hat{H}(t_2) + \int_{t_0}^{t} dt_2 \int_{t_0}^{t} dt_1 \Theta(t_2 - t_1)\hat{H}(t_2)\hat{H}(t_1)\right).$$

Here, we have used the Heaviside function:

$$\Theta(x) = \begin{cases} 1, & \text{for } x \geq 0 \\ 0, & \text{for } x < 0 \end{cases}$$

to extend the domains of integral so that they may be the same. Expressions $\int_{t_0}^{t} dt_1 \int_{t_0}^{t} dt_2 \Theta(t_1 - t_2) \cdots$ and $\int_{t_0}^{t} dt_1 \int_{t_0}^{t} dt_2 \Theta(t_2 - t_1) \cdots$ define the domains of integration. Now, we adopt another crucial approximation to preserve the Hermitian property of the scattering operator. The product of Hamilton operators distinguished by instant of time is non-Hermitian. Therefore, to keep the Hermiticity of the scattering operator, such non-Hermitian terms are neglected. On this line, we make an additional approximation that ensures the Hermitian property of the scattering operator. Thus, we obtain

$$\int_{t_0}^{t} dt_1 \int_{t_0}^{t} dt_2 \Theta(t_1 - t_2) \hat{H}(t_1) \hat{H}(t_2) = \frac{1}{2} \int_{t_0}^{t} dt_1 \int_{t_0}^{t} dt_2 \hat{H}(t_1) \hat{H}(t_2)$$

and

$$\int_{t_0}^{t} dt_2 \int_{t_0}^{t} dt_1 \Theta(t_2 - t_1) \hat{H}(t_2) \hat{H}(t_1) = \frac{1}{2} \int_{t_0}^{t} dt_2 \int_{t_0}^{t} dt_1 \hat{H}(t_2) \hat{H}(t_1).$$

From this, it follows that the Dyson series:

$$\frac{1}{n!} \int_{t_0}^{t} dt_1 \int_{t_0}^{t} dt_2 \ldots \int_{t_0}^{t} dt_n \mathcal{T}[\hat{H}(t_1) \hat{H}(t_2) \ldots \hat{H}(t_n)] \tag{16}$$

becomes

$$\frac{1}{n!} \prod_{k=1}^{n} \int_{t_0}^{t} \hat{H}(t_k) dt_k = \frac{1}{n!} \left[\int_{t_0}^{t} \hat{H}(t') dt' \right]^n. \tag{17}$$

Therefore, the time-ordering operator and Wick's theorem are needless. Naturally, the Feynman diagram which is framed topologically in terms of the Dyson series is unnecessary for the mathematical reason as well.

For the scattering operator to be consistent, it should satisfy the following requirements.

- The scattering operator should give the finite solution of the Schrödinger equation.
- It should be Hermitian.
- It should be unitary.
- It should satisfy the causality condition of time evolution.

Obviously, the Dyson series is not Hermitian and unitary and does not satisfy the causality condition of time evolution.

This conclusion is serious but provides the key to solving the infinity problem of scattering matrix. This procedure which is mathematically legitimate is sufficient to keep us from imaging higher-order scattering processes which the Feynman diagram dictates and overlapping divergence thereof. According to Eq. (17), Nth higher-order approximation is described by N successive applications of $\int_{t_0}^{t} \hat{H}(t_k) dt_k$ to the wave function.

Thus, the time-evolution operator in a series, Eq. (10) can be represented without using the time-ordering operator and the Feynman diagram as

$$S(t, t_0) = \sum_{n=0}^{\infty} \frac{1}{n!} \left(-\frac{\mathrm{i}}{\hbar}\right)^n \left[\int_{t_0}^{t} \hat{H}(t')dt'\right]^n . \tag{18}$$

Using the power series expansion $\exp(x) = 1 + x + \frac{1}{2}x^2 + \frac{1}{3!}x^3 + \cdots$, formula (18) is recast as

$$S(t, t_0) = \exp\left[-\frac{\mathrm{i}}{\hbar}\int_{t_0}^{t} \hat{H}(t')dt'\right]. \tag{19}$$

Equation (18) is different from the Dyson series (10). In the end, the time-ordering operator makes the difference. It is obvious that ignoring the time-ordering operator in the Dyson series yields Eq. (19). So far as the Hamiltonian $\bar{H}$ is correct, Eq. (19) ensures convergence.

The infinity problem of scattering matrix, in a sense, is an instance illustrative of the imperfection of the adopted mathematical language for quantum field theory. Purely from the point of view of mathematics, such a mathematical theory that one must separate a finite quantity from a given infinity cannot be legitimate. We can understand the points in question of renormalization theory by considering it based merely on mathematical logic. As the starting point, we should presuppose that the Schrödinger equation

$$\mathrm{i}\hbar\frac{\partial\Phi}{\partial t} = \hat{H}\Phi \tag{20}$$

gives an exact finite solution in its domain of definition.

Let us remember that the solution in terms of scattering operator is represented as

$$\Phi(t) = S(t, t_0)\Phi(t_0). \tag{21}$$

Obviously, Eqs. (20) and (21) should be considered to be identical from the mathematical viewpoint. If the solution to Eq. (21) becomes infinite, it gives nothing but the conclusion that just the scattering operator $S(t, t_0)$ is incorrect. In this case, we must examine the exactitude of the interaction Hamiltonian defined. In fact, the Hamiltonians of interaction in the scattering problem are provided by Feynman's rule assumed, so that they are not absolute.

There is neither necessity nor possibility of conceiving of every mathematical approximation carrying physical meaning. If renormalization is needed on firm physical ground, it indicates that the Schrödinger Eq. (20) too holds only on condition that renormalization is performed and the renormalization is necessary for all problems to which the fundamental equation of quantum mechanics is applied. Evidently, it is not logic because there are many cases where it is possible to obtain exact results without renormalization, even though in the presence of interaction.

3 Alternative Time-Evolution Operators and Convergence

In this section, we aim to show without using the procedure of second quantization that it is possible to conceive of scattering matrices which are always convergent. Therefore, the discussion below goes virtually at the level of quantum mechanics without using the field operators. If the conceived method ensures the convergence of scattering matrix, it implies that it is necessary to examine and correct the conventional approach.

3.1 Local Time-Evolution Operator and Analysis of Convergence

The description in this subsection aims to demonstrate the convergence of the scattering matrix in a mathematically intuitive way. It is worth noting that it is not reasonable to introduce the successive approximation in the form of Eq. (8). This is because in order that it is possible to represent

$$|\Phi(t)\rangle = |\Phi(t_0)\rangle - \frac{\mathrm{i}}{\hbar}\int_{t_0}^{t} \hat{H}(t')\Phi(t')dt' \tag{22}$$

as

$$|\Phi(t_1)\rangle \approx |\Phi(t_0)\rangle - \frac{\mathrm{i}}{\hbar}\int_{t_0}^{t_1} \hat{H}(t')dt'|\Phi(t_0)\rangle, \tag{23}$$

a close proximity of t_0 in general should comprise the limits of integration. Only if so, Eq. (23) in general is justified from the mathematical aspect. Actually, it is easy to understand that only for sufficiently short integration interval, it is possible to guarantee convergence of the above solution. The violation of this condition may lead to the divergence of solution. In this connection, it is necessary to improve the successive approximation considered above so as to guarantee convergence. What is best is to prevent the infinity problem.

Let us consider that it is possible to avoid the infinity problem by applying the successive approximation properly. If one first partitions the time interval properly and then makes phased evolution of solutions in every time interval rather than applies the successive approximation using the whole time interval, the desired convergence is possible.

With the help of Eq. (22), we take as the first time-evolution

$$|\Phi(t_1)\rangle \approx |\Phi(t_0)\rangle - \frac{\mathrm{i}}{\hbar}\int_{t_0}^{t_1} \hat{H}(t')dt'|\Phi(t_0)\rangle = \left[1 - \frac{\mathrm{i}}{\hbar}\int_{t_0}^{t_1} \hat{H}(t')dt'\right]|\Phi(t_0)\rangle, \tag{24}$$

where $t_1 \in [t_0, t]$ and as the second time-evolution

$$\begin{aligned} |\Phi(t_2)\rangle &\approx |\Phi(t_1)\rangle - \frac{\mathrm{i}}{\hbar}\int_{t_1}^{t_2} \hat{H}(t')dt'|\Phi(t_1)\rangle \\ &= \left[1 - \frac{\mathrm{i}}{\hbar}\int_{t_1}^{t_2} \hat{H}(t')dt'\right]|\Phi(t_1)\rangle \approx \prod_{k=1}^{2}\left[1 - \frac{\mathrm{i}}{\hbar}\int_{t_{k-1}}^{t_k} \hat{H}(t')dt'\right]|\Phi(t_0)\rangle, \end{aligned} \tag{25}$$

where $t_2 \in [t_1, t]$. Let us partition the interval, $[t_0, t]$ into n equal elementary subintervals defined by a set of points $\{t_0, t_1, t_2, \cdots t_n\}$ such that $t_0 = 0 < t_1 < t_2, \cdots < t_n = t$. Then, the state function, in general, is written as

$$|\Phi(t_n)\rangle \approx \prod_{k=1}^{n}\left[1 - \frac{\mathrm{i}}{\hbar}\int_{t_{k-1}}^{t_k} \hat{H}(t')dt'\right]|\Phi(t_0)\rangle. \tag{26}$$

With reference to the above equation, the time-evolved state function is represented as

$$|\Phi(t)\rangle = \lim_{n\to\infty}\prod_{k=1}^{n}\left[1 - \frac{\mathrm{i}}{\hbar}\int_{t_{k-1}}^{t_k} \hat{H}(t')dt'\right]|\Phi(t_0)\rangle. \tag{27}$$

From this, we should take the time-evolution operator as

$$S(t, t_0) = \lim_{n\to\infty}\prod_{k=1}^{n}\left[1 - \frac{\mathrm{i}}{\hbar}\int_{t_{k-1}}^{t_k} \hat{H}(t')dt'\right]. \tag{28}$$

We shall refer to this as the local time-evolution operator in the sense that it is represented by integrals in small time intervals. According to the theorem of mean value of integral calculus, we have

$$\int_{t_{k-1}}^{t_k} \hat{H}(t')dt' = \hat{H}(\bar{t}_k)(t_k - t_{k-1}) = \hat{H}(\bar{t}_k)\Delta t, \tag{29}$$

where $\bar{t}_k$ is between $[t_{k-1}, t_k]$. By Eq. (28), the nth order approximation of the time-evolution operator is taken as

$$S^{(n)}(t, t_0) = \prod_{k=1}^{n}\left[1 - \frac{\mathrm{i}}{\hbar}\int_{t_{k-1}}^{t_k} \hat{H}(t')dt'\right]. \tag{30}$$

Let us consider the convergence of the scattering matrix represented in terms of the local time-evolution operator. Of course, for a finite partition of the time interval, there is no infinity problem, since it gives the scattering matrix in a finite series. Let us consider the case of infinite partition. We begin with the fact that there exists $\bar{E}_k$ which satisfies

$$\int_{t_{k-1}}^{t_k} \hat{H}(t')dt'|\Phi(t_0)\rangle = \Delta t \hat{H}(\bar{t}_k)|\Phi(t_0)\rangle = \Delta t \bar{E}_k|\Phi(t_0)\rangle,$$

namely,

$$\hat{H}(\bar{t}_k)|\Phi(t_0)\rangle = \bar{E}_k|\Phi(t_0)\rangle,$$

where $\Delta t = (t - t_0)/n$. To clarify this, it is necessary to take into consideration that for an element of the scatering matrix, it holds that

$$\langle\Phi(\bar{t}_k)|\hat{H}(\bar{t}_k)|\Phi(t_0)\rangle = \langle\hat{H}(\bar{t}_k)\Phi(\bar{t}_k)|\Phi(t_0)\rangle = \bar{E}_k\langle\Phi(\bar{t}_k)|\Phi(t_0)\rangle,$$

where we took into consideration that $\hat{H}(\bar{t}_k)$ is Hermitian and the following eigenvalue equation for it should hold:

$$\hat{H}(\bar{t}_k)\Phi(\bar{t}_k) = E(\bar{t}_k)\Phi(\bar{t}_k).$$

Moreover, $|\Phi(\bar{t}_k)\rangle \approx |\Phi(t_0)\rangle$ was supposed. Thus, we identify that in the sense of the scattering matrix, $\hat{H}(\bar{t}_k)|\Phi(t_0)\rangle = \bar{E}_k|\Phi(t_0)\rangle$ holds true. In the end, Eq. (30) is represented as

$$S^{(n)}(t, t_0) = \prod_{k=1}^{n}\left(1 - \frac{\mathrm{i}}{\hbar}\bar{E}_k\Delta t\right). \tag{31}$$

Evidently, of $\{\bar{E}_k\}$, there exists E_{max} that satisfies

$$\left|\left(1 - \frac{\mathrm{i}}{\hbar}E_{\max}\Delta t\right)^n\right| \geq \left|\prod_{k=1}^{n}\left(1 - \frac{\mathrm{i}}{\hbar}\bar{E}_k\Delta t\right)\right|. \tag{32}$$

From Eq. 32), we should examine the convergence of

$$S_m = \lim_{n\to\infty}\left(1 - \frac{\mathrm{i}}{\hbar}E_{\max}\frac{t - t_0}{n}\right)^n. \tag{33}$$

Setting $A = -\frac{\mathrm{i}}{\hbar}E_{\max}(t - t_0)$, Eq. (33) is reduced to

$$S_m = \lim_{n\to\infty}\left(1 + \frac{A}{n}\right)^n. \tag{34}$$

In virtue of the binomial theorem, Eq. (34) is expanded into a power series:

$$S_m = \lim_{n \to \infty} \sum_{k=0}^{n} C_n^k \left(\frac{A}{n} \right)^k, \tag{35}$$

where C_n^k are the binomial coefficients. If S_m is absolutely convergent, namely,

$$S_{abs} = \lim_{n \to \infty} \sum_{k=0}^{n} \left| C_n^k \left(\frac{A}{n} \right)^k \right| \tag{36}$$

is convergent, S_m is absolutely convergent. After the manner of Eqs. (34) and (36) is recast as

$$S_{abs} = \lim_{n \to \infty} \left(1 + \frac{|A|}{n} \right)^n.$$

Setting $x = n/|A|$, we immediately obtain

$$S_{abs} = \lim_{x \to \infty} \left(1 + \frac{1}{x} \right)^{x|A|}.$$

Since

$$\lim_{x \to \infty} \left(1 + \frac{1}{x} \right)^{x|A|} = e^{|A|},$$

S_m is absolutely convergent.

Thus, it has been demonstrated that in the case of the infinite partition of a time interval, the local time-evolution operator ensures the convergence of the scattering matrix. This shows that it is possible to avoid the divergence of the scattering matrix by using time evolution based on the partition of a given time interval.

In the end, introducing Eq. (28) instead of the Dyson series, i.e., Eq. (10) enables us to prevent the infinity problem of the scattering matrix.

3.2 *Global Time-Evolution Operator and Analysis of Convergence*

The description of this subsection aims to show that the improved Heisenberg picture is qualified as the exact scattering operator preserving its convergence. The Heisenberg picture is an important mathematical formulation for investigating the time evolution of quantum states together with the Schrödinger formulation. In particular, the Heisenberg picture plays a key role in the case of the investigation of the scattering problem. The scattering matrix is closely related to the Heisenberg picture.

It is important to review whether this formulation is possessed of mathematical generality. To begin with, let us consider how the Heisenberg picture is derived. The state function for a microscopic system in the Schödinger picture is determined by

$$i\hbar\frac{\partial\Phi_S}{\partial t} = \hat{H}\Phi_S. \tag{37}$$

The formal solution to this equation is supposed to be

$$\Phi_S(t) = e^{-\frac{i}{\hbar}\hat{H}t}\Phi_H, \tag{38}$$

where Φ_H as a time-independent function is defined as the wave function in the Heisenberg picture [47, 48]. Here, subscript H refers to the Heisenberg picture. Operator $\hat{S} = e^{-\frac{i}{\hbar}\hat{H}t}$ is considered to be the unitary operator making transformation between the two pictures. By Eq. (38), the matrix elements of a certain operator $\hat{F}$ are represented as

$$F_{mn}(t) = \left\langle \Phi_{Sm}(t)\,|\hat{F}_S|\Phi_{Sn}(t)\right\rangle = \left\langle \Phi_{Hm}(t)\,\|e^{\frac{i}{\hbar}\hat{H}t}\hat{F}_S e^{-\frac{i}{\hbar}\hat{H}t}\|\Phi_{Hn}(t)\right\rangle. \tag{39}$$

The last expression in Eq. (39) can be interpreted as matrix elements of the operator with respect to the wave function in the Heisenberg picture.

Naturally, the expression in Eq. (39):

$$\hat{F}_H(t) = e^{\frac{i}{\hbar}\hat{H}t}\hat{F}_S e^{-\frac{i}{\hbar}\hat{H}t} \tag{40}$$

can be considered the operator in the Heisenberg picture, which is time-dependent unlike that in the Schrödinger picture.

From Eq. (40), it follows that the equation of motion in the Heisenberg picture is

$$i\hbar\frac{\partial\hat{F}_H(t)}{\partial t} = \left[\hat{F}_H(t), \hat{H}\right]. \tag{41}$$

In order to examine whether the Heisenberg picture is correct, it is necessary to obtain the formal solution of the Schrödinger equation in a rigorous way.

It is always possible that the wave function is set as

$$\Phi_S(\mathbf{q}, t) = \hat{f}(\mathbf{q}, t)\varphi(\mathbf{q}), \tag{42}$$

where $\hat{f}(\mathbf{q}, t)$ is an operator dependent on time and position. Thus, the Schrödinger Eq. (37) is represented as

$$i\hbar\frac{\partial\left[\hat{f}(\mathbf{q},t)\varphi(\mathbf{q})\right]}{\partial t}=\hat{H}\left[\hat{f}(\mathbf{q},t)\varphi(\mathbf{q})\right]. \tag{43}$$

Since operator $i\hbar\frac{\partial}{\partial t}$ is applied only to $\hat{f}(\mathbf{q},t)$, we have

$$i\hbar\frac{\partial\hat{f}(\mathbf{q},t)}{\partial t}\varphi(\mathbf{q})=\hat{H}\hat{f}(\mathbf{q},t)\varphi(\mathbf{q}). \tag{44}$$

Since φ is an arbitrary function, the following operator equation should hold.

$$i\hbar\frac{\partial\hat{f}(\mathbf{q},t)}{\partial t}=\hat{H}(\mathbf{q},t)\hat{f}(\mathbf{q},t). \tag{45}$$

To begin with, we assume $\hat{f}(\mathbf{q},t)$ to be represented as a function f of an unknown operator $\hat{x}(\mathbf{q};t,t_0)$, namely,

$$\hat{f}(\mathbf{q},t)=f(\hat{x}(\mathbf{q};t,t_0)).$$

Then Eq. (45) becomes

$$i\hbar\frac{\partial f(\hat{x}(\mathbf{q};t,t_0))}{\partial t}=\hat{H}(\mathbf{q},t)f(\hat{x}(\mathbf{q};t,t_0)).$$

Adopting the formal derivative with respect to $\hat{x}(\mathbf{q};t,t_0)$, $\frac{\partial}{\partial\hat{x}(\mathbf{q};t,t_0)}$ and the chain rule $\frac{\partial}{\partial t}=\frac{\partial\hat{x}(\mathbf{q};t,t_0)}{\partial t}\frac{\partial}{\partial\hat{x}(\mathbf{q};t,t_0)}$, we rewrite (45) as

$$i\hbar\frac{\partial\hat{x}(\mathbf{q};t,t_0)}{\partial t}\frac{\partial f(\hat{x}(\mathbf{q};t,t_0))}{\partial\hat{x}(\mathbf{q};t,t_0)}=\hat{H}(\mathbf{q},t)f(\hat{x}(\mathbf{q};t,t_0)).$$

Assuming that

$$\frac{\partial f(\hat{x}(\mathbf{q};t,t_0))}{\partial\hat{x}(\mathbf{q};t,t_0)}=f(\hat{x}(\mathbf{q};t,t_0)),$$

we have

$$i\hbar\frac{\partial\hat{x}(\mathbf{q};t,t_0)}{\partial t}=\hat{H}(\mathbf{q},t).$$

From this, we get

$$f(\hat{x}(\mathbf{q};t,t_0))=\exp(\hat{x}(\mathbf{q};t,t_0))$$

and

$$\hat{x}(\mathbf{q}; t, t_0) = -\frac{\mathrm{i}}{\hbar}\int_{t_0}^{t} \hat{H}(\mathbf{q}, t')dt'.$$

Finally, we obtain

$$\hat{f}(\mathbf{q}, t) = \exp\left[-\frac{\mathrm{i}}{\hbar}\int_{t_0}^{t} \hat{H}(\mathbf{q}, t')dt'\right]. \tag{46}$$

where $\hat{f}(\mathbf{q}, t)$ is treated formally as an algebraic function, while operating on $\varphi(\mathbf{q})$. Evidently, Eq. (46) becomes the improved time-evolution operator which supersedes that in the Heisenberg pcture. Accordingly, we arrive at the conclusion that the wave function should be represented formally as

$$\Phi_S(\mathbf{q}, t) = \exp\left[-\frac{\mathrm{i}}{\hbar}\int_{t_0}^{t} \hat{H}(\mathbf{q}, t')dt'\right]\varphi(\mathbf{q}). \tag{47}$$

For initial condition $t = t_0$, Eq. (47) reduces to

$$\Phi_S(\mathbf{q}, t_0) = \varphi(\mathbf{q}) = \Phi_H(\mathbf{q}). \tag{48}$$

Hereafter, $\Phi_0(\mathbf{q}, t_0)$ is represented briefly as $\Phi_0(\mathbf{q})$. Then, the wave function is written as

$$\Phi_S(\mathbf{q}, t) = \exp\left[-\frac{\mathrm{i}}{\hbar}\int_{t_0}^{t} \hat{H}(\mathbf{q}, t')dt'\right]\Phi_0(\mathbf{q}) \tag{49}$$

or

$$\Phi_S(\mathbf{q}, t) = \exp\left[-\frac{\mathrm{i}}{\hbar}\int_{t_0}^{t} \hat{H}(\mathbf{q}, t')dt'\right]\Phi_H(\mathbf{q}). \tag{50}$$

Hence, it follows that the time-evolution operator, i.e., the unitary operator is

$$U(t, t_0) = \exp\left[-\frac{\mathrm{i}}{\hbar}\int_{t_0}^{t} \hat{H}(\mathbf{q}, t')dt'\right]. \tag{51}$$

This operator is represented by the Maclaurin expansion as

$$U(t, t_0) = \sum_{n=0}^{\infty}\frac{1}{n!}\left[-\frac{\mathrm{i}}{\hbar}\int_{t_0}^{t} \hat{H}(\mathbf{q}, t')dt'\right]^n. \tag{52}$$

We shall refer to $U(t, t_0)$ as the global time-evolution operator in the improved Heisenberg picture, since it gives an analytical representation of time evolution from an initial time to a final time, i.e., on the whole interval of time, $[t_0, t]$. Evidently, our formal solution, i.e., Eq. (50) is clearly distinguished from Eq. (38).

Interestingly, the global time-evolution operator coincides with Eq. (18) as an expression of the Dyson series in the case of neglecting the time-ordering operator. This fact tells us that if Eq. (52) is right, it means that the time-ordering operator and the Feynman diagram are meaningless and unnecessary.

For the scattering problem, the global time-evolution operator becomes the scattering operator. Thus, we write Eq. (49) as

$$\Phi_S(\mathbf{q}, t) = S(t, t_0)\Phi_S(\mathbf{q}, t_0). \tag{53}$$

The substitution of Eq. (50) into the Schrödinger equation yields

$$\mathrm{i}\hbar\frac{\partial}{\partial t}\Phi_S(\mathbf{q}, t) = \mathrm{i}\hbar\frac{\partial}{\partial t}\exp\left[-\frac{\mathrm{i}}{\hbar}\int_{t_0}^{t}\hat{H}(\mathbf{q}, t')dt'\right]\varphi(\mathbf{q}) = \hat{H}(\mathbf{q}, t)\Phi_S(\mathbf{q}, t).$$

Therefore, we have

$$\mathrm{i}\hbar\frac{\partial}{\partial t}\Phi_S(\mathbf{q}, t) = \hat{H}(\mathbf{q}, t)\Phi_S(\mathbf{q}, t).$$

Here, we adopted the formal derivative with respect to $\hat{x}(\mathbf{q}; t, t_0) = -\frac{\mathrm{i}}{\hbar}\int_{t_0}^{t}\hat{H}(\mathbf{q}, t')dt'$, i.e., $\frac{\partial}{\partial\hat{x}(\mathbf{q};t,t_0)}$ and applied the chain rule $\frac{\partial}{\partial t} = \frac{\partial\hat{x}(\mathbf{q};t,t_0)}{\partial t}\frac{\partial}{\partial\hat{x}(\mathbf{q};t,t_0)}$.

This confirms that the representation of state function in terms of the global time-evolution operator satisfies the Schrödinger equation. Conclusively, it should be emphasized that the correct time-evolution operator is $\exp\left[-\frac{\mathrm{i}}{\hbar}\int_{t_0}^{t}\hat{H}(\mathbf{q}, t')dt'\right]$ instead of $\exp\left(-\frac{\mathrm{i}}{\hbar}\hat{H}t\right)$.

If $\exp\left(-\frac{\mathrm{i}}{\hbar}\hat{H}t\right)\Phi_0(\mathbf{q})$ does not satisfy the Schrödinger equation, this is sufficient to confirm that the Heisenberg picture is not correct.

Substituting $\exp\left(-\frac{\mathrm{i}}{\hbar}\hat{H}t\right)\Phi_0(\mathbf{q})$ into the Schrödinger equation proves that this formal solution does not satisfy the equation.

In fact, in view of the time dependence of $\hat{H}$, we get

$$\mathrm{i}\hbar\frac{\partial}{\partial t}\left[\exp\left(-\frac{\mathrm{i}}{\hbar}\hat{H}t\right)\Phi_0(\mathbf{q})\right] = \left[\hat{H} + t\frac{\partial\hat{H}}{\partial t}\right]\Phi_S(\mathbf{q}, t) \neq \hat{H}\Phi_S(\mathbf{q}, t). \tag{54}$$

Obviously, the formal solution of the Schrödinger equation $\exp\left(-\frac{\mathrm{i}}{\hbar}\hat{H}t\right)\Phi_0(\mathbf{q})$ which is really supposed is not mathematically correct, since it does not satisfy the wave equation itself.

Therefore, we should adopt the improved representation of operator in place of the Heisenberg picture as

$$\hat{F}_H(t) = \exp\left[\frac{i}{\hbar}\int_{t_0}^{t}\hat{H}(\mathbf{q},t')dt'\right]\hat{F}_S\exp\left[-\frac{i}{\hbar}\int_{t_0}^{t}\hat{H}(\mathbf{q},t')dt'\right]. \tag{55}$$

Obviously, unlike the Dyson series, the global time-evolution operator is Hermitian and unitary, and does satisfy the causality condition of time evolution and the requirement for convergence.

Let us consider that the scattering matrix in terms of the global time-evolution operator is always convergent. The global time-evolution operator expanded in a Maclaurin expansion, namely

$$S(t,t_0) = \sum_{n=0}^{\infty}\frac{1}{n!}\left[-\frac{i}{\hbar}\int_{t_0}^{t}\hat{H}(\mathbf{q},t')dt'\right]^n. \tag{56}$$

purports that the identical operator, $-\frac{i}{\hbar}\int_{t_0}^{t}\hat{H}(\mathbf{q},t')dt'$ operates n times successively on the wave function. The expansion makes sense only if it is convergent. Evidently, the number of terms of the series determines the order of approximation. For example, n-th approximation is denoted by

$$S^{(n)}(t,t_0) = \sum_{k=0}^{n}\frac{1}{k!}\left[-\frac{i}{\hbar}\int_{t_0}^{t}\hat{H}(\mathbf{q},t')dt'\right]^k. \tag{57}$$

It is obvious that a finite order approximation converges.

Now, let us verify that the infinite series of the global time-evolution operator ensures convergence. First, according to the mean value theorem of integral calculus, we set

$$\int_{t_0}^{t}\hat{H}(\mathbf{q},t')dt' = \hat{H}(\mathbf{q},\bar{t})(t-t_0), \tag{58}$$

where $\bar{t}$ becomes a parameter. Equation (58) helps us eliminate the integral symbol from the time-evolution operator to get the following representation:

$$\exp\left[-\frac{i}{\hbar}\int_{t_0}^{t}\hat{H}(\mathbf{q},t')dt'\right] = \exp\left[-\frac{i}{\hbar}(t-t_0)\hat{H}(\mathbf{q},\bar{t})\right]. \tag{59}$$

Next, we expand the time-evolution operator as the following Maclaurin series:

$$S(\mathbf{q},t) = \sum_{n=0}^{\infty}\frac{1}{n!}\left(-\frac{i}{\hbar}(t-t_0)\hat{H}(\mathbf{q},\bar{t})\right)^n. \tag{60}$$

The solution of the time-dependent Schrödinger equation is represented with the help of the time-evolution operator as

$$\Phi_S(\mathbf{q}, t) = \sum_{n=0}^{\infty} \frac{1}{n!} \left(-\frac{\mathrm{i}}{\hbar}(t - t_0)\hat{H}(\mathbf{q}, \bar{t}) \right)^n \Phi_0(\mathbf{q}). \tag{61}$$

It is possible to verify that in any case, interaction does not give rise to infinity. We represent the Hamilton operator $\hat{H}(\mathbf{q}, \bar{t})$ in Eq. (61) as the sum of a free part, $\hat{H}_0$ and an interaction part, $\hat{H}_I$, namely,

$$\hat{H}(\mathbf{q}, \bar{t}) = \hat{H}_0 + \hat{H}_I(\mathbf{q}, \bar{t}). \tag{62}$$

Taking into consideration that for operators $\hat{H}_0$ and $\hat{H}(\mathbf{q}, \bar{t})$ the eigenvalue equations hold, namely,

$$\hat{H}_0 \Phi_0(\mathbf{q}) = E_0 \Phi_0(\mathbf{q}) \tag{63}$$

and

$$\hat{H}(\mathbf{q}, \bar{t})\Phi_S(\mathbf{q}, \bar{t}) = E(\bar{t})\Phi_S(\mathbf{q}, \bar{t}), \tag{64}$$

we deal with the following integrals:

$$\langle \Phi_S(\mathbf{q}, \bar{t}) | \hat{H}_0 \Phi_0(\mathbf{q}) \rangle = E_0 \langle \Phi_S(\mathbf{q}, \bar{t}) | \Phi_0(\mathbf{q}) \rangle \tag{65}$$

and

$$\langle \Phi_0(\mathbf{q}) | \hat{H}(\mathbf{q}, \bar{t}) \Phi_S(\mathbf{q}, \bar{t}) \rangle = E(\bar{t}) \langle \Phi_0(\mathbf{q}) | \Phi_S(\mathbf{q}, \bar{t}) \rangle. \tag{66}$$

Since $\hat{H}(\mathbf{q}, \bar{t})$ is Hermitian, Eq. (66) becomes

$$\langle \hat{H}(\mathbf{q}, \bar{t}) \Phi_0(\mathbf{q}) | \Phi_S(\mathbf{q}, \bar{t}) \rangle = E(\bar{t}) \langle \Phi_0(\mathbf{q}) | \Phi_S(\mathbf{q}, \bar{t}) \rangle. \tag{67}$$

The complex conjugate of Eq. (67) is

$$\langle \Phi_S(\mathbf{q}, \bar{t}) | \hat{H}(\mathbf{q}, \bar{t}) \Phi_0(\mathbf{q}) \rangle = E(\bar{t}) \langle \Phi_S(\mathbf{q}, \bar{t}) | \Phi_0(\mathbf{q}) \rangle. \tag{68}$$

Then, the subtraction of Eq. (65) from Eq. (68) yields

$$\langle | \Phi_S(\mathbf{q}, \bar{t}) | \left[\hat{H}(\mathbf{q}, \bar{t}) - \hat{H}_0(\mathbf{q}) \right] \Phi_0(\mathbf{q}) \rangle = (E(\bar{t}) - E_0) \langle \Phi_S(\mathbf{q}, \bar{t}) | \Phi_0(\mathbf{q}) \rangle.$$

Taking Eq. (62) into consideration, we have

$$\langle \Phi_S(\mathbf{q}, \bar{t}) | \hat{H}_I(\mathbf{q}, \bar{t}) \Phi_0(\mathbf{q}) \rangle = \langle \Phi_S(\mathbf{q}, \bar{t}) | (E(\bar{t}) - E_0) \Phi_0(\mathbf{q}) \rangle.$$

Hence, we obtain

$$\hat{H}_I(\mathbf{q}, \bar{t}) \Phi_0(\mathbf{q}) = (E(\bar{t}) - E_0) \Phi_0(\mathbf{q}). \tag{69}$$

In the end, we reach the conclusion that since $(E(\bar{t}) - E_0)$ is finite, $\hat{H}_I(\mathbf{q}, \bar{t})\Phi_0(\mathbf{q})$ is finite as well. Thus, it is obvious that $\hat{H}_I(\mathbf{q}, \bar{t})$ is impossible to carry infinity. Even if $\hat{H}_I(\mathbf{q}, \bar{t})$ is represented in the language of quantum field theory, it should give a finite result.

From Eqs. (61) and (64), the solution of the time-dependent Schrödinger equation is represented as

$$\Phi_S(\mathbf{q}, t) = \sum_n \frac{1}{n!}\left[-\frac{\mathrm{i}}{\hbar}(t-t_0)\right]^n \left[\hat{H}(\mathbf{q}, \bar{t})\right]^n \Phi_0(\mathbf{q})$$
$$= \sum_n \frac{1}{n!}\left(-\frac{\mathrm{i}}{\hbar}\right)^n \left[(t-t_0)E(\bar{t})\right]^n \Phi_0(\mathbf{q}). \quad (70)$$

Hence, we can examine the convergence based on the relation:

$$S(t, t_0) = \sum_n \frac{1}{n!}\left(-\frac{\mathrm{i}}{\hbar}\right)^n [(t-t_0)E(\bar{t})]^n. \quad (71)$$

The following consideration leads to the conclusion that the scattering operator, Eq. (71) is always convergent. The n-th and $n+1$-th terms in the series $\sum_n \frac{1}{n!}\left(-\frac{\mathrm{i}}{\hbar}\right)^n [(t-t_0)E(\bar{t})]^n$ read as

$$a(n+1) = \frac{1}{(n+1)!}\left[-\frac{\mathrm{i}}{\hbar}(t-t_0)E(\bar{t})\right]^{n+1}, \quad (72)$$

$$a(n) = \frac{1}{n!}\left[-\frac{\mathrm{i}}{\hbar}(t-t_0)E(\bar{t})\right]^n. \quad (73)$$

The ratio between them is

$$\frac{a(n+1)}{a(n)} = \frac{1}{(n+1)}\left[-\frac{\mathrm{i}}{\hbar}(t-t_0)E(\bar{t})\right]. \quad (74)$$

Since its limit is zero, namely,

$$\lim_{n\to\infty} \frac{a(n+1)}{a(n)} = 0, \quad (75)$$

according to the convergence condition of infinite series, it follows that the series S converges to a finite value.

Equation (45) is different from the Dyson series (42). In the end, the time ordering operator makes the difference. It is obvious that ignoring the time ordering operator in the Dyson series yields Eq. (46). So far as the Hamiltonian $\hat{H}$ is correct, Eq. (46) ensures convergence. Obviously, unlike the Dyson series, the global time-evolution operator is Hermitian and unitary, and does satisfy the causality condition of time evolution and the requirement for convergence.

Let us consider the case where the Hamiltonian is represented in the language of quantum field theory. Then the Hamiltonian is expressed by the Hamiltonian density $\mathcal{H}$ as $\hat{H} = \int \mathcal{H}(x)d^3x$. As an example, in the case of the interaction between electron-positron field and electromagnetic field, the Hamiltonian density of interaction is assumed to be

$$\mathcal{H} = -j_\mu(x)A^\mu(x) = -\frac{e}{2}\left[\bar{\psi}\gamma_\mu\psi\right]A^\mu = eN\left(\bar{\psi}\gamma_\mu\psi\right)A^\mu. \tag{76}$$

Of course, this should ensure finiteness of the scattering matrix. By use of Eq. (76), the scattering operator is given as

$$S(t, t_0) = \sum_{n=0}^{\infty}\frac{1}{n!}\left(-\frac{i}{\hbar}\right)^n\left[\int eN\left(\bar{\psi}\gamma_\mu\psi\right)A^\mu d^4x\right]^n. \tag{77}$$

Based on this, it is possible to assess the convergence of the scattering matrix in a general way. Obviously, the first approximation $eN\int\left(\bar{\psi}\gamma_\mu\psi\right)A^\mu d^4x$, in general, is convergent. Therefore, the Taylor expansion, Eq. (77) is always convergent and thus the transition probability from an initial state $|i\rangle$ to a final state $|f\rangle$:

$$W_{i\to f} = |S_{i\to f}|^2 = |\langle f|S(t, t_0)|i\rangle|^2$$

is the same too. Thus, the use of the global time-evolution operator keeps the scattering matrix from divergence. Hence, it is concluded that within our formulation, there is no physical world causing the infinity problem. This implies that if we would encounter the infinity problem, it might be due to the mathematical formulation of scattering theory.

4 Conclusion

The main aim of this work has been to present an alternative mathematical formulation capable of solving the divergence problem of the scattering matrix without relying on renormalization theory. Should it be true, it undoubtedly would be a major advance in the final solution to the infinity problem and a clear indication that there does exist the infinity problem.

We have demonstrated that it is to avoid the infinity problem by an appropriate mathematical representation of the scattering operator. Our work has shown that the representation of scattering matrix by means of the method of successive approximation based on the Dyson series and the Feynman diagram is not perfect from the mathematical and physical views.

An alternative mathematical representation of scattering matrix in terms of the local and global time-evolution operators that we have proposed is independent of the

Dyson series and the Feynman diagram. Using a mathematically rigorous method, the local time-evolution operator Eq. (28) and the global time-evolution operator Eq. (51) have been derived, which can supersede the Dyson series and the Heisenberg picture, and importantly, does not need the Feynman diagram. These operators are essentially distinguished from the Dyson series by being free of the time-ordering operator. The global time-evolution operator which provides the exact solution of the wave equation improves the Heisenberg picture.

Within the confines of our formulation, there does not exist the divergence problem of the scattering matrix, since the convergence of the scattering operators is proved as shown in Sects. 3.1 and 3.2. The fact that despite the long-term studies of the infinity problem, we have not yet found a promising route to renormalization ensuring generality and completeness of the solution needs innovative perspectives which are based on consideration from a new angle. Complying with such a requirement, we have explored the local and global time-evolution operators which always guarantee convergence of the scattering matrix. Our formulation is simple in the mathematical aspect but gives clear explanations of the divergence problems of scattering matrix. Our work has revealed that the Dyson series is not unique selection, and it is possible to find more consistent scattering operator. Since there is no paper associated with the research direction like ours, we believe that our work will contribute significantly to developing the mathematical formalism of quantum scattering theory in a new direction.

Acknowledgements I would like to thank the organizers of the 12th Int'l Conference on Mathematical Modeling in the Physical Sciences for the invitation to participate in a stimulating meeting. In particular, I thank Professors Ji-Min Pak and Il-Yong Kang for help and encouragement.

References

1. Das, A.: Lectures on Quantum Field Theory, Chap. 6, 8. World Scientific (2008)
2. Dyson, F.: The radiation theories of Tomonaga, Schwinger, and Feynman. Phys. Rev. **75**, 406–502 (1949)
3. Dyson, F.: The S-matrix in quantum electrodynamics. Phys. Rev. **75**, 1736–1755 (1949)
4. Ward, J.: On the renormalization of quantum electrodynamics. Proc. Phys. Soc. London A **64**, 54–56 (1951)
5. Salam, A.: Overlapping divergences and the S-matrix. Phys. Rev. **82**, 217–227 (1951)
6. Weinberg, S.: High energy behavior in quantum field theory. Phys. Rev. **118**, 838–849 (1960)
7. Stueckelberg, E., Green, T.: Elimination des constantes arbitraires dans la theéorie relativiste des quanta. Helv. Phys. Acta **24**, 153–174 (1951)
8. Bogoliubov, N., Parasiuk, O.: On the multiplication of propagators in quantum field theory. Acta Math. **97**, 227–326 (1957). (in German)
9. Hepp, K.: Proof of the Bogoliubov-Parasiuk theorem on renormalization. Commun. Math. Phys. **2**, 301–326 (1966)
10. Zimmermann, W.: Convergence of Bogoliubov's method of renormalization in momentum space. Commun. Math. Phys. **15**, 208–234 (1969)
11. Gelfand, I., Yaglom, A.: Integration in functional spaces and its applications in quantum physics. J. Math. Phys. **1**, 48–69 (1960)

12. Cameron, R.: A family of integrals serving to connect the Wiener and Feynman integrals. J. Math. Phys. Sci. MIT **39**, 126–140 (1960)
13. Helset, A., et al.: Geometry in Scattering Amplitudes. arXiv:2210.08000 [hep-ph] (2022)
14. Abe, Y., Fukuma, M.: Gradient flow and the renormalization group. Prog. Theor. Exp. Phys. 083B02 (2018)
15. Makino, H., et al.: Gradient flow and the Wilsonian renormalization group flow. Prog. Theor. Exp. Phys. 053B02 (2018)
16. Pagani, C., Sonoda, H.: Products of composite operators in the exact renormalization group formalism. Prog. Theor. Exp. Phys. 023B02 (2018)
17. Corianò, C., Maglio, M.M.: Renormalization, conformal ward identities and the origin of a conformal anomaly pole. Phys. Lett. B **781**, 283–289 (2018)
18. Dawid, S.M., et al.: Renormalization group procedure for potential $-g/r^2$. Phys. Lett. B **777**, 260–264 (2018)
19. Lippoldt, S.: Renormalized functional renormalization group. Phys. Lett. B **782**, 275–279 (2018)
20. Maiezza, A., Vasquez, J.C.: Renormalons in a general Quantum Field Theory. Ann. Phys. **394**, 84–97 (2018)
21. Cen, L.-X.: Revisiting numerical real-space renormalization group for quantum lattice systems. Ann. Phys. **397**, 151–158 (2018)
22. Bindera, M., Schmidt, I.: Functional renormalization group flow of massive gravity. Eur. Phys. J. C **80**, 271 (2020)
23. Gracey, J.A.: Renormalization of scalar field theories in rational spacetime dimensions. Eur. Phys. J. C **80**, 604 (2020)
24. Kuwahara, T., Tanaka, G., Tsuchiya, A., Yamashiro, K.: Exact renormalization group for wave functionals. arXiv:2211.05534 [hep-th] (2022)
25. Bellon, M.P., Clavier, P.J.: Alien calculus and a Schwinger-Dyson equation: two-point function with a nonperturbative mass scale. Lett. Math. Phys. **108**, 391–412 (2018)
26. Ohta, N., Rachwal, L.: Effective action from the functional renormalization group. Eur. Phys. J. C **80**, 877 (2020)
27. Ema, Y., et al.: Renormalization group equations of Higgs-R^2 inflation. JHEP **02**, 109 (2021)
28. Giuliani, A., et al.: Gentle introduction to rigorous Renormalization Group: a worked fermionic example. JHEP **01**, 026 (2021)
29. Herren, F., et al.: On ambiguities and divergences in perturbative renormalization group functions. JHEP **06**, 116 (2021)
30. Buccio, D., et al.: Renormalization group flows between Gaussian fixed. JHEP **10**, 113 (2022). points
31. Burke, M.D., et al.: Renormalized Perturbation Theory for Fast Evaluation of Feynman Diagrams on the Real Frequency Axis. arXiv:2211.02453 [cond-mat.str-el] (2022)
32. Prinz, D.: Renormalization of Gauge Theories and Gravity. arXiv:2210.17510 [hep-th] (2022)
33. Fraboulet, K.: Path-integral Approaches to Strongly-coupled Quantum Many-body Systems. arXiv:2210.16676 [cond-mat.str-el] (2022)
34. Pottel, S.: BPHZ renormalization in configuration space for the $\mathcal{A}^4$-model. Nucl. Phys. B **927**, 274–293 (2018)
35. Helset, A., et al.: Renormalization of the standard model effective field theory from geometry. JHEP **02**, 063 (2023)
36. Moreno-Pulido, C., Peracaula, J.S.: Renormalizing the vacuum energy in cosmological spacetime: implications for the cosmological constant problem. Eur. Phys. J. C **82**, 551 (2022)
37. Karki, S., Altschul, B.: Renormalization scheme dependence of β-functions in Lorentz-violating quantum field theory. Eur. Phys. J. C **82**, 676 (2022)
38. Casadio, R., Kuntz, I., Paci, G.: Quantum fields in teleparallel gravity: renormalization at one-loop. Eur. Phys. J. C **82**, 186 (2022)
39. Huang, K.: A critical history of renormalization. Int. J. Mod. Phys. **28**(29), 1330050 (2013)
40. Gavrilov, S.P., Gitman, D.M.: Regularization, renormalization and consistency conditions in QED with x-electric potential steps. Eur. Phys. J. C **80**, 820 (2020)

41. Kreimer, D.: On the Hopf algebra structure of perturbative quantum eld theories. Adv. Theor. Math. Phys. **2**, 303 (1998)
42. Connes, A., Kreimer, D.: Hopf algebras, renormalization and noncommutative geometry. Comm. Math. Phys. **199**, 203 (1998)
43. Kazakov, D.I.: Kinematically dependent renormalization. Phys. Lett. B **786**, 327 (2018)
44. Curtright, T.L., Zachos, C.K.: Renormalization group functional equations. Phys. Rev. D **83**, 065019 (2011)
45. Hooft, G.T.: Renormalization without infinities. Int. J. Mod. Phys. A **20**(6), 1336 (2005)
46. Slavnov, D.A.: The hybrid renormalization. Theo. Math. Phys. **122**(3), 335 (2000)
47. Landau, L.D., Lifshitz, E.M.: Quantum Mechanics, Non-relativistic Theory, Chap. 2. Pergamon Press (1991)
48. Greiner, W.: Quantum Mechanics, An Introduction, Chap. 10. Springer (2001)

Nonlinear Loading and Damping of a Single Degree of Freedom Oscillator

Ove Tobias Gudmestad

Abstract A single degree of freedom oscillator system subject to regular sinusoidal loading does exhibit a well-known linear solution. In real situations, we shall add damping as there in any physical system will be a damping mechanism that will limit the oscillations. In case of forced oscillations in a fluid, the loading and the damping may be nonlinear and the response may then exhibit nonlinear characteristics. A simple drag loading term proportional to the square of the oscillator's velocity is studied as an example of a system exhibiting limit cycles: When the external loading and the oscillator are in higher order resonances, where the natural period of the loading is an integer of the natural period of the system, resonances will occur. The resulting response of the system is a limit cycle oscillation. Furthermore, for oscillators with non-negligible motions, the motion will act as a nonlinear damping term, that for some situations may be interpreted as "negative damping." In case of irregular wave loading, the system can get into higher order resonances occasionally (when the periods of the main waves are an integer of the oscillator's periods), triggering large temporary limit cycle motions. A review of other nonlinear damping models is also given with a brief discussion of the performance of systems exposed to these damping models.

Keywords Single degree of freedom oscillator system · Nonlinear loading and damping mechanisms · Fourier decomposition of the loading · Limit cycle motion · Nonlinear higher order motions · Irregular loading input

O. T. Gudmestad (✉)
Western Norway University of Applied Science, Haugesund, Norway
e-mail: otgudmestad@gmail.com

D. Vlachos (ed.), *Mathematical Modeling in Physical Sciences*, Springer Proceedings in Mathematics & Statistics 446, https://doi.org/10.1007/978-3-031-52965-8_34

1 Introduction

1.1 The Dynamic Equation for a Single Degree of Oscillator with Forcing Function

We notice the Newton's law for a system of particles that have non-variable masses m_i (in kg):

$$\boldsymbol{F}_i^T = m_i \boldsymbol{a}_i \tag{1}$$

Here, $\boldsymbol{F}_i^T$ represents the total forces (in N) acting on the system's particles and $m_i \boldsymbol{a}_i$ represents the inertial forces that result from the action of the total forces; a_i being the acceleration $\frac{m}{s^2}$ of the particles. For the case of a single degree of oscillator, $i = 1$. The total forces are:

- the ***external forces***, for example the force of gravity, the force generated by the environment (for example by wind or waves) or forces generated by humans or machinery. The external forces could be written in the form of Fourier series, where a representative time series must be selected for an irregular loading case
- the ***restoring forces*** acting in the direction opposite to the displacement in order to bring the system back to equilibrium. For a spring-mass system, the restoring force can be written as: $F_R = -kx$ where k is the spring-constant $(\frac{N}{m})$ that represents the stiffness of the spring and x (m) is the displacement of the oscillator.

Rearranging the equation, we obtain:

$$\frac{d^2x}{dt^2} = -\frac{k}{m}x \tag{2}$$

The solution to this equation is a regular harmonic motion with amplitude depending on the initial conditions and with angular frequency ω_0, expressed as $\omega_0^2 = \frac{k}{m}$.

In reality, no physical system will continue to oscillate with harmonic motions as there will be internal or external damping acting on the system. Therefore, ***damping forces*** must be introduced in the Eq. (2). Normally, and for smaller values of the displacements, linear damping is introduced, expressed as proportional to the velocity of the displacement: $F_d = -c\frac{dx}{dt}$, where c is a damping constant. Other models for damping are viable in many situations and will be investigated in the further.

For an oscillator being exposed to external varying forces, it is sufficient to consider harmonic forcing due to the possibility to expand regular and irregular forcing functions in Fourier Series. We can therefore study forcing functions of the form: $F_0\sin(\omega_f t)$ where F_0 is the amplitude of the forcing function and ω_f is the frequency of the forcing function:

$$m\frac{d^2x}{dt^2} = \mathrm{F_0 sin(\omega_f t) - kx} \tag{3}$$

A resonant motion is obtained when the energy of the forcing function with frequency ω_f is equal to ω_0. In case of no-damping, the resulting displacement is unlimited. This result, being non-physical, requires that damping be introduced in all engineering calculations.

$$m\frac{d^2x}{dt^2} = \mathrm{F_0 sin(\omega_f t) - kx - c\frac{dx}{dt}} \tag{4}$$

However, the resonant motion can be very large, in fact the amplitude of the particular solution to this equation can be equal to $x(t) = \left(\frac{F_0}{k}\right)\left(\frac{1}{2\lambda}\right)$. Here, the term $\left(\frac{F_0}{k}\right)$ represents the displacement in case of a static system (no motions) and the term $\left(\frac{1}{2\lambda}\right)$ represents the dynamic amplification. Note that λ is the damping ratio; $\lambda = \frac{c}{2m\omega_0}$. Note also that the amplitude is at maximum at the resonant frequency $\omega_r = \omega_0\sqrt{1-2\lambda^2}$.

The generalized form of the dynamic equation for a single degree of oscillator with forcing function is normally given in the format:

$$\frac{d^2x}{dt^2} + 2\lambda\omega_0\frac{dx}{dt} + \omega_0^2 kx = \left(\frac{F_0}{m}\right)\sin(\omega_f t) \tag{5}$$

This introduction to the problem represents the basics if all dynamic analysis and is possible too simplistic for a scientific paper. As will be demonstrated below, however, this introduction is necessary for the continued discussion.

1.2 A Non-linear Forcing Function Could Represent a Damping Term

The Morison equation is normally used for analyzing hydrodynamic motion of slender members, that is, members with member diameters no larger than 1/5 of the wave length. In this discussion, such a member will be treated as an oscillator.

The horizontal component of the forcing function is given as the sum of an ***inertia term*** (proportional to the acceleration a_x of the water particles) and a ***drag term*** (proportional to the square of the velocity v_x of the water particles normal to the axis of the member and in the direction of the water particle flow). The force per unit length of a cylindrical member with diameter D is given by the nonlinear expression:

$$F = \rho_w C_m\left(\frac{\pi}{4}\pi D^2\right)a_x + \frac{D}{2}\rho_w C_d v_x|v_x| \tag{6}$$

Here,

ρ_w is the density of the fluid.

C_m is the hydrodynamic mass (inertia) coefficient.

C_d is the hydrodynamic drag coefficient.

$|v_x|$ is the numerical value of v_x. Note that the drag term is a nonlinear function of the velocity of the water particles v_x. For a linearization of the drag forcing term, see [1].

To estimate the total force on the member, the force components per unit lengths must be integrated (summed) from the seafloor to the top of the wave surface, this being a nonlinear summation as the water surface varies with the wave crest. Furthermore, the velocities and accelerations of the water particles are complex nonlinear functions, determined by the surface conditions of the flow, however, they could be written in the form of sums of harmonic sinus functions of the form $\sin(j\omega t)$. It should be noticed that the term $\sin(\omega t)|\sin(\omega t)|$ can be linearized using Fourier series transformation:

$$\sin(\omega t)|\sin(\omega t)| = c_0 + c_1\sin(\omega t) + c_2\cos(2\omega t) + c_3\sin(3\omega t) + \cdots \quad (7)$$

In the case of a constant current, U_0, the drag term will read

$$F_D = \frac{D}{2}\rho_w C_d(U_0 + v_x)|U_0 + v_x|\frac{D}{2}\rho_w C_d \quad (8)$$

And in the case of flexible members, which has non-negligible movements in line with the waves, the drag loading will be adjusted by the movements of the structure and could be written as

$$F_D = \frac{D}{2}\rho_w C_d(b\frac{dx}{dt} + v_x)|b\frac{dx}{dt} + v_x| \quad (9)$$

The member's velocity $\frac{dx}{dt}$ appears in the forcing function and could then represent a damping term. This situation has been considered by the author, see [2], and the solutions are discussed in Sect. 2 below.

2 Limit Cycle Presentation of the Resulting Motion

The solution of the equation of motion could be presented in a phase plane where the member's position and velocity (the member's trajectory) are plotted for an increasing time, t. In the phase plane, a closed trajectory towards to which the trajectories spiral either towards or away, is termed a ***limit cycle***, [3, 4]. When the surrounding trajectories all spiral towards the limit cycle, it is said to be a stable limit cycle. If all the near trajectories spiral away, it is unstable, and if some spiral towards while other spiral away, the limit cycle is half-stable. Systems with a stable limit cycle will,

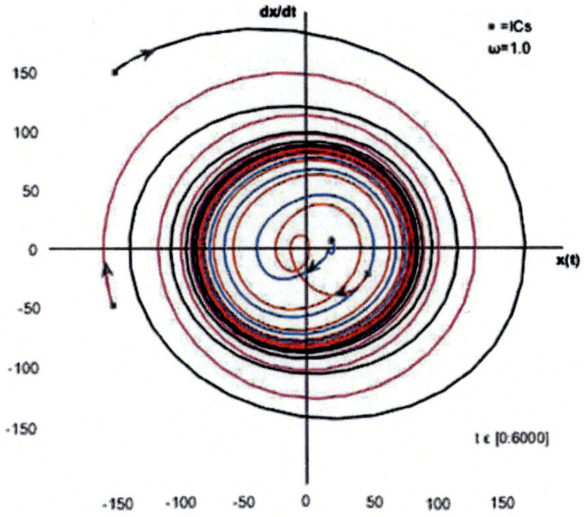

Fig. 1 Phase plane diagram for a system with nonlinear drag force at resonance, $\beta = 1$, Four different trajectories (with different initial conditions) are shown, which all spiral towards the limit cycle marked in red. Notice that systems started with initial conditions inside the limit cycle, evolve outwards towards the limit cycle (Copied from [2])

independent of the starting conditions, settle into a steady trajectory in the phase plane, where there is a balance between generation and dissipation of energy.

Limit cycles only occur in systems with nonlinear terms. A linear system may have a closed trajectory, but it is then surrounded by other closed trajectories when we select other initial conditions. Nonlinear systems may be described by ordinary differential equations, containing nonlinearity in the stiffness, damping, or loading terms.

As shown in [2], for forcing functions given by (6) and (8), trajectories starting at different initial conditions, all are spiraling towards the limit cycle, see Fig. 1.

Some of the trajectories have initial conditions making them start outside the limit cycle, spiraling inwards towards the limit cycle. However, some trajectories start on the inside of the limit cycle, experiencing a growth in amplitudes for both position and velocity until they reach the limit cycle. These trajectories are of interest as they experience an ***apparent negative damping***, increasing the member's displacement and velocity until the generation and dissipation of energy in the system is balanced. Limit cycles are also demonstrated in the case of currents, forcing function given by (9).

For higher order resonances, $\beta = \frac{\omega_f}{\omega_0} = 1/2$ and $\beta = 1/3$, similar behavior is observed, thereby energy from waves with higher periods (lower frequency ω_f), for example 15 s and 10 respectively, are fed into a system with natural period of 5 s, respectively. In storm situations, the energy spectrum has high values for higher periodic waves and considerable displacements and velocities of the member can be expected.

Regarding the softer member, represented by Eq. 8, an increased value of the motions of the member when the value of b increases, results in larger amplitude limit cycles as might be expected.

3 Consequences of the Nonlinear Drag Loading on the Motion of Structures Exposed to Wave Loading

3.1 Case Examples, Large Cylindrical Structure with Diameter in the Order of 15 m and Slender Space-Frame Structure with Member Diameters in the Order of 2 m

The consequences of nonlinear drag loading on the motion of offshore structures, for example fixed offshore platforms for production of oil and gas and fixed wind turbines are discussed in [5, 6].

A case, referred to in [5], is where a *fixed large one tower offshore unit*, the Draugen platform, located on Haltenbanken offshore central Norway at 64.35 degrees North and 7.77 degrees East was hit by a huge wave on 12th March 1995, rather shortly after its installation on site during the summer of 1993.

The conical shaft of the Draugen Concrete Gravity Structure (GBS) has its narrowest diameter at the sea surface, where it measures just over 15 m increasing to 22 m down at the multiple storage cells located approximately 50 m above the bottom of the sea. The top of the shaft is designed as a box structure with a square cross-section measuring 22 m to a side, Fig. 2 [7].

Fig. 2 The Draugen platform. Photo: Heine Schjølberg/A/S Norske Shell

The wave conditions at the site are very rough with a design significant wave height of 17 m and spectral peak period of 18 s [8]. Using the dispersion relation for deep water (the water depth is 280 m), $L = \frac{gT^2}{2\pi}$, gives a wave length L of more than 500 m so the structure will act as a drag dominant structure when exposed to design waves.

The platform manager was warned by a meteorology center that a huge wave was moving in the direction of the platform and when the wave hit the platform, it was "shaken fiercely." The situation is described in details in [5]. The response was reported as a "ringing event." Such an event was identified during the wave tank model tests carried out at Marintek, Trondheim during the detailed design phase of the platform: "Resonant ringing vibrations were triggered in steep irregular waves, in the form of a resonant build-up during a wave period, at frequency being 3–4 times the fundamental wave frequency", [9]. Noting that the natural period of vibration of the structure is in the range of 5 s, the higher harmonics are in the range of 10 s, 15 s, respectively. Based on the above discussion, it could be suggested that the ringing effects experienced for the Draugen platform were caused by resonance effects as described above for oscillators exposed to nonlinear drag loading. It should be noted that the nonlinearities in the total loading and the effects on the response as caused by integrating the force up to the actual nonlinear surface have not been included in the discussion above, further analysis is required to include all nonlinearities in the analysis. For a thorough discussion of the wave conditions at the day of the event, reference is made to [6, 9].

In the case of a *steel-frame structure*, a specific study has been reported by [6]. The dimensions of the structural members are in the range of diameters of 2 m. The conclusions of the paper [6] are that "drag dominated structure may experience a significant dynamic amplification at the 3ω load frequency. This is the most important effect for the investigated extreme sea-state." The slender space-frame structure (the Kvitebjørn jacket for production of gas, Fig. 3) has been located in the Northern North Sea at a depth of 190 m. Note that the effects of the higher harmonics were pointed out in [10].

3.2 Additional Nonlinearities to Be Considered for Single Degree of Freedom Systems with Drag Loading

As discussed above, the nonlinearity investigated in this paper is a nonlinear drag forcing function that could exhibit negative damping terms in case of non-negligible oscillations of the system under consideration.

The force per unit length of members have been investigated while the nonlinear surface conditions must be explored for members exposed to waves for a full analysis of the motions of the structures. Furthermore, the velocities and accelerations of water particles influencing the forces, are of nonlinear character.

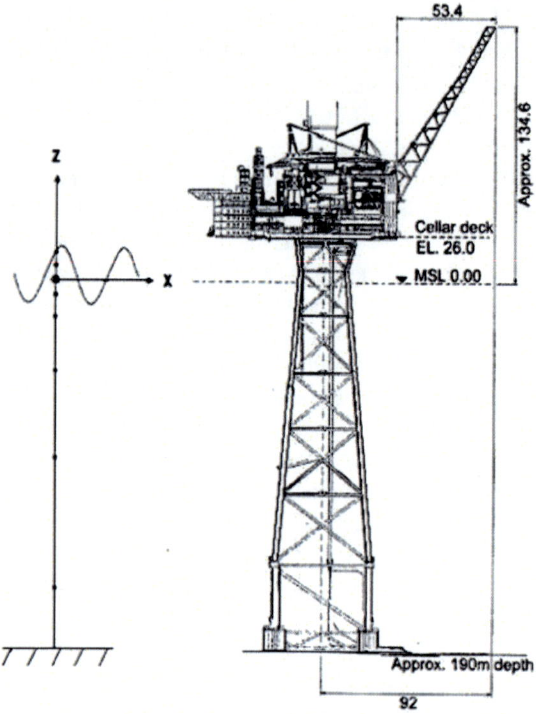

Fig. 3 Slender space-frame structure for Northern North Sea, "The Kvitebjørn jacket"

4 Other Relevant Non-linear Models of Interest

The focus has been on drag loading models for members placed in the wave environment; thus, on fluid–structure interaction. Negative damping has been identified as a possible phenomenon in case of members with oscillatory motions exposed to drag loading, causing large amplification of the response until a limit cycle is reached.

A relevant ***non-linear damping*** model is the van der Pol Eq. (10) where the damping term for some time will work as negative damping, pumping energy into the system [2]. Note that all trajectories spiral towards the limit cycle.

$$\frac{d^2x}{dt^2} - (a_1 - a_2x^2)\frac{dx}{dt} + a_3x = Q_0 cos\Omega t \tag{10}$$

For a more elaborate discussion, see [4] in case of no external force, $Q_0 = 0$, and [11] in case of external loading.

The Rayleigh equation is a further development of the van der Pol equation, where an additional damping term is introduced, see [12]:

$$\frac{d^2x}{dt^2} - \left(a_1 - a_2x^2\right)\frac{dx}{dt} + a_3\left(\frac{dx}{dt}\right)^3 + a_4x = Q_0 cos\Omega t \tag{11}$$

The relevance is the case where the damping is developed in a Taylor series.

The term "negative damping" is also discussed in Chap. 10.1 and App. D5 of [12].

In case of a ***nonlinear restoring force***, the Duffing equation represents a much-investigated model [4, 11]:

$$\frac{d^2x}{dt^2} + a_1\frac{dx}{dt} + (a_2x + a_3x^3) = Q_0 cos\Omega t \tag{12}$$

Here, a_1 is the linear damping coefficient and a_2 and a_3 are liner and cubic spring coefficients, respectively. Resonances are present in the system as discussed in [4]. It should be noticed that the motion of a simple pendulum with (mass-less) length L and mass m hanging from the pendulum is:

$$\frac{d^2\theta}{dt^2} + \frac{\mathrm{g}}{\mathrm{L}}\sin\theta \tag{13}$$

When expanding the $\sin\theta$ term in a Taylor series, adding damping and an external force, we obtain the Duffing equation. For a discussion of the example of a pendulum with variable length, see [13].

Furthermore, a quadratic nonlinear oscillator varying proportionally to the displacement is discussed in [11]:

$$\frac{d^2x}{dt^2} + a_1\frac{dx}{dt} + (a_2 + Q_0 cos\Omega t)x + a_3x^2) = 0 \tag{14}$$

5 Application of Nonlinear Models to Other Branches of Science, Conclusions

In the electronic industry, oscillators can be tuned with nonlinear stiffness and damping components. The loading can be made sinusoidal, nonlinear, or chaotic. The response can thus be tuned to the wishes of the manufacturer and the user. Filters could possibly be designed to pick out electronic malware.

In [14] a discussion of ice forces on vertical offshore structures is discussed. The term "negative damping" to explain the motions is introduced.

In medicine, the flow of blood in veins could be modeled with the use of the terminology of fluid flow. The drag term could represent interaction of the flow with obstacles like blood clogs, and the external loading is the pumping of the blood by the heart. The equation for the dynamics of a drag dominated flow could possibly explain the movements of the blood clogs. It might be important to keep the frequency of the heartbeat at a rate where we avoid that the blood clogs move quickly in the veins, to avoid the clogs are moving to the heart or the brain.

The conclusion is that nonlinear models are of interest in several branches of research and that studies of such models are very useful and should be promoted.

References

1. Gudmestad, O.T., Connor, J.J.: Linearization methods and the influence of current on the nonlinear hydrodynamic drag force. J. Appl. Ocean Res. **5**(4) (1983). Also published in Spanos, P.O. (eds.) Probabilistic offshore mechanics. CML Publications, Southampton, (1985)
2. Hellevik, K., Gudmestad, O.T.: Limit cycle oscillations at resonances for systems subjected to nonlinear damping or external forces. IOP Conf. Ser.: Mater. Sci. Eng. **276**, 012020 (2017). https://doi.org/10.1088/1757-899X/276/1/012020
3. Struble, R.A., Martin, W.T.: Nonlinear Differential Equations. McGraw-Hill Book Company (1962)
4. Bogoliubov, N.N., Mitropolsky, Y.A.: Asymptotic Methods in the Theory of Non-Linear Oscillations. Hindustan Publishing Corporation (1961)
5. Gudmestad, O.T.: Modelling of waves for the design of offshore structures. J. Mar. Sci. Eng. **8**, 293 (2020). https://doi.org/10.3390/jmse8040293
6. Horn, J.-T.H., Amdahl, J., Haver, S.K.: Dynamic amplification of drag dominated structures in irregular seas. OCEANS 2015-MTS/IEEE (2015). https://core.ac.uk/download/pdf/154668304.pdf
7. Sandberg, F.H.: Squaring the circle, industrial heritage DRAUGEN. https://draugen.industriminne.no/en/2018/07/02/squaring-the-circle/
8. Standards Norway: Norsok N-003, Actions and Action Effects. Lysaker, Norway (2016)
9. Grue, J.: Two phenomena: Honji instability and ringing of offshore structures. Theor. Appl. Mech. Lett. **1**, 062001 (2011)
10. Gudmestad, O.T., Poumbouras, G.A.: Time and frequency domain wave forces on offshore structures. J. Appl. Ocean Res. **10**(1) (1988)
11. Luo, A.C.J.: Analytical Routes to Chaos in Nonlinear Engr. John Wiley & Sons Ltd. (2014)
12. Yabuno, H.: Linear and Nonlinear Instabilities in Mechanical Systems. John Wiley & Sons Ltd. (2021)
13. Dysthe, K.B., Gudmestad, O.T.: On resonance and stability of conservative systems. J. Math. Phys. **16**(1) (1973)
14. Kärnä, T., Qu, Y., Bi, X, Yue, Q., Kuehnlein, W.: A spectral model for forces due to ice crushing. J. Offshore Mech. Arctic Eng. **129**, 138–146 (2005). https://doi.org/10.1115/1.2426997

Correction for the Classical Conditions for a Collision in Three-Body System Using General Relativity and Machine Learning

Hadi Salloum, Manuel Mazzara, and Mohammad Reza Bahrami

Abstract This paper comprehensively investigates collision conditions in three-body problems, incorporating General Relativity (GR) effects. The study analyzes the initial values of the bodies to determine the collision possibility and develops a high-accuracy machine learning model for classifying collision events. The study introduces the concept of GR-effective potential energy derived from the Einstein Field Equations and solves the equation using the Schwarzschild solution for spherically symmetric gravity fields. Additionally, a code is developed to examine collisions using the GR-effective potential energy, and a machine learning model is trained accordingly. The study provides a modified equation that accurately describes the collision condition in three-body problems, accounting for GR effects. The results have significant implications in astrophysics and contribute to advancing knowledge in understanding celestial body dynamics in collision scenarios.

Keywords Three-body problem · Collision conditions · General Relativity · Machine learning · Astrophysics

1 Introduction

The history of geometry dates back to ancient times, with its origins found in the endeavors of ancient Babylonians and Egyptians to establish their civilizations. However, the development of geometry as a formal discipline began in the seventh century BC with the Greek schools. Many of the fundamental principles were discovered in the sixth and fifth centuries BC, and were further refined in the concept of theory

H. Salloum (✉) · M. Mazzara
Department of Computer Science and Engineering, Innopolis University, Innopolis, Russia
e-mail: h.salloum@innopolis.university

M. Mazzara
e-mail: m.mazzara@innopolis.ru

M. R. Bahrami
Lab of CyberPhysical Systems, Innopolis University, Innopolis, Russia
e-mail: mo.bahrami@innopolis.ru

D. Vlachos (ed.), *Mathematical Modeling in Physical Sciences*, Springer Proceedings in Mathematics & Statistics 446, https://doi.org/10.1007/978-3-031-52965-8_35

and proof. In the third century BC, geometry reached a significant milestone with the emergence of the mathematician Euclid, whose work was based on postulates, hypotheses, and definitions [2]. The postulates of Euclidean geometry include:

1. For any two points, there is a unique straight line that passes through them.
2. Every line segment can be extended uniquely to a straight line.
3. It is possible to draw a circle given any center and radius.
4. All right angles are equal.
5. If a straight line intersects two other lines and a1 and a2 are the two internal angles of intersection on one side of a, and they add up to less than two right angles, then a1 and a2 intersect only on that side of a [3].

Furthermore, the fifth postulate, also known as the parallel postulate, plays a crucial role in studying geometric shapes. It forms the basis of parallelism theory, similarity of shapes, methods for matching triangles, perpendicularity of lines, and inclination of straight lines. The parallel postulate can be reformulated as follows: "Two straight lines are parallel if they do not contain any common point." Therefore, the fifth postulate is fundamental to Euclidean geometry. Despite many unsuccessful attempts to prove this postulate from the other postulates, it remained unproven until the development of non-Euclidean geometries in the 19th century [2, 3]. In 1826, Lobachevsky's groundbreaking work provided proof of the futility of attempts to reconcile Euclid's fifth postulate. Lobachevsky formulated a consistent theory of geometry that did not adhere to this assumption, while independently, Bolyai arrived at similar conclusions. Lobachevsky published the first account of his non-Euclidean geometry in 1829, marking the beginning of a new era in the field of geometry. Subsequently, in the late nineteenth century, the consistency of Lobachevsky's geometry was formally established, validating the theory based on the counter-intuitive claim:

> From a point outside a straight line in a plane, at least two straight lines that don't intersect the line pass through.

In other words, assuming the existence of a straight line a in a plane, and a point A not lying on it, at least two straight lines a_1 and a_2 pass through point A without intersecting the line a (as shown in Fig. 1) [2–4]. In the early 20th century, our understanding of the universe was primarily based on Euclidean geometry [2].

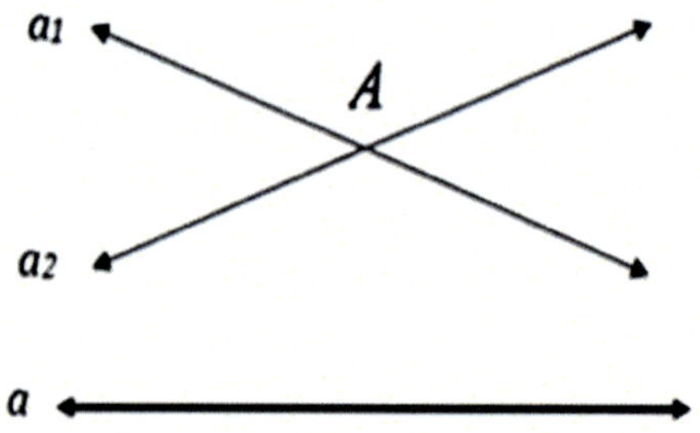

Fig. 1 Lobachevsky geometry postulate

However, the Einstein Equivalence Principle (EEP) showed that space-time signals a breakdown of Euclidean geometry. The EEP states that in small enough regions of space-time, the laws of physics reduce to those of special relativity, making it impossible to detect the existence of a gravitational field by means of local experiments. One of the celebrated predictions of the EEP is the gravitational redshift [1, 5, 6].

Consider two boxes, a distance z apart, each moving with some constant acceleration a in a region far away from any gravitational fields. At time t_0, the trailing box emits a photon of wavelength λ_0. The boxes remain a constant distance apart, so the photon reaches the leading box after a time $\Delta t = z/c$ in our background reference frame (assuming $\Delta v/c$ is small, so we work to first order). In this time, the boxes will have picked up an additional velocity $\Delta v = a\Delta t = az/c$. Therefore, the photon reaching the leading box will be red-shifted by the conventional Doppler effect by an amount

$$\Delta\lambda/\lambda_0 = az/(c^2). \tag{1}$$

According to the EEP, the same thing should happen in a uniform gravitational field. Therefore, a photon emitted from the ground with wavelength λ_0 will be red-shifted by an amount

$$\Delta\lambda/\lambda_0 = a_g z/(c^2) \tag{2}$$

Where g is the acceleration due to gravity. The formula for the red-shift is more often stated in terms of the Newtonian potential Φ, where $\Delta\Phi = ag$. We then have

$$\Delta\lambda/\lambda_0 = (1/c)\int \Delta\Phi\, dt = (1/c^2)\int \partial_z\Phi\, dz = \Delta\Phi \tag{3}$$

(setting $c = 1$).

Now, we should think of space-time as curved. Consider a person on the ground who emits a beam of light with wavelength λ_0 from a height z_0, which travels to the top of a tower at height z_1. The time between when the beginning of any single wavelength of the light is emitted and the end of that same wavelength is $\Delta t_0 = \lambda_0/c$, and the same time interval for the absorption is $\Delta t_1 = \lambda_1/c$, where time is measured by clocks located at the respective elevations. Simple geometry seems to imply that the times Δt_0 and Δt_1 must be the same. But of course, they are not; the gravitational red-shift implies that the elevated observers observe fewer wavelengths per second, so that $\Delta t_1 > \Delta t_0$ as shown in Fig. 2. We can interpret this roughly as "the clock on the tower appears to run more quickly." Therefore, the space-time through which the photons traveled was curved, signaling a breakdown of Euclidean geometry [1]. The Structure of Space-Time depends on the Lorenz transformations of the coordinates of events which have the form:

$$x' = \gamma(x - \beta ct) \tag{4}$$

$$x = \gamma(x' + \beta ct') \tag{5}$$

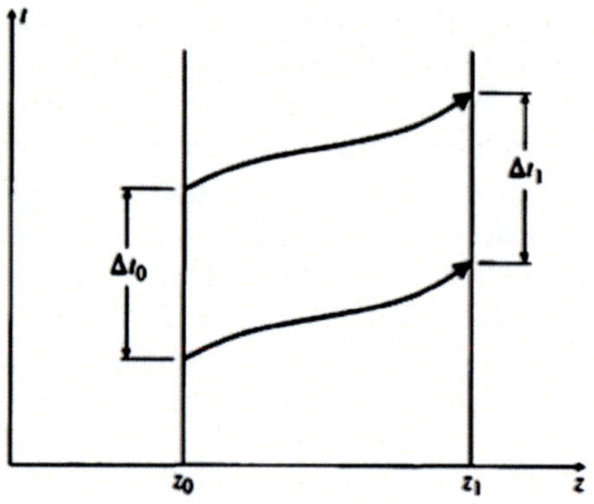

Fig. 2 Spacetime diagram of the gravitational-redshift experiment [1]

$$ct' = \gamma(ct - \beta x) \tag{6}$$

$$ct = \gamma(ct' - \beta x') \tag{7}$$

Let us define the interval Δs of the pair of events E_1 and E_2 as:

$$\Delta s = c^2 \Delta t^2 - \Delta x^2 - \Delta y^2 - \Delta z^2 \tag{8}$$

The in-variance of the interval allows an absolute classification of pairs of events. A pair of events is said to be:

- time-like separated if $\Delta s > 0$
- space-like separated if $\Delta s < 0$
- Light-like separated if $\Delta s = 0$

Figure 3 shows a light cone of event E. The events belonging to the interior of the cone, such as event A, are time-like separated from E. The events outside the cone, such as event B, are space-like separated from E. Moreover, the interval for time-like separated events is the measure of the proper time between them:

$$\Delta\tau = \frac{\Delta s}{c} \tag{9}$$

Any pair of events belonging to a world line of a particle is time-like separated.

Let's assume that there is a system frame S and two events E_1 and E_2 such that the coordinates of E_1 and E_2 in this reference system are (t, x, y, z) and $(t + dt, x + dx, y + dy, z + dz)$ respectively. So, the interval has the value:

$$ds^2 = c^2 dt^2 - |dr|^2 \tag{10}$$

where dr in the equation is the particle displacement during the time dt.

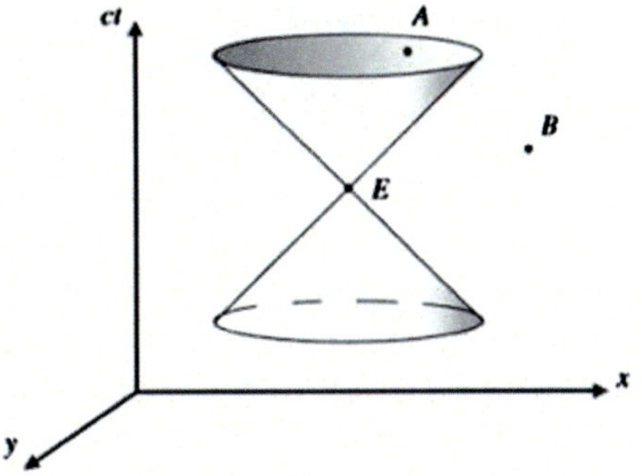

Fig. 3 The light cone of event E [7]

$$d\tau = \frac{ds}{c} = \frac{1}{c}\sqrt{c^2dt^2 - |dr|^2} = \sqrt{1 - \left(\frac{dr}{dt}\right)^2 c^2 dt} \tag{11}$$

We can get the time read by a clock fixed to the particle (proper time) by integrating the equation [1, 7, 15]:

$$\Delta\tau = \int dt\sqrt{1 - \left(\frac{dr}{dt}\right)^2 c^2} \tag{12}$$

The structure of space-time has been profoundly shaped by the mathematical framework of Riemannian geometry, which provides a powerful tool for describing the curvature and geometry of space-time. Riemannian geometry, as a non-Euclidean geometry, rejects the validity of Euclid's fifth postulate and modifies his second postulate, leading to intriguing implications in physics.

Euclid's fifth postulate states that through a point not on a given line, there is only one line parallel to the given line. In Riemannian geometry, this postulate is abandoned, and as a result, there are no lines that are strictly parallel to a given line. This departure from Euclidean geometry has significant consequences for our understanding of space-time. Furthermore, Riemannian geometry also modifies Euclid's second postulate, which states that a straight line of finite length can be extended continuously without bounds. In Riemannian geometry, the concept of straightness becomes more subtle due to the curvature of space-time. In the context of physics, Riemannian geometry has found profound applications in describing the geometry of space-time in classical physics and Special Relativity. Classical physics' Euclidean space and Special Relativity's pseudo-Euclidean space-time are special cases of Riemannian geometries. In Riemannian geometry, the distance between two infinitesimally separated points is described by a quadratic form of the coordinate differences, denoted as

$$s^2 = g_{ij}(x)dx^i dx^j \tag{13}$$

Where g_{ij} are the covariant components of a symmetric second-order tensor known as the metric tensor. The metric tensor, which encodes information about the curvature and geometry of space-time, can be expressed in terms of the coordinate-system scale factors h_i, $i = 1, 2, 3$, as [8]

$$g_{ij} = \begin{cases} h_i^2 & \text{if } i = j \\ 0 & \text{if } i \neq j \end{cases}$$

In 1915, Einstein published the General Theory of Relativity (GR), by incorporating Special Relativity and Riemannian geometry. The resulting Einstein Field Equations (EFE) explain how the curvature of space-time is caused by the presence of energy-momentum. In the case of celestial bodies, where the gravity field is spherically symmetric, the equation can be solved using the Schwarzschild solution. The GR-Effective Potential Energy for celestial bodies can be derived from the EFE. However, the concepts of potential and kinetic energy can only be defined in the context of the Schwarzschild metric. By using the GR-Effective Potential Energy, it is possible to correct classical conditions for the the collision for celestial bodies in the three-body problem.

2 Methodology

2.1 Einstein Field Equation (EFE)

The gravitational field equation in Newton's theory of gravity can be expressed in an alternative form as follows:

$$\Delta\Phi = 4\pi G\rho_m, \tag{14}$$

where ρ_m represents the mass density acting as the source of the gravitational field, $\Delta = \delta^{ij}\partial_i\partial_j$ is the Laplacian operator in three-dimensional space, and Φ is the gravitational potential. In the framework of general relativity, the geometry of space-time is described by the metric tensor, and the curvature of space-time is determined by the Riemann tensor. The Riemann tensor, denoted by $R^l{}_{jki}$, can be expressed in terms of the Christoffel symbols $\Gamma^l{}_{ji}$ as follows:

$$R^l{}_{jki} = \Gamma^l{}_{jk,i} - \Gamma^l{}_{ji,k} + \Gamma^e{}_{jk}\Gamma^l{}_{ei} - \Gamma^e{}_{ji}\Gamma^l{}_{ek}, \tag{15}$$

where comma denotes partial derivative, and the indices i, j, k, and l take values from 0 to 3. The Ricci tensor R_{ij}, which describes the curvature of space-time in a contracted form, can be defined in terms of the Riemann tensor as:

$$R_{ij} = R^l{}_{jli}. \tag{16}$$

The curvature scalar R is defined as the contraction of the indices of the Ricci tensor:

$$R = g^{ij} R_{ij}, \tag{17}$$

where g^{ij} represents the contra-variant components of the metric tensor. With the mathematical formalism of curved space-time established, Einstein derived the field equation that describes the dynamics of gravity in the presence of matter and energy. In his general theory of relativity, Einstein introduced an energy-momentum tensor $T_{\mu\nu}$ that describes the distribution of matter and energy in space-time, analogous to the energy components of the gravitational field. The Einstein field equation (EFE) is given by [1, 7, 9]:

$$R_{\mu\nu} - \frac{1}{2} g_{\mu\nu} R + \Lambda g_{\mu\nu} = 8\pi G T_{\mu\nu}, \tag{18}$$

where μ and ν represent the indices of the space-time coordinates, $g_{\mu\nu}$ is the covariant components of the metric tensor, and Λ is the cosmological constant. The left-hand side of the equation describes the curvature of space-time, while the right-hand side describes the distribution of matter and energy. Einstein initially introduced the cosmological constant Λ to obtain a static cosmological model, which was consistent with the astronomical observations of that time. However, subsequent astronomical observations indicated that the universe is expanding, and the cosmological constant was deemed unnecessary. Nevertheless, modern cosmological models, such as the Lambda-CDM model, incorporate a non-zero cosmological constant to explain the observed accelerated expansion of the universe [10, 11, 16]. The Einstein field equation is a set of ten coupled partial differential equations that describe the behavior of gravity in the presence of matter and energy. Solving these equations requires knowledge of the metric tensor and the energy-momentum tensor, which in turn depend on the distribution of matter and energy in space-time. The field equations are highly non-linear and their solutions can exhibit a wide range of phenomena, such as black holes, gravitational waves, and cosmological expansion [12].

2.2 *GR Effective Potential Energy for the Celestial Bodies Using EFE (Geodesics of the Schwarzschild Metric)*

The geodesic equation, which describes the motion of particles in curved space-time, can be derived from the Einstein field equations (EFE). For the Schwarzschild metric, the geodesic equation implies that the quantity ϵ is constant, where $\epsilon = -g_{uv} \frac{dx^u}{d\lambda} \frac{dx^v}{d\lambda}$ is the energy per unit mass of the particle along its trajectory. Expanding this expression, we obtain:

$$-(1 - 2GM/r) \left(\frac{dt}{d\lambda} \right)^2 + \frac{1}{1 - 2GM/r} \left(\frac{dr}{d\lambda} \right)^2 + r^2 \left(\frac{d\phi}{d\lambda} \right)^2 = -\epsilon \tag{19}$$

Together with the conserved quantities E (energy) and L (angular momentum), this equation provides a convenient way to understand the orbits of particles in the Schwarzschild geometry. Multiplying both sides of the equation by $(d\phi/dr)^{-2} = r^4 L^2$, we obtain:

$$\left(\frac{dr}{d\phi}\right)^2 + r^4 L^2 - 2GML^2/r^3 + r^2 - 2GM/r = \frac{E^2}{L^2 r^4} \tag{20}$$

We can define a new variable $\psi = L^2 GM/r$, and rewrite the equation as:

$$\left(\frac{d\psi}{d\phi}\right)^2 + \frac{L^2 G^2 M^2}{r^2} - 2\psi + \psi^2 - \frac{2G^2 M^2}{L^2 r^3}\psi^3 = \frac{E^2 G^2 M^2}{L^2} \tag{21}$$

From this equation, we can assume an effective potential energy for celestial bodies in general relativity (GR):

$$V_{\text{eff}} = -\frac{GM}{r} + \frac{L^2}{2mr^2} - \frac{GML^2c^2}{r^3} \tag{22}$$

where c is the speed of light, and the term $-\frac{GML^2c^2}{r^3}$ is the difference between the effective potential energy in classical mechanics and GR [1, 8, 13].

2.3 *Mathematical Description for the Three-Body Problem*

The three-body problem in Classical Mechanics. Let K_i represent the three bodies with masses m_i, distances $K_i K_j = r_{ij}$, and coordinates y_{ij} in an inertial coordinate system. The equations of motion are

$$\ddot{y}_{1i} = G\left(\frac{m_2 \cdot (y_{2i} - y_{1i})}{r_{12}^3} + \frac{m_3 \cdot (y_{3i} - y_{1i})}{r_{13}^3}\right) \tag{23}$$

$$\ddot{y}_{2i} = G\left(\frac{m_1 \cdot (y_{1i} - y_{2i})}{r_{12}^3} + \frac{m_3 \cdot (y_{3i} - y_{2i})}{r_{23}^3}\right) \tag{24}$$

$$\ddot{y}_{3i} = G\left(\frac{m_1 \cdot (y_{1i} - y_{3i})}{r_{13}^3} + \frac{m_2 \cdot (y_{2i} - y_{3i})}{r_{23}^3}\right) \tag{25}$$

where G is the gravitational constant.

The three-body problem in General Relativity. By using the GR-effective potential energy, we can write the equations of motion in General Relativity as follows:

$$\ddot{y}_{1i} = G\left(\frac{m_2 \cdot (y_{2i} - y_{1i})}{r_{12}^3} + \frac{m_3 \cdot (y_{3i} - y_{1i})}{r_{13}^3} + \frac{4}{m_1^2 \cdot c^2}\left(\frac{m_2 \cdot L_1^2 \cdot (y_{3i} - y_{1i})}{r_{12}^5} + \frac{m_3 \cdot L_1^2 \cdot (y_{3i} - y_{1i})}{r_{13}^5}\right)\right) \tag{26}$$

$$\ddot{y}_{2i} = G\left(\frac{m_1 \cdot (y_{1i} - y_{2i})}{r_{12}^3} + \frac{m_3 \cdot (y_{3i} - y_{2i})}{r_{23}^3} + \frac{4}{m_2^2 \cdot c^2}\left(\frac{m_1 \cdot L_2^2 \cdot (y_{3i} - y_{2i})}{r_{12}^5} + \frac{m_3 \cdot L_2^2 \cdot (y_{3i} - y_{2i})}{r_{23}^5}\right)\right) \tag{27}$$

$$\ddot{y}_{3i} = G\left(\frac{m_1 \cdot (y_{1i} - y_{3i})}{r_{13}^3} + \frac{m_2 \cdot (y_{2i} - y_{3i})}{r_{23}^3} + \frac{4}{m_3^2 \cdot c^2}\left(\frac{m_1 \cdot L_3^2 \cdot (y_{3i} - y_{1i})}{r_{13}^5} + \frac{m_2 \cdot L_3^2 \cdot (y_{3i} - y_{1i})}{r_{23}^5}\right)\right) \tag{28}$$

where L_1, L_2, and L_3 are the angular momentum of the three bodies, and c is the speed of light [13, 14].

2.4 Data Generation and Collision Checking

In the context of celestial mechanics research, the generation of comprehensive and diverse datasets plays a crucial role in understanding the complex dynamics of three-body systems. To achieve this, we employed a rigorous data generation process to explore a wide range of scenarios and capture the various characteristics and initial conditions exhibited by celestial bodies. For this purpose, multiple sets of random data were generated using numpy, where each dataset consisted of 21500 rows for the classical collision and 20500 for GR collision. The datasets were meticulously designed to encompass different bounds for the masses, initial positions (represented by x and y coordinates), radii of the three celestial bodies, and angular momentum for the GR collision dataset. By incorporating varying bounds, we ensured a broad spectrum of possibilities, thus facilitating a more thorough investigation of the system's behavior. Through merging these distinct datasets, we constructed a larger and more comprehensive dataset. The merging process allowed us to conduct comparative analyses and assess the impact of diverse factors, such as varying masses and initial positions, on the motion of celestial bodies and potential collision events. The core aspect of our analysis involved the implementation of a classical three-body collision checking function. This function is of paramount importance, as it enables the detection of critical events that may lead to collisions within the three-body system. To effectively model the long-term behavior of the system, we set the time span (tspan) to a significantly large value (e.g., 3.154e+9), which essentially simulates an infinite time horizon. This approach provides valuable insights into the system's dynamics over extended periods, contributing to a more comprehensive understanding of its behavior. Leveraging the SciPy library's solve_ivp function, we performed numerical integration of the equations of motion governing the three-body system. Consequently, we calculated the distances between the celestial bodies at discrete time steps and scrutinized potential collision scenarios. Whenever the minimum distance between any two bodies fell below the sum of their radii, the function identified a collision and returned a True value; otherwise, it returned False, indicating the absence of any collision. Upon completion of the collision checking process, we recorded the results in a new column named 'output' within the DataFrame. This column served as an indicator of collision occurrence for each specific set of celestial bodies. Finally, since we set the time span (tspan) to simulate an infinite time horizon, we no longer need the 'tspan' column. Thus, we remove the 'tspan' column from the DataFrame and save the updated DataFrame, including the collision checking results, as a new CSV file for further analysis as shown in Figs. 4 and 5 [17, 18].

	A	B	C	D	E	F	G	H	I	J	K	L	M
1	m1	m2	m3	r1_x	r1_y	r2_x	r2_y	r3_x	r3_y	r01	r02	r03	output
2	1.44E+24	1.89E+24	7.37E+24	2.8E+11	1E+11	2.9E+10	2.6E+11	3.5E+11	1E+11	7.4E+10	3.3E+10	5.6E+10	TRUE
3	1.00E+24	9.84E+24	1.01E+24	3.8E+10	2.3E+10	1.5E+11	3.1E+11	6.5E+10	2.5E+11	6.3E+10	2.7E+10	9.3E+10	TRUE
4	7.09E+24	5.20E+24	4.41E+24	4.4E+10	2.3E+11	3.4E+11	3.3E+11	3.1E+11	6.8E+10	4.3E+10	2E+10	1.7E+09	FALSE
5	8.92E+24	8.79E+24	3.51E+24	3.7E+11	2.1E+11	2.3E+11	2.9E+11	2.3E+11	2.2E+11	9E+09	4.5E+10	8.3E+10	TRUE
6	1.42E+24	5.02E+24	9.26E+24	8E+10	2.3E+11	2.6E+11	3.7E+11	3.5E+11	2.9E+11	5.2E+09	4.8E+10	6.2E+09	FALSE
7	3.65E+24	4.86E+24	3.15E+24	5.7E+10	1.9E+11	1.2E+11	3.5E+10	2.4E+11	1.2E+11	7.5E+10	5E+10	6.6E+09	FALSE
8	6.69E+24	6.62E+24	4.75E+24	2E+11	2.5E+11	2.4E+10	2.5E+11	1.8E+11	9.1E+10	8.1E+10	7E+10	5.5E+09	FALSE
9	3.57E+24	6.36E+24	4.14E+24	3.4E+11	1.8E+11	1.9E+11	2.7E+11	1.8E+11	2.7E+10	9.1E+10	2.4E+10	2.2E+10	TRUE
10	5.80E+23	3.80E+24	4.69E+24	2.4E+11	4.1E+10	1.1E+11	1.9E+11	3.2E+11	1.3E+11	6.5E+10	3.1E+10	6.1E+10	TRUE
11	5.21E+24	3.45E+24	9.15E+24	2.1E+11	3.3E+11	2.3E+11	1.3E+11	3.7E+11	3.6E+11	6.2E+10	2.1E+10	4.7E+10	FALSE
12	9.40E+24	8.64E+24	7.46E+24	1.5E+11	2.7E+11	1.3E+11	1.1E+11	2.5E+11	1.6E+11	6.4E+10	2.7E+10	3.1E+10	FALSE
13	5.85E+24	3.10E+24	2.52E+24	7.1E+10	1.4E+11	3.2E+11	2.7E+11	3.1E+11	1.9E+11	6.8E+09	9.4E+10	5.2E+10	TRUE
14	8.98E+24	5.77E+23	5.26E+24	3.8E+11	1.1E+11	1.3E+11	3.7E+11	3E+11	1.8E+11	5.7E+10	3.9E+10	3.2E+10	TRUE
15	8.40E+24	5.91E+24	7.20E+24	2.7E+11	7.8E+10	3.4E+11	2E+11	3.4E+11	2.1E+11	6.8E+10	7.1E+10	9.2E+10	TRUE
16	2.61E+24	5.46E+24	3.01E+24	2.9E+11	1E+11	1.3E+11	8.8E+10	3.8E+11	3.5E+11	8E+10	4.2E+10	7.8E+10	FALSE
17	2.49E+24	5.19E+24	5.40E+24	3.1E+11	1.5E+11	2E+11	1.7E+11	1.1E+11	1.1E+11	5.6E+10	1.6E+10	4.7E+10	FALSE
18	6.12E+24	9.10E+24	5.61E+24	2.4E+11	4.1E+10	3.1E+11	1.9E+11	4.7E+10	1E+11	8.7E+10	5.2E+09	6E+10	FALSE
19	2.56E+24	8.34E+23	4.41E+24	2.2E+11	3E+11	2.8E+11	1.9E+11	2.2E+10	2.1E+10	3.3E+10	5.7E+10	2.1E+10	FALSE
20	4.72E+24	4.80E+24	6.50E+24	1.5E+11	3.4E+11	3E+11	1.2E+11	1.2E+11	3.4E+11	5.5E+10	4.7E+10	1.9E+10	TRUE
21	3.74E+24	1.15E+24	8.06E+24	6.8E+10	3.5E+11	2.4E+11	9.3E+10	1.5E+11	2.7E+11	4.9E+10	8.5E+10	8.8E+09	FALSE
22	6.47E+24	2.16E+24	8.46E+23	2.9E+11	1.3E+11	2.1E+11	2.7E+10	5.1E+10	3.8E+11	3.3E+10	4.2E+10	8E+10	FALSE
23	4.58E+24	4.17E+24	9.05E+24	1.3E+11	1.3E+11	2.5E+11	3E+11	1.6E+11	1.3E+11	7.9E+10	9.2E+09	9.7E+10	TRUE
24	8.59E+24	9.22E+24	6.06E+24	1.8E+11	2.1E+10	2.4E+11	3.2E+11	1.9E+11	1E+11	1.3E+10	5.4E+10	6E+10	TRUE
25	5.07E+24	5.09E+24	6.32E+24	2.4E+11	2.7E+11	3.5E+11	1.4E+11	3.4E+11	6.7E+10	4E+10	2.9E+10	3.6E+10	FALSE
26	9.93E+24	7.88E+24	6.88E+24	4.4E+10	1.4E+11	2.7E+10	9E+10	1.4E+10	1.3E+11	2.9E+10	5E+10	4.4E+10	TRUE
27	7.42E+24	2.69E+24	1.90E+24	2E+11	3.8E+11	2.1E+11	3.8E+11	1.9E+11	2.4E+11	6.7E+10	9.5E+10	5.3E+10	TRUE
28	7.20E+24	3.09E+24	8.22E+24	2.1E+11	2.5E+11	1.9E+11	8E+10	1.7E+11	1.1E+10	8.6E+10	7.3E+09	5.1E+10	FALSE
29	2.77E+24	1.57E+23	6.74E+24	3E+10	3.6E+11	2.4E+11	3.7E+11	3.6E+11	2.1E+11	5.7E+10	8.6E+09	7.9E+10	FALSE
30	7.47E+24	3.94E+24	5.62E+24	3.8E+11	2.6E+11	1.1E+11	2E+11	6.3E+10	1.9E+11	2E+09	1.3E+10	4.2E+10	TRUE
31	9.58E+24	9.72E+24	3.33E+24	2.6E+11	2.8E+11	3E+11	8.6E+10	2.5E+11	3.5E+10	9.9E+10	8.7E+10	3.8E+10	TRUE
32	1.08E+24	7.63E+24	8.96E+24	3.2E+11	3E+11	1.5E+11	4.3E+10	2.5E+11	1.2E+11	8.4E+10	8.6E+10	6.1E+10	TRUE
33	5.94E+24	4.77E+24	4.67E+24	8.4E+10	1.7E+11	8.5E+10	3.4E+11	2.2E+11	1.4E+11	4.5E+10	7.1E+10	2.5E+09	FALSE
34	5.81E+24	9.67E+24	1.34E+24	2.7E+11	3.2E+11	2E+11	3.1E+10	3.3E+11	4.1E+10	7.3E+10	2.2E+10	5.7E+10	TRUE
35	9.42E+24	6.26E+23	3.06E+24	2.6E+11	1.4E+11	1.9E+11	5.4E+10	4.8E+10	2.3E+11	6.5E+10	3E+10	1.6E+10	FALSE
36	3.07E+24	2.15E+24	6.42E+24	3.8E+11	3.3E+11	3.2E+11	1.5E+11	2.4E+11	3E+11	8.9E+10	8.4E+10	1.6E+10	TRUE
37	8.51E+24	7.19E+24	4.56E+24	2.3E+11	3E+11	1.4E+11	1.2E+10	3.6E+11	1.8E+11	1.9E+10	5.3E+09	8.3E+10	FALSE
38	2.34E+24	9.13E+24	3.22E+24	3.1E+11	4.7E+10	5.3E+10	3.3E+11	3.3E+11	4.9E+10	1E+11	2.6E+10	4.2E+10	TRUE
39	4.27E+24	9.74E+24	5.99E+24	2.9E+10	1.9E+11	1.9E+11	1.1E+11	3.3E+11	3.6E+11	8.1E+10	4.8E+10	8.9E+10	FALSE
40	8.52E+24	4.40E+24	6.22E+24	2.3E+11	2.1E+11	6.7E+10	3.2E+11	3.7E+11	1.3E+11	5.7E+10	1.1E+10	5.8E+10	FALSE
41	6.06E+24	2.57E+24	4.91E+24	1.3E+11	1.1E+11	1.2E+11	5E+10	1.4E+11	8.4E+10	9.1E+10	7.1E+10	8.8E+10	TRUE
42	7.99E+24	5.24E+24	6.62E+24	1.1E+11	2.1E+11	2.2E+11	1.1E+11	2.7E+11	1.6E+11	7.7E+10	5.1E+09	4.3E+10	FALSE

Fig. 4 The classical collision dataset

	A	B	C	D	E	F	G	H	I	J	K	L	M	N	O	P
1	m1	m2	m3	r1_x	r1_y	r2_x	r2_y	r3_x	r3_y	r01	r02	r03	L1	L2	L3	output
2	4.04E+24	6.31E+24	9.24E+24	3.42E+10	1.84E+11	1.16E+10	1.92E+11	3.29E+11	3.66E+11	5.55E+10	2.24E+10	9.59E+10	1.96E+44	9.17E+44	9.85E+44	TRUE
3	6.98E+24	5.79E+24	3.27E+24	1.4E+11	1.54E+11	9.85E+10	1.47E+11	2.31E+11	3.22E+11	1.18E+10	3.01E+10	4.88E+10	3.81E+43	4.67E+44	6.24E+44	TRUE
4	7.28E+24	6.39E+24	4.71E+24	9.63E+10	2.78E+11	6.3E+10	2.36E+11	1.97E+11	1.51E+11	3.92E+09	2.05E+10	4.45E+10	7.14E+44	7.98E+44	8.12E+44	TRUE
5	1.92E+23	2.72E+24	5.55E+24	2.51E+11	4.57E+10	2.14E+11	3.3E+11	3.6E+11	6.05E+10	9.17E+10	2.2E+10	3.2E+10	3.75E+44	9.85E+44	7.98E+44	TRUE
6	6.29E+24	8.73E+24	6.28E+24	3.76E+11	2.95E+11	3.62E+11	1.71E+10	3.29E+11	3.76E+11	3.27E+10	1.03E+10	7.27E+10	2.37E+44	5.57E+44	9.13E+44	FALSE
7	8.13E+24	5.85E+24	3.63E+24	2.85E+11	1.78E+10	3.11E+11	2.08E+11	2.91E+11	1.5E+11	1.23E+10	9.06E+10	9.61E+10	8.97E+44	9.11E+44	7.26E+44	TRUE
8	5.68E+24	7.90E+24	6.54E+24	4.17E+10	3.74E+11	7.21E+10	1.44E+11	1.07E+10	8.77E+10	8.37E+10	6.24E+10	7.77E+10	3.99E+44	7.01E+44	7.51E+44	TRUE
9	3.74E+24	9.45E+24	6.14E+24	3.56E+11	4.25E+10	2.19E+11	3.11E+11	6.86E+10	2.16E+11	6.27E+10	7.89E+10	2.95E+10	9.49E+44	9.95E+44	4.95E+44	TRUE
10	1.82E+23	4.91E+24	5.09E+24	3.79E+11	3.09E+11	3.58E+11	2.74E+11	7.71E+10	1.89E+11	9.23E+10	7.69E+10	6.52E+10	8.61E+44	2.10E+44	7.86E+44	TRUE
11	2.87E+24	4.79E+23	3.51E+24	2.33E+11	2.18E+11	3.59E+11	7.8E+10	2.77E+11	2.52E+11	1.2E+10	7.06E+10	3.61E+10	5.72E+44	9.63E+44	2.24E+44	FALSE
12	4.65E+24	8.01E+24	4.54E+24	2.38E+11	5.81E+10	2.89E+10	3.06E+11	2.47E+10	1.68E+11	7.97E+10	6.25E+10	1.14E+10	2.61E+44	3.97E+44	8.56E+44	TRUE
13	5.32E+24	1.45E+24	4.20E+24	9.45E+10	2.17E+10	3.15E+11	1.97E+11	2.43E+11	3.61E+11	1.26E+10	7.96E+10	6.87E+10	6.03E+44	8.13E+43	4.67E+44	TRUE
14	9.66E+24	7.82E+24	1.21E+24	1.7E+11	3.77E+11	2.75E+11	2.58E+11	3.03E+11	2.9E+10	8.64E+10	5.55E+10	1.15E+10	5.80E+44	1.35E+44	4.41E+44	FALSE
15	6.11E+24	4.37E+24	9.76E+24	8.83E+10	1.18E+11	3.31E+11	6.57E+10	2.5E+11	1.53E+11	9.93E+10	8.4E+10	5.01E+10	8.26E+44	1.48E+44	4.34E+44	TRUE
16	5.73E+24	3.81E+23	2.32E+24	2.03E+10	1.44E+11	2.78E+11	3.58E+11	2.21E+11	2.85E+11	5.09E+10	3.27E+10	8.64E+10	1.11E+44	4.67E+44	1.95E+44	TRUE
17	3.27E+24	1.71E+24	5.83E+24	1.25E+11	1.72E+11	1.55E+11	9.71E+10	1.84E+11	1.94E+11	2.11E+10	7.4E+10	2.82E+09	4.90E+44	7.38E+44	2.66E+44	TRUE
18	7.38E+24	7.83E+24	4.22E+24	2.8E+11	3.14E+11	7.83E+10	1.64E+10	3.61E+11	7.37E+10	5.84E+10	9.94E+10	2.65E+10	2.62E+44	1.98E+44	8.59E+43	TRUE
19	3.21E+24	6.98E+24	7.52E+24	4.06E+10	1.38E+11	1.95E+11	3.67E+11	1.7E+11	2.07E+11	4.33E+10	2.47E+10	1.79E+10	3.02E+44	4.74E+44	1.42E+44	TRUE
20	3.65E+24	6.56E+24	4.78E+24	1.4E+11	3.37E+11	3.61E+11	1.2E+11	3.21E+11	3.65E+11	9.94E+10	3.82E+10	7.19E+10	6.79E+44	1.82E+44	5.83E+44	TRUE
21	8.11E+24	9.45E+24	4.67E+24	9.64E+10	5.51E+10	1.96E+11	1.96E+11	1.26E+11	1.63E+11	3.16E+10	4.46E+09	2.98E+10	5.67E+44	4.69E+44	8.21E+44	TRUE
22	5.04E+23	5.40E+23	5.86E+24	2.99E+11	2.97E+10	3.64E+11	2.24E+11	2.84E+11	2.22E+11	1.53E+10	8.22E+10	6.56E+10	2.08E+44	1.53E+44	9.97E+44	TRUE
23	2.75E+24	6.40E+23	8.87E+24	1.26E+11	3.36E+11	1.29E+11	1.73E+11	2.54E+11	2.78E+11	2.26E+10	6.78E+09	9.08E+10	7.69E+44	4.31E+44	1.98E+44	TRUE
24	2.41E+23	9.83E+24	5.51E+23	3.34E+11	3.8E+11	1.45E+11	1.02E+11	2.04E+11	1.25E+11	7.15E+10	8.67E+09	8.14E+10	4.21E+44	1.32E+44	2.86E+44	TRUE
25	8.02E+24	8.71E+23	5.34E+24	1.22E+11	3.6E+11	4.86E+10	3.12E+11	1.71E+11	1.69E+11	9.04E+10	8.34E+10	1.14E+10	2.02E+44	9.92E+44	9.77E+44	TRUE
26	3.30E+24	5.31E+24	1.67E+24	2E+11	2.2E+11	3.2E+11	2.76E+11	2.65E+11	1.18E+11	6.08E+10	7.19E+10	4.15E+10	9.53E+44	3.88E+44	1.09E+44	TRUE
27	8.08E+23	4.30E+24	5.99E+24	1.99E+11	3.12E+11	2.19E+11	3.84E+11	2.28E+11	9.33E+10	2.83E+10	9.51E+10	6.45E+10	5.75E+44	8.19E+44	4.54E+43	FALSE
28	6.52E+24	8.71E+24	6.18E+24	1.59E+11	3.66E+11	7.63E+10	2.66E+11	2.79E+11	3.64E+11	9.78E+09	9.3E+10	5.1E+10	6.45E+44	1.07E+42	8.74E+43	TRUE
29	5.68E+24	9.96E+24	4.50E+24	2.08E+11	3.02E+11	1.89E+11	2.18E+11	9.85E+10	2.22E+10	3.24E+10	9.71E+10	5.42E+09	1.86E+44	7.42E+44	9.52E+44	FALSE
30	9.42E+23	4.47E+24	3.71E+24	3.2E+10	2.59E+11	2.46E+11	4.47E+10	3.29E+10	2.08E+11	6.19E+10	5.03E+10	9.53E+10	6.89E+44	7.84E+44	4.20E+44	TRUE
31	5.61E+24	2.95E+24	2.60E+24	5.89E+10	1.82E+11	2.78E+11	9.5E+10	3.77E+11	1.39E+11	8.36E+10	6.46E+10	2.41E+10	9.91E+44	3.91E+43	3.42E+44	TRUE
32	1.29E+24	7.33E+24	3.21E+24	2.85E+10	3.84E+11	3.16E+11	2.83E+11	5.92E+10	2.01E+11	4.16E+10	7.63E+10	3.85E+10	2.73E+44	9.48E+44	6.34E+44	TRUE
33	5.66E+24	3.12E+24	1.44E+24	5.27E+10	3.75E+11	1.35E+11	3.71E+11	3.41E+11	1.51E+11	4.88E+10	1.78E+10	8.84E+10	4.78E+44	3.40E+44	6.95E+44	FALSE
34	7.71E+24	8.96E+24	2.78E+24	2.83E+11	2.67E+11	2.02E+11	1.26E+11	5.84E+10	8.05E+10	7.94E+10	4.68E+10	2.86E+09	9.44E+44	4.97E+44	6.86E+44	FALSE
35	5.07E+24	9.11E+24	7.42E+24	1.54E+11	2.77E+11	3.46E+11	3.8E+11	3.61E+11	1.43E+11	7.11E+10	3.44E+10	4.29E+10	8.16E+44	1.86E+44	8.51E+44	TRUE
36	6.85E+24	8.87E+24	9.52E+24	1.83E+11	1.85E+11	2.91E+11	2.18E+10	1.41E+11	3.02E+11	4.92E+10	8.6E+10	8.24E+10	7.29E+44	6.22E+44	3.31E+44	TRUE
37	1.97E+24	8.88E+24	1.93E+24	1.5E+11	2.33E+11	2.21E+11	1.82E+11	4.6E+10	1.89E+11	2.61E+10	1.99E+10	6.42E+10	3.63E+44	8.50E+44	7.61E+44	FALSE
38	1.96E+24	7.86E+24	1.58E+24	2.37E+11	2.27E+11	1.68E+11	3.45E+10	4.33E+10	2.97E+11	6.62E+10	2.3E+10	8.45E+10	4.35E+44	1.69E+44	3.33E+44	FALSE

Fig. 5 The GR collision dataset

2.5 *Machine Learning Model*

Polynomial Regression and Model Representation. In order to capture nonlinear relationships between the input features and the output variable, a polynomial regression model was employed. Polynomial regression extends the concept of linear regression by introducing polynomial terms as additional features. The model allows for a more flexible representation of the data, enabling the exploration of complex relationships. The polynomial regression model transforms the input features by adding polynomial terms up to a specified degree. The degree represents the highest power of the input features in the polynomial equation. The general equation for a polynomial regression model with a single input feature can be written as follows:

$$y = c0 + c1 * x + c2 * x^2 + ... + cn * x^n \tag{29}$$

Here, y represents the predicted output, x is the input feature, c0, c1, ..., cn are the coefficients of the polynomial terms, and n is the degree of the polynomial [19].

Choosing the Polynomial Degree. The selection of the polynomial degree is a critical aspect in polynomial regression as it plays a vital role in avoiding overfitting or underfitting the data. The degree of the polynomial determines the model's complexity and its ability to capture the underlying relationships accurately. Higher-degree polynomials provide increased flexibility in capturing complex relationships between the input features and the output variable. They can effectively represent curved or nonlinear patterns in the data. However, higher-degree polynomials are more prone to overfitting the training data. Overfitting occurs when the model excessively fits the noise or peculiarities in the training data, leading to poor performance on new, unseen data. Conversely, lower-degree polynomials may underfit the data by oversimplifying the relationship between the features and the output. Underfitting occurs when the model is too simple to capture the underlying patterns, resulting in inadequate predictive performance. To strike an appropriate balance between model complexity and generalization to unseen data, a careful evaluation of different polynomial degrees is necessary. This evaluation involves comparing the model's performance with varying degrees of polynomials. Performance metrics such as mean squared error, cross-validation, or information criteria can help assess how well the model fits the training data and its ability to generalize to new data. In our study, we have chosen a second-degree polynomial as it offers the best accuracy. The Fig. 6a, b present the training and testing Accuracy verse the degree of polynomial features for the classical model and GR model, and the accuracy for each degree was relativity close [18, 19].

Prediction and Evaluation. Upon training the polynomial regression model, it becomes applicable for making predictions on new data by leveraging the learned coefficients within the polynomial equation. The performance evaluation of the GR model and the Classical model for the collision task yielded insightful findings. The metrics used to assess the models' effectiveness included training accuracy, testing accuracy, precision, recall, and F1 score. The confusion matrices provided further details on the models' predictive capabilities. The GR model achieved a training accuracy of 0.88 and a testing accuracy of 0.89, while the Classical model exhibited slightly higher performance with a training accuracy of 0.92 and a testing accuracy

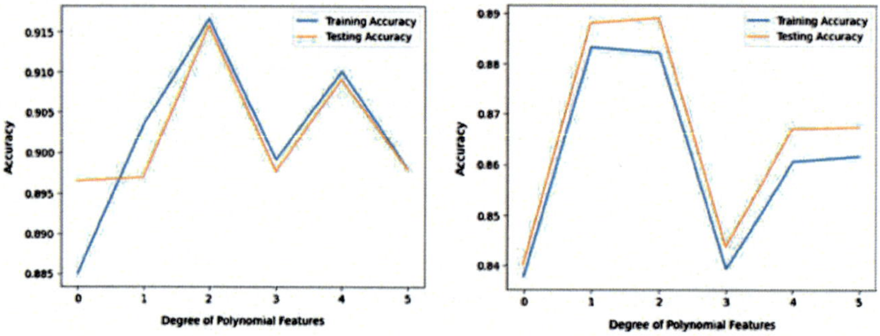

(a) Classical model (b) GR model

Fig. 6 Training and testing accuracy versus degree of polynomial features

Table 1 Performance comparison of GR model and classical model

Model	Training accuracy	Testing accuracy	Precision	Recall	F1 score
GR model	0.88	0.89	0.7016	0.5313	0.6047
Classical model	0.92	0.93	0.7251	0.5210	0.6064

Table 2 Confusion matrix for GR model and classical model

	GR model			Classical model		
	True class	False class	Total	True class	False class	Total
True positive	348	–	348	248	–	248
True negative	3297	–	3297	3730	–	3730
False positive	–	148	148	–	94	94
False negative	–	307	307	–	228	228
Total	3645	455	4100	3978	322	4300

of 0.93. Both models demonstrated comparable precision and F1 scores, with the Classical model marginally surpassing the GR model in these aspects. Regarding recall, the GR model displayed a value of 0.5313, indicating its capacity to correctly identify positive instances from the dataset. In contrast, the Classical model demonstrated a slightly lower recall of 0.5210, suggesting a relatively lower ability to capture true positive instances as shown in Table 1. The confusion matrices provided valuable insights into the models' predictive strengths and weaknesses. The GR model correctly predicted 3297 instances of the negative class (true negatives) and 348 instances of the positive class (true positives). However, it also misclassified 148 instances of the negative class (false positives) and 307 instances of the positive class (false negatives). On the other hand, the Classical model achieved a higher true negative count of 3730, indicating its proficiency in identifying negative instances correctly. However, it exhibited a lower true positive count of 248 and misclassified 94 instances of the negative class and 228 instances of the positive class as shown in Table 2.

Mathematical Formulation for Collision Detection in a Three-Body System. In our pursuit of investigating collision conditions within a three-body system, a robust polynomial regression model has been diligently trained. The model now offers a formal mathematical equation represented as $Ax + b = y$, where each constituent element holds pivotal significance in the collision detection process.

Coefficient Matrix (A) and Intercept (b). Through extensive training of the machine learning model, we have derived the coefficient matrix A and the intercept b, which facilitate the collision prediction process. The coefficient matrix A of size 91 (as

Fig. 7 Coefficient Matrix (A) and Intercept (b) for the models

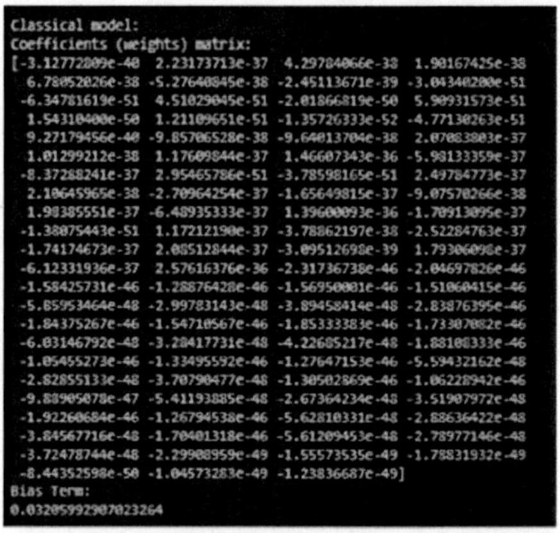

(a) Classical model

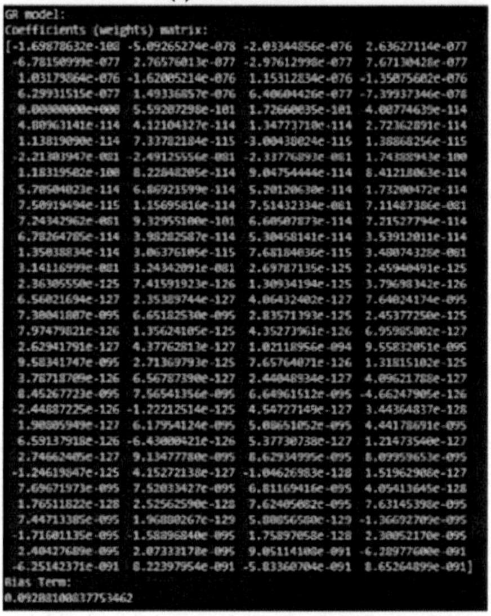

(b) GR model

shown in Fig. 7a) captures intricate relationships between the input features (x) and the predicted collision outcome (y). These weighted coefficients within A play a crucial role in quantifying the influence of each input feature, thereby providing valuable insights for accurate predictions. The intercept b complements the predictive equation by enabling necessary adjustments to the output (y) or decision boundary. This adaptability ensures predictions remain feasible even when all input features assume zero values.

Correction for Collision Condition Using General Relativity (GR). Recognizing the potential limitations of considering classical conditions alone, we have made enhancements to the model by accounting for the influence of General Relativity (GR). Consequently, additional input parameters have been incorporated to effectively capture the subtle effects of GR on the three-body system. Specifically, the augmented model now includes input columns for the angular momentum of each body (L_1, L_2, L_3). These angular momentum values play a critical role in characterizing the influence of General Relativity (GR) on the dynamics of the system, thereby providing a more comprehensive understanding of potential collision scenarios. As a result, the coefficient matrix A has expanded to a size of 136 elements as shown in Fig. 7b, reflecting the increased complexity and coverage of input-feature relationships. The revised matrix elements encapsulate the intricate interplay between classical conditions and the effects of General Relativity (GR), further enhancing the model's predictive capabilities.

Collision Outcome (y). The outcome variable y remains fundamental to the polynomial regression model, serving as the predicted probability of collision occurrence within the three-body system. As established earlier, when y exceeds or equals the threshold of 0.5, it indicates an anticipated collision event.

3 Discussion

Upon training and evaluating the polynomial regression model for collision detection in a three-body system, we obtained insightful findings concerning the performance and effectiveness of both the General Relativity (GR) and Classical models. The Classical model exhibited slightly higher performance than the GR model, achieving a training accuracy of 0.92 and a testing accuracy of 0.93, compared to the GR model's training accuracy of 0.88 and testing accuracy of 0.89. These results indicate the Classical model's superior ability to generalize to unseen data. Both models demonstrated comparable precision and F1 scores, indicating a balanced trade-off between the number of true positives and false positives. The Classical model marginally outperformed the GR model in these aspects. However, when considering recall, the GR model displayed a higher value (0.5313) compared to the Classical model (0.5210), signifying the GR model's capacity to correctly identify positive instances from the dataset. The confusion matrices provided valuable insights into the models' predictive strengths and weaknesses. The GR model correctly predicted 348 instances of the positive class (true positives) but also misclassified 307 instances of the positive class (false negatives). Additionally, it misclassified 148 instances of the negative class (false positives). On the other hand, the Classical model exhibited a higher true negative count of 3730, indicating its proficiency in identifying negative instances correctly. However, it misclassified 94 instances of the negative class and 228 instances of the positive class.

4 Limitations

Despite the promising performance of both models, there are several limitations that warrant consideration.

Linear Assumption of Polynomial Regression Model. The polynomial regression model assumes a linear relationship between the input features, which may not fully capture the complexity of interactions in the three-body system [20]. More sophisticated modeling techniques, such as neural networks or non-linear regression, could be explored to address this limitation.

Incomplete Representation of Gravitational Interactions. Although the incorporation of General Relativity (GR) effects through angular momentum enhances the model's predictive capabilities [21, 22], it may still overlook certain subtle gravitational interactions in the three-body system. More comprehensive models that consider additional gravitational factors could potentially yield further improvements.

Data Limitations. The performance of the model heavily relies on the quality and diversity of the data used for training and testing. The dataset used in this study might be limited in its representation of the vast range of possible scenarios in three-body systems, leading to potential biases and generalization issues.

5 Conclusions

In conclusion, the trained polynomial regression model, considering both Classical conditions and General Relativity (GR) effects, showcases promising predictive capabilities for collision detection in a three-body system. While the Classical model exhibits higher accuracy, the GR model demonstrates better recall in identifying positive instances due to the inclusion of angular momentum. Nonetheless, both models have their limitations, and further research is warranted to explore more sophisticated models and diverse datasets to enhance collision detection accuracy in complex celestial systems. By addressing these limitations, we can move closer to a more comprehensive and accurate understanding of collision dynamics in celestial bodies, which is crucial for space exploration and astronomical studies.

Acknowledgements We thank Innopolis University for generously funding this endeavor.

References

1. Carroll, S.M.: Spacetime and Geometry: An Introduction to General Relativity. Cambridge University Press (2019). https://doi.org/10.1017/9781108292607
2. Tabak, J.: Geometry: The Language of Space and Form. Infobase Publishing (2014)
3. Heiberg, J.L., Fitzpatrick, R.: Euclid's Elements of Geometry: Edited, and Provided with a Modern English Translation, by Richard Fitzpatrick. Independently Published (2008)
4. Braver, S.P.: Lobachevski Illuminated: Content, Methods, and Context of the "Theory of Parallels". University of Montana (2007)
5. Tino, G.M., Cacciapuoti, L., Capozziello, S., Lambiase, G., Sorrentino, F.: Precision gravity tests and the Einstein equivalence principle. Prog. Part. Nucl. Phys. **112**, 103772 (2020). https://doi.org/10.1016/j.ppnp.2020.103772
6. Nobili, A.M., Lucchesi, D.M., Crosta, M.T., Shao, M., Turyshev, S.G., Peron, R., Catastini, G., Anselmi, A., Zavattini, G.: On the universality of free fall, the equivalence principle, and the gravitational redshift. Am. J. Phys. **81**(7), 527–536 (2013). https://doi.org/10.1119/1.4793595
7. Ferraro, R.: Einstein's Space-Time: An Introduction to Special and General Relativity. Springer (2007)
8. Riley, K.F., Hobson, M.P., Bence, S.J.: Mathematical Methods for Physics and Engineering. American Association of Physics Teachers (1999)
9. Misner, C.W., Thorne, K.S., Wheeler, J.A.: Gravitation. Macmillan (1973)
10. Mitra, A., Bhattacharyya, S., Bhatt, N.: ΛcDM cosmology through the lens of Einstein's static universe, the mother of Λ. Int. J. Modern Phys. D **22**(03), 1350012 (2013). https://doi.org/10.1142/S0218271813500122
11. Oriti, D.: Approaches to Quantum Gravity: Toward a New Understanding of Space, Time and Matter. Cambridge University Press (2009)
12. Einstein, A., Davis, F.A.: The Principle of Relativity. Courier Corporation (2013)
13. Morin, D.: Introduction to Classical Mechanics: With Problems and Solutions. Cambridge University Press (2008)
14. Barrow-Green, J.: Poincaré and the Three Body Problem. American Mathematical Society (1997)
15. Einstein, A.: Relativity: The Special and the General Theory-100th Anniversary Edition. Princeton University Press (2019)
16. Cheng, T.P.: A College Course on Relativity and Cosmology. Oxford University Press (2015)
17. Bressert, E.: SciPy and NumPy. O'Reilly, Germany (2012)
18. Müller, A.C., Guido, S.: Introduction to Machine Learning with Python: A Guide for Data Scientists. O'Reilly Media, United States (2016)
19. Hackeling, G.: Mastering Machine Learning with Scikit-learn. Packt Publishing, United Kingdom (2017)
20. Bahrami, M.: Modeling of Physical Systems Through Bond Graphs. St. Petersburg Polytechnical University, St. Petersburg (2018)
21. Bahrami, M.: Mechanical challenges of inspection robot moving along the electrical line: effect of flexural rigidity. Advances in Mechanical Engineering: Selected Contributions from the Conference "Modern Engineering: Science and Education", Saint Petersburg, Russia, June 2020, pp. 30–37 (2021)
22. Bahrami, M., Ramezani, A., Osquie, K.: Modeling and simulation of non-contact atomic force microscope. Eng. Syst. Des. Anal. **49194**, 565–569 (2010)

A Novel Way of Calculating Scattering Integrals

Alfredo Takashi Suzuki and Timothy Suzuki

Abstract The technique coined as NDIM—Negative Dimensional Integration Method by their discoverers, relies on a three-pronged basis: Gaussian integration, series expansion and analytic continuation. The technique has been successfully applied to the calculation of covariant and non covariant Feynman integrals in a generic dimensional regularization space, i.e., D-dimensional space-time for D including the negative domain values. Since the dimensionality is general, we can use specifically for one-dimensional integrals. In this work we show how this technique can be applied to tackle certain improper integrals and give an example of a particular improper integral that appears in quantum mechanical scattering process. Traditionally, improper integrals are ascribed certain values through the limiting approach or as is known, by the Cauchy principal value via residues concept technique. Here we use the NDIM approach to do the calculations and show it works fine for the improper integrals. This novel approach we believe is more straightforward and does not require to handle poles, residues, or difficult closed contours as in the traditional approach.

Keywords Scattering theory · Integration techniques

1 Introduction

NDIM (Negative Dimensional Integration Method) [1] was born within the context of perturbative quantum field theories where Feynman diagrams that give a pictorial view of physical processes are calculated. When the perturbative series entails loop calculations, these often lead to diverging Feynman integrals, for which a regularization procedure is called for. Among several regularization procedures, dimensional

A. T. Suzuki (✉)
Department of Physics, La Sierra University, 4500 Riverwalk Pkw., Riverside, CA 92505, USA
e-mail: suzuki@ift.unesp.br

T. Suzuki
Department of Physics and Astronomy, Michigan State University, 567 Wilson Rd., East Lansing, MI 48824, USA

D. Vlachos (ed.), *Mathematical Modeling in Physical Sciences*, Springer Proceedings in Mathematics & Statistics 446, https://doi.org/10.1007/978-3-031-52965-8_36

regularization [2] is a popular choice, where a four-dimensional integration measure is analytically extended to a D-dimensional one. In essence, what NDIM does is to allow this analytic continuation to span even into the negative dimensionality region. With the help of a D-dimensional Gaussian integration and series expansion, an integrand with poles are transformed into polynomials that are integrated in negative dimensions. After the polynomial integration in negative dimensions is carried out, analytic continue the result back into positive dimensions. [3].

This D-dimensional NDIM procedure can be recast into a one-dimensional integration measure without difficulty [4, 5]. In our previous works, we considered integrands with pure imaginary poles, so the calculation entails no closed circuit with real poles standing on the integration path. In the present work we consider improper integrals with real poles, and show that even for this case NDIM can be applied successfully.

2 Improper Integrals

Let us consider the following improper integral:

$$I = \int_{-\infty}^{+\infty} \frac{dx}{\left(x^2 - a^2\right)} = \int_{-\infty}^{+\infty} \frac{dx}{(x-a)(x+a)}; \quad a \in \mathbb{R}. \tag{1}$$

Since the integrand is not well-defined (diverges) at $x = \pm a$, we may try to ascribe a value to this integral by the limiting procedure as follows:

$$I = \frac{1}{2a}\left\{ \lim_{\substack{\epsilon\to 0^+;\\ R\to\infty}} \left[\int_{-R}^{a-\epsilon} \frac{dx}{(x-a)} + \int_{a+\epsilon}^{+R} \frac{dx}{(x-a)} \right] + \lim_{\substack{\epsilon'\to 0^+;\\ R\to\infty}} \left[\int_{-R}^{-a-\epsilon'} \frac{dx}{(x+a)} + \int_{-a+\epsilon'}^{+R} \frac{dx}{(x+a)} \right] \right\}. \tag{2}$$

We can ascribe a principal value (PV) to it in the sense of Cauchy by using the same infinitesimal, that is, $\epsilon = \epsilon' \to 0^+$, in the limiting analysis

$$I_{\text{PV}} = \frac{1}{2a}\left\{ \lim_{\substack{\epsilon\to 0^+;\\ R\to\infty}} \left[\int_{-R}^{a-\epsilon} \frac{dx}{(x-a)} + \int_{a+\epsilon}^{+R} \frac{dx}{(x-a)} \right] + \lim_{\substack{\epsilon\to 0^+;\\ R\to\infty}} \left[\int_{-R}^{-a-\epsilon} \frac{dx}{(x+a)} + \int_{-a+\epsilon}^{+R} \frac{dx}{(x+a)} \right] \right\} = 0. \tag{3}$$

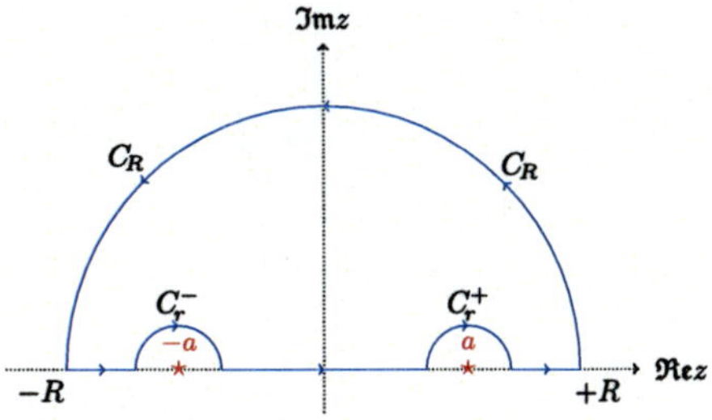

Fig. 1 Contour excluding both poles

We could also use the closed contour integration analytically continuing the integration variable into complex variables, $x \to z$,

$$I_{\rm PV} = \oint \frac{dz}{\left(z^2 - a^2\right)} = \oint \frac{dz}{(z-a)\,(z+a)}; \quad z \in \mathbb{C}; \;\; a \in \mathbb{R}. \tag{4}$$

Here, because we have two poles on the real axis, $x = \pm a$, we need to distort the contour around them by a semi-circle of infinitesimal radius ($r \to 0^+$) in order to avoid the poles. We can do this in several manners, closing the contour either on the upper or lower hemispheres:

1. Excluding both poles;
2. Including both poles;
3. Excluding the pole $x = -a$ and including the pole $x = a$;
4. Including the pole $x = -a$ and excluding the pole $x = a$.

All theses choices are totally equivalent, so, if we choose, for instance, the following contour, excluding both poles from the closed upper hemisphere circuit (Fig. 1).

Cauchy residue theorem yields this closed loop integration as, with $f(z) \equiv (z^2 - a^2)^{-1} = [(z-a)(z+a)]^{-1}$

$$I_{\rm PV} = \oint \frac{dz}{\left(z^2 - a^2\right)} = 2i\pi \sum \text{Residues}\, f(z) = 0, \tag{5}$$

since the chosen closed contour does not contain any poles.

The contour integration, on the other hand, can be broken down into

$$\begin{aligned} I_{\rm PV} &= \oint f(z)dz \\ &= \int_{C_R} f(z)dz + \int_{-R}^{-a-r} f(z)dz + \int_{C_r^-} f(z)dz \\ &+ \int_{-a+r}^{a-r} f(z)dz + \int_{C_r^+} f(z)dz + \int_{a+r}^{+R} f(z)dz \end{aligned} \tag{6}$$

Let us look at in detail these integrals. First, let us consider

$$C_R \equiv \int_{C_R} f(z)dz = \int_{C_R} \frac{dz}{(z-a)(z+a)}. \tag{7}$$

For this integral, along the upper hemisphere counterclockwise semi-circle, we have

$$\begin{aligned} z &= Re^{i\theta}, \quad 0 \le \theta \le \pi; \\ dz &= iRe^{i\theta}d\theta. \end{aligned} \tag{8}$$

Therefore,

$$C_R = \int_0^{\pi} \frac{iRe^{i\theta}d\theta}{(Re^{i\theta}-a)(Re^{i\theta}+a)}. \tag{9}$$

For the relevant limit $R \to \infty$, the integrand's leading term is

$$\lim_{R\to\infty} C_R \approx \frac{i}{R}\int_0^{\pi} e^{-i\theta}d\theta \to 0. \tag{10}$$

Next, let us examine

$$C_r^- \equiv \int_{C_r^-} f(z)dz = \int_{C_r^-} \frac{dz}{(z-a)(z+a)}. \tag{11}$$

For this integral, along the upper hemisphere clockwise semi-circle around $x = -a$, we have

$$\begin{aligned} z &= -a + re^{i\varphi}, \quad \pi \le \varphi \le 0; \\ dz &= ire^{i\varphi}d\varphi. \end{aligned} \tag{12}$$

Therefore,

$$\begin{aligned} C_r^- &= \int_{\pi}^{0} \frac{ire^{i\varphi}d\varphi}{(-2a+re^{i\varphi})(re^{i\varphi})} \\ &= i\int_{\pi}^{0} \frac{d\varphi}{(-2a+re^{i\varphi})}. \end{aligned} \tag{13}$$

Here, for the relevant limit $r \to 0^+$, the integrand's leading term is

$$\lim_{r\to 0^+} C_r^- \approx -\frac{i}{2a}\int_{\pi}^{0} d\varphi \to \frac{i\pi}{2a}. \tag{14}$$

Now, for the

$$C_r^+ \equiv \int_{C_r^+} f(z)dz = \int_{C_r^+} \frac{dz}{(z-a)(z+a)}. \tag{15}$$

For this integral, along the upper hemisphere clockwise semi-circle around $x = a$, we have

$$\begin{aligned} z &= a + re^{i\varphi}, \quad \pi \leq \varphi \leq 0; \\ dz &= ire^{i\varphi}d\varphi. \end{aligned} \tag{16}$$

Therefore,

$$\begin{aligned} C_r^+ &= \int_\pi^0 \frac{ire^{i\varphi}d\varphi}{(re^{i\varphi})(2a + re^{i\varphi})} \\ &= i\int_\pi^0 \frac{d\varphi}{(2a + re^{i\varphi})}. \end{aligned} \tag{17}$$

Here, for the relevant limit $r \to 0^+$, the integrand's leading term is

$$\lim_{r\to 0^+} C_r^+ \approx \frac{i}{2a}\int_\pi^0 d\varphi \to \frac{-i\pi}{2a}. \tag{18}$$

Finally, plugging the results for C_R, C_r^- and C_r^+ into Eq. (6) with the relevant limits taken, $R \to \infty$; $r \to 0^+$, we have

$$\begin{aligned} I_{\text{PV}} &= \oint f(z)dz \\ &= 0 + \int_{-\infty}^{-a} f(x)dx + \frac{i\pi}{2a} + \int_{-a}^{a} f(x)dx - \frac{i\pi}{2a} + \int_{a}^{+\infty} f(x)dx \\ &= \int_{-\infty}^{+\infty} f(x)dx \equiv \int_{-\infty}^{+\infty} \frac{dx}{(x^2 - a^2)}. \end{aligned} \tag{19}$$

Comparing with Eq. (5), we get, in accordance with Eq. (3)

$$\int_{-\infty}^{+\infty} \frac{dx}{(x^2 - a^2)} = 0. \tag{20}$$

Since the integral in Eq. (1) is an improper one, we can in principle ascribe other values to it, other than the PV value just calculated. It is possible to consider the original integral, Eq. (1), with $a \in \mathbb{C}$, by letting the real pole be dislocated by an infinitesimal imaginary part, i.e., either $a_+ = a + i\varepsilon$ or $a_- = a - i\varepsilon$.

Fig. 2 Contour including right pole

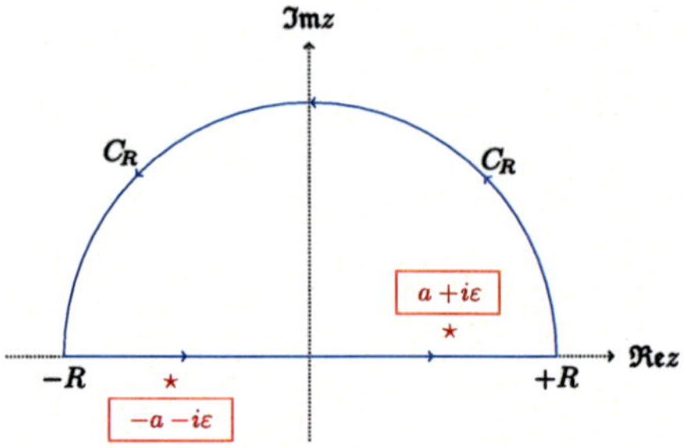

2.1 Improper Integral for a_+ shift

Let us consider now the following integral,

$$I_+ = \lim_{\varepsilon\to 0^+} I_+(\varepsilon), \tag{21}$$

where

$$\begin{aligned} I_+(\varepsilon) &\equiv \int_{-\infty}^{+\infty} \frac{dz}{\left(z^2 - a_+^2\right)}; \quad a_+ \in \mathbb{C} \\ &= \int_{-\infty}^{+\infty} \frac{dz}{(z-a-i\varepsilon)(z+a+i\varepsilon)}, \quad a \in \mathbb{R}. \end{aligned} \tag{22}$$

For this, we consider the following closed circuit (Fig. 2).
Then, we have

$$\begin{aligned} I_+(\varepsilon) &\equiv \oint \frac{dz}{\left(z^2 - a_+^2\right)} = \oint \frac{dz}{(z-a-i\varepsilon)(z+a+i\varepsilon)} \\ &= 2i\pi \sum \text{Residues}\, f_+(z)\,, \quad f_+(z) \equiv \frac{1}{(z-a-i\varepsilon)(z+a+i\varepsilon)} \\ &= 2i\pi \left\{ \lim_{z\to a+i\varepsilon} (z-a-i\varepsilon) f_+(z) \right\} \\ &= \frac{i\pi}{a+i\varepsilon}. \end{aligned} \tag{23}$$

So, remembering that the large counterclockwise contour does not contribute, i.e., $C_R \to 0$ for $R \to \infty$, we have

$$I_+ = \lim_{\varepsilon\to 0^+} \frac{i\pi}{a+i\varepsilon} = \frac{i\pi}{a} \tag{24}$$

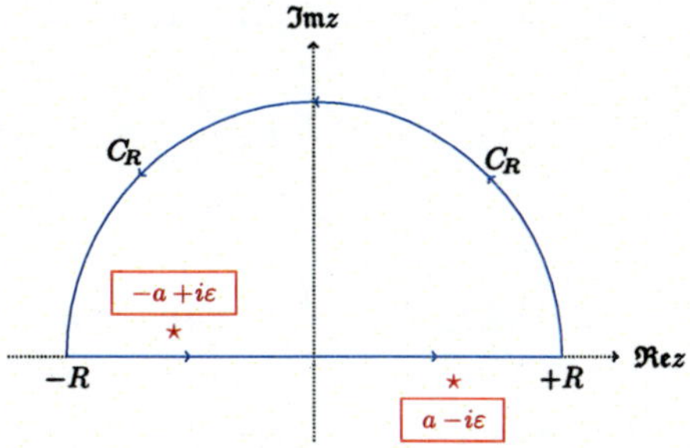

Fig. 3 Contour including left pole

2.2 *Improper Integral for a_- shift*

For this integral, we have

$$I_- = \lim_{\varepsilon \to 0^+} I_-(\varepsilon), \tag{25}$$

where

$$\begin{aligned} I_-(\varepsilon) &\equiv \int_{-\infty}^{+\infty} \frac{dz}{\left(z^2 - a_-^2\right)}; \quad a_- \in \mathbb{C} \\ &= \int_{-\infty}^{+\infty} \frac{dz}{(z - a + i\varepsilon)(z + a - i\varepsilon)}, \quad a \in \mathbb{R}. \end{aligned} \tag{26}$$

The same previous closed circuit now encloses the pole in the second quadrant (Fig. 3).

Then, we have

$$\begin{aligned} I_-(\varepsilon) &\equiv \oint \frac{dz}{\left(z^2 - a_-^2\right)} = \oint \frac{dz}{(z - a + i\varepsilon)(z + a - i\varepsilon)} \\ &= 2i\pi \sum \text{Residues} f_-(z), \quad f_-(z) \equiv \frac{1}{(z - a + i\varepsilon)(z + a - i\varepsilon)} \\ &= 2i\pi \left\{ \lim_{z \to -a + i\varepsilon} (z + a - i\varepsilon) f_-(z) \right\} \\ &= \frac{i\pi}{-a + i\varepsilon}. \end{aligned} \tag{27}$$

So, finally,

$$I_- = \lim_{\varepsilon \to 0^+} \frac{i\pi}{-a + i\varepsilon} = -\frac{i\pi}{a}. \tag{28}$$

We note that, as the PV is also defined by

$$I_{\mathrm{PV}} = \frac{1}{2} \lim_{\varepsilon \to 0^+} \{I_+(\varepsilon) + I_-(\varepsilon)\} = 0. \tag{29}$$

3 NDIM and the Improper Integrals

NDIM has already been applied for some definite integrals [4, 5]. Here we are going to employ the NDIM to evaluate Eq. (1). To implement the NDIM technique for the sought integral, let us introduce the generating functional Gaussian integral that is pertinent to our calculation:

$$G_a = \int_{-\infty}^{+\infty} dx\; e^{-\lambda(x^2-a^2)}. \tag{30}$$

The Gaussian integration can be performed without difficulty, yielding

$$\begin{aligned} G_a &= e^{\lambda a^2} \int_{-\infty}^{+\infty} e^{-\lambda x^2} \\ &= e^{\lambda a^2} \sqrt{\frac{\pi}{\lambda}}. \end{aligned} \tag{31}$$

Next, expanding the exponential function in the result above in power series we get

$$G_a = \pi^{1/2} \sum_{n=0}^{\infty} \frac{a^{2n}}{n!} \lambda^{n-1/2}. \tag{32}$$

On the other hand, expanding in power series the original Gaussian integral (30), we have

$$G_a = \int_{-\infty}^{+\infty} dx \sum_{k=0}^{\infty} (-1)^k \frac{\lambda^k}{k!} (x^2 - a^2)^k. \tag{33}$$

As the above integral has a polynomial integrand equivalent to the desired integral with poles, we have the negative dimensional integration characterized by this procedure, i.e., introducing the notation

$$I_{\text{NDIM}}(k) = \int_{-\infty}^{+\infty} \hat{d}x \left(x^2 - a^2\right)^k, \tag{34}$$

with $\hat{d}x$ to remind that we are in the negative dimension measure continuation. Then, the later (33) can be written as

$$G_a = \sum_{k=0}^{\infty} (-1)^k \frac{\lambda^k}{k!} I_{\text{NDIM}}(k). \tag{35}$$

Comparing the two series expansion for G_a, term by term, we have that $n = k + 1/2$ and $I_{\text{NDIM}}(k)$ can be obtained as

$$\begin{aligned} I_{\text{NDIM}}(k) &= (-1)^{-k} k! \pi^{1/2} \frac{a^{2k+1}}{(k+1/2)!} \\ &= (-1)^{-k} \pi^{1/2} a^{2k+1} \frac{\Gamma(1+k)}{\Gamma(1+k+1/2)} \\ &= (-1)^{-k} \pi^{1/2} a^{2k+1} \frac{1}{(1+k)_{1/2}}, \end{aligned} \tag{36}$$

where in the last line above we have introduced the Pochhammer's symbol

$$(\alpha)_\beta \equiv \frac{\Gamma(\alpha+\beta)}{\Gamma(\alpha)}. \tag{37}$$

Since we want (34) analytically continued (AC) to allow for negative values of k (in positive dimensional measure) we proceed by using in (36) the following Pochhammer's identity

$$(1-\alpha)_\beta = (-1)^\beta \frac{1}{(\alpha)_{-\beta}}. \tag{38}$$

Then,

$$I_{\text{NDIM}}^{\text{AC}}(k) = (-1)^{-k-1/2} \pi^{1/2} a^{2k+1} (-k)_{-1/2}, \tag{39}$$

In (34), we want $k = -1$ to reproduce the original integral in positive dimensions, so we are interested in

$$\begin{aligned} I_{\text{NDIM}}^{\text{AC}}(k=-1) &= (-1)^{1/2} \pi^{1/2} a^{-1} (1)_{-1/2} \\ &= \frac{i\pi}{a} = I_+. \end{aligned} \tag{40}$$

Since the original integral is, of course invariant under the symmetry $a \to -a$, it follows that we also have

$$I_{\text{NDIM}}^{\text{AC}}(k=-1) = (-1)^{1/2}\pi^{1/2}(-a)^{-1}(1)_{-1/2}$$
$$= -\frac{i\pi}{a} = I_{-}. \tag{41}$$

And once again, we can calculate the PV value,

$$I_{\text{PV}} = \frac{1}{2}\lim_{\varepsilon\to 0^+}\{I_+(\varepsilon) + I_-(\varepsilon)\} = 0. \tag{42}$$

4 NDIM in Quantum Mechanics Scattering Calculation

In quantum mechanics scattering problems, there is an improper integral of the following form to be calculated:

$$S(\sigma) = \int_{-\infty}^{+\infty} dx \frac{x \sin x}{(x^2 - \sigma^2)}, \quad \sigma \in \mathbb{R} \tag{43}$$

In order to perform this integration using NDIM, we first need to express it in terms of integrands that would be fitting for applying the technique. So the 'road preparation' for it is done by considering the integral

$$E(a,\sigma) = \int_{-\infty}^{+\infty} dx \frac{e^{iax}}{(x^2 - \sigma^2)}, \quad a \in \mathbb{R}. \tag{44}$$
$$= \sum_{m=0}^{\infty} \frac{(ia)^m}{m!} I^m(\sigma),$$

where we have defined

$$I^m(\sigma) = \int_{-\infty}^{+\infty} dx \frac{x^m}{(x^2 - \sigma^2)}. \tag{45}$$

We introduce then the relevant generating functional Gaussian integration as follows

$$G(\alpha, \beta) = \int_{-\infty}^{+\infty} dx e^{\alpha x - \beta(x^2 - \sigma^2)} \tag{46}$$

$$= e^{\frac{\alpha^2}{4\beta} + \beta\sigma^2} \sqrt{\frac{\pi}{\beta}}$$

$$= \sqrt{\frac{\pi}{\beta}} \sum_{k,l=0}^{\infty} \frac{\left(\beta\sigma^2\right)^k}{k!} \frac{\alpha^{2l}}{4^l \beta^l l!}. \tag{47}$$

On the other hand, series expansion of integrand in (46) gives

$$G(\alpha, \beta) = \sum_{r,s=0}^{\infty} (-1)^s \frac{\alpha^r \beta^s}{r!s!} \int_{-\infty}^{+\infty} dx\, x^r \left(x^2 - \sigma^2\right)^s. \tag{48}$$

Comparing (47) and (48) we have that

$$\sum_{r,s=0}^{\infty} (-1)^s \frac{\alpha^r \beta^s}{r!s!} \int_{-\infty}^{+\infty} dx\, x^r \left(x^2 - \sigma^2\right)^s = \pi^{1/2} \sum_{k,l=0}^{\infty} \frac{\sigma^{2k}}{4^l} \frac{\beta^{k-l-1/2}}{k!} \frac{\alpha^{2l}}{l!}. \tag{49}$$

The term by term equality entails

$$\begin{aligned} r &= 2l \\ s &= k - l - 1/2. \end{aligned} \tag{50}$$

Then, we have that $k = s + r/2 + 1/2$ and $l = r/2$, leading to

$$\int_{-\infty}^{+\infty} dx\, x^r \left(x^2 - \sigma^2\right)^s = (-1)^{-s} r!s!\pi^{1/2} \frac{\beta^{2s+r+1}}{4^{r/2}} \frac{1}{(s + r/2 + 1/2)!} \frac{1}{(r/2)!}$$

$$= (-1)^{-s} \pi^{1/2} \frac{\sigma^{r+2s+1}}{4^{r/2}} \frac{(1 + r/2)_{r/2}}{(1 + s)_{r/2+1/2}}. \tag{51}$$

Note that the integral

$$I_{\text{NDIM}}(r, s) \equiv \int_{-\infty}^{+\infty} dx\, x^r \left(x^2 - \sigma^2\right)^s, \tag{52}$$

will be equivalent to (45) when $r = m$ and $s = -1$. Note also that henceforth we are going to be omitting the reminder sign over $\hat{d}x$ for the NDIM integration measure. Since we need s to assume negative values in $I_{\text{NDIM}}(r, s)$, we analytic continue the Pochhammer's symbol on the righthandside of (51) that contains s, i.e.,

$$\frac{1}{(1+s)_{r/2+1/2}} \xrightarrow{AC} \frac{(-s)_{-r/2-1/2}}{(-1)^{r/2+1/2}}. \tag{53}$$

Then,

$$I_{\text{NDIM}}^{\text{AC}}(r,s) = (-1)^{-s-r/2-1/2}\pi^{1/2}\frac{\sigma^{r+2s+1}}{4^{r/2}}(1+r/2)_{r/2}(-s)_{-r/2-1/2}. \tag{54}$$

Comparing (45) with (52) we see that we need to have $r = m$ and $s = -1$ in (52) to get the relevant result for (45). So

$$\begin{aligned} I_{\text{NDIM}}^{\text{AC}}(m,-1) &= (-1)^{1/2-m/2}\pi^{1/2}\frac{\sigma^{m-1}}{4^{m/2}}(1+m/2)_{m/2}(1)_{-m/2-1/2} \\ &= i\pi^{1/2}\frac{\sigma^m}{\sigma}\frac{1}{(-4)^{m/2}}\frac{\Gamma(1+m)\Gamma(1/2-m/2)}{\Gamma(1+m/2)}. \end{aligned} \tag{55}$$

Now, using the gamma function relation

$$\Gamma(1/2-m/2) = \pi^{1/2}(-4)^{m/2}\frac{\Gamma(1+m/2)}{\Gamma(1+m)}. \tag{56}$$

Then

$$I_{\text{NDIM}}^{\text{AC}}(m,-1) \equiv I^m(\sigma) = \frac{i\pi\sigma^m}{\sigma}. \tag{57}$$

Substituting this last result into (44) we get

$$\begin{aligned} E(a,\sigma) &= \sum_{m=0}^{\infty}\frac{(ia)^m}{m!}\frac{i\pi\sigma^m}{\sigma} \\ &= \frac{i\pi}{\sigma}\sum_{m=0}^{\infty}\frac{(ia\sigma)^m}{m!}. \end{aligned} \tag{58}$$

Thus, finally:

$$\int_{-\infty}^{+\infty} dx \frac{e^{iax}}{(x^2-\sigma^2)} = \frac{i\pi}{\sigma}\, e^{ia\sigma}. \tag{59}$$

Next, we need to evaluate

$$E_x(a,\sigma) = \int_{-\infty}^{+\infty} dx \frac{x\, e^{iax}}{(x^2-\sigma^2)}. \tag{60}$$

We do not need to recalculate from scratch; we use the identity

$$x\, e^{iax} = -i\frac{d}{da}e^{iax}, \tag{61}$$

in the result (59). Doing this, we get

$$E_x(a,\sigma) \equiv \int_{-\infty}^{+\infty} dx \frac{x\, e^{iax}}{(x^2-\sigma^2)} = i\pi e^{ia\sigma}. \tag{62}$$

From the result (62) we have that

$$\int_{-\infty}^{+\infty} dx \frac{x\cos(ax)}{(x^2-\sigma^2)} = 0 \tag{63}$$

$$\int_{-\infty}^{+\infty} dx \frac{x\sin(ax)}{(x^2-\sigma^2)} = \pi\, e^{ia\sigma}. \tag{64}$$

Our original quantum mechanics integral $S(\sigma)$ is our result (64) for $a = 1$, so

$$\begin{aligned} E_x^+(1,\sigma) &\equiv S(\sigma) \\ &= \int_{-\infty}^{+\infty} dx \frac{x\sin(x)}{(x^2-\sigma^2)} = \pi\, e^{i\sigma}. \end{aligned} \tag{65}$$

The above result for quantum mechanical scattering problem corresponds to an outgoing wave. This result is equivalent to letting $\sigma \to \sigma + i\varepsilon$ in the Cauchy contour technique calculation. Noting that the integral on the left handside is invariant under $\sigma \leftrightarrow -\sigma$, we have the result for quantum mechanical scattering problem corresponding to an incoming wave.

$$\begin{aligned} E_x^-(1,-\sigma) &\equiv S(-\sigma) \\ &= \int_{-\infty}^{+\infty} dx \frac{x\sin(x)}{(x^2-\sigma^2)} = \pi\, e^{-i\sigma}. \end{aligned} \tag{66}$$

This corresponds to letting $\sigma \to \sigma - i\varepsilon$ in the Cauchy residue calculation. And the PV result for the integral is, as before,

$$\begin{aligned} S_{PV}(\sigma) &= \frac{1}{2}\left\{E_x^+(1,\sigma) + E_x^-(1,-\sigma)\right\} = \frac{1}{2}\left\{S(\sigma) + S(-\sigma)\right\} \\ &= \pi\left\{\frac{e^{i\sigma}+e^{-i\sigma}}{2}\right\} = \pi\cos\sigma. \end{aligned} \tag{67}$$

Since the original integral $S(\pm\sigma)$ is an improper integral, we can see that the value we ascribe to it will depend on which type of physical boundary conditions we want: incoming or outgoing waves or even the average between those boundary conditions (PV).

5 Conclusions

We have shown that using the NDIM technique, it was possible to evaluate some improper integrals and more specifically, the one integral that is relevant for quantum mechanical scattering problems. We have shown how to do the calculations in the NDIM technique and shown that the results it provides are equivalent to the results obtained using the Cauchy residue procedure, in which poles in the real axis is given an infinitesimal shift into the complex plane, either $\sigma + i\varepsilon$ or $\sigma - i\varepsilon$. In the NDIM procedure, each pole residue gives a distinct answer [4]. In our present calculation for the quantum mechanical scattering problem, there are two poles, and so two residues; and the PV result can be calculated by the simple average between the two residue calculations.

References

1. Halliday, I.G., Ricotta, R.M.: Phys. Lett. B **193**, 241–246 (1987)
2. 't Hooft, G., Veltman, M.I.G.: Nucl. Phys. B **44**, 189-213 (1972); Bollini, C.G., Giambiagi, J.J.: Il Nuovo Cim. B **12**, 20–26 (1972); Ashmore, J.F.: Lett. Nuovo Cim. **4**, 289–290 (1972)
3. Suzuki, A.T., Schmidt, A.G.M., Bentin, R.: Nucl. Phys. B **537**, 549 (1999); Suzuki, A.T., Schmidt, A.G.M.: Prog. Theor. Phys. **103**, 1011 (2000); Eur. Phys. J. C **12**, 361 (2000); Suzuki, A.T., Schmidt, A.G.M.: Phys. Lett. B **494**, 332 (2000); Suzuki, A.T., Suzuki, T.: Phys. Rev. D **106**, 085007 (2022)
4. Suzuki, A.T.: Acta Phys. Polonica B **37**, 2767–2779 (2006)
5. Suzuki, A.T.: arXiv:math-ph/0407032v1 (2004); Suzuki, A.T.: arXiv:0806.3216v1 (2008)
6. Capper, D.M., Dulwich, J.J., Litvak, M.J.: Nucl. Phys. B **241**(2), 463 (1984)
7. Dirac, P.A.M.: Rev. Mod. Phys. **21**, 392 (1949)
8. Gradshteyn, I.S., Ryzhik, I.M.: Table of Integrals, Series and Products, 8th edn, Edited by Daniel Zwillinger, Academic Press, Oxford, UK (2015)

Study of Linear Stability for Cylindrically Symmetrical States of Dynamic Equilibrium of Two-Component Vlasov–Poisson Plasma

Yuriy G. Gubarev and Jingyue Luo

Abstract We consider the linear stability problem for dynamic equilibria of two-component Vlasov–Poisson plasma in cylindrically symmetrical statement. The hydrodynamic substitution of independent variables is performed in order to transform the Vlasov–Poisson equations to an infinite system of gas-dynamic equations. It is important that exact stationary solutions to gas-dynamic equations are equivalent to exact stationary solutions to the Vlasov–Poisson equations. The sufficient condition of linear stability for exact stationary solutions to the Vlasov–Poisson equations is studied. Previously, this condition was not reversed either for small or, especially, for finite perturbations. To fulfill such reversion in the linear approximation, these gas-dynamic equations are linearized near their exact stationary solutions. The a priori exponential estimate from below is constructed for a subclass of small cylindrically symmetrical perturbations of exact stationary solutions to gas-dynamic equations, which grow over time and are described by the field of Lagrangian displacements. The countable set of sufficient conditions for linear practical instability is obtained. Thus, the Newcomb-Gardner-Rosenbluth sufficient condition for linear stability of exact stationary solutions to the Vlasov–Poisson equations is reversed. Moreover, a formal nature of this condition is revealed with respect to the considered small perturbations. As a result, by the direct Lyapunov method, an absolute instability for exact stationary solutions to the mathematical model of two-component Vlasov–Poisson plasma in relation to small cylindrically symmetrical perturbations is proved.

Keywords Vlasov–Poisson plasma · Cylindrically symmetrical dynamic equilibria · Absolute linear instability

Y. G. Gubarev
Lavrentyev Institute for Hydrodynamics, Lavrentyev Av. 15, 630090 Novosibirsk, Russian Federation

Y. G. Gubarev · J. Luo (✉)
Novosibirsk State University, Pirogova Str. 2, 630090 Novosibirsk, Russian Federation
e-mail: 644603544@qq.com

D. Vlachos (ed.), *Mathematical Modeling in Physical Sciences*, Springer Proceedings in Mathematics & Statistics 446, https://doi.org/10.1007/978-3-031-52965-8_37

1 Introduction

In the electrostatic approximation, when the electric field of electrons and ions is self-consistent, the plasma dynamics is described by kinetic Vlasov–Poisson equations [1]. In this case, such equations are used to study the collisionless motion of electrons, which interact with each other through the Coulomb repulsive forces, against the background of a homogeneous distribution of resting ions in the whole physical continuum. In this paper, the mathematical model of two-component Vlasov–Poisson plasma that we study is the following in three-dimensional statement:

$$\frac{\partial f}{\partial t} + \vec{v}\nabla_{\vec{x}} f - \nabla_{\vec{x}}\phi\nabla_{\vec{v}} f = 0, \quad \Delta_{\vec{x}}\phi = 4\pi\left(1 - \int_{\Re^3} f(\vec{x}, \vec{v}, t)\, d\vec{v}\right)$$
$$f = f(\vec{x}, \vec{v}, t) \geq 0, \quad f(\vec{x}, \vec{v}, 0) = f_0(\vec{x}, \vec{v}) \tag{1}$$

Here, f denotes the distribution function of electrons; t is time; $\vec{x}$, $\vec{v}$ denote electron coordinates and velocities; φ is the potential of self-consistent electric field; $\Re^3$ denotes the space of velocities; $d\vec{v}$ is a volume element in the velocity continuum; f_0 denotes the initial data. It is assumed that f and φ are the smooth functions, and the integral from right side of the Poisson equation exists.

In [2–4], a sufficient condition for linear stability of dynamic equilibrium states of the Vlasov–Poisson plasma was discovered. It was called as the Newcomb–Gardner–Rosenbluth's stability condition. Also, there was a series of results for dynamic equilibrium states of this plasma. For instance, in [2, 3], it was demonstrated that small perturbations in the form of normal waves, which are growing with time, don't exist for dynamic equilibrium states of the Vlasov–Poisson plasma with stationary distribution functions of charged particles, which are monotonically decreasing in the velocity space. Besides, in [4], it has shown that stationary distribution functions can be not only monotonically decreasing, but non-monotonically decreasing too.

In this paper, by studying the linear stability for exact stationary solutions to cylindrically symmetrical Vlasov–Poisson equations, it will be demonstrated that the corresponding dynamic equilibria of two-component Vlasov–Poisson plasma are absolutely unstable with respect to small cylindrically symmetrical perturbations. Additionally, the previously known sufficient Newcomb-Gardner-Rosenbluth's condition [5] for linear stability will be reversed. However, it is valid only for small cylindrically symmetrical perturbations from some incomplete unclosed subclass. Nevertheless, this sufficient condition for linear stability of cylindrically symmetrical states of dynamic equilibria of two-component Vlasov–Poisson plasma is also necessary. At last, it will be obtained the sufficient conditions for linear practical instability, and then the a priori exponential lower estimate for small cylindrically symmetrical perturbations growing over time will be built.

2 The Kinetic Vlasov–Poisson Equations

The mathematical model (1) of a boundless electrically neutral collisionless fully ionized plasma of electrons and ions in the electrostatic approximation (the two-component Vlasov–Poisson plasma) is considered further in cylindrically symmetrical formulation:

$$\frac{\partial f}{\partial t}+v\frac{\partial f}{\partial r}-\frac{\partial \phi}{\partial r}\frac{\partial f}{\partial v}=0,\quad \frac{1}{r}\frac{\partial}{\partial r}(r\frac{\partial \phi}{\partial r})=4\pi\left(1-2\pi\int_0^\infty f(r,v,t)\,v\,dv\right) \tag{2}$$
$$f=f(r,v,t)\geq 0,\quad f(r,v,0)=f_0(r,v)$$

where r, v denote electron radial coordinates and velocities. The boundary conditions: $f\to 0$ when $v\to\infty$; $f,\varphi\to 0$ when $r\to\infty$, or they are periodic when $r\to\infty$. Besides, we have the divergence form for the Vlasov equation:

$$\frac{\partial f}{\partial t}+\frac{1}{r}\frac{\partial}{\partial r}(vrf)-\frac{1}{v}\frac{\partial}{\partial v}\left(vf\frac{\partial \phi}{\partial r}\right)=0 \tag{2'}$$

System (2), (2') has the following functional of full energy and integral of motion which conserve with time (proven, not shown here):

$$E\equiv 2\pi^2\int_0^\infty\int_0^\infty v^3 fr\,drdv+\pi\int_0^\infty \phi(2\pi\int_0^\infty f\,vdv-1)\,rdr=const$$
$$C\equiv 4\pi^2\int_0^\infty\int_0^\infty \Phi(f)\,vr\,drdv=const \tag{3}$$

where $\Phi=\Phi(f)$ is an arbitrary function of its argument f.

The mathematical model (2), (2') of two-component Vlasov–Poisson plasma has the following exact stationary solutions

$$f=f^0(r,v),\quad \phi=\phi^0(r) \tag{4}$$

which satisfy the following system of stationary equations

$$v\frac{\partial f^0}{\partial r}-\frac{d\phi^0}{dr}\frac{\partial f^0}{\partial v}=0,\quad \frac{1}{r}\frac{d}{dr}(r\frac{d\phi^0}{dr})=4\pi(1-2\pi\int_0^\infty f^0 v\,dv) \tag{5}$$

and the divergence form for stationary Vlasov equation

$$\frac{1}{r}\frac{\partial}{\partial r}(vrf^0)-\frac{1}{v}\frac{\partial}{\partial v}(vf^0\frac{d\phi^0}{dr})=0 \tag{5'}$$

Next, we will study the stability of exact stationary solutions (4)–(5') against small cylindrically symmetrical perturbations. For this purpose, we take the desired functions in the following form (f', φ' denote small cylindrically symmetrical perturbations): $f = f^0(r, v) + f'(r, v, t)$, $\varphi = \varphi^0(r) + \varphi'(r, t)$. Then we substitute these functions into kinetic system (2) and relation (2'). Further, we consider stationary Eqs. (5) and (5'). Since cylindrically symmetrical perturbations f', φ' are small, the terms quadratic with respect to these perturbations are neglected. Thus, system of Eq. (2) linearized in the vicinity of exact stationary solutions (4) is transformed into the following relations:

$$\begin{gathered}\frac{\partial f'}{\partial t} + v\frac{\partial f'}{\partial r} - \frac{d\phi^0}{dr}\frac{\partial f'}{\partial v} - \frac{\partial \phi'}{\partial r}\frac{\partial f^0}{\partial v} = 0 \\ \frac{1}{r}\frac{\partial}{\partial r}(r\frac{\partial \phi'}{\partial r}) = -8\pi^2 \int_0^\infty f' v\, dv\end{gathered} \tag{6}$$

Analogously, Eq. (2') takes the form:

$$\frac{\partial f'}{\partial t} + \frac{1}{r}\frac{\partial}{\partial r}(vrf') - \frac{1}{v}\frac{\partial}{\partial v}(vf^0\frac{\partial \phi'}{\partial r}) - \frac{1}{v}\frac{\partial}{\partial v}(vf'\frac{d\phi^0}{dr}) = 0 \tag{6'}$$

Now, we'll compose the sum $E + C$ from integrals E and C (3), which is defined as I and is conserved over time. In order for the first variation δI of functional I to become equal to zero, we have such expression:

$$\frac{d\Phi}{df}(f^0) = -\frac{v^2}{2} - \phi^0$$

So, we have obtained that the integral I reaches its extreme values on exact stationary solutions (4)–(5'). The second variation $\delta^2 I$ of functional I has the following form:

$$\delta^2 I = 2\pi^2 \int_0^\infty \int_0^\infty \frac{d^2\Phi}{df^2}(f^0)(\delta\, f)^2 vr\, dv dr + \frac{1}{4}\int_0^\infty (\frac{\partial \delta\phi}{\partial r})^2\, r\, dr$$

From this variation by replacing the first variation δf with small cylindrically symmetrical perturbations f' and the first variation $\delta\phi$ with small cylindrically symmetrical perturbations φ', we have such linear analog of full energy functional

$$E_1 \equiv 2\pi^2 \int_{0^3}^\infty \int_0^\infty \frac{d^2\Phi}{df^2}(f^0) f'^2 vr\, dv dr + \frac{1}{4}\int_0^\infty (\frac{\partial \phi'}{\partial r})^2 r\, dr = const \tag{7}$$

It is not hard to check that the integral E_1 is conserved over time.

The functional (7) is non-negative definite when it meets the inequality:

$$\frac{d^2\Phi}{df^2}(f^0) \geq 0 \;\Leftrightarrow\; \frac{d(\frac{v^2}{2} + \phi^0)}{df^0} \leq 0 \tag{8}$$

Therefore, if condition (8) is true, exact stationary solutions (4) will be stable with respect to small cylindrically symmetrical perturbations (6), (6'). According to the variational method, inequality (8) is the sufficient condition of linear stability for dynamic equilibrium states (4)–(5') of the Vlasov–Poisson plasma. Our further goal is to prove that inequality (8) is not only sufficient, but also a necessary condition of linear stability for exact stationary solutions (4) to the Vlasov–Poisson Eqs. (2), (2'). It is not yet known how to do this in the kinetic case. However, it is known how to do this in the gas-dynamic case. So, we will replace the independent variables through a hydrodynamic substitution [6] and, thereby, will convert kinetic system of Eq. (2) into the gas-dynamic one.

3 The System of Vortex Shallow Water Equations

In order to establish a linear stability criterion for exact stationary solutions (4)–(5') in relation to small cylindrically symmetrical perturbations (6), (6'), we are transitioning from old independent variables (r, v, t) to new independent ones (r, λ, t) :

$$g = vrf, \;\; v = u(r, \lambda, t), \;\; \lambda \in [0, +\infty), \;\; u_\lambda \neq 0; \;\; g(r, \lambda, t) = \frac{\rho(r, \lambda, t)}{u_\lambda(r, \lambda, t)} \tag{9}$$

Here, λ is the Lagrangian coordinates of electrons ($d\lambda/dt \equiv 0$); u, ρ denote new sought functions.

As a result, kinetic Eq. (2) are transformed to the following gas-dynamic equations:

$$\begin{gathered}
\frac{\partial u}{\partial t} + u\frac{\partial u}{\partial r} + \frac{\partial \phi}{\partial r} = 0, \quad \frac{\partial \rho}{\partial t} + \frac{\partial}{\partial r}(\rho u) = 0 \\
\frac{\partial}{\partial r}(r\frac{\partial \phi}{\partial r}) = 4\pi(r - 2\pi\int_0^\infty \rho(r, \lambda, t)\, d\lambda) \\
u(r, \lambda, 0) = u_0(r, \lambda), \quad \rho(r, \lambda, 0) = \rho_0(r, \lambda)
\end{gathered} \tag{10}$$

We assume that $u \to 0, \rho \to 0$ when $\lambda \to \infty$; $u \to 0, \rho \to 0, \varphi \to 0$ when $r \to \infty$, or they are periodic when $r \to \infty$.

For gas-dynamic system of Eq. (10), there are integrals of full energy and motion conserving over time:

$$\begin{gathered}
E_2 \equiv 2\pi^2 \int_0^\infty \int_0^\infty \rho u^2 d\lambda dr + \pi \int_0^\infty \phi(2\pi \int_0^\infty \rho d\lambda - r)dr = const \\
C_1 \equiv 4\pi^2 \int_0^\infty \int_0^\infty \Phi_1(\kappa) u_\lambda d\lambda dr = const
\end{gathered} \tag{11}$$

where $\kappa \equiv \rho/u_\lambda$ is the reverse vorticity, $\Phi_1 = \Phi_1(\kappa)$ denotes an arbitrary function of its argument κ. By the way, replacement of independent variables (9) is non-degenerate one due to the fact that the reverse vorticity κ is preserved in each fluid particle during their movement.

Furthermore, the exact stationary solutions to initial-boundary value problem (10) are

$$u = u^0(r, \lambda), \ \rho = \rho^0(r, \lambda), \ \phi = \phi^0(r) \tag{12}$$

What's more, they are satisfying the following system of stationary equations:

$$\begin{gathered} u^0 \frac{\partial u^0}{\partial r} = -\frac{d\phi^0}{dr}, \quad \frac{\partial}{\partial r}(\rho^0 u^0) = 0 \\ \frac{d}{dr}(r\frac{d\phi^0}{dr}) = 4\pi(r - 2\pi \int_0^\infty \rho^0 \, d\lambda) \end{gathered} \tag{13}$$

By highlighting at least only one growing small cylindrically symmetrical perturbation, it is possible to prove instability of exact stationary solutions (12), (13) to initial-boundary value problem (10) with respect to small cylindrically symmetrical perturbations u', ρ', φ'. Therefore, to reach our target, we will study farther a particular class of evolutionary solutions to linearized mixed problem, which is characterized by the property that small cylindrically symmetrical perturbations for it are deviations of electron flight trajectories from streamlines corresponding to exact stationary solutions (12), (13) to initial-boundary value problem (10). Thus, we linearize the equations of system (10) near exact stationary solutions (12), (13), and it takes the following form:

$$\begin{gathered} \frac{\partial u'}{\partial t} + u^0 \frac{\partial u'}{\partial r} + u' \frac{\partial u^0}{\partial r} + \frac{\partial \phi'}{\partial r} = 0, \quad \frac{\partial \rho'}{\partial t} + \frac{\partial}{\partial r}(\rho^0 u' + \rho' u^0) = 0 \\ \frac{\partial}{\partial r}(r\frac{\partial \phi'}{\partial r}) = -8\pi^2 \int_0^\infty \rho' \, d\lambda \\ u'(r, \lambda, 0) = u'_0(r, \lambda), \quad \rho'(r, \lambda, 0) = \rho'_0(r, \lambda) \end{gathered} \tag{14}$$

Like for the kinetic system, we have the integral $I_1 = E_2 + C_1$ (11). In order to put the first variation of this integral equal to zero, we accept the condition:

$$-\frac{u^{02}}{2} - \phi^0 = \frac{d\Phi_1}{d\kappa}(\kappa^0); \quad \kappa^0 = \frac{\rho^0}{u^0_\lambda}$$

Then we calculate the second variation of functional I_1

$$\delta^2 I_1 = 2\pi^2 \int_0^\infty \int_0^\infty [\rho^0(\delta u)^2 + 2u^0 \delta u \delta\rho + \frac{d^2\Phi_1}{d\kappa^2}(\kappa^0) u_\lambda^0 (\delta\kappa)^2] dr d\lambda + \\ + \frac{1}{4}\int_0^\infty (\frac{\partial \delta\phi}{\partial r})^2 r dr \tag{15}$$

and, after replacing the first variation δu with small cylindrically symmetrical perturbations u', the first variation $\delta\rho$ with small cylindrically symmetrical perturbations ρ', the first variation $\delta\kappa$ with small cylindrically symmetrical perturbations κ' and the first variation $\delta\phi$ with small cylindrically symmetrical perturbations φ' in integrands of functional $\delta^2 I_1$ (15), obtain the linear analog of full energy integral, which is conserved over time:

$$E_3 \equiv 2\pi^2 \int_0^\infty \int_0^\infty [\rho^0 u'^2 + 2u^0 u' \rho' + \frac{d^2\Phi_1}{d\kappa^2}(\kappa^0) u_\lambda^0 \kappa'^2] d\lambda dr + \\ + \frac{1}{4}\int_0^\infty (\frac{\partial \phi'}{\partial r})^2 r dr = const \tag{16}$$

By means of the Sylvester criterion [7], it is not difficult to find out that functional E_3 (16) has no definite or constant sign. Therefore, it can be deduced that there is no sufficient condition for linear stability of exact stationary solutions (12), (13) to initial-boundary value problem (10) in relation to small cylindrically symmetrical perturbations u', ρ', φ' (14).

Next, with the goal to establish such sufficient condition for linear stability or to prove absolute instability for exact stationary solutions (12), (13) to initial-boundary value problem (10), we think of using the Lagrangian displacements field [8] to characterize subclass of small cylindrically symmetrical perturbations (14) with the property that these perturbations are deviations of electron flight trajectories from streamlines corresponding to the studied stationary solutions.

4 The a Priori Exponential Lower Estimate for Small Cylindrically Symmetrical Perturbations Growing Over Time

Such perturbations can be described by the field of Lagrangian displacements ξ [8], which is determined by the following equation:

$$\xi = \xi(r, \lambda, t): \quad \frac{\partial \xi}{\partial t} = u' + \xi \frac{\partial u^0}{\partial r} - u^0 \frac{\partial \xi}{\partial r} \tag{17}$$

At the base of relation (17), mixed problem (14) is rewritten, so that we have new initial-boundary value problem in the form:

$$\begin{gathered}\frac{\partial^2\xi}{\partial t^2}+2u^0\frac{\partial^2\xi}{\partial t\partial r}+u^0\frac{\partial}{\partial r}(u^0\frac{\partial\xi}{\partial r})-\xi\frac{\partial}{\partial r}(u^0\frac{\partial u^0}{\partial r})=-\frac{\partial\phi'}{\partial r}\\ \frac{\partial}{\partial r}(r\frac{\partial\phi'}{\partial r})=8\pi^2\int_0^\infty\frac{\partial}{\partial r}(\xi\rho^0)d\lambda\\ \rho'=-\frac{\partial}{\partial r}(\xi\rho^0)\\ \xi(r,\lambda,\,0)=\xi_0(r,\lambda),\ \frac{\partial\xi}{\partial t}(r,\lambda,\,0)=(\frac{\partial\xi}{\partial t})_0(r,\lambda)\end{gathered} \tag{18}$$

Then linear analog of the full energy integral (16) takes the following form

$$\begin{gathered}E_3\equiv 2\pi^2\int_0^\infty\int_0^\infty\rho^0\left[\left(\frac{\partial\xi}{\partial t}\right)^2-\left(u^0\frac{\partial\xi}{\partial r}\right)^2+\xi^2\frac{d^2\phi^0}{dr^2}\right]d\lambda dr+\\ +\frac{1}{4}\int_0^\infty\left(\frac{\partial\phi'}{\partial r}\right)^2 rdr=const\end{gathered} \tag{19}$$

and is still conserved over time. According to the Sylvester criterion [7], it can be seen from expression (19) that the functional E_3 is not constant or definite by sign in principle.

This fact justifies again the choice of subclass (17), (18) of small cylindrically symmetrical perturbations (14) to prove absolute instability of exact stationary solutions (12), (13) to initial-boundary value problem (10). By the above analysis, we can make the hypothesis that absolute instability of exact stationary solutions (12), (13) can be proved whether sufficient condition (8) for linear stability holds or not. Subsequently, we introduce the Lyapunov functional [9]

$$M\equiv 4\pi^2\int_0^\infty\int_0^\infty\rho^0\xi^2 d\lambda dr \tag{20}$$

and, in accordance with mixed problem (17), (18), its first and second time derivatives are given by the following:

$$\begin{gathered}\frac{dM}{dt}=8\pi^2\int_0^\infty\int_0^\infty\rho^0\xi\frac{\partial\xi}{\partial t}d\lambda dr\\ \frac{d^2M}{dt^2}=8\pi^2\int_0^\infty\int_0^\infty\rho^0\left[\left(\frac{\partial\xi}{\partial t}\right)^2+\xi\frac{\partial^2\xi}{\partial t^2}\right]d\lambda dr\end{gathered} \tag{21}$$

Using initial-boundary value problem (17), (18) and relations (20), (21), we can construct an original differential inequality for the integral M [10]:

$$\frac{d^2M}{dt^2}-2\lambda\frac{dM}{dt}+2(\lambda^2+\gamma)M\geq 0 \tag{22}$$

Here, λ is an arbitrary constant, while γ denotes the known positive constant. Relation (22) is a second order linear differential inequality with constant coefficients. If constant λ is the positive one, then we can supplement relation (22) by a countable set of the conditions [11]

$$\begin{aligned}
&M(\pi n/2\sqrt{\lambda^2+2\gamma}) > 0;\ n = 0, 1, 2, \ldots \\
&\frac{dM}{dt}(\pi n/2\sqrt{\lambda^2+2\gamma}) \geq 2(\lambda+\gamma/\lambda)M(\pi n/2\sqrt{\lambda^2+2\gamma}) \\
&M(\pi n/2\sqrt{\lambda^2+2\gamma}) \equiv M(0)\exp(\pi n\gamma/2\sqrt{\lambda^2+2\gamma}) \\
&\frac{dM}{dt}(\pi n/2\sqrt{\lambda^2+2\gamma}) \equiv \frac{dM}{dt}(0)\exp(\pi n\gamma/2\sqrt{\lambda^2+2\gamma}) \\
&M(0) > 0,\ \frac{dM}{dt}(0) \geq 2(\lambda+\gamma/\lambda)M(0)
\end{aligned} \tag{23}$$

From relations (23), we summarize the a priori exponential lower estimate:

$$M(t) \geq C_2 \exp(\lambda t) \tag{24}$$

where C_2 denotes the known positive constant.

If conditions (23) are satisfied, then functional M (20) does not grow slower than exponentially with time. Thereby, a priori exponential lower estimate (24) established, and states (12), (13) (or, equivalently, (4), (5), and (5')) of dynamic equilibrium of two-component Vlasov–Poisson plasma are absolutely unstable.

First two inequalities in system (23) can be interpreted as the sufficient conditions for linear practical instability [12]. For application of these conditions, no additional information about two-component Vlasov–Poisson plasma under consideration is required. So, the obtained conditions for linear practical instability are constructive in their character.

5 Conclusion

We studied the linear stability problem for the states of dynamic equilibrium of two-component Vlasov–Poisson plasma in cylindrically symmetrical statement. The direct Lyapunov method [6, 9–11, 13] was applied to show that these dynamical equilibrium states of the considered Vlasov–Poisson plasma are absolutely unstable with respect to small cylindrically symmetrical perturbations when this plasma contains electrons with a stationary distribution function that is variable both over the physical continuum and the velocity space as well as ions uniformly distributed over the physical continuum. What's more, we obtained the sufficient conditions for linear practical instability of the studied equilibria of two-component Vlasov–Poisson plasma and constructed the a priori exponential lower estimate.

The results of our research are important for solving the problem of controlled thermonuclear fusion. Furthermore, they can be applied to the development and subsequent operation of devices designed to perform the controlled thermonuclear fusion.

For reliable operation of such plasma confinement devices, we need to ensure the practical stability for dynamic equilibrium states of plasma under all acceptable perturbations. In particular, these equilibrium states should be stable in a practical sense in relation to small cylindrically symmetric perturbations.

References

1. Krall, N., Trivelpiece, A.: Principles of Plasma Physics. McGraw-Hill Book Company Inc., NY (1973)
2. Bernstein, I.: Waves in a plasma in a magnetic field. Phys. Rev. **109**, 10–21 (1958)
3. Gardner, C.: Bound on the energy available from a plasma. Phys. Fluids **6**, 839–840 (1963)
4. Rosenbluth, M.: Topics in microinstabilities. In: Rosenbluth, M. (ed.) Advanced Plasma Theory, pp. 137–158. Academic, NY (1964)
5. Holm, D., Marsden, J., Ratiu, T., Weinstein, A.: Nonlinear stability of fluid and plasma equilibria. Phys. Reports **123**, 1–116 (1985)
6. Gubarev, Yu.: On instability of dynamic equilibrium states of Vlasov-Poisson plasma. Appl. Math. Phys. **4**(1), 1–8 (2016)
7. Yakubovich, V., Starzginskiy, V.: Linear Differential Equations with Periodical Coefficients and its Applications. Nauka, Moscow (1972). [in Russian]
8. Chandrasekhar, S.: Ellipsoidal Figures of Equilibrium. Yale University Press, New Haven (1969)
9. Gubarev, Yu.: The Direct Lyapunov Method. The Stability of Quiescent States and Steady-State Flows of Fluids and Gases. Palmarium Academic Publishing, Saarbrücken (2012). [in Russian]
10. Gubarev, Yu.: Linear stability criterion for steady screw magnetohydrodynamic flows of ideal fluid. T & A **16**, 407–418 (2009)
11. Gubarev, Yu.: Two-flow instability of particular class of one-dimensional dynamic equilibrium states of Vlasov-Poisson plasma. Plasma Res. Express **1**(4), 045008-1–045008-13 (2019)
12. La Salle, J., Lefschetz, S.: Stability by Liapunov's Direct Method with Applications. Academic, NY (1961)
13. Chetaev, N.: Stability of Motion. Nauka, Moscow (1990). [in Russian]

Ranking of Linear Spaces in Generating a Hierarchy Between Concepts

Yehuda Roth

Abstract Human perception is composed of fundamental concepts. If we associate these with states in Hilbert spaces such that the fundamental concepts are associated with the spanning sets of the spaces, a composed concept can be defined through the superposition of states within the selected spanning sets. Under these definitions, no particular spanning set can be defined as being fundamental. Therefore, in this article, we describe the concept of a hierarchy of spaces.

Keywords Hierarchy of Spaces · Rank of a spaces · Fundamental concepts

1 Introduction

This paper presents a mathematical framework that associates concepts with states. Fundamental concepts are described as space-spanning sets, and the superposition of fundamental concepts results in complex concepts. Spaces with different ranks define the differences between fundamental and complex concepts. Our analysis is based on a quantum-like approach, i.e., systems that refer to physical, mathematical, or conceptual frameworks that share specific characteristics with quantum mechanics (see, for example, Ref. [1]).

We suggest that implementing concepts is by using mathematical tools borrowed from quantum mechanics as in Ref. [2].

To describe the term "conceptualizing", we will associate the the superposition process with the creation of new concepts.

In the Merriam-Webster dictionary, the term concept used as a noun, is defined as "something conceived in the mind" or "an abstract or generic idea generalized from particular instances." [3]. The word conceptualization refers to the way of generating a concept [3]. Consider for example some space with two items: We can locate an item in on of two sides to define the *sides*-concept; *right* denoted by $|R\rangle$ and *left* described by the state $|L\rangle$. Conceptualizing is related to a unitary transformation

Y. Roth (✉)
Oranim Academic College of Education, 3600600 Tivon, Israel
e-mail: yudroth@gmail.com

D. Vlachos (ed.), *Mathematical Modeling in Physical Sciences*, Springer Proceedings in Mathematics & Statistics 446, https://doi.org/10.1007/978-3-031-52965-8_38

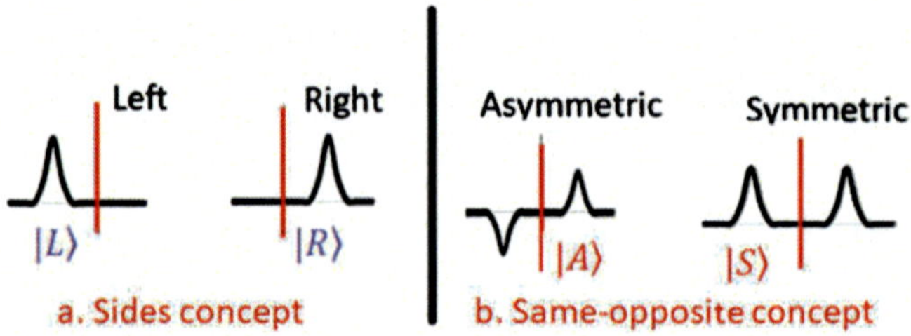

Fig. 1 Two alternatives to describe the condition of the bumps: **a** a description using the right and left sides, and **b** using symmetric or asymmetric states

transforming the *sides* states into a new basis $\left|\begin{matrix} S \\ A \end{matrix}\right\rangle = \frac{1}{\sqrt{2}}\{|L\rangle \pm |R\rangle\}$ to define the two items symmetrically or asymmetrically.

Defining a linear space corresponds with a basis of states, sometimes referred as to spanning sets. Superposition of these states generates the items within the space and can form a new basis of states to generate new set of concepts.

2 Concept Generation in a Bump Example

As a visual example, suppose bumps with a Gaussian shape are described in two fixed locations along a horizontal axis.

For example the composed shape,  can be described by two alternatives (see states symbols in Fig. 1):

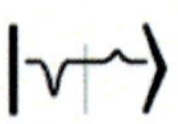 $= 0.3\,|R\rangle - \sqrt{0.92}\,|L\rangle$

Or,

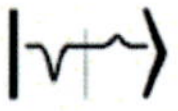 $= \frac{0.3-\sqrt{0.92}}{\sqrt{2}}\,|S\rangle + \frac{0.3+\sqrt{0.92}}{\sqrt{2}}\,|A\rangle$ (1)

These two fundamental concepts describe the same shape. Which of them is the most fundamental? Mathematically, there are no significant differences [4]. However human perception does have preferences for one approach over another. For this purpose, we define the concept of a hierarchy of spaces.

3 Hierarchy of Spaces

Consider N items represented by states.

We rank spaces as follows:

1. A space of rank N is an N-dimensional space spanned by all states.It allows all types of superposition.
2. Single-rank spaces are N 1D spaces having only one state, each disallowing linear combinations. This turns off the ability to generate new concepts. These definitions define what we refer to as fundamental concepts.
3. Generally, a ρ rank item defines a space with ρ dimensions.

This is shown in Fig. 2.

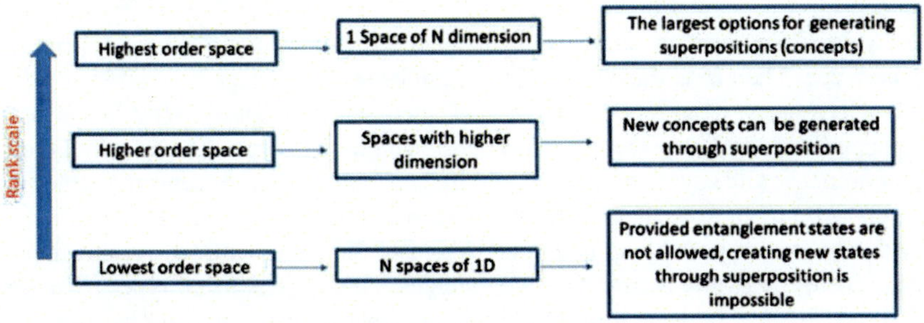

Fig. 2 Diagram illustrating the ranking of the spaces

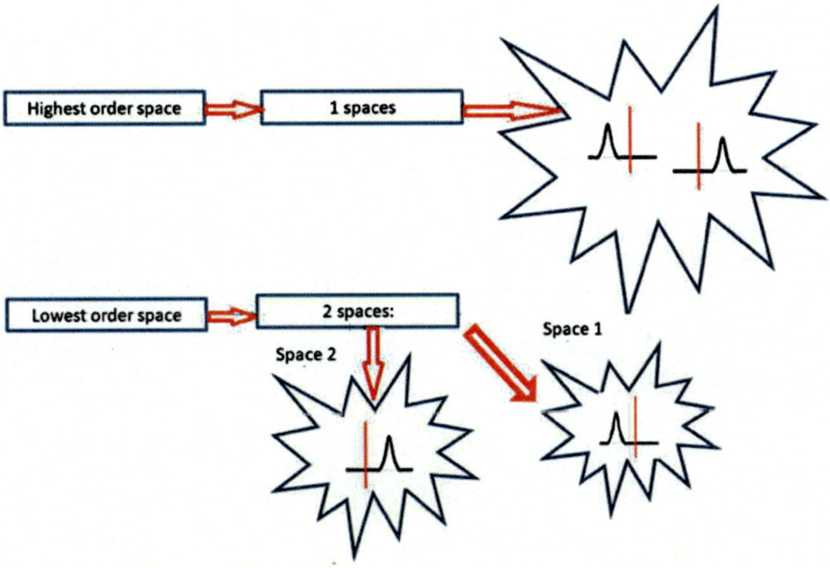

Fig. 3 Diagram illustrating the ranking of the spaces in the bump example

The bump example expressed by the space hierarchy is illustrated in Fig. 3 (we selected the right-left basis of states).

4 Human Recognition of Concepts

Most people prefer specific concepts at the lowest scale, whereas mathematically, there is no difference between spanning sets. In the following discussion, we use the example of bombers to demonstrate a possible debate on how to define the fundamental concepts to describe space. Let us present two speakers, **Mr. Sides** and **Miss Symmetry**, who had the following discussion.
Miss. Symmetry: "I think that the symmetric and asymmetric are the fundamental concepts."
Mr. Sides: "No, most people will tell you that the locations of objects are of primary importance. Therefore, the object side is a fundamental concept."
Miss. Symmetry: "In any case, you have two states in space. Why invent new concepts, *right* and *left* where it is simpler to compare the two. This is why I think the symmetric and asymmetric terms are fundamental." "The concepts the same and the opposite seem to be more appropriate as basic concepts."
Evidently, it is impossible to determine which of the two is correct. As previously mentioned, the two sets have the same weight. This applies to all types of space. However, if most observers use the same fundamental concepts, an observer who uses a different hierarchy can be considered unorthodox.

5 Discussion

This paper presents a mathematical procedure for generating ranks between linear spaces. Our study was based on a quantum-like approach, allowing us to believe it can be implemented on quantum computers. Designing computers with unconventional fundamental concepts and using quantum tools such as logic gates (unitary transformations in quantum terminology) may introduce alternative world descriptions. Moreover, the possibility that non-orthodox people have different interpretive systems logically correct for describing reality cannot be excluded.

Acknowledgements

In memory of my son Lior Roth (1987–2022).

Lior was an individual with special needs who had unconventional interpretations of reality.
I was told that he is programmed differently.
He was the inspiration for this study.

References

1. Basieva, I., Khrennikov, A., Ozawa, M.: Quantum-like modeling in biology with open quantum systems and instruments. Biosystems **201**, 104328 (2021). https://doi.org/10.1016/j.biosystems.2020.104328, https://www.sciencedirect.com/science/article/pii/S0303264720301994
2. Svensson, B.: Pedagogical review of quantum measurement theory with an emphasis on weak measurements. Quanta **2**(1), 18–49 (2013). https://doi.org/10.12743/quanta.v2i1.12
3. Carleo, G., et al.: Machine learning and the physical sciences. Rev. Mod. Phys. **91**(4), 045002–045002 (2019). https://doi.org/10.1103/RevModPhys.91.045002
4. Sarfaraj, M.N., Mukhopadhyay, S.: All-optical scheme for implementation of tri-state pauli-x, y and z quantum gates using phase encoding. Optoelectron. Lett. **17**(12), 746–750 (2021). https://doi.org/10.1007/s11801-021-1037-y

Physical Modeling Using Stochastic Differential Equations

Monte Carlo Solution of Semi-linear Helmholtz Boundary Value Problem

Abdujabbor Rasulov and Gulnora Raimova

Abstract In these work we will study a probabilistic representation of the solution of the Helmholtz boundary problem for the non-linear problem

$$-\Delta u(x) + cu(x) = g \cdot f(u), \quad x \in D, \quad u|_\Gamma = \psi$$

where, $f(u)$ in our case could be hyperbolic functions $sh(u)$ or $ch(u)$. Under the assumption of the existence of a solution, an unbiased estimator is constructed on the trajectories of the proposed branching process "walk on spheres". To do this, using Green's formula, a special integral equation is written that connects the value of the function with its integrals over a ball and a sphere of maximum radius centered at a point and entirely contained in the region under consideration. A probabilistic representation of the solution of the problem in the form of the mathematical expectation of some random variable is obtained. In accordance with the probabilistic representation, a branching process of walk on spheres is constructed and an unbiased estimator of the solution of the problem with finite variance is constructed on its trajectories.

Keywords Helmholtz equation · Dirichlet problem · Monte Carlo method · Branching random process · Walk on spheres · Martingale · Unbiased estimator

1 Introduction

As is known, there are many applied problems, the solution of which is associated with boundary value problems for nonlinear elliptic equations containing hyperbolic functions of unknown functions. For example, the equation $\Delta u(x) = k^2 \cdot ch(u)$ arises in problems of constructing surfaces of constant mean curvature in the hyperbolic space H^3 [1]. The equations $\Delta u(x) = k^2 \cdot sh(u)$ arise when solving prob-

WWW home page: https://www.uwed.uz/en/faculties/5/departments?department=33.

A. Rasulov (✉) · G. Raimova
University of World Economy and Diplomacy, Tashkent, Uzbekistan
e-mail: asrasulov@gmail.com

D. Vlachos (ed.), *Mathematical Modeling in Physical Sciences*, Springer Proceedings in Mathematics & Statistics 446, https://doi.org/10.1007/978-3-031-52965-8_39

lems of bio-molecular electrostatic theory [2]. This article considers a probabilistic approach to solving the first boundary value problem for the following two below equations. Let D be a bounded domain in R^3 with regular boundary Γ, $\varphi(x) \in C(\bar{\Gamma})$, $\psi(x) \in C(\bar{\Gamma})$. Consider the following Dirichlet problems:

$$-\Delta u(x) + c \cdot u(x) = g \cdot sh(u), \quad x \in D, \quad u|_{\Gamma} = \varphi;$$

$$-\Delta u(x) + c \cdot u(x) = g \cdot ch(u), \quad x \in D, \quad u|_{\Gamma} = \psi.$$

It is assumed that the functions $\varphi(x)$, $\psi(x)$ and the coefficients c, g are such that there exists a unique continuous solution to these semilinear problems ([3, 4]). Assuming the existence of a solution for the problems, an unbiased estimator is constructed on the trajectories of the branching process walk on spheres.

Unbiased and biased estimators for the solution of boundary value problems for the linear Helmholtz equation $\Delta u - cu = -g(x)$ were considered for $c(x) = const$ by Mikhailov and Elepov in [5, 6], for the variable case $c(x)$ in [7, 8] for the Dirichlet problem, Simonov in [9] for the mixed problem and the problem Neumann, Sipin in [10] for the Dirichlet problem for the equation

$$\Delta u + \sum_{i=1}^{n} a_i \frac{\partial u}{\partial x_i} + au = -g.$$

In the papers [11, 12] by Mikhailov and Makarov, estimates for the solution of boundary value problems for the linear Helmholtz equation are built on the basis of a special integral-difference equation using the process of walking on spheres with reflection from the border. In the works [13, 14] the Monte Carlo solution of one applied problem of biomolecular electrostatic theory for the linearized equation $\Delta u(x) = k^2 \cdot u$ was considered. The Monte Carlo solution of the Dirichlet problem for a nonlinear equation of the form

$$\Delta u(x) = \sum_{i=1}^{n} a_i(x) u^{2i}(x) + a_0(x)$$

was proposed by Rasulov in his work [15, 16]. Mikhailov [7] studied special case the equation $\Delta u + u^n = 0$ and G.M.Raimova applied for the equation

$$\Delta u(x) + c\, u(x) = \sum_{i=0}^{\infty} a_i(x) u^i(x)$$

in [17, 18].

2 Formulation of the Problem

Let D be a bounded domain in R^3 with regular boundary Γ, $\varphi(x) \in C(\bar{\Gamma})$, $\psi(x) \in C(\bar{\Gamma})$. Consider the following Dirichlet problems:

$$-\Delta u(x) + c \cdot u(x) = g \cdot sh(u), \quad x \in D, \quad u|_\Gamma = \varphi. \tag{1}$$

$$-\Delta u(x) + c \cdot u(x) = g \cdot ch(u), \quad x \in D, \quad u|_\Gamma = \psi. \tag{2}$$

Considering that for the hyperbolic sine and cosine functions for arbitrary values of the argument, the following expansions take place

$$sh(u) = u + \frac{u^3}{3!} + \cdots + \frac{u^{2m-1}}{(2m-1)!} + \cdots, \quad ch(u) = 1 + \frac{u^2}{2!} + \cdots + \frac{u^{2m}}{(2m)!} + \cdots,$$

the considered problems (1) and (2) can be written as:

$$-\Delta u(x) + cu(x) = f(u), \quad x \in D, \quad u|_\Gamma = \varphi, \quad f(u) = g \cdot \sum_{n=0}^{\infty} \frac{u^{2n+1}}{(2n+1)!} \tag{1'}$$

$$-\Delta u(x) + cu(x) = f(u), \quad x \in D, \quad u|_\Gamma = \psi, \quad f(u) = g \cdot \sum_{n=0}^{\infty} \frac{u^{2n}}{(2n)!} \tag{2'}$$

First of all let us present some well-known results from [5], for the solution of the Dirichlet problem in the three-dimensional case for the linear Helmholtz equation $\Delta u - c\,u = -f$, where $c > 0$, The function $f(x)$ satisfies the Hölder condition in the domain D bounded by the regular boundary Γ, on which the boundary condition $u|_\Gamma = \varphi(x)$ is given:

$$\Delta u(x) - cu(x) = -f, \quad x \in D, \quad u|_\Gamma = \varphi \tag{3}$$

Let's introduce the notation: $\overline{D}$—closure of the region D; function $d(x) = \min_{y \in \Gamma} |x - y|$ is the distance from the point x to the boundary Γ; $\Gamma_\varepsilon = \{x \in \overline{D} : d(x) < \varepsilon\}$, ε- neighborhood of Γ; the sphere $S(x)$ is the largest of the spheres centered at the point x that lie entirely in $\overline{D}$, i.e. $S(x) = \{y \in \overline{D} : |y - x| = d(x)\}$. Let it be necessary to estimate the solution $u(x_0)$ of the stated problem at a given point $x_0 \in D$. It is known from the theory of fundamental solutions [20] that $u(x_0)$ can be represented as

$$u(x_0) = \frac{d_0\sqrt{c}}{4\pi d_0^2 sh(d_0\sqrt{c})} \int_{S(x_0)} u(s)ds + \int_{|x-x_0|<d_0} \frac{sh[(d_0 - |x - x_0|)\sqrt{c}]}{4\pi |x - x_0| sh(d_0\sqrt{c})} f(x)dx, \tag{4}$$

where $d_0 = d(x_0)$. The relation (4) must be supplemented with the following equality, which takes into account the boundary condition of the problem:

$$u(x_0) = \varphi(x_0), \quad x_0 \in \Gamma. \tag{5}$$

For the function $u(x)$, we can write the following integral equation:

$$u(x) = \int_D k(x, y)u(y)dy + \phi(y), \tag{6}$$

where

$$k(x, y) = \begin{cases} \frac{\sqrt{c}}{sh(d\sqrt{c})} \delta_r(y), & x \notin \Gamma_\varepsilon, \\ 0, & x \in \Gamma_\varepsilon, \end{cases}$$

$$\phi(x) = \begin{cases} \frac{1}{4\pi} \int\limits_{|x-y|<d} \frac{sh[(d-|x-y|)\sqrt{c}]}{|x-y|sh(d\sqrt{c})} f(y)d(y), & x \notin \Gamma_\varepsilon, \\ u(x), & x \in \Gamma_\varepsilon. \end{cases}$$

Here $d = d(x)$, $\delta_x(y)$ is the generalized density corresponding to the uniform probability distribution on the sphere $S(x)$. The probabilistic approach to solving the problem is associated with the Markov chain, which is called a walk in spheres or a spherical process.

We will consider a probabilistic approach to solving the problem (2). For the problem (1), by analogy, we can get the same results.

3 Probabilistic Representation of the Solution

Applying the relation (4) to Eq. (1') $-\Delta u(x) + cu(x) = f(u)$ we get the following integral representation:

$$u(x) = \frac{R\sqrt{c}}{4\pi R^2 sh(R\sqrt{c})} \int_{S_R} u(y)dy + \int_{K_R} \frac{sh[(R-|x-y|)\sqrt{c}]}{4\pi|x-y|sh(R\sqrt{c})} f(u(y))dy \tag{7}$$

Here $R(x) = \min_{y \in \Gamma} |x - y|$ is the distance from the point x to the boundary, K_R is the ball of radius R centered at the point x, S_R - corresponding sphere. The first integral in (7) is the integral over the surface of the sphere S_R. Since the surface area of the sphere S_R is equal to $4\pi R^2$, after introducing this expression under the sign of the first integral, it will represent a uniform distribution on the sphere:

$$u(x) = \frac{R\sqrt{c}}{sh(R\sqrt{c})} \int\limits_{S_R} u(x + R\omega)d\omega +$$

$$+ \frac{sh(R\sqrt{c}) - R\sqrt{c}}{sh(R\sqrt{c})} \int\limits_{K_R} \frac{sh\,[(R - |x - y|)\sqrt{c}]}{4\pi |x - y|(sh(R\sqrt{c}) - R\sqrt{c})} \cdot \frac{f(u(y))}{c} dy.$$

Let's rewrite this expression as follows

$$u(x) = q \int\limits_{S_R} u(y)d\omega + (1 - q) \int\limits_{K_R} p(x, y) \frac{f(u(y))}{c} dy, \tag{8}$$

the following notation was used here: $q = \frac{R\sqrt{c}}{sh(R\sqrt{c})}$, ω is the uniform distribution on S_R,

$$p(x, y) = \frac{sh[(R - |x - y|)\sqrt{c}]}{4\pi |x - y|(sh(R\sqrt{c}) - R\, sqrtc)}$$

transition density from x to y ($x, y \in K_R$).

Before proceeding to the description of a random branching process consistent with the probabilistic representation (8), we give a general definition of the process of walk on spheres with branching and some of its properties from [21].

The process walk on sphere with branching is defined as follows: in a bounded domain $D \subset R^3$ with boundary Γ particles n of types T_1, T_2, ..., T_n wander. A particle of type T_i located at point x moves to point y per unit time with probability $q_i(x)$, distributed with density $p_i(x, y)$, and generates there α_1 particles of T_1 type, α_2 particles of T_2, ..., α_n - type T_n . With probability $1 - q_i(x)$ no transformation occurs, and the particle passes to the point y, distributed uniformly on the sphere $S_R(x)$, $R(x)$ is the distance from the point x to the boundary Γ

$$1 - q_i(x) = \frac{R(x)\sqrt{c_i}}{sh(R(x)\sqrt{c_i})}, \quad r = |x - y|,$$

$$p_i(x, y) = \begin{cases} \frac{1}{4\pi r} \cdot \frac{c_i sh(\sqrt{c_i}(R - r))}{sh(R\sqrt{c_i})q_i(x)}, & y \in K_R(x); \\ 0, & y \notin K_R(x). \end{cases}$$

The new particles move similarly to the considered one, independently of each other. If the number of branches N of the described branching Markov chain is finite, then, starting from some random moment of time, the chain is N independent processes of walk on spheres and, therefore, has a limit belonging to Γ. The number N is naturally also random. Let us present a theorem giving necessary and sufficient conditions for the average number of branches of the process of wandering over spheres with branching to be finite. Let b^i_j be the average number of particles of type T_j

obtained from a particle of type T_i in one transformation, $I = \left\|\delta^i_j\right\|$ - identity matrix of order n.

Theorem (see: [21]) The average number of branches for a process of wandering over spheres with branching is finite if and only if the Peron root of the matrix $\left\|c_i(b^i_j - \delta^i_j)\right\|$ is strictly less than the first eigenvalue of the boundary problem $\Delta u + \lambda u = 0, u|_\Gamma = 0$.

Corollary If only particles of the same type participate in the walk, then the average number of branches in a walk over branched spheres is finite if and only if $c(b-1) < \lambda_1$. Here b is the average number of particles resulting from the fission of one particle, and λ_1 is the first eigenvalue of the boundary value problem $\Delta u + \lambda u = 0, u|_\Gamma = 0$.

This facts we will be used in next step. Further the formula (8) can be rewritten as follows:

$$u(x) = q \int\limits_{S_R(x)} u(y)d\omega + (1-q) \int\limits_{K_R(x)} p(x,y) \sum_{n=0}^{\infty} \frac{g}{c(2n)!} u^{2n}(y)dy. \qquad (9)$$

In accordance with (8), it is natural to construct unbiased estimates for solving problem (2) on the trajectories of the process of wandering over branched spheres, in which particles of the same type participate.

Now we will construct random branching process consistent with a probabilistic representation. In accordance with the representation (9), we construct a branching random walk process in the phase space D, in which particles of the same type participate. Let $\alpha < 1$ be some parameter. About this parameter and the condition of choosing it we will discourse when we prove our theorems. Suppose at the initial moment there is one particle at the point $x_0 = x$. Let $n > 0$ and x_n be known. For a unit of time, a particle with probability $q = q(x) = \frac{R\sqrt{c}}{sh(R\sqrt{c})}$ goes to the point x_{n+1} , uniformly distributed on the sphere $S_R(x)$ and with probability $1-q$ goes to the point x_{n+1} distributed in the ball $K_R(x)$ with density $p(x, y)$. In the second case, with probability π_n $(n = 1, 2, \cdots)$, the particle is divided into n new particles or dies with probability π_0. Branching probabilities are defined as follows:

$$\pi_{2n} = \frac{\alpha}{sh(1)\cdot(2n)!}, \quad \pi_{2n-1} = 0, (n = 1, 2, \ldots).$$

The absorption probability is equal to:

$$\pi_0 = 1 - \sum_{n=1}^{\infty} \pi_n = 1 - \frac{\alpha}{sh(1)} \sum_{n=0}^{\infty} \frac{1}{(2n)!} = 1 - \alpha\,\frac{ch(1)-1}{sh(1)}.$$

The new particles behave independently of each other in a similar way to the considered particle. The process terminates if all particles in $\overline{D}$ die, or if all particles fall into Γ. The α parameter allows you to control the number of branches.

Let us consider the structure of the probability space on which the process of wandering over spheres with branching is specified. In this regard, we need recall some definitions from the theory of general branching processes (see: [22]).

Definition 1 A point distribution on $\overline{D}$ is a vector $w = (x_1, n_1; x_2, n_2; \ldots; x_k, n_k)$, where k is an arbitrary non-negative number, $n_1, \ldots, n_k$ are integers positive numbers, $x_1, x_2, \ldots, x_k$ - distinct points of $\overline{D}$.

The case $k = 0$ corresponds to the absence of particles in $\overline{D}$. Such a point distribution is called zero and is denoted by θ. The point distribution w can be interpreted as follows: for $i = 1, \ldots, k$ there are n_i particles at the point x_i. The set of all point distributions on $\overline{D}$ is denoted by $\Omega^0_{\overline{D}}$. For a point distribution w, the set function $\widetilde{w}$ is introduced: $\widetilde{w}(A) = \sum\limits_{x_i \in A} n_i$, which is a measure on the measurable space $(\overline{D}, \mathcal{F}(\overline{\mathcal{D}}))$, where $\mathcal{F}(\overline{D})$ is the σ-algebra of all subsets of $\overline{D}$. The correspondence between w and $\widetilde{w}$ is one-to-one, so in what follows w and $\widetilde{w}$ are identified. A separable σ-algebra C^* is distinguished in $\mathcal{F}(\overline{D})$.

Definition 2 Random integral of a bounded function $h(x)$ over a point distribution $w = (x_1, n_1; x_2, n_2; \ldots; x_k, n_k)$ called a number

$$\int h\, dw = \int\limits_{\overline{D}} h(x)\, dw(x) = \sum_{i=1}^{k} n_i\, h(x_i).$$

If $w = \theta$, then $\int h\, dw = 0$. Let $\mathbf{P}$ be a probability measure on $\left(\Omega^0_{\overline{D}}, C^*\right)$, $0 \leq s(x)$ is a bounded function on $\overline{D}$.

Definition 3 Functional $\Phi(s) = \int\limits_{\Omega_{\overline{D}}} \exp\left(-\int s dw\right) d\mathbf{P}(w)$ is called the generating functional of the measure $\mathbf{P}$.

Definition 4 A branching process on $\overline{D}$ is a homogeneous Markov stochastic process with phase space $\left(\Omega_{\overline{D}}, C^*\right)$, transition function $P(w, B)$ of which satisfies the condition $P(w, B) = P\left\{Z_{n+1} \in B \,|Z_n = w\right\}$.

As is known, $\Phi(w, s) = \prod\limits_{i=1}^{k} \left[\Phi_{x_i}(s)\right]^{n_i}$, where $\Phi(w, s)$, $\Phi_{x_i}(s)$ are the generating functionals of the measures $P(w, \cdot)$, $P((x_i, 1), \cdot)$, respectively. To define a process, it is necessary to define the transition function $P((x, 1), B)$, $x \in \overline{D}$, $B \in C^*$. It follows from the definition of C^* that it suffices to define $P((x, 1), B)$ on the algebra of cylindrical sets [22], which have the form

$$B = \{w(A_1) = r_1,\ w(A_2) = r_2,\ \ldots\ , w(A_n) = r_n\},\quad n = 1, 2, \ldots,$$

where $A_1, A_2, \ldots$ is a countable family of $\overline{D}$ subsets generating a Borel σ-algebra in it. We define $P((x, 1), B)$ by the equalities

$$P((x,1),B) = P\left((x,1), B\bigcap\left(\bigcup_{i=0}^{\infty}\{w(\overline{D})=i\}\right)\right) =$$

$$= \sum_{i=0}^{\infty} P\left((x,1), B\bigcap\{w(\overline{D})=i\}\right),$$

where

$$P((x,1), B\cap\{\theta\}) = I_B(\theta)\cdot\begin{cases}\pi_0(1-q(x)), & x\in D;\\ 0, & x\in\Gamma;\end{cases}$$

$$P((x,1), B\cap\{w(\overline{D})=1\}) =$$

$$= \begin{cases} q(x)\ P\left(I_{A_1}(x_1)=r_1,\ldots,I_{A_n}(x_1)=r_n\right) + \\ +\pi_1(1-q(x))\ P\left(I_{A_1}(x_2)=r_1,\ldots,I_{A_n}(x_2)=r_n\right), & x\in D;\\ I_B((x,1)), & x\in\Gamma; \end{cases} \tag{10}$$

$$[P((x,1), B\cap\{w(\overline{D})=1\}) =$$

$$= \begin{cases} q(x)\ P\left(I_{A_1}(x_1)=r_1,\ldots,I_{A_n}(x_1)=r_n\right) + \\ +\pi_1(1-q(x))\ P\left(I_{A_1}(x_2)=r_1,\ldots,I_{A_n}(x_2)=r_n\right), & x\in D;\\ I_B((x,1)), & x\in\Gamma; \end{cases} \tag{11}$$

$$P((x,1), B\cap\{w(\overline{D})=i\}) =$$

$$\begin{cases} \pi_i(1-q(x))\ P\left(iI_{A_1}(x_2)=r_1,\ldots,iI_{A_n}(x_2)=r_n\right), & x\in D;\\ 0, & x\in\Gamma. \end{cases} \tag{12}$$

Here $I_A(x)$ is the indicator of the set A, x_1 is a point uniformly distributed on the sphere $S_R(x)$, x_2 is a point distributed in $K_R(x)$ with density $p(x, y)$, $i = 2, 3, \ldots$. Next, we give a method for modeling the transition $x_n \to x_{n+1}$. If a point x_{n+1} is uniformly distributed on the sphere $S_R(x_n)$, then $x_{n+1} = x_n + R\omega_n$, where ω_n is an isotropic vector. If the distribution density of the point x_{n+1} is equal to $p(x_n, y)$ for a fixed x_n, then it is advisable to use the [19] elimination method for modeling:

I. $\alpha_n^0, \alpha_n^1, \alpha_n^2$ are simulated - random variables uniformly distributed on the segment (0; 1); the value $\xi = -\ln\left(\alpha_n^1\alpha_n^2\right)/\sqrt{c}$ is determined; if $\xi > R$, then the same thing is repeated, etc., otherwise the value $\zeta = \alpha_n^0\ \xi \exp(-\sqrt{c}\xi)$ is chosen;
II. if $\zeta \geq \xi \frac{sh\left[(R-\xi)\sqrt{c}\right]}{sh(R\sqrt{c})}$ then I is executed, etc., otherwise $x_{n+1} = \xi$.

Let $Z_0 = (x, 1),\ Z_1,\ Z_2,\ \ldots,\ Z_n,\ \ldots$ be the trajectory of the branching process, where $Z_i = (x_i^1, n_i^1; x_i^2, n_i^2; \ldots; x_i^{l_i}, n_i^{l_i})$ is the point distribution at time i. Let us prove the following lemma.

Lemma 1 *A random branching process $\{Z_n\}$ with probability one either degenerates in the domain D or converges to a point distribution $Z = (x^1, n^1; x^2, n^2; \ldots; x^k, n^k)$, where $x^i \in \Gamma$, $i = 1, 2, \ldots k$.*

Proof The process can be considered as a branching process for particles diffusing in a limited region $\overline{D}$ with absorbing boundaries. If the average number of particles K resulting from the division of one particle is $K = K(x, 1) \leq 1$, then the constructed process terminates with probability 1, and the condition $K < 1$ is necessary and sufficient for boundedness of the average number of process forks. Let us show that, $K < 1$:

$$K = q(x)\int\limits_{\overline{D}} d\omega + (1 - q(x))\int\limits_{\overline{D}} p(x, y)\sum_{n=0}^{\infty}(n\pi_n)dy = q(x) + (1 - q(x)) \cdot \alpha < 1.$$

Consequently, this process degenerates and the total number of particles involved in the process is finite. If the process does not terminate inside D, then, starting from some moment, all particles reach the boundary Γ, i.e., there exists $n = n_0$, starting from which Z_n has the form $(x_n^1, 1;\ x_n^2, 1;\ \ldots;\ x_n^k, 1)$, where k does not depend on n. In what follows, all k particles perform independent walks over the spheres. It follows from the results of [21] that $x_n^i \to x^i \in \Gamma$ for $n \to \infty$ a.s. The lemma is proven.

Note that for the process of walk on spheres with branching, by consequence of the corollary to above given Theorem , the average number of branches will be finite even in the case when K is the average number of particles resulting from the division of one particle is equal to 1 (i.e., $b = 1$). Therefore, the value of the parameter $\alpha = 1$ for which $K = 1$ is valid. Next, on the trajectory of walk on spheres in branching, we construct unbiased and ε-biased estimates for the solution.

4 Construction of an Unbiased Estimators of the Solutions

To construct an estimators of Monte Carlo method for problem (2), the process of walk on spheres with branching from the point $x_0 = x$ is implemented, i.e. at the beginning there is one particle at the point x. Per unit time, the particle with probability $q = \frac{R\sqrt{c}}{sh(R\sqrt{c})}$ passes to the point $x_1 = y_1$, which is chosen uniformly over the sphere $S_R(x)$ and with probability $1 - q$ to the point $x_1 = y_2$, which is chosen inside the ball $K_R(x)$ according to the density $p(x, y)$. In the second case, branching occurs with probability π_n ($n \neq 0$), i.e. the particle is divided into n particles and the estimator is multiplied by $W = \frac{g \cdot sh(1)}{c \cdot \alpha}$. In the case of $n = 0$, the particle is absorbed at the point x_1 and the estimator is multiplied by $W_0 = \frac{g}{c\pi_0}$. n independent walk trajectories are constructed from the point x_1. The new particles behave independently of each other in a similar way to the considered particle. The process terminates either if all particles in $\overline{D}$ die, or if all particles reach the ε neighbourhood of the boundary.

Let us define an estimator in a recursive way. Let $\zeta_0(x) = u(x)$, $\zeta_k(x) = \Psi(\zeta_{k-1}(x))$, where

$$\Psi(\zeta(x)) = \begin{cases} \zeta(y_1), & \text{with probability } q(x); \\ W \cdot \prod_{i=1}^{n} \zeta^{(i)}(y_2), & \text{with probability } (1 - q(x))\pi_n, \quad n \neq 0; \\ W_0, & \text{with probability } (1 - q(x))\pi_0. \end{cases}$$

Here $\zeta^{(i)}(y)$ are independent realizations of the random variable $\zeta(y)$. Let $\mathfrak{R}_k$ be the σ-algebra generated by the sequences $\{\omega_i\}_{i=0}^{k-1}$, $\{\alpha_i^0\}_{i=0}^{k-1}$, $\{\alpha_i^1\}_{i=0}^{k-1}$, $\{\alpha_i^2\}_{i=0}^{k-1}$. The following assertion is true.

Theorem 1 *The sequence* $\{\zeta_k(x)\}_{k=0}^{\infty}$ *forms a martingale with respect to* $\{\mathfrak{R}_k\}_{k=0}^{\infty}$. *If* $g \cdot sh(1) < c$, *then* $\zeta_k(x, t)$ *is a uniformly integrable martingale.*

Proof It is obvious from the definition of $\mathfrak{R}_k$ that $\zeta_k(x)$ is $\mathfrak{R}_k$-measurable. From the properties of conditional mathematical expectations it follows that:

$$\mathbf{E}(\zeta_{n+1}(x)/\mathfrak{R}_n) = \mathbf{E}\left(\Psi(\zeta_n(x))/\mathfrak{R}_n\right) = q(x)\mathbf{E}\left(\zeta_n(y_1)/\mathfrak{R}_n\right) +$$

$$+ (1 - q(x)) \sum_{i=1}^{\infty} \pi_n \, \mathbf{E}\left(W \cdot \prod_{j=1}^{i} \zeta_n^{(j)}(y_2)/\mathfrak{R}_n \right) + (1 - q(x))\pi_0 \mathbf{E}\left(W_0/\mathfrak{R}_n\right).$$

In view of the validity of the probabilistic representation (9), we get:

$$\mathbf{E}(\zeta_{n+1}(x)/\mathfrak{R}_n) = q(x) \int_{S_R(x)} \zeta_n(y) d\omega +$$

$$+ (1 - q(x)) \int_{K_R(x)} p(x, y) \left[\sum_{i=1}^{\infty} \pi_n \frac{g \cdot sh(1)}{c\alpha} \zeta_n^i(y) + \pi_0 \frac{g}{c\pi_0} \right] = \zeta_n(x).$$

Thus the sequence $\{\zeta_k(x)\}_{k=0}^{\infty}$ forms a martingale with respect to $\{\mathfrak{R}_k\}_{k=0}^{\infty}$. To prove the uniform integrability of $\zeta_k(x)$, it suffices to show that $|\zeta_k(x)| \leq C$, ($C = const$). Since the parameter α is chosen from the condition $g \cdot sh(1)/c \leq \alpha < 1$.

Since $u(x) \in C(\overline{D}) \cap C^2(\overline{D})$ and $\overline{D}$ is a bounded domain, then $|u(x)| \leq const$ for $(x) \in \overline{D}$. Further, under the conditions of the theorem, $W \leq 1$, and hence $|\zeta_n(x)| \leq C < \infty$, $(C = const)$. Whence it follows that the sequence $\{\zeta_n(x, t)\}$ is uniformly integrable. The theorem has been proven.

In order to reduce the time of obtaining one implementation of the estimator for the solution of problem (2'), we consider a slightly modified process with a smaller number of branches and estimates on it with an arbitrarily small bias. Let us take ε- small enough and consider the inner ε- neighbourhood of the boundary: Γ_ε. Let N_1 be the moment of termination of the process inside the region, and N_ε be the moment of the first hit of all particles in Γ_ε. $N = \min\{N_1, N_\varepsilon\}$—process stop time. Then the probability of breaking the trajectory at the point x_n will be equal to:

$$g(x_n) = \begin{cases} 1, & x_n \in \Gamma_\varepsilon, \\ (1 - q(x_{n-1}))\pi_0, & x_n \in \overline{D} \backslash \Gamma_\varepsilon. \end{cases}$$

It follows from Lemma 1 and above Corollary, that $N < \infty$.

Theorem 2 *If $g \cdot sh(1) < c$, then $\zeta_N(x)$ is an unbiased estimator for $u(x)$ with finite variance.*

Proof Since $\zeta_n(x)$ is a uniformly integrable martingale and N is a Markov moment, then according to Doob's free choice transformation theorem [23] for the martingale $\{\zeta_k(x)\}_{k=0}^{\infty}$, we get $\mathbf{E}\zeta_N(x) = \mathbf{E}\zeta_1(x)$. From the definition of $\zeta_1(x)$ and the probabilistic representation (8) it follows that

$$\mathbf{E}(\zeta_1(x)) = q(x)\mathbf{E}u(y_1) + (1 - q(x))\mathbf{E}f(y_2, u(y_2))/c = u(x).$$

By consequence of the conditions of Theorem 1, $\mathbf{E}(\zeta_N(x))^2 < \infty$ and hence its variance is finite. The theorem has been proven.

Next, from $\zeta_N(x)$ a biased but practically realizable estimator $\zeta_N^*(x)$ is constructed in the standard way. Let x^* be the boundary point closest to x Γ. $\zeta_N^*(x)$ is obtained by replacing $u(x_N)$ in $\zeta_N(x)$ with $\varphi(x_N^*)$. Let us estimate the displacement $\zeta_N^*(x)$. It's clear that $|\mathbf{E}\zeta_N^*(x) - u(x)| \leq \mathbf{E}|\zeta_N^*(x) - \zeta_N(x)|$. If $N = N_1$, then $T_N = \{\theta\}$ and the process terminates without hitting Γ_ε. In this case $\zeta_N^*(x) = \zeta_N(x)$ and the offset is zero. If $N = N_\varepsilon$, then $Z_N = (x_N^1, n_N^1; x_N^2, n_N^2; \ldots; x_N^k, n_N^k)$, where $x_N^i \in \Gamma$, $i = 1, 2, \ldots, k$ and the number k does not depend on N. Given that $|W_n(y)| \leq 1$ for arbitrary n and $y \in \overline{D}$, we have

$$\zeta_N^*(x) - \zeta_N(x) \leq \left(\prod_{i=1}^{k} [\varphi(x_N^i*)]^{n_N^i} - \prod_{i=1}^{k} [u(x_N^i*)]^{n_N^i} \right).$$

Let $L(\varepsilon)$ be the modulus of continuity of the function $u(x)$; then

$$|\mathbf{E}\zeta_N^*(x) - u(x)| \leq L(\varepsilon)\mathbf{E}(n_N^1 + n_N^2 + \cdots + n_N^k).$$

Considering that the average number of particles in the Nth generation

$$\mathbf{E}\left(n_N^1 + n_N^2 + \cdots + n_N^k\right) \leq K^N < 1,$$

we find that the displacement does not exceed $L(\varepsilon)$. The finiteness of the variance $\zeta_N^*(x)$ follows from $|W_n(y)| \leq 1$.

5 The Numerical Experiment

The results of numerical experiment are given in this section. The following problems were considered:
Problem 1:

$$\begin{cases} -\Delta u(x) + 4u(x) = \mathrm{sh}(u) + f_0, & x \in D = \{(x_1, x_2, x_3) : x_1^2 + x_2^2 + x_3^2 \leq 1\}, \\ u(x) = 0.4\exp(-x_1 - x_2 - x_3), & x \in \partial D, \end{cases}$$

where $f_0(x) = 0.4\exp(-x_1 - x_2 - x_3) - sh(0.4\exp(-x_1 - x_2 - x_3))$.
Problem 2:

$$\begin{cases} -\Delta u(x) + 4u(x) = \mathrm{ch}(u) + f_0, & x \in D = \{(x_1, x_2, x_3) : x_1^2 + x_2^2 + x_3^2 \leq 1\}, \\ u(x) = 0.4\exp(-x_1 - x_2 - x_3), & x \in \partial D, \end{cases}$$

where $f_0(x) = 0.4\exp(-x_1 - x_2 - x_3) - ch(0.4\exp(-x_1 - x_2 - x_3))$.

Denote (x_0) as the point where the problem is solved. α is a parameter, which is used in π_n and regulating the quantity of the branches. Let's designate the average quantity of the particles of the tree as AQP and the average number of branches as AQB. U_{ex} is the exact solution in point x_0, that is $U_{ex} = u(x_0)$. MS is the estimator of $MS = \bar{\xi} = \frac{1}{N_t}\left(\xi_1 + \cdots + \xi_{N_t}\right)$, where $\xi = \xi_N(x0)$ and ξ_i are independent realizations of the random estimate for the solution ξ. σ is the statistical estimator of the value $\sqrt{D\xi/N}$, where $D\xi$ is the variance of the estimator of the solution of our given problem. 3σ is the confidence interval for the estimator. Let's designate err as the difference between the exact solution and estimator, $err = |u(x_0) - MS|$. For the computational experiment, the values $N = 5000$ (the number of simulated trajectories) and $\varepsilon = 0.005$ (the neighborhood of the boundary) were used The results of the computational experiment are given in the following table (Table 1).

The results of computational experiment show that with the discussed algorithms we can obtain practically efficient estimators. The parameter α, which is used in calculating the probabilities of the branching, gives us a chance to control the average quantity of the branches and average number of the particles in the tree. In our case, with the probability 0.997, the exact solution of the problem will be in the confi-

Table 1 The results of computational experiments

Problem	α	x_0	U_{ex}	MS	err	3σ	AQB	AQP
Pr1	0,9	(0,25;0,75;0,5)	0,08925	0,08858	0,00067	0,00301	1,02	11,17
Pr1	0,9	(0,5;0,5;0,5)	0,08925	0,09135	0,00210	0,00577	1,04	12,26
Pr1	0,9	(-0,5;0,1;0,6)	0,32749	0,32742	0,00007	0,01178	1,06	12,90
Pr1	0,9	(-0,5;0,3;-0,7)	0,98384	0,98397	0,00013	0,01564	1,03	11,90
Pr2	0,9	(-0,5;0,3;-0,7)	0,98384	0,97398	0,00987	0,01541	0,96	11,42
Rr2	0,9	(0,5;0,5;0,5)	0,08925	0,08906	0,00019	0,00470	0,95	11,94
Pr2	0,9	(0,1;0,7;-0,5)	0,29633	0,29438	0,00195	0,00869	0,95	11,94
Pr2	0,9	(0,2;-0,3;-0,6)	0,80550	0,79775	0,00775	0,02367	0,89	12,51
Pr1	0,9	(0,2;-0,3;-0,6)	0,80550	0,81204	0,00653	0,02602	1,08	13,13

dence interval $\left(\bar{\xi} - 3\sigma, \bar{\xi} + 3\sigma\right)$. Since in the given examples, the exact solutions are known, we can make sure that all the estimators are in the confidence interval (see table).

6 Conclusion

The application of Monte Carlo methods in various fields is constantly growing due to increases in computer capabilities. Increasing speed and memory, and the wide availability of multiprocessor computers, allows us to solve many problems using the Monte Carlo method. Error estimates can be empirically obtained for all types of problems in parallel way and ease of efficient parallel implementation.

In this paper we proposed Monte Carlo algorithms for the solution of the interior Dirichlet boundary value problem for the Helmholtz operator with a polynomial nonlinearity on the right-hand side.

Partial differential equations and Monte Carlo method, which have been used to model many phenomena in the natural sciences and engineering. In some cases, our method could be applied in the area reaction diffusion equations, homogenization of PDEs, stochastic dynamics, interacting particle systems, non-linear Poisson-Boltzmann equations, electrostatics in bio-molecular structure and dynamics and etc.

The preliminary results of on computational experiment show, that with the discussed algorithms we can obtain practically efficient estimators. The parameter, which is used in calculating the probabilities of the branching, gives us the opportunity to control the average quantity of branches and the average number of the particles in the constructed tree. Thus the estimators were constructed on trees with a minimal number of branches. In our approach we have used an estimator via absorption. In the future we could built an estimator via collision and hope will get a new results. The results of numerical experiments and applications to concert problems we are going to give in our next publications.

References

1. Bobenko, A.I. All constant mean curvature tori in R^3, S^3, H^3 in terms of theta-functions. Mathematische Annalen **290**(2), 209–246 (1991). http://eudml.org/doc/164816
2. Davis, M., McCammon, A.: Electrostatics in-molecular bio-structure and dynamics. Chem. Rev. **90**, 509–521 (1990)
3. Courant R.: Methods of Mathematical Physics: Partial Differential Equations, Vol. 2, p.852 (2008)
4. Dubinsky, Y.A.: Quasilinear elliptic and parabolic equations of any order. Russian Math. Surv. Moscow N.23(1)(139), 45–90 (1968)
5. Elepov, B.S., Mikhailov, G.A., Algorithm of "wandering over spheres" for the equation. Doklady Math. Moscow T.212. **1**, 15–18 (1973)
6. Elepov, B.S., Mikhailov, G.A.: Using fundamental solutions of elliptic equations to construct algorithms for the Monte Carlo method. Comput. Math. Math. Phys. - Moscow **14**(3), 728–736 (1974)
7. Mikhailov, G.A.: Solving the Dirichlet problem for nonlinear elliptic equations by the Monte Carlo method. Siberian Math. J. - Novosibirsk,: T.35. **5**, 1085–1093 (1994)
8. Sipin, A.S.: Solution of the Dirichlet Problem for the Equation - $\Delta u(x)$ + a(x)u = f(x) by Monte Carlo Methods, Vestnik of Leningrad State University. Ser. Math. Mech. Astr. **1**, 60–63 (1976). (in Russian)
9. Simonov N.A.: Algorithms for random walk over spheres for solving a mixed boundary value problem and the Neumann problem. Numerical Analysis and Applications, Novosibirsk. - T. 10, No.2, 209–220 (2009)
10. Sipin A.S.: Solution of two Dirichlet boundary value problems by the Monte Carlo method. Comput. Math. Math. Phys. Moscow **19**(2), 388–401 (1979)
11. Mikhailov G.A., Makarov R.N.: Solving boundary value problems of the second and third kind by Monte Carlo methods. Sib. Math. J. Novosibirsk **38**(3), 603–614 (1997)
12. Mikhailov G.A., Makarov R.N.: Solution of boundary value problems by the method of "walking on spheres" with reflection from the boundary. Doklady Math. Moscow T.353, No.6, 720–722 (1997)
13. Fenley M., Mascagni M., McClain J., Silalahi A.A., Simonov N.: Using correlated Monte Carlo sampling for efficiently solving the linearized Poisson-Boltzmann equation over a broad range of salt concentration. J. Chem. Theory Comput. **6**, 300–314 (2009)
14. Fleming, C., Mascagni, M., Simonov, N.: An efficient Monte Carlo approach for solving linear problems in bio-molecular electrostatics, Computational Science-ICCS. Springer **2005**, 760–765 (2005)
15. Rasulov A.S.: Monte-Carlo Method for Solving Nonlinear Problems, Monograph, Tashkent, Fan, (1992), p. 105 (in Russian)
16. Rasulov, A.S.: Monte Carlo algorithms for the solution Quasi-Linear Dirichlet boundary value problems of elliptical type. Math. Stat. Horizon Res. Publishing corporation **11**(2), 200–205 (2023)
17. Raimova G.M.: Probabilistic models for solving the Dirichlet problem for nonlinear elliptic equations. Uzbek Math. J. Tashkent No. 1, 114–123 (2017)
18. Raimova, G.M.: Probabilistic approach to solution of the Neumann problem for some nonlinear equation. Commun. Stat. - Simul. Comput. **45**(8), 2981–2990 (2016)
19. Ermakov S.M., Mikhailov G.A. Statistical modeling, 296 P., 2nd edn. Moscow: Nauka. (in Russian)
20. Fridman A.: Partial Differential Equations of Parabolic Type,Dover Publications p. 368 (2008)
21. Ermakov S.M., Nekrutkin V.V., Sipin A.S.: Random Processes for Classical Equations of Mathematical Physics. Kluwer Academic Publisher, p. 282 (1989)
22. Harris, T.E.: The Theory of Branching Processes. Springer, Berlin (1963)
23. Shiryaev, A.N.: Probability, p. 624. Springer, New York, NY (1996)

Strongly Entangling Neural Network: Quantum-Classical Hybrid Model for Quantum Natural Language Processing

J. Ismael Díaz-Ortiz, Axel Villanueva, and Francisco Delgado

Abstract One of the most used techniques to improve a Machine Learning model is to gather more data. An interesting field in Machine Learning is Sequence Modelling, having Natural Language Processing as the peak of the field. The capabilities of Quantum Computing have been growing recently entering the novel field of Quantum Machine Learning. In this paper, we propose a Quantum Natural Language Processing classification model named Strongly Entangling Neural Network. This model leverages the quantum advantage to imitate part of the behavior of a Recurrent Neural Network to process text data into the circuit and perform the classification task. This is accomplished by representing our data in a quantum circuit that relies heavily on the entanglement property of qubits. The results of our model have very favorable metrics, particularly obtaining a 97.70% of accuracy.

Keywords Quantum machine learning · Quantum natural processing · Quantum information

1 Introduction

Classical Machine Learning (CML) algorithms have proven to be powerful at finding the hidden patterns underneath large amounts of data. Nevertheless, the time and hardware requirements needed to train complex state-of-the-art models have been increasing in recent years. Together, the complexity, abstractness of the data, and desired tasks have also increased. With the rise in popularity and capabilities of quantum technologies and their counterintuitive algorithms of Quantum Machine Learning (QML) when compared with classical computing, it can be expected that QML algorithms running on quantum computers can outperform CML algorithms for classical computers [1]. A field recently growing at a rapid pace within CML is Natural Language Processing (NLP). The motivation behind NLP is to provide a computer with the ability to analyze, process, and find meaning inside large amounts

J. I. Díaz-Ortiz · A. Villanueva · F. Delgado (✉)
School of Engineering and Sciences, Tecnologico de Monterrey, Monterrey, Mexico
e-mail: fdelgado@tec.mx

D. Vlachos (ed.), *Mathematical Modeling in Physical Sciences*, Springer Proceedings in Mathematics & Statistics 446, https://doi.org/10.1007/978-3-031-52965-8_40

of Natural Language Data (NLD). Once abstracted from the human perspective about how language is seen, the phrases or sentences can be understood just as a sequence of data, being then NLP a problem for Sequence Modelling (SM) [2].

In classical computing there exist several NLP algorithms capable of extracting the desired information from the text depending on the desired task: classification, text segmentation, Named Entity Recognition, etc. [3–5]. In the field of Deep Learning, the most popular models are based on the use of Convolutional Neural Networks, Recurrent Neural Networks (RNN) [6] and Transformers/Attention-based models [7, 8]. Some state-of-the-art models in NLP using these previous base architectures are Generative Pre-trained Transformer [9], Bidirectional Encoder Representations from Transformers [10], and Hugging Face's BLOOM [11]. The performance and possibilities given by QML algorithms opened a new field for NLP: Quantum Natural Language Processing (QNLP) based on the modeling of language sequences taking the advantage of quantum systems, which is the main topic of this research centered particularly on QNLP for classification tasks. In Fig. 1a it can be seen in better detail

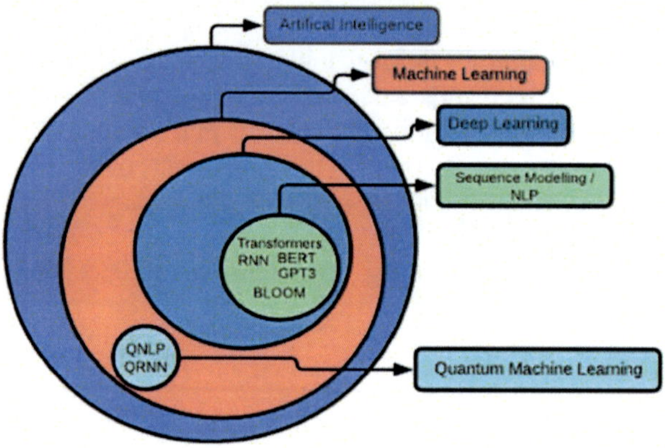

(a)

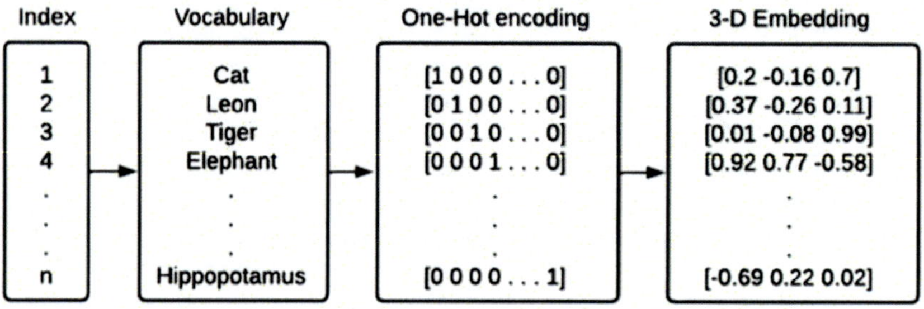

(b)

Fig. 1 **a** Interactions inside Artificial Intelligence with Deep Learning, Machine Learning, and Quantum Machine Learning. **b** One-hot encoding and Word embedding are illustrated. Closer words under certain criteria exhibit characteristic measures of similitude (inner product, distance measure, or another norm) among the corresponding vectors how CML and QML are a subcategory derived from Artificial Intelligence, and how the different fields and techniques are categorized in general.

In this paper is proposed an architecture for SM made specifically for QNLP: *Strongly Entangling Neural Network*. It is based on the use of Variational Quantum Circuits (VQC) from which are generated circuit neurons to mimic the general structure of an RNN, without some explicit connections. The skeleton of the model is based on the Circuit-Centric Quantum classifier proposed in [12]. The structure of the article is as follows. The Fundamentals section presents the general theory and background regarding the model explained in deeper detail. Their corresponding subsections will work as building blocks for the proposed model. Then, the model is explained in detail in the section Strongly Entangling Neural Network. Experiment outcomes are shown in the Results section. Finally, conclusions and future work are presented in the last section.

2 Fundamentals

The aim of this section is to describe the different elements and techniques used, as well as the mathematical foundations related to Quantum information and Quantum processing underlying our proposed model. Also, the subsections correspond to the elements used in the final model being presented in the order that they appear: Word Embedding (text to vector representation), Quantum Embedding (classical to quantum data), VQC (Quantum circuit with trainable parameters), Measurements and Loss function for the deployment and optimization of the model.

2.1 Words Embedding

When talking about CML, it is needed to understand that a computer is not capable of understanding anything but numbers. Depending on the desired task for the NLP model, the architecture may vary considering the inputs and outputs. For this research, it is sought that computers understand and perform classification. The inputs are a sequence of data in the form of text, which should be transformed into a numeric representation. This process is known as Text Vectorization (TV) [13].

One way of doing TV is to generate a one-hot-encoding for the whole vocabulary. That is, to generate a vector representation considering a vocabulary of length n, then obtain a vector representation for each word with the same length n. There, just one value in the vector is 1, otherwise, 0 [13]. The problem with this representation is that the size of the vector increases directly with the size of the vocabulary. It represents a memory consuming, not properly convenient for efficient calculations. In addition, there is no relation between the words in the vocabulary.

Despite this problem, another way to generate the TV is Word Embedding (WE). It consists of a dense vector representation of n dimensions (chosen by the practitioner depending on the data set), with floating-point values. Here, the values in the vector for every word are trainable parameters, so similar words (according to their meaning and/or context), once trained, have a similar or near vector representation. Figure 1b illustrates the one-hot-encoding and WE. It shows the transition from a set of words into a set of quantum states during the classical and quantum encoding (3D as an instance). Considering the benefits, WE was chosen as the TV algorithm for the proposed model.

2.2 Encoding Layer and Angle Embedding

The output of WE is still in the form of classical information, which naturally does not fit into a quantum circuit by itself. Therefore, an extra step called Quantum Embedding (QE) is needed. There, classical data is encoded into a quantum circuit. There are several methods of QE, some most used are: Amplitude Embedding, Basis Embedding, and Displacement Embedding, among others [14].

The method selected for the current model was Angle Embedding. Here, n features are encoded as the angles of rotations applied to n qubits after they are initialized in the state $|0\rangle^{\otimes n}$. X, Y or Z rotations can be used. A circuit example using $R_x(x_i)_i$ rotations on each qubit i is shown in Fig. 2. By introducing the classical information into the circuit in the form of rotations angles, the qubits initialized as $|0\rangle^{\otimes n}$ are now representatives of classical data as well as quantum data. In (1), it is shown how the classical data is encoded and introduced to the quantum circuit using the corresponding operation, considering an array of input features $X = [x_1, \ldots, x_n]$, where $S(X) = \bigotimes_{i=1}^{n} R_x(x_i)_i, x_i \in [-\pi, \pi]$) will describe the whole encoding block and $|\psi(x)\rangle$ is the state of the whole circuit after the encoding layer or encoding block. Then, it reads:

$$\begin{array}{c}\text{Word binary}\\ \text{representation}\end{array} \rightarrow [0, 1, \ldots, 0, 0] \rightarrow [x_1, \ldots, x_n] \rightarrow |\psi(x)\rangle = S(X)|0\rangle^{\otimes n} \quad (1)$$

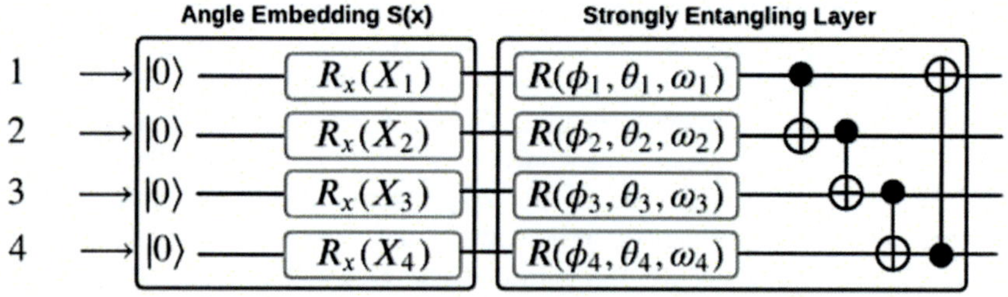

Fig. 2 EB $S(X)$ for the angle embedding and the Strongly Entangling VQC. This whole circuit corresponds to the Circuit Neuron used in the model

2.3 Variational Quantum Circuits and Ansatz

Analogous to how CML works, particularly in the field of Deep Learning, QML makes use of VQC, which as the name mentions, contains Variational parameters which are optimized during training depending on the architecture and task. With the use of VQC, a QML model can be described as follows (2). By considering a trainable set of parameters θ and input features X, $f(X, \theta)$ is the expectation value of some observable in the circuit while M is the observable. Then, $U(X, \theta)$ is VQC whose output depends on the input X and the trainable parameters set θ. $|\mathbf{0}\rangle$ corresponds to the initial state (there, $|\mathbf{0}\rangle = |0\rangle^{\otimes n}$). This equation considers that the encoding block described in the previous section is part of the VQC, $U(X, \theta)$:

$$f(X, \theta) = \langle \mathbf{0}|S(X)^{\dagger}U^{\dagger}(X, \theta)MU(X, \theta)S(X)|\mathbf{0}\rangle \tag{2}$$

There, the encoding layer will be referred as the Encoding Block (EB) $S(X)$. While the VQC having all trainable parameters doing the processing to get the proper output will be referred as the Training Block (TB) $U(\theta)$. Then, we define the initial state $|\psi(x)\rangle = S(X)|\mathbf{0}\rangle$ for the QML model after applying the EB.

$U(\theta)$ can be conformed by different layers depending on the whole circuit depth. In VQC, commonly several layers are used, each one conformed similarly but including its own trainable parameters. Thus, $U(\theta)$ can be described considering L layers as:

$$U(\theta) = U_L(\theta_L) \ldots U_2(\theta_2)U_1(\theta_1) \tag{3}$$

For this paper, the basis VQC used as a neuron for the whole model will only have one layer conformed by single qubit rotations $R(\phi, \theta, \omega)$ and 2-qubit unitary gates called *imprimitives*, which are not local gates, but controlled gates referred as entangling gates (as an instance, a controlled-NOT gate C^aNOT_b). As mentioned, in general, the single-qubit rotation depends on 3 parameters, which are the trainable parameters for that specific gate. The 3 parameter rotation is the product of other Pauli rotation gates. It is (4):

$$R(\phi, \theta, \omega) = R_z(\omega)R_y(\theta)R_z(\phi) = \begin{pmatrix} e^{-\frac{i}{2}(\phi+\omega)}\cos(\frac{\theta}{2}) & -e^{+\frac{i}{2}(\phi-\omega)}\sin(\frac{\theta}{2}) \\ e^{-\frac{i}{2}(\phi-\omega)}\sin(\frac{\theta}{2}) & e^{+\frac{i}{2}(\phi+\omega)}\cos(\frac{\theta}{2}) \end{pmatrix} \tag{4}$$

while, the controlled gate has the form (G^0 is the identity):

$$C^iG_j|i\rangle \otimes |j\rangle = \sum_{i=0}^{1} |i\rangle \otimes G^i|j\rangle \rightarrow C^iNOT_j|i\rangle \otimes |j\rangle = \sum_{i=0}^{1} |i\rangle \otimes NOT^i|j\rangle \tag{5}$$

The 1-layer VQC working as the basis neuron in this model is called *Strongly Entangling Circuit* (SEC). It gives the idea to prepare quantum states generating a strong entanglement inside each circuit neuron. It prepares quantum states generating a strong entanglement using the $C^a NOT_b$ gates inside each circuit neuron. The SEC $U(W)$ is described in (6), then we define the following indexes and quantities: (a) n = number of wires in the circuit; (b) r = range for the imprimitives, range to select the control wire, (c) c_j are the nodes for the controls, $(jr - r)$ in total; (d) $t_j = jr$ are the nodes for the controlled qubits, jr in total; and (e) $W_j = (\phi_j, \theta_j, \omega_j)$, the array with the trainable parameters (for each wire in the circuit). Based on this, the dimensions of the total array of trainable parameters, $W = [W_1 W_2 \ldots W_n]$, is $n \times 3$. Then:

$$U(W) = \prod_{k=1}^{n} C^{c_k} NOT_{t_k} \prod_{j=1}^{n} R(W_j) \tag{6}$$

2.4 Measurements and Loss Function

To generate this hybrid quantum-classical model there are needed certain interactions between both parts to build the whole model. The classical information was introduced in the Quantum Circuit after the angle embedding, and then the quantum states generated after embedding are processed in the circuit. In the QML model the output of the model can be taken directly from the circuit as a measurement in one of the states or certain measurements can be taken and processed afterward in a classical model so it ends up as a classical-quantum-classical model.

In the field of Quantum Information, a measurement is an operation made on a qubit, which is in superposition in the basis vectors, so a classical bit or classical result no longer in superposition is obtained. During the development of this research, we are not working with a mere measurement, but with the expectation value, which is the probabilistic expected value of a measurement on a qubit. So in the VQC to get the quantum data into classical data and continue with another stage of the model, there are needed to take the expectation values of different wires (qubits, states) in the circuit, which may be continuously vary depending on the rest of the circuit, particularly in the change of the trainable parameters during the training.

As mentioned before, the trainable parameters in the circuit are optimized during the training depending on two main elements: Optimization algorithm and Loss Function (Cost Function). The Loss Function takes the prediction-output $\hat{y}$ value of the whole circuit and calculates depending on the function the error with the expected value y. Depending on the model's task the Loss Function is selected, in this case as it will be explained in the following section, the selected task is sequence modeling for classification. Considering this fact, the selected Loss Function is called Binary Cross-entropy [15] which is used for models in which output is the probability that an input belongs to a particular class. The Loss function is presented in the following equation considering N training examples:

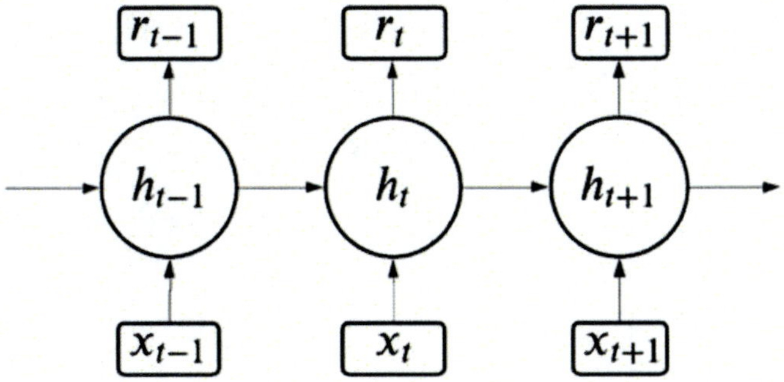

Fig. 3 Representation of some steps of the Recurrent Neural Network

$$L = -\frac{1}{N}\sum_{i=1}^{N} y_i \cdot \log(\hat{y}) + (1 - y_i) \cdot \log(1 - \hat{y}) \tag{7}$$

The Optimization algorithm, using the result of the Loss Function optimizes the weights or trainable parameters in the model by performing a modified version of a normal Gradient Descent, in which by back-propagating the error obtained by the Loss Function considering a differentiable model (function model), the trainable parameters get update till finding (ideally) a global minimum in which the values of the trainable parameters obtain the global minimum error. Specifically, we perform different tests using the following Optimization Algorithms: Stochastic Gradient Descent (SGD) [16] and Adam [17].

3 Strongly Entangling Neural Network

In the previous section, different elements were introduced and explained in detail to use them as construction blocks for the model presented in this section. The proposed model architecture has been called *Strongly Entangling Neural Network* (SENN). The Circuit Neuron presented in Fig. 2 is used as the basis for the creation of a model intending to imitate part of the behavior of an RNN. It is a type of neural network, where a closed loop between the units inside a layer generates a Forward Propagation among those hidden units. It causes the output of a unit can affect a subsequent one, then allowing a temporal dynamic behavior. The output of the unit $t-1$ affects the unit t, whose output then affects unit $t+1$ and so on, as illustrated in Fig. 3. RNNs are used for sequenced inputs with temporal behavior so that the behavior of previous inputs influences subsequent units.

The purpose is that SENN can be used for any kind of sequence modeling. However, the architecture will be explained for a specific use case: QNLP Classification Task. The model consists of 6 layers which can be either Classical Layers (CL) or

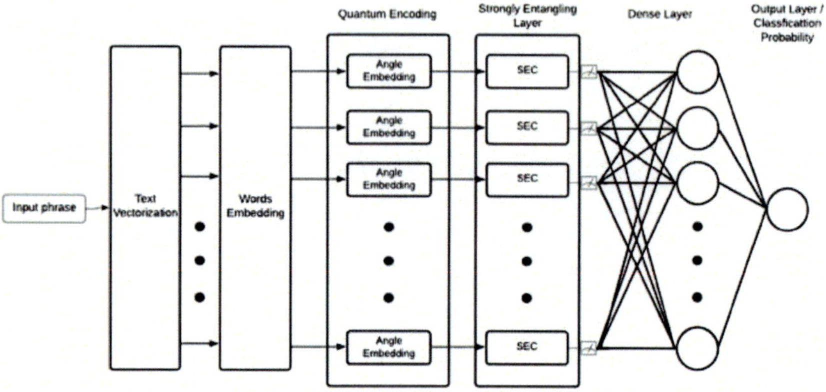

Fig. 4 Entire diagram for the Strongly Entangling Neural Network showing its several layers

Quantum Layers (QL): (1) TV Layer (CL), (2) Embedding Layer (CL), (3) Quantum Encoding/Angle Embedding (QL), (4) Strongly Entangling Layers (QL), (5) Dense Layer (CL), and (6) Output Dense Layer (CL). The whole model can be seen in Fig. 4. The TV layer is a preparing step for the Embedding layer which simply maps text features into integer sequences [18].

The data flow in the architecture is as follows: A phrase enters the TV Layer and gets converted into an integer sequence. It then enters in the Embedding Layer, where every word gets converted into a vector representation of size d (considering $d = 3$ in the experiment explained in the following section). Also, in Fig. 2, the same value for d was used. Considering n maximum words in each phrase entering into the circuit, then, the Embedding Layer outputs the same n number of vectors which now enters n Circuit Neurons as explained in the previous section (consisting of the Angle Embedding and the SEC as it can be seen in Fig. 2, a circuit neuron). The workflow generates each word converted into an embedding that will be processed by a Circuit Neuron. It was tried to generate a forced Forward Propagation between circuit neurons using some two-qubit entangling gates such as C^aNOT_b gates, but it was found that this decreased the accuracy of the model and increased the loss. Decisions about this point are discussed in the following section. Considering the length of the embedding, then: $Total\ number\ of\ qubits = embedding\ length \times words\ per\ phrase.$ To extract the information from the quantum circuits (n circuit neurons), it is needed to measure the qubits according to the explanation given in the next section. There, taking as mentioned, the expectation value using a Z Pauli Gate.

The measurements enter into a Dense Layer with a number of hidden units equal to the number of qubits using a *ReLU* (rectified linear unit) activation function [19]. Finally, the last layer is a Dense Layer with a single hidden neuron using *Sigmoid* activation function [19] considering a classification task having then a stochastic output. The whole architecture is by itself very similar to some other NLP architectures, just replacing the RNN, a Long Short-Term Memory layer for example, with the QE and the Strongly Entangling Layer, getting the model completely into the field of QML and giving the possibility of treat sequential data in the form of NLD in the context of Quantum Computing.

For the experimentation, a dataset provided by [20] was used, which contains phrases translated from English to Spanish. This dataset was originally used for Deep Learning models with translation tasks, but it was modified for its use in the model test by generating a labeled dataset of phrases in both, English and Spanish, having a 1−label for phrases in English and a 0−label for phrases in Spanish. With this dataset in mind, the experiment consisted of using the model to create a text sorter for English and Spanish. The dataset has more than 230, 000 training examples but considering the computational cost for the Proof of Concept, a reduced version from the original dataset was used. The model was trained for just 25 epochs, to see the performance and evolution considering a small training time. The number of epochs is defined as the number of times that the model sees the whole training dataset, which means that is the number of times that the model iterates and optimizes the trainable variables over the dataset.

All the programming for the model and experimentation was done in Python environment using two main libraries: (a) Tensorflow [18] to generate the model and for the training; and (b) PennyLane [21] for the generation of all the Quantum layers. PennyLane has integrated a tool allowing to convert a circuit into a Tensorflow Layer, letting the integration of the Classical Layers and the Quantum Layers into a single model. PennyLane has different devices available for the simulation of quantum circuits. In this case, the configuration *Default.Qubit* [21] was used. The results of the experimentation performed are shown and discussed in the following section.

4 Results

The experiments were done considering the details explained in the last section. After 25 epochs the model achieved a Loss of 0.0768, Validation Loss of 0.1641, Accuracy of 0.9770, and Validation Accuracy of 0.9100. The validation of the model is an extra step in each epoch in which a Validation Dataset different from the training dataset is passed to the model to evaluate how the model performs with data that it has never seen before. Weights are not optimized during validation. The Losses and Accuracy changes during training can be seen in Fig. 5.

To get these results different decisions were taken related to the number of qubits used, embedding length, and used gates in circuits, among other parameters of the model. As a summary of the model, in this case, the following values were selected for all these parameters: (a) Embedding Length: 3, (b) Imprimitive 2-qubit gate: C^aNOT_b, (c) Maximum length for the phrases: 4 words, (d) Pauli gate used for the expectation values: Z, and using the previous values we can calculate (e) $Total\ number\ of\ qubits = embedding\, length\ \times\ words\ per\ phrase = 12\, qubits$. As it can be seen in Fig. 5, Loss and Validation loss decrease properly in every epoch. Similarly, Accuracy and Validation Accuracy increases properly in

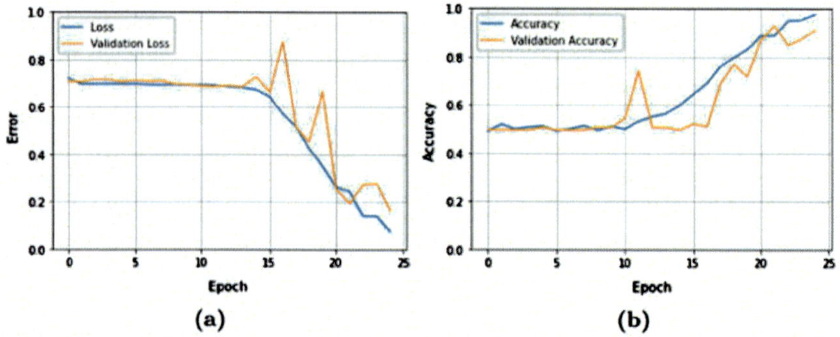

Fig. 5 **a** Loss and Validation Loss during training. **b** Accuracy and Validation Accuracy during training

every epoch. As can be appreciated in the same figure, there are no major changes in Loss or in Accuracy during the first 10 epochs. This is since the fact that in the model the weights in the Quantum circuits were initialized randomly, which correspond to the rotation angles in the circuits, so the model needs to learn completely by itself the ranges of the values for these angles. But, once that finds the proper pattern, it starts to improve rapidly, increasing the accuracy to around 0.47 in less than 15 epochs.

As mentioned before and as it can be seen in Fig. 4, the model still does not use an explicit Forward propagation between the circuit neurons, but instead takes the expectation value in every qubit and uses an extra dense layer, where connections between expectation values of different qubits are inferred by the model during training. This was expected to affect the model performance, but that is not the case as can be seen in the results, having then a quantum-classical hybrid model that is able to encode text and learn from it to perform a specific task. This forward propagation is needed to generate more complex models based on this one. Details about this point are discussed in the following section where conclusions and further work are mentioned.

5 Conclusions

During research and experimentation, we were able to encode NLD in a Quantum-Classical hybrid model, which can learn from the data and perform properly a classification task. At first instance, it was desired to have a model analog to an RNN with an explicit forward propagation using C^aNOT_b gates between circuit neurons. But during experimentation, it was found that since the circuit neurons use a Strongly Entangling Circuit with C^aNOT_b gates, the use of extra C^aNOT_b between circuit neurons will generate an entanglement of all the qubits in the model, then complicating more the learning task and affecting the performance. Due to these reasons, it was discarded to use a 2-qubit entangling gate to generate this forward propagation. It was chosen to leave the model without this step since it got proper results in this simpler version. Despite this, it is left to future research to find some mechanism to generate forward propagation. In fact, it is possible to use this base to build more complex models, for example applying SENN as a base to generate a Quantum Attention model or a Quantum Transformer model.

For this research, the focus has been put almost uniquely on QNLP, leaving behind other Sequence Modelling problems. In future work, it is planned to implement SENN with different kinds of sequence data to get a complete generalization. For the current proposal, the adequate use of quantum circuits and the proper interactions between quantum and classical parts have allowed taking advantage by exploiting the features of quantum processing reaching a high accuracy in just a few epochs. The classification task solved for the experimentation can be considered simple, but it still can be grown to more complex problems and architectures in different fields of SM and particularly in QNLP.

Acknowledgements J. Ismael Díaz-Ortiz acknowledges Emmanuel Díaz the unconditional support given at every time. The authors would like to acknowledge the financial support of NOVUS (Grant number: 214), Institute for the Future of Education, Tecnologico de Monterrey, Mexico, in the production of this work.

References

1. Biamonte, J., Wittek, P., Pancotti, N., Rebentrost, P., Wiebe, N., Lloyd, S.: Quantum machine learning. Nature **549**(7671), 195–202 (2017)
2. Yousuf, H., Lahzi, M., Salloum, S.A., Shaalan, K.: A systematic review on sequence-to-sequence learning with neural network and its models. Int. J. Elec. Comp. Eng. **11**(3), 2315 (2021)
3. Manning, C.D., Schutze, H.: Foundations of Statistical Natural Language Processing (Chap. 16), pp. 575–608. MIT Press, England (1999)
4. Koshorek, O., Cohen, A., Mor, N., Rotman, M., Berant, J.: Text segmentation as a supervised learning task. In: Proceedings of the 2018 Conference of the North American Chapter of the Association for Computational Linguistics: Human Language Technologies, vol. 2 (Short Papers), pp. 469–473. Association for Computational Linguistics, New Orleans, Louisiana (2018)
5. Tjong Kim Sang, E.F., De Meulder, F.: Introduction to the CoNLL-2003 shared task: Language-independent named entity recognition. In: Proceedings of Seventh Conference on Natural Language Learning at HLT-NAACL 2003, pp. 142–147 (2003)
6. Goodfellow, I., Bengio, Y., Courville, A.: Deep Learning (Chaps. 9–10), pp. 326–415. MIT Press (2016)
7. Vaswani, A., Shazeer, N., Parmar, N., Uszkoreit, J., Jones, L., Gomez, A.N., Kaiser, L., Polosukhin, I.: Attention is all you need (2017)
8. Tunstall, L., von Werra, L., Wolf, T.: Natural Language Processing with Transformers: Building Language Applications with Hugging Face. O'Reilly Media Inc. (2022)
9. Brown, T.B., Mann, B., Ryder, N., Subbiah, M., Kaplan, J., Dhariwal, P., Neelakantan, A., Shyam, P., Sastry, G., Askell, A., Agarwal, S., Herbert-Voss, A., Krueger, G., Henighan, T., Child, R., Ramesh, A., Ziegler, D.M., Wu, J., Winter, C., Hesse, C., Chen, M., Sigler, E., Litwin, M., Gray, S., Chess, B., Clark, J., Berner, C., McCandlish, S., Radford, A., Sutskever, I., Amodei, D.: Language models are few-shot learners (2020)

10. Devlin, J., Chang, M.W., Lee, K., Toutanova, K.: BERT: Pre-training of deep bidirectional transformers for language understanding. In: Proceedings of the 2019 Conference of the North American Chapter of the Association for Computational Linguistics: Human Language Technologies, vol. 1 (Long and Short Papers), pp. 4171–4186. Association for Computational Linguistics, Minneapolis, Minnesota (2019)
11. Mitchell, M., Pistilli, G., Jernite, Y., Ozoani, E., Gerchick, M., Rajani, N., Luccioni, S., Solaiman, I., Masoud, M., Nikpoor, S., Ferrandis, C.M., Bekman, S., Akiki, C., Contractor, D., Lansky, D., McMillan-Major, A., Thrush, T., Ilić, S., Dupont, G., Longpre, S., Dey, M., Biderman, S., Kiela, D., Baylor, E., Scao, T.L., Gokaslan, A., Launay, J., Muennighoff, N.: Large open-science open-access multilingual language model (2022). https://huggingface.co/bigscience/bloom
12. Schuld, M., Bocharov, A., Svore, K.M., Wiebe, N.: Circuit-centric quantum classifiers. Phys. Rev. A **101**(3) (2020)
13. Vajjala, S., Majumder, B., Gupta, A., Surana, H.: Practical Natural Language Processing. O'Reilly Media, Sebastopol, CA (2020)
14. Schuld, M., Petruccione, F.: Supervised Learning with Quantum Computers. Springer International Publishing (2018)
15. Ruby, U., Yendapalli, V.: Binary cross entropy with deep learning technique for image classification. Int. J. Adv. Trends Comput. Sci. Eng. **9**(10) (2020)
16. Ruder, S.: An overview of gradient descent optimization algorithms (2017)
17. Kingma, D.P., Ba, J.: Adam: a method for stochastic optimization (2017)
18. Abadi, M., Agarwal, A., Barham, P., Brevdo, E., Chen, Z., Citro, C., Corrado, G.S., Davis, A., Dean, J., Devin, M., Ghemawat, S., Goodfellow, I., Harp, A., Irving, G., Isard, M., Jia, Y., Jozefowicz, R., Kaiser, L., Kudlur, M., Levenberg, J., Mané, D., Monga, R., Moore, S., Murray, D., Olah, C., Schuster, M., Shlens, J., Steiner, B., Sutskever, I., Talwar, K., Tucker, P., Vanhoucke, V., Vasudevan, V., Viégas, F., Vinyals, O., Warden, P., Wattenberg, M., Wicke, M., Yu, Y., Zheng, X.: TensorFlow: Large-scale machine learning on heterogeneous systems (2015). https://www.tensorflow.org/
19. Nwankpa, C., Ijomah, W., Gachagan, A., Marshall, S.: Activation functions: comparison of trends in practice and research for deep learning (2018)
20. Kelly, C.: Tab-delimited bilingual sentence pairs (2022). https://www.manythings.org/anki/
21. Bergholm, V., Izaac, J., Schuld, M., Gogolin, C., Ahmed, S., Ajith, V., Alam, M.S., Alonso-Linaje, G., AkashNarayanan, B., Asadi, A., Arrazola, J.M., Azad, U., Banning, S., Blank, C., Bromley, T.R., Cordier, B.A., Ceroni, J., Delgado, A., Matteo, O.D., Dusko, A., Garg, T., Guala, D., Hayes, A., Hill, R., Ijaz, A., Isacsson, T., Ittah, D., Jahangiri, S., Jain, P., Jiang, E., Khandelwal, A., Kottmann, K., Lang, R.A., Lee, C., Loke, T., Lowe, A., McKiernan, K., Meyer, J.J., Montañez-Barrera, J.A., Moyard, R., Niu, Z., O'Riordan, L.J., Oud, S., Panigrahi, A., Park, C.Y., Polatajko, D., Quesada, N., Roberts, C., Sá, N., Schoch, I., Shi, B., Shu, S., Sim, S., Singh, A., Strandberg, I., Soni, J., Száva, A., Thabet, S., Vargas-Hernández, R.A., Vincent, T., Vitucci, N., Weber, M., Wierichs, D., Wiersema, R., Willmann, M., Wong, V., Zhang, S., Killoran, N.: Pennylane: automatic differentiation of hybrid quantum-classical computations (2022)

COVID-19 and Virus Spreading Mathematical Modeling

Mathematical Analysis and Optimal Strategy for a COVID-19 Pandemic Model with Intervention

Padma Bhushan Borah and Hemanta Kumar Sarmah

Abstract The COVID-19 pandemic has spread to every corner of the globe. The virus was first spotted in Wuhan, China, in December 2019 and has since spread worldwide. It has impacted every one of us in the hardest way possible. In this paper, a mathematical model based on differential equations is proposed. This model depicts the infection patterns of COVID-19 transmission, taking asymptomatic individuals and hospitalization into account. The basic reproduction number is computed using the next-generation matrix method. This is found to be a critical signal describing the dynamics of the COVID-19 transmission. The local stability of the steady states has been investigated. The model's global stability is demonstrated using the second method of Lyapunov and the LaSalle invariance principle. In addition, an optimal control problem is formulated to reduce fatality by considering pharmaceutical intervention options as control functions. COVID-19 transmission dynamics change when an intervention is introduced. In order to limit the number of infected individuals and reduce control costs, an appropriate objective functional has been developed and solved using Pontryagin's maximal principle. Furthermore, extensive simulations have been conducted for various initial conditions and parameter values in order to validate the theoretical aspects.

Keywords COVID-19 pandemic · Lyapunov function · Basic reproduction number · Hospitalisation · Optimal control · Pontryagin maximum principle

2020 AMS Subject Classification 34D05 · 34D20 · 34D23 · 92-10 · 92D30

1 Introduction

The COVID-19 outbreak has been one of the most devastating pandemics in history. COVID-19 was classified as a pandemic by the World Health Organization (WHO) on March 11, 2020. Human-to-human contact has been the primary mode of trans-

P. B. Borah (✉) · H. K. Sarmah
Department of Mathematics, Gauhati University, Guwahati 781014, Assam, India
e-mail: padmabhushanborah@gmail.com

D. Vlachos (ed.), *Mathematical Modeling in Physical Sciences*, Springer Proceedings in Mathematics & Statistics 446, https://doi.org/10.1007/978-3-031-52965-8_41

mission of COVID-19. Direct contact with an infected person transmits the virus. There have been over 600 million instances of COVID-19 infection and over 6 million fatalities across the globe [1–3]. Several reasons, including a paucity of diagnoses and asymptomatic cases, contribute to the underestimation of the number of cases [4–12]. Wuhan, Hubei province, China, was the first place where COVID-19 was reported [13–17]. Subsequently, the virus expanded globally, impacting practically every nation.

Epidemiological models have contributed significantly to our understanding of infectious disease transmission dynamics [18–21]. A 1760 study of smallpox dynamics initiated the study of epidemic dynamics, which has since become an indispensable tool for examining the transmission and management of infectious diseases [22]. In a ground breaking study, Kermack and McKendrick developed the Susceptible-Infected-Removed (SIR) model of ordinary differential equations [23]. By adding an additional compartment of exposed class E(t), Anderson and May [24] modified the well-established SIR model developed by Kermack and McKendrick. In the 1950s and early 1960s, Bailey and Bartlett developed stochastic models of disease dynamics [25, 26]. Bartlett was also a pioneer in the use of Monte Carlo simulations to study epidemics [27, 28].

Since the appearance of COVID-19, several authors have contributed to the field of epidemiological modeling; for examples, see [29–55]. According to the classical SIR compartmental model, the entire population is divided into three compartments: S(t), I(t), and R(t), indicating the susceptible individuals, infected individuals, and removed individuals (recovered or dead) [23, 56]. However, for most infectious diseases, a latent period precedes the onset of symptoms. This needs an extra compartment in the model, namely the exposed population E(t), resulting in four ODEs. [24, 57]. In order to understand the early dynamics of the COVID-19 epidemic and to evaluate the effectiveness of various post-outbreak interventions, SEIR was effectively used [58–62]. Many delay models [63–67] have also been studied extending the traditional SEIR model to suit the COVID-19 dynamics. In [68–70], optimal control techniques have been used in the study of COVID-19 disease transmission, and it has been demonstrated that optimal control strategies can be used to prevent the spread of novel coronavirus disease.

In this study, we enhance the SEIR model by including the class $A(t)$ of asymptomatic but infected people and the class $H(t)$ of infected individuals receiving therapy. The compartment $H(t)$ is not contagious since they are quarantined properly. As demonstrated in [71], deterministic models with fewer compartments, such as SIR or SEIR, cannot accurately characterize the development of COVID-19; so this work tries to analyze the dynamics of the COVID-19 outbreak using dynamical system theory and the SEIAHR model. Although the discrete model is equally accurate, we prefer the continuous deterministic model because it is easier to process. The local stability of the model will be evaluated under both disease-free and endemic equilibrium circumstances before the global stability of the model is examined. We also formulate a suitable optimal control problem for our model by incorporating a pharmaceutical control strategy for the symptomatic infected individuals, and characterize the resulting control solution completely.

The remainder of the paper is structured as follows. In Sect. 2, we establish the mathematical model for the transmission of COVID-19 infection. In Sect. 3, we describe the model's fundamental features, including positivity and boundedness. The subsequent part is devoted to determining the basic reproduction number using the next generation matrix method. Section 5 comprises of finding the steady state solutions of the model. In Sect. 6, we investigate the model's stability, which includes both local and global stability. In Sect. 7, we study an optimal control version of our model and in Sect. 8, we examine numerical simulations of our model to validate the conclusions obtained in the preceding sections. The final portion contains comments and conclusions.

2 Formulation of the Model

To represent COVID-19 spread, we build a compartmental model [72, 73] based on a nonlinear differential equation system. The entire population is divided into six compartments for the purposes of the model: those who are susceptible to the illness S(t), those who have been exposed to it but are not yet contagious E(t), those who are infected and contagious I(t), those who are contagious but need to be hospitalized because of the illness H(t), and the group of people R(t) who survived COVID-19. It is assumed that each compartment is homogenous. As a result, the hazards are the same for everyone in the same compartment. We also assume that a person develops immunity after he has recovered from the illness. Furthermore, it is considered that all babies are vulnerable. Total population is provided by

$$N(t) = S(t) + E(t) + I(t) + A(t) + H(t) + R(t),$$

at any given time t. With the above assumptions, our model constitutes of the following system of nonlinear ordinary differential equations:

$$\begin{cases} \dfrac{dS(t)}{dt} = b - \mu S - (\beta_1 I + \beta_2 A)S, \\ \dfrac{dE(t)}{dt} = (\beta_1 I + \beta_2 A)S - (\mu + \alpha)E, \\ \dfrac{dI(t)}{dt} = (1-p)\alpha E - (\mu + h + \gamma)I, \\ \dfrac{dA(t)}{dt} = p\alpha E - (\mu + \gamma)A, \\ \dfrac{dH(t)}{dt} = hI - (\mu + \delta + \rho)H, \\ \dfrac{dR(t)}{dt} = \gamma(I + A) + \rho H - \mu R, \end{cases} \tag{1}$$

with initial conditions

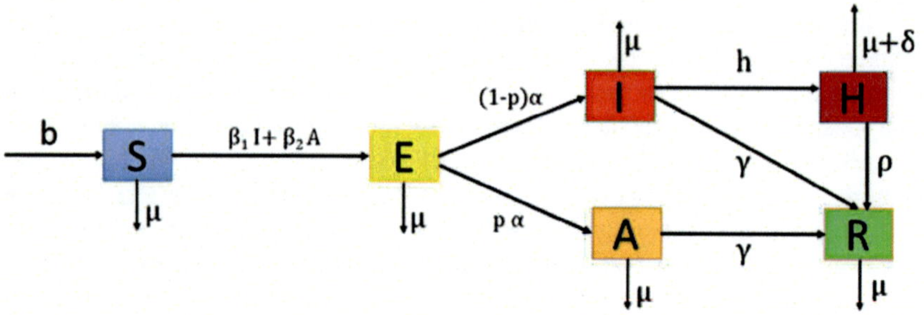

Fig. 1 Model's graphical depiction

$$S(0) \geq 0, E(0) \geq 0, I(0) \geq 0, A(0) \geq 0, H(0) \geq 0, R(0) \geq 0. \tag{2}$$

Here, b stands for the birth rate, μ for the natural death rate, β_1 for the rate at which people become exposed to the virus due to contact with infected people, β_2 for the effective contact rate with asymptomatic people, p for the probability of being asymptomatic, α for the proportion of exposed class that becomes infectious, h for the rate of hospitalization, γ for the natural rate of recovery, δ for the disease-induced death rate, and ρ for the rate of recovery caused by therapy.

Figure 1 is a schematic representation of the mathematical model (1).

3 Basic Properties of the Model

According to the fundamental theory of differential equations [74, 75], there exists a unique solution of the system (1) with the initial conditions (2) for all $t \geq 0$. To be realistic, the model's solutions must be nonnegative and uniformly bounded. In this part, we investigate it.

3.1 *Positivity of Solutions*

Theorem 1 *Under initial conditions (2), the solutions of the system of Eq. (1), namely $S(t)$, $E(t)$, $I(t)$, $A(t)$, $H(t)$, and $R(t)$ are all positive for all $t \geq 0$ if all parameters are positive.*

Proof Let

$$\kappa = \sup\{\phi > 0 | \forall t \in [0, \phi], S(t) \geq 0, E(t) \geq 0, I(t) \geq 0, A(t) \geq 0, H(t) \geq 0, R(t) \geq 0\}.$$

It suffices to show $\kappa = +\infty$.

If possible, $\kappa < \infty$ say. Consequently, the continuity of solutions implies that

$$S(\kappa) = 0, \text{ or } E(\kappa) = 0, \text{ or } I(\kappa) = 0, \text{ or } A(\kappa) = 0, \text{ or } H(\kappa) = 0, \text{ or } R(\kappa) = 0.$$

Now, if $S(\kappa) = 0$ before the other variables do, then

$$\left.\frac{dS(t)}{dt}\right|_{t=\kappa} = \lim_{t\to\kappa^-} \frac{S(\kappa) - S(t)}{\kappa - t} = -\lim_{t\to\kappa^-} \frac{S(t)}{\kappa - t} \le 0. \tag{3}$$

But, from the very first equation of the system of Eq. (1), we get

$$\left.\frac{dS(t)}{dt}\right|_{t=\kappa} = b - (\beta_1 I(\kappa) + \beta_2 A(\kappa))\, S(\kappa) - \mu S(\kappa) = b > 0,$$

which contradicts inequality (3).

Similarly, if $E(\kappa)$ equals zero over the other variables, then

$$\left.\frac{dE(t)}{dt}\right|_{t=\kappa} = \lim_{t\to\kappa^-} \frac{E(\kappa) - E(t)}{\kappa - t} = -\lim_{t\to\kappa^-} \frac{E(t)}{\kappa - t} \le 0, \tag{4}$$

but from the second equation of the system (1), we have

$$\begin{aligned}\left.\frac{dE(t)}{dt}\right|_{t=\kappa} &= (\beta_1 I(\kappa) + \beta_2 A(\kappa))\, S(\kappa) - (\mu + \alpha) E(\kappa)\\ &= (\beta_1 I(\kappa) + \beta_2 A(\kappa))\, S(\kappa) > 0,\end{aligned}$$

contradicting (4).

In a similar fashion, we may establish analogous contradictions with the remaining variables. For example: if $I(\kappa)$ is equal to zero before the other variables being zero, then

$$\left.\frac{dI(t)}{dt}\right|_{t=\kappa} = \lim_{t\to\kappa^-} \frac{I(\kappa) - I(t)}{\kappa - t} = -\lim_{t\to\kappa^-} \frac{I(t)}{\kappa - t} \le 0, \tag{5}$$

but the third equation of the system (1) gives us

$$\begin{aligned}\left.\frac{dI(t)}{dt}\right|_{t=\kappa} &= (1 - p)\alpha E(\kappa) - (\mu + h + \gamma) I(\kappa)\\ &= (1 - p)\alpha E(\kappa) > 0,\end{aligned}$$

contradicting inequality (5); etc. Therefore, κ cannot be finite.

As a result, all the solutions of the model are positive.

Theorem 1 is now proved.

3.2 *Boundedness of Solutions*

Theorem 2 *All possible solutions* $S(t)$, $E(t)$, $I(t)$, $A(t)$, $H(t)$, *and* $R(t)$ *of the model (1) are uniformly bounded on* $[0, \infty)$.

Proof By combining all of the model (1)'s equations, we get

$$\frac{dN(t)}{dt} = b - \mu N(t) - \delta H(t). \tag{6}$$

Since $\delta > 0$ and $H(t) \geq 0$, Eq. (6) gives

$$\frac{dN(t)}{dt} \leq b - \mu N(t). \tag{7}$$

From Gronwall's inequality [76] for the inequality (7) we have for $t \geq 0$,

$$N(t) \leq \frac{b}{\mu} + \left(N(0) - \frac{b}{\mu}\right) e^{-\mu t}, \tag{8}$$

where $N(0)$ is the total initial population. As a result, if $N(0) \leq \frac{b}{\mu}$, then $N(t) \leq \frac{b}{\mu}$. Therefore, the set defined by

$$\mathcal{R} = \left\{(S(t), E(t), I(t), A(t), H(t), R(t)) \in \mathbb{R}_+^6 : 0 \leq N(t) \leq \frac{b}{\mu}\right\}, \tag{9}$$

is positively invariant and the solutions of the model (1) remain bounded if $N(0) \leq \frac{b}{\mu}$. Further, Eq. (8) implies that

$$\lim_{t \to +\infty} \sup N(t) \leq \frac{b}{\mu}. \tag{10}$$

From inequality (10), we have for $N(0) > \frac{b}{\mu}$, $N(t)$ asymptotically approaches $\frac{b}{\mu}$. Therefore, all possible solutions of the model (1) are uniformly bounded on $\mathbb{R}^+$.

Due to Theorems 1 and 2, it is sufficient to consider the region $\mathcal{R}$ as provided by (9) in order to analyze model (1).

4 The Basic Reproduction Number

The basic reproduction number $\mathscr{R}_0$, is defined as the "expected number of secondary cases that would arise from a typical primary case in a completely susceptible population" [77–84]. The value of $\mathscr{R}_0$ makes it possible to determine the severity of the infection [81–84]. Utilizing the next generation matrix methods described in [83–85], we shall derive the algebraic formula of the basic reproduction number for model (1).

In the next generation matrix method, we first calculate the matrices $\mathscr{F}$ and $\mathscr{V}$, where $\mathscr{F}$ is the matrix of newly arising infections, and $\mathscr{V}$ is the matrix of net transition terms from the infected compartments of the model. Next, we compute the corresponding Jacobian matrices F and V of $\mathscr{F}$ and $\mathscr{V}$, respectively, at the disease free equilibrium. Then basic reproduction number is the spectral radius of the matrix FV^{-1}, i.e., $\mathscr{R}_0 = \rho(FV^{-1})$.

Following [78, 85], we deduce the formula of the basic reproduction number $\mathscr{R}_0$ for model (1) in the following steps:

Step 1. There are $four$ infected compartments, namely E, I, A and H.

Step 2. The 4×1 matrix $\mathscr{F}$ of new infections is given by

$$\mathscr{F} = \begin{bmatrix} (\beta_1 I + \beta_2 A)\, S \\ 0 \\ 0 \\ 0 \end{bmatrix}.$$

Step 3. The 4×1 matrix $\mathscr{V}^-$ that includes expressions going out of every infected compartment is

$$\mathscr{V}^- = \begin{bmatrix} (\mu + \alpha)E \\ (\mu + h + \gamma)I \\ (\mu + \gamma)A \\ (\mu + \delta + \rho)H \end{bmatrix}.$$

Step 4. The 4×1 matrix $\mathscr{V}^+$ that includes terms going into every infected compartment except new infections is

$$\mathscr{V}^+ = \begin{bmatrix} 0 \\ (1 - p)\alpha E \\ p\alpha E \\ hI \end{bmatrix}.$$

Step 5. The 4×1 matrix of net transfer amongst the infected compartments excluding new infections is $\mathscr{V} = \mathscr{V}^- - \mathscr{V}^+$

$$\mathscr{V} = \begin{bmatrix} (\mu + \alpha)E \\ (\mu + h + \gamma)I - (1 - p)\alpha E \\ (\mu + \gamma)A - p\alpha E \\ (\mu + \delta + \rho)H - hI \end{bmatrix}.$$

Step 6. The 4×4 Jacobian matrix F which is $\mathscr{F}$'s partial derivatives with regard to the vector $x = [E, I, A, H]^{\mathrm{T}}$ of the four infectious state variables evaluated at the disease free equilibrium point $\mathscr{E}_0$ of the model (1) is

$$F = \left[\frac{\partial \mathscr{F}}{\partial x}\right]_{\text{at } \mathscr{E}_0} = \begin{bmatrix} 0 & \frac{\beta_1 b}{\mu} & \frac{\beta_2 b}{\mu} & 0 \\ 0 & 0 & 0 & 0 \\ 0 & 0 & 0 & 0 \\ 0 & 0 & 0 & 0 \end{bmatrix}.$$

Step 7. The 4×4 Jacobian matrix V which is $\mathscr{V}$'s partial derivatives with regardto the vector $x = [E, I, A, H]^{\mathrm{T}}$ of the four infectious state variables evaluated at the disease free equilibrium point $\mathscr{E}_0\left(\frac{b}{\mu}, 0, 0, 0, 0, 0\right)$ of the model (1) is

$$V = \left[\frac{\partial \mathscr{V}}{\partial x}\right]_{\text{at } \mathscr{E}_0} = \begin{bmatrix} \mu+\alpha & 0 & 0 & 0 \\ -(1-p)\alpha & \mu+h+\gamma & 0 & 0 \\ -p\alpha & 0 & \mu+\gamma & 0 \\ 0 & -h & 0 & \mu+\delta+\rho \end{bmatrix}.$$

Step 8. The inverse of V is

$$V^{-1} = \begin{bmatrix} \frac{1}{\mu+\alpha} & 0 & 0 & 0 \\ \frac{(1-p)\alpha}{(\mu+\alpha)(\mu+h+\gamma)} & \frac{1}{\mu+h+\gamma} & 0 & 0 \\ \frac{p\alpha}{(\mu+\gamma)(\mu+\alpha)} & 0 & \frac{1}{\mu+\gamma} & 0 \\ \frac{(1-p)\alpha h}{(\mu+\delta+\rho)(\mu+h+\gamma)(\mu+\alpha)} & \frac{h}{(\mu+\delta+\rho)(\mu+h+\gamma)} & 0 & \frac{1}{\mu+\delta+\rho} \end{bmatrix}.$$

Step 9. We obtain the matrix FV^{-1} as

$$FV^{-1} = \begin{bmatrix} \frac{\alpha b}{\mu(\mu+\alpha)}\left(\frac{(1-p)\beta_1}{\mu+h+\gamma} + \frac{p\beta_2}{\mu+\gamma}\right) & \frac{\beta_1 b}{\mu(\mu+\gamma+h)} & \frac{\beta_2 b}{\mu(\mu+\gamma)} & 0 \\ 0 & 0 & 0 & 0 \\ 0 & 0 & 0 & 0 \\ 0 & 0 & 0 & 0 \end{bmatrix}.$$

Step 10. Finally, we derive the basic reproduction number as the largest eigenvalue $\rho(FV^{-1})$ of the matrix FV^{-1}. Hence, we have

$$\mathscr{R}_0 = \frac{(1-p)\alpha b\beta_1}{\mu(\mu+\alpha)(\mu+h+\gamma)} + \frac{p\alpha b\beta_2}{\mu(\mu+\alpha)(\mu+\gamma)}. \tag{11}$$

5 Steady State Solutions

We determine the model's steady states by setting all the time derivatives in (1) to zero. There are two alternative steady states, one of which corresponds to infection co-existing and the other to no infection. We find them below here.

5.1 *The Disease-Free Steady State*

The disease-free steady state $\mathscr{E}_0$ is achieved when there is no infection, or when $I = 0$. Solving system (1) for $I = 0$ and $\dot{S} = \dot{E} = \dot{I} = \dot{A} = \dot{H} = \dot{R} = 0$ gives us

$$\begin{aligned}\mathscr{E}_0 &= (S_0, E_0, I_0, A_0, H_0, R_0)\\ &= \left(\frac{b}{\mu}, 0, 0, 0, 0, 0\right).\end{aligned} \tag{12}$$

5.2 *The Endemic Steady State*

The endemic steady state $\mathscr{E}_1$ is achieved when there is an infection, or when $I \neq 0$. Setting the right hand side of system (1) to zero at $\mathscr{E}_1 = (S_1, E_1, I_1, A_1, H_1, R_1)$, we get

$$\begin{cases} 0 = b - \mu S_1 - (\beta_1 I_1 + \beta_2 A_1) S_1, \\ 0 = (\beta_1 I_1 + \beta_2 A_1) S_1 - (\mu + \alpha) E_1, \\ 0 = (1 - p)\alpha E_1 - (\mu + h + \gamma) I_1, \\ 0 = p\alpha E_1 - (\mu + \gamma) A_1, \\ 0 = h I_1 - (\mu + \delta + \rho) H_1, \\ 0 = \gamma (I_1 + A_1) + \rho H_1 - \mu R_1. \end{cases} \tag{13}$$

Solving the system of Eq. (13) along with $I_1 \neq 0$ yields the following endemic steady state $\mathscr{E}_1$:

$$\begin{cases} S_1 = \dfrac{b}{\mu \mathscr{R}_0}, \\ E_1 = \dfrac{b}{\mu+\alpha}\dfrac{\mathscr{R}_0 - 1}{\mathscr{R}_0}, \\ A_1 = \dfrac{p\alpha b}{(\mu+\alpha)(\mu+\gamma)}\dfrac{\mathscr{R}_0 - 1}{\mathscr{R}_0}, \\ I_1 = \dfrac{(1-p)\alpha b}{(\mu+\alpha)(\mu+h+\gamma)}\dfrac{\mathscr{R}_0 - 1}{\mathscr{R}_0}, \\ H_1 = \dfrac{h(1-p)\alpha b}{(\mu+\alpha)(\mu+h+\gamma)(\mu+\delta+\rho)}\dfrac{\mathscr{R}_0 - 1}{\mathscr{R}_0}, \\ R_1 = \dfrac{\alpha b}{\mu(\mu+\alpha)}\left(\dfrac{\gamma(1-p)}{\mu+h+\gamma} + \dfrac{\gamma p}{\mu+\gamma} + \dfrac{\rho h(1-p)}{(\mu+\delta+\rho)(\mu+h+\gamma)}\right)\dfrac{\mathscr{R}_0 - 1}{\mathscr{R}_0}. \end{cases} \tag{14}$$

6 Stability Analysis

In this section, we study the stability of the steady-state solutions of the model (1) obtained in (12) and (14). We will show that the value of the basic reproduction number $\mathscr{R}_0$ of model (1) is intimately tied to the stability of the steady states.

6.1 *Stability of* $\mathscr{E}_0$

We proceed to examine the local and global stability behavior of the disease-free steady state $\mathscr{E}_0$ derived in (12).

Local Stability

Theorem 3 *The disease-free steady state* $\mathscr{E}_0$ *provided by (12) is locally asymptotically stable if* $\mathscr{R}_0 < 1$ *and unstable if* $\mathscr{R}_0 > 1$.

Proof This result is a direct outcome of Theorem 2 in [83].

Global Stability

Theorem 4 *The disease-free steady-state* $\mathscr{E}_0$ *provided by (12) is globally asymptotically stable in* $\mathcal{R}$ *if* $\mathscr{R}_0 \leq 1$.

Proof We take the Lyapunov function as follows:

$$\mathscr{L}(Z(t)) = E(t) + \frac{S_0\beta_1}{\mu+h+\gamma}I(t) + \frac{S_0\beta_2}{\mu+\gamma}A(t), \tag{15}$$

where $Z(t) = (S(t), E(t), I(t), A(t)), H(t), R(t))$ is the vector of state variables, and $S_0 = \frac{b}{\mu}$ as obtained in (12). The function $\mathscr{L}$ satisfies:

$$\begin{cases} \mathscr{L}(\mathscr{E}_0) = 0, \\ \mathscr{L}(\mathrm{Z}(t)) > 0, \text{ for all } \mathrm{Z}(t) \neq \mathscr{E}_0, \text{ and} \\ \mathscr{L}(\mathrm{Z}(t)) \text{ is radially unbounded: } \mathscr{L}(\mathrm{Z}(t)) \to \infty \text{ as } \|\mathrm{Z}\| \to \infty. \end{cases} \tag{16}$$

Now, by considering the time derivative of $\mathscr{L}$ along the trajectory of model (1) in the region $\mathcal{R}$, we have

$$\begin{aligned} \frac{d\mathscr{L}}{dt} &= \dot{E} + \frac{S_0\beta_1}{\mu+h+\gamma}\dot{I} + \frac{S_0\beta_2}{\mu+\gamma}\dot{A} \\ &= (\beta_1 I + \beta_2 A)S - (\mu+\alpha)E + \frac{S_0\beta_2}{\mu+\gamma}(p\alpha E - (\mu+\gamma)A) + \frac{S_0\beta_1}{\mu+h+\gamma}((1-p)\alpha E - (\mu+h+\gamma)I) \\ &= (\beta_1 I + \beta_2 A)S - (\mu+\alpha)E - S_0\beta_1 I - S_0\beta_2 A + \frac{S_0\beta_1}{\mu+h+\gamma}(1-p)\alpha E + \frac{S_0\beta_2}{\mu+\gamma}p\alpha E \\ &= (\beta_1 I + \beta_2 A)(S - S_0) + \left(\frac{b\beta_1(1-p)\alpha}{\mu(\mu+h+\gamma)(\mu+\alpha)} + \frac{b\beta_2 p\alpha}{\mu(\mu+\gamma)(\mu+\alpha)} - 1\right)(\mu+\alpha)E \\ &\leq (\mathscr{R}_0 - 1)(\mu+\alpha)E. \end{aligned}$$

Hence, $\frac{d\mathscr{L}(\mathrm{Z}(t))}{dt} \leq 0$ for $\mathscr{R}_0 \leq 1$, and $\frac{d\mathscr{L}(\mathrm{Z}(t))}{dt} = 0$ iff $E = 0$ iff $I = 0$ iff $\mathrm{Z} = \mathscr{E}_0$. This indicates that the set

$$\mathscr{L}_{\mathscr{E}_0} = \left\{\mathrm{Z}(t) \in \mathcal{R} : \frac{d\mathscr{L}(\mathrm{Z}(t))}{dt} = 0\right\} = \{\mathscr{E}_0\}, \tag{17}$$

contains the unique point $\mathscr{E}_0$ derived in (12). According to Lasalle's invariance principle [86–89] then, the disease-free steady state $\mathscr{E}_0$ is globally asymptotically stable in $\mathcal{R}$ if $\mathscr{R}_0 \leq 1$.

6.2 *Stability of* $\mathscr{E}_1$

We now consider the local and global stability of the endemic steady state $\mathscr{E}_1$ obtained in (14).

Local Stability

Theorem 5 *The endemic steady state* $\mathscr{E}_1$ *provided by (14) is locally asymptotically stable if* $\mathscr{R}_0 > 1$ *and unstable if* $\mathscr{R}_0 < 1$.

Proof This result is achieved by applying Theorem 4 in [83].

Global Stability For the global stability of the endemic steady state, the following type of function will be utilized:

$$H(z) = z - 1 - \ln z, \tag{18}$$

which is non-negative for $z > 0$, and $H(z) = 0$ if and only if $z = 1$. Additionally,

$$H'(z) = 1 - 1/z. \tag{19}$$

Theorem 6 *The endemic steady-state $\mathscr{E}_1$ provided by (14) is globally asymptotically stable in $\mathcal{R}$ if $\mathscr{R}_0 > 1$.*

Proof We define the following Lyapunov function $\mathscr{U}$:

$$\mathscr{U}(Z(t)) = S_1 H\left(\frac{S}{S_1}\right) + E_1 H\left(\frac{E}{E_1}\right) + \frac{\beta_1 I_1 S_1}{(1-p)\alpha E_1} I_1 H\left(\frac{I}{I_1}\right) + \frac{\beta_2 A_1 S_1}{p\alpha E_1} A_1 H\left(\frac{A}{A_1}\right), \tag{20}$$

where $Z(t) = (S(t), E(t), I(t), A(t), H(t), R(t))$ is the state vector. The function $\mathscr{U}$ satisfies:

$$\begin{cases} \mathscr{U}(\mathscr{E}_1) = 0, \\ \mathscr{U}(Z(t)) > 0, \text{ for all } Z(t) \neq \mathscr{E}_1, \text{ and} \\ \mathscr{U}(Z(t)) \text{ is radially unbounded: } \mathscr{U}(Z(t)) \to \infty \text{ as } \|Z\| \to \infty. \end{cases} \tag{21}$$

Taking the time derivative of $\mathscr{U}$ along the trajectory of the model (1) in the region $\mathcal{R}$, we get

$$\frac{d\mathscr{U}(Z(t))}{dt} = H'\left(\frac{S}{S_1}\right)\dot{S} + H'\left(\frac{E}{E_1}\right)\dot{E} + \frac{\beta_1 I_1 S_1}{(1-p)\alpha E_1} H'\left(\frac{I}{I_1}\right)\dot{I} + \frac{\beta_2 A_1 S_1}{p\alpha E_1} H\left(\frac{A}{A_1}\right)\dot{A}. \tag{22}$$

From (22), using (1) and (19), we get

$$\begin{aligned} \dot{\mathscr{U}}(Z(t)) = {} & \left(1 - \frac{S_1}{S}\right)(b - \mu S - (\beta_1 I + \beta_2 A)S) + \left(1 - \frac{E_1}{E}\right)((\beta_1 I + \beta_2 A)S) - (\mu + \alpha)E) \\ & + \frac{\beta_1 I_1 S_1}{(1-p)\alpha E_1}\left(1 - \frac{I_1}{I}\right)((1-p)\alpha E - (\mu + h + \gamma)I) + \frac{\beta_2 A_1 S_1}{p\alpha E_1}\left(1 - \frac{A_1}{A}\right)(p\alpha E - (\mu+\gamma)A). \end{aligned} \tag{23}$$

From (13), we have

$$\begin{aligned} b &= \mu S_1 + (\beta_1 I_1 + \beta_2 A_1)S_1, \\ \mu + \alpha &= \frac{(\beta_1 I_1 + \beta_2 A_1)S_1}{E_1}, \\ \mu + h + \gamma &= \frac{(1-p)\alpha E_1}{I_1}, \\ \mu + \gamma &= \frac{p\alpha E}{A_1}. \end{aligned} \tag{24}$$

Substituting (24) in (23), and regrouping the terms, we get

$$\dot{\mathscr{U}}(Z) = -\frac{\mu(S - S_1)^2}{S} + \beta_1 I_1 S_1\left(3 - \frac{S_1}{S} - \frac{E_1 I S}{E I_1 S_1} - \frac{I_1 E}{I E_1}\right) + \beta_2 A_1 S_1\left(3 - \frac{S_1}{S} - \frac{E_1 A S}{E A_1 S_1} - \frac{A_1 E}{A E_1}\right) \tag{25}$$

From $AM \geq GM$, we get

$$\begin{aligned} \frac{S_1}{S} + \frac{E_1 I S}{E I_1 S_1} + \frac{I_1 E}{I E_1} \geq 3, \\ \frac{S_1}{S} + \frac{E_1 A S}{E A_1 S_1} + \frac{A_1 E}{A E_1} \geq 3. \end{aligned} \tag{26}$$

Therefore, from (25),

$$\dot{\mathscr{U}}(Z) \leq -\frac{\mu (S(t) - S_1)^2}{S(t)} \leq 0.$$

Moreover, for $\mathscr{R}_0 > 1, \dot{\mathscr{U}}(Z) = 0$ if and only if $Z(t) = (S_1, E_1, I_1, T_1, R_1)$. i.e., $\dot{\mathscr{U}}(Z) = 0$ if and only if $Z = \mathscr{E}_1$ when $\mathscr{R}_0 > 1$. This shows the set

$$\mathscr{U}_{\mathscr{E}_1} = \left\{ Z(t) \in \mathcal{R} : \dot{L}((Z(t)) = 0 \right\} = \{\mathscr{E}_1\}, \tag{27}$$

contains the unique point $\mathscr{E}_1$ derived in (14). According to Lasalle's invariance principle [86–89] then, the endemic steady-state $\mathscr{E}_1$ is therefore globally asymptotically stable in $\mathcal{R}$ if $\mathscr{R}_0 > 1$.

7 An Optimal Control Problem

In this section, we apply the theory of optimal control, a potent method for the analysis of infectious diseases. Using the most effective intervention options, we seek to reduce the number of affected people with minimum cost. We undertake a theoretical study and numerical simulations to demonstrate how these control measures influence the transmission of the COVID-19 virus. We introduce the control variable $u(t)$ in our model (1), which externally controls the number of symptomatic infectious cases over a given time period.

7.1 Improved Treatment Plan for Infected Patients

A better treatment program for symptomatic infected persons not only reduces the prevalence of the disease but also influences its progression. In the current scenario, we assume that medication is administered to the clinically ill population. As the availability of resources associated with medical treatments is limited, we introduce the medication in the form of $-\xi u(t) I$ in our model (1), where ξ represents the treatment rate with intensity $u(t)$. When medication is administered, there are a variety of expenses linked, including those for drugs, kits, diagnostics, and health protocols. As a result, we assume that the treatment intensity $u(t)$, which is our control variable, ranges between 0 and 1. Here, there is no response to the treatment for $u(t) = 0$ and and for $u(t) = 1$, the response is maximum after the treatment. The control $u(t)$ modulates the proportion of symptomatic infectious instances receiving antivirals per unit of time $\xi I(t)$.

The permissible set for the control variable $u(t)$ is therefore defined as follows:

$$\mathcal{U} = \{u(t) \text{ is a measurable function} : 0 \leq u(t) \leq 1, t \in [0, T]\}. \tag{28}$$

Using the aforementioned control procedure, the following optimum control problem is studied:

$$\begin{cases} \dfrac{dS(t)}{dt} = b - \mu S - (\beta_1 I + \beta_2 A)S, \\ \dfrac{dE(t)}{dt} = (\beta_1 I + \beta_2 A)S - (\mu + \alpha)E, \\ \dfrac{dI(t)}{dt} = (1 - p)\alpha E - (\mu + h + \gamma)I - \xi u(t)I, \\ \dfrac{dA(t)}{dt} = p\alpha E - (\mu + \gamma)A, \\ \dfrac{dH(t)}{dt} = hI - (\mu + \delta + \rho)H, \\ \dfrac{dR(t)}{dt} = \gamma(I + A) + \rho H - \mu R + \xi u(t)I, \end{cases} \tag{29}$$

subject to minimizing the objective functional

$$\mathcal{W}(u(t)) = \int_0^T \left[\varphi_1 I + \varphi_2 A + \frac{1}{2} m u^2(t)\right] dt \tag{30}$$

with the initial conditions as stated in the model Eq. (2). Here, the objective functional $\mathcal{W}$ reflects the total cost incurred, which is the sum of the expenses specified in the integrand: $\varphi_1 I + \varphi_2 A + \frac{1}{2} m u^2$. All the parameters, viz., φ_1, φ_2, and m are non-negative. These are weight constants indicating the relative costs of each associated term over $[0, T]$.

7.2 *Existence of Optimal Control*

This subsection examines the existence of an optimal control of (29), (30). The optimal control is then characterized by applying Pontryagin's-Maximum Principle in the subsequent subsection.

Theorem 7 *There exists optimal control $u^*(t) \in \mathcal{U}$ that minimizes the objective functional $\mathcal{W}(u(t))$ corresponding to the control system (29) and (30).*

Proof In Theorems 1 and 2 of Sect. 3, we have already shown that all the solutions of the system (1) under the initial conditions (2) are non-negative and uniformly bounded. This along with the boundedness of the control variable $u(t)$ implies the non-negativity and the boundedness of the objective functional. Therefore, sequence of controls $u_k(t)$ in $\mathcal{U}$ exists such that

$$\lim_{k\to\infty} \mathcal{W}(u_k(t)) = \inf_{u\in\mathcal{U}} \mathcal{W}(u(t)).$$

Also, the control $u(t) \in \mathcal{U}$ is uniformly bounded in L^∞ and hence in $L^2([0, T])$ as well. Since $\mathcal{U}$ is closed, and the space $L^2([0, T])$ is reflexive [90], there exists $u^*(t) \in \mathcal{U}$ and a subsequence $\left\{u_{k_n}(t)\right\}_{n\in\mathbb{N}}$ in $\mathcal{U}$ so that $u_{k_n}(t) \rightharpoonup u^*(t)$, weakly in $L^2([0, T])$ as $n \to \infty$.

Moreover, clearly the sequence $Y_k = (S_k, E_k, I_k, A_k, H_k, R_k)$ of state variables is uniformly bounded corresponding to $u_k(t)$ for all $k \in \mathbb{N}$. This entails the uniform boundedness of the derivatives of the state variables owing to (1), which in turn implies the equi-continuity of the corresponding state sequence Y_k. Then, by the Arzela-Ascoli Theorem, there exists a subsequence Y_{k_n} and a state solution $Y^* = (S^*, E^*, I^*, A^*, H^*, R^*)$ such that

$$Y_{k_n} \to Y^* \quad \text{uniformly on } [0, T].$$

Finally, the lower semi-continuity of the L^2-norm with respect to L^2 weak convergence gives

$$\inf_{u\in\mathcal{U}} \mathcal{W}(u(t)) = \lim_{k\to\infty} \mathcal{W}(u_k(t)) = \lim_{n\to\infty} \mathcal{W}\left(u_{k_n}(t)\right) \geq \mathcal{W}\left(u^*(t)\right),$$

where

$$\mathcal{W}\left(u^*(t)\right) = \int_0^T \left[\varphi_1 I^* + \varphi_2 A^* + \frac{1}{2} m u^{*2}(t)\right] dt.$$

Thus, $u^*(t)$ is an optimal control as desired.

7.3 Characterization of Optimal Control

We use Pontryagin's Maximum Principle [91–93] to determine the necessary conditions for our optimal control problem.

Theorem 8 *For optimal control $u^*(t) \in \mathcal{U}$ and the corresponding state solutions $Y^* = (S^*, E^*, I^*, A^*, H^*, R^*)$ of the system (29) that minimizes the objective functional (30), there exists co-state variables λ_S, λ_E, λ_I, λ_A, λ_H, and λ_R satisfying the following canonical equations:*

$$\lambda'_S = \lambda_S\mu + (\lambda_S - \lambda_E)(\beta_1 I^* + \beta_2 A^*),$$
$$\lambda'_E = \lambda_E(\mu + \alpha) + (\lambda_I - \lambda_A)p\alpha - \lambda_I\alpha,$$
$$\lambda'_I = -\left[\varphi_1 + (\lambda_E - \lambda_S)\,\beta_1 S^* - \lambda_I\left(\mu + h + \gamma + \xi u^*(t)\right) + \lambda_H h + \lambda_R(\gamma + \xi u^*(t))\right],$$
$$\lambda'_A = -\left[\varphi_2 + (\lambda_E - \lambda_S)\,\beta_2 S^* - \lambda_A\,(\mu + \gamma) + \lambda_R\gamma\right],$$
$$\lambda'_H = \lambda_H(\mu + \delta + \rho) - \lambda_R\rho,$$
$$\lambda'_R = \lambda_R\mu,$$

with the transversality conditions:

$$\lambda_S(T) = 0, \lambda_E(T) = 0, \lambda_I(T) = 0, \lambda_A(T) = 0, \lambda_H(T) = 0, \lambda_R(T) = 0.$$

Further, the corresponding optimal control $u^*(t)$ is given by

$$u^*(t) = \min\left\{\max\left\{0, \frac{(\lambda_I - \lambda_R)\,\xi I}{m}\right\}, 1\right\}.$$

Proof We apply Pontryagin's Maximum Principle for our system (29) and (30). To this end, we define the Hamiltonian $\mathcal{H}$ for all $t \in [0, T]$ as below:

$$\begin{aligned}\mathcal{H} =& \varphi_1 I + \varphi_2 A + \frac{1}{2}mu^2(t) + \lambda_S\dot{S} + \lambda_E\dot{E} + \lambda_I\dot{I} + \lambda_A\dot{A} + \lambda_H\dot{H} + \lambda_R\dot{R}\\ =& \varphi_1 I + \varphi_2 A + \frac{1}{2}mu^2(t)\\ &+ \lambda_S\,[b - \mu S - (\beta_1 I + \beta_2 A)S]\\ &+ \lambda_E\,[(\beta_1 I + \beta_2 A)S - (\mu + \alpha)E]\\ &+ \lambda_I\,[(1 - p)\alpha E - (\mu + h + \gamma)I - \xi u(t)I]\\ &+ \lambda_A\,[p\alpha E - (\mu + \gamma)A] + \lambda_H\,[hI - (\mu + \delta + \rho)H]\\ &+ \lambda_R\,[\gamma(I + A) + \rho H - \mu R + \xi u(t)I],\end{aligned}$$

where λ_S, λ_E, λ_I, λ_A, λ_H, and λ_R are the adjoints or co-state variables.

From Pontryagin's Maximum Principle, adjoint variables satisfy:

$$\lambda'_S = -\frac{\partial\mathcal{H}}{\partial S}, \quad \lambda'_E = -\frac{\partial\mathcal{H}}{\partial E}, \quad \lambda'_I = -\frac{\partial\mathcal{H}}{\partial I},$$
$$\lambda'_A = -\frac{\partial\mathcal{H}}{\partial A}, \quad \lambda'_H = -\frac{\partial\mathcal{H}}{\partial H}, \quad \lambda'_R = -\frac{\partial\mathcal{H}}{\partial R},$$

with the transversality conditions: $\lambda_S(T) = 0$, $\lambda_E(T) = 0$, $\lambda_I(T) = 0$, $\lambda_A(T) = 0$, $\lambda_H(T) = 0$, and $\lambda_R(T) = 0$.

Next, minimization of the Hamiltonian $\mathcal{H}$ at $u(t) = u^*(t)$ is provided by

$$\left.\frac{\partial\mathcal{H}}{\partial u}\right|_{u(t)=u^*(t)} = 0. \tag{31}$$

Solving (31) for $u^*(t)$, we get

$$u(t) = \frac{(\lambda_I - \lambda_R)\,\xi I}{m} \quad \text{at } u(t) = u^*(t) \text{ subject to } u^* \in \mathcal{U}. \tag{32}$$

Then, (32) along with (28), gives us the following characterization of optimal control

$$u^*(t) = \min\left\{\max\left\{0, \frac{(\lambda_I - \lambda_R)\,\xi I}{m}\right\}, 1\right\}. \tag{33}$$

Additionally, $\frac{\partial^2 \mathcal{H}}{\partial u^2(t)} = m > 0$, which means the optimal control u^* indeed minimizes the Hamiltonian $\mathcal{H}$.

Finally, from (29) and (33), we get the optimality system Y^* corresponding to the optimal control $u^*(t)$ as

$$\begin{cases} \dfrac{dS^*(t)}{dt} = b - \mu S^* - (\beta_1 I^* + \beta_2 A^*)S^*, \\ \dfrac{dE^*(t)}{dt} = (\beta_1 I^* + \beta_2 A^*)S^* - (\mu + \alpha)E^*, \\ \dfrac{dI^*(t)}{dt} = (1-p)\alpha E^* - (\mu + h + \gamma)I^* - \xi u^*(t) I^*, \\ \dfrac{dA^*(t)}{dt} = p\alpha E^* - (\mu + \gamma)A^*, \\ \dfrac{dH^*(t)}{dt} = hI^* - (\mu + \delta + \rho)H^*, \\ \dfrac{dR^*(t)}{dt} = \gamma(A^* + I^*) - \mu R^* + \rho H^* + \xi u^*(t) I^*, \end{cases} \tag{34}$$

with initial conditions: $S^*(0) \geq 0$, $E^*(0) \geq 0$, $A^*(0) \geq 0$, $I^*(0) \geq 0$, $H^*(0) \geq 0$, $R^*(0) \geq 0$, and the corresponding adjoint variables or co-states satisfying

$$\begin{aligned} \lambda'_S &= \lambda_S \mu + (\lambda_S - \lambda_E)(\beta_1 I^* + \beta_2 A^*), \\ \lambda'_E &= \lambda_E(\mu + \alpha) + (\lambda_I - \lambda_A)p\alpha - \lambda_I \alpha, \\ \lambda'_I &= -\left[\varphi_1 + (\lambda_E - \lambda_S)\,\beta_1 S^* - \lambda_I\left(\mu + h + \gamma + \xi u^*(t)\right) + \lambda_H h + \lambda_R(\gamma + \xi u^*(t))\right], \\ \lambda'_A &= -\left[\varphi_2 + (\lambda_E - \lambda_S)\,\beta_2 S^* - \lambda_A\,(\mu + \gamma) + \lambda_R \gamma\right], \\ \lambda'_H &= \lambda_H(\mu + \delta + \rho) - \lambda_R \rho, \\ \lambda'_R &= \lambda_R \mu, \end{aligned}$$

together with the transversality conditions: $\lambda_S(T) = 0$, $\lambda_E(T) = 0$, $\lambda_I(T) = 0$, $\lambda_A(T) = 0$, $\lambda_H(T) = 0$, and $\lambda_R(T) = 0$. This completes the proof.

8 Numerical Simulations

In this section, we use numerical simulations to illustrate our theoretical findings. This section is divided into two parts. In the first part, we have verified the analytical findings discussed in the earlier sections and the optimal control problem is solved numerically in the later part. We shall employ RK4 method for this purpose. To support our findings on local and global stability, we will change the initial sub-population values in the model's various compartments. The entire initial population in each situation is computed using the formula $N(0) = S(0) + E(0) + I(0) + A(0) + H(0) + R(0)$.

8.1 Without Control Strategy

We employ three different sets of parameter values with the intent of demonstrating the dependency of stability on the value of $\mathscr{R}_0$ in the ranges of $\mathscr{R}_0 < 1$, $\mathscr{R}_0 = 1$ and $\mathscr{R}_0 > 1$, respectively.

Assuming the parameter values $b = 200$, $\beta_1 = 0.005$, $\beta_2 = 0.003$, $p = 0.5$, $\alpha = 0.1$, $\rho = 0.25$, $\delta = 0.25$, $\mu = 0.4$, $h = 0.05$ and $\gamma = 0.10$, we get $\mathscr{R}_0 = 0.754545$. It is observed that the disease-free equilibrium is $\mathscr{E}_0 = (500, 0, 0, 0, 0)$ for these parameter values. Since $\mathscr{R}_0 = 0.754545 < 1$, so according to Theorem 5, the disease-free equilibrium $\mathscr{E}_0$ is globally asymptotically stable. Figure 2 demonstrates that for any initial sub-population, the final susceptible population approaches the value 500 asymptotically, whereas the other classes, namely the exposed class, the infective class, the recovered class, the asymptomatic class and the hospitalized class, approach zero; thus verifying Theorem 5.

For the subsequent set of numerical simulations, the following parameter values: $b = 40$, $\beta_1 = 0.0275$, $\beta_2 = 0.025$, $p = 0.5$, $\alpha = 0.1$, $\rho = 0.25$, $\delta = 0.25$, $\mu = 0.4$, $h = 0.05$, and $\gamma = 0.10$ are assumed. In this instance, $\mathscr{R}_0 = 1$ and the disease-free equilibrium is $\mathscr{E}_0 = (100, 0, 0, 0, 0)$. Theorem 4 states that $\mathscr{E}_0$ is globally asymptotically stable given that $\mathscr{R}_0 = 1$. Figure 3 demonstrates that the final susceptible population approaches the value 100 asymptotically for any initial sub-population, but the other classes, namely the exposed class, the infective class, the recovered class, the asymptomatic class, and the hospitalized class, converge to zero. This establishes Theorem 4.

In the last scenario, we assume the parameter values $b = 1000$, $\beta_1 = 0.002$, $\beta_2 = 0.0015$, $p = 0.5$, $\alpha = 0.001$, $\rho = 0.25$, $\delta = 0.25$, $\mu = 0.04$, $h = 0.3$, and $\gamma = 0.15$. In this situation, $\mathscr{R}_0 = 3.6513$ and $\mathscr{E}_1 = (6846, 17710, 18, 47, 10, 305)$ represents the endemic equilibrium. As $\mathscr{R}_0 > 1$, Theorem 6 states that $\mathscr{E}_1$ is globally asymptotically stable. In fact, as shown in Fig. 4, the state variables approach their respective endemic steady state, confirming Theorem 6.

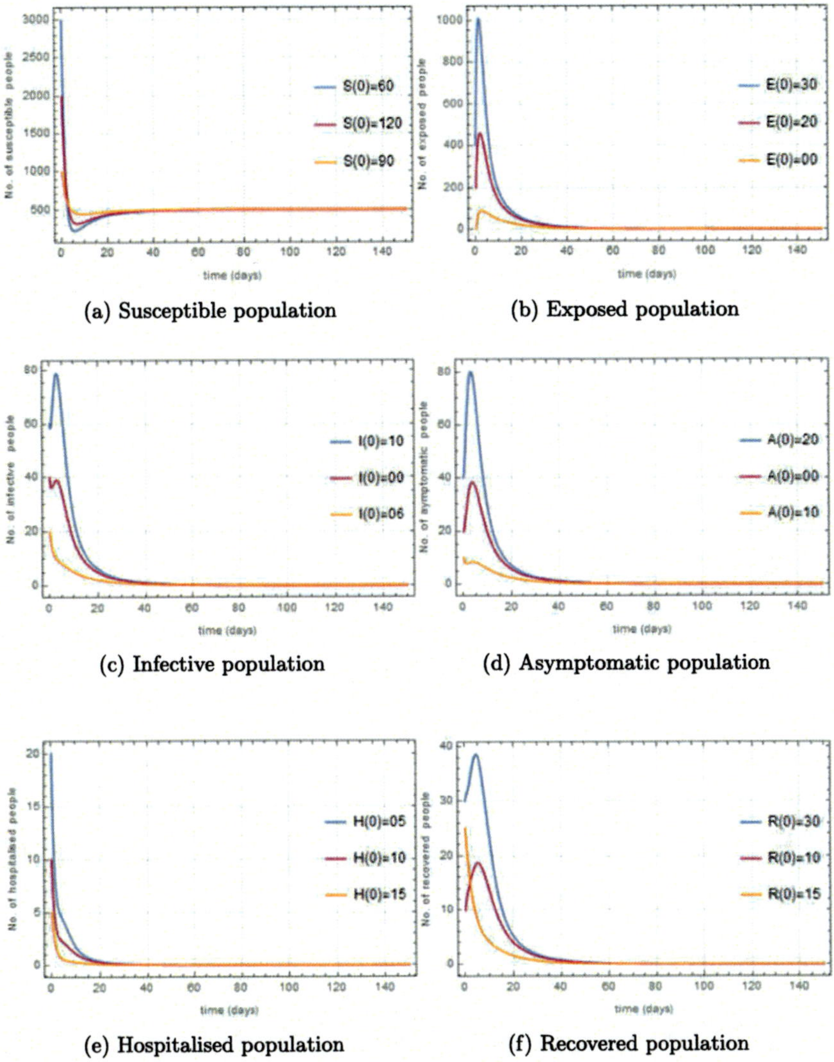

(a) Susceptible population

(b) Exposed population

(c) Infective population

(d) Asymptomatic population

(e) Hospitalised population

(f) Recovered population

Fig. 2 The system (1) approaches the disease-free equilibrium point $\mathscr{E}_0 = (500, 0, 0, 0, 0)$ asymptotically for $\mathscr{R}_0 < 1$

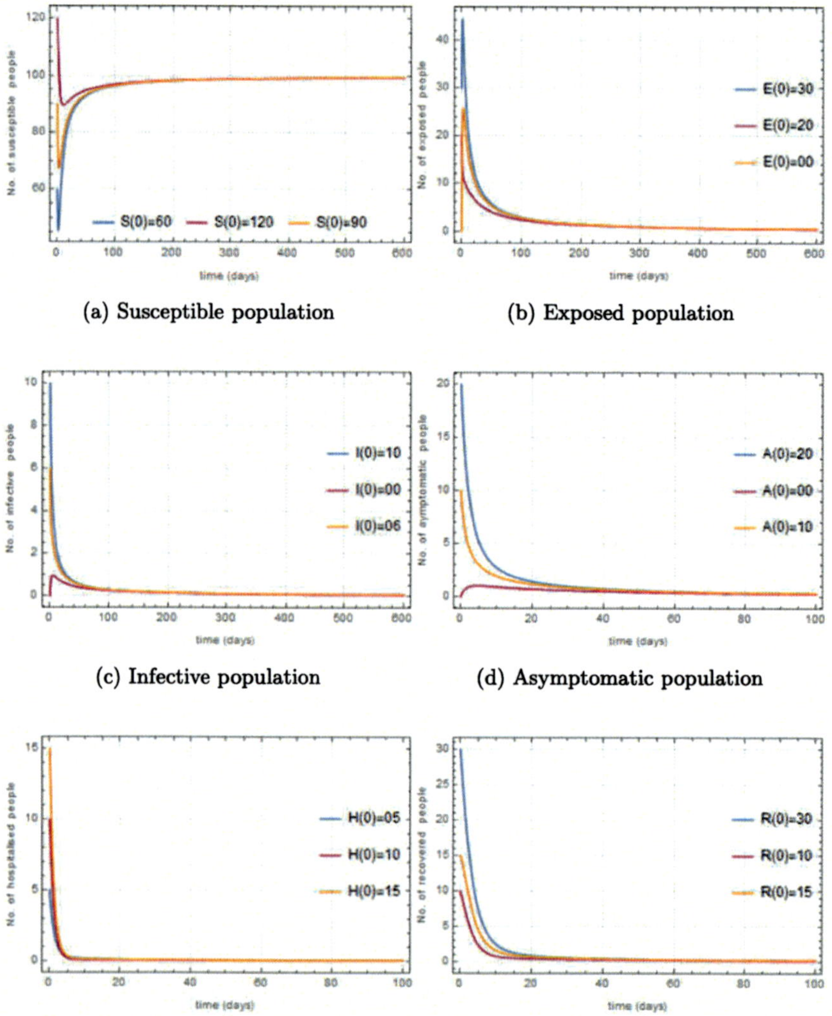

(a) Susceptible population

(b) Exposed population

(c) Infective population

(d) Asymptomatic population

(e) Hospitalised population

(f) Recovered population

Fig. 3 The system (1) approaches the disease-free equilibrium $\mathscr{E}_0 = (100, 0, 0, 0, 0)$ asymptotically for $\mathscr{R}_0 = 1$

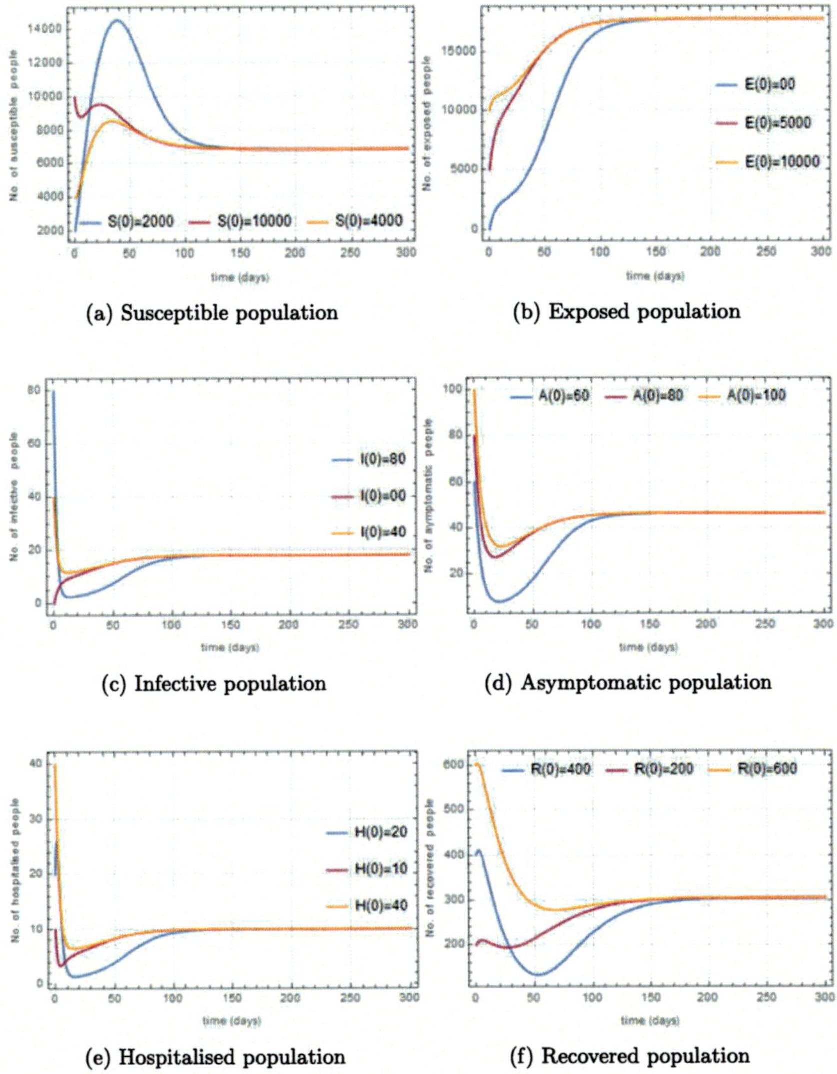

(a) Susceptible population

(b) Exposed population

(c) Infective population

(d) Asymptomatic population

(e) Hospitalised population

(f) Recovered population

Fig. 4 The system (1) approaches the endemic equilibrium point $\mathscr{E}_1 = (6846, 17710, 18, 47, 10, 305)$ asymptotically for $\mathscr{R}_0 > 1$

8.2 *With Control Strategy*

In order to propose the optimal control problem, we have introduced a control $u(t)$ resulting the additional recovery through the medication and aimed to minimize the cost associated with the medication. Here, we have performed the simulations to investigate the outcomes of the optimal strategy of the medication and then to compare it to the no control scenario. For solving the controlled system and the corresponding adjoint system, we employ following parameter values: $b = 500$, $\beta_1 = 0.002$, $\beta_2 = 0.0015$, $p = 0.5$, $\alpha = 0.001$, $\rho = 0.25$, $\delta = 0.25$, $\mu = 0.04$, $h = 0.3$, and $\gamma = 0.15$ along with the weight constants $\varphi_1 = 2$, $\varphi_2 = 1$, and $m = 40$. In this instance, $\mathscr{R}_0 = 1.82565 > 1$. In Fig. 5, the time series plots for the exposed, infective, asymptomatic and hospitalized populations are depicted in both cases: without control and with optimal control for a period of 300 days. Comparing the plots, we observe less damage in each case when control measure is applied. The variation of the control for this period of 300 days is shown in the Fig. 6.

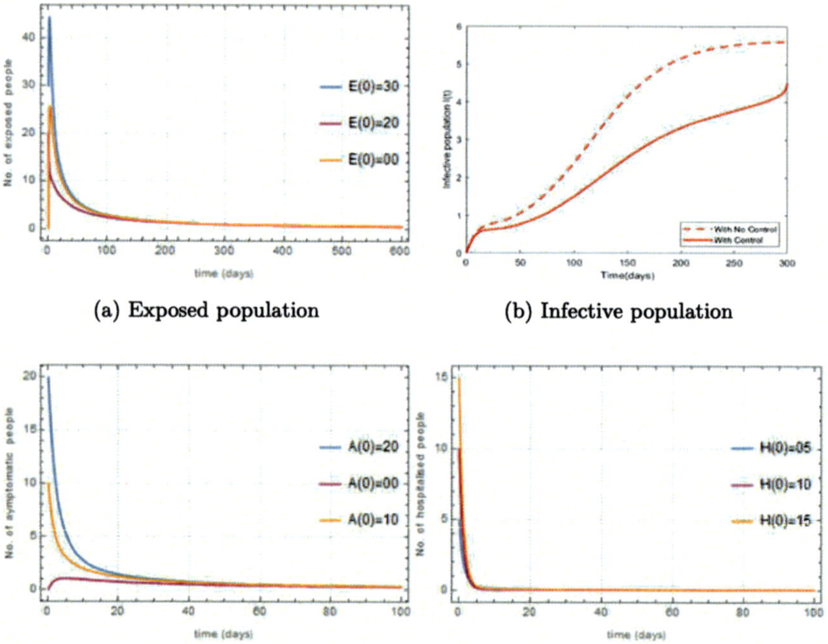

(a) Exposed population

(b) Infective population

(c) Asymptomatic population

(d) Hospitalized population

Fig. 5 Numerical simulation of model (1) for $\mathscr{R}_0 = 1.82565$ with and without control. The system predicts fewer infectious cases with control

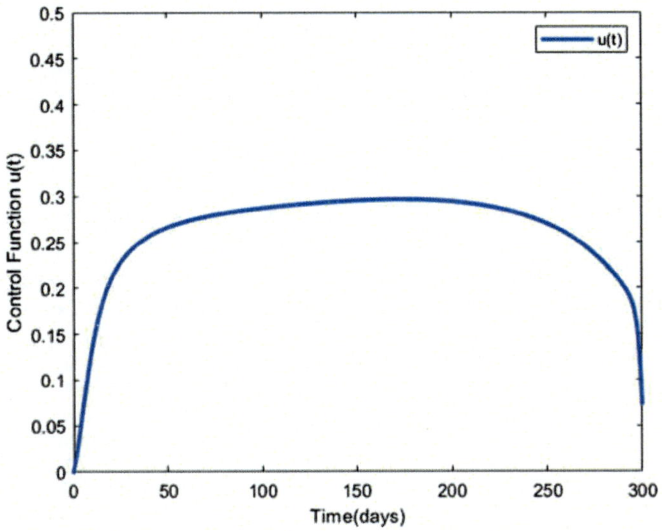

Fig. 6 Control function $u^*(t)$

9 Conclusion and Discussion

In this study, a system of ordinary differential equations is used to create a deterministic mathematical model for the propagation of the COVID-19 pandemic. We have investigated the fundamental features of the model (1), namely positivity and boundedness. The basic reproduction number has been calculated as $\mathscr{R}_0 = \frac{(1-p)\alpha b\beta_1}{\mu(\mu+\alpha)(\mu+h+\gamma)} + \frac{p\alpha b\beta_2}{\mu(\mu+\alpha)(\mu+\gamma)}$, and its relationship to the model's stability has been shown. We determined the steady-state solutions of the model and also found that the disease-free steady state is locally asymptotically stable for $\mathscr{R}_0 \leq 1$ while the endemic steady state is locally asymptotically stable for $\mathscr{R}_0 > 1$. Further, for $\mathscr{R}_0 \leq 1$, there is a unique globally stable disease-free steady state, and when $\mathscr{R}_0 > 1$, there is a unique globally stable endemic steady state. The global stability of the two steady states has been demonstrated by taking into account an appropriate Lyapunov functional for each scenario and applying the LaSalle invariance principle. Further, incorporating a control strategy for symptomatic infected individuals, we developed an appropriate optimal control problem for our model and solved it. We found that the prevalence of infection is less when control is applied. Lastly, we demonstrated our theoretical findings using a series of numerical simulations conducted under a variety of initial conditions. In this model, we assumed permanent immunity after recovery. This needs further study. We think this to have no effect on the model's qualitative behavior. Vaccination has not been considered either. We intend to include these elements in future research.

Declaration

- **Competing interests** The authors declare that there is no competing interests.
- **Funding** The authors declare that there are no funding available for this paper.

References

1. Centers for Disease Control and Prevention (2020). https://www.cdc.gov/coronavirus/2019-nCoV/index.html
2. Johns Hopkins University and Medicine (2020). https://coronavirus.jhu.edu
3. WHO official website. https://www.who.int/emergencies/diseases/novel-coronavirus-2019/situation-reports/
4. Azmon, A., Faes, C., Hens, N.: On the estimation of the reproduction number based on misreported epidemic data. Stat. Med. **33**, 1176–1192 (2014)
5. Burki, T.: COVID-19 in Latin America. Lancet Infect. Dis. **20**, 547–548 (2020)
6. Alves, T.H.E., de Souza, T.A., Silva de Almeida, S., Ramos, N.A., de Oliveira, S.V.: Underreporting of death by COVID-19 in Brazil's second most populous state. Front. Public Health **8** (2020)
7. Arvisais-Anhalt, S., Lehmann, C.U., Park, J.Y., Araj, E., Holcomb, M., Jamieson, A.R., McDonald, S., Medford, R.J., Perl, T.M., Toomay, S.M., et al.: (COVID-19) What the coronavirus disease, pandemic has reinforced: the need for accurate data. Clin. Infect. Dis. **72**(2020), 920–923 (2019)
8. do Prado, M.F., et al.: Analysis of COVID-19 under-reporting in Brazil. Rev. Bras. Ter. Intensiva **32**, 224–228 (2020)
9. Lau, H., et al.: Evaluating the massive underreporting and undertesting of COVID-19 cases in multiple global epicenters. Pulmonology **27**(2), 110–115 (2021)
10. Sarnaglia, A.J., Zamprogno, B., Molinares, F.A.F., de Godoi, L.G., Monroy, N.A.J.: Correcting notification delay and forecasting of COVID-19 data. J. Math. Anal. Appl. 125202 (2021)
11. Rasjid, Z.E., Setiawan, R., Effendi, A., Comparison, A.: Prediction of death and infected COVID-19 cases in Indonesia using time series smoothing and LSTM neural network. Procedia Comput. Sci. **179**, 982–988 (2021)
12. Saberi, M., et al.: Accounting for underreporting in mathematical modeling of transmission and control of COVID-19 in Iran. Front. Phys. **8**, 289 (2020)
13. Worobey, M., Levy, J.I., et al.: The Huanan market was the epicenter of SARS-CoV-2 emergence. Science (2022). https://doi.org/10.5281/zenodo.6299600
14. Gao, G., et al.: Surveillance of SARS-CoV-2 in the environment and animal samples of the Huanan seafood market. Preprint at Research Square (2022). https://doi.org/10.21203/rs.3.rs-1370392/v1
15. Zhou, P., et al.: A pneumonia outbreak associated with a new coronavirus of probable bat origin. Nature **579**, 270–273 (2020)
16. Pekar, J.E., Magee, A., Parker, E., Moshiri, N., Izhikevich, K., Havens, J.L., Wertheim, J.O., et al.: SARS-CoV-2 emergence very likely resulted from at least two zoonotic events (2022). https://doi.org/10.5281/zenodo.6291628
17. Maxmen, A.: Wuhan market was epicentre of pandemic's start, studies suggest. Nature **603**, 15–16 (2022)
18. Levin, S.A.: New Directions in the Mathematics of Infectious Disease. In: Castillo-Chavez, C., Blower, S., van den Driessche, P., Kirschner, D., Yakubu, A.-A. (eds.) Mathematical Approaches for Emerging and Reemerging Infectious Diseases: Models, Methods, and Theory, pp. 1–5. Springer, New York (2002)

19. Chowell, G., Castillo-Chavez, C., Fenimore, P.W., Kribs-Zaleta, C.M., Arriola, L., Hyman, J.M.: Model parameters and outbreak control for SARS. Emerg. Infect. Dis. **10**, 1258–1263 (2004)
20. Hethcote, H.W.: The mathematics of infectious diseases. SIAM Rev. **42**, 599–653 (2000)
21. Kretzschmar, M., Wallinga, J.: Mathematical models in infectious disease epidemiology. In: Kramer, A., Kretzschmar, M., Krickeberg, K. (eds.) Modern Infectious Disease Epidemiology, pp. 209–221. Springer, New York, NY (2009)
22. Brauer, F., Carlos, C.: Mathematical Models in Population Biology and Epidemiology, vol. 2. Springer, New York (2012)
23. Kermack, W.O., McKendrick, A.G.: A contribution to the mathematical theory of epidemics. Proc. R. Soc. Lond. A: Math. Phys. Eng. Sci. **115**(772), 700–721 (1927)
24. May, R.M., Anderson, R.M.: Population biology of infectious diseases: part II. Nature **280**(5722), 455–461 (1979)
25. Bartlett, M.S.: Stochastic Population Models in Ecology and Epidemiology. Wiley, New York (1960)
26. Bailey, N.T.J.: The Mathematical Theory of Epidemics. Griffin, London (1957)
27. Bartlett, M.S.: The critical community size for measles in the U.S. J. R. Stat. Soc. Ser. A **123**, 37–44 (1960)
28. Bartlett, M.: Deterministic and stochastic models for recurrent epidemics. In: Proceedings of the Third Berkeley Symposium on Mathematical Statistics and Probability, vol. 4, p. 109 (1956)
29. Hao, Y., et al.: Prediction and analysis of corona virus disease 2019. PloS one **15**, e0239960 (2020)
30. Asamoah, J.K.K., Bornaa, C.S., Seidu, B., Jin, Z.: Mathematical analysis of the effects of controls on transmission dynamics of SARS-CoV-2. Alex. Eng. J. **59**(6), 5069–5078 (2020)
31. Zhou, T., et al.: Preliminary prediction of the basic reproduction number of the Wuhan novel coronavirus 2019-nCoV. J. Evid. Based Med. **13**, 3–7 (2020)
32. Lin, Q., Zhao, S., Gao, D., et al.: A conceptual model for the coronavirus disease 2019 (COVID-19) outbreak in Wuhan, China with individual reaction and governmental action. Int. J. Infect. Dis. **93**, 211–216 (2020)
33. Kassa, S.M., Njagarah, J.B., Terefe, Y.A.: Analysis of the mitigation strategies for COVID-19: from mathematical modelling perspective. Chaos, Solitons, Fractals **138**, 109968 (2020)
34. Kucharski, A.J., Russell, T.W., Diamond, C., Liu, Y., Edmunds, J., Funk, S., Flasche, S., et al.: Early dynamics of transmission and control of COVID-19: a mathematical modelling study. Lancet Infect. Dis. **20**(5), 553–558 (2020)
35. Tang, B., Yia, F., Tang, S., Bragazzi, N.L., Li, Q., Sun, Y., Liang, J., Yiao, Y., Wu, J.: The effectiveness of quarantine and isolation determine the trend of the COVID-19 epidemics in the final phase of the current outbreak in China. Int. J. Infect. Dis. **95**, 288–293 (2020)
36. Batista, M.: Estimation of the final size of the COVID-19 epidemic, medRxiv 2020.02.16.20023606; https://doi.org/10.1101/2020.02.16.20023606
37. Bai, Z., Gong, Y., Tian, Y., Cao, Y., Liu, W., Li, J.: The rapid assessment and early warning models for COVID-19. Virol. Sin. **35**, 272–279 (2020)
38. Liu, Y., Hewings, G.J.D., Qin, M., Yiang, Y., Zheng, S., Li, Y., Wang, S.L.: Modelling the situation of COVID- 19 and effects of different containment strategies in china with dynamic differential equations and parameters estimation, SSRN (2020). https://doi.org/10.2139/ssrn.3551359
39. Anastassopoulou, C., Russo, L., Tsakris, A., Siettos, C.: Data based analysis, modelling and forecasting of the COVID-19 outbreak. PLoS ONE. **15**, e0230405 (2020)
40. Lopez, L., Rodo, Y.: A modified SEIR model to predict the COVID-19 outbreak in Spain and Italy: simulating control scenarios and multi-scale epidemics. Results Phys. **21**, 103746 (2021)
41. Radulescu, A., Williams, C., Cavanagh, K.: Management strategies in a SEIR-type model of COVID 19 community spread. Sci. Rep. **10**, 21256 (2020)
42. Jia, J., Ding, J., Liu, S., Liao, G., Li, J., Duan, B., Wang, G., Zhang, R.: Modeling the control of COVID-19: impact of policy interventions and meteorological factors (2020). arXiv:2003.0298

43. Liu, Z., Magal, P., Seydi, O., Webb, G.: Predicting the cumulative number of cases for the covid-19 epidemic in china from early data. Math. Biosci. Eng. **17**(4), 3040–3051 (2020)
44. Bastos, S.B., Cajueiro, D.O.: Modeling and forecasting the early evolution of the Covid-19 pandemic in Brazil. Sci. Rep. **10**, 19457 (2020)
45. Senapati, A., Rana, S., Das, T., Chattopadhyay, J.: Impact of intervention on the spread of COVID-19 in India: a model based study (2020). arXiv:2004.04950 [Math, q-Bio]
46. Ivorra, B., Ferrandez, M.R., Vela-Perez, M., Ramos, A.M.: Mathematical modeling of the spread of the coronavirus disease 2019 (COVID-19) taking into account the undetected infections. The case of China. Commun. Nonlinear Sci. Numer. Simul. **88**, 105303 (2020)
47. Pengpeng, S., Shengli, C., Peihua, F.: SEIR Transmission dynamics model of 2019 nCoV coronavirus with considering the weak infectious ability and changes in latency duration, Infectious Diseases (except HIV/AIDS), medRxiv 2020.02.16.20023655; https://doi.org/10.1101/2020.02.16.20023655
48. Tsay, C., Lejarza, F., Stadtherr, M.A., Baldea, M.: Modeling, state estimation, and optimal control for the US COVID-19 outbreak. Sci. Rep. **10**, 10711 (2020)
49. Hou, C., Chen, J., Zhou, Y., et al.: The effectiveness of quarantine of Wuhan city against the Corona Virus Disease 2019 (COVID-19): a well-mixed SEIR model analysis. J. Med. Virol. **92**, 841–848 (2020)
50. Cao, J., Jiang, Y., Zhao, B., et al.: Mathematical modeling and epidemic prediction of covid-19 and its significance to epidemic prevention and control measures. J. Biomed. Res. Innov. **1**(1), 103 (2020)
51. Chatterjee, A.N., Basir, F.A.: A model for 2019-nCoV infection with treatment, medRxiv 2020.04.24.20077958; https://doi.org/10.1101/2020.04.24.20077958
52. Zhao, S., Lin, Q., Ran, J., et al.: Preliminary estimation of the basic reproduction number of novel coronavirus (2019-nCoV) in China, from 2019 to 2020: a data-driven analysis in the early phase of the outbreak. Int. J. Infect. Dis. **92**, 214–217 (2020)
53. Liang, K.: Mathematical model of infection kinetics and its analysis for COVID-19, SARS and MERS. Infect. Genet. Evol. **82**, 104306 (2020)
54. Boudrioua, M.S., Boudrioua, A.: Predicting the COVID-19 epidemic in Algeria using the SIR model. medRxiv 2020.04.25.20079467; https://doi.org/10.1101/2020.04.25.20079467
55. Zhou, Y., Ma, Y., Hong, N., et al.: Forecasting the worldwide spread of COVID-19 based on logistic model and SEIR model. medRxiv 2020.03.26.20044289; https://doi.org/10.1101/2020.03.26.20044289
56. Anderson, R.M., Anderson, B., May, R.M.: Infectious Diseases of Humans: Dynamics and Control. Oxford University Press, Oxford (1992)
57. Hethcote, H.W.: The mathematics of infectious diseases. SIAM Rev. **42**(4), 599–653 (2000)
58. Tang, B., Bragazzi, N.L., Li, Q., Tang, S., Yiao, Y., Wu, J.: An updated estimation of the risk of transmission of the novel coronavirus (2019-ncov). Infect. Dis. Model. **5**, 248–255 (2020)
59. Yiong, H., Yan, H.: Simulating the infected population and spreadtrend of 2019-ncov under different policy by eir model. Available at SSRN 3537083 (2020)
60. Tang, B., Wang, X., Li, Q., Bragazzi, N.L., Tang, S., Xiao, Y., Wu, J.: Estimation of the transmission risk of the 2019-nCoV and its implication for public health interventions. J. Clin. Med. **9**(2), 462 (2020)
61. Hamzah, F.B., Lau, C., Nazri, H., Ligot, D.V., Lee, G., Tan, C.L., Chung, M.H.: CoronaTracker: worldwide COVID-19 outbreak data analysis and prediction. Bull. World Health Org. **1**(32), 1–32 (2020)
62. Clifford, S.J., Klepac, P., Van Zandvoort, K., Quilty, B.J., Eggo, R.M., Flasche, S., et al.: Interventions targeting air travellers early in the pandemic may delay local outbreaks of sars-cov-2. medRxiv (2020)
63. Chen, Y., Cheng, J., Jiang, Y., Liu, K.: A time delay dynamical model for outbreak of 2019-ncov and the parameter identification. J. Inverse Ill-Posed Probl. **28**(2), 243–250 (2020)
64. Rong, X., Yang, L., Chu, H., Fan, M.: Effect of delay in diagnosis on transmission of COVID-19. Math. Biosci. Eng. **17**(3), 2725–2740 (2020)
65. Menéndez, J.: Elementary time-delay dynamics of COVID-19 disease. medRxiv (2020)

66. Liu, X., Zheng, X., Balachandran, B.: COVID-19: data-driven dynamics, statistical and distributed delay models, and observations. Nonlinear Dyn. **101**(3), 1527–1543 (2020)
67. Zaitri, M.A., Silva, C.J., Torres, D.F.: Stability analysis of delayed COVID-19 models. Axioms **11**(8), 400 (2022)
68. Rohith, G., Devika, K.B.: Dynamics and control of COVID-19 pandemic with nonlinear incidence rates. Nonlinear Dyn. **101**(3), 2013–2026 (2020)
69. Huang, J., Qi, G.: Effects of control measures on the dynamics of COVID-19 and double-peak behavior in Spain. Nonlinear Dyn. **101**(3), 1889–1899 (2020)
70. Silva, C.J., Cruz, C., Torres, D.F.M., Munuzuri, A.P., Carballosa, A., Area, I., Nieto, J.J., et al.: Optimal control of the COVID-19 pandemic: controlled sanitary deconfinement in Portugal. Sci. Rep. **11**, 3451 (2021)
71. Niu, R., Wong, E.W., Chan, Y.C., Van Wyk, M.A., Chen, G.: Modeling the COVID-19 pandemic using an SEIHR model with human migration. IEEE Access **8**, 195503–195514 (2020)
72. Diekmann, O., Heesterbeek, J.A.P.: Mathematical Epidemiology of Infectious Diseases: Model Building, Analysis and Interpretation. Wiley, New York (2000)
73. Kot, M.: Elements of Mathematical Ecology. Cambridge University Press, Cambridge (2000)
74. Lambert, J.D.: Computational Methods in Ordinary Differential Equations. Wiley, New York, NY, USA (1973)
75. Fred Brauer, J.A.N.: The Qualitative Theory of Ordinary Differential Equations: An Introduction. Dover Publications, New York, NY, USA (1989)
76. Dragomir, S.S., City, M.: Some Gronwall type inequalities and applications (2002). http://rgmiavu.edu.au/SSDragomirWeb.html
77. Hethcote, H.W.: Mathematics of infectious diseases. SIAM Rev. **42**, 599–653 (2005)
78. Bjørnstad, O.N.: Epidemics: Models and Data Using R. Springer, New York (2018)
79. Bjørnstad, O.N., Shea, K., Krzywinski, M., et al.: The SEIRS model for infectious disease dynamics. Nat. Methods **17**, 557–558 (2020)
80. Roberts, M.G., Heesterbeek, J.A.: A new method for estimating the effort requiredto control an infectious disease. Proc. Biol. Sci. **270**, 1359–1364 (2003)
81. Ma, J., Earn, D.J.: Generality of the final size formula for an epidemic of a newly invading infectious disease. Bull. Math. Biol. **68**, 679–702 (2006)
82. Miller, J.C.: A note on the derivation of epidemic final sizes. Bull. Math. Biol. **74**, 2125–2141 (2012)
83. Van den Driessche, P., Watmough, J.: Reproduction numbers and sub-threshold endemic equilibria for compartmental models of disease transmission. Math. Biosci. **180**(1–2), 29–48 (2002)
84. Van den Driessche, P., Watmough, J.: Further notes on the basic reproduction number. Math. Epidemiol. 159–178 (2008)
85. Diekmann, O., Heesterbeek, J.A.P., Metz, J.A.: On the definition and the computation of the basic reproduction ratio R_0 in models for infectious diseases in heterogeneous populations. J. Math. Biol. **28**(4), 365–382 (1990)
86. LaSalle, J.P.: The Stability of Dynamical Systems. Society for Industrial and Applied Mathematics (1976)
87. LaSalle, J.: Some extensions of Liapunov's second method. IRE Trans. Circuit Theory **7**(4), 520–527 (1960). https://doi.org/10.1109/TCT.1960.1086720
88. Hale, J.K., Lunel, S.M.V.: Introduction to Functional Differential Equations, vol. 99. Springer Science *And* Business Media (2013)
89. Hale, J.K.: Functional differential equations. In: Analytic Theory of Differential Equations, pp. 9–22. Springer, Berlin, Heidelberg (1971)
90. Schechter, M.: Principles of Functional Analysis (No. 36). American Mathematical Soc (2001)
91. Pontryagin, L.S.: Mathematical Theory of Optimal Processes. CRC Press (1987)
92. Kamien, M.I., Schwartz, N.L.: Dynamic Optimization: The Calculus of Variations and Optimal Control in Economics and Management. Elsevier Science B.V, North-Holland (2012)
93. Lukes, D.L.: Differential Equations: Classical to Controlled. Academic, New York (1982)

Modeling Public Fear Under the Information Environment of Emergencies as COVID-19 and Wars

Teruaki Ohnishi

Abstract Under the recent pandemic of corona-virus disease COVID-19, people have repeatedly reacted to the information concerning the spread of the epidemic, which has flooded in our society throughout the whole time. By using the statistical data obtained in Japan, a mathematical model treating the relation between the information environment and the public reaction caused by fear is proposed. The input quantities for this model are the time-varying environment of information on the COVID-19 together with the amount of public communication or the amount of information search on the COVID-19 via the SNS. Between those two quantities, we introduce three variables such as the amount of information released by the news media every day, the novelty of information, and the extent of people's surprise to the information. Regarding the amount of information, we use the real quantity observed in Japan, whereas the novelty of information is related to the so-called oblivion function by which the extent people forget the information in the past is given. Two types of oblivion function, the power-law type and the exponential type, are introduced. On the other hand, the extent of people's surprise by the information on COVID-19, which is determined so as to make sure the consistency of the model as a whole, has revealed to have sharp peaks at the earliest time of the pandemic, and the beginning of every emergent-state declarations over eight times. Such a reaction around the people's surprise is due to the emotional contagion in the fearful field of information environment. To compare the time-dependent trend of those variables to the other cases of social events, we have introduced the case of Ukraine-Russia conflict reported in Japan since 1 January 2022. The people's surprise in this case has revealed that people astonished only at the first time of the conflict, without any surprise after that even the novelty of information has been high. Discussion was made on the sources about the difference regarding the people's surprise between the cases COVID-19 and Ukraine-Russia conflict.

Keywords COVID-19 · Mathematical model · Emotional contagion · Statistical quantity · People's surprise · Oblivion function · Novelty of information · Information environment · Public fear · Ukraine–Russia conflict

T. Ohnishi (✉)
Institute of Science and Technology for Society, Urayasu 2790012, Chiba, Japan
e-mail: teruakiohnishi@yahoo.co.jp

D. Vlachos (ed.), *Mathematical Modeling in Physical Sciences*, Springer Proceedings in Mathematics & Statistics 446, https://doi.org/10.1007/978-3-031-52965-8_42

1 Introduction

The author has proposed a mathematical model, which relates the amount of news released in the society to the public reaction such as the search for more reliable information under the circumstance of emergency of COVID-19, by using the preliminarily data up to September 2021 in Japan [1]. The public behavior of information seeking during the period from February to April 2020, when a sort of panic had prevailed among Japanese, clearly showed the abrupt increase of the amount of communication among people via the SNS.

The time-varying feature of the communication has revealed to be composed of three stages as (1) the sudden occurrence of communication among people caused by the insufficient information and mysterious fear, (2) the transient phase of communication and (3) the roughly-equilibrating phase of communication with the news reporting the emergency in the society. Here the public behavior as such was assumed to originate from the unknown and weird character of the disease and the fearfulness and disgust to the virus.

The strength of such emotion was also assumed to be proportional to the amount of communication via the SNS and the information retrieval via the Internet, which are considered as information-seeking behaviors. Such behaviors are originated from the compensation of anxiety and uneasiness about the lack of reliable information. In such a case the society can be treated as a single reactor immersed in a homogeneous environment where the emotion is oriented to a given direction.

The collective behavior of the public as a whole was, therefore, also considered as caused by the emotional contagion of fear and disgust from the information environment regarding COVID-19. We should note that situations as such are realized also in music halls and theaters, or some type of deeply impressed lectures which cause some moving emotion in the audience common to all [2, 3].

Our mathematical model for deriving emotional quantities under the condition of emergent climate is outlined in Sect. 2. In Sect. 3 applying the statistical data up to 31 March 2023, together with the data of time-varying public fear to COVID-19 derived by using the public opinion survey, we investigate the appropriateness of the form of so-called oblivion function and the extent of time–varying surprise of the Japanese during the whole period of COVID-19 event.

To compare the variables for COVID-19 to the other emergent event, we study the case of Ukraine-Russia conflict during 1 January to 31 March in 2023 in Sect. 4, and concluding remarks are made in Sect. 5.

From what type of news media we acquire the primary information regarding those social events depends on the circumstance of the society different from every country.

It also depends on the difference of society and public sensitivity on the matter how the every-day information diffuses into the society, and what reactions of people are resulted from the information. A model for the Japanese society is, therefore, proposed here by using real data in Japan, but it is well applicable for any society by adjusting values of several variables and constants.

2 Mathematical Model

Figure 1 shows the schematic figure of our model. The emotion induced strongly in all people by the contagion from the information environment is almost unpleasant and energetic [4]. We suppose it as a negative and fearful emotion so that, in what follows, we exemplify the public fear of a corona virus disease and the fright of a war. The strength of the fear people feel at time t from the information environment, $\xi(t)$, is assumed to be given by:

$$\frac{d\xi}{dt} = -c_1\frac{\xi}{\tau} + c_2 \cdot \varepsilon(t) \cdot I_{evn}(t) \cdot \Delta(t) \tag{1}$$

where c_1 and c_2 are constants, τ is the time scale for the variable ξ, I_{env} is the strength of information environment, ε is the novelty of the information and Δ is an extent of emotional contagion which is a factor connecting the fear ξ and the information environment.

We assume the extent of the fear proportional to the strength of the information environment which, in turn, is proportional to the strength of information seeking by people. Differentiating Eq. (1) and substituting the ξ by the number of people's access to the SNS or the Internet per unit time, which indicate the strength of information seeking by people, $x(t)$ [1], we obtain the quantity Δ as

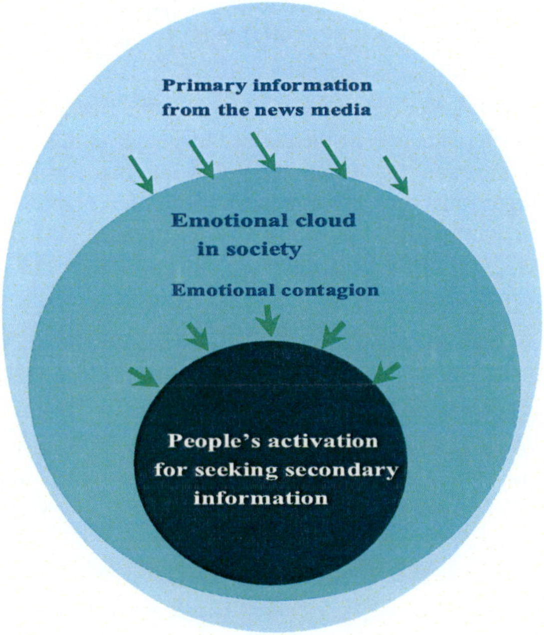

Fig. 1 Schematic figure for the diffusion of negative information

$$\Delta(t) = MAX\left[\frac{1}{c_3 \cdot \varepsilon(t) \cdot I_{env}(t)}\left\{x(t+1) - x(t-1) + \frac{2}{\tau}x(t)\right\}, 0\right] \quad (2)$$

where c_3 is a constant and MAX [*a*, *b*] is the maximum of either *a* or *b*. The quantity $\Delta(t)$ depends on the difference of the strengths of public's information seeking before and after the time *t*. Such dependence on the difference *x*(*after event*)–*x*(*before event*) indicates the extent of unexpectedness of the event which occurs at *t,* so that it resembles the dependence of a surprise function [5]. We therefore call the quantity $\Delta(t)$ as the surprise factor which represents the extent of surprise caused by the emotional contagion.

The second term on the right hand side of Eq. (1) is a source term for the variation of the ξ., which indicates the public emotion such as fear, being derived by the product of the strength of the novelty of information (which is given by the product of the extent of the novelty and the amount of information) and the extent of surprise by the information.

The extent of novelty of information $\varepsilon(t)$ is assumed to be given by a learning curve which is a function of the extent of public memory $Z(t)$ regarding the event considering. Approximating an exponential form, the $\varepsilon(t)$ is given by

$$\varepsilon(t) = \exp(-\gamma Z(t)) \quad (3)$$

where γ is a constant. On the other hand, the $Z(t)$ is given by:

$$Z(t) = c_4 \int_0^t I_{env}(t) \cdot \omega(t', t)dt' \quad (4)$$

where c_4 is a constant and ω (t', t) is the fraction of the information released at t', which remains to be memorized by people until *t*. The ω (t', t) is called the oblivion

Function for which the following two forms of a power-law type and an exponential type are introduced [6] for the comparison of its appropriateness:

$$(I) \quad \omega(t\prime, t) \propto g \cdot \left(\frac{t_0}{t_0 + -t\prime}\right)^k + (1-g) \quad (5)$$

$$(II) \quad \omega(t\prime, t) \propto g \cdot \exp\left(-\frac{(t - t\prime)}{t_0}\right) + (1-g) \quad (6)$$

where t_0 and k are constants so as to be determined as they reproduce some observational data, and *g* is the fraction by which people finally forget information.

More precise description of the model is given in a previous paper [1].

3 Public Emotion Relating to COVID-19

Here we use [day] as a unit of time and see the time-varying feature of variables on and after 1 January 2020 when we set $t = 1$. The secular variation of the observed fear people felt to COVID-19 is shown in Fig. 2 by circles. The ordinate shows an indicator of people's fearfulness P which was derived by using Likert number r_i and the data of public opinion surveys executed during the period of COVID-19 regarding the fearfulness and the risky feeling of people against the infection of the disease [7–9] as $P = \sum_i r_i \cdot p_i$, where p_i is the rate people select the i'th choice in the choices of question. Quantity $P = 0$ corresponds to the state people never feel fearfulness, whereas $P = 3$ to the sate people feel maximum fearfulness. According to Fig. 2 people's fear seemed rapidly increased up to $t = 100 \sim 200$, became minimum and maximum at around $t = 200 \sim 300$ and $400 \sim 500$, respectively, gradually decreased after that with several bumps of small maxima and minima.

Figure 3 (a) shows the secular variation of the intensity of information environment in Japan up to 31 March 2023 ($t = 1186$), the derivative method of which is given in the previous paper [1]. The interest in COVID-19 by the news media seems to have decreased in $t > 400$ although a wavy feature has appeared several times in $t > 400$ in the released amount of information. Figure 3(b) shows the frequency of information search on Google with a key word *corona*, which we consider as an indicator for people to compensate their anxiety due to the shortage of information.

When people are immersed in an environment of negative information, the quantity $Z(t)$ is considered to be proportional to the extent of negative emotion held in

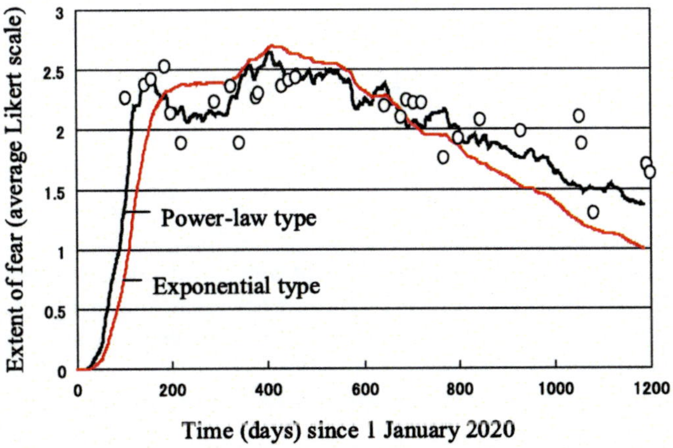

Fig. 2 Time variation of the extent of public fear against COVID-19 in Japan. Circles are average Likert number in the society derived by using public opinion surveys, and black and red solid lines are the time dependent memories of negative information people possessed, which were calculated by using the oblivion functions of the power-law type and of the exponential type, respectively

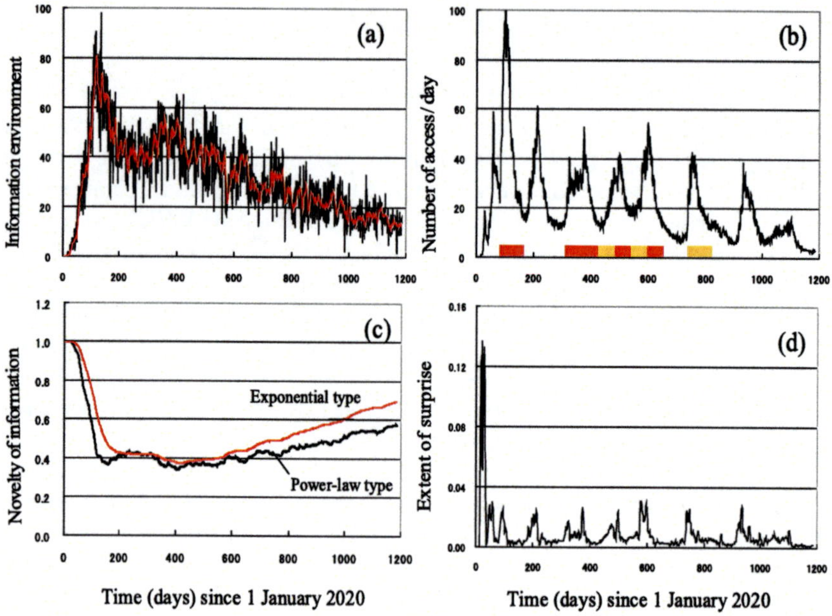

Fig. 3 **a** Time variation of the strength of information on COVID-19 released by the new media in Japan where a red line is the seven days moving average; **b** The number of access to Google search. Red and yellow bands indicate the periods of the declaration of the state of emergency and the semi-state of emergency in Tokyo, respectively; **c** The time behavior of the novelty of information newly released on COVID-19; and **d** The extent of public surprise by the information. This is the seven days moving average of the surprise

people [1, 6, 10]. Hence the time dependent behavior of $Z(t)$ for the case of COVID-19, for instance, might be resemble to the secular variation of people's fear shown in Fig. 2. We show two curves for the $Z(t)$ corresponding to the above respective functions (I) and (II) for oblivion in Fig. 2, for which constants were determined so as the index defined by $\Xi = \left\{\sum_{i=1}^{n} [Z(t_i) - P(t_i)]^2/n\right\}^{1/2}$ becomes minimum. The values of the Ξ thus determined are respectively Ξ (I) $= 0.275$ with $t_0 = 1.45$ and $k = 0.947$ for the type (I), and Ξ (II) $= 0.440$ with $t_0 = 16.84$ for the type (II) both for $g = 1.0$, which infers a favorable form of the power-law function (I) as compared to the exponential one (II) for the oblivion function. We therefore adopt the type (I) form in what follows.

By using the quantity $Z(t)$ with such a behavior we show in Fig. 3(c) the time-varying novelty of information by using Eq. (3). The novelty people felt decreased rapidly around $t \simeq 100$, after that it was almost constant or gradually increased with time. On the other hand Fig. 3 (d) shows the seven days moving average of the quantity Δ (t), which as noted before gives an indicator for people to be surprised by the information of COVID-19, that is the extent of the emotional contagion people received from the information environment from time to time. The Δ (t) became maxima at the very first time of the epidemic and every beginning of the spreading wave of the disease with much weaker values than that of the beginning at $t \simeq 100$.

4 The Case of Ukraine-Russia Conflict

Figure 4 (a) shows the variation of the amount of primary information released by the news media in Japan during 1 January 2022 (when it was set as $t = 1$ with a unit of [day]) to 31 March 2023 ($t = 455$), which was derived by the same statistical method as the COVID-19 case. The unit of the ordinate of this figure is the same as that in Fig. 3(a) so that the intensity of the information environment was weaker to about 20% than the case of COVID-19. Three tremendous peaks have appeared in the information, a, b and c; the peak a at 4 March 2022 (t = 63) corresponds to just one weak after the invasion of Russia, where the overseas evacuation of Ukrainian and various difficulties caused by this conflict were largely reported. The peak b is at 23 June (t = 174) corresponds just to the Okinawa memorial day in Japan and four months after the beginning of this conflict so that many reports appeared regarding the antiwar ideas and movements. On the other hand the peak c is at 24 February 2023 (t = 420), when it is just one year after the outbreak of the conflict, corresponds to the news reporting the state of agglutination of the war.

Figure 4 (b) is the distribution of the number of information search on Google with the key word "*Ukraine*". The maximum scale of this ordinate is assumed as (maximum amount of media release per day for Ukraine-Russia case)/(maximum amount of media release per day for COVID-19 case) × 100.

The primary information shown in Fig. 4 (a) has a maximum at 4 March ($t =$ 63), whereas the number of information search in (b) only had one sharp peak at 24 February ($t = 55$) when the first report on the outbreak of this conflict was released from the news media with an amount of relatively small scale. Although people sensitively reacted to this emergent news to result in seeking more precise and reliable information, which caused the early peak at $t = 55$ in Fig. 4 (b), they seemed rapidly to lose interest after that without requiring any information about it.

Figure 4 (c) shows the extent of information of this issue memorized by people $Z(t)$ with the use of the oblivion function of the type (I) where we have used constants as the same value as COVID-19, together with the novelty of information $\varepsilon(t)$. Whereas in Fig. 4 (d) shown is the emotional contagion factor $\Delta(t)$, that is an indicator by what extent people were surprised by the war. The unit of the ordinate of Fig. 4 (d) is also the same as Fig. 3 (d) so that this figure indicates the relatively weaker strength of the surprise caused by the war occurring in Europe far from Japan, even at its initial time around $t = 55$, than the case of COVID-19 in Japan.

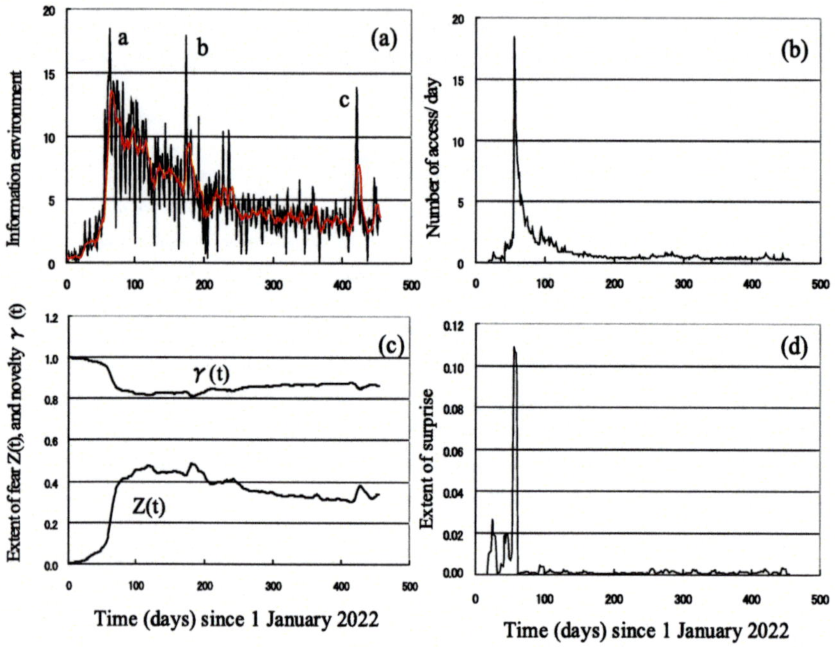

Fig. 4 **a** Time variation of the strength of information on Ukraine-Russia conflict released from 1 January 2022 by the new media in Japan where a red line is the seven days moving average. See the text about the peaks a, b and c; **b** The number of access to Google search with a key word *Ukraine*; **c** The time behavior of the novelty of information newly released and the extent of public memory on Ukraine–Russia conflict; and **d** The extent of public surprise by the information. This is the seven days moving average of the surprise

5 Concluding Remarks

Since January 2020 negative climate regarding an infectious disease of COVID-19 has been formed via the information released primarily by the traditional news media and Internet news sites. Emotions as anxiety and fear against corona virus have been brought in people who have searched more precise and reliable information via the SNS and the Internet search. In this paper a mathematical model has been proposed which describes the relation between the public reaction of information seeking as such and the strength of information environment.

With this model we have made clear that the oblivion function of the form of power-law type is much better than an exponential type to reproduce the extent by what fraction people forget the information obtained in the past. By using such an oblivion function, we have derived the extent of surprise people feels from the information on COVID-19 released every day. People's surprise thus derived has revealed to make a sharp maximum just at the beginning of the spread of disease and small local maxima at every occurrence of the waves of spread.

Such a model has extended to adopt another case of emergency, the Ukraine-Russia conflict which has been loudly reported since January 2022 in Japan, to compare the case of COVID-19. In the case of this conflict people were surprised by the information environment only just at the beginning of this conflict without any surprise after that even the novelty of information has been high. Such a behavior of people is quite different from the case of COVID-19, indicating that people's surprise is dependent on the extent of possible and direct risk of their life, which is imagined from the information environment.

For the possible matter which may cause some risk on the lives of their own, people sensitively react to it to result in the surprise, leading the action of information seeking. But on the other hand when the event occurs in far from themselves, they only make a surprise just at the beginning. People's surprise rapidly disappears and does not continue after that without holding their interest even the information is continuously released every day.

Although the COVID-19 was an unhappy event, it may become a happy event if the study newly finds relations between the public behavior and feeling and the negative information environment in emergency period by analyzing their relevant data.

References

1. Ohnishi, T.: By what extent the Japanese has been surprised by COVID-19 information? Emotional contagion and its mathematical model. Eur. J. Appl. Sci. **10**, 39–56 (2022)
2. Hallam, S., Cross, I., Thaut, M. (eds.): The Oxford Handbook of Music Psychology. Oxford Univ. Press, Oxford (2009)
3. Lewis, M., Haviland-Jones, J.M. (eds.): Handbook of Emotions. Guilford Press, NY (2008)
4. Barsade, S.G.: The ripple effect: Emotional contagion and its influence on group behavior. Admin Sci. Quart. **47**, 644–675 (2002)
5. Macedo, L., Cardoso, A.: A contrast-based computational model of surprise and its applications. Topics Cogn. Sci. **11**, 88–102 (2019)
6. Ohnishi, T., Shimano, K.: Public interest immersed in the field of information environment: How has Japanese interest in energy and environmental problems varied? Rep. Advances Phys. Sci. **2**, 1850005 (2018)
7. Hara, M.: How COVID-19 changed our lives and attitudes. NHK Mon. Rep. Br oadcast Res. (in Japanese) **71**(6), 2–30 (2021)
8. Kobayasi, T., Murata, H.: How the COVID-19 pandemic has changed people's lives and attitudes. NHK Mon. Rep. Br oadcast Res. (in Japanese) **72**(7), 52–87 (2022)
9. Japanese newspapers The Yomiuri, The Asahi, The Mainichi: Articles reporting public opinion surveys on COVID-19 in Japan (in Japanese). (2020–2023)
10. Ohnishi, T.: Can interactive dual fields of information explain the prevalent phenomena. Am. J. Phys. Appl. **7**, 144–155 (2019)

Chaotic Model for Development of HIV Virus

Mohammad Qabaja, Jihad Asad, and Rania Wannan

Abstract In this paper we focus on the numerical treatment of nonlinear system of differential equations arises from biological model. Three numerical methods namely, Runge Kutta method differential transform and multistep differential transform methods are applied to solve the system. Some numerical examples are solved to show the validity and applicability of these numerical methods. Numerical results show clearly that the multistep differential transform method is one of the most efficient and accurate method in comparison with its counterparts.

Keywords Differential transform method · Multi-step differential transform method

1 Introduction

Systems of nonlinear differential equations arise naturally in several application fields, for instance, in classical physics, fluids mechanics, electrical engineering, chemical reaction theory, mathematical biology, ... etc. [1–4]. Since most systems are nonlinear, many different sciences, including engineers, biologists, physicists, and mathematics, are interested in nonlinear problems.

This paper is organized as follows. In section two description of HIV virus model are discussed, and some basic analyses of the system are presented (uniqueness and stability). In section two numerical methods Runge Kutta method, differential transform, and multistep differential transform method techniques for solving nonlinear systems are discussed, however, in section four some results are simulated and discussed. Finally, we closed the paper by a conclusion.

M. Qabaja · R. Wannan (✉)
Department of Applied Mathematics, Faculty of Applied Science, Palestine Technical University-Kadoorie, Tulkarm P 305, Palestine
e-mail: r.wannan@ptuk.edu.ps

J. Asad
Department of Physics, Faculty of Applied Science, Palestine Technical University- Kadoorie, Tulkarm P 305, Palestine

D. Vlachos (ed.), *Mathematical Modeling in Physical Sciences*, Springer Proceedings in Mathematics & Statistics 446, https://doi.org/10.1007/978-3-031-52965-8_43

2 Mathematical Model

From a biological point of view, the HIV virus is a predatory virus, as it attacks healthy immune cells such as prey, and then healthy immune cells become infected, which causes the number to decline of healthy immune cells, and in order to keep the body healthy, immune cells that are functioning properly simultaneously assault and attempt to eradicate viruses. However, the infected immune cells begin to produce the virus due to changing their internal organization in a way that the spread of the virus, knowing that the virus is not able to reproduce by itself, so it exploits the prey for growth and thus multiplies more and spreads through it. The main problem is that the virus multiplies and spreads on the infected immune cells and their survival depends on them, moreover, immune cells are in charge of battling the virus (circular relationship). Therefore, a two-dimensional mathematical model was added to describe the method of spreading the virus and trying to control the virus [5].

$$\frac{dx}{dt} = a_1 x - a_2 xy$$

$$\frac{dy}{dt} = a_3 xy - a_4 y$$

where the prey population is $x(t)$ and the predator population is $y(t)$, over the time t. All parameters a_1, a_2, a_3 and a_4 are constants greater than zero. Parameter a_1 is the prey species' normal growth rate in the absence of the predator. Parameter a_2 it is the single component that lowers the prey population and indicates the effect of the predator on the prey population. Parameter a_3 shows the only element that raises the predator population is the predator's impact on the prey.

A species ability to reproduce depends on the number of resources available. Parameter a_4 indicates the percentage of a predator species natural deaths. But this model is not realistic, because the infinite increase in the number of preys in the absence of a predator is not realistic ($a_2 = 0$), and at the same time the infinite decrease in the number of preys in the absence of prey does not comply with the laws of nature ($a_3 = 0$), so it was developed. A three-dimensional mathematical model is able to make up for this shortcoming in the mathematical model.

$$\begin{aligned} \frac{dx}{dt} &= rx - \beta xz - dx^2 \\ \frac{dy}{dt} &= \beta xz - \alpha y - cy^2 \\ \frac{dz}{dt} &= \lambda y - bz - \sigma xz - ez^2 \end{aligned} \tag{1}$$

where the population of healthy immune cells that have been activated is the variable x. The population of the infected immune cells is represented by variable y. Variable z represents the virus' population. The proliferation of active and healthy T cells is represented by the parameter r, and symbolize βxz the rate when the HIV virus infects the activated T cells, the population of healthy T cells would decrease.

The healthy T cell saturation rate is the parameter d, and it sets a limit on the total number of healthy T cells, even in the absence of a virus. Parameters α and c are the morality rate of the infected T cells and saturation coefficient of their population, respectively.

The growth of the virus by infected T cells is the parameter λ. The constant morality rate and saturation rate of the virus are represented by the parameters b and e, respectively. The virus population would be ruined in a struggle with the healthy T cells at the rate σxz. By using a lot of computer search, the model's parameters were found in such a way that it displayed chaotic behavior. Where $e = 0.39$, $\beta = 0.9$, $d = 0.28$, $\alpha = 1.29$, $c = 0.65$, $\lambda = 0.79$, $b = 0.15$ and $\sigma = 1.24$ [6].

2.1 Existence and Uniqueness

Theorem 1 [7] If f and $\frac{\partial f}{\partial y}$ are continuous in a rectangle $R : |t| \leq a, |y| \leq b$, then there is some interval $|t| \leq h \leq a$ in which there exists a unique solution $y = \varphi(t)$ of the initial value problem.

Applied Theorem 1 in (1), we have

$$f_1(x, y, z) = \frac{dx}{dt} = rx - \beta xz - dx^2$$

$$f_2(x, y, z) = \frac{dy}{dt} = \beta xz - \alpha y - cy^2$$

$$f_3(x, y, z) = \frac{dz}{dt} = \lambda y - bz - \sigma xz - ez^2,$$

f_1, f_2 and f_3 are continuous in any closed cube in $\mathbb{R}^3$.

Now, we will find the derivative about f_1, f_2 and f_3.

$$\frac{\partial f_1}{\partial x} = r - \beta z - 2dx, \frac{\partial f_1}{\partial y} = 0, \frac{\partial f_1}{\partial z} = -\beta x,$$

$$\frac{\partial f_2}{\partial x} = \beta z, \frac{\partial f_2}{\partial y} = -\alpha - 2cy, \frac{\partial f_2}{\partial z} = \beta x$$

$$\partial 3x\partial t\frac{\partial f_3}{\partial x} = -\sigma Z, \frac{\partial f_3}{\partial y} = \lambda, \frac{\partial f_3}{\partial z} = -b - \sigma x - 2ez,$$

all derivatives are continuous. From Theorem 1, we guarantee the existence and uniqueness of the solution on $\mathbb{R}^3$.

2.2 Equilibrium Points

For the purpose of calculating the model's equilibrium points, equations right-hand sides are assumed to be zero.

Let

$$f_1(x, y, z) = \frac{dx}{dt} = rx - \beta xz - dx^2 = 0$$

$$f_2(x, y, z) = \frac{dy}{dt} = \beta xz - \alpha y - cy^2 = 0$$

$$f_3(x, y, z) = \frac{dz}{dt} = \lambda y - bz - \sigma xz - ez^2 = 0$$

If we solve these three equations, we get three fixed points as:

$$E_1 = (0, 0, 0), E_2 = \left(0, 0, -{}^{5}/_{13}\right), E_3 = \left({}^{25r}/_{7}, 0, 0\right).$$

We have to find the eigenvalues of the Jacobian matrix at the fixed points. The form of a Jacobian matrix containing three equations is:

$$\begin{bmatrix} \frac{\partial f_1}{\partial x} & \frac{\partial f_1}{\partial y} & \frac{\partial f_1}{\partial z} \\ \frac{\partial f_2}{\partial x} & \frac{\partial f_2}{\partial y} & \frac{\partial f_2}{\partial z} \\ \frac{\partial f_3}{\partial x} & \frac{\partial f_3}{\partial y} & \frac{\partial f_3}{\partial xz} \end{bmatrix}.$$

Applying Jacobian matrix in f_1, f_2 and f_3

$$\begin{bmatrix} r - \beta z - 2dx & 0 & -\beta x \\ \beta z & -\alpha - 2cy & \beta x \\ -\sigma z & \lambda & -b - \sigma x - 2ez \end{bmatrix}.$$

Applying the first fixed point $E_1 = (0, 0, 0)$ in Jacobian matrix and find eigen values, we have

$$A_1 = \begin{bmatrix} r & 0 & 0 \\ 0 & -\alpha & 0 \\ 0 & \lambda & -b \end{bmatrix},$$

to find eigenvalues from A_1 take the determinant

$$|A_1| = \begin{vmatrix} r - k & 0 & 0 \\ 0 & -\alpha - k & 0 \\ 0 & \lambda & -b - k \end{vmatrix}, \text{wherek : eigenvalues.}$$

The eigenvalues are: [r, −0.15, −1.29].

Applying the second fixed point $E_2 = (0, 0, -5/13)$ in Jacobian matrix and find eigenvalues, we have

$$A_2 = \begin{bmatrix} r + \frac{5}{13}\beta & 0 & 0 \\ \frac{-5}{13}\beta & -\alpha & 0 \\ \frac{5}{13}\sigma & \lambda & -b + \frac{10}{13}e \end{bmatrix}.$$

To find eigenvalues from A_2 take the determinant

$$|A_2| = \begin{vmatrix} r + \frac{5}{13}\beta - k & 0 & 0 \\ \frac{-5}{13}\beta & -\alpha - k & 0 \\ \frac{5}{13}\sigma & \lambda & -b + \frac{10}{13}e - k \end{vmatrix}.$$

The eigenvalues are: [0.35 + r, 0.15, −1.29].

Applying the last fixed point $E_3 = (25r/70, 0)$ in Jacobian matrix and find eigenvalues, we have

$$A_3 = \begin{bmatrix} r - \frac{50r}{7}d & 0 & \frac{-25r}{7}\beta \\ 0 & -\alpha & \frac{25r}{7}\beta \\ 0 & \lambda & -b - \frac{25r}{7}\sigma \end{bmatrix}.$$

To find eigenvalues from A_3 take the determinant

$$|A_3| = \begin{vmatrix} r - \frac{50r}{7}d - k & 0 & \frac{-25r}{7}\beta \\ 0 & -\alpha - k & \frac{25r}{7}\beta \\ 0 & \lambda & -b - \frac{25r}{7}\sigma - k \end{vmatrix}.$$

The eigenvalues are:

$$[-r, -2.21r + \sqrt{4.9r^2 + 0.015r + 0.32} - 0.72, -2.21r - \sqrt{4.9r^2 + 0.015r + 0.32} - 0.72].$$

The fixed point is stable if all the eigenvalues of the Jacobian matrix for the fixed point are negative. The first fixed point is not defined, the second fixed point has nonnegative eigenvalue, because $r \epsilon [0, \infty)$, then the second point is unstable for any value of r, however, all of the eigenvalues of the Jacobian matrix for the third fixed point will be negative if the change of r is taken into account at the interval $[0.01, 0.04]$, and as a result, the last fixed point is stable.

3 Numerical Methods

Numerical methods are methods for approximating mathematical operations. Because we are unable to solve the system analytically, we need to use approximations.

In this section, we will study the system of equations through three methods and a simple description of the methods, in Sect. 3.1 by Runge Kutta method, in Sect. 3.2 by Differential Transform method and in Sect. 3.3 by Multistep Differential Transform method.

3.1 Runge Kutta Method

A collection of iterative implicit, and procedural methods, including the Euler method, that are used to approximate the solutions of many simultaneous nonlinear equations. Runge and Kutta [8], two German mathematicians, created these techniques in the early 1900s, to solve the following system by Runge Kutta method,

$$\begin{aligned} \tfrac{dx}{dt} &= \mathrm{f}_1(t, x, y, z) \\ \tfrac{dy}{dt} &= \mathrm{f}_2(t, x, y, z) \\ \tfrac{dz}{dt} &= \mathrm{f}_3(t, x, y, z), \end{aligned} \tag{2}$$

in the initial circumstances

$$\mathrm{x}(\mathrm{t}_0) = \mathrm{x}_0, \mathrm{y}(\mathrm{t}_0) = \mathrm{y}_0, \mathrm{z}(\mathrm{t}_0) = \mathrm{z}_0 \tag{3}$$

We must establish approximations of values. $x_1, x_2, \ldots, x_n, y_1, y_2, \ldots, y_n$ and $z_1, z_2, \ldots, z_n$ of the solution $x = \varnothing(\mathrm{t}), y = \varphi(\mathrm{t})$ and $z = \omega(\mathrm{t})$ at the points $\mathrm{t_n} = \mathrm{t}_0 + \mathrm{nh}$ with $\mathrm{n} = 1.2.\ldots.$

The initial value issue (2) and (3) can be expressed in vector notation as

$$\mathrm{x}' = \mathrm{f}(\mathrm{t} \cdot \mathrm{x}) \cdot \mathrm{x}(\mathrm{t}_0) = \mathrm{x}_0.$$

where the vector with three components *x*, *y* and *z* is called **x**, and **f** is the vector function with three components $\mathrm{f}_1, \mathrm{f}_2\mathrm{and}\mathrm{f}_3$, and $\mathbf{x}_0$ is the vector with three components $\mathrm{x}_0 \cdot \mathrm{y}_0\mathrm{and}\mathrm{z}_0$.

It is possible to use a system for the step from $\mathrm{t_n}$ to t_{n+1} using the Runge–Kutta method.

We have

$$\mathbf{x}_{\mathrm{n}+1} = \mathbf{x}_{\mathrm{n}} + \left(\frac{\mathrm{h}}{6}\right)(\mathbf{k}_{\mathrm{n}1} + 2\mathbf{k}_{\mathrm{n}2} + 2\mathbf{k}_{\mathrm{n}3} + \mathbf{k}_{\mathrm{n}4}).$$

In another form

$$\begin{pmatrix} \mathrm{x_{n+1}} \\ \mathrm{y_{n+1}} \\ \mathrm{z_{n+1}} \end{pmatrix} = \begin{pmatrix} \mathrm{x_n} \\ \mathrm{y_n} \\ \mathrm{z_n} \end{pmatrix} + \left(\frac{\mathrm{h}}{6}\right)(\mathbf{k}_{\mathrm{n1}} + 2\mathbf{k}_{\mathrm{n2}} + 2\mathbf{k}_{\mathrm{n3}} + \mathbf{k}_{\mathrm{n4}}).$$

where

$$\mathbf{k}_{\mathrm{n1}} = \mathbf{f}(\mathrm{t_n} \cdot \mathbf{x}_{\mathrm{n}}).$$

$$\mathbf{k}_{\mathrm{n2}} = \mathbf{f}\left[\mathrm{t_n} + \left(\frac{\mathrm{h}}{2}\right).\mathbf{x}_{\mathrm{n}} + \left(\frac{\mathrm{h}}{2}\right)\mathbf{k}_{\mathrm{n1}}\right].$$

$$\mathbf{k}_{\mathrm{n3}} = \mathbf{f}\left[\mathrm{t_n} + \left(\frac{\mathrm{h}}{2}\right).\mathbf{x}_{\mathrm{n}} + \left(\frac{\mathrm{h}}{2}\right)\mathbf{k}_{\mathrm{n2}}\right].$$

$$\mathbf{k}_{\mathrm{n4}} = \mathbf{f}[\mathrm{t_n} + \mathrm{h}.\mathbf{x}_{\mathrm{n}} + \mathrm{h}\mathbf{k}_{\mathrm{n3}}]$$

where h is the time step of moving any point to the next point.

By applying Runge Kutta method in (1) with $\mathrm{h} = 0.01$, and the initial conditions

$$\mathrm{x(t_0 = 0) = -0.6.}\quad \mathrm{y(t_0 = 0) = -0.66.}\quad \mathrm{z(t_0 = 0) = 1.03}$$

we have

$$\mathbf{k}_{\mathrm{n1}} = \mathbf{f}(\mathrm{t_n} \cdot \mathbf{x}_{\mathrm{n}}) = \begin{pmatrix} rx_n - \beta x_n z_n - d{x_n}^2 \\ \beta x_n z_n - \alpha y_n - c{y_n}^2 \\ \lambda y_n - bz_n - \sigma x_n z_n - e{z_n}^2 \end{pmatrix}.$$

$$\mathbf{k}_{\mathrm{n2}} = \mathbf{f}\left[\mathrm{t_n} + \left(\frac{\mathrm{h}}{2}\right).\begin{pmatrix} \mathrm{x_n} \\ \mathrm{y_n} \\ \mathrm{z_n} \end{pmatrix} + \left(\frac{\mathrm{h}}{2}\right)\mathbf{k}_{\mathrm{n1}}\right].$$

$$\mathbf{k}_{\mathrm{n3}} = \mathbf{f}\left[\mathrm{t_n} + \left(\frac{\mathrm{h}}{2}\right).\begin{pmatrix} \mathrm{x_n} \\ \mathrm{y_n} \\ \mathrm{z_n} \end{pmatrix} + \left(\frac{\mathrm{h}}{2}\right)\mathbf{k}_{\mathrm{n2}}\right].$$

$$\mathbf{k}_{\mathrm{n4}} = \mathbf{f}\left[\mathrm{t_n} + \mathrm{h}.\begin{pmatrix} \mathrm{x_n} \\ \mathrm{y_n} \\ \mathrm{z_n} \end{pmatrix} + \mathrm{h}\mathbf{k}_{\mathrm{n3}}\right]$$

By using MATLAB, we were able to get a solution to this system, and get a plot to describe the behavior of this system.

3.2 Differential Transform Method

A numerical and analytical method for resolving integral equations, ordinary and partial differential equations is the differential transform method (DTM). The approach offers a convergent series solution with components that are simple to compute. The concept of the differential transform was first proposed by Pukhov in 1982 [9], then Zhou in 1986 [10] used this method to solve linear and nonlinear initial value problems in electrical circuit analysis.

This method constructs, for differential equations, an analytical solution in the form of a polynomial. It is different from the traditional high order Taylor series method, which requires symbolic computations of the necessary derivatives of the data functions. The Taylor series method is computationally taken long time for large orders. The DTM is an iterative procedure for obtaining analytic Taylor series solutions of differential equations. The DTM provides a promising method with numerous applications in a variety of scientific fields [11, 12].

Definition: The differential transform of a function f(t) is defined as follows

$$F(k) = \frac{1}{k!}\left[\frac{d^k f(t)}{dt^k}\right]_{t=t_0}$$

where f(t) is the original function and F(k) is the transformed function.

The inverse transform of F(k) is defined by,

$$f(t) = \sum_{k=0}^{\infty}\left(\frac{(t-t_0)^k}{k!}\frac{d^k f(t)}{dt^k}\right)|_{t=t_0} \quad (4)$$

The following Table 1 can be deduced using the above definition.

By applying DTM in (1), we have

Table 1 Transformed function and basic operation [13]

Original function	Transformed function
$f(t) = u(t) \pm v(t)$	$\mathbf{F(k) = U(k) \pm V(k)}$
$f(t) = \alpha u(t)$	$\mathbf{F(k) = \alpha U(k)}$ **where** $\propto$ **is constant**
$f(t) = u(t)v(t)$	$\mathbf{F(k) = \sum_{l=0}^{k} U(l)V(k-l)}$
$f(t) = \frac{du(t)}{dt}$	$\mathbf{F(k) = (k+1)U(k+1)}$
$f(t) = t^m$	$\mathbf{F(k) = \delta(k-m)}$

$$x(t) = \sum_{k=0}^{\infty} X(k)t^k = X(0) + X(1)t + X(2)t^2 + \ldots$$

$$y(t) = \sum_{k=0}^{\infty} Y(k)t^k = Y(0) + Y(1)t + Y(2)t^2 + \ldots$$

$$z(t) = \sum_{k=0}^{\infty} Z(k)t^k = Z(0) + Z(1)t + Z(2)t^2 + \ldots$$

where X(k) and Y(k) and Z(k), satisfy the following recurrence relation,

$$X(k+1) = \frac{1}{k+1}\left(rX(k) - \beta\sum_{l=0}^{k} X(l)Z(k-l) - d\sum_{l=0}^{k} X(l)X(k-l)\right)$$

$$Y(k+1) = \frac{1}{k+1}\left(\beta\sum_{l=0}^{k} X(l)Z(k-l) - \alpha Y(k) - c\sum_{l=0}^{k} Y(l)Y(k-l)\right)$$

$$Z(k+1) = \frac{1}{k+1}\left(\lambda Y(k) - bZ(k) - \sigma\sum_{l=0}^{k} X(l)Z(k-l) - e\sum_{l=0}^{k} Z(l)Z(k-l)\right)$$

with the initial condition $X(0) = -0.6$, $Y(0) = -0.66$, $Z(0) = 1.03$.

The sequential solution we obtain via DTM is actually a scattered sequential solution, which is one of its limitations. This sequential approach provides a good approximation of the genuine solution in a very small interval but does not depict the problem's true behaviors, so we will turn to the multistep differential transformation method (Ms-DTM).

3.3 *Multistep Differential Transform Method*

DTM displays the problem's actual behavior while providing an accurate prediction of the right answer in a short amount of time, in contrast to the Ms-DTM, it gives series solutions over large time intervals and improve accuracy [12].

Definition: Let $[0.T]$ be the interval for nonlinear initial value problem

$$f\left(t \cdot u \cdot u' \cdots u^{(p)}\right) = 0. \tag{5}$$

Subject to the initial condition $u^{(k)}(0) = c_k$ for $k = 0.1. \cdots .p - 1$.
Can be expressed by the finite series

$$\mathrm{u(t)} = \sum_{n=0}^{N} a_n t^n.$$

Assume that the interval $[0, T]$ is divided into M subintervals $\left[\mathrm{t_{m-1}}, \mathrm{t_m}\right]$, $\mathrm{m} = 1, 2, \cdots .\mathrm{M}$, with step size $\mathrm{h} = \frac{\mathrm{T}}{\mathrm{M}}$, by using the nodes $\mathrm{t_m} = \mathrm{mh}$.

The main idea of the Ms-DTM are as follows. first, over the first interval $[0, t_1]$, we consider initial conditions $u_1^{(k)}(0) = c_k$, at this term the DTM is applied to (5), and the solution is given as

$$\mathrm{u_1(t)} = \sum_{\mathrm{n=0}}^{\mathrm{K}} \mathrm{a_{1n}t^n.t} \in [0, \mathrm{t_1}]. \tag{6}$$

At each subinterval $[\mathrm{t_{m-1}} \cdot \mathrm{t_m}]$ and $\mathrm{m} \geq 2$, the initial conditions.

$\mathrm{u_m}^{(\mathrm{k})}(\mathrm{t_{m-1}}) = \mathrm{u_{m-1}}^{(\mathrm{k})}(\mathrm{t_{m-1}})$, and apply the DTM to Eq. (5) over the interval $[\mathrm{t_{m-1}} \cdot \mathrm{t_m}]$, where $\mathrm{t_0}$ in Eq. (4) is altered with t_{m-1}.

The operation is repeated again, a sequence of approximate solutions $\mathrm{u_m(t)}$, $\mathrm{m} = 1, 2, \cdots, \mathrm{M}$. is generated as follows:

$$u_m(t) = \sum_{n=0}^{K} a_{mn}(t - t_{m-1})^n.\ t \in \left[t_m, t_{m+1}\right]$$

where $N = K \cdot M$. In actual fact, the multi-step DTM suggests the following solution,

$$\mathrm{u(t)} = \begin{cases} \mathrm{u_1(t)} \cdot t \in [0.\mathrm{t_1}] \\ \mathrm{u_2(t)} \cdot t \in [\mathrm{t_1} \cdot \mathrm{t_2}] \\ \vdots \\ \mathrm{u_M(t)} \cdot t \in \left[\mathrm{t_{M-1}} \cdot \mathrm{t_M}\right] \end{cases}$$

Actually, there is no difference between the classic DTM and the Ms-DTM algorithm, However, Ms-DTM reduces the step size if we consider h = T, However, the reason for using Ms-DTM to gives a more accurate solution in a long period of time.

4 Simulation Results and Discussion

In this section, we will find the absolute difference between the values of $x(t)$, $y(t)$ and $z(t)$ in the three ways through which solutions are found for the system (Tables 2, 3, 4 and Figs. 1, 2, 3, 4, 5 and 6).

Table 2 The absolute error rate between Runge–Kutta method and Ms-DTM very few compared to Runge Kutta method and DTM, Ms-DTM and DTM Where the percentage of error was very large with the increase of time on x

t	RK	Ms-DTM	DTM	|RK-Ms-DTM|	|RK-DTM|	|Ms-DTM-DTM|
	$x(t)$	$x(t)$	$x(t)$			
0	−0.6	−0.6	−0.6	0	0	**0**
0.5	−0.4249	−0.4249	−0.4251	0	0.0002	**0.0002**
⋮	⋮	⋮	⋮	⋮	⋮	⋮
3	−0.1878	−0.1884	−6.6151	0.0006	6.4273	**6.4267**
3.5	−0.1752	−0.1758	−4.9999	0.0006	4.8247	**4.8241**
4	−0.1653	−0.1659	30.538-	0.0006	30.3727	**30.3721**
4.5	−0.1572	−0.1578	−57.0637	0.0006	56.9065	**50.9059**
5	0.1503-	−0.1510	−99.5887	0.0007	99.4384	**99.4377**

Table 3 The absolute error rate between Runge–Kutta method and Ms-DTM very few compared to Runge–Kutta method and DTM, and with Ms-DTM and DTM Where the percentage of error was very large with the increase of time on *y*

t	RK	Ms-DTM	DTM	|RK-Ms-DTM|	|RK-DTM|	|Ms-DTM-DTM|
	$y(t)$	$y(t)$	$y(t)$			
0	−0.66	−0.66	−0.66	0	0	**0**
0.5	−0.5950	−0.5950	−0.5948	0	0.0002	**0.0002**
⋮	⋮	⋮	⋮	⋮	⋮	⋮
3	−0.0985	−0.0985	5.0525	0	5.151	**5.151**
3.5	−0.0663	−0.0664	11.8473	0.0001	11.9136	**11.9137**
4	−0.0464	−0.0464	24.4234	0	24.4698	**24.4698**
4.5	−0.0339	−0.0339	45.905	0	45.9389	**45.9389**
5	−0.0260	−0.0260	80.3661	0	80.3921	**80.3921**

5 Conclusion

In this study, the chaotic behavior of the HIV virus was studied numerically using Runge–Kutta method, DTM, and Ms-DTM methods respectively. It was obvious from the tables that the DTM displays the problem's actual behavior while providing an accurate prediction of the right answer in a short period of time. However, the Ms-DTM gives a more accurate solution in a long period of time.

Table 4 The absolute error rate between Runge–Kutta method and Ms-DTM very few compared to Runge–Kutta method and DTM, and with Ms-DTM and DTM where the percentage of error was very large with the increase of time on z

T	RK	Ms-DTM	DTM	|RK-Ms-DTM|	|RK-DTM|	|Ms-DTM-DTM|
	$z(t)$	$z(t)$	$z(t)$			
0	1.03	1.03	1.03	0	0	**0**
0.5	0.8318	0.8318	0.8313	0	0.0005	**0.0005**
⋮	⋮	⋮	⋮	⋮	⋮	⋮
3	0.2391	0.2384	−10.5654	0.0007	10.8045	**10.8038**
3.5	0.2057	0.2052	−23.8054	0.0005	24.0111	**24.0106**
4	0.1824	0.1819	−47.5024	0.0005	47.6848	**47.6843**
4.5	0.1652	0.1648	−86.8398	0.0004	87.005	**87.0046**
5	0.1518	0.1514	−148.4447	0.0004	148.5965	**148.5961**

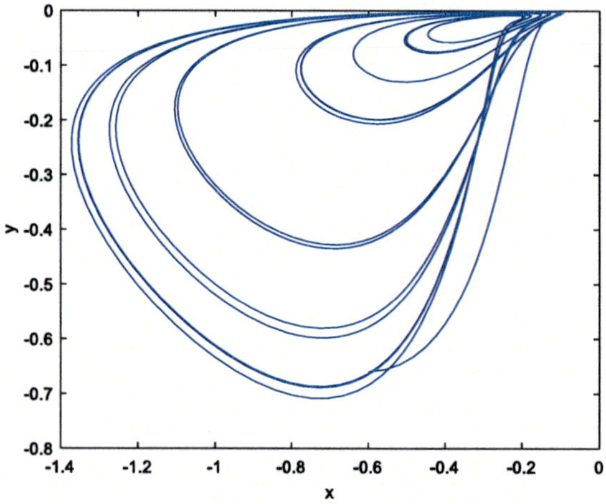

Fig. 1 The relationship between $x(t)$ and $y(t)$ when $h = 0.1$ and $r = 0.01$ using Ms-DTM

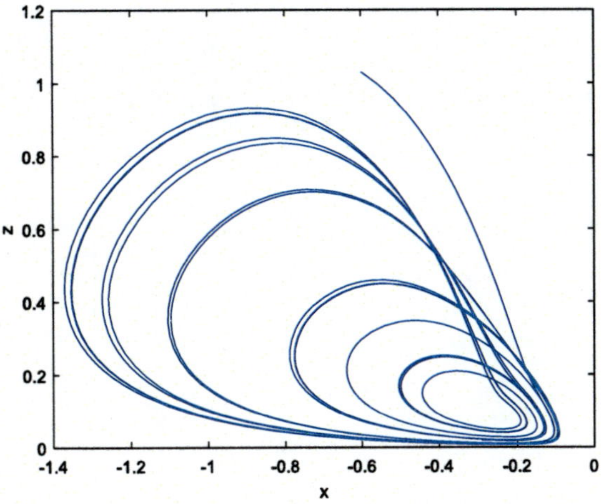

Fig. 2 The relationship between $x(t)$ and $z(t)$ when $h = 0.1, r = 0.01$ using Ms-DTM method

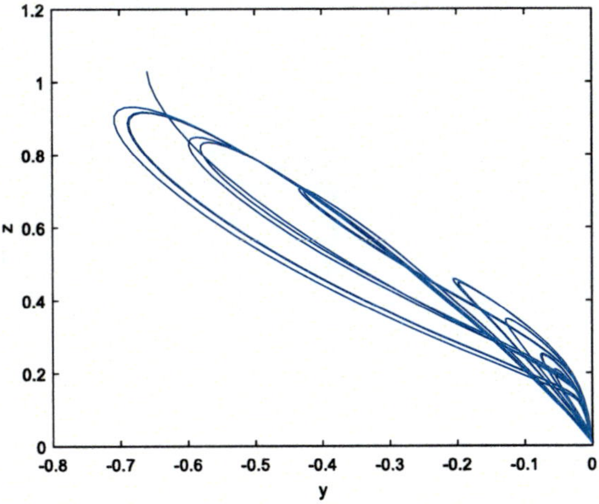

Fig. 3 The relationship between $y(t)$ and $z(t)$ when $h = 0.1, r = 0.01$ to using Ms-DTM method

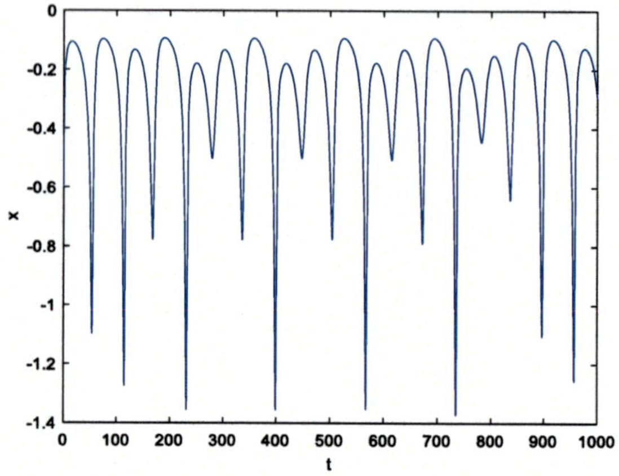

Fig. 4 The relationship between t and $x(t)$ when $h = 0.1, r = 0.01$ to using Ms-DTM method

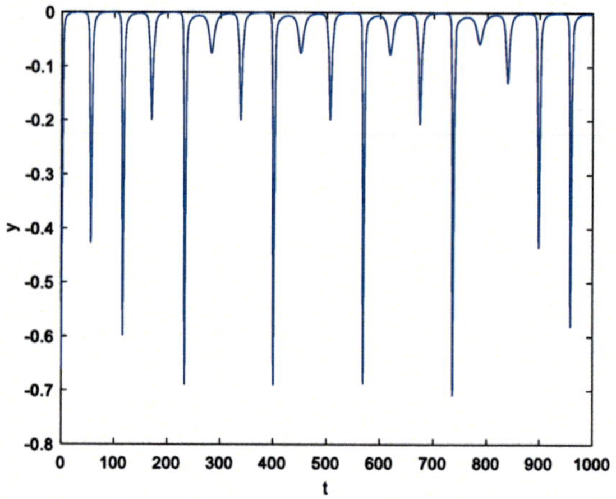

Fig. 5 The relationship between t and $y(t)$ when $h = 0.1, r = 0.01$ to using Ms-DTM method

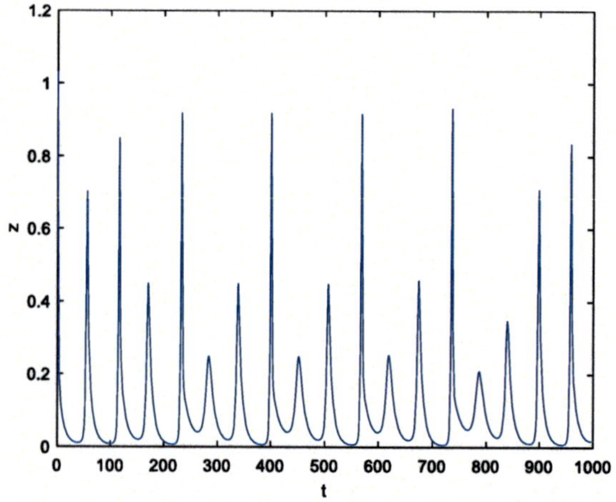

Fig. 6 The relationship between t and $z(t)$ when $h = 0.1, r = 0.01$ to using Ms-DTM method

Acknowledgements The authors would like to thank Palestine Technical University-Kadoorie for their financial support to conduct this research.

References

1. Scott, A.: Nonlinear Biology, The Nonlinear Universe, The Frontiers Collection. Springer, Heidelberg, pp. 181–276 (2007)
2. Korenberg, M.J., Hunter, I.W.: The identification of nonlinear biological systems: Volterra kernel approaches. Ann. Biomed. Eng. **24**(2), 250–268 (1996). Springer
3. Mosconi, F., Julou, T., Desprat, N., Sinha, D.K., Allemand, J.F., Vincent, C., Bensimon, D.: Some nonlinear challenges in biology. Nonlinearity **21**(8) (2008)
4. Gintautas, V.: Resonant forcing of nonlinear systems of differential equations. Chaos **18**(3) (2008)
5. Parastesha, F., Arama, Z., Namazib, H., Sprottc, J.C., Jafari, S.: A simple chaotic model for development of HIV virus. Sci. Iran. D **28**(3), 1643–1652 (2021)
6. Rahmani Doust, M., Gholizade, S.: The Lotka-Volterra predator-prey equation. Casp. J. Math. Sci. (CJMS) **3**(2), 221–225 (2014)
7. Devries, P.L., Hasbun, J.E.: A first course in computational physics, Second edn. Jones and Bartlett Publishers, p. 215 (2011)
8. Odibat, Z.: Differential transform method for solving Volterra integral equation with separable kernels. Math. Comput. Model. **48**(7–8) (2008)
9. Pukhov, G.E.: Differential transforms and circuit theory. J. Circuit Theory Appl. **10**, 265–276 (1982)
10. Zhou, J.K.: Differential Transformation and Its Applications for Electronic Circuits. Huazhong Science & Technology University Press, China (1986)
11. Hassan, I.H.: Comparison differential transformation technique with Adomian decomposition method for linear and nonlinear initial value problems. Chaos Solitons Fractals **36**(1), 53–56 (2008)
12. Odibat, Z.: Comput. Math. Appl. **59**, 1462–1472 (2010)
13. Nourifar, M., Sani, A.A., Keyhani, A.: Efficient multi-step differential transform method: Theory and its application to nonlinear oscillators. Commun. Nonlinear Sci. Numer. Simul. **53**, 154–183 (2017)

Stability Analysis of a COVID-19 SEIQR Model with Switching Constant Transmission Rates

Timothy Robin Y. Teng, Destiny S. Lutero, and Mark Anthony C. Tolentino

Abstract In this paper, we analyze threshold conditions for a COVID-19 susceptible-exposed-infectious-quarantined-recovered (SEIQR) model with a constant recruitment rate and a piecewise constant disease transmission rate using results from the theory of switching systems. We establish that if the basic reproduction number of the system under each mode p is less than 1, then the corresponding solutions tend to the disease-free equilibrium. Further, this condition is sufficient but is not necessary to guarantee disease eradication. Simulations show that it is possible to keep the transmission of infection at bay even if the basic reproduction numbers under some modes are not less than 1.

Keywords Switching transmission rate · Non-autonomous · Infectious disease · COVID-19 · SEIQR

1 Introduction

The World Health Organization (WHO) declared that COVID-19 is no longer a public health emergency of international concern as of May 5, 2023 [10, 11]. The announcement came in more than three years after the agency's declaration of COVID-19 as a global pandemic. Over that span of time, COVID-19 placed a massive strain on health, economic, and social systems worldwide. The disease claimed close to 7 million lives and infected over 760 million people as of early July 2023 [12].

Despite WHO's pronouncement, COVID-19 remains an ongoing public health concern. While many societies have transitioned back to pre-pandemic normalcy, the evolving nature of the SARS-COV-2 has led to the emergence of new variants and subvariants that are capable of causing waves of COVID-19 infections and deaths [10].

T. R. Y. Teng (✉) · D. S. Lutero · M. A. C. Tolentino
Ateneo de Manila University, 1101 Loyola Heights, Quezon City, Philippines
e-mail: tteng@ateneo.edu

D. S. Lutero
University of the Philippines Los Baños, 4031 Los Baños, Laguna, Philippines

D. Vlachos (ed.), *Mathematical Modeling in Physical Sciences*, Springer Proceedings in Mathematics & Statistics 446, https://doi.org/10.1007/978-3-031-52965-8_44

Amidst all this, mathematical modeling has emerged as a vital tool in understanding COVID-19 disease transmission dynamics in a community. These models have played a crucial role for the past three years in disease surveillance and have guided health and government officials in formulating policy decisions for disease control. A key aspect considered by such models is the variability of disease transmissibility over time, resulting from factors such as mutations in virus structure and changes in policy intervention strategies. This variability influences the size and severity of the pandemic and emphasizes the need for adaptive models to capture accurately the ever-evolving COVID-19 situation.

To better understand these aforementioned models, it is important to study them from a theoretical perspective. Represented as systems of differential equations, compartmental models for infectious diseases with time-dependent parameters have been discussed in literature. Some of these discussions revolve around exploring the properties of the system's solutions and/or analyzing their long-term behavior. Ma and Ma, for instance, studied a seasonally forced SEIR model [8]. Liu and Stechlinksi analyzed several compartmental models with time-dependent parameters; these included an SIS model [7] and an SIR model [4] with switching and pulse control, an SIR model with a switched nonlinear incidence rate [5], and a switched multi-city SIR model [6]. Sottile and Liu, in addition, analyzed an SEIR model for measles with a time-varying transmission rate [9].

This paper aims to add to the existing literature on compartmental models with time-varying parameters by considering the SEIQR model for COVID-19 in [3] and assuming that the transmission rate is piecewise constant instead of constant. After deriving the basic reproduction numbers, the stability of the disease-free equilibrium is analyzed using results from switching theory as presented in [1]. Finally, numerical simulations are presented to illustrate the established stability results.

2 The Mathematical Model

2.1 Model Formulation

The proposed COVID-19 model provides a time-dependent extension of the compartmental model formulated in [3]. Similar to the previous model, the new model subdivides the human population into five compartments. Individuals who are vulnerable to COVID-19 infection comprise the susceptible compartment S. Once they interact with infectious individuals, susceptible individuals may become exposed to the virus and, in effect, transition towards the exposed compartment E. Individuals in the E compartment remain there until they become capable of transmitting the virus to others, at which point they transition to the infectious compartment I. One possibility for these infectious individuals is for them to be detected and clinically diagnosed with COVID-19. This detection marks their transition to the quarantined compartment Q, where they are assumed to be isolated from the general population and receiving treatment. Another possibility is for infectious individuals to remain undetected and eventually recover from the disease; once they have recovered, they are placed in the recovered compartment R.

Aside from the five compartments, the model also has parameters that govern the movement of individuals across these states. The transmission parameter $\beta(t)$ indicates the rate at which susceptible individuals get infected upon contact with individuals from the I compartment. In this work, this parameter is made time-dependent, whereas it is assumed constant in [3], to capture the changes in disease transmissibility over time due to factors such as emergence of new variants, behavioral changes (such as mask-wearing), and policy interventions (such as lockdowns). The transition rate α is dependent on the COVID-19 latency period, which refers to the time during which an infected individual becomes contagious. Subsequently, infectious individuals can either be detected at a rate δ towards Q, or naturally recover without isolating at a rate θ towards R. The recovery rate ρ is applied to the Q compartment to capture the recovery of those who have been confirmed to have COVID-19.

Other model parameters include the parameter A that corresponds to the birthrate of the population. The model also includes the natural death rate μ and the COVID-19-attributed deaths rates ϵ_I and ϵ_Q that affect compartments I and Q, respectively.

The proposed model is illustrated in Fig. 1, and its transmission dynamics is represented by the following system of ordinary differential equations:

$$S'(t) = A - \mu S(t) - \beta(t)S(t)I(t), \tag{1}$$

$$E'(t) = \beta(t)S(t)I(t) - (\alpha + \mu)E(t), \tag{2}$$

$$I'(t) = \alpha E(t) - (\theta + \epsilon_I + \delta + \mu)I(t), \tag{3}$$

$$Q'(t) = \delta I(t) - (\epsilon_Q + \rho + \mu)Q(t), \tag{4}$$

$$R'(t) = \rho Q(t) + \theta I(t) - \mu R(t), \tag{5}$$

where

$$N(t) = S(t) + E(t) + I(t) + Q(t) + R(t). \tag{6}$$

The parameters of the model are assumed to be positive. In particular, the time-varying disease transmission rate $\beta(t)$ is assumed to be a bounded positive piecewise constant function on the t-interval $[0, \infty)$. We shall refer to the system of ordinary differential equations defined by Eqs. (1) to (5) as system Π.

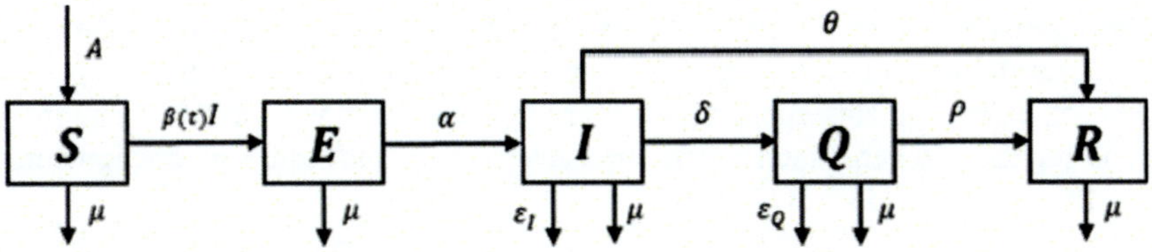

Fig. 1 The flow diagram illustrating COVID-19 transmission for the SEIQR compartmental model

Let $t_0 = 0$. Given that $\beta(t)$ is piecewise constant function on $[0, \infty)$, we may assume that $\beta(t) = b_i$ on $[t_{i-1}, t_i)$ for each $i \in \mathbb{N}$ and that $\beta(t)$ has a jump discontinuity at $t = t_i$ for each i. Furthermore, we assume $\inf_i (t_{i+1} - t_i) \geq h$ for some $h > 0$. The range of $\beta(t)$ is also assumed to be finite; that is, for each $i \in \mathbb{N}$, we must have $b_i = \beta_p$, where p belongs to a finite set P. Following from this, it can be observed that the system goes through different phases as the value of $\beta(t)$ changes over time. We say then that system Π is in *mode* p when $\beta(t)$ assumes the value β_p.

Let $\mathcal{D}$ be the biologically feasible region defined as

$$\mathcal{D} = \left\{ (S, E, I, Q, R) \in \mathbb{R}^5_{0+} : S + E + I + Q + R \leq \frac{A}{\mu} \right\}. \tag{7}$$

The following proposition states the fundamental properties of solutions to system Π.

Proposition 1 *For any initial condition $(S_0, E_0, I_0, Q_0, R_0) \in \mathcal{D}$, there exists a unique, nonnegative, and bounded solution $X : [0, +\infty) \to \mathbb{R}^5_{0+}$ to system Π.*

We note that the above result follows from the properties of solutions with initial condition in $\mathcal{D}$ and satisfying the corresponding system when $\beta(t)$ is equal to a positive constant on $[0, +\infty)$. These properties are presented in [3].

2.2 Equilibrium Point and Basic Reproduction Number

A constant solution $(S(t), E(t), I(t), Q(t), R(t)) = (S, E, I, Q, R)$ of system Π satisfies $[S'(t)\ E'(t)\ I'(t)\ Q'(t)\ R'(t)] = [0\,0\,0\,0\,0]$ for all $t \geq 0$. This implies that for each mode p, we have

$$I = 0 \quad \text{or} \quad \beta_p S - (\alpha + \mu)\left(\frac{\delta + \epsilon_I + \mu + \theta}{\alpha}\right) = 0. \tag{8}$$

The first equation yields the disease-free equilibrium (DFE)

$$X_{\text{dfe}} = \left(\frac{A}{\mu}, 0, 0, 0, 0\right),$$

which satisfies system Π across all modes p. The latter equation in (8), on the other hand, results in a value of S that is dependent on β_p. Since the values of β_p are distinct, there is no corresponding equilibrium point resulting from this equation.

Using the next generation matrix approach from [2], the basic reproduction number R_0^p for each mode p is given by

$$R_0^p = \frac{A\alpha\beta_p}{\mu(\alpha + \mu)(\delta + \epsilon_I + \mu + \theta)}. \tag{9}$$

It should be noted that system Π only admits a single equilibrium point X_{dfe}. Moreover, X_{dfe} exists for any value of R_0^p and is contained in the feasible region $\mathscr{D}$.

3 Stability Analysis

The stability analysis of the DFE begins by considering the following non-autonomous subsystem Π_s of system Π

$$S'(t) = A - \mu S(t) - \beta(t)S(t)I(t), \tag{10}$$
$$E'(t) = \beta(t)S(t)I(t) - (\alpha + \mu)E(t), \tag{11}$$
$$I'(t) = \alpha E(t) - (\theta + \epsilon_I + \delta + \mu)I(t). \tag{12}$$

We note that this subsystem has an associated feasible region given by

$$\Omega = \left\{(S, E, I) \in \mathbb{R}^3_{0+} : S + E + I \leq \frac{A}{\mu}\right\}.$$

Now, let us perform the following change of variables:

$$\bar{S}(t) = \frac{A}{\mu} - S(t),$$
$$\bar{I}(t) = I(t),$$
$$\bar{E}(t) = E(t).$$

We express subsystem Π_s in terms of $\bar{S}(t)$, $\bar{I}(t)$, and $\bar{E}(t)$, and we denote the converted subsystem as system $\bar{\Pi}_s$:

$$\bar{S}'(t) = -\mu\bar{S}(t) + \beta(t)\left(\frac{A}{\mu} - \bar{S}(t)\right)\bar{I}(t), \tag{13}$$
$$\bar{E}'(t) = \beta(t)\left(\frac{A}{\mu} - \bar{S}(t))\right)\bar{I}(t) - (\mu + \alpha)\bar{E}'(t), \tag{14}$$
$$\bar{I}'(t) = \alpha\bar{E}(t) - (\mu + \delta + \epsilon_I + \theta)\bar{I}(t). \tag{15}$$

As defined in the preceding section, we also say that system $\bar{\Pi}_s$ is in *mode p* when $\beta(t)$ assumes the value β_p. We also note that the feasible region for the converted system is given by

$$\bar{\Omega} = \left\{(\bar{S}, \bar{E}, \bar{I}) \in \mathbb{R}^3_{0+} : \bar{S} \leq \frac{A}{\mu} \text{ and } \bar{E} + \bar{I} \leq \bar{S}\right\}.$$

Since $X_{\text{dfe}} = \left(\dfrac{A}{\mu}, 0, 0, 0, 0\right)$ is the disease-free equilibrium of system Π, it follows that $\bar{X}_{\text{dfe}} := (0, 0, 0)$ is the corresponding disease-free equilibrium of system $\bar{\Pi}_s$. Moreover, the set $\bar{\Omega}$ contains $\bar{X}_{\text{dfe}}$, and that for any initial condition in $\bar{\Omega}$, there exists a unique, nonnegative, and bounded solution for system $\bar{\Pi}_s$.

Let $x : [0, +\infty) \to \bar{\Omega}$, where $x(t) = \left(\bar{S}(t), \bar{E}(t), \bar{I}(t)\right)$ and define $f_p(x) : \bar{\Omega} \to \mathbb{R}^3_{0+}$ as

$$f_p(x) = \begin{pmatrix} -\mu\bar{S} - \beta_p\left(\dfrac{A}{\mu} - \bar{S}\right)\bar{I} \\ \beta_p\left(\dfrac{A}{\mu} - \bar{S}\right)\bar{I} - (\mu + \alpha)\bar{E} \\ \alpha\bar{E} - (\mu + \delta + \epsilon_I + \theta)\bar{I} \end{pmatrix},$$

where p belongs to the finite index set P. Observe that $f_p(x)$ is a continuous vector field on $\mathbb{R}^3_{0+}$ and $f_p((0, 0, 0)) = (0, 0, 0)$. Moreover, we can represent system $\bar{\Pi}_s$ as $x' = f_p(x)$ whenever the system is in mode p.

We proceed following the definitions on switched systems from [1]. Let $\mathscr{F}_s = \left\{f_p(x), p \in P\right\}$ and note that this finite family $\mathscr{F}_s$ defines a switching system. Moreover, given the assumption that $\inf_i (t_{i+1} - t_i) \geq h$, where $h > 0$, we remark that solutions of system $\bar{\Pi}_s$ with initial condition in $\bar{\Omega}$ are elements of $\mathcal{S}_{\text{dwell}}$, which is defined as the set of switched solutions of $\mathscr{F}_s$ with nonvanishing dwell time.

Let $R_{\max} = \max\{R_0^p : p \in P\}$. Define the function $V : \bar{\Omega} \to \mathbb{R}$ as

$$V(\bar{S}, \bar{E}, \bar{I}) = \frac{1 - R_{\max}}{R_{\max}} \cdot \alpha\bar{S} + \alpha\bar{E} + (\alpha + \mu)\bar{I}.$$

We establish some properties of V in the succeeding lemmas.

Lemma 1 *If $R_0^p < 1$ for all $p \in P$, then the function V is positive definite and continuously differentiable.*

Proof It is clear that $V(0, 0, 0) = 0$. Since $R_0^p < 1$ for all $p \in P$, we have $R_{\max} < 1$ and $\frac{1-R_{\max}}{R_{\max}} > 0$. This implies that all the coefficients of V are positive. Thus, $V(\bar{S}, \bar{E}, \bar{I}) > 0$ for every $\left(\bar{S}, \bar{E}, \bar{I}\right) \in \bar{\Omega} \setminus \{(0, 0, 0)\}$. Clearly, V is also continuously differentiable on $\bar{\Omega}$. □

Lemma 2 *If $R_0^p < 1$ for all $p \in P$, then $\nabla V(x) \cdot f_p(x) \leq 0$ for all $p \in P$ and for all $x = \left(\bar{S}, \bar{E}, \bar{I}\right) \in \bar{\Omega}$.*

Proof Given the assumption that $R_0^p < 1$ for all $p \in P$, note that $R_{\max} < 1$ as well. Now, fix an arbitrary $p \in P$ and an arbitrary $x = (\bar{S}, \bar{E}, \bar{I}) \in \bar{\Omega}$. We have

$$
\begin{aligned}
\nabla V(x) \cdot f_p(x) &= \left[\frac{\partial V}{\partial \bar{S}} \; \frac{\partial V}{\partial \bar{E}} \; \frac{\partial V}{\partial \bar{I}}\right] \cdot f_p(x) \\
&= -\mu\alpha \frac{1 - R_{\max}}{R_{\max}} \cdot \bar{S} - \alpha\beta_p \left[\frac{1 - R_{\max}}{R_{\max}} + 1\right] \bar{S}\bar{I} \\
&\quad + \left[\frac{1 - R_{\max}}{R_{\max}} \cdot \frac{\alpha\beta_p A}{\mu} + \frac{\alpha\beta_p A}{\mu} - (\alpha + \mu)(\theta + \epsilon_I + \delta + \mu)\right] \bar{I} \\
&= -\mu\alpha \frac{1 - R_{\max}}{R_{\max}} \cdot \bar{S} - \alpha\beta_p \left[\frac{1}{R_{\max}}\right] \bar{S}\bar{I} \\
&\quad + (\alpha + \mu)(\theta + \epsilon_I + \delta + \mu) \left[\frac{1 - R_{\max}}{R_{\max}} \cdot R_0^p + R_0^p - 1\right] \bar{I}.
\end{aligned}
$$

Recall that $R_0^p \leq R_{\max} < 1$. Consequently, $1 - R_{\max} > 0$ and $\frac{1-R_{\max}}{R_{\max}} \cdot R_0^p + R_0^p - 1 = \frac{R_0^p}{R_{\max}} - 1 \leq 0$. Therefore, $\nabla V(x) \cdot f_p(x) \leq 0$ for all $p \in P$ and for all $(\bar{S}, \bar{E}, \bar{I}) \in \bar{\Omega}$. □

Lemmas 1 and 2 imply that V is a common weak Lyapunov function for $\mathscr{F}_s = \{f_p(x), p \in P\}$ if $R_0^p < 1$ for all $p \in P$. We are now ready to state the following convergence result.

Theorem 1 *Suppose $R_0^p < 1$ for all $p \in P$. Let $\phi_s(t) = \left(\bar{S}(t), \bar{E}(t), \bar{I}(t)\right)$ be any solution to system $\bar{\Pi}_s$ with initial condition in $\bar{\Omega}$. Then $\phi_s(t) \to (0, 0, 0)$ as $t \to \infty$.*

Establishing the above theorem requires the following result of Baccioti and Mazzi [1]:

Theorem 2 *Let $V(x) : \mathcal{O} \to [0, +\infty)$ be a weak common Lyapunov function, where $\mathcal{O}$ is an open subset of $\bar{\Omega}$. Let $l > 0$ and let $\mathcal{O}_l$ be the connected component of the level set $\{x \in \mathcal{O} : V(x) < l\}$ such that $(0, 0, 0) \in \mathcal{O}_l$. Assume that $\mathcal{O}_l$ is bounded and let*

$$Z = \{x \in \mathcal{O} : \exists p \in P \text{ such that } \nabla V(x) \cdot f_p(x) = 0\}.$$

Finally let M be the union of all compact, weakly invariant sets which are contained in $Z \cap \mathcal{O}_l$. Then every solution $\phi(t) \in \mathcal{S}_{dwell}$ such that $\phi(0) \in \mathcal{O}_l$ is attracted by M.

Proof (of Theorem 1) Recall that our common weak Lyapunov function V is defined over the entire feasible region $\bar{\Omega}$, which is open in itself.

From the proof of Lemma 2, we have

$$
\begin{aligned}
\nabla V(x) \cdot f_p(x) = -\mu\alpha \frac{1 - R_{\max}}{R_{\max}} \cdot \bar{S} - \frac{\alpha\beta_p}{R_{\max}} \cdot \bar{S}\bar{I} \\
+ (\alpha + \mu)(\theta + \epsilon_I + \delta + \mu) \left(\frac{R_0^p}{R_{\max}} - 1\right) \cdot \bar{I}
\end{aligned}
$$

for all $p \in P$ and $x = (\bar{S}, \bar{E}, \bar{I}) \in \bar{\Omega}$.

Since $R_0^p < 1$ for all $p \in P$, it follows that $R_{max} < 1$ as well. Then

$$-\mu\alpha\frac{1-R_{\max}}{R_{\max}} < 0 \quad \text{and} \quad -\frac{\alpha\beta_p}{R_{\max}} < 0$$

for all $p \in P$. Moreover, note that

$$(\alpha+\mu)(\theta+\epsilon_I+\delta+\mu)\left(\frac{R_0^p}{R_{\max}}-1\right)$$

is less than 0 when $R_0^p < R_{\max}$, and is equal to 0 when $R_0^p = R_{\max}$. Thus, for all $p \in P$ for which $R_0^p < R_{\max}$, we have $\nabla V(x) \cdot f_p(x) = 0$ if and only if $\bar{S} = \bar{I} = 0$.

Now, let $p \in P$ for which $R_0^p = R_{\max}$. Then in this case, we have

$$\begin{aligned}\nabla V(x) \cdot f_p(x) &= -\mu\alpha\frac{1-R_{\max}}{R_{\max}} \cdot \bar{S} - \frac{\alpha\beta_p}{R_{\max}}\bar{S}\bar{I} \\ &= -\frac{\alpha\mu}{R_{\max}}\bar{S}\left[1-R_{\max}+\frac{\beta_p}{\mu}\bar{I}\right]\end{aligned}$$

and $\nabla V(x) \cdot f_p(x) = 0$ if and only if $\bar{S} = 0$ or $\bar{I} = \frac{\mu}{\beta_p}(R_{\max}-1)$. But $\frac{\mu}{\beta_p}(R_{\max}-1) < 0$ so $\bar{I}$ cannot attain this value in $\bar{\Omega}$. Thus, $\nabla V(x) \cdot f_p(x) = 0$ in $\bar{\Omega}$ if and only if $\bar{S} = 0$.

Let $Z = \{x \in \bar{\Omega} : \exists p \in P \text{ such that } \nabla V(X) \cdot f_p(x) = 0\}$. Given the preceding discussion, we must have $Z = \{x \in \bar{\Omega} : \bar{S} = 0 \text{ or } \bar{S} = \bar{I} = 0\}$. By the definition of $\bar{\Omega}$, we must have $\bar{E} \geq 0$, $\bar{I} \geq 0$, and $\bar{E} + \bar{I} \leq \bar{S}$; thus, Z must be the singleton $\{(0, 0, 0)\}$.

Now, let $l > 0$ and let $\bar{\Omega}_l$ be the connected component of the level set $\{x \in \bar{\Omega} : V(x) < l\}$. We can choose l large enough so that $\bar{\Omega}_l = \bar{\Omega}$. Clearly, $\bar{\Omega}_l$ is bounded for this choice of l. Moreover, $(0, 0, 0) \in \bar{\Omega}_l$. Then it follows that $Z \cap \bar{\Omega}_l = \{(0, 0, 0)\}$.

Let M be the union of all compact, weakly invariant sets which are contained in $Z \cap \bar{\Omega}_l$; then $M = \{(0, 0, 0)\}$ as well. Therefore, by Theorem 2, if $\phi_s(t)$ is a solution to system $\bar{\Pi}_s$ with initial condition in $\bar{\Omega}_l = \bar{\Omega}$, then $\phi_s(t)$ is attracted by $M = \{(0, 0, 0)\}$, as desired. □

From Theorem 1, it follows that if $X_s(t) = (S(t), E(t), I(t))$ is a solution of Π_s with initial condition in Ω, then $S(t) \to \frac{A}{\mu}$, $E(t) \to 0$ and $I(t) \to 0$ as $t \to \infty$. We can then extend this conclusion to any solution $X(t) = (S(t), E(t), I(t), Q(t), R(t))$ of system Π with $X(0) \in \mathcal{D}$, in which $S(t)$ converges to $\frac{A}{\mu}$ and both $E(t)$ and $I(t)$ converge to 0.

We can now present the following convergence result involving solutions to system Π.

Theorem 3 *If $R_0^p < 1$ for all $p \in P$, then solutions of system Π with initial condition in $\mathscr{D}$ converge to the disease-free equilibrium $(A/\mu, 0, 0, 0, 0)$; that is, the disease-free equilibrium is globally attractive in $\mathscr{D}$.*

Proof Let $X(t) = (S(t), E(t), I(t), Q(t), R(t))$ be a solution of system Π with $X(0) \in \mathscr{D}$. As remarked earlier, we have $\lim_{t\to\infty} S(t) = \frac{A}{\mu}$ and $\lim_{t\to\infty} E(t) = \lim_{t\to\infty} I(t) = 0$.

We can express $Q(t)$ using (4) as

$$Q(t) = \frac{Q(0)}{e^{(\epsilon_Q+\rho+\mu)t}} + \frac{\int_0^t e^{(\epsilon_Q+\rho+\mu)\xi}(\delta I(\xi))}{e^{(\epsilon_Q+\rho+\mu)t}} d\xi.$$

Observe that $\lim_{t\to\infty} \frac{Q(0)}{e^{(\epsilon_Q+\rho+\mu)t}} = 0$, and

$$\lim_{t\to\infty} \frac{\int_0^t e^{(\epsilon_Q+\rho+\mu)\xi}(\delta I(\xi))}{e^{(\epsilon_Q+\rho+\mu)t}} d\xi = \lim_{t\to\infty} \frac{e^{(\epsilon_Q+\rho+\mu)t}\delta I(t)}{(\epsilon_Q+\rho+\mu)e^{(\epsilon_Q+\rho+\mu)t}}$$
$$= \lim_{t\to\infty} \frac{\delta}{\epsilon_Q+\rho+\mu} I(t) = 0.$$

Hence, $\lim_{t\to\infty} Q(t) = 0$.

From (5), we derive an expression for $R(t)$ as

$$R(t) = \frac{R(0)}{e^{\mu t}} + \frac{\int_0^t e^{\mu\xi}(\theta I(\xi) + \rho Q(\xi))\, d\xi}{e^{\mu t}}.$$

As presented above, we can also similarly show that $\lim_{t\to\infty} R(t) = 0$. The desired conclusion then follows. □

3.1 Simulations

We now illustrate how the solutions of system Π behave in relation to the disease-free equilibrium X_{dfe}, including the stability result established in the previous discussion.

Periodic switching is assumed for the transmission parameter $\beta(t)$, with period length of 200 (time units or days). In particular,

$$\beta(t) = \begin{cases} \beta_1 & \text{if } t \in \bigcup_{k=0}^{\infty} [200k, 30 + 200k) \\ \beta_2 & \text{if } t \in \bigcup_{k=0}^{\infty} [30 + 200k, 60 + 200k) \\ \beta_3 & \text{if } t \in \bigcup_{k=0}^{\infty} [60 + 200k, 90 + 200k) \\ \beta_4 & \text{if } t \in \bigcup_{k=0}^{\infty} [90 + 200k, 120 + 200k) \\ \beta_5 & \text{if } t \in \bigcup_{k=0}^{\infty} [120 + 200k, 150 + 200k) \\ \beta_6 & \text{if } t \in \bigcup_{k=0}^{\infty} [150 + 200k, 200 + 200k) \end{cases}.$$

The values for $\beta(t)$ as well as other parameters values utilized for the simulations are summarized in Tables 1 and 2, respectively. We also use the following form of the initial condition for all simulation runs:

$$(S(0), E(0), I(0), Q(0), R(0)) = \left(0.899\frac{A}{\mu}, 0.1\frac{A}{\mu}, 0.001\frac{A}{\mu}, 0, 0\right).$$

The outputs for all simulations are presented in Figs. 2 and 3. Figure 2 shows some examples where $X(t) \to X_{\text{dfe}}$ as t tends to infinity. Figure 2a and c have $R_0^p < 1$ for all p while Fig. 2b and d have $R_0^p > 1$ for some values of p. These outputs thus illustrate that the condition $R_0^p < 1$ for all p is sufficient but not necessary for the eradication of the disease.

Table 1 Transmission rate values used for generating outputs in Figs. 2 and 3

Figure Nos.	β_1	β_2	β_3	β_4	β_5	β_6
2a and c	0.2725	0.02	0.0167	0.0125	0.0167	0.02
2b and d	0.0375	0.0275	0.025	0.0167	0.025	0.03125
3a and c	0.0375	0.03125	0.025	0.0167	0.0275	0.03125
3b and d	3×10^{-4}	2.7×10^{-4}	3×10^{-4}	3.76×10^{-4}	3.45×10^{-4}	3.3×10^{-4}

Table 2 Other parameter values used for generating outputs in Figs. 2 and 3

Figure Nos.	A	μ	α	δ	ϵ_I	θ	ρ	ϵ_T
2a and c	10	0.2	0.9	0.9	0.01	0.01	0.01	0.01
2b and d	10	0.2	0.9	0.9	0.01	0.01	0.01	0.01
3a and c	10	0.2	0.9	0.9	0.01	0.01	0.01	0.01
3b and d	1	0.0005	1.25	0.2	0.02	0.08	0.04	0.03

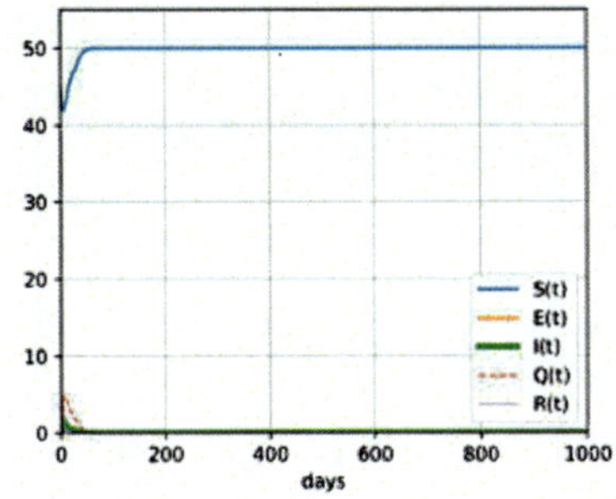

(a) $R_0^p < 1$ for all p, $X(t) \to X_{\text{dfe}}$

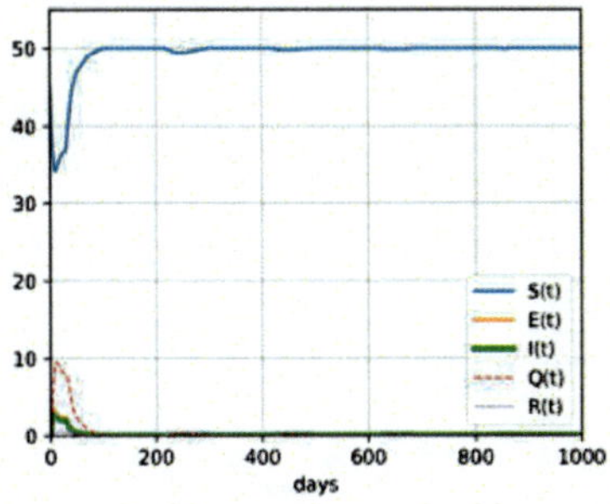

(b) $R_0^p > 1$ for some p, $X(t) \to X_{\text{dfe}}$

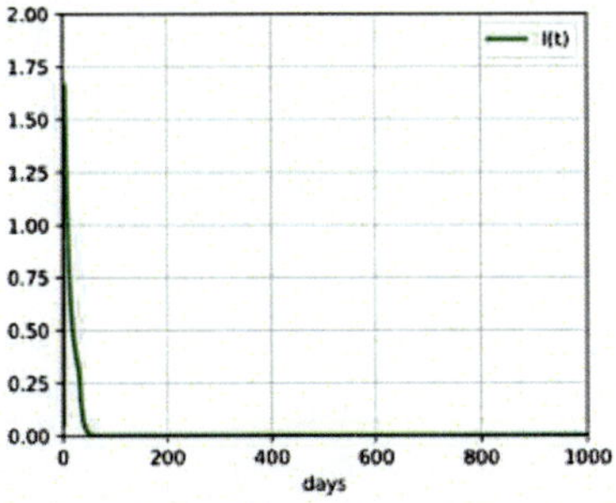

(c) $R_0^p < 1$ for all p, $I(t) \to 0$

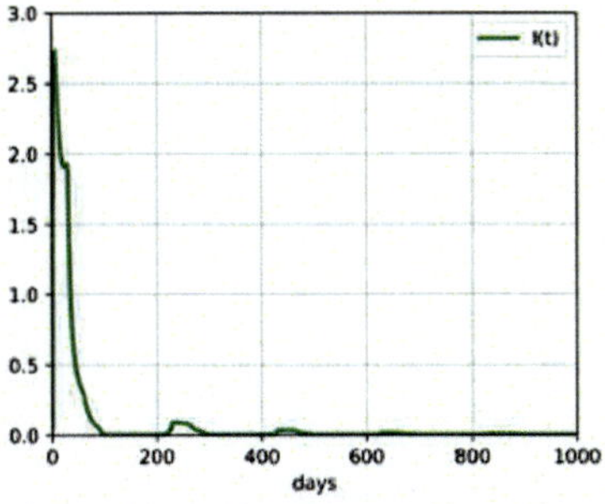

(d) $R_0^p > 1$ for some p, $I(t) \to 0$

Fig. 2 $R_0^p < 1$ for all p is sufficient but not necessary condition for $X(t) \to X_{\text{dfe}}$

In contrast, Fig. 3 shows other examples where $X(t) \not\to X_{\text{dfe}}$ as t tends to infinity, making the disease-free equilibrium unstable. Figure 3a and c have $R_0^p > 1$ for some values of p while Fig. 3b and d have $R_0^p > 1$ for all p.

The corresponding values for R_0^p and how they change over time as system Π goes through different modes are summarized in Table 3.

4 Conclusion

We analyzed a COVID-19 SEIQR model with a piecewise constant disease transmission rate. The stability of an equilibrium point was established by using the invariance-principle type theorem formulated by Bacciotti and Mazzi [1]. More specifically, if the basic reproduction number in each mode p of system Π is less than 1, then the disease-free equilibrium X_{dfe} is globally asymptotically stable. Using simulations, we also showed that the condition $R_0^p < 1$ for all p is sufficient but not necessary for a solution to converge to the DFE.

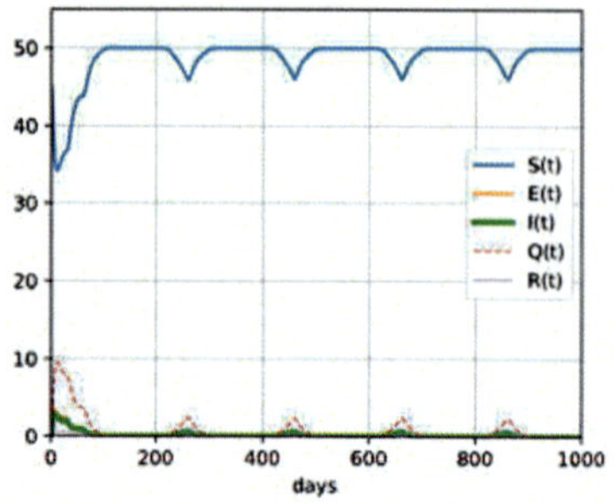

(a) $R_0^p > 1$ for some p, $X(t) \not\to X_{\mathrm{dfe}}$

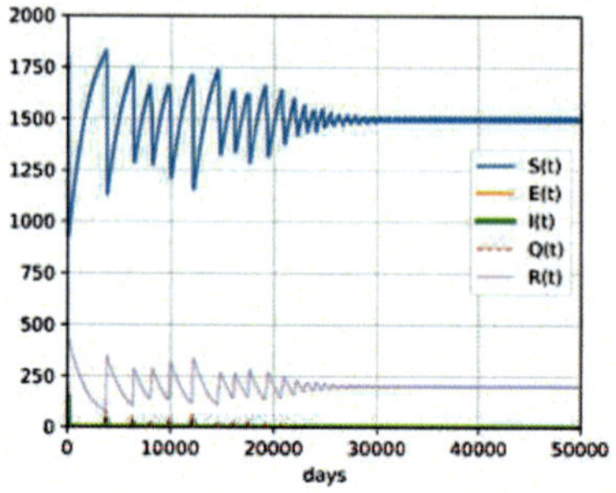

(b) $R_0^p > 1$ for all p, $X(t) \not\to X_{\mathrm{dfe}}$

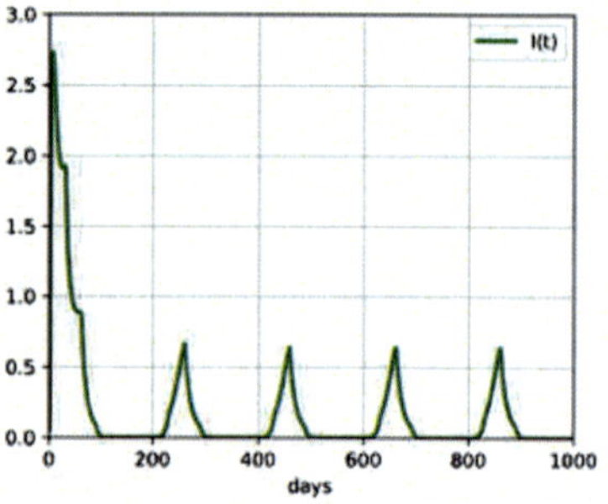

(c) $R_0^p > 1$ for some p, $I(t) \not\to 0$

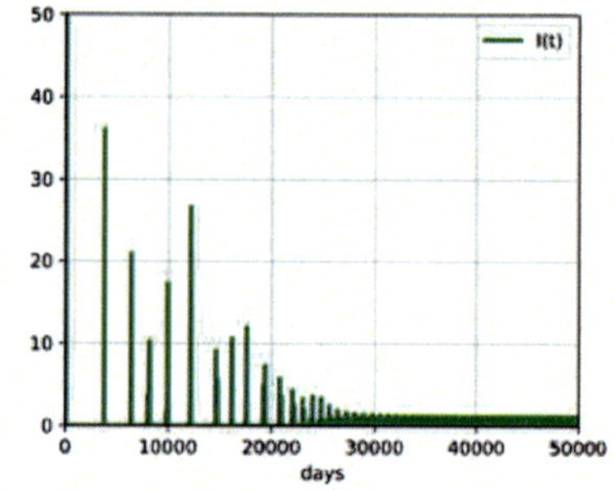

(d) $R_0^p > 1$ for all p, $I(t) \not\to 0$

Fig. 3 Other examples where $X(t) \not\to X_{\mathrm{dfe}}$

Table 3 Values of R_0^p across time, associated with outputs in Figs. 2 and 3

Figure Nos.	$t \in [0, 30)$	$t \in [30, 60)$	$t \in [60, 90)$	$t \in [90, 120)$	$t \in [120, 150)$	$t \in [150, 200)$	...
2a and c	0.995	0.731	0.609	0.457	0.609	0.731	...
2b and d	1.370	1.004	0.913	0.609	0.913	1.141	...
3a and c	1.370	1.141	0.913	0.609	1.004	1.141	...
3b and d	1.25	1.125	1.25	1.5625	1.4375	1.375	...

While our results are theoretical in nature, they provide some guidance for COVID-19 disease control. First, the results indicate that sufficient measures in reducing community transmission should be in place to ensure that the basic reproduction number is less than 1 through all phases/modes of the COVID-19 pandemic. However, it is possible to achieve the disease-free state even when the basic reproduction number exceeds the value of 1 over certain periods. Such may occur, for instance, because stricter mobility restrictions are not always in place or because a new variant emerges with heightened transmissibility. This opens up for consideration that in future outbreaks, economic and social activities may not need to undergo prolonged and severe restrictions to control the spread of the disease.

Acknowledgements The first author thanks the Ateneo de Manila University for its support through the RCW Faculty Grant. The second author recognizes the financial support provided by the Department of Science and Technology and University of the Philippines Los Baños. Finally, the authors thank the organizers and committees of the 12th International Conference on Mathematical Modeling in Physical Sciences.

References

1. Bacciotti, A., Mazzi, L.: An invariance principle for nonlinear switched systems. Syst. & Control Lett. **54**(11), 1109–1119 (2005)
2. Diekmann, O., Heesterbeek, J., Roberts, M.: The construction of next-generation matrices for compartmental epidemic models. J. R. Soc. Interface **7**, 873–885 (2010)
3. Lara-Tuprio, E.P.d., Estadilla, C.D.S., Teng, T.R.Y., Uyheng, J., Estuar, M.R.J.E., Espina, K.E., Macalalag, J.M.R., Sarmiento, R.F.R.: Mathematical analysis of a covid-19 compartmental model with interventions. In: AIP Conference Proceedings, vol. 2423, p. 020025. AIP Publishing LLC (2021)
4. Liu, X., Stechlinski, P.: Pulse and constant control schemes for epidemic models with seasonality. Nonlinear Anal.: Real World Appl. **12**(2), 931–946 (2011)
5. Liu, X., Stechlinski, P.: Infectious disease models with time-varying parameters and general nonlinear incidence rate. Appl. Math. Modell. **36**(5), 1974–1994 (2012)
6. Liu, X., Stechlinski, P.: Transmission dynamics of a switched multi-city model with transport-related infections. Nonlinear Anal.: Real World Appl. **14**(1), 264–279 (2013)
7. Liu, X., Stechlinski, P.: Sis models with switching and pulse control. Appl. Math. Comput. **232**, 727–742 (2014)
8. Ma, J., Ma, Z.: Epidemic threshold conditions for seasonally forced seir models. Math. Biosci. Eng. **3**(1), 161 (2006)
9. Sottile, S., Liu, X.: Time-varying epidemic transmission in heterogeneous networks and applications to measles. J. Biol. Syst. **28**(04), 901–926 (2020)
10. UN News: WHO chief declares end to COVID-19 as a global health emergency (2023). https://news.un.org/en/story/2023/05/1136367
11. World Health Organization: Statement on the fifteenth meeting of the IHR (2005) emergency committee on the COVID-19 pandemic (2023). www.who.int/news/item/05-05-2023-statement-on-the-fifteenth-meeting-of-the-international-health-regulations-(2005)-emergency-committee-regarding-the-coronavirus-disease-(covid-19)-pandemic
12. World Health Organization: Weekly epidemiological update on COVID-19–13 July 2023: Edition 151 (2023). https://www.who.int/publications/m/item/weekly-epidemiological-update-on-covid-19---13-july-2023

Mathematical Modeling for Sustainable Development

Modeling Air Pollution Data Using a Generalized Birnbaum-Saunders Distribution with Different Estimation Procedures

Bushra Saad Alosaimi, Farouq Mohammad Alam, and Hanan Mohammed Baaqeel

Abstract As the detrimental impact of air pollution becomes more prevalent; it is crucial to accurately model the distribution of air contamination levels. In practice, the Birnbaum-Saunders distribution is a well-known lifetime model for modeling positively skewed phenomena. Due to its relationship to the normal distribution and other desirable properties, different generalizations for the Birnbaum-Saunders distribution have been extensively studied to improve its flexibility. This article discusses inferential procedures of a specific generalization developed previously but has not received much consideration regarding this specific research area. The considered generalization has an extra shape parameter, known as Type-II generalized Birnbaum-Saunders distribution. The generalized family is more flexible than the Birnbaum-Saunders classical model, as it shares similar properties and can display both unimodality and bimodality. Eight frequentist estimation procedures are considered in this article, including the maximum likelihood and the maximum product of spacing estimation, several regression-based estimations, and goodness-of-fit estimations. Monte Carlo simulations are performed to investigate the estimation efficiency of the methods under various combinations of shape parameters, some conclusions are presented. Furthermore, air pollution concentration data are analyzed using the considered methods to illustrate their practical application. Overall, the analysis results favor the least-squares estimators according to the goodness-of-fit criteria.

Keywords Birnbaum-Saunders distribution · Maximum likelihood estimation · Maximum produce of spacings estimation · Least-square-based estimation · Goodness-of-fit estimation

---

B. S. Alosaimi (✉) · F. M. Alam · H. M. Baaqeel
Department of Statistics, Faculty of Science, King Abdulaziz University, Jeddah 21589, Kingdom of Saudi Arabia
e-mail: balaotibi0001@stu.kau.edu.sa

D. Vlachos (ed.), *Mathematical Modeling in Physical Sciences*, Springer Proceedings in Mathematics & Statistics 446, https://doi.org/10.1007/978-3-031-52965-8_45

1 Introduction

Various world phenomena can be described through the statistical distribution, identifying the best lifetime distribution is crucial for accurate data analysis. Current distributions fit the central part of the data well, but poorly in the tails due to lack of observations. Thus, research aims to improve distributions' ability to fit lifetime data with varying degrees of skewness and kurtosis [1]. Air pollution, a persistent global phenomenon with profound implications for public health, ecosystems, and the overall quality of life. Accurate modeling is crucial for understanding and assessing its environmental effects. Among the diverse statistical distributions available for modeling such data, the generalized Birnbaum-Saunders (GBS) distribution has garnered increasing attention due to its adaptability to capture the intricate statistical nuances associated with air pollution concentrations. This distribution provides a flexible framework for analyzing non-negative continuous data, accommodating varying degrees of kurtosis and asymmetry, and exhibiting characteristics of both unimodality and bimodality. Its ability to accommodate both central tendencies and tail behavior is of particular importance when dealing with pollution datasets that may exhibit extreme values and asymmetry. Among other contributions, this study broadens the scope of modeling air pollution concentrations using a generalization previously devised by [3, 4]. This particular generalization introduces an additional shape parameter, effectively enhancing the distribution's flexibility, and is commonly referred to as the Type-II generalized Birnbaum-Saunders (GBS-II) distribution. Similar to the GBS, the GBS-II distribution is a promising candidate for accurately representing the statistical properties of air pollution concentrations, and consequently providing valuable insights into the risks associated with high pollution levels. For additional information on modeling air pollution concentrations of various particulate matters with different asymmetric distributions, including the GBS distribution, see [2].

The statistical probability distribution known as the Birnbaum-Saunders (BS) distribution, named after its creators, Birnbaum and Saunders [5, 6], it is commonly recognized as the fatigue life distribution. It finds extensive applications across diverse fields, including environmental, engineering, and reliability analysis, due to its flexibility in accurately modeling practical fatigue scenarios where conventional distributions fall short. Furthermore, it exhibits strong connections with several significant distributions, such as the Gaussian, sinh-normal (S-N), and inverse Gaussian (I-G) distributions. This distribution originates from a physical argument involving cumulative damage, particularly in materials exposed to cyclic stress factors like pressure and repeated loading. The BS distribution generally provides a versatile framework capable of accommodating cumulative effects that eventually lead to failure when the damage exceeds a certain strength threshold [7]. A non-negative continuous random variable T is said to follow the BS distribution with shape parameter $\alpha > 0$, scale parameter $\beta > 0$, if the corresponding cumulative distribution function (CDF) is given by:

$$F(t) = \Phi\left(\frac{1}{\alpha}\left[\sqrt{\frac{t}{\beta}} - \sqrt{\frac{\beta}{t}}\right]\right), t > 0, \alpha, \beta > 0, \quad (1)$$

Where $\Phi()$ is the CDF of the standard Gaussian distribution. Numerous studies, including hundreds of papers have extensively covered this distribution's properties and advancements. For more details, consult [8]'s comprehensive review.

Over the past decade, various approaches have emerged for extending the scope of the BS distribution. This discussion provides an overview of certain related generalizations. The GBS distribution represents an expansive family of flexible sub-models, which was introduced by [9, 10]. It aims to broaden the BS distribution by relaxing the assumption of normality in Eq. (1) to a set of elliptical kernels, including Student's t, Cauchy, Logistic, and Laplace distributions. The classical BS distribution is regarded as a special case within this broader GBS family. Hence, this family of models is distinguished by similar favorable characteristics and can be applied to situations related to cumulative damage. For more in-depth information regarding these GBS sub-models, refer to the following recent studies and the references therein [11–14].

A novel generalization known as the Type-I Birnbaum-Saunders (GBS-I) distribution was proposed by [15], with the primary objective of formulating a more empirically based generalization. However, a notable limitation is the inherent complexity of this model. This approach deviated from the earlier generalization presented by [9, 10], as it extended the original theoretical framework initially laid out by Birnbaum and Saunders [9], while relaxing some of its underlying physical assumptions. For comprehensive insights into the GBS-I model, for further information consult recent studies, such as, [16, 17].

The resulting GBS-I distribution is obtained by transforming the standard normal variate in Eq. (1), incorporating an additional rate parameter that modifies the exponent of the BS variate T from (1/2) to $(1 - \kappa)$. This modification is represented by the equation $Z = (T^{1-\kappa}/\sqrt{\beta} - \sqrt{\beta}/T^{1-\kappa})/\alpha$. This particular model is referred to as GBS-I, distinguishing it from a prior generalization introduced by the same author [3], which was denoted as GBS-II. This study places specific emphasis on the latter generalization. As previously mentioned, the primary objective of this generalization is to enhance the flexibility of the Birnbaum-Saunders distribution by introducing an additional shape parameter. A similar extension is achieved by altering the standard normal variate in Eq. (1), where the exponent of the BS variate T is generalized from (1/2) to θ, with $\theta > 0$, such that $Z = ((T/\beta)^{\theta} - (\beta/T)^{\theta})/\alpha$.

In a recent academic investigation carried out by [18], significant emphasis was placed on highlighting the significance of this particular generalization. This study extended the valuable connection between the classical distribution and several prominent distributions to encompass the GBS-II distribution. Owen's research delved into examining the intricate relationship between GBS-II and other consequential distributions, including the I-G and the S-N distributions. It is important to note that analytical expressions for the moments of the GBS-II distribution are unattainable. Nevertheless, this study pioneered the derivation of numerical moments for the GBS-II distribution. This was accomplished by leveraging its advantageous relationship with the three parameter S-N distribution. This numerical approach facilitates the approximation of GBS-II distribution moments, thereby contributing significantly to a more profound comprehension of its statistical characteristics.

Furthermore, it is worth noting that [4]'s study stands as one of the pioneering inquiries into the GBS-II distribution. The study primarily focused on investigating the GBS-II distribution from both univariate and multivariate perspectives. Additionally, their contribution extended to exploring various extensions of the GBS-II model, which were derived from its intriguing association with the S-N distribution. To delve further into this area, readers are encouraged to examine recent research, such as the study introducing the bivariate GBS-II distribution and its relationships with the bivariate BS and bivariate S-N distributions [19], along with subsequent studies in this specialized field.

In the context of properties associated with the GBS-II distribution, scholars have undertaken a comprehensive investigation into its distributional and statistical attributes. This exploration encompasses crucial functions such as reliability and hazard rate functions, in addition to the utilization of diverse numerical methodologies for the examination of moments. These inquiries have been pursued in studies like the one presented in the thesis by [20], among other notable contributions. Moreover, [21] provides a mathematical perspective on the behavior of the probability density function that is intrinsic to the GBS-II distribution.

In practical terms, the process of estimating distributional properties of a probability distribution, including key measures like reliability and hazard rate, involves obtaining appropriate parameter estimates through the use of suitable estimation techniques. Furthermore, these parameters offer valuable insights into the characteristics of the lifetime distribution, covering aspects such as its shape, skewness, and kurtosis. Ensuring precise parameter estimation is vital to effectively represent the underlying information of the distribution, making the development of inferential procedures that yield optimal parameter estimates a central objective in the field of statistical research. Researchers have undertaken extensive comparative investigations on certain lifetime distributions, appraising the performance of identical estimation methods advocated in this paper. You can reference pertinent research in the publications authored by [11, 12, 14, 22–27] and the references therein.

The investigation into estimation procedures for the GBS-II distribution has been relatively limited in its extent, with a predominant emphasis on Maximum Likelihood Estimation (MLE) method, as elucidated in the works of [18, 20, 21]. Furthermore, the latter study has notably contributed to this field by introducing an innovative numerical approach. This approach is predicated on utilizing moments derived from the GBS-II distribution's intrinsic relationship with the sinh-normal distribution. In contrast to prior investigations, this study examines several frequentist estimators that diverge from those previously explored in the existing scholarly literature. Specifically, it delves into the performance evaluation of eight distinct frequentist estimation methodologies utilized for the estimation of the three parameters associated with the GBS-II distribution, while concurrently assessing their robustness. The estimators derived within this study for the model parameters encompass Maximum

Likelihood Estimators (MLEs), Least-Squares Estimators (LSEs), Weighted Least-Squares Estimators (WLSEs), Percentile Estimators (PCEs), Maximum Product of Spacings Estimators (MPSEs), Cramér-von Mises Estimators (CVMEs), Anderson-Darling Estimators (ADEs), and right-tailed Anderson-Darling Estimators (RADEs). Notably, as previously highlighted, obtaining analytical expressions for the moments of the GBS-II distribution proves unattainable. Consequently, estimation methodologies predicated on moment-based approaches are excluded from consideration in this study.

The central objective of this research paper is to address the challenge of estimating parameters in the GBS-II distribution by conducting a comprehensive numerical comparison of various estimators to identify the most efficient and robust performer. This novel contribution aims to enhance the understanding of GBS-II parameter estimation through a combination of Monte Carlo simulations and the analysis of real data. The research's originality lies in its comparative approach, as prior studies have not undertaken such a thorough examination of different estimation methods for GBS-II parameters. The research encompasses two critical dimensions: an extensive numerical evaluation, employing well-defined criteria for estimation efficiency and goodness-of-fit assessment through Monte Carlo simulations, and the practical application of the GBS-II distribution to analyze real-life data, illustrated through the air pollution concentrations dataset, thereby reinforcing the practical relevance of the research outcomes.

The subsequent sections of this paper are structured as follows: Sect. 2 delves into the distributional characteristics of the GBS-II distribution, while Section elaborates on eight frequentist estimation procedures. Section 4 presents the results of extensive Monte Carlo simulations, facilitating a comparative analysis of each estimator's performance across various criteria. Section 5 provides a practical demonstration of the discussed estimators, utilizing a real dataset concerning air pollution concentrations. The paper culminates in Sect. 6 with conclusions and discussions of potential avenues of future research.

2 Generalization Description

A non-negative continuous random variable T is said to be following the GBS-II distribution with first shape parameter $\alpha > 0$, scale parameter $\beta > 0$, and second shape parameter $\theta > 0$, i.e., $T \sim \text{GBS-II}(\alpha, \beta, \theta)$ if the corresponding cumulative distribution function (CDF), probability density function (PDF) of T are given by:

$$F(t) = \Phi\left[\frac{1}{\alpha}\varepsilon(t)\right], \quad (2)$$

and

$$f(t; \alpha, \beta, \theta) = \frac{\theta}{\alpha t}\delta(t)\phi\left[\frac{1}{\alpha}\varepsilon(t)\right], \quad (3)$$

Respectively, such that $\Phi(.)$ and $\phi(.)$ are the CDF and PDF of the standard Gaussian distribution. respectively, where

$$\varepsilon(t;\beta,\theta) = \left(\frac{t}{\beta}\right)^{\theta} - \left(\frac{\beta}{t}\right)^{\theta}, \tag{4}$$

and

$$\delta(t;\beta,\theta) = \left(\frac{t}{\beta}\right)^{\theta} + \left(\frac{\beta}{t}\right)^{\theta}. \tag{5}$$

note that, $\varepsilon(t;\beta,\theta)$, $\delta(t;\beta,\theta)$ are abbreviated as $\varepsilon(t)$, $\delta(t)$, respectively, throughout the study.

Consequently, the hazard function (HF) of the GBS-II distribution is obtained as:

$$h(t;\alpha,\beta,\theta) = \frac{f(t;\alpha,\beta,\theta)}{1-F(t;\alpha,\beta,\theta)} = \frac{\theta\delta(t)\phi\left[\frac{1}{\alpha}\varepsilon(t)\right]}{\alpha t\left(1-\Phi\left[\frac{1}{\alpha}\varepsilon(t)\right]\right)} \tag{6}$$

Figures 1, 2 illustrate the PDF and HF curves for the GBS-II distribution with various parameter values, respectively. However, a detailed mathematical analysis of the underlying behavior of these distributional features falls outside the scope of this study.

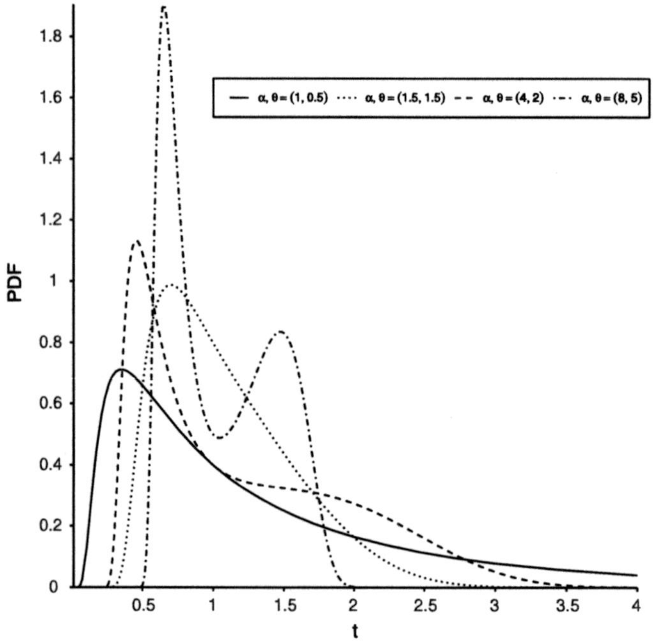

Fig. 1 The PDF curves of the GBS-II for various parameter values

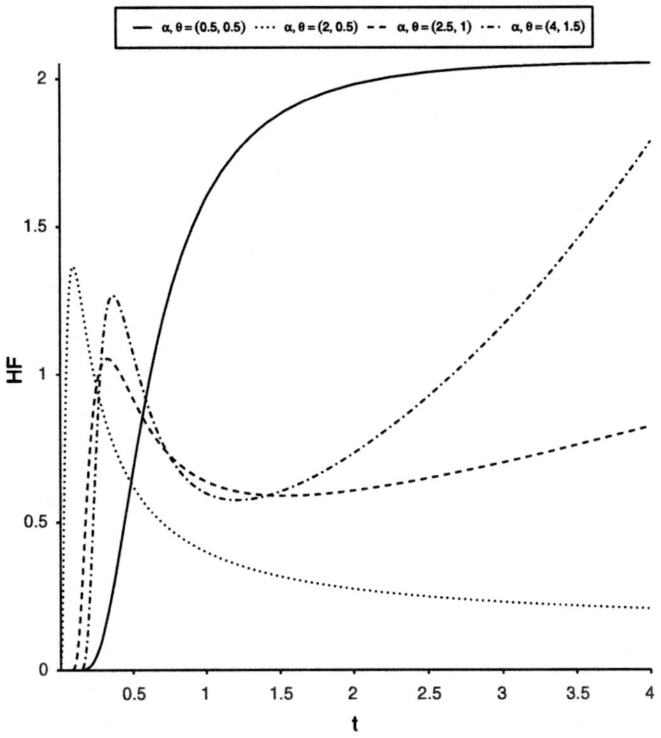

Fig. 2 The HF curves of the GBS-II for various parameter values

Property 1. *Suppose that* $T \sim$ GBS-II(α, β, θ), *pth percentile* $t_p = F^{-1}(p)$ *is obtained by applying the inverse probability integral transform on the CDF of the distribution, then the quantile function(QF) can be obtained as:*

$$t_p = Q(z; \alpha, \beta, \theta) = \beta \left(\frac{\alpha z_p + \sqrt{[\alpha z_p]^2 + 4}}{2} \right)^{\frac{1}{\theta}} \tag{7}$$

such that z_p *is the pth percentile of* $Z \sim N(0, 1)$. Where z_p is the quantile of the standard normal distribution. The QF is essential in generating random varieties from the GBS-II distribution.

Property 2.

$$T = \beta \left(\frac{\alpha Z + \sqrt{[\alpha Z]^2 + 4}}{2} \right)^{\frac{1}{\theta}} \sim \text{GBS-II}(\alpha, \beta, \theta). \tag{8}$$

Property 3.

$$E\left(T^r\right)=\frac{\beta^r \exp\left(\alpha^{-2}\right)}{\alpha\sqrt{2\pi}}\left[K_{(r/\theta+1)/2}\left(\alpha^{-2}\right)+K_{(r/\theta-1)/2}\left(\alpha^{-2}\right)\right], \tag{9}$$

where $K_\omega(x)$ is the third kind modified Bessel function of order ω.
Property 4. *if* $T \sim$ GBS-II(α, β, θ), *then:*

1. $\delta(t) \sim N(0, \alpha)$, *such that* $\delta(t)$ *is given by* (5)*;*
2. $cT \sim$ GBS-II$(\alpha, c\beta, \theta)$, *i.e., closed under scale transformations, given* $c > 0$;
3. $T^{-1} \sim$ GBS-II$(\alpha, \beta^{-1}, \theta)$, *i.e., closed under reciprocation.*

3 Estimation Procedures

In this section, we discuss eight frequentist estimation methods, to estimate the three parameters of GBS-II distribution:

3.1 Maximum Likelihood Estimation

The maximum likelihood method is unquestionably the most widely used method in statistical inference, mostly due to its numerous interesting properties [28]. Let $\mathbf{t}=(t_1, ..., t_n)$ be the realization of n independent and identically distributed random variables (i.e., an observed random sample with n observations) from the GBS-II distribution with PDF Eq. (3). The corresponding MLEs of the model parameters α, β and θ; denoted as, $\hat{\alpha}_{\text{MLE}}$, $\hat{\beta}_{\text{MLE}}$ and $\hat{\theta}_{\text{MLE}}$, respectively, are obtained by maximizing the objective function:

$$\ell(\alpha, \beta, \theta|\mathbf{t}) = n \log\theta - n\log\alpha + \sum_{i=1}^{n}\log\delta(t_i) - \frac{1}{2\alpha^2}\sum_{i=1}^{n}\varepsilon^2(t_i) \tag{10}$$

Where $\ell(\alpha, \beta, \theta|\mathbf{t})$ is the log-likelihood of the GBS-II distribution (without the constant term), such that $\varepsilon(t_i)$ and $\delta(t_i)$ are given in Eq. (4, 5), respectively, where $i = 1, ..., n$. Using a suitable numerical optimization method, the MLEs can be obtained by solving the following non-linear equations, simultaneously.

$$\frac{\partial\ell}{\partial\alpha} = -\frac{n}{\alpha} + \frac{1}{\alpha^3}\sum_{i=1}^{n}\varepsilon^2(t_i),$$

and

$$\frac{\partial \ell}{\partial \beta} = -\frac{\theta}{\beta}\left[\sum_{i}^{n}\frac{\varepsilon(t_i)}{\delta(t_i)} - \frac{1}{\alpha^2}\sum_{i}^{n}\varepsilon(t_i)\delta(t_i)\right],$$

and

$$\frac{\partial \ell}{\partial \theta} = \frac{n}{\theta} + \sum_{i=1}^{n} log(\beta^{-1}t_i)\frac{\varepsilon(t_i)}{\delta(t_i)} - \frac{1}{\alpha^2}\sum_{i=1}^{n} log(\beta^{-1}t_i)\varepsilon(t_i)\delta(t_i).$$

3.2 Maximum Product of Spacings Estimation

The maximum product of spacings method was pioneered by [29, 30] as a rival to the MLE method in terms of consistency and asymptotic efficiency, for further insights into spacings-based inferential methods; (refer to [31]). Recall that $t_{1:n} < ... < t_{n:n}$ are the observed order statistics based on a random sample n, the MPSEs of the model parameters, denoted as; $\hat{\alpha}_{\mathrm{MLE}}$, $\hat{\beta}_{\mathrm{MLE}}$ and $\hat{\theta}_{\mathrm{MLE}}$ are the values that maximize the logarithm of the geometric mean of spacings, given as:

$$P = \frac{1}{n+1}\sum_{i=1}^{n+1} log\nabla_i \tag{11}$$

Such that,

$$\nabla_i = \begin{cases} \Phi\left(\frac{\varepsilon(t_{1:n})}{\alpha}\right) & \text{if } i = 1 \\ \Phi\left(\frac{\varepsilon(t_{i:n})}{\alpha}\right) - \Phi\left(\frac{\varepsilon(t_{i-1:n})}{\alpha}\right) & \text{if } 1 < i \le n \\ 1 - \Phi\left(\frac{\varepsilon(t_{n:n})}{\alpha}\right) & \text{if } i = n+1 \end{cases} \tag{12}$$

The first-order derivatives of the objective function Eq. (11) w.r.t to the model parameters α, β and θ, respectively, given as:

$$\frac{\partial P}{\partial \alpha} = \frac{1}{n+1}\sum_{i=1}^{n+1}\frac{\varpi_\alpha \nabla_i}{\nabla_i},$$

and

$$\frac{\partial P}{\partial \beta} = \frac{1}{n+1}\sum_{i=1}^{n+1}\frac{\varpi_\beta \nabla_i}{\nabla_i},$$

and

$$\frac{\partial P}{\partial \theta} = \frac{1}{n+1}\sum_{i=1}^{n+1}\frac{\varpi_\theta \nabla_i}{\nabla_i},$$

Where $\varpi_\alpha \nabla_i$, $\varpi_\beta \nabla_i$ and $\varpi_\theta \nabla_i$ are the first-order derivatives of the function ∇_i w.r.t to the model parameters are as follows:

$$\varpi_\alpha \nabla_i = \begin{cases} \varpi_\alpha(t_{1:n}) \\ \varpi_\alpha(t_{i:n}) - \varpi_\alpha(t_{i-1:n}) \\ -\varpi_\alpha(t_{n:n}) \end{cases},$$

and

$$\varpi_\beta \nabla_i = \begin{cases} \varpi_\beta(t_{1:n}) \\ \varpi_\beta(t_{i:n}) - \varpi_\beta(t_{i-1:n}) \\ -\varpi_\beta(t_{n:n}) \end{cases},$$

and

$$\varpi_\theta \nabla_i = \begin{cases} \varpi_\theta(t_{1:n}) \\ \varpi_\theta(t_{i:n}) - \varpi_\theta(t_{i-1:n}) \\ -\varpi_\theta(t_{n:n}) \end{cases}. \tag{13}$$

Where $\varpi_\alpha(t)$, $\varpi_\beta(t)$ and $\varpi_\theta(t)$ are the first-order derivatives of the CDF of the GBS-II distribution w.r.t to the model parameters, consider that these expressions are used throughout the study when obtaining the gradient of the remaining estimation methods, given by:

$$\varpi_\alpha(t) = -\frac{\varepsilon(t)}{\alpha^2}\phi\left[\frac{\varepsilon(t)}{\alpha}\right],$$

and

$$\varpi_\beta(t) = -\frac{\theta}{\alpha\beta}\delta(t)\phi\left[\frac{1}{\alpha}\varepsilon(t)\right],$$

and

$$\varpi_\theta(t) = \frac{log(\beta^{-1}t)}{\alpha}\delta(t)\phi\left[\frac{1}{\alpha}\varepsilon(t)\right].$$

3.3 Least-Squares Estimation

Since [32] proposed the least-squares estimation method to estimate the parameters of the beta distribution, the two variations of the least-squares method, also known as, the ordinary least-squares and the weighted extension of the least-squares have been widely discussed in the literature. Suppose that $t_{1:n} < ... < t_{n:n}$ are the observed order statistics from an observed random sample **t** of size n from a distribution function $F(t)$. It is known that the expectation and variance of the associated order statistic $F(t_{1:n}), ..., F(t_{n:n})$ which are form a standard uniform distribution are given as:

$$E[F(t_{i:n})] = \frac{i}{n+1}, \ Var[F(t_{i:n})] = \frac{i(n-i+1)}{(n+1)^2(n+1)}$$

Consequently, the two variates of the least-squares the LSEs and WLSEs can be acquired by minimizing their associated objective functions, which are discussed in this sub-section and the succeeding one, respectively. The LSEs of α, β and θ are denoted as $\hat{\alpha}_{\text{LSE}}$, $\hat{\beta}_{\text{LSE}}$ and $\hat{\theta}_{\text{LSE}}$ of the GBS-II distribution, can be obtained by minimizing the objective function:

$$S = \sum_{i=1}^{n} \left(\Phi \left[\frac{\varepsilon(t_{i:n})}{\alpha} \right] - \frac{i}{n+1} \right)^2, \tag{14}$$

The first-order derivatives of the LSE objective function w.r.t the model parameters are given by:

$$\frac{\partial S}{\partial \alpha} = 2 \sum_{i=1}^{n} \left(\Phi \left[\frac{\varepsilon(t_{i:n})}{\alpha} \right] - \frac{i}{n+1} \right) \varpi_{\alpha} (t_{i:n}),$$

and

$$\frac{\partial S}{\partial \beta} = 2 \sum_{i=1}^{n} \left(\Phi \left[\frac{\varepsilon(t_{i:n})}{\alpha} \right] - \frac{i}{n+1} \right) \varpi_{\beta} (t_{i:n}),$$

and

$$\frac{\partial S}{\partial \theta} = 2 \sum_{i=1}^{n} \left(\Phi \left[\frac{\varepsilon(t_{i:n})}{\alpha} \right] - \frac{i}{n+1} \right) \varpi_{\theta} (t_{i:n}).$$

3.4 Weighted Least-Squares Estimation

The WLSEs of α, β and θ are denoted as, $\hat{\alpha}_{\text{WLSE}}$, $\hat{\beta}_{\text{WLSE}}$ and $\hat{\theta}_{\text{WLSE}}$, respectively, and can be obtained by minimizing the weighted version of the LSE objective function, given by:

$$S_w(\alpha, \beta, \theta) = \sum_{i=1}^{n} w_{i:n} \left(\Phi \left[\frac{\varepsilon(t_{i:n})}{\alpha} \right] - \frac{i}{n+1} \right)^2$$

where $w_{1:n}$, ..., $w_{n:n}$ represents the correction factor which are the weights of $x_{1:n} < \ldots < x_{n:n}$, respectively, such that:

$$w_{i:n} = \frac{(n+1)(n+1)^2}{i(n-i+1)}$$

The first-order derivatives of the WLSE objective function w.r.t the model parameters are given by:

$$\frac{\partial S_w}{\partial \alpha} = 2\sum_{i=1}^{n} w_{i:n}\left(\Phi\left[\frac{\varepsilon(t_{i:n})}{\alpha}\right] - \frac{i}{n+1}\right)\varpi_{\alpha}\left(t_{i:n}\right),$$

and

$$\frac{\partial S_w}{\partial \beta} = 2\sum_{i=1}^{n} w_{i:n}\left(\Phi\left[\frac{\varepsilon(t_{i:n})}{\alpha}\right] - \frac{i}{n+1}\right)\varpi_{\beta}\left(t_{i:n}\right),$$

and

$$\frac{\partial S_w}{\partial \theta} = 2\sum_{i=1}^{n} w_{i:n}\left(\Phi\left[\frac{\varepsilon(t_{i:n})}{\alpha}\right] - \frac{i}{n+1}\right)\varpi_{\theta}\left(t_{i:n}\right).$$

3.5 Percentile Estimation

The percentile estimation method is an effective method when data are obtained from a distribution function with an analytical form, then the unknown parameters can be estimated by fitting a straight line to the theoretical points derived from the quantile function and the sample percentile points. The PCE method was introduced by [35] to estimate the parameters of the Weibull distribution. Since the GBS-II distribution has an explicit CDF and QF, thus the PCE method is applicable to estimate the parameters the parameters α, β and θ denoted as $\hat{\alpha}_{\text{PCE}}$, $\hat{\beta}_{\text{PCE}}$ and $\hat{\theta}_{\text{PCE}}$. Recall that $t_{1:n} < ... < t_{n:n}$ are the observed order statistics, If $F(t_{i:n}; \alpha, \beta, \theta)$ are estimated by $p_{i:n} = E[F(t_{i:n})]$ where $p_{i:n} = \frac{i}{n+1}$ for $i = 1, 2, ..., n$. Then PCEs of the model parameters are obtained by minimizing the objective function:

$$S_p = \sum_{i=1}^{n}\left[t_{i:n} - \beta\left(\frac{\alpha\, Q(p_{i:n}) + \sqrt{[\alpha\, Q(p_{i:n})]^2 + 4}}{2}\right)^{\frac{1}{\theta}}\right]^2 \tag{15}$$

The first-order derivatives of the PCE objective function w.r.t the model parameters are given by:

$$\frac{\partial S_p}{\partial \alpha} = 2\sum_{i=1}^{n}\left[t_{i:n} - \beta\left(\frac{\alpha\, Q(p_{i:n}) + \sqrt{[\alpha\, Q(p_{i:n})]^2 + 4}}{2}\right)^{\frac{1}{\theta}}\right]\varpi_{\alpha}\left(t_{i:n}\right),$$

and

$$\frac{\partial S_p}{\partial \beta} = 2\sum_{i=1}^{n}\left[t_{i:n} - \beta\left(\frac{\alpha\ Q(p_{i:n}) + \sqrt{[\alpha\ Q(p_{i:n})]^2 + 4}}{2}\right)^{\frac{1}{\theta}}\right]\varpi_{\beta}\left(t_{i:n}\right),$$

and

$$\frac{\partial S_p}{\partial \theta} = 2\sum_{i=1}^{n}\left[t_{i:n} - \beta\left(\frac{\alpha\ Q(p_{i:n}) + \sqrt{[\alpha\ Q(p_{i:n})]^2 + 4}}{2}\right)^{\frac{1}{\theta}}\right]\varpi_{\theta}\left(t_{i:n}\right).$$

3.6 *Cramér-von Mises Estimation*

The goodness-of-fit statistics are used for inferential purposes, in the forthcoming sub-sections, some goodness-of-fit estimation methods are discussed. Based on the observed order statistics $t_{1:n} < \ldots < t_{n:n}$ from a random sample of size n, then the cramér-von mises estimators of the parameters α, β and θ denoted as$\hat{\alpha}_{\text{CVME}}$, $\hat{\beta}_{\text{CVME}}$ and $\hat{\theta}_{\text{CVME}}$, respectively, can be obtained by minimizing the objective function:

$$C = \frac{1}{12n} + \sum_{i=1}^{n}\left(\Phi\left[\frac{\varepsilon(t_{i:n})}{\alpha}\right] - \frac{2i-1}{2n}\right)^2, \tag{16}$$

The first-order derivatives of the objective function w.r.t the model parameters are given as follows:

$$\frac{\partial C}{\partial \alpha} = 2\sum_{i=1}^{n}\left(\Phi\left[\frac{\varepsilon(t_{i:n})}{\alpha}\right] - \frac{2i-1}{2n}\right)\varpi_{\alpha}(t_{i:n}),$$

and

$$\frac{\partial C}{\partial \beta} = 2\sum_{i=1}^{n}\left(\Phi\left[\frac{\varepsilon(t_{i:n})}{\alpha}\right] - \frac{2i-1}{2n}\right)\varpi_{\beta}(t_{i:n}),$$

and

$$\frac{\partial C}{\partial \theta} = 2\sum_{i=1}^{n}\left(\Phi\left[\frac{\varepsilon(t_{i:n})}{\alpha}\right] - \frac{2i-1}{2n}\right)\varpi_{\theta}(t_{i:n}).$$

3.7 Anderson-Darling Estimation

Based on the observed order statistics $t_{1:n} < \ldots < t_{n:n}$ from a random sample of size n, then the Anderson-Darling estimators Of the parameters α, β and θ denoted as $\hat{\alpha}_{\text{ADE}}$, $\hat{\beta}_{\text{ADE}}$ and $\hat{\theta}_{\text{ADE}}$, respectively, can be obtained by minimizing the objective function:

$$A = -n - \frac{1}{n}\sum_{i=1}^{n}(2i-1)\, log\, \Phi\left[-\frac{\varepsilon(t_{n-i+1:n})}{\alpha}\right] - \frac{1}{n}\sum_{i=1}^{n}(2i-1)\, log\, \Phi\left[\frac{\varepsilon(t_{i:n})}{\alpha}\right] \tag{17}$$

The first-order derivatives of the AD objective function w.r.t the model parameters are given by:

$$\frac{\partial A}{\partial \alpha} = \frac{1}{n}\sum_{i=1}^{n}(2i-1)\frac{\varpi_{\alpha}(t_{n-i+1:n})}{\Phi\left[-\frac{\varepsilon(t_{n-i+1:n})}{\alpha}\right]} - \frac{1}{n}\sum_{i=1}^{n}(2i-1)\frac{\varpi_{\alpha}(t_{i:n})}{\Phi\left[\frac{\varepsilon(t_{i:n})}{\alpha}\right]},$$

and

$$\frac{\partial A}{\partial \beta} = \frac{1}{n}\sum_{i=1}^{n}(2i-1)\frac{\varpi_{\beta}(t_{n-i+1:n})}{\Phi\left[-\frac{\varepsilon(t_{n-i+1:n})}{\alpha}\right]} - \frac{1}{n}\sum_{i=1}^{n}(2i-1)\frac{\varpi_{\beta}(t_{i:n})}{\Phi\left[\frac{\varepsilon(t_{i:n})}{\alpha}\right]},$$

and

$$\frac{\partial A}{\partial \theta} = \frac{1}{n}\sum_{i=1}^{n}(2i-1)\frac{\varpi_{\theta}(t_{n-i+1:n})}{\Phi\left[-\frac{\varepsilon(t_{n-i+1:n})}{\alpha}\right]} - \frac{1}{n}\sum_{i=1}^{n}(2i-1)\frac{\varpi_{\theta}(t_{i:n})}{\Phi\left[\frac{\varepsilon(t_{i:n})}{\alpha}\right]}.$$

3.8 Right-Tailed Anderson-Darling Estimation

Based on the observed order statistics $t_{1:n} < \ldots < t_{n:n}$ from a random sample of size n, then the corresponding right-tailed Anderson-Darling estimators of the model parameters α, β and θ denoted as $\hat{\alpha}_{\text{RADE}}$, $\hat{\beta}_{\text{RADE}}$ and $\hat{\theta}_{\text{RADE}}$ respectively, can be acquired by minimizing the objective function:

$$A_R = \frac{n}{2} - \frac{1}{n}\sum_{i=1}^{n}(2i-1)\log \Phi\left[-\frac{\varepsilon(t_{n-i+1:n})}{\alpha}\right] - 2\sum_{i=1}^{n}\Phi\left[\frac{\varepsilon(t_{i:n})}{\alpha}\right] \tag{18}$$

The first-order derivatives of the RAD objective function w.r.t the model parameters:

$$\frac{\partial A_R}{\partial \alpha} = \frac{1}{n}\sum_{i=1}^{n}(2i-1)\frac{\varpi_{\alpha}(t_{n-i+1:n})}{\Phi\left[-\frac{\varepsilon(t_{n-i+1:n})}{\alpha}\right]} - 2\sum_{i=1}^{n}\varpi_{\alpha}(t_{i:n}),$$

and

$$\frac{\partial A_R}{\partial \beta} = \frac{1}{n}\sum_{i=1}^{n}(2i-1)\frac{\varpi_\beta(t_{n-i+1:n})}{\Phi\left[-\frac{\varepsilon(t_{n-i+1:n})}{\alpha}\right]} - 2\sum_{i=1}^{n}\varpi_\beta(t_{i:n}),$$

and

$$\frac{\partial A_R}{\partial \theta} = \frac{1}{n}\sum_{i=1}^{n}(2i-1)\frac{\varpi_\theta(t_{n-i+1:n})}{\Phi\left[-\frac{\varepsilon(t_{n-i+1:n})}{\alpha}\right]} - 2\sum_{i=1}^{n}\varpi_\theta(t_{i:n}).$$

4 Numerical Results and Discussions

This section contains relevant numerical results to assess the performance of the discussed estimation methods, and consists of two sub-sections; firstly, a numerical example to illustrate the performance of the methods based on one simulation run. Secondly, the conclusions made based on one sample are strengthened by a comprehensive Monte Carlo simulation study alongside discussions of the performance of the methods under multiple criteria.

4.1 *Illustrative Numerical Example*

In this sub-section, a random sample with size $n = 100$ is simulated from GBS-II(2,1,2), and the model parameters are estimated, accordingly. Mathematically proving the existence and uniqueness of the estimators is beyond the scope of this study. Therefore, an extensive Monte Carlo numerical investigation of the existence and uniqueness of these eight estimators are conducted based on three-dimensional (4D) visualizations of the profile objective functions of the estimation methods.

The outcomes of the illustrative example are presented as follows; Fig. 3 shows the established 4D surface plots, the values of the objective functions evaluated at a massive number of suitable initial values for the model's shape parameters α and θ represented by the x-axis and y-axis, respectively, whereas, the values of the scale parameters β is represented by the z-axis. The 4D visualizations indicate that global extrema exist and are unique for the considered methods. Whereas the profile of the PCE method's objective function, indicates that its optimization might be challenging. Regarding the eight estimators of the model parameters α, β and θ are reported in Table 2, based on the simulated data reported in Table 1, the results indicated that MPSEs and MLEs provided the closest estimators for the true values of the model shape parameters, respectively. In the case of the scale parameter, all methods generated highly accurate estimates of the parameter's true value.

Table 1 Simulated data from GBS-II(2,1,2)

0.381033	0.662139	0.824820	1.149991	1.532865
0.471370	0.662196	0.827182	1.157471	1.555793
0.492859	0.670139	0.832450	1.158149	1.562576
0.513774	0.671045	0.858571	1.161033	1.581635
0.531509	0.678847	0.883626	1.182545	1.596464
0.572486	0.688349	0.895456	1.185097	1.605072
0.584799	0.693003	0.919046	1.199885	1.654608
0.595102	0.693260	0.923000	1.232197	1.668021
0.596361	0.697517	0.934579	1.234134	1.696823
0.596683	0.698570	0.948627	1.239646	1.752017
0.599898	0.701407	0.977045	1.247900	1.754714
0.610107	0.742524	0.982693	1.254516	1.771307
0.612193	0.743289	0.985200	1.302343	1.778116
0.615495	0.751050	1.031548	1.320662	1.818152
0.617002	0.751411	1.051541	1.331086	1.824825
0.618193	0.759483	1.068790	1.433394	1.883645
0.642846	0.764109	1.109901	1.450591	1.977557
0.645893	0.771013	1.121226	1.455121	1.987324
0.649232	0.775233	1.126250	1.458902	1.994245
0.658431	0.802326	1.144298	1.471789	2.200421

Table 2 Estimates for the model parameters using eight methods based on data in Table 1

Method	$\hat{\alpha}$	$\hat{\beta}$	$\hat{\theta}$
MLE	2.5694590	0.9851577	2.3242629
MPSE	2.1497993	0.9834604	2.0120684
LSE	3.6391635	0.9789142	2.9112840
WLSE	3.3825023	0.9892785	2.7844917
PCE	2.9310926	0.9742943	2.4373446
CVME	3.7208910	0.9788101	2.9854021
ADE	3.1393409	0.9855018	2.6531823
RADE	3.0474166	0.9780505	2.5416120

4.2 Monte Carlo Simulation Outcomes

In this sub-section, a numerical investigation of the performance of the proposed estimation methods is conducted through a comprehensive Monte Carlo simulation study and the associated results are presented. The study is established as a two-fold simulation process according to the criteria used for the evaluation of the methods. Firstly, an evaluation of the performance based on some estimation efficiency criteria.

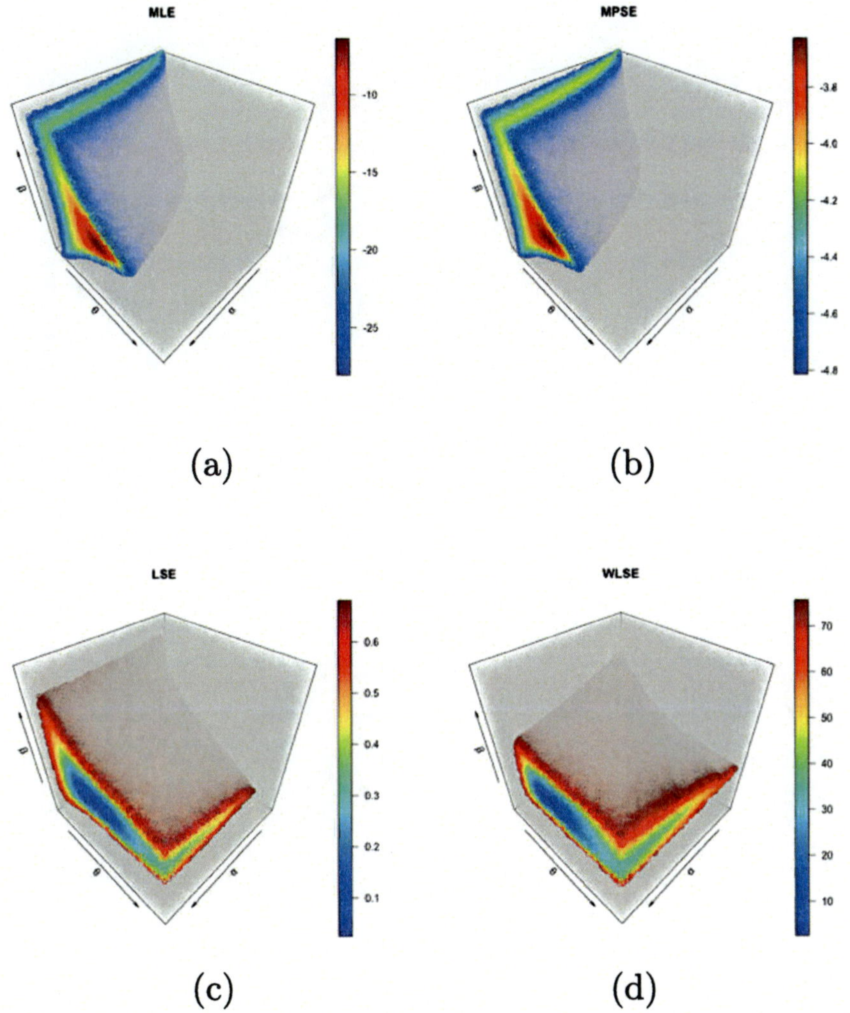

Fig. 3 4D plots of the objective functions based on data in Table 1

Secondly, the performance is analyzed based on some goodness-of-fit criteria. For each part, the performance is assessed under different variations of sample sizes and parameter values; as previously mentioned, the scale parameter is known to be the median of the distribution, thus, its value is fixed at $\beta = 1$, whereas the two shape parameters α and θ are assigned different values, given as follows (Table 3).

Based on each parameter setting the sample sizes are taken to be $n = 10(10)100$. This results in 160 different combinations of GBS-II(α, $\beta = 1$, θ), where the simulation outcomes are based on $M = 10,000$ simulation runs, and the results are detailed as follows:

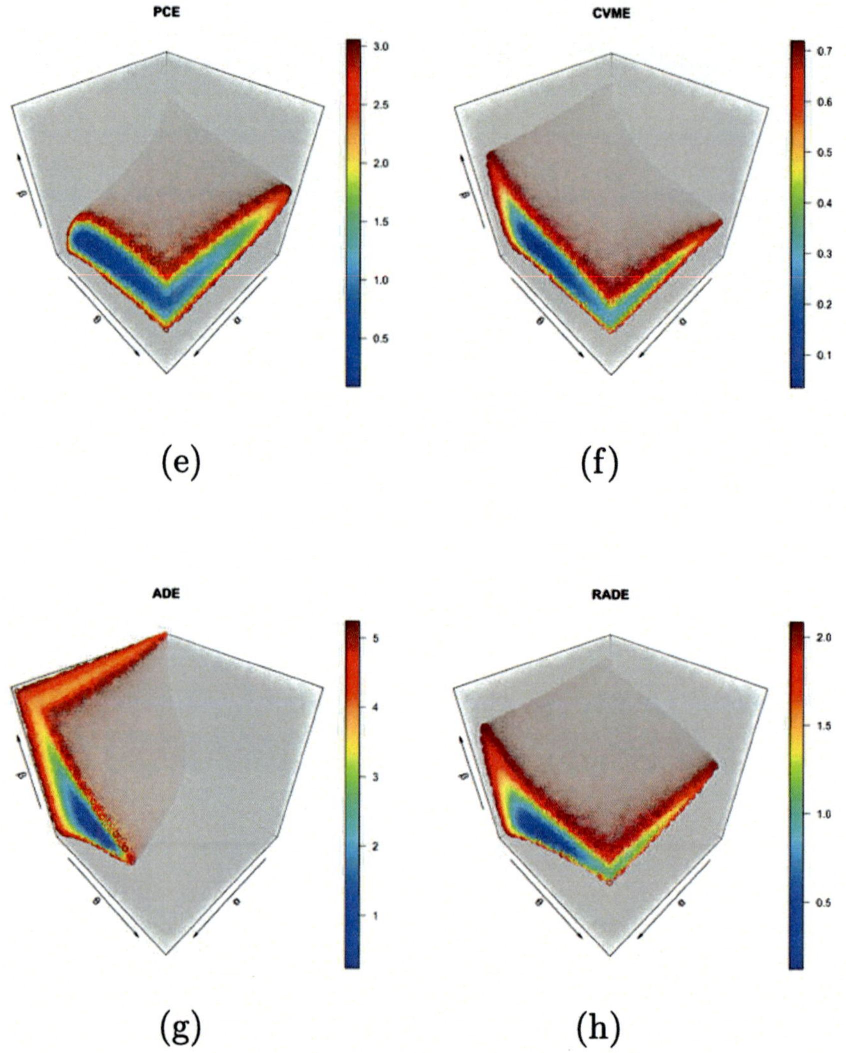

Fig. 3 (continued)

Table 3 Shape parameter variations for MC Simulation

1st Variation		2nd Variation		3rd Variation		4th Variation	
α	θ	α	θ	α	θ	α	θ
0.5	0.5	0.5	1	0.5	2	0.5	4
1	0.5	1	1	1	2	1	4
2	0.5	2	1	2	2	2	4
4	0.5	4	1	4	2	4	4

Estimation efficiency

The first set of criteria used to compare the efficiency of the eight estimators are, the simulated bias and the simulated root mean-squared-error(RMSE), computed by:

$$\text{Bias}(\hat{\alpha}) = \frac{1}{M}\sum_{i=1}^{M}(\hat{\alpha} - \alpha), \text{Bias}(\hat{\theta}) = \frac{1}{M}\sum_{i=1}^{M}\left(\hat{\theta} - \theta\right),$$

$$\text{RMSE}(\hat{\alpha}) = \frac{1}{M}\sum_{i=1}^{M}(\hat{\alpha} - \alpha)^2, \text{RMSE}(\hat{\theta}) = \frac{1}{M}\sum_{i=1}^{M}\left(\hat{\theta} - \theta\right)^2,$$

The outcomes of this part of the simulation study is discussed as follows, Overall, all the estimators' performance was satisfactory since their biases and RMSEs start to decrease as the sample size increases to the point that they become similar. The simulation outcomes of the efficiency of the shape parameters estimators are summarized in Figs. 4, 5, 6, 7. From these results, one can notice the following observations:

Regarding the biases of the estimators of α: In the case of $\theta \leq 1$, under large and extremely small sample sizes i.e. $n \geq 60$ and $n = 10$, MPSEs consistently provided the smallest bias regardless of the first shape parameter values α. An exception is made at the combination of shape parameter values $\theta = 1$ and $0.5 \leq \alpha \leq 1$ under sample size $n = 10$ then the smallest bias values were yielded by PCEs. As for moderate sample sizes $20 \leq n \leq 50$ and first shape parameter values $0.5 \leq \alpha \leq 2$ the following observation is noted regarding the performance of the MPSE and ADE methods, where the ADEs progressively compete with the MPSEs in estimating the first shape parameter α under larger sample sizes as the values of α increase. For instance, at $\alpha = 0.5$ the smallest bias values are provided by MPSEs except under small sample size $n = 20$, where ADEs yielded better results, then as α values increase and become $\alpha = 1$ ADEs gradually show better performance regarding larger sample sizes in the range $20 \leq n \leq 40$, finally, at $\alpha = 2$ ADEs became more effective than MPSEs in estimating the first shape parameter by providing the least bias values under all the specified combinations i.e. $20 \leq n \leq 50$ and $0.5 \leq \alpha \leq 2$.

On the other hand, in case of $2 \leq \theta \leq 4$, under extremely small sample sizes $n = 10$ PCEs provided the smallest bias values regardless of the values of α, except for the combination of shape parameter values $\alpha = 0.5$ and $\theta = 4$ WLSEs provided the best results. Overall, MPSEs produced the smallest bias values. Although some exceptions are made, firstly, in case of the shape parameter combination $\alpha = 0.5$ and $\theta = 4$ the OLSEs indicators yielded better results under large sample sizes $n \geq 50$, whereas the PCEs provided better results for the shape parameter combination $\alpha = 4$ and $\theta = 4$ under relatively small sample size $n = 20$. Last but not least, the ADEs outperformed the MPSEs at some combinations specified as follows, in case of $\theta = 4$, when $\alpha = 0.5$ and under sample sizes $20 \leq n \leq 40$ as well as $\alpha = 1$ and $20 \leq n \leq 30$ in addition to the combination $\alpha = 2$ and $20 \leq n \leq 60$. As for $\theta = 2$,

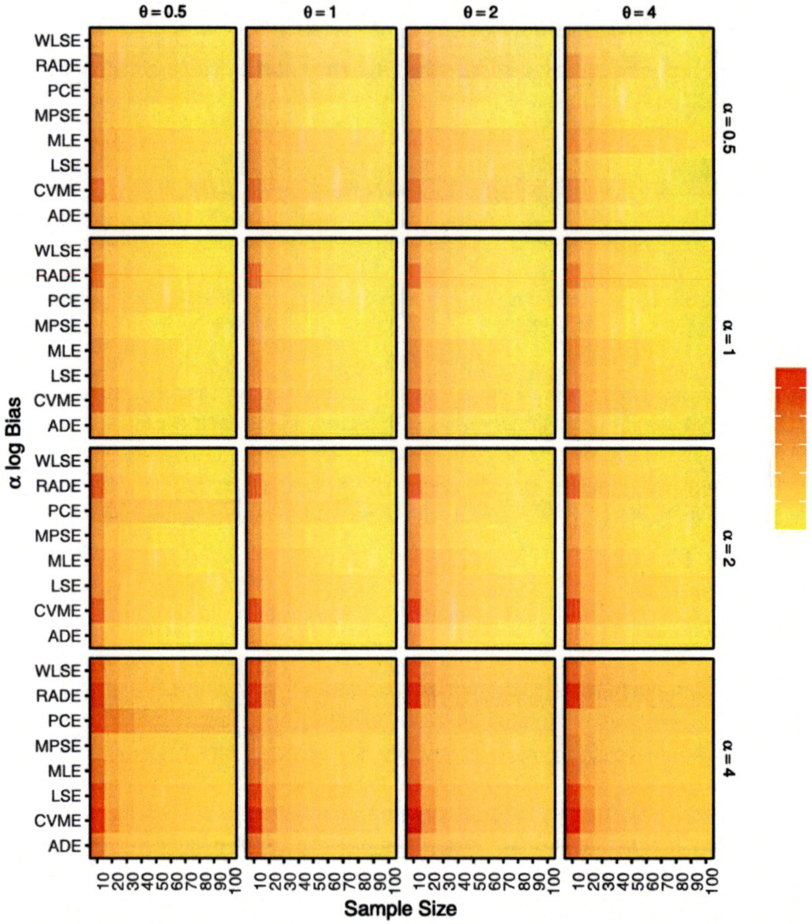

Fig. 4 The simulated biases of the estimators of the 1st shape parameter α

the previous observation made under $\theta \leq 1$ regarding the performance of the MPSE and ADE methods is also true in this case, where the ADEs begin to progressively outperform the MPSEs as the values of the first shape parameter α and sample sizes increase, under sample sizes $20 \leq n \leq 50$ and for $0.5 \leq \alpha \leq 2$.

Regarding the biases of the estimators of θ: Regarding the cases of $\alpha \leq 1$ and $\alpha = 4$, the overall performance of the MPSEs indicators suggests that the MPS method is the best at estimating the second shape parameter θ regardless of different sample sizes and θ parameter values. Some exceptions are made where some methods outperformed the MPSEs, specified as follows, firstly, the MLEs outperformed the MPSEs regarding large sample sizes at the following combinations, when $\theta = 1$ and under sample sizes $n \geq 60$, as well as $2 \leq \theta \leq 4$ and sample sizes $n \geq 50$. Secondly,

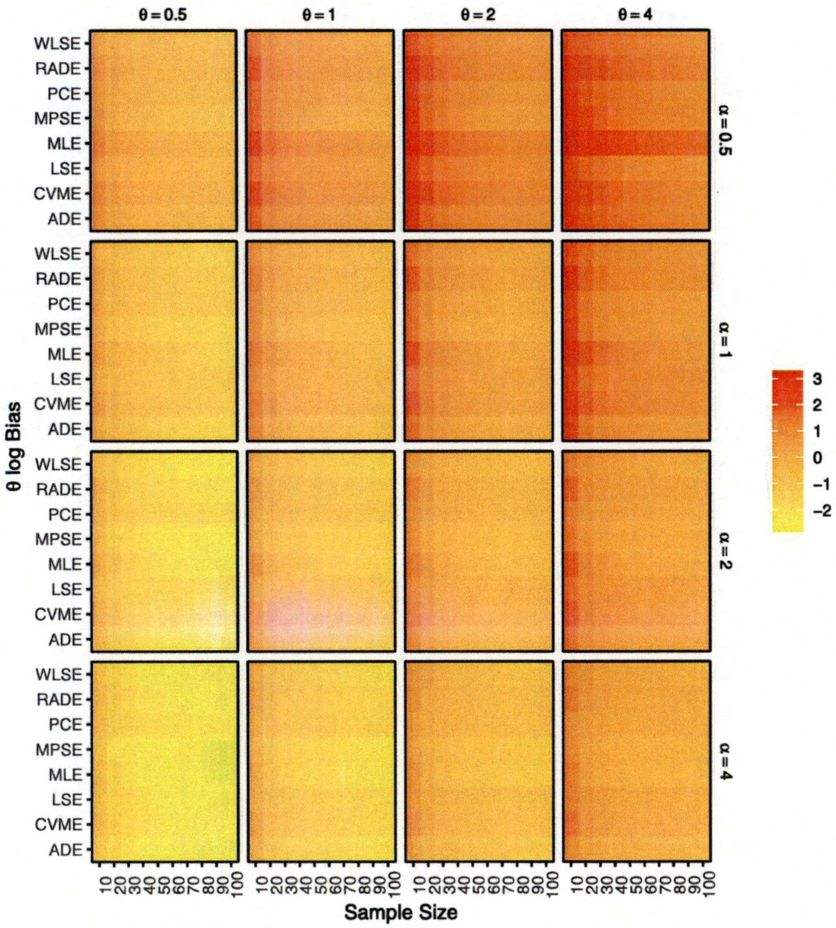

Fig. 5 The simulated biases of the estimators of the 2nd shape parameter θ

the ADEs provided better results at shape parameter combination $\alpha = 1$ and $\theta = 0.5$ and under relatively small sample size $n = 20$. Finally, OLSEs produced the smallest bias values at shape parameter combination $\alpha = 0.5$ and $\theta = 4$ regardless of sample sizes.

In case of $\alpha = 2$ and under sample sizes $20 \leq n \leq 60$, an observation can be made regarding the performance of the ADEs indicators, as the second shape parameter increases the ADEs progressively show better performance at estimating the second shape parameter θ as the sample sizes increase. Whereas for large sample sizes $70 \leq n \leq 100$ the MLEs provided the best results, except at $\theta = 0.5$ and $n = 70$ where MPSEs provided better results. Moreover, the smallest bias values were produced by MPSEs at $\theta = 1$ and under sample size $n = 60$.

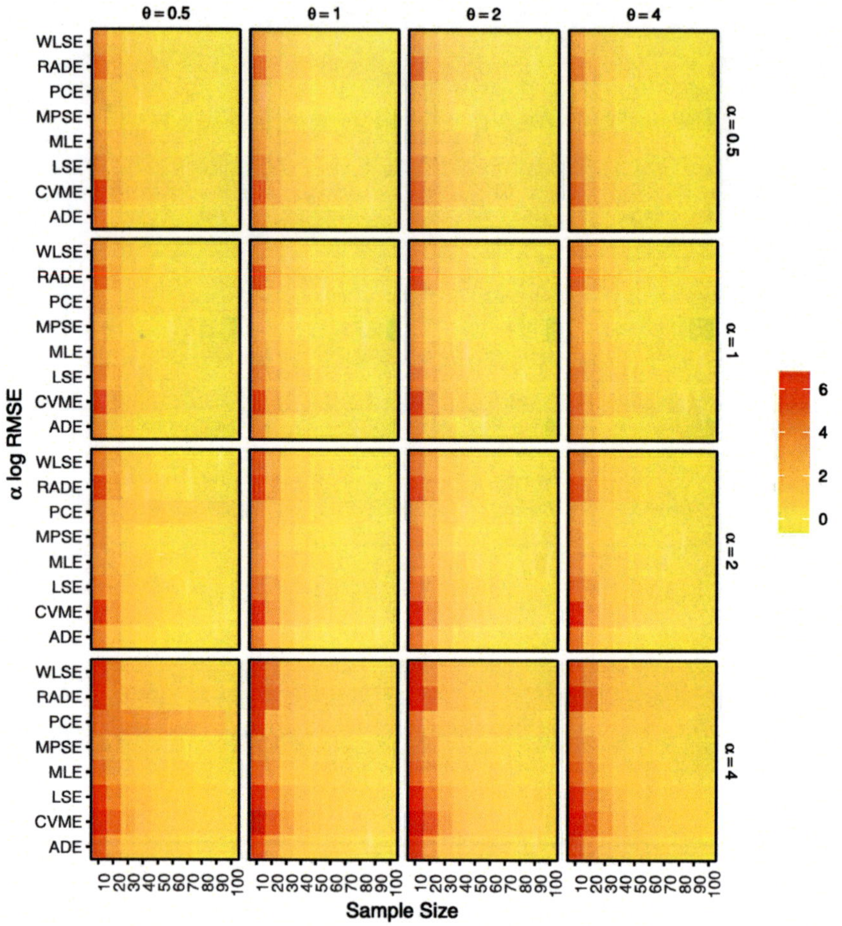

Fig. 6 The simulated RMSEs of the estimators of the 1st shape parameter α

Regarding extremely small sample sizes $n = 10$ the following observations can be made, at the first shape parameter values $0.5 \leq \alpha \leq 2$ the smallest biases were provided by WLSEs regardless of second shape parameter values, some exceptions are made, at shape parameter combination $\theta = 4$ with $\alpha = 2$ and $\alpha = 4$ the PCEs provided the smallest bias values. On the other hand, for the rest of combinations under the case $\alpha = 4$ the smallest bias values for small sample sizes $n = 10$ were yielded by MPSEs regardless of θ values.

Regarding the RMSEs of the estimators of α: Overall, the MPSEs provided the smallest RMSE values regardless of different sample sizes and first shape parameter values α. Although at some scenarios a few methods outperformed the MPSEs. For instance, the ADEs provided better results at the following combinations, under moderate sam-

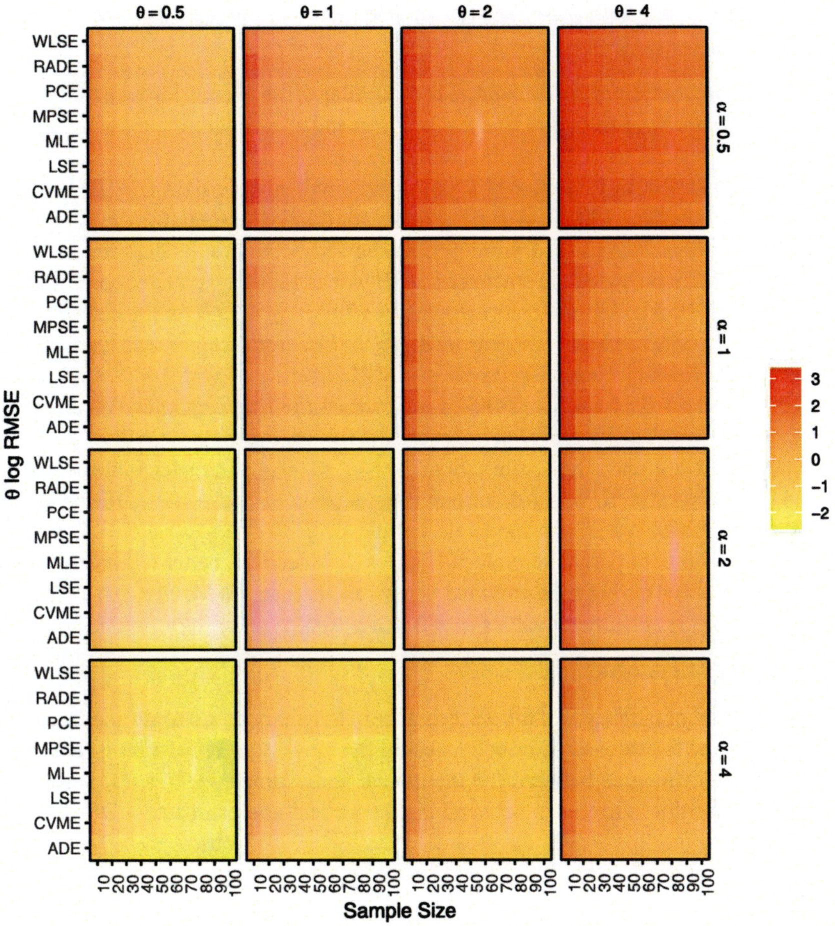

Fig. 7 The simulated RMSEs of the estimators of the 2nd shape parameter θ

ple sizes $20 \leq n \leq 30$ and shape parameter combinations $\alpha \leq 1$ and $\theta \leq 2$. Additional cases where ADEs provided the smallest RMSEs, under sample size $n = 30$ at shape parameter combinations $\theta = 0.5$ and $\alpha = 2$ as well as $\theta = 4$ and $\alpha = 1$ also $\theta = 4$ and $\alpha = 2$ and under sample sizes $20 \leq n \leq 40$ and $\theta = 4$ and $\alpha = 0.5$. The MLEs yielded the smallest RMSE values under extremely small sample sizes $n = 10$ and shape parameter combinations $\theta \leq 1$ with $\alpha = 2$. Another method that provided better performance than the MPSEs are the PCEs, whose indicators show better results under small to moderate sample sizes in the following scenarios, for sample size $n = 10$, in the case of $\theta = 2$ with first shape parameter values $\alpha \leq 2$, also in case of $\theta = 4$ with $\alpha \leq 4$, whereas under sample size $n = 20$, and $\theta = 1$ with $\alpha = 4$, as well as, in the case of $\theta = 2$ with $\alpha = 2$ and $\alpha = 4$, and finally in the case of $\theta = 4$ with first shape parameter values $1 \leq \alpha \leq 4$.

Regarding the RMSEs of the estimators of θ: Similarly, the overall performance of the MPSE indicators suggests that it is the best method at estimating the second shape parameter θ regardless of different sample sizes and θ values. Some exceptions are made, for instance, in case of $\alpha \leq 1$ and under small sample sizes $n = 10$ WLSEs provided the smallest RMSE values regardless of θ values except for the parameter combination $\theta = 2$ and $\alpha = 1$ where PCEs showed better results than the MPSEs. Moreover, PCEs provided the smallest RMSE values for the parameter combination $\theta = 4$ and $\alpha = 0.5$ under moderate sample sizes $20 \leq n \leq 40$. As for the cases of $\alpha = 2$ and $\alpha = 4$ some methods outperformed the MPSEs, for instance, MLEs provided better results for large sample sizes $60 \leq n \leq 100$ for $\alpha = 4$ with $1 \leq \theta \leq 4$ except under $n = 60$ for $\theta = 1$ where MPSEs maintained their performance. Whereas, in the case of $\alpha = 2$, MLEs provided better results for large sample sizes $90 \leq n \leq 100$ at $\theta = 0.5$ in addition to the combinations $70 \leq n \leq 100$ at parameter values $1 \leq \theta \leq 2$ and $80 \leq n \leq 100$ at $\theta = 4$. Additionally, under the $\alpha = 2$ case, the ADEs provided the smallest RMSE values for moderate sample sizes $30 \leq n \leq 70$ at $\theta = 4$, as well as $40 \leq n \leq 50$ with parameter values $1 \leq \theta \leq 2$. Moreover, under extremely small sample size $n = 10$ the PCEs provided better results than the MPSEs at $\theta = 4$ with both cases of first parameter values $\alpha = 2$ and $\alpha = 4$.

Goodness-of-fit analysis

The second set of indicators of the performance of the methods are some goodness-of-fit criteria used to evaluate the effectiveness of the estimators by calculating the average absolute difference between the theoretical and estimated CDFs (D_{abs}) and the maximum absolute difference between the theoretical and estimated CDFs (D_{max}), given as:

$$D_{abs} = \frac{1}{n \times M} \sum_{i=1}^{M} \sum_{j=1}^{n} \left| F(t_j; \alpha, \beta, \theta) - F(t_j; \hat{\alpha}_i, \hat{\beta}_i, \hat{\theta}_i) \right|$$

$$D_{max} = \frac{1}{M} \sum_{i=1}^{M} \max_{j=1,\ldots,n} \left| F(t_j; \alpha, \beta, \theta) - F(t_j; \hat{\alpha}_i, \hat{\beta}_i, \hat{\theta}_i) \right|$$

The outcomes of this part of the simulation study is discussed as follows. For each combinations of GBS-II($\alpha, \beta = 1, \theta$), the above goodness-of-fit criteria are obtained for the eight estimators as shown in Figs. 8, 9, respectively. The ADEs provided the smallest values for D_{abs} in case of $\alpha \leq 0.5$ regardless of sample sizes and across different values of the second shape parameter θ. Similarly, ADEs provided the smallest values for D_{max} when $\alpha \leq 0.5$ for moderate sizes $20 \leq n \leq 80$ regardless of θ parameter values. For extremely small sample sizes, $n = 10$ WLSEs provided the smallest D_{max} values regardless of θ values, as for large sample sizes $n \geq 90$ MPSEs produced the smallest values except for the combinations of $\theta = 0.5$ and

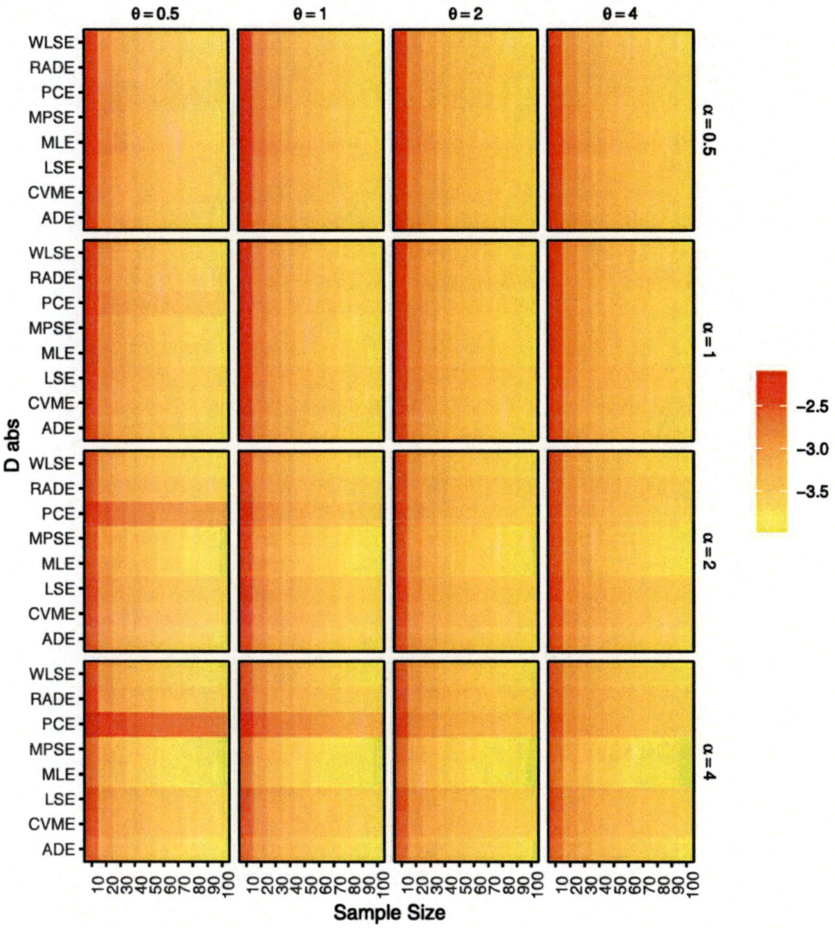

Fig. 8 The simulated average absolute difference between the true and estimated CDFs

$\theta = 4$ and under sample size $n = 90$. In case of $\alpha = 1$, MPSEs provided the smallest values for D_{abs} under large sample sizes $n \geq 70$ regardless of different θ parameter values, except for the combination where $n = 70$ and $\theta = 2$ where ADEs yielded the smallest values. As for moderate to small sample sizes $n \leq 60$ ADEs yielded the smallest D_{abs} values regardless of θ parameter values. Similarly, MPSEs provided the smallest values for D_{max} when $\alpha = 1$ under large sample sizes $n \geq 80$ regardless of θ values, whereas moderate to small sample sizes n ≤ 70 ADEs yielded the smallest values regardless of θ values, except for the combination of $\theta = 4$ and under $n = 10$ where WLSEs yielded the smallest D_{max} values.

The MLEs provided the smallest values for D_{abs} when $\alpha = 4$ regardless of sample sizes and across different values of the second shape parameter θ . The same

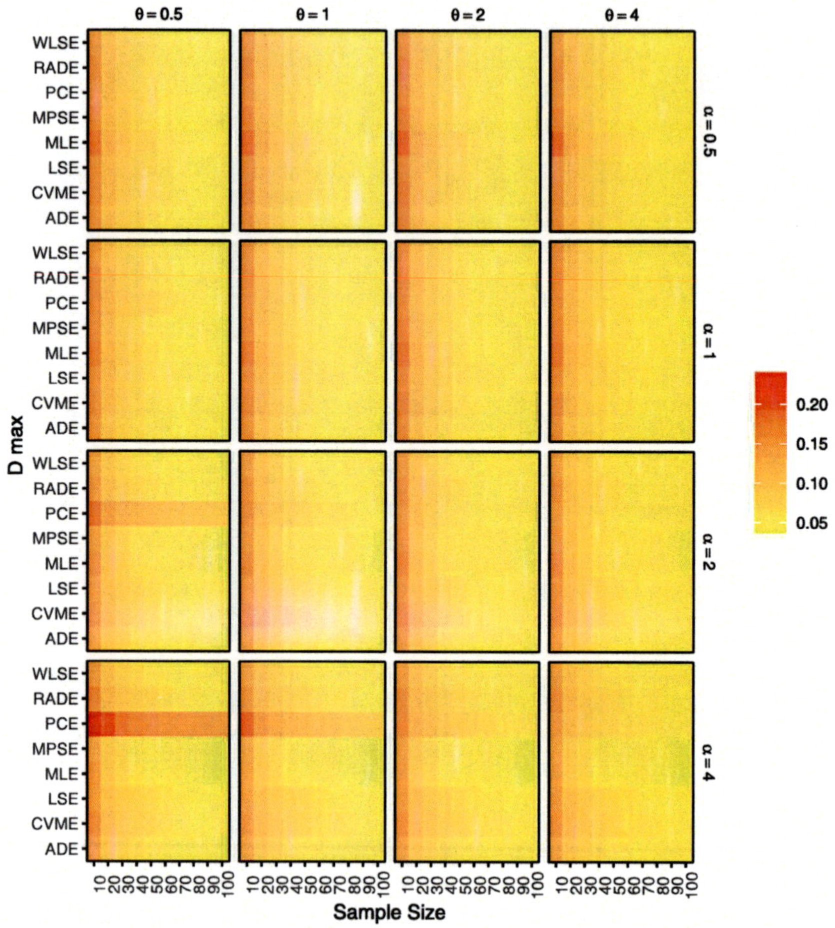

Fig. 9 The simulated maximum absolute difference between the true and estimated CDFs

performance of MLEs can be observed in case of $\alpha = 2$ under moderate to large sample sizes $n \geq 20$ regardless of θ parameter values, except for the combination of $\theta = 0.5$ and $n = 20$ where the smallest values were produced by ADEs. Regarding small sample sizes $n \leq 10$ ADEs yielded the smallest D_{abs} values across different θ parameter values. As for D_{max} indicator results, the MPSEs provided the smallest values in case of $\alpha = 4$ regardless of sample sizes and second shape parameter values. Similarly, when $\alpha = 2$ the smallest values for D_{max} were produced by MPSEs for moderate to large sample sizes $n \geq 40$ regardless of θ parameter values, whereas ADEs yielded the smallest values under small sample sizes $n \leq 30$ regardless of θ values. Similarly, the MPSEs provided the smallest values for D_{max} when $\alpha = 2$ under moderate to large sample sizes $n \geq 40$ regardless of θ parameter values, in case

Table 4 Daily ozone concentrations (in ppb)

41	7	30	37	12	97	79	64	122	31	118	20	13	7	80
36	16	1	29	13	85	63	59	89	44	84	23	46	14	28
12	11	11	71	49	10	16	39	110	21	85	21	18	30	73
18	14	4	39	32	27	108	9	44	9	96	24	13	14	13
28	18	32	23	64	7	20	16	65	45	78	44	24	18	
23	14	23	21	40	48	52	78	22	168	91	21	16	20	
19	34	45	37	77	35	82	35	59	73	47	28	23	11	
8	6	115	20	97	61	50	66	23	76	32	9	36	135	

of small sample sizes $n \leq 30$ the ADEs yielded the smallest values for D_{max} across different values of the second shape parameter θ. From the overall performance of the D_{abs} and D_{max} indicators, one can observer that as the sample sizes increase, all methods produce decreasing values (Table 4).

5 Data Analysis

In this sub-section, an analysis of the O_3 concentrations data is conducted in a similar approach to what has been done in the preceding subsection, this dataset has been reported and analysed by [34] [?] using a an extension of the BS distribution, the dataset is given as follows; The values represents the ozone O_3 contamination levels present in the air and measured in an hourly basis (ppb) in New York. The outcomes of the the analysis are reported in Tables 5, 6 and 7, which represent the estimates of the model parameters, and the KS distance statistics and their corresponding bootstrap p-values, in addition to the estimated theoretical percentiles and their sample counterparts, respectively. While the graph of the fitted PDFs placed over the observed histogram of the O_3 dataset is presented in Fig. 10.

From the outcomes of the analysis, all methods have close biases and RMSEs. Nevertheless, based on each KS test and the associated bootstrap p-values, the LSEs and WLSEs provided the best-fitted models, respectively. Here, CVMEs and LSEs provided the closest estimated theoretical percentiles compared to the sample percentiles. Regarding the bootstrap bias and RMSE results, the RADEs followed by WLSEs, and PCEs provided competitive performances as the best-fitted models.

Note that the bootstrap RMSE $RMSE_B(.)$ is abbreviated as $R_B(.)$ in the Table 5.

Table 5 Outcomes of fitted models according to different estimators based on O_3 concentration data

Method	$\hat{\alpha}$	$\text{Bias}_B(\hat{\alpha})$	$R_B(\hat{\alpha})$	$\hat{\beta}$	$\text{Bias}_B(\hat{\beta})$	$R_B(\hat{\beta})$	$\hat{\theta}$	$\text{Bias}_B(\hat{\theta})$	$R_B(\hat{\theta})$
MLE	0.00001	−10.16451	17.61426	30.52406	20.32902	24.91032	0.00000	−10.16451	17.61426
MPSE	0.00001	−12.37918	21.45211	37.17474	24.75837	30.33786	0.00000	−12.37919	21.45211
LSE	1.66174	−9.50743	17.15183	31.41997	20.29524	24.83169	0.84106	−10.44482	17.63314
WLSE	1.70638	−9.38702	16.92300	31.19994	20.09650	24.58096	0.86584	−10.35059	17.49978
PCE	2.22918	−9.22711	16.76498	31.33864	20.14074	24.65678	1.07037	−10.48172	17.50360
CVME	1.72757	−9.46460	17.06124	31.42796	20.27880	24.81849	0.87885	−10.42966	17.62424
ADE	0.57995	−9.72267	17.45366	31.22485	20.61751	25.30310	0.32696	−10.18877	17.73027
RADE	2.47088	−9.12812	16.70541	31.51951	20.09131	24.54093	1.17208	−10.52646	17.52409

Table 6 Goodness-of-fit statistics based on O_3 concentration data

Method	KS	p-value
MLE	0.06228	0.68931
MPSE	0.10951	0.11089
LSE	0.03939	0.77822
WLSE	0.03973	0.81019
PCE	0.04490	0.81618
CVME	0.04011	0.72727
ADE	0.04954	0.42258
RADE	0.04278	0.75025

Table 7 Sample percentiles versus theoretical percentiles estimated using different estimators based on O_3 concentration data

	P_{10}	P_{25}	P_{50}	P_{75}	P_{90}
Sample	0.99145	0.72650	0.50427	0.25641	0.00855
MLE	0.99996	0.73002	0.48543	0.19892	0.02390
MPSE	0.99984	0.76485	0.56549	0.29838	0.06661
LSE	1.00000	0.72065	0.49897	0.22668	0.01022
WLSE	1.00000	0.71889	0.49613	0.22281	0.00863
PCE	1.00000	0.71372	0.49803	0.22974	0.00424
CVME	1.00000	0.72240	0.49907	0.22445	0.00836
ADE	1.00000	0.73384	0.49605	0.21099	0.02304
RADE	1.00000	0.71584	0.50023	0.23067	0.00240

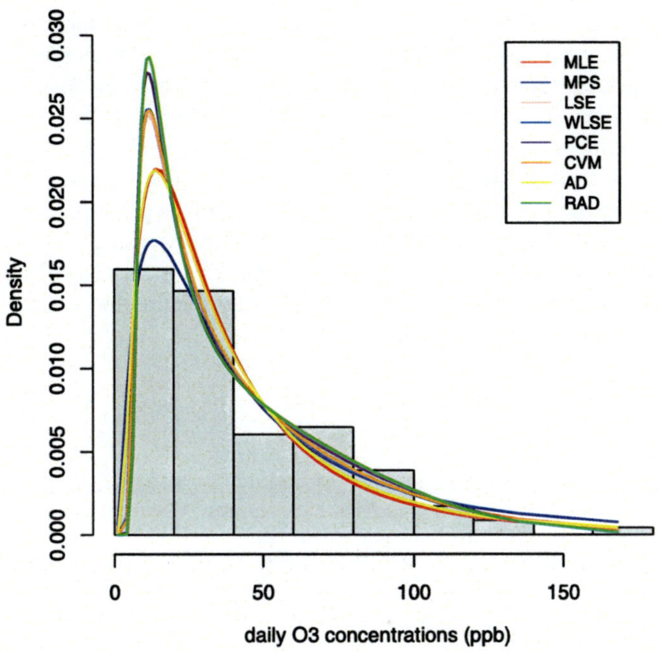

Fig. 10 Data histogram versus fitted PDFs using different estimators based on O_3 concentration data

6 Conclusions

In this paper, we discussed eight frequentist estimation methods for estimating the three parameters of the GBS-II distribution. Including the maximum likelihood and the maximum product of spacing estimation, several regression-based estimations, and goodness-of-fit estimations. An investigation of the performance of the estimators was conducted in two main areas; firstly, a numerical evaluation of the performance via extensive Monte Carlo simulations, where the results were examined based on some estimation efficiency criteria as well as some goodness-of-fit criteria. Secondly, two applications related to the environmental sciences were considered for the purpose of illustrating the utility of the distribution, additionally, to determining which method best fits such data.

The numerical results and Monte Carlo simulation outcomes indicated that the choice of the estimator depends on the true value of the shape parameters and the sample size. All the methods performed well, although the MPSEs outperformed the rest of the methods regarding the considered criteria. The utility of the considered estimators has been illustrated by analyzing air pollution concentration data. Regarding bootstrap bias and RMSE, the PCEs and RADEs provided the best-fitted models. On the other hand, according to the chosen goodness-of-fit procedure the

LSEs provided the best-fitted model for the considered data according to the chosen goodness-of-fit procedure. The analysis also indicated that LSEs provided the closest theoretical percentiles to their sample counterparts concentrations data.

Some potential directions for future research regarding the GBS-II distribution are suggested, firstly, the considered estimators are known as frequentist estimates and comparing their performances to those of Bayesian counterparts is an important research direction to be addressed. Additionally, due to the prevalence of censoring when performing reliability and survival analysis in real-world applications, it is of interest to discuss the considered estimation methods for the GBS-II distribution under the assumption of censored samples and assess their performances.

References

1. Sanhueza, A., Leiva, V., Balakrishnan, N.: The generalized Birnbaum-Saunders distribution and its theory, methodology, and application. Commun. Stat. Theory Methods **37**(5), 645–670 (2008). https://doi.org/10.1080/03610920701541174
2. Marchant, C., Leiva, V., Cavieres, M.F., Sanhueza, A.: Air contaminant statistical distributions with application to PM10 in Santiago, Chile. In: Whitacre, D.M. (ed.) Reviews of Environmental Contamination and Toxicology, vol. 223, pp. 1–31. Springer, New York (2013). https://doi.org/10.1007/978-1-4614-5577-6_1
3. Owen, W.J.: Another look at the Birnbaum-Saunders distribution (2004). http://www.stat.lanl.gov/MMR2004/Extended%20Abstract/WOwnn.pdf
4. Díaz-García, J.A., Domínguez-Molina, J.R.: Some generalizations of Birnbaum-Saunders and sinh-normal distributions. Int. Math. Forum 1(33–36), 1709–1727 (2006). https://doi.org/10.12988/imf.2006.06146
5. Birnbaum, Z., Saunders, S.: A new family of life distributions. J. Appl. Probab. **6**(2), 319–327 (1969). https://doi.org/10.2307/3212003
6. Birnbaum, Z., Saunders, S.: Estimation for a family of life distributions with applications to fatigue. J. Appl. Probab. **6**(2), 328–347 (1969). https://doi.org/10.2307/3212004
7. Leiva, V.: Chapter 1-Genesis of the Birnbaum-Saunders distribution. In: Leiva, V. (ed.) The Birnbaum-Saunders Distribution, pp. 1–15. Academic Press (2016). https://doi.org/10.1016/B978-0-12-803769-0.00001-7
8. Balakrishnan, N., Kundu, D.: Birnbaum-Saunders distribution: a review of models, analysis, and applications. Appl. Stoch. Models Bus. Ind. **35**(1), 4–49 (2019). https://doi.org/10.1002/asmb.2348
9. Díaz-García, J.A., Leiva-Sánchez, V.: A new family of life distributions based on the elliptically contoured distributions. J. Stat. Planning Inference **128**(2), 445–457 (2005). https://doi.org/10.1016/j.jspi.2003.11.007
10. Díaz-García, J.A., Leiva-Sánchez, V.: Erratum to "A new family of life distributions based on the elliptically contoured distributions": [J. Stat. Plann. Inference **128**(2), 445–457 (2005)]. J. Stat. Plann. Inference **137**(4), 1512–1513 (2007). https://doi.org/10.1016/j.jspi.2006.06.040
11. Balakrishnan, N., Alam, F.M.A.: Maximum likelihood estimation of the parameters of student's t Birnbaum-Saunders distribution: a comparative study. Commun. Stat. Simul. Comput. **51**(3), 793–822 (2020). https://doi.org/10.1080/03610918.2019.1659359
12. Alam, F.M.A.: On comparing different methods of estimation for the parameters of a pathological distribution with application to climate data. Austrian J. Stat. **51**(4), 120–147 (2022). https://doi.org/10.17713/ajs.v51i4.1331

13. Alam, F.M.A., Almalki, A.M.: On modeling cancer and tuberculosis data using the Birnbaum-Saunders lifetime model established on a logistic kernel. Appl. Sci. **12**(10), 5000 (2022). https://doi.org/10.3390/app12105000
14. Alam, F.M.A., Nassar, M.: On modeling concrete compressive strength data using Laplace Birnbaum-Saunders distribution assuming contaminated information. Crystals **11**(7), 830 (2021). https://doi.org/10.3390/cryst11070830
15. Owen, W.J.: A new three-parameter extension to the Birnbaum-Saunders distribution. IEEE Trans. Reliab. **55**(3), 475–479 (2006). https://doi.org/10.1109/TR.2006.879646
16. Wang, R., Sha, N., Xu, X.: Statistical inference for type-I generalized Birnbaum-Saunders distribution. J. Indian Soc. Probab. Stat. **19**, 469–487 (2018). https://doi.org/10.1007/s41096-018-0044-1
17. Wang, R., Sha, N., Xu, X.: Correction to: statistical inference for Type-I generalized Birnbaum-Saunders distribution. J. Indian Soc. Probab. Stat. **19**, 489 (2018). https://doi.org/10.1007/s41096-018-0046-z
18. Owen, W.J., Tony Ng, H.K.: Revisit of relationships and models for the Birnbaum-Saunders and inverse-Gaussian distributions. J. Stat. Distrib. App. **2**, 11 (2015). https://doi.org/10.1186/s40488-015-0034-8
19. Kundu, D.: Bivariate sinh-normal distribution and a related model. Brazilian J. Probab. Stat. **29**(3), 590–607 (2015). https://www.jstor.org/stable/26359001
20. Cheng Y.: A general Birnbaum-Saunders model and its application in ALT models. Ph.D. Thesis, Rutgers University-Graduate School-New Brunswick, USA (2013). https://rb.gy/r7m7g
21. Sha, N.: A new inference approach for Type-II generalized Birnbaum-Saunders distribution. Stats **2**(1), 148–163 (2019). https://doi.org/10.3390/stats2010011
22. Gupta, R.D., Kundu, D.: Generalized exponential distribution: different method of estimations. J. Stat. Comput. Simul. **69**(4), 315–337 (2001). https://doi.org/10.1080/00949650108812098
23. do Espirito Santo, A.P.J., Mazucheli, J.: Comparison of estimation methods for the Marshall-Olkin extended Lindley distribution. J. Stat. Comput. Simul. **85**(17), 3437–3450 (2015). https://doi.org/10.1080/00949655.2014.977904
24. Menezes, A.F.B., Mazucheli, J., Dey, S.: The unit-logistic distribution: Different methods of estimation. Pesquisa Operacional **38**(3), 555–578 (2018). https://doi.org/10.1590/0101-7438.2018.038.03.0555
25. Qoshja, A., Hoxha, F.: Different methods of estimation for generalized inverse Lindley distribution. Math. Theory Model. **6**(5), 132–141 (2016)
26. Dey, S., Raheem, E., Mukherjee, S.: Statistical properties and different methods of estimation of transmuted Rayleigh distribution. Revista Colombiana de Estadística **40**(1), 165–203 (2017). https://doi.org/10.15446/rce.v40n1.56153
27. Nassar, M., Afify, A.Z., Dey, S., Kumar, D.: A new extension of Weibull distribution: properties and different methods of estimation. J. Comput. Appl. Math. **336**, 439–457 (2018). https://doi.org/10.1016/j.cam.2017.12.001
28. Casella, G., Berger, R.L.: Statistical Inference, 2nd edn. Duxbury Press, Pacific Grove (2002)
29. Cheng, R., Amin, N.: Maximum product-of-spacings estimation with applications to the log-normal distribution. Math. Report **791** (1979)
30. Cheng, R.C.H., Amin, N.A.K.: Estimating parameters in continuous univariate distributions with a shifted origin. J. R. Stat. Soc. Ser. B (Methodol.) **45**, 394–403 (1983). https://doi.org/10.1111/j.2517-6161.1983.tb01268.x
31. Ghosh, K., Jammalamadaka, S.R.: A general estimation method using spacings. J. Stat. Plann. Inference **93**(1), 71–82 (2001). https://doi.org/10.1016/S0378-3758(00)00160-9
32. Swain, J.J., Venkatraman, S., Wilson, J.R.: Least-squares estimation of distribution functions in Johnson's translation system. J. Stat. Comput. Simul. **29**(4), 271–297 (1988). https://doi.org/10.1080/00949658808811068
33. Ferreira , M., Gomes , M.I., Leiva , V.: On an extreme value version of the Birnbaum-Saunders distribution. REVSTAT-Stat. J. **10**(2), 181–210 (2012). https://doi.org/10.57805/revstat.v10i2.116

34. Saulo, H., Leiva, V., Ziegelmann, F.A., Marchant, C.: A nonparametric method for estimating asymmetric densities based on skewed Birnbaum-Saunders distributions applied to environmental data. Stoch. Environ. Res. Risk Assess. **27**(6), 1479–1491 (2013). https://doi.org/10.1007/s00477-012-0684-8
35. Dubey, S.D.: Some percentile estimators for Weibull parameters. Technometrics **9**(1), 119–129 (1967). https://doi.org/10.1080/00401706.1967.10490445

Improved Routing for the In-Band Network Telemetry Problem

Thierno Bocar N'Diaye, Marcelo Caggiani Luizelli, Jules Degila, and Luciana Salete Buriol

Abstract The continuous growth of networks imposes on managers to find new technologies for better handling of traffic measurement for network wide-visibility (i.e., the ability to access most desired flow-level information per hop for all flows), performance monitoring requirements and network robustness (i.e., the ability to survive partial network failures). In-band network telemetry has emerged as one of the latest network monitoring paradigms for the mega-scale network. Previous research has focused on orchestrating the processes of collecting in-band network telemetry data from the network and addressed the scalability. Yet little has been done to route network flows efficiently to address the low network coverage and monitoring applications performance in this new network monitoring paradigm. This paper analyses two mathematical models for in-band network telemetry. The first model, proposed by [1], exploits shortest path flow routes as inputs. The second model proposed in this work, extends the first to compute flow route paths efficiently, providing a comprehensive solution to the problem.

Keywords Network monitoring · Monitoring application · Mathematical model · In-band network telemetry · Network flow route · MILP

T. B. N'Diaye (✉) · J. Degila
Institut de Mathématiques et Sciences Physiques (IMSP), Université d'Abomey-Calavi (UAC), Abomey Calavi, Benin
e-mail: thierno.ndiaye@imsp-uac.org

J. Degila
e-mail: jules.degila@imsp-uac.org

M. C. Luizelli
Graduate Program in Electrical Engineering, Universidade Federal do Pampa (UNIPAMPA), Bagé, Brazil
e-mail: marceloluizelli@unipampa.edu.br

L. S. Buriol
Graduate Program in Computing, Universidade Federal do Rio Grande do Sul (UFRGS), Porto Alegre, Brazil
e-mail: buriol@inf.ufrgs.br

D. Vlachos (ed.), *Mathematical Modeling in Physical Sciences*, Springer Proceedings in Mathematics & Statistics 446, https://doi.org/10.1007/978-3-031-52965-8_46

1 Introduction

Network monitoring is essential in providing high network-wide visibility in several sectors. The process consists of monitoring the network usage and performance, detecting failures, and reporting it as much as possible in real-time. The network monitoring mechanism started with adding taps in the network for collecting monitoring packets, then evolved to recording the internal states of the switches and nodes into the data packet as the packets go through the network. The last mechanism, the in-band network telemetry (INT), has recently emerged as a promising alternative to traditional network monitoring. The INT is a framework designed to monitor, collect, and report flows and network states, by the data plane, without requiring intervention or work by the control plane. Most of the recent progress in the INT has been driven by advances in programmable network devices and high-level domain-specific networking descriptions and query languages [2].

Using programmable devices, in-band data plane telemetry allows us to collect and encapsulate low-level telemetry information into production network traffic. Packets contain header fields interpreted as telemetry instructions by network devices [1]. However, the INT detection of failures in the network is limited because telemetry paths and metrics must be preassigned by network operators (or control plane applications). Consequently, INT can only monitor certain telemetry data of specific paths [3]. Thus the INT coverage is most likely limited to particular devices and links in the network. Recent efforts that discuss INT orchestration include [1, 3–7]. Liu and al. [3] proposed NetVision to provide higher network-wide visibility to network operators. Their approach is based on customizing the probe path by changing the segment routing labels at run-time. However, inserting an active probe can be expensive. It may lead to queue buildups, buffer-bloat, packet drops, network congestion, and delays, mainly if performed in an unplanned ad-hoc manner [8].

In addition, the probe path must be carefully designed to achieve comprehensive network visibility. Marques et al. [5] provided the first effort towards in-band network orchestration. They proposed two heuristics showing that the network visibility may degrade in coverage, consistency, and freshness depending on which telemetry items are collected. However, the proposed heuristics do not optimally assign telemetry items to network flows due to the local decisions taken by these heuristics. Hohemberger et al. [1] formulated a mathematical model for the in-band network telemetry orchestration plan problem as a mixed integer linear programming (MILP). They proposed a machine-learning mechanism to dynamically coordinate the collection of telemetry items from the network. Recently, Hohemberger et al. [4] tackled the scalability limitation of the INT by proposing an iterated local search procedure.

Despite the effort to deliver the best out of the in-band telemetry to improve network coverage and scalability, little has been done to optimize route flows through the network. Typically, network managers employ static routing or distance vector routing, such as the shortest path, as the primary method for routing within the network. There are several issues with these methods. First, static routing is a manual process that requires network administrators to configure and update routing tables manually. Second, manually configuring network routes becomes very challenging in large and complex networks, as it can be difficult to keep track of all the routes and ensure that they are up to date. Third, the shortest path algorithm for routing network flows to interconnect forwarding devices does not provide full network coverage and is most likely to miss some forwarding devices within the network.

This paper proposes an automated approach for routing network flows for INT while collecting telemetry data from the network. This approach optimally assigns network flows to telemetry items and provides higher network-wide visibility with the network. The main idea involves introducing new flow variables to the model proposed by Hohemberger et al. [1]. These variables ensure more flexibility in routing network flows and allow control the longest path length. For modelling the flow paths the model requires the number of network flows, the maximum number of hops a path can have, and the endpoints (i.e., ingress and egress forwarding devices).

2 Related Work

This section focuses on previous works investigating in-band network telemetry mechanisms to monitor networks and services.

The earliest studies investigating in-band network telemetry began with the concept of programmable data planes in early 2014. Programmable data planes were the cornerstone for evolving from the software-defined networking (SDN) paradigm [9]. Data plane programmability has made network measurement much more manageable through domain-specific programming languages such as P4 [2]. With the development of software-defined networks and programmable data plane technology, Mazieres et al. [10] laid the foundation for in-band network telemetry by introducing the concept of tiny packet programs (TPP). The combination of the software-defined network and programmable data planes has profited the in-band network telemetry by allowing the network's owner to define how packets are processed, insert information about the network state, and control and manage various network incidents rapidly and intuitively [11]. Everflow [12] proposed a scalable packet-level telemetry system for large data center networks based on the property of one of the main switches: 'match and mirror'. By this property, packages can be traced and selected satisfying specific patterns focusing on in-band tracing. PathDump [13, 14] is a system in which a switch extracts telemetry data and reports it to a monitoring sink/analyzer. At the same time, the packet's route is kept on track for subsequent analysis of the recipient hosts.

Recently, much research has been conducted on in-band network telemetry in various aspects. Network telemetry has emerged as the newest way of collecting data from the network due to its automated process for remotely collecting and processing network information [15]. Pan et al. [6] proposed an INT-path, a network-wide telemetry framework. INT-path embeds source routing into INT-probe to track the probe packet route through the network. This mechanism is based on the Euler tracking algorithm. Later, Pan et al. [16] proposed INT-label, a lightweight network-wide telemetry architecture without explicitly relying on probe packets. Instead of embedding source routing into the INT probe, device states are adaptively periodically labelled onto sampled packets.

As visibility is essential for network measurement, INT has stepped in to offer high-precision device and traffic state visibility. However, INT is an underlying technique. Each INT instance covers only one monitoring path, and the number of probes to deploy in the network to trace INT paths is constrained [17]. Thus, a high-level orchestration is required to guarantee better network coverage. Castro et al. [18] proposed Patcher—fault—tolerant probing planning for INT. Patcher optimizes using the INT probe while considering the possibility of device failures (e.g., power failure, hardware failure). Qianchen Yuan et al. [19] proposed INT-react, an improvement of INT-path to solve the optimal path planning for a megascale network in a relatively short time. Marques et al. [5], and Hohemberger et al. [1, 4] presented the first mathematical formulation for the in-band network telemetry problem as a mixed-integer linear problem. They provided the first effort towards in-band network telemetry orchestration. Marques et al. [5] proposal is based on heuristic approaches that orchestrate how network telemetry data is collected into packets. Hohemberger et al. [1] orchestration approach is based on a machine learning-based model that wisely collects telemetry data based on its importance. Most previous mechanisms for collecting network telemetry data from network infrastructure have focused on maximizing the amount of data collected while neglecting or giving little attention to monitoring applications requirements. And this may considerably decrease the network measurement performance. Hohemberger et al. [4] dressed the scalability limitation of existing approaches for solving the in-band network telemetry problem by proposing a heuristic-based approach for large-scale networks. Pan et al. [20] proposed probabilistic in-band network telemetry (PINT), a probabilistic version of INT. PINT analyzes the in-band network telemetry on its terms by considering some key telemetry factors and proposing approaches that produce solutions with maximum network visibility.

Tan et al. [21] conducted an exhaustive survey on developing the in-band network telemetry and network measurement from classical to the most recent. However, most studies on the orchestration of the in-band network telemetry problem [6, 16–18] have focused on network wide-visibility based on the amount of network telemetry data collected from the network relying on probes packets.

3 The In-Band Network Telemetry Problem

In-band network telemetry (INT) is a paradigm used in computer networks to collect accurate real-time telemetry data about traffic flowing through the network. It provides wide-network visibility of network traffic. INT embeds telemetry data into data packets as they travel across the network.

The in-band network telemetry problem aims to coordinate the collection of network telemetry data to maximize the efficiency of network monitoring applications by satisfying as many network monitoring requirements as possible while minimizing the impact on production network flows.

The in-band network telemetry problem considers a physical network infrastructure $G = (D, L)$, a set of telemetry items V, and a set of monitoring applications M. Set D in the infrastructure network G represents programmable forwarding devices while set L contains the links interconnecting pairs of devices $(d_1, d_2) \in (D \times D)$. Each forwarding device $d \in D$ can embed a subset of items $V_d \subseteq V$ into packets of flow $f \in F$. Each telemetry item $v \in V$ has a size defined as a positive integer number.

Each network flow has two endpoint forwarding devices. The path between these two endpoints is simple (without loops) and is given or determined by the model. Network flows F are encapsulated in a forwarding protocol (e.g., NSH [12], IPv4). Therefore, the amount of space to embed telemetry items into packets is bounded by a constant $K_f \in \mathbb{N}^+$.

A monitoring application $m \in M$ requires a subset of telemetry items $R_m \subseteq V$ to operate appropriately. These telemetry items might have spatial and temporal dependencies. We say two or more telemetry items have spatial dependence if and only if they must be collected from the same forwarding device $d \in D$. The set of sets $R_m^s \subseteq \mathcal{P}(R_m)$ represents spatial dependencies. We say that a group of telemetry items have temporal dependence if and only if they must be collected within a given deadline. The set $R_m^t \subseteq \mathcal{P}(R_m)$ represents temporal dependencies; the required deadline is expressed as a function $\xi : R_m^t \to \mathbb{N}^+$. The model keeps track of the last time an item $P \in R_m^t$ was collected using a function $\zeta : R_m^t \to \mathbb{N}^+$. When $(\forall P \in R_m^t) : \zeta(P) > \xi(P)$, item P is outdated (i.e., the deadline has expired). Table 1 summarizes the notation used in the mathematical models.

Next, we present the first variation of the in-band network telemetry problem introduced by [1], which we label here as the in-band network telemetry basic problem (INTB). Then we propose a second variation of the problem, an extension of the INTB problem. We label it as the in-band network telemetry extended problem (INTE).

3.1 In-Band Network Telemetry Basic Problem

This subsection presents the first variation of the in-band network telemetry problem. In this variation the route of all network flows is pre-defined, computed as shortest paths, and given as input. We start by illustrating the problem and presenting the corresponding mathematical model.

3.1.1 Problem Illustration

To illustrate the INTB problem, let's consider the example depicted in Fig. 1, where we have four monitoring applications (M_1, M_2, M_3, and M_4) each demanding a

Table 1 Description of the notations used in the mathematical models

Symbols	Description
Sets:	
D	Set of forwarding devices
L	Set of links in the network
V	Set of telemetry items
V_d	Set of telemetry items embedded by device $d \in D$
M	Set of monitoring applications
F	Set of network flows
GF	Subset of network flows with pre-defined path
S_f	Set of starting points of network flows, $\forall f \in F$
E_f	Set of ending points of network flows, $\forall f \in F$
R_m	Set of monitoring requirements of monitoring application $m \in M$
$R_m^s \subseteq \mathcal{P}(R_m)$	Set of spatial dependencies
$R_m^t \subseteq R_m^s$	Set of temporal dependencies
Parameters:	
$\delta(v)$	Size of telemetry item $v \in V$
$\mathcal{P}(f)$	Routing path of flow $f \in GF$
$K(f)$	Capacity of network flow $f \in F$
$\mathcal{N}^+(d)$	Nodes reached from $d \in D$
$\mathcal{N}^-(d)$	Nodes that point to $d \in D$
max_route	Upper bound to the number of hops in a route
Binary variables:	
s_{mdp}	Equal 1 if spatial dependency $p \in P \in R_m^s$ is met on device d through monitoring application m; 0 otherwise
t_{mp}	Equal 1 if temporal dependency $p \in P \in R_m^t$ is met on monitoring application m; 0 otherwise
y_{dvf}	Equal 1 if telemetry item v is collected from device d by flow f; 0 otherwise
x_{ijf}	Equal 1 if flow f traverses arc $(i, j) \in L$; 0 otherwise
Integer variables:	
a_{mdp}	Counts the number of met spacial dependencies
b_{mp}	Counts the number of met temporal dependencies

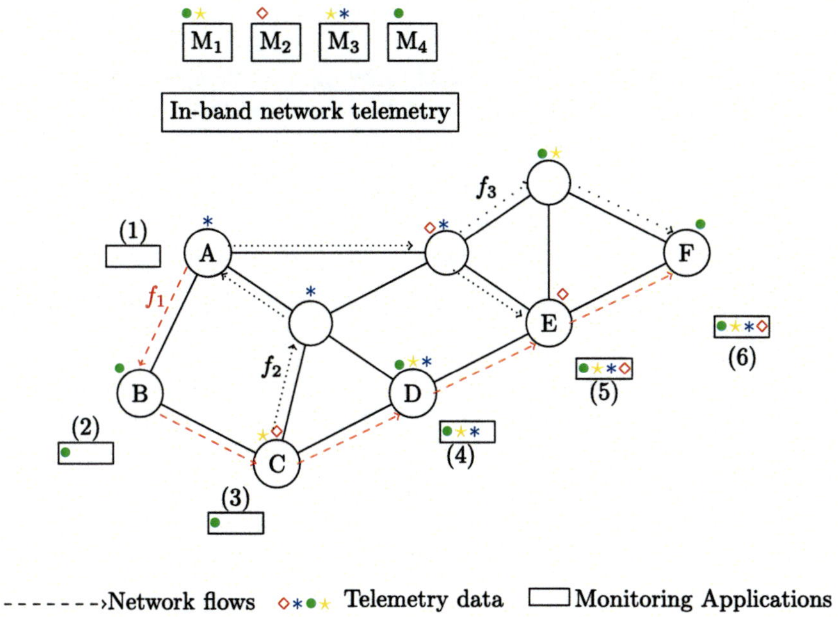

Fig. 1 A network infrastructure of eight forwarding devices (nodes) where each device embeds some telemetry items (red (◇), green (•), yellow (⋆), and blue (∗)). Three network flows are used to collect telemetry items from devices in their routes. The dashed arrows indicate the path of network flow f_1, the dotted arrows represent the route of network flow f_2, and the loosely dotted arrows represent the route of network flow f_3. Four monitoring applications (M_1, M_2, M_3, and M_4) are used and on top of each monitoring application, there's a subset of telemetry items representing the monitoring application's requirements

different subset of telemetry items (i.e., monitoring requirements) and three network flow packets (f_1, f_2, and f_3) with a limited capacity of four telemetry items at a time unit. In this example, the requirements of monitoring application M_4 are satisfied by routing the network flow packet f_1 from device A to device F and collecting the green (•) telemetry item in (2). To meet monitoring application M_3 requirements, the packet is sent to device C in (3) and the yellow (⋆) and blue (∗) items are collected from the same forwarding device in (4). However, the packet failed to meet the monitoring requirements of M_1, which demanded both the green (•) and yellow (⋆) items be collected from the same device. In (5), the packet is forwarded to device E, where it collects the red (◇) telemetry item and meets the requirements of monitoring application M_2.

Finally, the flow packet reaches F in (6), and all telemetry items are extracted and reported to an INT collector before the original packet is sent to its final destination. This solution satisfies the maximum number of monitoring application requirements and is optimal.

3.1.2 INTB Mathematical Model

We now present the in-band network telemetry basic problem as a mixed-integer linear programming problem. The mathematical model inputs include:

- A physical network infrastructure $G = (D, L)$.
- V_d, Set of telemetry items embedded by each device d.
- $\delta(v)$, Size of each telemetry item $v \in V_d$.
- A set of active network flows F and their capacity K_f.

The outputs for the INTB mathematical model are denoted by the 3-tuple $\mathcal{X} = \{Y, S, T\}$ where,

1. $Y = \{y_{dvf}, \quad \forall d \in D, v \in V, f \in F\}$ is the set of collected telemetry items.
2. $S = \{s_{mdP}, \quad \forall m \in M, d \in D, p \in P\}$ is the set of satisfied spatial dependencies.
3. $T = \{t_{mP}, \quad \forall m \in M, p \in P\}$ is the set of satisfied temporal dependencies.

Variables s_{mdp} indicate that a spatial dependency $p \in P \in R^s_m$ from monitoring application m is satisfied in network device d. Conversely, variable t_{mp} indicates that temporal dependency $p \in P \in R^t_m$ from monitoring application m is satisfied.

$$\text{Maximize} \sum_{m \in M} \sum_{d \in D} \sum_{p \in P \in R^s_m} s_{mdp} + \sum_{m \in M} \sum_{p \in P \in R^t_m} t_{mp} \tag{1}$$

Subject to:

$$\sum_{d \in \mathcal{P}(f)} \sum_{v \in V_d} \delta(v).y_{dvf} \leqslant K(f) \qquad \forall f \in F \tag{2}$$

$$\sum_{f \in F} y_{dvf} \leqslant 1 \qquad \forall d \in D, v \in V_d \tag{3}$$

$$a_{mdp} = \sum_{v \in P} \sum_{f \in F} y_{dvf} \qquad \forall m \in M, p \in P \in R^s_m, d \in D \tag{4}$$

$$s_{mdp} \leqslant \frac{a_{mdp}}{|P|} \qquad \forall m \in M, p \in P \in R^s_m, d \in D \tag{5}$$

$$b_{mp} = \sum_{d \in D} \sum_{v \in P} \sum_{f \in F} y_{dvf} \qquad \forall m \in M, p \in P \in R^t_m : \zeta(P) > \xi(P) \tag{6}$$

$$t_{mp} \leqslant \frac{b_{mp}}{|P|} \qquad \forall m \in M, p \in P \in R^t_m \tag{7}$$

$$y_{dvf} \in \{0, 1\} \qquad \forall d \in D, v \in V_d, f \in F \tag{8}$$

$$s_{mdp} \in \{0, 1\} \qquad \forall m \in M, d \in D, p \in P \in R^s_m \tag{9}$$

$$t_{mp} \in \{0, 1\} \qquad \forall m \in M, p \in P \in R^t_m \tag{10}$$

$$a_{mdp} \in \mathbb{N} \qquad \forall m \in M, d \in D, p \in P \in R^s_m \tag{11}$$

$$b_{mp} \in \mathbb{N} \qquad \forall m \in M, p \in P \in R^t_m \tag{12}$$

The objective function (1) is to maximize the number of monitoring requirements satisfied (i.e. spatial and temporal dependencies). Constraint (2) ensures that the capacities of the active network flows are respected. Constraint (3) provides that a single telemetry item $v \in V_d$ is collected at most by a single network flow $f \in F$ from device $d \in D$. Constraints (4) and (5) count the number of telemetry items v collected from device d and verify if spatial dependencies are satisfied. Similarly, constraints (6) and (7) count the items collected from device d and verify if temporal dependencies are satisfied. The constraints (8)–(12) define the variables' output domains.

3.2 In-Band Network Telemetry Extended Problem

This subsection presents the second variation of the in-band network telemetry problem. In this new variation the path of network flows is not given as input but computed by the model while solving the problem. For that, we have introduced a new decision variable x_{ijf} to track the route of the active network flows. We also introduced a set of new constraints to ensure the consistency of the mathematical model. These constraints include a constraint for cycle elimination, a constraint for flow conservation, and a constraint for limiting the number of hops in a route. First, we illustrate the problem and then give the corresponding mathematical model.

3.2.1 Problem Illustration

To illustrate the INTE problem, we consider the network infrastructure depicted in Fig. 2, where we have four monitoring applications (M_1, M_2, M_3, and M_4), each demanding a different subset of telemetry items (i.e. monitoring requirements). We also consider one network flow packet f_1 with source node A and destination node F. In this example network flow packet f_1 is routed following the shortest path in Fig. 2b (3 hops) and with a computed path of 4 hops in Fig. 2a. From Fig. 2a, the requirements of the four monitoring applications are satisfied by routing the network flow packet f_1 from A to F. The requirements of monitoring applications M_1 have been met by collecting the yellow (⋆) and green (•) items from the same forwarding device in (2). The requirements of monitoring application M_3 also are satisfied by collecting both the yellow (⋆) and blue (∗) items from the same forwarding device in (3). The monitoring requirements of monitoring applications M_2 and M_4 are met by collecting the red (◇) item from forwarding device E in (4) and the green (•) item from forwarding device F in (5). In Fig. 2b, the requirements of monitoring application M_1 are satisfied as both yellow (⋆) and green (•) items have been collected from the forwarding device G by the packet in (3). The requirements of monitoring application M_4 have been satisfied in (5) as the green (•) item is collected. The requirements of monitoring application M_2 have been satisfied in (2). However, the packet failed to meet the requirements of monitoring application M_3 at devices A

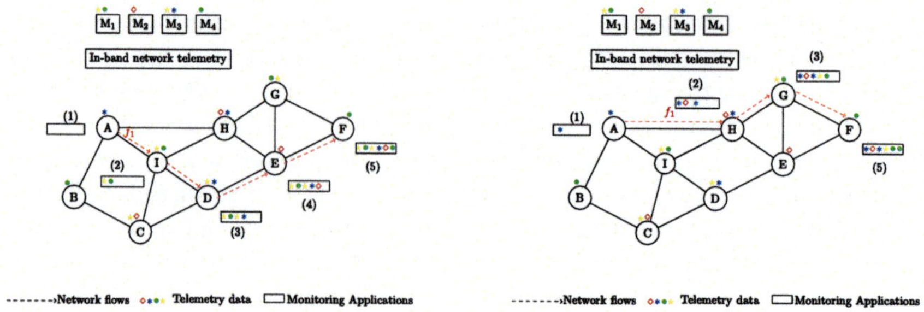

(a) Network flow packet routed from A to F following the computed path.

(b) Network flow packet routed from A to F following the shortest path.

Fig. 2 A network infrastructure of eight forwarding devices (nodes) where each device embeds some telemetry items (red (◇), green (•), yellow (⋆), and blue (∗)) and a network flow packet f_1 to collect telemetry items from forwarding devices in its routes. The network flow f_1 is routed from A to F following two routes. Four monitoring applications (M_1, M_2, M_3, and M_4) are used, and on top of each monitoring application, there's a subset of telemetry items representing the monitoring application's requirements

and H, which demanded both the blue (∗) and yellow (⋆) items be collected from the same device. The solution in Fig. 2a with the longer route is optimal. It satisfies the maximum number of monitoring application requirements since it has satisfied the monitoring requirements of all four monitoring applications. The solution in Fig. 2b with a shorter route failed in satisfying the monitoring requirements of monitoring application M_3 because blue (∗) and yellow (⋆) items were not collected from the same forwarding device.

3.2.2 INTE Mathematical Model

This section presents an extended problem of the INTB problem by introducing a new set of variables of flows that allows some active network flows paths to be computed. In this problem, some active network flow paths can be given while others are calculated within the model. In addition to the inputs of the INTB mathematical model, the INTE model requires:

- percentage of active network flows to be given with pre-defined routes GF. The model will compute the remaining routes of active network flows.
- max_route is an integer limiting the maximum number of hops in a route.
- two endpoints, S_f and E_f for each active network flow.

The outputs for the INTE model are denoted by the 4-tuple $\mathcal{X} = \{X, Y, S, T\}$ where $X = \{x_{ijf}, \quad \forall i \in D, j \in D, f \in F \setminus GF\}$ is the set of routes for non-defined routes of active network flows.

$$\text{Maximize} \sum_{m \in M} \sum_{d \in D} \sum_{p \in P \in R_m^s} s_{mdp} + \sum_{m \in M} \sum_{p \in P \in R_m^t} t_{mp} \tag{13}$$

Subject to:

$$\sum_{d \in \mathcal{P}(f)} \sum_{v \in V_d} \delta(v).y_{dvf} \leqslant K(f) \qquad \forall f \in GF \tag{14}$$

$$\sum_{j \in \mathcal{N}^+(i)} x_{ijf} = 1 \qquad i = S_f, \forall f \in F \setminus GF \tag{15}$$

$$\sum_{i \in \mathcal{N}^-(j)} x_{ijf} = 1 \qquad j = E_f, \forall f \in F \setminus GF \tag{16}$$

$$x_{ijf} + x_{jif} \leqslant 1 \qquad \forall (i,j) \in L, f \in F \setminus GF \tag{17}$$

$$u_{jf} \geqslant u_{if} + 1 - |D|(1 - x_{ijf}) \qquad \forall (i,j) \in L, f \in F \tag{18}$$

$$\sum_{j \in \mathcal{N}(i)} x_{jif} - \sum_{j \in \mathcal{N}(i)} x_{ijf} = 0 \qquad \forall i \in D \setminus \{S_f, E_f\}, f \in F \setminus GF \tag{19}$$

$$\sum_{(i,j) \in L} x_{ijf} \leqslant max\_route \qquad \forall f \in F \setminus GF \tag{20}$$

$$y_{dvf} \leqslant \sum_{j \in \mathcal{N}^+(d)} x_{djf} \qquad \forall d \in D, v \in V_d, f \in F \setminus GF \tag{21}$$

$$\sum_{f \in F} y_{dvf} \leqslant 1 \qquad \forall d \in D, v \in V_d \tag{22}$$

$$\sum_{d \in D} \sum_{v \in V_d} \delta(v).y_{dvf} \leqslant K(f) \qquad \forall f \in F \setminus GF \tag{23}$$

$$a_{mdp} = \sum_{v \in P} \sum_{f \in F} y_{dvf} \qquad \forall m \in M, p \in P \in R_m^s, d \in D \tag{24}$$

$$s_{mdp} \leqslant \frac{a_{mdp}}{|P|} \qquad \forall m \in M, p \in P \in R_m^s, d \in D \tag{25}$$

$$b_{mp} = \sum_{d \in D} \sum_{v \in P} \sum_{f \in F} y_{dvf} \qquad \forall m \in M, p \in P \in R_m^t : \zeta(P) > \xi(P) \tag{26}$$

$$t_{mp} \leqslant \frac{b_{mp}}{|P|} \qquad \forall m \in M, p \in P \in R_m^t \tag{27}$$

$$x_{ijf} \in \{0,1\} \qquad \forall (i,j) \in L, f \in F \tag{28}$$

$$y_{dvf} \in \{0,1\} \qquad \forall d \in D, v \in V_d, f \in F \tag{29}$$

$$s_{mdp} \in \{0,1\} \qquad \forall m \in M, d \in D, p \in P \in R_m^s \tag{30}$$

$$t_{mp} \in \{0,1\} \qquad \forall m \in M, p \in P \in R_m^t \tag{31}$$

$$a_{mdp} \in \mathbb{N} \qquad \forall m \in M, d \in D, p \in P \in R_m^s \tag{32}$$

$$b_{mp} \in \mathbb{N} \qquad \forall m \in M, p \in P \in R_m^t \tag{33}$$

The objective function (13) is to maximize the number of spatial and temporal dependencies. Constraint (14) ensures no active network flow with a pre-defined path exceeds capacity. Constraints (15) and (16) ensure that each flow will start at its origin and end at its destination. Constraint (19) is the flow conservation constraint (i.e. for every node d in the path of flow $f \in F$ different from the origin and destination of the flow, there is precisely one arc entering d and one arc leaving d). Constraints (18) and (17) are for sub-tours elimination. Constraint (17) eliminates sub-tours of size two, and constraint (18) is the Miller–Tucker–Zemlin (MTZ) [22] constraint and it eliminates longer sub-tours. Constraint (20) limits the maximum route a flow can have; the *max_route* is an upper limit to the total number of hops an active network flow can achieve. Constraint (21) ensures that a telemetry item v is collected by an active network flow only if the device d that collects it belongs to the active network flow f path. Constraint (22) ensures that a single telemetry item is collected at most by a one network flow $f \in F$ in a given device d. Constraint (23) ensures that no flow f exceeds its capacity. Constraints (24) and (25) aim to count whether or not a spatial dependency is met. Given a spatial dependency, constraint (24) counts the number of telemetry items collected from a device d. Constraint (25) verifies whether or not the dependency is met (i.e., checking if all items were collected). Similarly, constraints (26) and (27) count and verify the number of satisfied temporal dependencies. Last, the constraints (28)–(33) define the variables' output domains.

Both INTB and INTE problems are NP-Hard problems. INTB was shown to be NP-Hard in [1] since it is a generalization of the well-known NP-Hard Bin-Packing problem [23]. INTE problem is a generalization of [1].

4 Computational Experiments

This section presents the computational experiments with the mathematical models introduced in Sects. 3.1.2 and 3.2.2. We first describe the experimental setup and the dataset used, and then evaluate three sets of experiments. First, we assess the network-wide visibility regarding network telemetry coverage (i.e., the percentage of telemetry items collected) concerning an increasing number of active network flows with up to 40% of the active network flows with a non-pre-defined route. Then, we evaluate the monitoring applications' performance (i.e. in terms of the number of satisfied monitoring application requirements), considering an increasing number of monitoring applications. Last, we evaluate the impact of the length of the monitoring application on the network monitoring applications' performance.

We have cited various articles in this paper that explore and maximize the potential of in-band network telemetry. Some articles take a non-mathematical approach to implement and extract the most value from this network telemetry. Other articles employ mathematical methods, such as modeling the problem as a mixed integer linear programming problem. Given that our paper introduces a new mixed integer linear programming model, we compare our results to the latest article proposing a mathematical model for resolving the in-band network telemetry problem.

4.1 Setup

All the experiments presented in this paper were conducted on a machine with forty-seven Intel(R) Xeon(R) E5-2697 processors running at 2.70 GHz and with 64GB of RAM, using Ubuntu GNU/Linux Server 18.04.5 LTS x86_64 operating system. Each mathematical model was coded and implemented in Python 3.7 using the Docplex package and solved by Cplex 12.10 under an academic license and using the default parameter values[1] The maximum computation time limit to find a solution was 14400 s (4h).

4.2 Dataset

The dataset used in the experiments to evaluate the in-band network telemetry problem comprises a physical network infrastructure generated following the Barabasi-Albert model [24]. We consider a physical network infrastructure of 50 nodes (forwarding devices) and an average of 200 physical links. On top of each network infrastructure instance, we considered a set F of network flows and varied the size of $|F|$ from 50 to 200. Each network flow has a limited capacity K_f ranging from 2 to 20 bytes. Forwarding devices have 8 possible telemetry items of size $S(v)$ varying from 2 to 10 bytes to export [6].

We generated multiple instances of network infrastructure and named each instance DxVyMz. The letters D, V and M respectively refer to the forwarding device, telemetry item and monitoring application while the letters x, y and z respectively indicate the number of forwarding devices, the number of telemetry items embedded by each forwarding device and the number of monitoring applications in the instance. For example, the sample $D50T8M2$ corresponds to a model of network infrastructure of 50 forwarding devices, each forwarding device embedding 8 telemetry items and 2 monitoring applications are used.

The forwarding devices, telemetry items, monitoring applications and network flows are used to solve the INTB and INTE optimization models. First, we solve the INTB optimization model where all network flows routing is pre-defined and follows the shortest path. Next, we solve the INTE optimization model, where 80% of the network flow route is pre-defined and follows the shortest path (INTE-80%). The optimization model computes the routes for the remaining 20% of network flows. Finally, we solve the INTE optimization model where only 60% of network flow routes are pre-defined and follow the shortest path (INTE-60%). The optimization model computes the remaining 40% of network flow routes. The maximum number of hops (number of links) for network flow routes that are not pre-defined is set to the length of the longest route in the shortest path of the pre-defined routes (4 hops for the experiments presented in this section).

[1] Reproducible material available on https://github.com/thiernoimsp/IRINT2023.

4.3 Results

In this subsection, we present the results of our experiments using the instances described in the previous subsection. The results obtained by different network infrastructures are summarized in Table 2. Each of the models INTB, INTE-80% and INTE-60% was run with the same instances with increasing network flows, $|F|$. The column "Instance" in Table 2 corresponds to the name of the used instance, the column "Items" corresponds to the amount of collected telemetry items, the column "Obj" corresponds to the best value of the objective function found, the column "UB" corresponds to the best objective bound found by the solver, the column "Gap" corresponds to the relative gap computed as follows: $Gap = frac|Obj - UB||Obj|$ and the column "CPU(s)" corresponds to the required time in seconds for obtaining the solution.

We base the evaluation of the results obtained from the different experiments on three aspects of the network monitoring: i) the network wide-visibility, ii) satisfied monitoring application requirements and iii) the impact of the monitoring application requirements length $|R_m|$ on the network coverage and the monitoring application's performance.

To evaluate the network wide-visibility for INTB and INTE models, consider the number of collected telemetry items for increasing network flows represented in the column "Items" from Table 2. We observe that for all the instances, the number of collected telemetry items increases with the number of network flows. We also observe that across all the instances, the minimal average values of collected telemetry items were obtained with the INTB model, and the maximal average values of collected telemetry items were obtained by the INTE-60% model. Notice that although the routes of network flow in INTE-80% are slightly different from the routing of network flows in INTB, INTE-80% gave better network visibility. This result shows that network flow routing impacts the network wide-visibility. Thus, the INTE model wisely computes the network flow routes that are not pre-defined and provides better visibility than the routes following the shortest path.

To evaluate the network monitoring application's performance, we consider Fig. 3 which illustrates the number of satisfied monitoring application requirements considering an increasing number of monitoring applications. Figure 3 represents the average value of satisfied monitoring application requirements from each instance from Table 2. The Figure shows that the average value of satisfied monitoring application requirements increases as the number of monitoring applications increases. We also notice that the performance of the monitoring applications has grown with the models INTE-80% and INTE-60%, which indicates the routing of network flows impacts both the network coverage and the monitoring application's performance.

To evaluate the impact of the length of the monitoring application requirements on the network coverage and the monitoring application's performance, let's consider the results in Table 3. This table summarizes the overall average value from different network infrastructure instances DxVyMz of 50 forwarding devices each embedding 8 telemetry items. The number of monitoring applications used varies from 2 to 10,

Table 2 Comparative Analysis of the optimization models considering different network infrastructures with an increasing number of monitoring applications and an increasing number of network flows

Instance	$\|F\|$	INTB					INTE-80%					INTE-60%				
		Items	Obj	UB	Gap(%)	CPU(s)	Items	Obj	UB	Gap(%)	CPU(s)	Items	Obj	UB	Gap(%)	CPU(s)
D50V8M2	50	100	590*	590	0.00	3.79	113	696*	696	0.00	7960.91	123	740	752	1.64	14400.78
	100	190	1164	1165	0.11	14402.86	233	1378	1398	1.51	14401.77	242	1468	1530	4.22	14400.70
	150	243	1448*	1448	0.00	73.99	250	1530*	1530	0.00	92.53	250	1530*	1530	0.00	794.44
	200	246	1474*	1474	0.00	120.79	250	1530*	1530	0.00	30.40	250	1530*	1530	0.00	365.16
	Mean	194	1169	1169	0.03	3650.36	211	1283	1288	0.38	5621.40	216	1317	1335	1.46	7490.27
D50V8M4	50	101	770*	770	0.00	10.37	121	864*	864	0.00	8051.44	141	936	959	2.47	14406.27
	100	188	1366*	1366	0.00	15.77	233	1579	1602	1.50	14400.93	259	1767	1836	3.91	14402.26
	150	282	1914*	1914	0.00	15.04	318	2143	2187	2.08	14400.93	343	2272	2365	4.10	14401.57
	200	346	2278*	2278	0.00	133.03	385	2418	2428	0.43	14405.17	400	2448*	2448	0.00	14401.86
	Mean	229	1582	1582	0.00	43.55	264	1751	1770	1.00	12814.62	285	1855	1902	2.62	14402.99
D50V8M6	50	97	853*	853	0.00	211.07	126	977	981	0.44	14401.08	144	1057	1084	2.59	14400.99
	100	196	1515*	1515	0.00	1055.00	231	1750	1779	1.68	14410.25	260	1968	2015	2.41	14402.63
	150	274	2034	2035	0.07	14407.46	316	2334	2346	0.56	14401.90	336	2402	2541	5.81	14401.28
	200	337	2458*	2458	0.00	1657.51	383	2686	2712	1.00	14405.45	399	2746	2754	0.29	14401.92
	Mean	226	1715	1715	0.02	4332.76	264	1936	1954	0.92	14404.67	284	2043	2098	2.77	14401.70
D50V8M8	50	93	806*	806	0.00	366.13	110	957	961	0.49	14430.66	130	1064	1092	2.65	14401.40
	100	177	1530	1531	0.07	14401.01	214	1753	1775	1.29	14401.07	241	1994	2042	2.42	14401.11
	150	260	2222*	2222	0.00	697.40	309	2582	2638	2.21	14400.62	344	2900	2977	2.69	14403.65
	200	341	2867*	2867	0.00	703.47	378	3074	3105	1.01	14401.19	391	3112	3162	1.61	14402.19
	Mean	217	1856	1856	0.02	4042.00	252	2091	2119	1.25	14408.38	276	2267	2318	2.34	14402.09
D50V8M10	50	95	882	882	0.00	590.88	113	1075	1104	2.71	14406.39	131	1227	1274	3.90	14400.55
	100	179	1783	1788	0.34	14518.89	215	2038	2083	2.24	14401.21	248	2227	2380	6.88	14401.31
	150	260	2583*	2583	0.00	2340.35	305	2989	3050	2.05	14401.03	342	3242	3472	7.13	14401.28
	200	318	3268	3269	0.04	14491.11	380	3614	3693	2.19	14401.27	400	3774*	3774	0.00	14402.06
	Mean	213	2129	2130	0.10	7985.31	253	2429	2482	2.30	14402.47	280	2617	2725	4.48	14401.30
	Overall mean	216	1690	1690	0.03	4010.80	249	1898	1923	1.17	12330.31	268	2020	2076	2.73	13019.67

*Optimal solution found

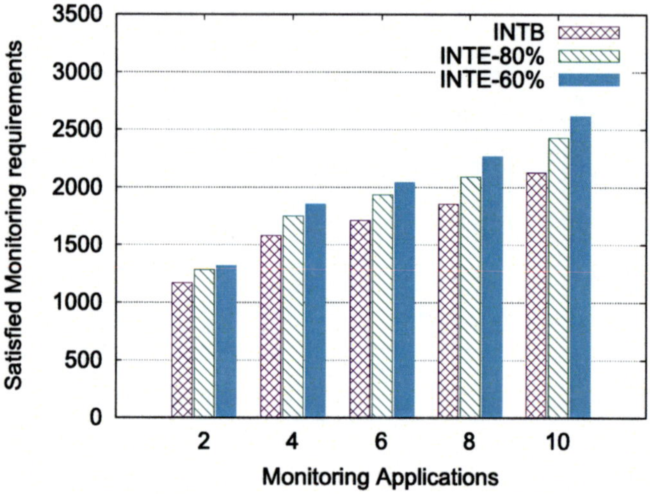

Fig. 3 Satisfied monitoring requirements for an increasing number of monitoring applications

Table 3 Overall average performance of monitoring applications and network coverage for monitoring applications of different sizes

	Objective function			Collected items		
	$\|R_m\| = 2$	$\|R_m\| = 3$	$\|R_m\| = 4$	$\|R_m\| = 2$	$\|R_m\| = 3$	$\|R_m\| = 4$
INTB	524	666	1690	220	234	216
INTE-80%	580	743	1898	243	264	249
INTE-60%	602	783	2020	253	280	268

and the length of the monitoring applications ranges from 2 to 4. The results in Table 2 correspond to an $|R_m| = 4$, and the values of the third column of the objective function in Table 3 are the same as the values of the last row of Table 2 (Overall average row).

Table 3 shows that the number of satisfied monitoring requirements rapidly increases as the length of monitoring applications increase. We observe that the average value of the objective function for both models INTB and INTE triple as the length of the monitoring applications doubles. For example, notice that the values of the third column of the objective function ($|R_m| = 4$) for INTB, INTE-80% and INTE-60% are respectively 3.23, 3.27 and 3.36 times the values of the first column of the "Objective Function" ($|R_m| = 2$). We also observe that both models, INTE-80% and INTE-60%, satisfied more monitoring application requirements than INTB, which shows the impact routing network flows on the performance of the monitoring applications.

The results also show that both models INTB and INTE prioritize satisfying monitoring requirements over the network coverage regarding the number of collected telemetry items. Notice that the average values of the number of collected telemetry items from the third column of the "Collected items" ($|R_m| = 4$) are lesser than the average values of the third column ($|R_m| = 3$). However, the table also shows that the number of satisfied monitoring application requirements of length $|R_m| = 4$ is superior to that of monitoring applications of length $|R_m| = 3$.

5 Conclusion

This paper presented two variations of the in-band network telemetry problem. The first variation of the problem, INTB, aims to maximize the number of monitoring requirements satisfied considering pre-defined network flow routes following the shortest path. The second variation, INTE, seeks to maximize the number of monitoring requirements satisfied considering an unknown network flows routing with the desired number of maximum hops. We tested the INTE problem with two different sets of instances. Initially, 80% of flow routes were pre-defined, while the model determined the remaining 20% (INTE-80%). In the second, 60% of routes were pre-defined, with the model computing the other 40% (INTE-60%). Our results show that the INTE model can effectively collect telemetry items and maximize the number of satisfied monitoring application requirements better than the INTB model in the worst case by up to 15%. However, this improvement comes at the expense of a higher computational time. In future work, we intend to implement a heuristicapproximation approach for the INTE problem to reduce computational time and use actual network infrastructure.

Acknowledgements The authors are grateful for Olinto Araujo for providing the access to the super calculator for carrying on the experimental results. They thank the African Center of Excellence in Mathematical Science, Informatics, and Applications (CEA-SMIA) and the German Academic Exchange Service (DAAD) for their support. Theirno thanks Henrique Becker, Gabriel Langeloh, Alberto Kummer and the Algorithms and Optimization group team of the Institute of Informatics Federal University of Rio Grande do Sul (INF-UFRGS) for research assistance. Thierno also thanks Sanni Mustapha for research guidance.

References

1. Hohemberger, R., Castro, A.G., Vogt, F.G., Mansilha, R.B., Lorenzon, A.F., Rossi, F.D., Luizelli, M.C.: Orchestrating in-band data plane telemetry with machine learning. IEEE Commun. Lett. **23**(12), 2247–2251 (2019)
2. Bosshart, P., Daly, D., Gibb, G., Izzard, M., McKeown, N., Rexford, J., Schlesinger, C., Talayco, D., Vahdat, A., Varghese, G., et al.: P4: Programming protocol-independent packet processors. ACM SIGCOMM Comput. Commun. Rev. **44**(3), 87–95 (2014)
3. Liu, Z., Bi, J., Zhou, Y., Wang, Y., Lin, Y.: Netvision: towards network telemetry as a service. In: 2018 IEEE 26th International Conference on Network Protocols (ICNP), pp. 247–248 (2018)
4. Hohemberger, R., Lorenzon, A.F., Rossi, F.D., Luizelli, M.C.: A heuristic approach for large-scale orchestration of the in-band data plane telemetry problem. In: International Conference on Advanced Information Networking and Applications, pp. 381–392 (2020)
5. Marques, J.A., Luizelli, M.C., da Costa Filho, R.I.T., Gaspary, L.P.: An optimization-based approach for efficient network monitoring using in-band network telemetry. J. Internet Serv. Appl. **10**(1), 12 (2019)
6. Pan, T., Song, E., Bian, Z., Lin, X., Peng, X., Zhang, J., Huang, T., Liu, B., Liu, Y.: Int-path: Towards optimal path planning for in-band network-wide telemetry. In: IEEE INFOCOM 2019-IEEE Conference on Computer Communications, pp. 487–495 (2019)

7. Dallanora, L.M., Castro, A.G., da Costa Filho, R.I., Rossi, F.D., Lorenzon, A.F., Luizelli, M.C.: Dypro: Dynamic probing planning for in-band network telemetry. In: 2022 IEEE Symposium on Computers and Communications (ISCC), pp. 1–6 (2022)
8. Bhamare, D., Kassler, A., Vestin, J., Khoshkholghi, M.A., Taheri, J.: Intopt: In-band network telemetry optimization for nfv service chain monitoring. In: ICC 2019-2019 IEEE International Conference on Communications (ICC), pp. 1–7 (2019)
9. Bosshart, P., Gibb, G., Kim, H.S., Varghese, G., McKeown, N., Izzard, M., Mujica, F., Horowitz, M.: Forwarding metamorphosis: fast programmable match-action processing in hardware for sdn. ACM SIGCOMM Comput. Commun. Rev. **43**(4), 99–110 (2013)
10. Jeyakumar, V., Alizadeh, M., Geng, Y., Kim, C., Mazières, D.: Millions of little minions: using packets for low latency network programming and visibility. ACM SIGCOMM Comput. Commun. Rev. **44**(4), 3–14 (2014)
11. Kim, C., Sivaraman, A., Katta, N., Bas, A., Dixit, A., Wobker, L.J.: In-band network telemetry via programmable dataplanes. In: ACM SIGCOMM, vol. 15 (2015)
12. Zhu, Y., Kang, N., Cao, J., Greenberg, A., Lu, G., Mahajan, R., Maltz, D., Yuan, L., Zhang, M., Zhao, B.Y., et al.: Packet-level telemetry in large datacenter networks. In: Proceedings of the 2015 ACM Conference on Special Interest Group on Data Communication, pp. 479–491 (2015)
13. Tammana, P., Agarwal, R., Lee, M.: Simplifying datacenter network debugging with pathdump. In: 12th USENIX Symposium on Operating Systems Design and Implementation (OSDI 16), pp. 233–248 (2016)
14. Thomas, J., Laupkhov, P.: Tracking packets' paths and latency via int (in-band network telemetry). In: 3rd P4 Workshop (2016)
15. Yu, M.: Network telemetry: towards a top-down approach. ACM SIGCOMM Comput. Commun. Rev. **49**(1), 11–17 (2019)
16. Pan, T., Song, E., Jia, C., Cao, W., Huang, T., Liu, B.: Lightweight network-wide telemetry without explicitly using probe packets. In: IEEE INFOCOM 2020-IEEE Conference on Computer Communications Workshops (INFOCOM WKSHPS), pp. 1354–1355 (2020)
17. Pan, T., Lin, X., Song, H., Song, E., Bian, Z., Li, H., Zhang, J., Li, F., Huang, T., Jia, C., et al.: Int-probe: Lightweight in-band network-wide telemetry with stationary probes. In: 2021 IEEE 41st International Conference on Distributed Computing Systems (ICDCS), pp. 898–909 (2021)
18. Castro, A.G., Lopes, V.H.S., Vogt, F.G., Rossi, F.D., Lorenzon, A.F., Luizelli, M.C.: Patcher: Towards fault - tolerant probing planning for in-band network telemetry. In: 2020 IEEE Latin-American Conference on Communications (LAT-INCOM), pp. 1–6 (2020). https://doi.org/10.1109/LATINCOM50620.2020.9282348
19. Yuan, Q., Li, F., Pan, T., Xu, Y., Wang, X.: Int-react: An o (e) path planner for resilient network-wide telemetry over megascale networks. In: 2022 IEEE 30th International Conference on Network Protocols (ICNP), pp. 1–11 (2022)
20. Pan, J., Wei, Z., Chen, W., Tian, Y.: Orchestrating probabilistic in-band network telemetry for network monitoring. In: 2021 7th International Conference on Computer and Communications (ICCC), pp. 441–446 (2021). https://doi.org/10.1109/ICCC54389.2021.9674307
21. Tan, L., Su, W., Zhang, W., Lv, J., Zhang, Z., Miao, J., Liu, X., Li, N.: In-band network telemetry: a survey. Comput. Netw. **186**, 107,763 (2021)
22. Miller, C., Tucker, A., Zemlin, R.: Integer programming formulations and traveling salesman problems. acm **7**, 326–329 (1960). Google Scholar Google Scholar Digital Library Digital Library
23. Hartmanis, J.: Computers and intractability: a guide to the theory of np-completeness (michael r. garey and david s. johnson). Siam Rev. **24**(1), 90 (1982)
24. Albert, R., Barabási, A.L.: Topology of evolving networks: local events and universality. Phys. Rev. Lett. **85**(24), 5234 (2000)

House Price Prediction Using XG-Boost Grid Search and Cross-Validation Methods

Tianlei Zhu, Yingke Yang, Shu Bao, and Hassan Raza

Abstract The primary factors influencing the variation in housing prices in different locations are the characteristics of the housing and the conditions of the area. This study will utilize the popular machine learning technique, XGBoost, along with grid-search and cross-validation techniques in deep learning. The chosen data set is appropriate as it provides adequate representation of both the range of house prices and the diversity.

Keywords House pricing · XG-boost · Boston · USA

1 Introduction

Economic development depends upon many factors, one of which is house prices-a crucial factor in determining whether people from urban areas can live better lives [1–3].

House pricing is essential in determining the market's overall price level, which further signifies the social economy's level of mass consumption and prosperity. As an essential element of the price system, house price, to some extent, specifies the overall price level of the market, which further signifies the level of mass consumption and prosperity of the social economy. As real estate is a basic requirement of production and a significant part of labor reproduction costs, companies include house price in the production costs to ensure the allocation of space for the production can

T. Zhu · Y. Yang · S. Bao · H. Raza (✉)
School of Mathematical Sciences, College of Science and Technology, Wenzhou-Kean University, Wenzhou 325060, China
e-mail: hraza@kean.edu

T. Zhu
e-mail: zhut@kean.edu

Y. Yang
e-mail: yangyin@kean.edu

S. Bao
e-mail: baos@kean.edu

D. Vlachos (ed.), *Mathematical Modeling in Physical Sciences*, Springer Proceedings in Mathematics & Statistics 446, https://doi.org/10.1007/978-3-031-52965-8_47

be compensated and generate revenue, resulting in a condition where real estate is a component of all commodity prices. From this perspective, the reasonability of house prices not only determines market prices' reliability but also regulates consumption prices. Many residents are anxious about house prices as one can never evaluate the price of houses according to the area or offices accessible [4]. Consequently, verifying and ranking the factors participating in house price calculation in order of relevance is essential.

Many aspects influence house prices. The factors(internal) affecting house prices incorporate the type of houses, age(what year they are built), community, and facilities, including nearest shops, parks, property services, etc. [5]. Other factors(external) may include access to public transport, distance to schools, supermarkets, safety, etc. There can be numerous factors that can influence people's willingness to spend money to purchase houses; a few of them can be the community environment, how fast the development can occur, greenness, playgrounds, a sense of security [6–9]. In this paper, (XGBoost) is employed to recognize the spatial variations in house prices with better accuracy. The reason for choosing (XGBoost) is the superiority in selecting the influential factors that can significantly influence the accurate outcome of the model. The model's probability of overfitting decreases with a better understanding of variable significance [10].

However, the essay uses data from an online database of Boston housing prices, assuming idealistic conditions to discuss the problem mathematically by neglecting the possible changes in social indicators. We focus on micro-level factors rather than macroscopic factors.

Ranking the factors affecting house prices is of significant importance for the government to finance the allocations on real estate and land agents to adjust supplies. This paper uses XGBoost and Grid Search to analyze factors affecting house prices.

2 Literature Review

Instead of evaluating the influence factors of house price in the specialty, previous researchers prefer to analyze the prediction models of the price as prediction models should be able to handle nonlinear problems [11]. Encountering the problem of choosing relevant arguments, they were inclined to look to the independent variables investigated by predecessors. For instance, a study [12] used support vector machine learning to predict house prices based on variables such as overall quality, living area, year built, and garage area. As house price is affected by plentiful attributes, it is tough to distinguish to what extent the independent variables are worth analyzing. Nevertheless, research investigating elements influencing the price of housing is rare and limited. In the commonly used hedonic model [13] for predicting house prices, common variables include type of house, age of property, and area of residence [14]. Preliminary research conducted in the early stages set sights on location, neighborhood, and architecture, and later studies incorporated additional factors such as accessibility and land-use planning [15]. Other studies analyzed some relevant considerations like race, pollution, distance to CBD, transportation and retail on housing prices [16]. In a study of the factors that influence house prices in Indonesia [17] researchers collected data through a questionnaire survey and used factor analysis, correlation analysis, and Cronbach's alpha test for processing. As the subjective factors are inevitable in their data source, for instance, the preference and class of the people under investigation, readers cannot ensure a valid relationship between possible influence components and house price fluctuation. The data utilized in this paper is presumed to contribute to high accuracy in two aspects. First, the geographic density of our samples is high, which can provide stable circumstances for house prices and a unified policy environment. Secondly, the data being of a single property type can simplify the codification or consolidation process, ultimately improving the investigation's accuracy [18]. Previous studies used hedonic price and quantile regression models to analyze micro-level aspects' average and distributional effects on housing prices and rents [15]. However, the hedonic model has limitations like multicollinearity and not fully accounting for market supply and demand's impact on real estate prices, reducing reliability. In our study, every parameter is estimated and optimized through cross-validation, which provides a highly reliable outcome.

3 Methodology

XGBoost is an open-source software library that accompanies an optimized distributed gradient lifting machine learning algorithm with the framework of Gradient Boosting. It goes through cycles that build new models and incorporates them into an ensemble model again and again, and the goal is to test the error between the old model and the new one, predicting the errors. Based on this, this model can accurately make some predictions than the Random Forest model [19].

4 Experiment

4.1 Selection of Sample Data

This paper uses Kaggle datasets to predict house prices, with 80 attributes and almost 3000 records. The data is then split into two parts: training and test sets. The missing values are removed through data processing. After creating the data frames, the following two figure shows the correlation between features and targets on the heatmap (Fig. 1).

Algorithm 1 Sample Algorithm

Input: Training set
$\{x_i, y_i\}$, where $\{i = 1, \ldots, N\}$
1. Initialize model with a constant value:
$\widehat{f_0}(x) - argmin_\alpha \sum_{i=1}^{N} L(y_I, \alpha)$
2. For $m = 1$ to M
1) Compute the "gradients" and "hessians":
$\widehat{g}(x_i) = [\frac{\alpha L(y_i, f(x))}{\alpha f(x_i)}]_{f(x)=\widehat{f_{m-1}}(x)}$
$\widehat{h}(x_i) = [\frac{\alpha^2 L(y_i, f(x))}{\alpha f(x_i)^2}]_{f(x)=\widehat{f_{m-1}}(x)}$
2) Fit a base leaner using the training set $x_i - \frac{\widehat{g_m}(x_i)}{\widehat{h_m}(x_i)}$ by solving the optimization problem below:
$\widehat{\phi'_m} = argmin_{\phi'=\varphi} \sum_{i=1}^{N} \frac{1}{2}\widehat{h_m}(x_i)[-\frac{\widehat{g_m}(x_i)}{\widehat{h_m}(x_i)} - \phi'(x_i)]^2$
$\widehat{f_m}(x) = \alpha\widehat{\phi}'(x)$
Update the Model:
$\widehat{f_m}(x) = \widehat{f_{m-1}}(x) + \widehat{f_m}(x)$
Output:
$\widehat{f}(x) = \widehat{f_m}(x) = \sum_{m=0}^{N} \widehat{f_m}(x)$

Through the thermal map, we observed the relationship between each factor, and the depth of the color also reflects the coefficient of the degree of relationship. We then visualized the relationship between the selling price and the other factors, as shown in the figure below (Figs. 2 and 3).

4.2 *Experimental Procedure*

After data processing, most of the columns with only one classification were removed, and some feature engineering was carried out. First, we created some new time data features and the "ExterQual" list of observation objects to get the following figure (Fig. 4).

We mainly create new Boolean features and use mathematical transformations; count, group transforms, feature interactions, and other ways to complete the feature engineering. Now, we will create a function to automate plotting three types of plots for a single numerical variable (Fig. 5).

In feature selection, two parameters, MI scores and Lin Correlation, measure the relationship between each factor and the housing price to know the ordering of factors affecting the housing price for model training or fitting. Deep learning

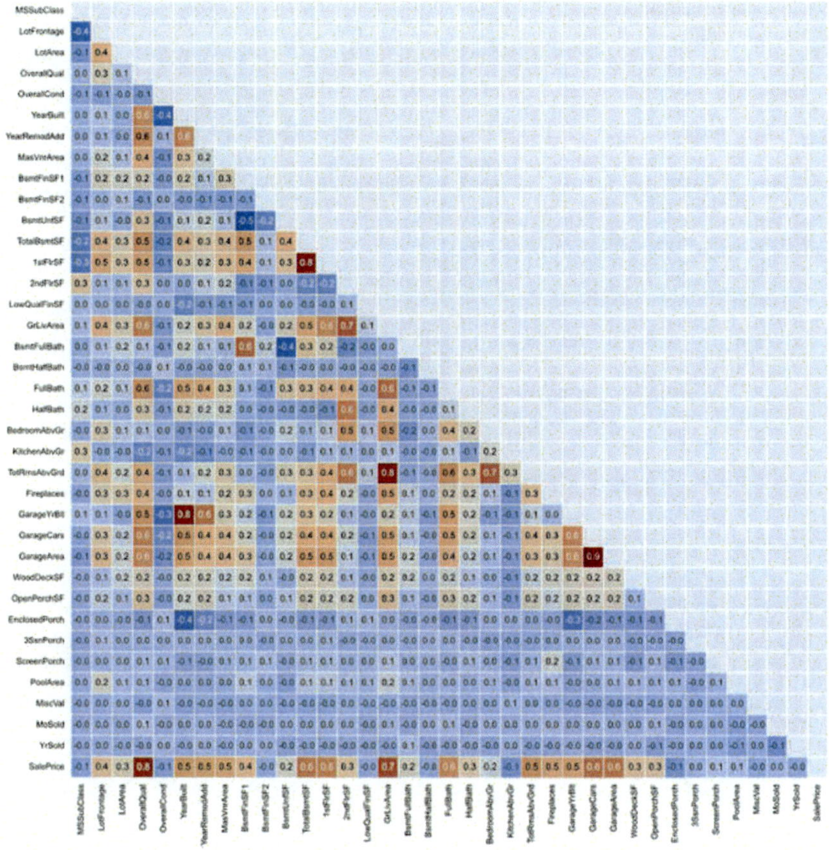

Fig. 1 Correlation between features and target on heatmap

algorithms can improve house price prediction and reduce test error rate [19]. The XGBoost algorithm increases the accuracy of estate choices and reduces investment risks for customers, satisfying their needs. It is also incorporating Grid Search and Cross-Validation for the experiment.

Fig. 2 Data visualization (categorical data)

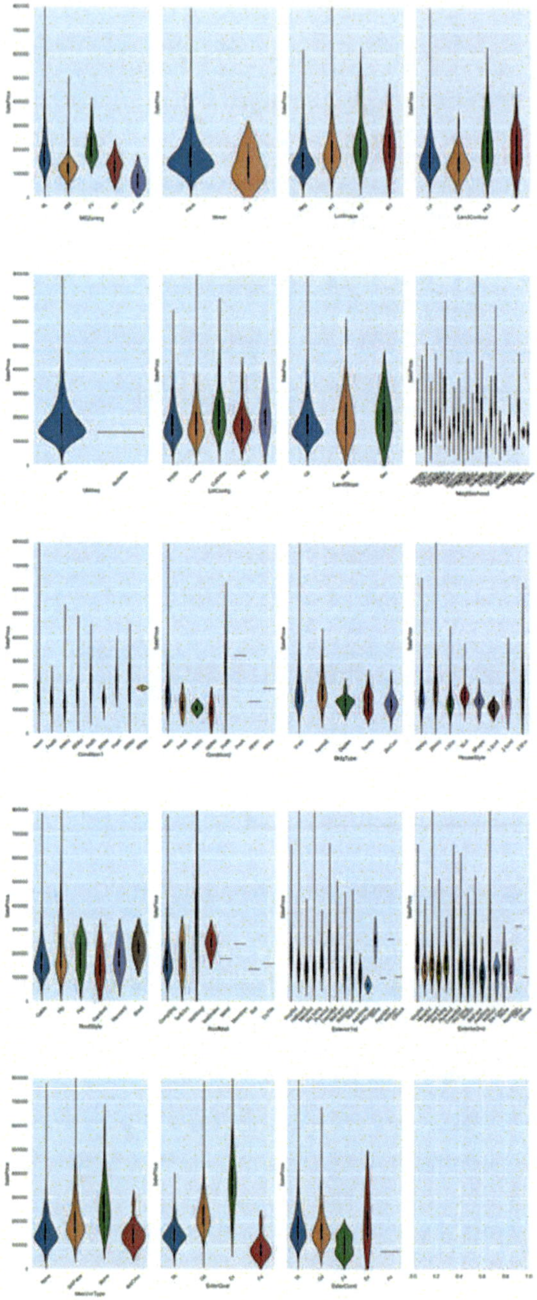

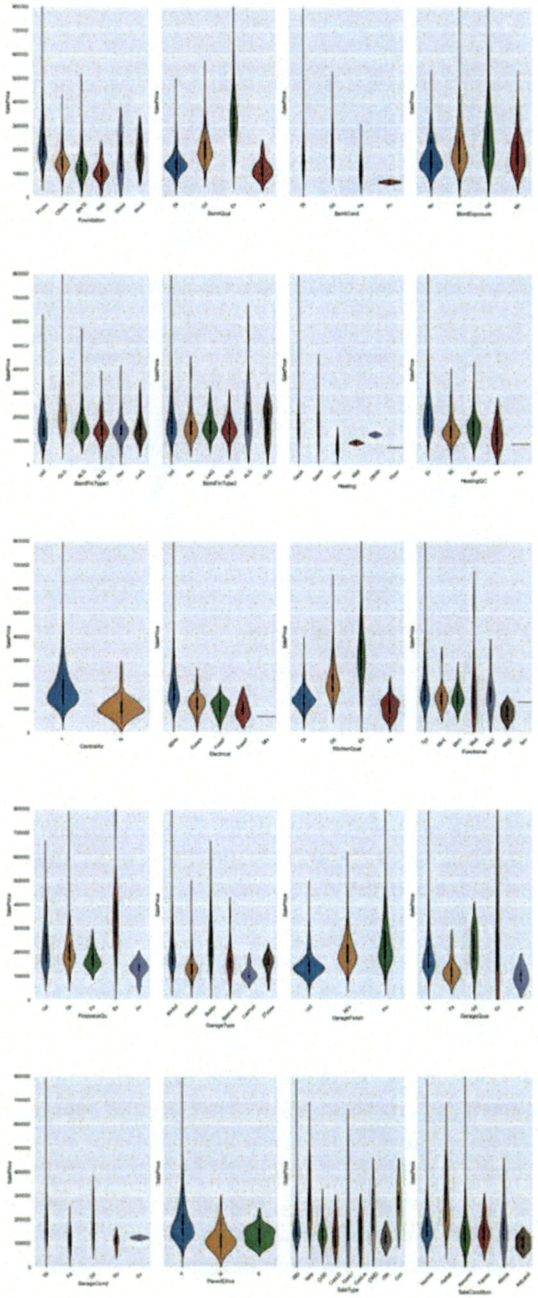

Fig. 3 Data visualization (categorical data)

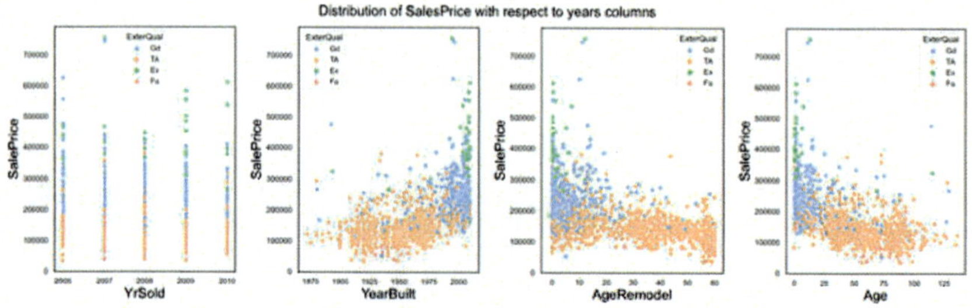

Fig. 4 Distribution of sales price with respect to years columns

Fig. 5 Distribution of sales price

5 Results and Discussion

In the process of debugging and training the model, we found that RMSLE on training data is 0.0719, RMSLE on validation data is 0.1226, MAE on validation data is 0.12262027279136213, best XGBoost on whole trained data's MAE: 197572.41698585558, so these errors are enough (Fig. 6).

6 Conclusion

In this study, we employed the XGBoost algorithm in machine learning. We used grid-search and cross-validation to analyze the impact of influencing factors on housing prices and identified the most significant ones. After all, understanding the factors that impact housing prices is crucial for informing urban planning and housing policies [20]. Furthermore, using the tuning method can boost the accuracy and reduce the error of XGBoost. However, it is also easy to find that using these two methods is very time-consuming, resulting in many computational costs, especially for processing

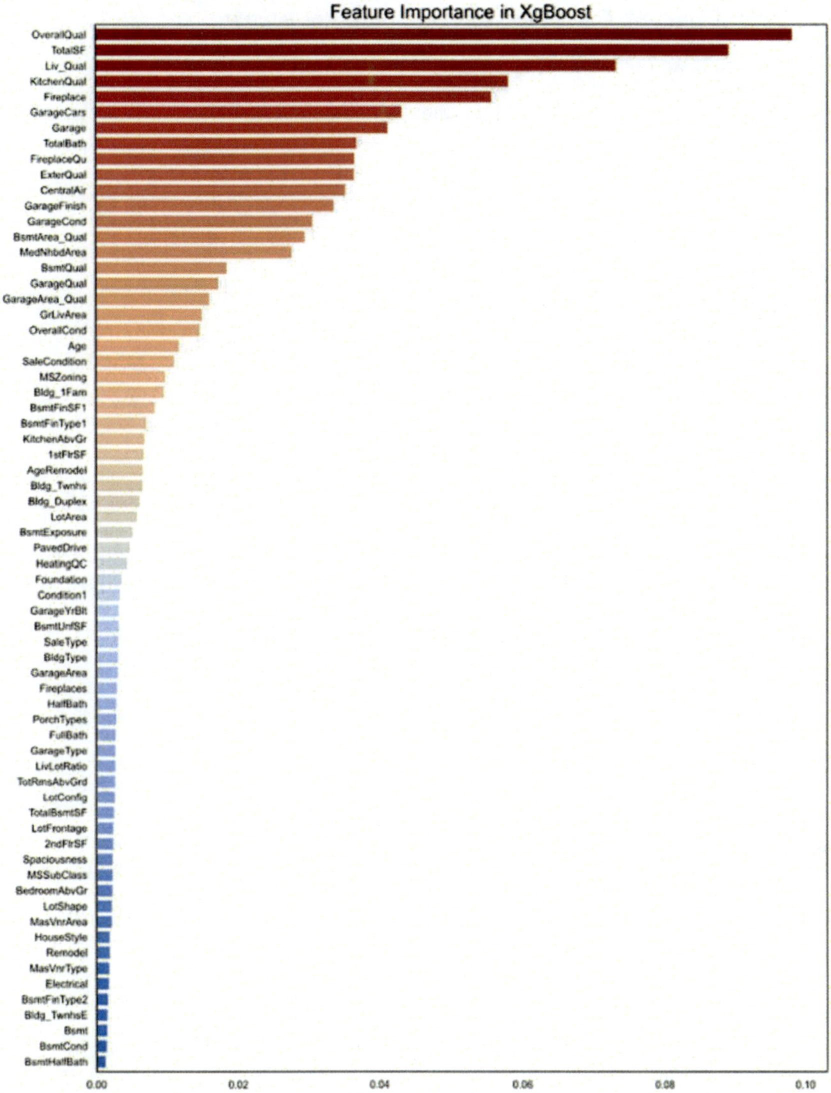

Fig. 6 Feature importance in XgBoost

large-scale and high-dimensional data. In the future, we will combine deep learning and other models to optimize this experiment and reduce the time complexity.

Funding The research is supported by the Student partnering with faculty Research programme grant number WKUSPF2023038.

Availability of Data and Material Data sharing is not applicable to this article as no data set were generated or analyzed during the current study.

Conflicts of Interest The authors have no conflict of interest.

Code There is no code which can shared at this time.

Authors' Contributions All authors contributed equally.

References

1. Yao, Y., Zhang, J., Hong, Y., Liang, H., He, J.: Mapping fine-scale urban housing prices by fusing remotely sensed imagery and social media data. Trans. GIS **22**(2), 561–81 (2018)
2. Wen, H., Goodman, A.C.: Relationship between urban land price and housing price: evidence from 21 provincial capitals in China. Habitat Int. **1**(40), 9–17 (2013)
3. Wu, C., Ren, F., Hu, W., Du, Q.: Multiscale geographically and temporally weighted regression: exploring the spatiotemporal determinants of housing prices. Int. J. Geogr. Inf. Sci. **33**(3), 489–511 (2019)
4. Jain, M., Rajput, H., Garg, N.,Chawla, P.: Prediction of house pricing using machine learning with python. IEEE (2022). Accessed 1 Dec 2022. https://ieeexplore.ieee.org/abstract/document/9155839
5. Hong, J., Choi, H., Kim, W.-S.: A house price valuation based on the random forest approach: the mass appraisal of residential property in South Korea. Int. J. Strateg. Prop. Manag. Accessed 20 Dec 2022
6. Yin, L., Wang, Z.: Measuring visual enclosure for street walkability: using machine learning algorithms and google street view imagery. Appl. Geogr. **1**(76), 147–53 (2016)
7. Helbich, M., Yao, Y., Liu, Y., Zhang, J., Liu, P., Wang, R.: Using deep learning to examine street view green and blue spaces and their associations with geriatric depression in Beijing. China Environ. Int. **1**(126), 107–17 (2019)
8. Zhang, F., Zhou, B., Liu, L., Liu, Y., Fung, H.H., Lin, H., Ratti, C.: Measuring human perceptions of a large-scale urban region using machine learning. Landscape Urban Plan. **1**(180), 148–60 (2018)
9. Li, S., Jiang, Y., Ke, S., Nie, K., Wu, C.: Understanding the effects of influential factors on housing prices by combining extreme gradient boosting and a hedonic price model (XGBoost-HPM). Land **10**(5), 533 (2021)
10. Kang, Y., Zhang, F., Peng, W., Gao, S., Rao, J., Duarte, F., Ratti, C.: Understanding house price appreciation using multi-source big geo-data and machine learning. Land Use Policy. **1**(111), 104919 (2021)
11. Zhu, T., Liao, Y., Tao, Z.: Predicting google's stock price with LSTM model. Proc. Bus. Econ. Stud. **5**(5), 82–87 (2022)
12. Vasquez, C., Chellamuthu, V.: House price prediction with statistical analysis in support vector machine learning for regression estimation. Curiosity: Interdiscip. J. Res. Innov. **1**(2), 22311 (2021)
13. Rosen, S.: Hedonic prices and implicit markets: product differentiation in pure competition. J. Polit. Econ. **82**(1), 34–55 (1974)
14. Chau, K.W., Chin, T.L.: A critical review of literature on the hedonic price model. Int. J. Hous. Sci. Appl. **27**(2), 145–165 (2003)
15. Cui, N., Gu, H., Shen, T., Feng, C.: The impact of micro-level influencing factors on home value: a housing price-rent comparison. Sustainability **10**(12), 4343 (2018)
16. Palmquist, R.B.: Valuing localized externalities. J. Urban Econ. **31**, 59–68 (1992)
17. Rahadi, R.A., Wiryono, S.K., Koesrindartoto, D.P., Syamwil, I.B.: Factors influencing the price of housing in Indonesia. Int. J. Hous. Markets Anal. Accessed 1 June 2015

18. Hong, J., Choi, H., Woo-sung, K.I.M.: A house price valuation based on the random forest approach: the mass appraisal of residential property in South Korea. Int. J. Strat. Prop. Manag. **24**(3), 140 (2020)
19. Avanijaa, J.: Prediction of house price using xgboost regression algorithm. Turk. J. Comput. Math. Educ. (TURCOMAT) **12**(2), 2151–2155 (2021)
20. Li, S., et al.: Understanding the effects of influential factors on housing prices by combining extreme gradient boosting and a hedonic price model (XGBoost-HPM). Land **10**(5), 533 (2021)

The Effectiveness of Submerged- Emerged Breakwaters: An Analytical and Numerical Study

Ikha Magdalena and Vinsensia Ferren

Abstract In this study, we investigate the wave attenuation by mangroves and submerged rigid breakwaters using a modified shallow water model. We aim to capture the interaction of waves with vegetation and breakwaters and quantify the wave height reduction. We derive the wave transmission coefficient analytically and numerically and examine the dependence of every parameter involved, from the structure and vegetation to the wave attenuation. We will show that both mangroves and submerged rigid breakwaters can effectively attenuate wave energy, with the degree of attenuation depending on various factors such as the width and length of the vegetation, the height and spacing of the breakwater, and the incident wave characteristics. Our findings have important implications for coastal management and the design of coastal protection structures.

Keywords Vegetation · Breakwater · Wave attenuation · Wave transmission coefficient

1 Introduction

Over the past few years, the use of sustainable and eco-friendly methods for coastline protection has been on the rise. One such method gaining popularity is the use of mangroves. According to [31], mangroves have a lifespan of 20–100 years and are highly effective in protecting coastal areas and their inhabitants from natural disasters due to their ability to reduce wind and swell waves [15, 23, 30], lower current velocity [27, 30], and mitigate storm surges. They can act as natural barriers against small to medium cyclones and low-intensity tsunamis [4, 14, 24], while also controlling waves and currents [9, 12, 35], reducing erosion, and promoting sedimentation in the forests [26]. This, in turn, contributes to the vertical growth of mangrove tidal flats. Moreover, [2] highlights that some mangrove forests are currently exceeding

I. Magdalena (✉) · V. Ferren
Faculty of Mathematics and Natural Sciences, Bandung Institute of Technology, Bandung 40132, Indonesia
e-mail: ikha.magdalena@itb.ac.id

D. Vlachos (ed.), *Mathematical Modeling in Physical Sciences*, Springer Proceedings in Mathematics & Statistics 446, https://doi.org/10.1007/978-3-031-52965-8_48

or keeping pace with the average sea-level rise, emphasizing their crucial role in maintaining the ecological balance of coastal areas.

Therefore, the restoration and plantation of mangroves are increasingly advocated in [3, 5, 10, 16] to reduce mangrove loss, which is crucial given that more than 35% of the original mangrove cover was reported to have been lost globally by the end of the 20th century [8, 34]. Various interventions aimed at reducing mangrove forest loss have been investigated by [1, 7, 13, 25, 29], with studies emphasizing the importance of protecting mangroves themselves to conserve the numerous ecosystem services that they provide [6].

Despite their effectiveness, mangroves are susceptible to damage caused by storms, water currents, waves, and strong winds, leading to tree mortality [32]. Additionally, flooding and sedimentation can indirectly affect mangrove survival by limiting the oxygen supply to their roots [28, 33]. To protect young mangroves from these effects, we are considering the installation of a protective barrier to dampen incoming waves before they reach the mangroves, allowing them to grow and thrive on deep-water shorelines. It is important to note that mangroves alone may not provide adequate protection, and additional coastal protection methods may be necessary.

Several experiments have been conducted to study the interaction between waves and mangrove forests. For instance, research by [22] demonstrated that Rhizophora mangrove forests can attenuate waves and offer protection. Additionally, [11] studied wave attenuation in mangroves along the Andaman coast of Trang Province in southern Thailand. However, it is believed that combining mangrove planting with structural construction methods, such as submerged breakwaters, may be more successful in significantly reducing the wave height. Nonetheless, there is still a lack of knowledge regarding the effectiveness of combining mangroves and submerged breakwaters to reduce wave amplitude, which is the focus of our study. Our study aims to investigate the potential of two submerged breakwaters to safeguard young mangroves from the disruptive effects of water waves. To simulate wave reduction, we will use a mathematical model, as it is a cost-effective and flexible alternative to experimental research. In order to accurately model the interaction between waves, breakwaters, and mangroves, we will employ shallow water equations. The equations are frequently used to model various fluid-related issues, as can be seen by their widespread application in academic research, which is detailed in sources such as [18, 21].

To account for the region where mangroves are present, we will modify the SWE by adding a friction force of the Dupuit-Forcheimer type, while no changes are necessary for the location of the submerged breakwater. The new analytical solution in terms of wave transmission coefficient will be derived using the separation of variables method, while the numerical solution will be obtained using the finite volume on a staggered grid approach. Our study will examine the effectiveness of both mangroves and submerged rigid breakwaters to reduce wave energy and how the reduction depends on various factors such as vegetation width and length, breakwater height and spacing, and incident wave characteristics.

2 Mathematical Model

In this section, we will go over the mathematical model used to investigate wave damping generated by Mangrove and Breakwater. First, we recall the well-known Shallow Water Equations, which are briefly described in [19, 20]. The Shallow Water Equations consist of two partial differential equations namely mass conservation and momentum.

We will use two variables: η for wave elevation and u for wave velocity. Assuming we consider linear wave and flat bottom topography ($d(x) = d$), then our model can be written as

$$\eta_t + du_x = 0, \tag{1}$$
$$u_t + g\eta_x = 0. \tag{2}$$

Notation d represents water depth while g denotes the gravitational acceleration.

To address our problem, we divide the observation domain into seven subdomains, denoted by $\Omega_i, i = 1, 2, \ldots, 7$, which can be seen in Fig. 1. Equations (1) and (2) describe the corresponding equations for each domain when there are no other structures present. However, when it comes to the mangrove area, we add an extra component that accounts for friction, which represents the impact of the fluid and porous medium interaction. This leads to the equations:

$$\eta_t + du_x = 0, \tag{3}$$
$$u_t + g\eta_x + f\omega u = 0, \tag{4}$$

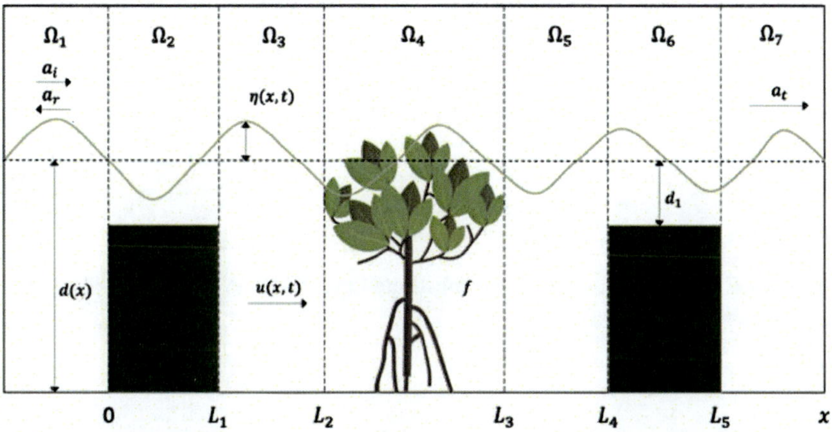

Fig. 1 Sketch of water waves passing through submerged rectangular breakwaters and mangroves

with f denoting the friction coefficient. To involve the presence of a submerged breakwater, we modify the bottom topography (d_1), resulting in the following equations for domains Ω_2–Ω_6, as depicted in Fig. 1:

$$\eta_t + (d_1 u)_x = 0, \tag{5}$$

$$u_t + g\eta_x = 0. \tag{6}$$

Here, d_1 represents the height of the submerged breakwater.

3 Analytical Wave Transmission Coefficient

This section aims to discuss the mathematical model utilized for the determination of the wave transmission coefficient, denoted by K_t. To commence the process, we will introduce the definitions of $\eta(x,t)$ and $u(x,t)$ for each domain, as elucidated in the first subsection. The subsequent step entails the application of continuity conditions to these equations, thereby resulting in the derivation of K_t. The details concerning this process are provided in the second subsection.

3.1 *Wave Elevation and Horizontal Velocity at the Corresponding Domain*

Here, the separation of variables method will be utilized to define the values of $\eta(x,t)$ and $u(x,t)$. It is assumed that the waves being considered are monochromatic, possessing a specific frequency denoted by ω. This assumption enables the following definition of an ansatz to be used in the calculations:

$$\eta(x,t) = F(x)e^{i\omega t}, \tag{7}$$

$$u(x,t) = G(x)e^{i\omega t}, \tag{8}$$

where $F(x)$ and $G(x)$ are arbitrary functions of x.

Domain $\boldsymbol{\Omega_1}$, $\boldsymbol{\Omega_3}$, $\boldsymbol{\Omega_5}$, and $\boldsymbol{\Omega_7}$ First, substitute Eqs. (7)–(8) to (1)–(2), we obtain two equations relating F to G:

$$F(x) = -\frac{d}{i\omega}G_x, \tag{9}$$

$$G(x) = -\frac{g}{i\omega}F_x. \tag{10}$$

By differentiating Eq. (10) with respect to x and substituting the result into Eq. (9), a second-order ordinary differential equation, Eq. (11), is obtained:

$$F_{xx} + \frac{\omega^2}{dg} F = 0. \tag{11}$$

The solution to this differential equation is then found by specifying wave amplitudes a_i, a_r, a_3, a_4, a_7, a_8, and a_t

$$F(x) = \begin{cases} a_i e^{-i\frac{\omega}{\sqrt{gd}}x} + a_r e^{i\frac{\omega}{\sqrt{gd}}x}, & x \in \Omega_1 \\ a_3 e^{-i\frac{\omega}{\sqrt{gd}}x} + a_4 e^{i\frac{\omega}{\sqrt{gd}}x}, & x \in \Omega_3 \\ a_7 e^{-i\frac{\omega}{\sqrt{gd}}x} + a_8 e^{i\frac{\omega}{\sqrt{gd}}x}, & x \in \Omega_5 \\ a_t e^{-i\frac{\omega}{\sqrt{gd}}(x)}, & x \in \Omega_7. \end{cases} \tag{12}$$

Thereafter, Eq. (12) is substituted into Eq. (10) to find $G(x)$. Finally, the wave elevation and fluid velocity are determined by substituting Eqs. (7)–(8) into the derived expressions for $F(x)$ and $G(x)$. The solution of the above equation is

$$\eta(x,t) = \begin{cases} a_i e^{i(\omega t - \frac{\omega}{\sqrt{gd}}x)} + a_r e^{i(\omega t + \frac{\omega}{\sqrt{gd}}x)}, & x \in \Omega_1 \\ a_3 e^{i(\omega t - \frac{\omega}{\sqrt{gd}}x)} + a_4 e^{i(\omega t + \frac{\omega}{\sqrt{gd}}x)}, & x \in \Omega_3 \\ a_7 e^{i(\omega t - \frac{\omega}{\sqrt{gd}}x)} + a_8 e^{i(\omega t + \frac{\omega}{\sqrt{gd}}x)}, & x \in \Omega_5 \\ a_t e^{i(\omega t - \frac{\omega}{\sqrt{gd}}x)}, & x \in \Omega_7. \end{cases} \tag{13}$$

$$u(x,t) = \begin{cases} \sqrt{\frac{g}{d}}(a_i e^{i(\omega t - \frac{\omega}{\sqrt{gd}}x)} - a_r e^{i(\omega t + \frac{\omega}{\sqrt{gd}}x)}), & x \in \Omega_1 \\ \sqrt{\frac{g}{d}}(a_3 e^{i(\omega t - \frac{\omega}{\sqrt{gd}}x)} - a_4 e^{i(\omega t + \frac{\omega}{\sqrt{gd}}x)}), & x \in \Omega_3 \\ \sqrt{\frac{g}{d}}(a_7 e^{i(\omega t - \frac{\omega}{\sqrt{gd}}x)} - a_8 e^{i(\omega t + \frac{\omega}{\sqrt{gd}}x)}), & x \in \Omega_5 \\ \sqrt{\frac{g}{d}}(a_t e^{i(\omega t - \frac{\omega}{\sqrt{gd}}x)}), & x \in \Omega_7. \end{cases} \tag{14}$$

Domain Ω_2, and Ω_6 To obtain the wave elevation $\eta(x,t)$ and fluid velocity $u(x,t)$ at Ω_2 and Ω_6, we can use a similar procedure as described above. However, since the water depth is different, we need to modify the water depth into d_1. The new second-order ordinary differential equation that relates F to G is given by

$$F_{xx} + \frac{\omega^2}{d_1 g} F = 0, \tag{15}$$

which has a slightly different form than Eq. (11). The solution to this equation is given by

$$F(x) = \begin{cases} a_1 e^{-i\frac{\omega}{\sqrt{gd_1}}x} + a_2 e^{i\frac{\omega}{\sqrt{gd_1}}x}, & x \in \Omega_2 \\ a_9 e^{-i\frac{\omega}{\sqrt{gd_1}}x} + a_{10} e^{i\frac{\omega}{\sqrt{gd_1}}x}, & x \in \Omega_6, \end{cases} \tag{16}$$

where the coefficients a_1, a_2, a_9, and a_{10} represent the wave amplitudes at each corresponding domain.

Using the solutions from Eq. (16), we can obtain the wave elevation and fluid velocity at Ω_2 and Ω_6

$$\eta(x,t) = \begin{cases} a_1 e^{i(\omega t - \frac{\omega}{\sqrt{gd_1}} x)} + a_2 e^{i(\omega t + \frac{\omega}{\sqrt{gd_1}} x)}, & x \in \Omega_2 \\ a_9 e^{i(\omega t - \frac{\omega}{\sqrt{gd_1}} x)} + a_{10} e^{i(\omega t + \frac{\omega}{\sqrt{gd_1}} x)}, & x \in \Omega_6 \end{cases} \tag{17}$$

$$u(x,t) = \begin{cases} \sqrt{\frac{g}{d_1}} (a_1 e^{i(\omega t - \frac{\omega}{\sqrt{gd_1}} x)} - a_2 e^{i(\omega t + \frac{\omega}{\sqrt{gd_1}} x)}), & x \in \Omega_2 \\ \sqrt{\frac{g}{d_1}} (a_9 e^{i(\omega t - \frac{\omega}{\sqrt{gd}} x)} - a_{10} e^{i(\omega t + \frac{\omega}{\sqrt{gd}} x)}), & x \in \Omega_6. \end{cases} \tag{18}$$

These equations provide the necessary information to model the wave dynamics in the given system and can be used to study the behavior of the waves under different conditions.

Domain Ω_4 We now turn to the final sub-domain, which consists of the porous medium. In this domain, the governing equations differ slightly from those in the other domains. However, the procedure for finding $\eta(x,t)$ and $u(x,t)$ is similar and explained below.

After substituting Eqs. (7)–(8) into (3)–(4), we have

$$F(x) = -\frac{d}{i\omega} G_x, \tag{19}$$

$$G(x) = -\frac{g}{\omega(i+f)} F_x. \tag{20}$$

Through elimination, we can reduce the above equations to the following second-order ordinary differential equation:

$$F_{xx} + \frac{\omega^2}{gd}(1 - if)F = 0. \tag{21}$$

The solution to Eq. (21) is,

$$F(x) = a_5 e^{-ikx} + a_6 e^{ikx}, \tag{22}$$

where $k = (\omega/\sqrt{gd})\sqrt{(1-if)}$. The amplitudes of the wave passing through the porous medium are represented by a_9 and a_{10}. Upon substituting Eq. (22) into Eqs. (7)–(8) yields

$$\eta(x,t) = a_5 e^{i(\omega t - kx)} + a_6 e^{i(\omega t + kx)}, \tag{23}$$

$$u(x,t) = \sqrt{\frac{g}{d}} \frac{1}{\sqrt{(1-if)}} (a_5 e^{i(\omega t - kx)} - a_6 e^{i(\omega t + kx)}), \tag{24}$$

which are the wave elevation and the fluid velocity at the porous domain.

3.2 Wave Transmission Coefficient

To determine the wave transmission coefficient, several matching conditions must be fulfilled at the discontinue points ($x = 0, L_1, L_2, L_3, L_4$, and L_5). Therefore, the following conditions are put in place to ensure the continuity of both wave height and horizontal momentum

$$\eta(x^-, t) = \eta(x^+, t), \tag{25}$$
$$d(x^-)u(x^-, t) = d(x^+)u(x^+, t), \tag{26}$$

where x^- denotes the left-handed limit and x^+ for the right-handed limit. Next, the equations derived in the previous subsection for $\eta(x, t)$ and $u(x, t)$ will be analyzed for each respective interface. From the matching conditions at $x = L_5$, we acquired

$$a_9 e^{-ik_1 L_5} + a_{10} e^{ik_1 L_5} = a_t e^{-ik_2 L_5}, \tag{27}$$
$$\sqrt{gd_1}(a_9 e^{-ik_1 L_5} + a_{10} e^{ik_1 L_5}) = \sqrt{gd}(a_t e^{-ik_2 L_5}), \tag{28}$$

where the wave numbers satisfy the following dispersion relations $k_1 = \frac{\omega}{\sqrt{gd_1}}$, and $k_2 = \frac{\omega}{\sqrt{gd}}$.

Applying the same routine for $x = L_4, L_3, L_2$, and L_1 we have

$$a_7 e^{-ik_2 L_4} + a_8 e^{ik_2 L_4} = a_9 e^{-ik_1 L_4} + a_{10} e^{ik_1 L_4}, \tag{29}$$
$$\sqrt{gd}(a_7 e^{-ik_2 L_4} - a_8 e^{ik_2 L_4}) = \sqrt{gd_1}(a_9 e^{-ik_1 L_4} - a_{10} e^{ik_1 L_4}), \tag{30}$$

$$a_5 e^{-ik_2 L_3 \sqrt{1-if}} + a_6 e^{ik_2 L_3 \sqrt{1-if}} = a_7 e^{-ik_2 L_3} + a_8 e^{ik_2 L_3}, \tag{31}$$
$$\sqrt{gd\epsilon}(a_5 e^{-ik_2 L_3 \sqrt{1-if}} - a_6 e^{ik_2 L_3 \sqrt{1-if}}) = \sqrt{gd}(a_7 e^{-ik_2 L_3} - a_8 e^{ik_2 L_3}), \tag{32}$$

$$a_3 e^{-ik_2 L_2} + a_4 e^{ik_2 L_2} = a_5 e^{-ik_2 L_2 \sqrt{1-if}} + a_6 e^{ik_2 L_2 \sqrt{1-if}}, \tag{33}$$
$$\sqrt{gd}(a_3 e^{-ik_2 L_2} - a_4 e^{ik_2 L_2}) = \sqrt{gd\epsilon}(a_5 e^{-ik_2 L_2 \sqrt{1-if}} - a_6 e^{ik_2 L_2 \sqrt{1-if}}), \tag{34}$$

$$a_1 e^{-ik_1 L_1} + a_2 e^{ik_1 L_1} = a_3 e^{-ik_2 L_1} + a_4 e^{ik_2 L_1}, \tag{35}$$
$$\sqrt{gd_1}(a_1 e^{-ik_1 L_1} - a_2 e^{ik_1 L_1}) = \sqrt{gd\epsilon}(a_3 e^{-ik_2 L_1} - a_4 e^{ik_2 L_1}), \tag{36}$$

$$a_i + a_r = a_1 + a_2, \tag{37}$$
$$\sqrt{gd}(a_i - a_r) = \sqrt{gd_1}(a_1 - a_2). \tag{38}$$

where $\epsilon = \frac{1}{\sqrt{1-if}}$. To simplify the process of getting the wave amplitudes for each domain, let us write:

$A = e^{-ik_2L_5}$	$B = e^{-ik_1L_5}$	$C = e^{ik_1L_5}$
$D = e^{-ik_2L_4}$	$E = e^{ik_2L_4}$	$F = e^{-ik_1L_4}$
$G = e^{ik_1L_4}$	$H = e^{-ik_2L_3}$	$I = e^{ik_2L_3}$
$J = e^{-ik_2L_3\sqrt{1-if}}$	$K = e^{ik_2L_3\sqrt{1-if}}$	$L = e^{-ik_2L_2\sqrt{1-if}}$
$M = e^{ik_2L_2\sqrt{1-if}}$	$N = e^{-ik_2L_2}$	$O = e^{ik_2L_2}$
$P = \sqrt{gd}$	$Q = \sqrt{gd_1}$	$W = e^{-ik_2L_1}$
$X = e^{ik_2L_1}$	$Y = e^{-ik_1L_1}$	$Z = e^{ik_1L_1}$

Therefore, by substituting the variables above, and solving Eqs. (27)–(38) we can express $a_{10}, a_9, \ldots, a_r, a_i$ as

$$a_{10} = \frac{(P-Q)Aa_7}{2PC}, \tag{39}$$

$$a_9 = \frac{(P+Q)Aa_7}{2PB}, \tag{40}$$

$$a_8 = \frac{(P-Q)Da_9 + (P+Q)Ea_{10}}{2PG}, \tag{41}$$

$$a_7 = \frac{(P+Q)Da_9 + (P-Q)Ea_{10}}{2PF}, \tag{42}$$

$$a_6 = \frac{P(\epsilon-1)Ha_7 + P(\epsilon+1)Ia_8}{2PK\epsilon}, \tag{43}$$

$$a_5 = \frac{P(\epsilon+1)Ha_7 + P(\epsilon-1)Ia_8}{2PJ\epsilon}, \tag{44}$$

$$a_4 = \frac{P(1-\epsilon)La_5 + P(1+\epsilon)Ma_6}{2PO}, \tag{45}$$

$$a_3 = \frac{P(1+\epsilon)La_5 + P(1-\epsilon)Ma_6}{2PN}, \tag{46}$$

$$a_2 = \frac{(Q-P)Wa_3 + (Q+P)Xa_4}{2QZ}, \tag{47}$$

$$a_1 = \frac{(Q+P)Wa_3 + (Q-P)Xa_4}{2QY}, \tag{48}$$

$$a_r = \frac{(Q-P)a_1 + (Q+P)a_2}{2Q}, \tag{49}$$

$$a_i = \frac{(Q+P)a_1 + (Q-P)a_2}{2Q}, \tag{50}$$

As the main objective of this paper is to determine the wave transmission coefficient K_t, we will only focus on solving Eq. (50) to obtain the value of a_i, rather than evaluating both a_i and a_r. Hence, the solution for wave transmission coefficient is given by,

$$K_t = \left| \frac{2Q}{(Q+P)a_1 + (Q-P)a_2} \right|. \tag{51}$$

4 A Staggered Finite Volume Method

The numerical solution to the mathematical model will be obtained through the implementation of a finite volume method. This particular method has been deemed suitable for this investigation as it is devoid of damping errors. We consider the computational domain from 0 to L with the observation time ranging from 0 to T. To facilitate this computation, a staggered grid will be employed, whereby the spatial domain is partitioned into a grid with half-partitions represented by $x_{1/2} = 0, x_1, \ldots, x_{j-1/2}, x_j, x_{j+1/2}, \ldots, x_{Nx}, x_{Nx+1/2} = L$. The observation time domain, on the other hand, will be divided into $N_t = \lfloor \frac{T}{\Delta t} \rfloor + 1$ with interval Δt

The mass conservation is represented by Eqs. (1), (3), and (5), and they are discretized at a cell-centered at x_j. In contrast, the momentum balance is represented by Eqs. (2) and (4), and they are discretized at a cell-centered at $x_j + 1/2$. Additionally, the wave elevation η and water depth d is evaluated in the full grid x_j, while wave velocity u is evaluated in the half grid $x_{j+1/2}$. This arrangement has been shown in Fig. 2.

The numerical approximation obtained by applying the Forward in Time and Center in Space (FTCS) method can be expressed as follows:

$$\eta_j^{n+1} = \eta_j^n - \frac{\Delta t}{\Delta x}(q_{j+1/2}^n - q_{j-1/2}^n), \tag{52}$$

$$u_{j+1/2}^{n+1} = u_{j+1/2}^n - g\frac{\Delta t}{\Delta x}(\eta_{j+1}^{n+1} - \eta_j^{n+1}). \tag{53}$$

where $q_{j+1/2}^n = {}^*d_{j+1/2}^n u_{j+1/2}^n$, $n = 1, 2, \ldots, N_t$. η_j^n denotes the approximation of η at point x_j and time t^n where $u_{j+1/2}^n$ denotes the approximation of u at point $x_{j+1/2}$ and time t^n.

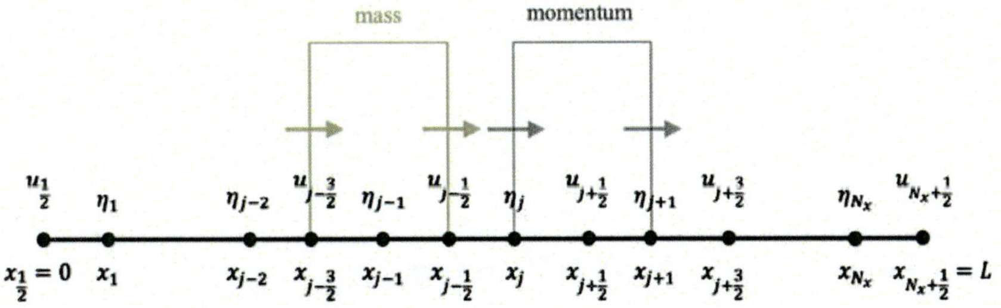

Fig. 2 Discretization of staggered grid scheme

To ensure the numerical scheme remains stable, we adjust the time at which η is evaluated in the momentum equation to t^{n_1+1} instead of t^n. This approach overcomes the limitation of the FTCS method, which is known to be inherently unstable for any transport equations. Furthermore, the stability condition for Eqs. (52) and (53) can be derived by implementing Von Neumann stability analysis. The condition is given by $0 \leq \sqrt{gd}\frac{\Delta t}{\Delta x} \leq 1$ with d denotes the flat bottom depth, as mentioned in [17].

It is imperative to acknowledge that in order to compute $q^n_{j+1/2}$, the value of $d^n_{j+1/2}$ is necessary. Nonetheless, as mentioned earlier, d is evaluated in the full grid. Therefore, we utilize the upwind method to approximate $d^n_{j+1/2}$, which is dependent on the wave velocity. This is accomplished by implementing the following conditions:

$$ {}^*d^n_{j+1/2} = \begin{cases} d^n_j, & u_{j+1/2} \geq 0 \\ d^n_{j+1}, & u_{j+1/2} < 0 \end{cases} \tag{54} $$

In the case where $u_{j+1/2} \geq 0$, the wave is moving to the right, then ${}^*d^n_{j+1/2} = d^n_j$. When the wave is moving to the left, where $u_{j+1/2} < 0$, then ${}^*d^n_{j+1/2} = d^n_{j+1}$. The numerical approximation for the mass conservation and momentum equation in the mangrove domain is presented as follows:

$$ \eta^{n+1}_j = \eta^n_j - \frac{\Delta t}{\Delta x}(q^n_{j+1/2} - q^n_{j-1/2}) \tag{55} $$

$$ u^{n+1}_{j+1/2} = \frac{1}{1 + f\Delta t\omega}\left(u^n_{j+1/2} - g\frac{\Delta t}{\Delta x}(\eta^{n+1}_{j+1} - \eta^{n+1}_j)\right) \tag{56} $$

Here f denotes the mangrove's friction coefficient.

5 Results and Discussion

This part of the paper will cover the implementation of our numerical solution to simulate the interaction among the incoming waves, mangroves, and Submerged-Emerged Breakwaters. To begin with, we will verify the accuracy of our model by examining wave propagation solely through mangroves. A comparable validation procedure has already been carried out by [18]. Once we have confirmed the model's validity, we will proceed to evaluate our numerical approach using the selected coastal protection of mangroves, along with the inclusion of breakwaters.

5.1 *Waves Passing Through Mangroves*

In this section, we verify the accuracy of both our numerical and analytical methods for modeling wave propagation through mangroves. Specifically, we investigate the performance of our numerical model in a scenario where there is no breakwater and the seabed is flat with a depth of d. Our computational domain spans from $x = -5$ to $x = 35$, and we set the water depth to 10 m. Within this domain, we include a section of mangroves with a length of 5 m, positioned between 10 and 15 m.

The initial condition is calm water, and the incoming wave is a monochromatic wave with amplitude 0.1 m and wave frequency $\omega = 2\pi$. On the right side, we implement a fully absorbing boundary. The wave is initially characterized by the following conditions:

$$\eta(x, 0) = 0 \quad u(x, 0) = 0. \tag{57}$$

The left boundary is subject to the following equation:

$$\eta(0, t) = A \sin(\omega t), \tag{58}$$

while the right boundary can be expressed as:

$$u(L, t) = \sqrt{\frac{g}{d}}\eta(L, t), \tag{59}$$

where $g = 9.81$ m/s^2.

To account for the force resulting from fluid-vegetation interaction in the mangrove area, a friction coefficient of $c_f = 0.1$ will be incorporated. During the simulation of wave propagation, a spatial partition $\Delta x = 0.1$ is used, along with a time partition determined by $\Delta t = \Delta x/\sqrt{gd}$, to ensure stability. The simulation results shown below display green lines to indicate the mangrove locations, red dashed lines representing the wave elevation from our simulation using 7 domains, and the blue line representing the wave elevation from the simulation conducted by [18], which used only 3 domains.

The presented Fig. 3 confirms the validity of our model since it is almost identical to the one developed by [18]. The graph demonstrates a reduction in wave amplitude as it passes through the porous media. To examine this reduction more closely, we calculated the numerical transmission coefficient (K_t) using the maximum η values in the first and final domains, similar to the analytical K_t. The numerical simulation resulted in a K_t value of 0.8594, whereas the analytical K_t produced a value of 0.8537.

5.2 Waves Passing Through Mangroves and Two Submerged Breakwaters

In this section, we aim to evaluate the accuracy of our numerical and analytical methods in simulating wave propagation through mangroves, taking into account the presence of two submerged breakwaters. These breakwaters are added before and

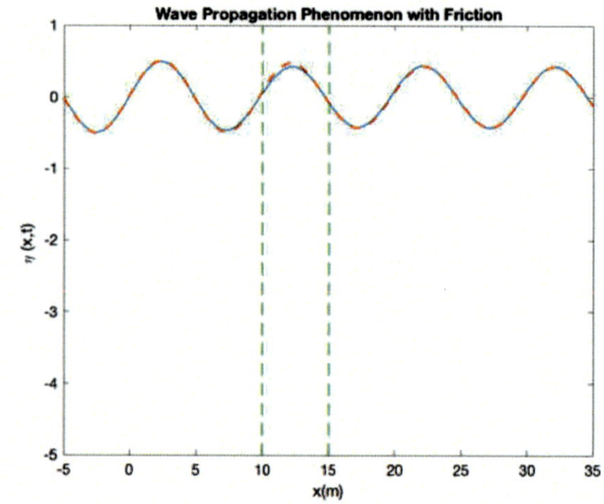

Fig. 3 Evolution of wave amplitude passing through mangroves

after the mangroves. The spatial domain spans from $-5 < x < 35$, and it is divided into several subdomains. Subdomain Ω_2 covers the range from 0 to 10 m, Ω_4 spans from 10 to 15 m, and Ω_6 spans from 20 to 25 m. The two submerged breakwaters are placed in both Ω_2 and Ω_6, while the mangroves are located in Ω_4. We set the water depth to $d = 10$ m for subdomains Ω_1, Ω_3, Ω_4, Ω_5, and Ω_7, which do not have submerged breakwaters. For subdomains Ω_2 and Ω_6, we set the water depth from the still water condition to the upper base of the submerged breakwater as $d_1 = 7$ m.

Here, we utilized the same set of parameters as in the previous subsection, with the exception of the water depth due to the addition of two submerged breakwaters. We maintained the same configuration as Eqs. (57)–(59) for the initial and left-right boundary conditions. The spatial domain partition length Δx, the time partition Δt, and the friction coefficient value c_f were also kept constant. The outcome of the numerical simulation is illustrated in Fig. 4.

The simulation results depict the locations of submerged breakwaters with red lines, the sub-domain with mangroves with green lines, and the wave elevation with a blue line. Figure 4 shows that the wave amplitude decreases after passing through the submerged breakwaters and mangroves. The numerical transmission coefficient K_t, which was computed from the simulation, provides a clearer indication of the reduction of the wave amplitude. The simulation yields a numerical K_t value of 0.8451, while the analytical K_t value is 0.8443. indicating that the submerged breakwaters and mangroves decrease the wave amplitude by more than 15% of the initial amplitude.

On the other hand, when considering the previous scenario in which solely mangroves were present, the computed K_t value was comparatively higher. In the current scenario, where two additional breakwaters were added, the numerical value of K_t is reduced by approximately 1.4%. This suggests that the combination of both mangroves and breakwaters as coastal protections leads to a more substantial reduction in wave amplitude.

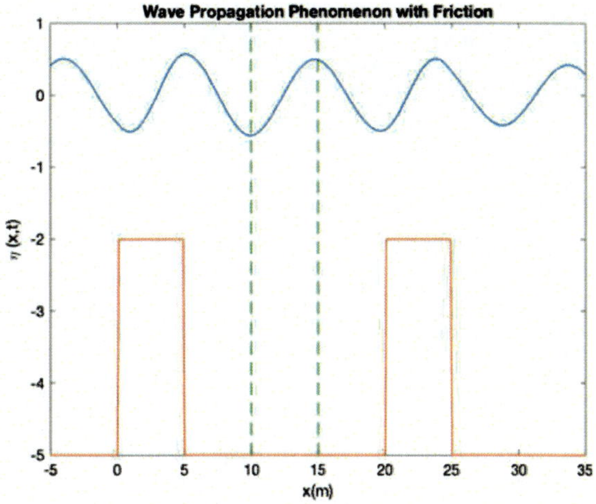

Fig. 4 Evolution of wave amplitude passing through mangroves and breakwaters

5.3 *Comparison Between Analytical and Numerical Simulations*

Afterward, we will examine the accuracy of our numerical model. We compare the numerical to the analytical values of the transmission coefficient K_t. To do so, we will vary the parameters related to the porous media and submerged breakwater, such as the width of the mangroves, the height and length of the breakwater, and the incident wave characteristics. By doing this, we will not only be assessing the accuracy of our numerical model, but we will also be able to better understand the relationship between mangroves, submerged breakwaters, and the reduction of wave amplitude.

First, we will conduct several numerical simulations using different c_f values from 0 to 1. The observational domain remains unchanged, with the simulation time set to $t = 4.2$. The results obtained from the numerical scheme provide a good approximation of our analytical solutions when it comes to describing the relationship between K_t and c_f. Moreover, as the parameters c_f increase, the K_t will gradually decrease. This is illustrated in Fig. 5.

Next, we validate our numerical results by comparing the relationship between the transmission coefficient K_t and mangrove length obtained from both numerical simulations and analytical solutions. We're varying the values of mangrove length within the range of 0–10 m with $c_f = 0.1$. From the result presented in Fig. 6, we can see that there is a hardly noticeable discrepancy between the numerical and analytical transmission coefficients. This implies that our numerical solution successfully approximates the analytical solution. Additionally, as the mangrove length increases, there is a gradual decrease in K_t.

To investigate the effect of submerged breakwaters height on K_t, we alter the water depth from the still water condition to the submerged breakwaters d_1 ranging from 1 to 9 m. The result of the simulation is presented in Fig. 7. Our numerical

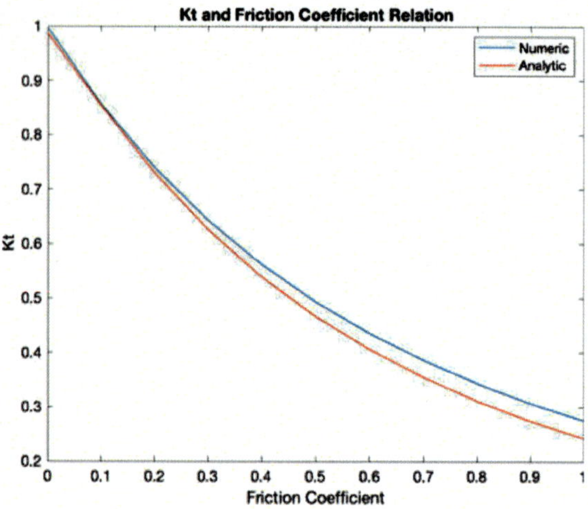

Fig. 5 The effect of friction coefficient on K_t

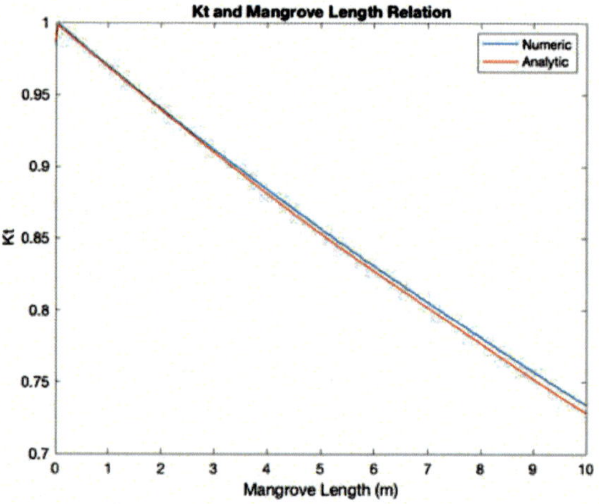

Fig. 6 The effect of mangrove length on K_t

simulations indicate that the height of the rectangular submerged breakwater has quite a significant effect on K_t as it decreases with an increase in d_1, which is consistent with the conclusions drawn from our analytical solution.

Finally, we will explore the analytical and numerical values of K_t for different breakwater lengths ranging from 0 to 10 m. The simulation result is illustrated in Fig. 8. Our numerical simulations suggest that the length of the rectangular submerged breakwater has a negligible effect on K_t. As the breakwater length (L_4) increases, the numerical K_t remains constant without any significant decrease or increase. This observation is in agreement with the findings obtained from our analytical solution.

Figures 5, 6, 7 and 8 demonstrate that the numerical and analytical transmission coefficients, denoted by K_t, are nearly identical, indicating that the numerical solution

Fig. 7 The effect of d_1/d on K_t

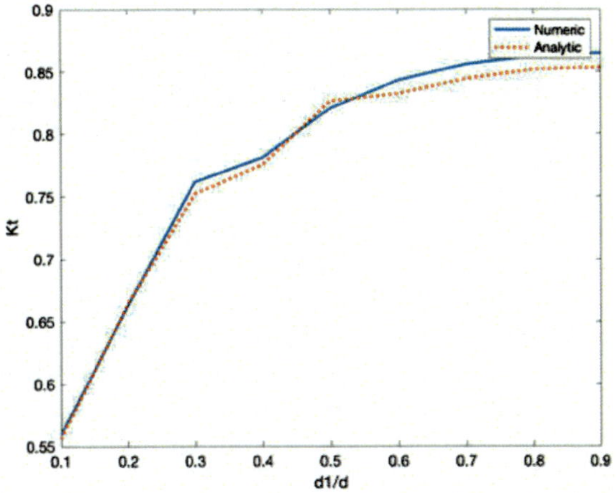

Fig. 8 The effect of breakwater length on K_t

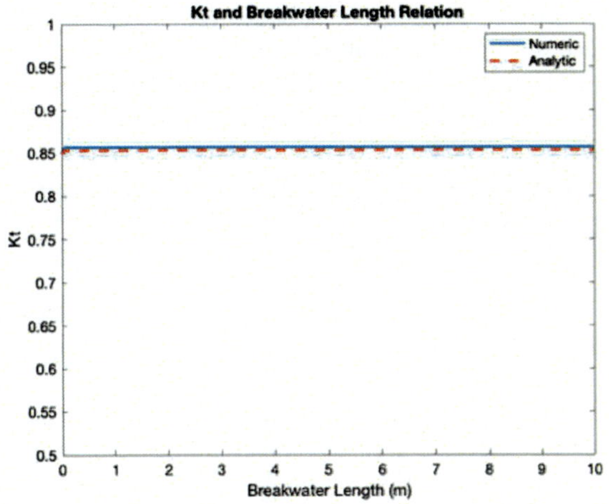

is accurately approximated. Furthermore, Figs. 5, 6 and 7 provide evidence of a robust relationship between K_t and both mangrove and breakwater parameters. Specifically, an increase in the mangrove friction coefficient, mangrove length, or breakwater height leads to a gradual decrease in the transmission coefficient. However, according to Fig. 8, there is no evident relationship between the transmission coefficient (K_t) and the length of the rectangular submerged breakwater.

6 Conclusion

A modified shallow water model has demonstrated its efficacy in capturing the interaction between waves, vegetation, and breakwater. By utilizing the method of separation of variables, an analytical solution for the wave transmission coefficient has been derived. The propagation and scattering processes of waves passing through breakwaters and vegetation have been simulated exceptionally well by employing a staggered finite volume method. The wave transmission coefficient obtained through analytical and numerical means enables us to quantify the reduction in wave height. Furthermore, it allows us to investigate the impact of each parameter, including structure, vegetation, and wave attenuation, on the wave transmission coefficient. Our findings reveal that both mangroves and submerged rigid breakwaters can effectively attenuate wave energy, with the degree of attenuation being dependent on several factors such as vegetation width and length, breakwater height and spacing, and incident wave characteristics. We have identified that the mangrove friction coefficient, mangrove-occupied area, and breakwater height play a significant role in reducing wave amplitude, whereas breakwater length has a minor influence. The practical implications of these findings are essential for designing effective coastal protection measures against wave damage by highlighting the crucial factors for wave energy attenuation. Overall, this study provides a significant contribution to the field of coastal engineering and can assist in developing sustainable coastal management practices.

References

1. Al Amin, M.A., Mulyana, D., Damar, A., Budiman, M.A.K.: Effectiveness and impact studies of mangrove rehabilitation in the northern coast of west java: a case study in karawang regency. OP Conf. Ser. Earth Environ. Sci. **744** (2021)
2. Alongi, D.M.: Mangrove forests: resilience, protection from tsunamis, and responses to global climate change. Estuar. Coast. Shelf Scie. **76**, 1–13 (2008)
3. Andradi-Brown, D.A., Howe, C., Mace, G.M., Knight, A.T.: Do mangrove forest restoration or rehabilitation activities return biodiversity to pre-impact levels? Environ. Evid. **2**, 20 (2013)
4. Badola, R., Hussain, S.A.: Valuing ecosystem functions: an empirical study on the storm protection function of bhitarkanika mangrove ecosystem, India. Environ. Conserv. **32**, 85–92 (2005)
5. Dale, P.E.R., Knight, J.M., Dwyer, P.G.: Mangrove rehabilitation: a review focusing on ecological and institutional issues. Environ. Evid. **22**, 587–604 (2014)
6. Drakou, E.G.: The Importance of Mangroves to People: A Call to Action. United Nations Environment Programme (UNEP) and World Conservation Monitoring Centre (WCMC) (2014)
7. Ellison, A.M., Felson, A.J., Friess, D.A.: Mangrove rehabilitation and restoration as experimental adaptive management. Front. Mar. Sci. **7**, 05 (2020)
8. Feller, I.C., Friess, D.A., Krauss, K.W., Lewis, R.R.: The state of the world's mangroves in the 21st century under climate change. Hydrobiologia **803**, 1–12 (2017)
9. Gedan, K.B., Kirwan, M.L., Wolanski, E., Barbier, E.B., Silliman, B.R.: The present and future role of coastal wetland vegetation in protecting shorelines: answering recent challenges to the paradigm. Clim. Change **106**, 7–29 (2010)

10. Hamilton, S.E., Casey, D.: Creation of a high spatio-temporal resolution global database of continuous mangrove forest cover for the 21st century (cgmfc-21). Glob. Ecol. Biogeogr. **25**, 729–738 (2016)
11. Horstman, E., Dohmen-Janssen, M., Narra, P., van den Berg, N.-J., Siemerink, M., Balke, T., Bouma, T., Hulscher, S.: Experimental investigation of wave attenuation and bulk drag coefficient in mangrove forest with complex root morphology. Appl. Ocean Res. **118**, 01 (2012)
12. Horstman, E.M., Dohmen-Janssen, C.M., Bouma, T.J., Hulscher, S.J.M.H.: Tidal-scale flow routing and sedimentation in mangrove forests: combining field data and numerical modelling. Geomorphology **228**, 244–262 (2015)
13. Kodikara, K.A.S., Mukherjee, N., Jayatissa, L.P., Dahdouh-Guebas, F., Koedam, N.: Have mangrove restoration projects worked? An in-depth study in Sri Lanka. Ecol. Econ. **25**, 705–716 (2017)
14. Koh, H.L., Teh, S.Y., Liu, P.L.F., Ismail, A.I.M., Lee, H.: Simulation of Andaman 2004 tsunami for assessing impact on Malaysia. J. Asian Earth Sci. **36**, 74–83 (2009)
15. Lee, W.K., Tay, S., Ooi, S.K., Friess, D.: Potential short wave attenuation function of disturbed mangroves. Estuar. Coast. Shelf Sci. **248** (2021)
16. Macintosh, D., Ashton, E., Havanon, S.: Mangrove rehabilitation and intertidal biodiversity: a study in the ranong mangrove ecosystem, Thailand. Estuar. Coast. Shelf Sci. **55**, 331–345 (2002)
17. Magdalena, I., Pudjaprasetya, S.R., Wiryanto, L.H.: Wave interaction with an emerged porous media. Adv. Appl. Math. Mech. **6**(5), 680–692 (2014)
18. Magdalena, I., Atras, M.F., Sembiring, L., Nugroho, M.A., Labay, R.S.B., Roque, M.P.: Wave transmission by rectangular submerged breakwaters. Computation **8**(2) (2020)
19. Magdalena, I., Kusnowo, V., Azis, M., Widowati: 1d–2d numerical model for wave attenuation by mangroves as a porous structure. Computation **9**, 66 (2021)
20. Magdalena, I., La'lang, R., Mendoza, R.: Quantification of wave attenuation in mangroves in manila bay using nonlinear shallow water equations. Resul. Appl. Math. **12**, 100–191 (2021)
21. Magdalena, I., Suhardi, J.L.E., Adityawan, M.B.: Analytical and numerical studies for the reduction of wave run-up height by a submerged breakwater. Int. J. GEOMATE **20**(77), 1–9 (2021)
22. Maza, M., Lara, J.L., Losada, I.J.: Experimental analysis of wave attenuation and drag forces in a realistic fringe rhizophora mangrove forest. Adv. Water Resour. **131** (2019)
23. Mazda, Y., Magi, M., Ikeda, Y., Kurokawa, T., Asano, T.: Wave reduction in a mangrove forest dominated by sonneratia sp. Wetlands Ecol. Manag. **248**, 365–378 (2006)
24. Mazda, Y., Magi, M., Kogo, M., Hong, P.N.: Mangroves as a coastal protection from waves in the tong king delta. Vietnam. Mangroves Salt Marshes **1**, 127–135 (1997)
25. Miteva, D.A., Murray, B.C., Pattanayak, S.K.: Do protected areas reduce blue carbon emissions? A quasi-experimental evaluation of mangroves in Indonesia. Ecolog. Econ. **119**, 127–135 (2015)
26. Mitra, A.: Mangrove Forests in India: Exploring Ecosystem Services. Springer, Cham (2020)
27. Mullarney, J., Henderson, S., Reyns, J., Norris, B., Bryan, K.: Spatially varying drag within a wave-exposed mangrove forest and on the adjacent tidal flat. Cont. Shelf Resear. **147**, 102–113 (2017)
28. Paling, E., Kobryn, H., Humphreys, G.: Assessing the extent of mangrove change caused by cyclone vance in the eastern exmouth gulf, Northwestern Australia. Estuar. Coast. Shelf Sci. **77**, 603–613 (2008)
29. Primavera, J.H., Esteban, J.M.A.: A review of mangrove rehabilitation in the Philippines: successes, failures and future prospects. Wetlands Ecol. Manag. **16**, 345–358 (2008)
30. Quartel, S., Kroon, A., Augustinus, P.G.E.F., van Santen, P., Tri, N.H.: Wave attenuation in coastal mangroves in the red river delta, Vietnam. J. Asian Earth Sci. **29**, 576–584 (2007)
31. Raju, R.D., Arockiasamy, M.: Coastal protection using integration of mangroves with floating barges: an innovative concept. Coast. Eng. **10** (2022)

32. Sippo, J., Lovelock, C., Santos, I., Sanders, C., Maher, D.: Mangrove mortality in a changing climate: an overview. Estuar. Coast. Shelf Sci. **215**, 241–249 (2018)
33. Spencer, T., Möller, I.: Mangrove systems. Treat. Geomorphol. **10**, 360–391 (2013)
34. Valiela, I., Bowen, J.L., York, J.K.: Mangrove forests: one of the world's threatened major tropical environments. BioScience **51**, 807–815 (2001)
35. Van Santen, P., Augustinus, P.G.E.F., Janssen-Stelder, B.M., Quartel, S., Tri, N.H.: Sedimentation in an estuarine mangrove system. J. Asian Earth Sci. **29**, 566–575 (2007)

Validation and Numerical Simulation of a Parabolic Trough Solar Collector Plant Using an Implicit Finite-Difference Scheme

Richard Rangel, Lesme Corredor, and Marco Sanjuan

Abstract The evolving energy landscape demands reliable integration of renewable resources to ensure security, sustainability, and reduced carbon emissions. Then, a high level of understanding about the behavior of the renewable energy sources, the technologies, and their interaction is needed. Particularly, parabolic trough collector (PTC) solar plants can positively affect power systems due to their dispatchability. Since PTC harvests direct normal irradiance (DNI), complex dynamic models are required to understand how PTC behaves. This work uses a 1-D dynamic model that captures the dynamic behavior of PTC plants and is solved numerically via a three-point backward finite-difference. Through two case studies, this research highlights the dynamic behavior and performance of the PTC rows and total solar fields. The row response to varying DNI is showcased in the first case study. The model predicts heat transfer fluid (HTF) temperature profiles in 0.7 s, paving the way for effective control strategies to prevent overheating. The second case study extends the analysis to disturbances caused by clouds and shadows, reaffirming the model's robustness. With a computation time of 2.7 s for temperature profiles over a day, the model offers both accuracy and computational efficiency for real-time applications.

Keywords Parabolic trough collector · Dynamic response · Numerical solution · Measured disturbances · Implicit finite-difference scheme

R. Rangel (✉) · L. Corredor
Department of Mechanical Engineering, Universidad del Norte, Barranquilla, Colombia
e-mail: rerangel@uninorte.edu.co

L. Corredor
e-mail: lcorredo@uninorte.edu.co

M. Sanjuan
Centro de Investigación e Innovación en Energía y Gas (CIIEG), Promigas S.A E.S.P., Barranquilla, Colombia
e-mail: marco.sanjuan@promigas.com

D. Vlachos (ed.), *Mathematical Modeling in Physical Sciences*, Springer Proceedings in Mathematics & Statistics 446, https://doi.org/10.1007/978-3-031-52965-8_49

1 Introduction

The current trend in the electricity sector looks to integrate different generation technologies as an alternative to ensure reliability, security, and low CO_2 emissions. An example of these integrations is the coupling of wind farms with photovoltaic parks, as solar and wind resources become complementary in some regions. In other words, the abundance of one resource balances the deficiency of the other in terms of generation during specific seasons. However, since both resources are intermittent, countries' power systems face significant challenges in terms of security and reliability because their infrastructure is not designed for highly variable systems. Furthermore, the uncertainty of the non-conventional renewable resource propagates throughout the system, often causing volatility in electricity prices. Therefore, generation portfolios will require dispatchable technologies capable of addressing the technical and economic difficulties introduced by variable renewable energy sources (VRE), ensuring optimal response to demand peaks, guaranteeing ancillary services, and reliability in electrical systems [1, 2]. Accordingly, concentrated solar power plants (CSPs) have an important role in the energy transition since it is a cross-cutting technology to the VRE and dispatchable technology concepts. Firstly, its primary energy source is Direct Normal Irradiance (DNI), which is uncertain, but when it is coupled with a thermal storage system, it acquires the ability to control the power dispatch [2–7] Therefore, dispatchable technologies like CSP are expected to have the flexibility to respond to known and unknown changes inherent to the electricity power system in different time scales in order to balance supply and demand cost-effectively [8, 9]. The parabolic trough collector (PTC) plant is a mature technology within the range of CSP technologies. To study in detail the capabilities of PTC plants in terms of dynamic behavior, interaction with thermal energy storage and power blocks, response to abrupt market signals, and DNI uncertainty, it is necessary to develop mathematical models that accurately explain the behavior of PTC systems in different timescales, which can be in the order of seconds or minutes to handle DNI variation or minutes to hours to support the grid system. The goal of this study is to simulate and validate the performance of parabolic trough solar collector assembly, modeled as a system of a mixed partial differential equation with two nonlinear ordinary differential equations and a group of nonlinear algebraic equations, using the three-point backward implicit finite-difference scheme proposed in [10].

2 Reference Systems

Parabolic trough solar power plants convert the solar radiation, specifically the DNI, into useful thermal energy, and then, the thermal energy can be converted to electricity if coupled with a power block. A typical scheme of a PTC power plant is depicted

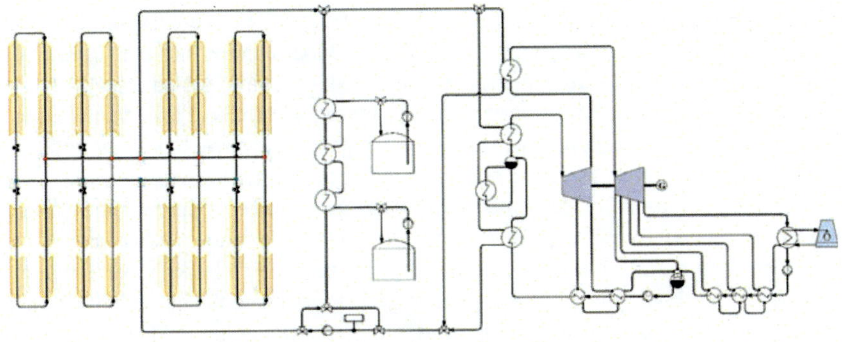

Fig. 1 Conventional scheme of a PTC solar plant

in Fig. 1. The main components of a PTC power plant are: (i) solar field, (ii) thermal energy storage, (iii) solar steam generator, and (iv) the power block.

The solar field consists of multiple rows of single-axis-tracking parabolic trough solar collectors. These solar collectors are modular and typically aligned on a north–south horizontal axis. Each collector features a linear parabolic-shaped reflector that concentrates the DNI from the sun onto a linear receiver positioned at the focal point of the parabola (See Fig. 2). Collectors track the sun's movement from east to west to ensure continuous focus on the linear receiver throughout the day [11].

A heat transfer fluid (HTF) circulates through the receiver pipe, absorbing the concentrated solar energy and reaching temperatures as high as 393 °C. The HTF then goes to the solar steam generator to produce high-pressure superheated steam (100 bar, 371 °C). This superheated steam is directed to the power block to generate electricity. In this work, the focus is centered on the solar field.

The core unit of the solar field is the solar collector assembly (SCA). Each SCA is an independent tracking group of PTCs including parabolic reflectors or mirrors,

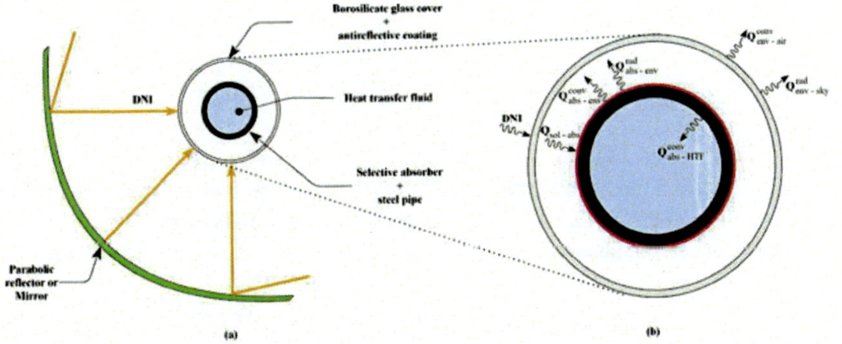

Fig. 2 **a** Basic design elements of parabolic trough concentrators and **b** one-dimensional energy balance for cross-section of a heat collecting element

heat collecting elements (HCE), a metal support structure, and a continuous tracking system [12]. Thus, the solar field in a PTC plant includes hundreds or even thousands of SCAs. The HCE includes a stainless-steel receiver pipe, which usually has a borosilicate glass cover tube around to reduce the convective heat loss, then the heat loss coefficient is reduced. A disadvantage of using the glass cover tube is the addition of transmittance losses due to the reflected light that pass through the glass to reach the receiver. Additionally, deposited coating on the absorber reduces heat loss and improves absorptivity and optical performance. One way to further reduce convective heat loss from the receiver tube and thereby increase the collector's performance, particularly for high-temperature applications, is to evacuate the space between the glass cover tube and the receiver and fill it with hydrogen or argon [13].

2.1 Mathematical Model and Numerical Method

Due to the nature of the DNI transformation towards useful thermal energy, the transient energy balance equations for the HTF, the absorber pipe, and the glass envelope must be computed. In this work, the one-dimensional transient energy balance for the cross-section of a HCE presented in Fig. 2 is used, which is based on the works of [13–16]. Most of the model variables presented below have the structure $X = f(t, x)$. For the sake of simplicity, the term (t, x) is omitted in the writing of equations. As a result, the following system of hyperbolic partial differential equation coupled with nonlinear ordinary differential equations (i.e., coupled PDE-ODEs system) must be solved.

For the HTF:

$$\rho_{HTF} C_{HTF} A_{abs,i} \frac{\partial T_{HTF}}{\partial t} = -\dot{m}_{HTF} C_{HTF} \frac{\partial T_{HTF}}{\partial x} + \dot{Q}^{conv}_{abs \to HTF} \tag{1}$$

Rearranging terms:

$$\frac{\partial T_{HTF}}{\partial t} = \Omega(x,t) \frac{\partial T_{HTF}}{\partial x} + \chi(x,t) \tag{2}$$

where $\Omega(x,t) = -\dot{m}_{HTF}/\rho_{HTF} A_{abs,i}$ and $\chi(x,t) = \dot{Q}^{conv}_{abs \to HTF}/\rho_{HTF} C_{HTF} A_{abs,i}$. Thus, ρ_{HTF}, C_{HTF}, and $\dot{m}_{HTF}$ are the density, specific heat, and mass flow rate of the HTF. $A_{abs,i}$ is the inner surface area of the steel pipe/absorber, T_{HTF} is the bulk temperature of the HTF. $\dot{Q}^{conv}_{abs \to HTF}$ is heat transfer between the absorber pipe and the HTF, which is calculated as follows:

$$\dot{Q}^{conv}_{abs \to HTF} = D_{abs,i} h_{abs-\mathrm{HTF}} [T_{abs} - T_{HTF}] \tag{3}$$

where $D_{abs,i}$ is the inner diameter of the absorber pipe, $h_{abs-\mathrm{HTF}}$ is the HTF convection heat transfer coefficient and T_{abs} is the temperature of the absorber pipe, which value is calculated with (4).

$$\rho_{abs} C_{abs} A_{abs} \frac{\partial T_{abs}}{\partial t} = -\dot{Q}^{conv}_{abs \to HTF} - \dot{Q}^{rad}_{abs \to env} - \dot{Q}^{conv}_{abs \to env} + \dot{Q}_{sol \to abs} \tag{4}$$

Rearranging terms:

$$\frac{\partial T_{abs}}{\partial t} = \alpha \left[\Theta^{losses}_{absorber}(x, t) + \dot{Q}_{sol \to abs} \right] \tag{5}$$

where $\alpha = 1/\rho_{abs} C_{abs} A_{abs}$ $\Theta^{losses}_{absorber}(x, t)$ groups the convection and radiation losses terms. Thus, ρ_{abs}, C_{abs} and A_{abs} are the density, specific heat, and cross-sectional area of the absorber pipe. These values stay constant. The functions $\dot{Q}^{rad}_{abs-env}$ and $\dot{Q}^{conv}_{abs-env}$ are the heat losses from the absorber to the glass envelope via radiation and convection, calculated with (6)–(7). The function $\dot{Q}_{sol-abs}$ is the solar irradiance, considered as a heat flux, absorbed and it is calculated according to (8). This flux is independent of the temperatures and can be calculated before solving the system.

$$\dot{Q}^{rad}_{abs \to env} = \frac{\sigma \pi D_{abs,o}}{\frac{1}{\varepsilon_{abs}} + \frac{1-\varepsilon_{env}}{\varepsilon_{env}} \left(\frac{D_{abs,o}}{D_{env,i}} \right)} \left[T^4_{abs} - T^4_{env} \right] \tag{6}$$

$$\dot{Q}^{conv}_{abs \to env} = D_{abs,o} h_{abs-env} [T_{abs} - T_{env}] \tag{7}$$

$$\dot{Q}_{sol \to abs} = G_{DNI} w_{PTC} \eta_0(\theta) k(\theta) \zeta(\theta) \eta_{shading}(\theta) \tag{8}$$

where σ is the Stefan–Boltzmann constant (5.67×10^{-8} W/m^{2}K^4), $D_{abs,o}$ is the outer diameter of the absorber pipe, $D_{env,i}$ is the inner diameter of the glass envelope, ε_{env} and ε_{abs} are the selective coating emissivity of the glass envelope and the absorber, respectively. $h_{abs-env}$ is the convection heat transfer coefficient between the absorber and the glass envelope. G_{DNI} is the direct normal irradiance, w_{PTC} is the width of the PTC, $\eta_0(\theta)$ is the peak optical efficiency, $k(\theta)$ is the incidence angle modifier, $\zeta(\theta)$ is the geometrical end effect and $\eta_{shading}(\theta)$ is the shading efficiency. All the variables with the form $X = f(\theta)$ are function of the angle of incidence, which is calculated according to the tracking mode and orientation of the PTC. T_{env} is the glass envelope temperature which is calculated as follows:

$$\rho_{env} C_{env} A_{env} \frac{\partial T_{env}}{\partial t} = \dot{Q}^{rad}_{abs \to env} + \dot{Q}^{conv}_{abs \to env} - \dot{Q}^{rad}_{env \to sky} - \dot{Q}^{conv}_{env \to air} \tag{9}$$

Rearranging terms:

$$\frac{\partial T_{env}}{\partial t} = \xi \left[\Phi^{\text{losses}}_{\text{abs}}(x, t) - \Psi^{\text{losses}}_{\text{env}}(x, t) \right] \tag{10}$$

where $\xi = 1/\rho_{env} C_{env} A_{env}$, $\Phi_{\text{abs}}^{\text{losses}}(x, t)$ groups the heat gain from the absorber to the glass cover and $\Psi_{\text{env}}^{\text{losses}}(x, t)$ groups the heat losses from the glass envelope external surface, Then, ρ_{env}, C_{env} and A_{env} are the density, specific heat, and cross-sectional area of the glass envelope. The remaining functions $\dot{Q}_{env-sky}^{rad}$ and $\dot{Q}_{env-air}^{conv}$ are the heat losses from the glass envelope via radiation and convection, calculated with (11)–(12).

$$\dot{Q}_{env \to sky}^{rad} = \pi D_{env,o} \sigma \varepsilon_{env} \left[T_{env}^4 - T_{sky}^4 \right] \tag{11}$$

$$\dot{Q}_{env \to air}^{conv} = \pi D_{env,o} h_{env-a} [T_{env} - T_{air}] \tag{12}$$

where $D_{env,o}$ is the outer diameter of the glass envelop, h_{env-a} is the heat transfer coefficient that considers the convective heat transfer between the glass envelope and its surrounding. T_{air} is the temperature of the air surrounding the HCE and T_{sky} is the effective temperature of the sky, assumed to be 8 °C lower than T_{air}.

Some assumptions are needed to solve the coupled PDE-ODEs system. Firstly, the convective heat transfer coefficients are evaluated at average temperatures. Secondly, the solar flux pattern on the HCE is considered uniform. In consequence, the HTF mass flow is heated uniformly. Third, negligible thermal resistance from absorber coating and glass envelope anti-reflection treatment.

The Eqs. (1)–(12) are the major expressions to describe the heat transfer phenomenon in PTC plants. However, most heat transfer coefficients, HTF thermal properties, SCA optical and geometrical properties require their own equations. Most of them are functions of the absorber, glass envelope or HTF temperature. The works of [13–16] explain in detail the different expressions used in this research.

The initial conditions are defined according to the proposed approximation [17]. Particularly for the HTF temperature initial condition $T_{HTF}(x, 0)$, it is assumed to be equal to the first value in the dataset. Regarding the boundary conditions at the fluid entrance, it is assumed that the HTF temperature is the same as the first value in the dataset, i.e., the boundary condition is of the first-type. The other BCs are of the second-type, i.e., Neumann conditions, representing adiabatic conditions for the pipe and the glass envelope.

Writing the final model in a more compact way, results in the following structure:

$$\begin{cases} \begin{array}{ll} \frac{\partial T_{HTF}}{\partial t} = \Omega(x,t) \frac{\partial T_{HTF}}{\partial x} + \chi(x,t) & \\ \frac{\partial T_{abs}}{\partial t} = \alpha \left[\Theta_{absorber}^{losses}(x,t) + \dot{Q}_{sol \to abs} \right] & 0 < x < l_{row} \\ \frac{\partial T_{env}}{\partial t} = \xi \left[\Phi_{\text{abs}}^{\text{losses}}(x,t) - \Psi_{\text{env}}^{\text{losses}}(x,t) \right] & \end{array} \\ \begin{cases} T_{HTF}(0,t) = T_{dataset}^{start} & \frac{\partial T_{HTF}(l_{row},t)}{\partial t} = 0 \\ \frac{\partial T_{abs}(0,t)}{\partial t} = 0 & \frac{\partial T_{abs}(l_{row},t)}{\partial t} = 0 \\ \frac{\partial T_{env}(0,t)}{\partial t} = 0 & \frac{\partial T_{env}(l_{row},t)}{\partial t} = 0 \end{cases} \quad t > 0 \\ \begin{cases} T_{HTF}(x,0) = T_{dataset}^{start} \\ T_{abs}(x,0) = T_f(x,0) + 4.5°\text{C} \\ T_{env}(x,0) = 0.2 \left[T_f(x,0) + 7°\text{C} \right] + 0.8 T_{air} - 1 \end{cases} \quad 0 \le x \le l_{row} \end{cases} \tag{13}$$

The terms $\chi(x,t)$, $\Theta_{absorber}^{losses}(x,t)$, $\Phi_{\text{abs}}^{\text{losses}}(x,t)$ and $\Psi_{\text{env}}^{\text{losses}}(x,t)$ are the so-called reaction terms. These terms are all nonlinear and are highly dependent on the temperature of the HTF, absorber, glass cover, thermal properties of the HTF, optical and geometrical properties of the SCA and HCE. In this work, the thermal properties of the absorber pipe, glass cover geometrical, and optical properties remain fixed as parameters while the HTF's thermal properties vary according to temperature changes.

The model presented is solved with the three-point backward implicit finite-difference scheme (TPB) proposed by [10]. The numerical method is specially developed for hyperbolic-parabolic PDEs. The accuracy of the spatial and temporal terms became of second order. Nevertheless, the resulting finite difference equation is not linear because of the nonlinear reaction terms. To manage it, the two-step expansion technique proposed in [10] is implemented to linearize the finite difference equation and uncouple it from the ODEs. Thus, the term A of the structure $\mathbf{Ax} = \mathbf{b}$ becomes tridiagonal, and then the iterative Generalized Minimum Residual Method (GMRES) is used via MATLAB function `gmres` [18]. The following algorithm describes the steps followed in this work.

Algorithm 1 three-points backward implicit finite-difference scheme

Inputs: $T_{amb}, DNI, \theta, \dot{m}_f, T_{HTF}^{inlet}$
Outputs: $T_{HTF}(x,t), T_{HTF}^{out}, T_{abs}(x,t), T_{env}(x,t)$
Parameters: Thermal, geometric, and optical properties of SCA and HCE
Discretize the spatial computational domain using $\delta_x = l_{row}/(n_x - 1)$.
Discretize the temporal computational domain using δ_t equal to dataset sampling time.
Initialization: Temperature distributions of $T_{HTF}(x,0), T_{abs}(x,0), T_{env}(x,0)$ and $\dot{Q}_{sol\rightarrow abs}(t)$ for the complete domain
While $\varepsilon = 10^{-4}$, convergence criterion
  For $k = 1{:}\,n_t$, number of time values in dataset
    For $j = 1{:}\,n_x$, number of nodes in axial direction
      Compute two-step linearization of $\chi(x,t)$, $\Theta_{absorber}^{losses}(x,t)$, $\Phi_{\text{abs}}^{\text{losses}}(x,t)$, $\Psi_{\text{env}}^{\text{losses}}(x,t)$
      Apply TPB structure of [10] to uncouple the nonlinear reaction terms.
      Build matrices **A** and **b**.
    End
  End
End

2.2 *Case Study 1: HTF-PTC Test Loop at the Plataforma Solar De Almería*

The Plataforma Solar de Almería (PSA) is Europe's largest concentrating solar technology R&D and test center. It is located in the Tabernas desert, Spain (Latitude:

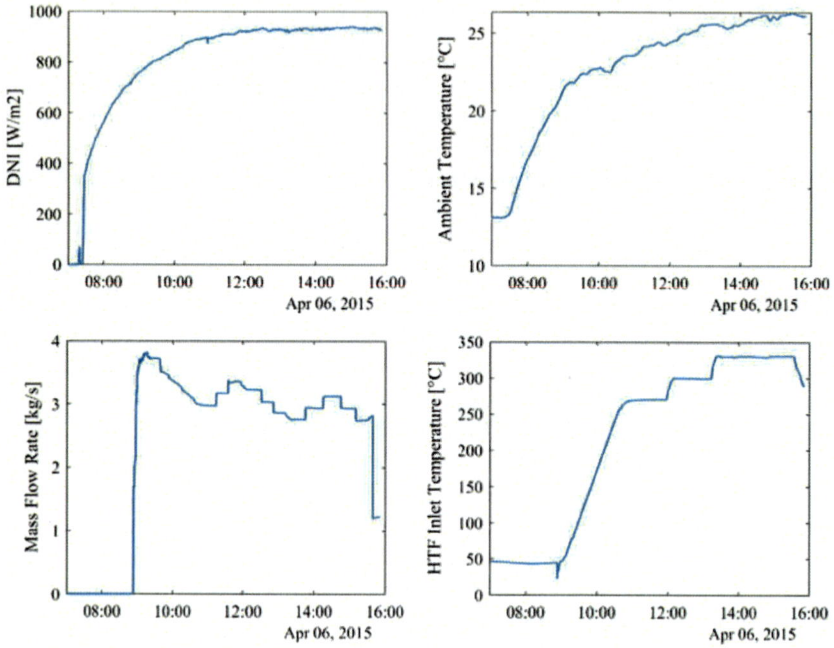

Fig. 3 Input variables for case study 1

37°06′00″ N and Longitude: 2°21′54″ W). The center has the HTF-PTC test bench dedicated to studying new PTC technologies, heat transfer fluids, and auxiliary components under real solar conditions. Extensive information about this facility can be found in [19, 20]. The variables depicted in Fig. 3 are the measured inputs from experiments in which the mass flow rate is varied on a clear-sky day and captures the HTF outlet temperature evolution.

The objective is to predict the HTF outlet temperature of the HTF-Loop with a 75 m collector EuroTrough with the model presented in the Sect. 2.1. Then, the simulated results are compared with the experimental data, and conclusions are elaborated about the usefulness of the model. Table 1 resumes the principal parameters considered in the simulation.

2.3 *Case Study 2: Commercial 50 MW PTC Power Plant*

As stated in the Introduction, the solar field in a PTC plant consists of multiple SCAs. These SCAs are connected in series and form multiple parallel rows. In this case, the complete solar field of a commercial PTC solar power plant is simulated. The plant's capacity is 50 MW and is located in Spain. The specific location is not provided for confidentiality reasons, and the input variables illustrated in Fig. 4 were rescaled

Table 1 Principal parameters for simulations in the case study 1 and 2

Parameter	Unit	Case study 1	Case study 2
Collector technology	–	EuroTrough 150	
HCE technology	–	Schott PTR70	
Heat transfer fluid	–	Syltherm 800	Therminol VP1
Collector orientation	–	East–West	North–South
Collector width	m	5.750	
SCA length	m	150	
Number of SCA per row	–	0.5	4
Number of rows	–	0	160
Absorber density	kg/m^3	8027	
Absorber heat capacity	kJ/kg–°C	0.502	
Absorber outer diameter	m	0.070	
Absorber inner diameter	m	0.066	
Glass envelope density	kg/m^3	2400	
Glass envelope heat capacity	kJ/kg–°C	0.840	
Glass envelope outer diameter	m	0.120	
Glass envelope inner diameter	m	0.115	

without compromising the variables' nature. Similar to Case Study 1, the aim is to predict the solar field HTF outlet temperature based on the input variables of Fig. 4. Additionally, due to the strong transient behavior of the DNI, the model is assessed to evaluate if it is capable of handling abrupt disturbances. Then, the simulated results are compared with the experimental data, and conclusions are elaborated about the usefulness of the model to study large PTC solar plants. Table 1 resumes the principal parameters considered in the simulation.

3 Simulation Results

The results presented in this section are focused on the HTF temperature evolution, even though the algorithm can predict the temperature of the absorber pipe and the glass envelope, these temperatures may not be useful from a control point of view. However, the thermal evolution of the absorber could be of interest if the aim were asset material performance due to thermal and mechanical stresses.

For case study 1, the HTF test bench was activated in the morning, around 8:00 a.m. local time, when the irradiance and ambient temperature increased. The experiment lasted until 4 p.m., during which the flow rate was varied, as illustrated in Fig. 3. The DNI plot shows clear-sky conditions throughout the test period; thus, no disturbances influenced the dynamics. In terms of the simulation, this was conducted with a sampling time of 5 s with a 10^{-4} convergence criterion. As a result, the algorithm

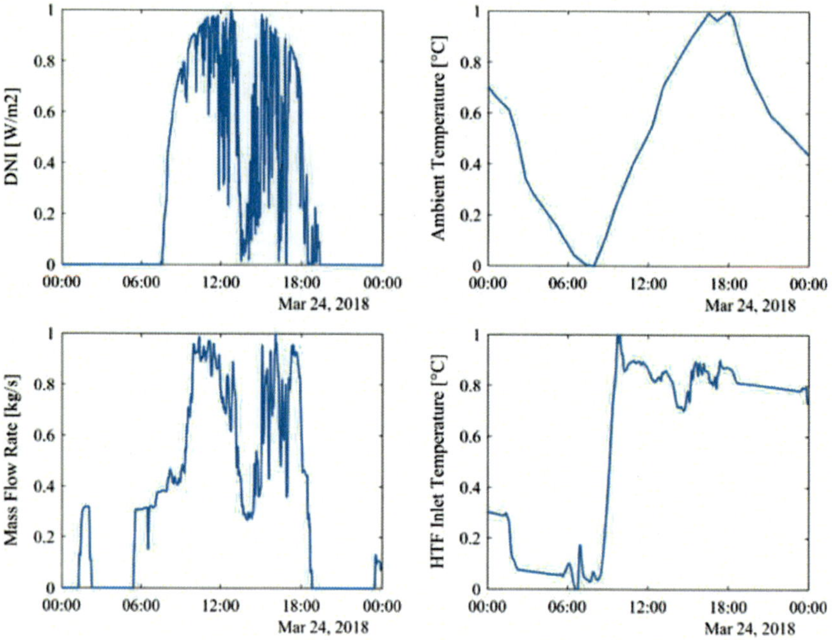

Fig. 4 Input variables for case study 2

computed the temperature profiles in 0.7 s for the whole test period, and the mean absolute percentage error (MAPE) reached was around 1.5%. For instance, Fig. 5 depicts the evolution of the HTF temperature along the 75 m SCA at each time step. This kind of HTF temperature profile prediction in such a short time allows designing efficient defocusing control strategies to avoid overheating the fluid. On the other hand, Fig. 6 compares the reached HTF outlet temperature via the model and the experimental data from the PSA test bench. The output temperature model is comparable to the test bench, ensuring accurate forecasts for actual PTC outlet temperatures.

For case study 2, the data in Fig. 4 presents the input variables that affected the regular operation of a PTC solar plant in Spain. The solar plant was operating from 08:15 a.m. to 11:50 p.m. As shown in Fig. 4, the flow rate is the manipulated variable, and the DNI and ambient temperature changes are measured. In this case, the multiple clouds and shadows in the solar field caused abrupt changes in the DNI, disturbing the system. Regarding the numerical performance, this case was conducted with a sampling time of 1 min with a 10^{-4} convergence criterion. As a result, the algorithm computed the temperature profiles in 2.7 s for the whole day, and the MAPE was around 1.5%. These results show an adequate level of accuracy while the computation speed is not compromised. For instance, Fig. 7 depicts the evolution of the HTF temperature along the 600 m row at each time step. Similar to case study 1, having this kind of HTF temperature profile prediction allows the design of

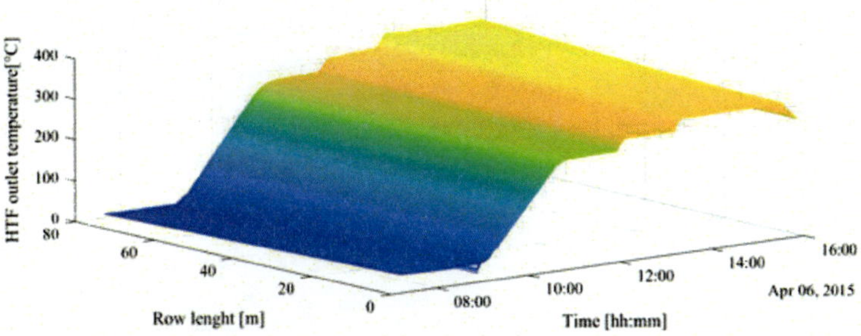

Fig. 5 HTF temperature evolution along the HCEs and time

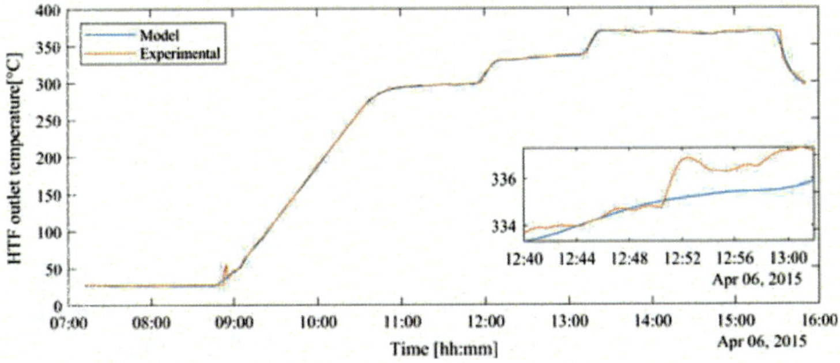

Fig. 6 Comparison between the HTF outlet temperature achieved by the model and the PSA test bench during a clear-sky day

automatic control strategies and dispatch energy strategies for wholesale electricity markets. On the other hand, Fig. 8 compares the reached HTF outlet temperature via model and the measured lar data from the real plant. The output temperature model is similar to the real plant, ensuring accurate forecasts for real PTC solar fields.

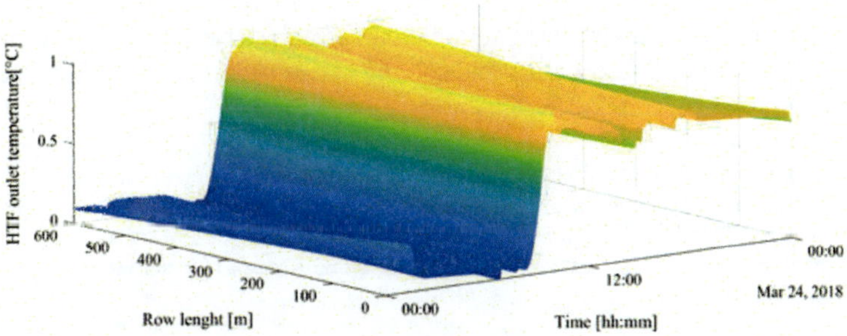

Fig. 7 HTF temperature evolution along the row and time

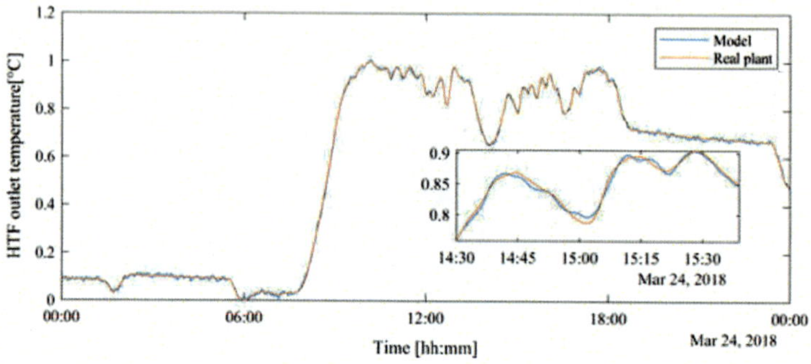

Fig. 8 Comparison between the HTF outlet temperature achieved by the model and an actual PTC solar field during a day with clouds

4 Conclusions and Future Work

This paper implements a three-point backward implicit finite-difference scheme in MATLAB to solve the coupled PDE-ODEs system that represents the dynamic behavior of parabolic trough solar plants.

The results from two case studies underline the viability and significance of numerical implementation to further applications in automatic control and market-based dispatch strategies for concentrating solar power plants. The first case study showcases the SCA response to varying direct normal irradiance. The model's remarkable speed in predicting temperature profiles, as short as 0.7 s, highlights its efficacy in enabling efficient automatic control strategies, a critical aspect in preventing fluid overheating. The second case study accentuates the PTC solar field resilience in the face disturbances caused by clouds and shadows. Even under such abrupt DNI fluctuations, the model's accurate predictions emphasize its robustness. A computation time of 2.7 s for temperature profiles over an entire day demonstrates both accuracy and computational efficiency, reinforcing the model's practicality for real-time applications.

In futures works, the model will be used to develop automatic control strategies for defocusing of SCA, balancing the solar field, thermal energy production, electricity production under DNI uncertainty and solving self-scheduling problems to allocate optimal generation.

Acknowledgements Richard Rangel acknowledges the financial support from the Colombian Ministry of Sciences, Technology, and Innovation through the National Doctoral Program 785-2018 and Universidad del Norte through the scholarship contract N° UN-OJ-2018-41048.

We thank Eduardo Zarza (CIEMAT-Plataforma Solar de Almería) for sharing the experimental data to validate the model.

References

1. IRENA: Electricity storage and renewables: costs and markets to 2030. Abu Dhabi (2017). http://irena.org/publications/2017/Oct/Electricity-storage-and-renewables-costs-and-markets, https://www.irena.org/-/media/Files/IRENA/Agency/Publication/2017/Oct/IRENA_Electricity_Storage_Costs_2017.pdf
2. Lovegrove, K., et al.: Comparison of dispatchable renewable electricity options. Australia (2018). https://arena.gov.au/knowledge-bank/comparison-of-dispatchable-renewable-electricity-options/.
3. Burghi, A.C.D.A., Hirsch, T., Pitz-Paal, R.: CSP dispatch optimization considering forecast uncertainties. In: AIP Conference Proceedings, vol. 2033, November 2018 (2018). https://doi.org/10.1063/1.5067206
4. Denholm, P., Mehos, M.: Enabling greater penetration of solar power via the use of CSP with thermal energy storage (2011)
5. Helman, U., Jacobowitz, D.: The economic and reliability benefits of CSP with thermal energy storage: literature review and research needs (2014). https://www.inship.eu/docs/TES4the_economic_and_reliability_benefits_of_csp_with_thermal_storage_2014_09_09_final.pdf
6. Sioshansi, R., Denholm, P.: The value of concentrating solar power and thermal energy storage. IEEE Trans. Sustain. Energy **1**(3), 173–183 (2010). https://doi.org/10.1109/TSTE.2010.2052078
7. Wittmann, M., Eck, M., Pitz-Paal, R., Müller-Steinhagen, H.: Methodology for optimized operation strategies of solar thermal power plants with integrated heat storage. Sol. Energy **85**(4), 653–659 (2011). https://doi.org/10.1016/j.solener.2010.11.024
8. Ela, E., Milligan, M., Bloom, A., Botterud, A., Townsend, A., Levin, T.: Evolution of wholesale electricity market design with increasing levels of renewable generation. Golden, CO (United States), September 2014 (2014). https://doi.org/10.2172/1159375
9. IEA: Technology roadmap: solar thermal electricity 2014, Paris (2014). https://webstore.iea.org/technology-roadmap-solar-thermal-electricity-2014
10. Wu, J.C., Fan, L.T., Erickson, L.E.: Three-point backward finite-difference method for solving a system of mixed hyperbolic—parabolic partial differential equations. Comput. Chem. Eng. **14**(6), 679–685 (1990). https://doi.org/10.1016/0098-1354(90)87036-O
11. Kalogirou, S.: Solar Energy Engineering. Elsevier (2009)
12. Blanco, M., Ramirez, L. (eds.): Advances in Concentrating Solar Thermal Research and Technology, vol. 91. Elsevier (2017)
13. Kalogirou, S.A.: A detailed thermal model of a parabolic trough collector receiver. Energy **48**(1), 298–306 (2012). https://doi.org/10.1016/j.energy.2012.06.023
14. Forristall, R.: Heat transfer analysis and modeling of a parabolic trough solar receiver implemented in engineering equation solver. Golden, CO (United States), October 2003. https://doi.org/10.2172/15004820
15. Fasquelle, T., Falcoz, Q., Neveu, P., Lecat, F., Flamant, G.: A thermal model to predict the dynamic performances of parabolic trough lines. Energy **141**, 1187–1203 (2017). https://doi.org/10.1016/j.energy.2017.09.063
16. Powell, K.M., Edgar, T.F.: Modeling and control of a solar thermal power plant with thermal energy storage. Chem. Eng. Sci. **71**, 138–145 (2012). https://doi.org/10.1016/j.ces.2011.12.009
17. Wagner, M. J., Gilman, P.: Technical manual for the SAM physical trough model. Golden, CO (United States), (2011). [Online]. http://www.nrel.gov/docs/fy11osti/51825.pdf
18. MathWorks: GMRES: solve system of linear equations (2023). https://www.mathworks.com/help/matlab/ref/gmres.html. Accessed 04 May 2022
19. Sallaberry, F., Valenzuela, L., Palacin, L.G.: On-site parabolic-trough collector testing in solar thermal power plants: experimental validation of a new approach developed for the IEC 62862-3-2 standard. Sol. Energy **155**, 398–409 (2017). https://doi.org/10.1016/j.solener.2017.06.045
20. Valenzuela, L., López-Martín, R., Zarza, E.: Optical and thermal performance of large-size parabolic-trough solar collectors from outdoor experiments: a test method and a case study. Energy **70**, 456–464 (2014). https://doi.org/10.1016/j.energy.2014.04.016

Transition to Rational Models of Production and Consumption in the Framework of the UN Sustainable Development Goals

S. Absalyamova, L. Zulfakarova, G. Shafigullina, R. Sakhapov, and M. Makhmutov

Abstract The article is devoted to the problems of transition to rational models of production and consumption as the basis for the formation of a circular economy. The authors consider the evolution of theoretical approaches to the formation of the concept of circular economy. The paper determines the factors of transition to rational models of production and consumption, and proposes an economic and mathematical model for a comprehensive assessment of the impact of economic, social, demographic factors on the development of a circular economy. The national initiatives of individual countries in the field of regulating the development of the circular economy for the implementation of the UN Sustainable Development Goals were studied in the article. Indicators are proposed that evaluate the effectiveness of achieving the UN SDGs. The authors analyse best practices of implementing the circular economy model by individual companies and countries. Finally, the research identifies the problems of Russia's transition to a circular economy and proposes ways to solve them.

Keywords Circular economy · Sustainable development · Rational models of production and consumption · Sharing economy

1 Introduction

Today, the transition to sustainable patterns of production and consumption is one of the seventeen UN Sustainable Development Goals (SDGs), which were adopted on September 25, 2015 by UN member states to combat poverty and to preserve the planet's resources and ensure the well-being of future generations [1].

S. Absalyamova (✉) · L. Zulfakarova · G. Shafigullina
Kazan Federal (Volga Region) University, 18 Kremlyovskaya Street, 420008 Kazan, Russia
e-mail: svetlana.absalyamova@kpfu.ru

R. Sakhapov · M. Makhmutov
Kazan State University of Architecture and Engineering, Zelenaya Str., 1, 420043 Kazan, Russia

D. Vlachos (ed.), *Mathematical Modeling in Physical Sciences*, Springer Proceedings in Mathematics & Statistics 446, https://doi.org/10.1007/978-3-031-52965-8_50

Calls for a shift to sustainable production and consumption patterns have repeatedly been heard before, but now they are officially enshrined in the 12th UN Sustainable Development Goal, which aims to replace the outdated model of production and consumption with a new sustainable model—the circular economy.

The term "circular economy" has several meanings. On the one hand, this is the antonym of the linear economy, the functioning of which is based on the transformation of resources into waste as a result of the production process, which leads to a reduction in natural resources, pollution of the soil and atmosphere, and environmental crises.

On the other hand, the circular economy's key concept is "cycle", namely, biogeochemical cycles and cycles of reuse of products and waste. The biogeochemical cycle can be attributed to the water cycle: water evaporating from the oceans forms rain clouds, falls on land in the form of rain, flows into rivers and returns back to the ocean.

The idea of recycling products is related to the cycle of resources. The concept of "waste as food", in which the waste from one production process is used as a raw material in another production process, reducing consumption through the sharing or reuse of goods, as well as recycling, is the core of the circular economy.

There are the following types of business models that meet the principles of a circular economy [2]:

- circular value chains, in which limited resources are replaced by fully renewable ones;
- recovery and recycling—a model in which technological innovation creates the possibility of recovery and recycling of natural resources;
- extending the life cycle of a product through repair, refurbishment or remarketing. This business model is based on the transition from selling things to providing services for their use;
- joint consumption and use—a model based on providing access to goods or assets that have a low utilization rate;
- product as a service (servicing)—a model in which customers pay for the rent of goods or property with payment upon use.

The circular economy is characterized by "3Rs"—Reduce, Reuse and Recycle: optimization of the production process, sharing or reuse of goods, recycling of waste [3]. Today, the circular economy is capable of transforming into a "4R", where the fourth principle is the restoration of resources.

2 Material and Method

The current approach to the concept of a circular economy has been developed over several decades. Originally born at the intersection of two sciences: ecology and economics, it gradually went beyond them, and at present, it is thanks to the development of the sharing economy that the circular economy has received a social

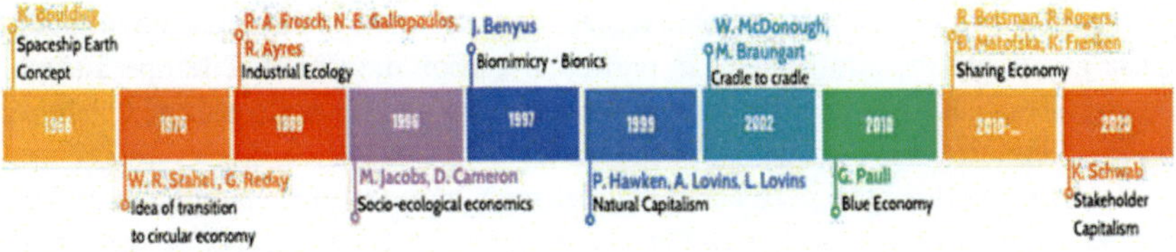

Fig. 1 Stages of formation of the concept of circular economy [3–16]

component. The milestones shown on Fig. 1 confirm the commonality of the idea that connects all the above teachings.

The milestones shown at the Fig. 1 confirm the commonality of the idea that connects all the above teachings.

Building on concepts such as "cradle to cradle"; biomimetics; industrial ecology; sharing economy; stakeholder capitalism—the circular economy today focuses on the optimization and sustainable development of the entire economic system, rather than its individual components.

Currently, the concepts of "weak" and "strong" sustainable consumption are distinguished, which is reflected on the UN website.

The principles of "weak" sustainable consumption include:

- choosing products that cause less harm to the environment;
- choosing products that cause the least harm to human health;
- choosing products which production requires less resources;
- choosing products taking into account the benefits for the buyer, seller and manufacturer.

This type of sustainable consumption creates demand for greener alternatives in various industries, stimulates the development of new technological solutions, which gradually leads to changes in the economic model itself.

In "strong" sustainable consumption, the focus is shifting to a change in the consumption model itself, therefore, within its framework, they talk about dematerialization, changes in supply and demand for goods and services, conscious choice and use of goods, as well as a critical assessment of consumption drivers.

3 Results

The circular economy would not be able to develop so quickly if there were no appropriate prerequisites for its formation, which originated both in economic theory and in the bowels of the traditional economic model itself.

First of all, the environmental factor has become a serious reason for the transition to a circular economy. As Stephen Cohen writes in his article in the Huffington Post (2015), people today are much more aware of the environmental crisis, they are increasingly aware that they can contribute to solving environmental problems [17].

The main source of anthropogenic impact on the climate is the emission of greenhouse gases into the atmosphere, in particular carbon dioxide. The temperature of the world's oceans has become higher, eternal snow and ice are melting. As shown in Fig. 2, emissions are growing at a high rate.

If no action is taken to mitigate climate change, temperatures could increase by 5 degrees by 2100. The world community has realized the significance of the threats and the need to move to a new environmental policy.

No less significant impact on the development of the circular economy was the technological factor, the emergence and rapid spread of information technologies that can unite owners of underutilized assets or resources on one platform.

Before the advent of IT technologies, the costs of finding participants in the sharing of goods, property, and waste significantly exceeded the growth in total utility from a more rational use of existing assets. It was online platforms that contributed to the increase in the efficiency of the implementation of the accumulated production potential, accelerated the transition to a circular economy.

The transition to a circular economy was influenced by economic factors too. At the beginning of the 21st century, crisis phenomena began to manifest themselves more and more clearly in the economies of developed countries.

This was partly due to the formation of a new technological order, the formation of which took place just in the first decades of the new century.

Social factors also contributed to the successful development of the circular economy—this is a return to social values, awareness of the importance of social interaction for the preservation of the environment (Fig. 3).

Thus, the factor model of the circular economy can be represented as follows:

CE = f (TF, EF, SF, EnvF), where.

TF—technological factor;

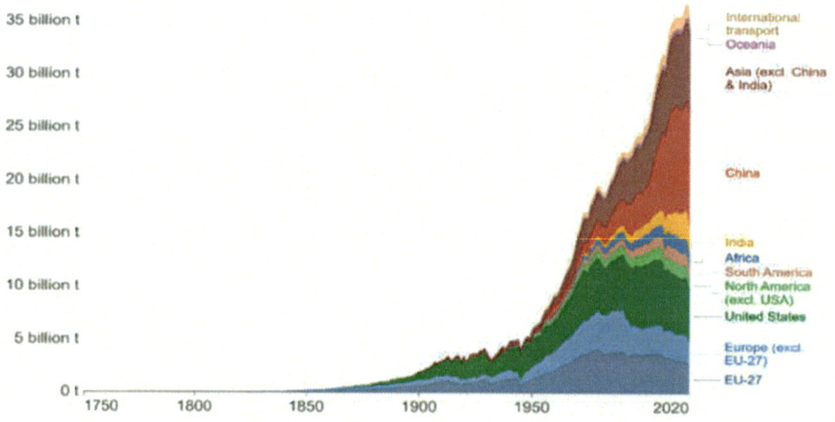

Fig. 2 Annual CO_2 emissions, tons [18]

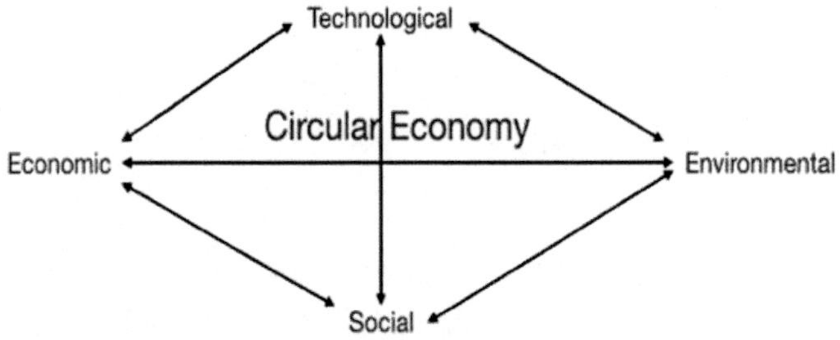

Fig. 3 Model of interaction of factors in the formation of a circular economy

EF—economic factor;
SF—social factor;
EnvF—an environmental factor.

New models of production and consumption are reflected not only in academic works, but also in the priorities of the practical activities of many states and private businesses.

Thus, the European Community adopted programs for the development of the "green", circular, and bio-economy for 2030–2050. The Paris climate agreement is aimed at the transition of all states to a low-carbon economy.

Developed countries are undergoing colossal structural and technological shifts aimed at reducing the consumption of traditional hydrocarbons; the share of renewable energy sources (RES), supported by a variety of economic instruments: taxes, loans, subsidies, etc., is sharply increasing.

The actions of many organizations in the world are aimed at the transition from a linear economy to a cyclical one. For example, the initiatives of the Ellen MacArthur Foundation: New Plastics Economy initiatives on a new approach to handling plastic so that it does not become garbage; Food Initiative—to rethink the food system and apply regenerative practices; Make Fashion Circular—on the restructuring of the fashion industry. The Rockefeller Foundation runs challenges and initiatives to transform the food system; IDEO has created a cycle design guide.

Global multinationals such as BP and Shell are now moving away from oil and gas and into integrated energy companies. Their new strategy includes finding ways to dramatically increase the production of low-carbon fuels such as biofuels, developing green energy. In Russia, Tatneft announced the transition to carbon neutrality by 2050.

The Board of Directors adopted a new environmental policy taking into account climate change and approved a roadmap. Targets have been set for a phased reduction in greenhouse gas emissions by 10% by 2025 and by 20% by 2030 from the base year of 2016 [19]. In developing the greenhouse gas emissions management system, Tatneft adheres to international standards and guidelines: GreenHouseGas (GHG) protocol—for measuring, controlling and disclosing emissions indicators, including supply chains; TCFD guidelines—carbon footprint reporting, data for the CDP climate rating.

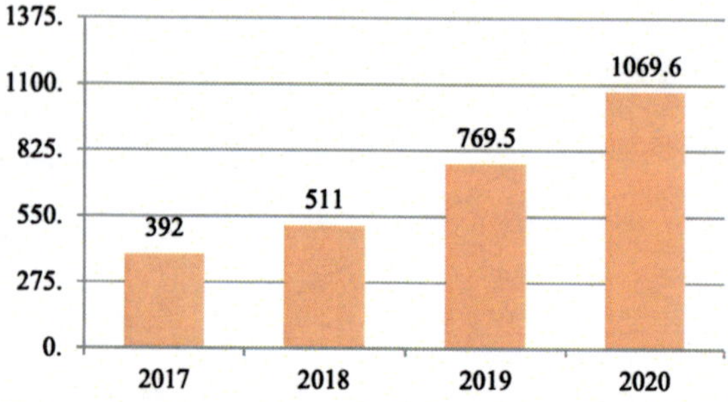

Fig. 4 Dynamics of the volume of sharing economy in Russia in 2017–2020, bln rub [20]

In assessing the scope of emission accounting, we intend to expand the application of the Corporate Standard for Accounting and Reporting of the Greenhouse Gas Protocol of the World Business Council for Sustainable Development (WBCSD) and the World Resources Institute (WRI).

The transition to new rational production models is accompanied by the rapid development of new consumption patterns.

As seen on Fig. 4, in recent years, the sharing economy in Russia has grown almost 3 times. The top 5 sectors of the sharing economy in 2022 included C2C commerce, P2P services, carsharing, carpooling, short-term rental housing—this is 98.7% of the entire Russian sharing market.

The sharing economy is understood as a socio-economic model of relations between its participants, based on changing the motives of consumer behavior, the transition to rational production and consumption, social interaction in the process of using underutilized assets through appropriate online platforms, contributing to the maximization of overall utility, more efficient distribution of resources and implementation of the principles of sustainable development [20, 21].

It assumes the minimization of costs while maximizing the level of utility of the good, achieving a higher level of overall utility within the available resource capabilities.

4 Conclusion

Thus, we observe two simultaneously occurring processes—the rationalization of the behavior of producers and consumers, which ultimately leads to a synergistic effect in the implementation of the idea of transition to a circular economy. A wide range of subjects is involved in the process of changing production and consumption patterns—states, international organizations, firms, and population groups.

Therefore, the transition to these models and the expansion of their use presupposes the adoption of a whole range of measures on a systematic basis, the development and implementation of which should account for different stakeholders.

Experts ambiguously evaluate the consequences of the development of new models, their impact on the functioning of traditional industries. The possible problems can be unequal competitive conditions between traditional and "new" types of entrepreneurship, the possibility of monopolizing certain industries using online platforms, etc.

The International Organization for Standardization (ISO) has established a technical committee—ISO/TC 323 "Circular Economy", whose task is to develop fundamental standards, requirements, principles, guidelines for the implementation of circular economy projects.

Today, many countries have accumulated some experience in the transition to a circular economy, have developed a certain mechanism for its regulation, which is of undoubted interest. Russia is also consistently stepping up measures aimed at implementing the concept of sustainable development and achieving its goals, developing targets with which Rosstat monitors the implementation of the roadmap to achieve the SDGs.

Thus, there were adopted the Decree of the President of the Russian Federation dated April 19, 2017 No. 176 "On the Strategy for the Environmental Safety of the Russian Federation for the period up to 2025", the Decree of the Government of the Russian Federation dated January 25, 2018 No. 84-r "On Approval of the Strategy for the Development of Industry on processing, recycling and neutralization of production and consumption waste for the period up to 2030".

The practical implementation of these strategies requires the adoption of a set of measures covering the most diverse areas of human life, one way or another related to the functioning of the circular economy.

These measures should contribute to the realization of the advantages of the circular economy, its potential in achieving sustainable development goals, and be aimed at involving an increasing number of participants in the process of transition to rational models of production and consumption.

Acknowledgements This paper has been supported by the Kazan Federal University Strategic Academic Leadership Program.

References

1. UN Sustainable Development Goals. (2015). Available from: https://www.un.org/sustainabledevelopment/ru/sustainable-development-goals/
2. Rutqvist, J., Lacy, P.S.: Waste to Wealth: The circular economy advantage [Internet] (2015). Available from: https://openlibrary.org/books/OL28262157M/Waste_to_Wealth
3. Ghisellini, P., Cialani, C., Ulgiati, S.: A review on circular economy: the expected transition to a balanced interplay of environmental and economic systems. J. Clean. Prod. [Internet] **114**, 11–32 (2016). Available from: https://doi.org/10.1016/j.jclepro.2015.09.007
4. Boulding, K.E.: The economics of the coming spaceship earth. Environmental Quality in a Growing Economy: Essays from the Sixth RFF Forum. John Hopkins University, Baltimore (1966)

5. Reday, G., Stahel, W.R.: Potential for Substituting Manpower for Energy. Battelle Memorial Institute (1976)
6. Ayres, R.U., Simonis, U.E.: Industrial Metabolism: Restructuring for Sustainable Development (1994)
7. Frosch, R.J., Gallopoulos, N.E.: Strategies for manufacturing. Sci. Am. [Internet] **261**(3), 144–152 (1989)
8. Benyus, J.M.: Biomimicry: Innovation Inspired by Nature [Internet]. (1997) Available from: https://ci.nii.ac.jp/ncid/BB03290291
9. Jacobs, M.J.: What is socio-ecological economics?. Ecol. Econ. Bull. [Internet] (1996). Available from: https://eprints.lancs.ac.uk/id/eprint/22089
10. Hawken, P., Lovins, A.B., Lovins, L.H.: Natural capitalism: The next industrial revolution. Earthscan (1999)
11. McDonough, W., Braungart, M.: Cradle to Cradle: Remaking the Way We Make Things. North Point Press (2010)
12. Pais, I., Provasi, G.: Sharing economy: A step towards the re-embeddedness of the economy? Stato e mercato (2015)
13. Botsman, R., Rogers, R.: What's Mine is Yours: How Collaborative Consumption is Changing the Way We Live. HarperCollins, UK (2011)
14. Frenken, K., Meelen, T., Arets, M., Van de Glind, P.: Smarter regulation for the sharing economy. The Guardian (2015)
15. Matofska, B.: What is the sharing economy. The people who share (2016)
16. Davos manifest, (2020). Available from: https://reputationcapital.blog/tag/davosskij-manifest/
17. Cohen, S.: The growing level of environmental awareness. Huffington Post (2014)
18. The Global Carbon Budget dataset https://doi.org/10.18160/gcp-2021
19. Tatneft report on sustainability, (2021). Available from: https://tatneft.ru/press-tsentr/press-relizi/more/8328?lang=ru.
20. RAEC, TIAR-CENTER: Sharing Economy (2020). https://tiarcenter.com/sharing-2020-key-results/
21. Absalyamov, T., Absalyamova, S.: Ekonomika sovmestnogo potreblenia:razlichiya v podhodah k opredeleniyu sushnosti ekonomicheskoi modeli. Kazansky ekonomicheskiy vestnik **3**(47) (2020)

The Chemical Elements with Their Applications in Fuzzy Delta-Algebraic Systems

Manal Al-Labad, Shuker Khalil, and Ahmed Naji Hassan

Abstract The preponderance of algebraic system applications has lately been generalized to many types of fuzzy algebras. We created the use of a spatial class of algebras called Delta − algebras. Using Delta − algebra, we explored a new form of algebraic semi groups including, Delta − semigroup, Delta − subsemigroup, Delta − ideal semigroup, Delta − ideal semigroup homomorphism, and Delta − semigroup homomorphism. Some application about the atomic numbers and valency of the chemical elements are studied in Delta − algebras. Next, fuzzy logic (FL) terms include fuzzy Delta − subsemigroup, fuzzy Delta − ideal semigroup, fuzzy Delta − ideal semigroup, bipolar fuzzy Delta − subsemigroup, bipolar fuzzy Delta − ideal semigroup, and bipolar fuzzy Delta − ideal semigroup are introduced. In addition, specific basic aspects of our notions are studied and addressed. Any Delta − ideal in a classical collection is not required to be a Delta par − ideal. Therefore, in this paper, we show this matter is holed after they are generalized in non − classical sets like fuzzy sets and bipolar fuzzy sets, and they also take a novel form in algebraic semi groups. This article investigates the Delta − semigroup homomorphism image, translations, and product characteristics of bipolar fuzzy (Delta/Deltapar) − ideals semigroups, and their applications are shown.

Keywords Chemical elements · Atomic numbers · Valency · Fuzzy sets · Algebraic semigroup · Homomorphism image · Bipolar fuzzy set · Product characteristics · Translations

M. Al-Labad
Department of Mathematics, Faculty of Arts and Sciences, University of Petra, Amman, Jordan

S. Khalil (✉) · A. N. Hassan
Department of Mathematics, College of Science, University Basrah, Basrah 61004, Iraq
e-mail: shuker.khalil@uobasrah.edu.iq

D. Vlachos (ed.), *Mathematical Modeling in Physical Sciences*, Springer Proceedings in Mathematics & Statistics 446, https://doi.org/10.1007/978-3-031-52965-8_51

1 Introduction

According to fuzzy set theory, [1] objects can pertain to a classes or a relation to a marked degree. It partially enables an object to belong to multiple classes, as well as taking into account the value of each companion in terms of the test cases. Following that, fuzzy set (FS) expansions such as intuitionistic fuzzy (IF), soft fuzzy (SF), bipolar fuzzy (BF), and others are defined and studied for their applications in various areas in our lives see [2–6]. Bipolar fuzzy sets $(BFSs)$ are an expansion of fuzzy sets (FSs) with a membership degree (MD) spectrum of $[-1, 1]$.

In a bipolar fuzzy set, the (MD) 0 of an element suggests that it is not relevant to the compliant property, the (MD) (0, 1) of an element suggests that it is somewhat important to the property, and the (MD) $[-1, 0]$ of an element suggests that it is somewhat important to the implicit refute, see [7, 8].

In this research, we discovered some new structures of bipolar fuzzy ideals in algebraic semigroups. (FL) implications include fuzzy Delta – subsemigroups, fuzzy Delta – ideal semigroups, bipolar fuzzy Delta – subsemigroups, bipolar fuzzy Delta – ideal semigroups, and bipolar fuzzy Delta – ideal semigroups. Further to that, specific fundamental aspects of our ideas are being researched and debated. Any Delta – ideal in a classical collection does not have to be Delta – ideal.

However, in this study, they can be extended to non – classical sets like fuzzy sets and bipolar fuzzy sets, in furthermore to a novel form in algebraic semigroups. Some application about the atomic numbers and valency of the chemical elements are shown in Delta – algebras. This paper discusses the image of the Delta – semigroup homomorphism, translations, and the product of bipolar fuzzy (Delta/Deltapar) – ideals semigroups, as well as their applications.

2 Preliminaries

This section reviews the fundamental notions of algebras.

Definition 2.1 (*Maji et al.* [9]) Assume that $\nu \in \mho \neq \varnothing$, we say $(\mho, \triangle, \nu)$ is a Delta-algebra (for short, $\delta - A$) if $\triangle : \mho \times \mho \to \mho$ such that:

(1) $\vartheta \triangle \vartheta = \nu \ \forall \vartheta \in \mho$,
(2) $\nu \triangle \vartheta = \nu \ \forall \vartheta \in \mho$,
(3) $\vartheta \triangle \theta = \nu$ and $\theta \triangle \vartheta = \nu \to \vartheta = \theta$, for all $\theta, \vartheta \in \mho$,
(4) $\vartheta \triangle \theta = \theta \triangle \vartheta \neq \mho, \forall \vartheta \neq \theta \in \mho - \{\nu\}$,
(5) $\vartheta \triangle \theta = \nu$ and $\theta \triangle \vartheta = \nu \to \vartheta = \theta$, for all $\theta, \vartheta \in \mho$,

Also, any $\varnothing \neq S \subseteq \mho$ is called $\delta -$ subalgebra $(\delta - SA)$ of $\mho$ if $\vartheta \triangle \vartheta \in S$, $\forall \vartheta, \theta \in S$, the relation $\leq$ on $\mho$ is defined by $\vartheta \leq \theta \Leftrightarrow \vartheta \triangle \theta = \nu$.

Definition 2.2 (*Maji et al.* [9]) Suppos that $(\mho, \triangle, \nu)$ is a $(\delta - A)$ and $\varphi \neq Z \subseteq \mho$. The set Z is called a Delta-Ideal $(\delta -$ ideal$)$ of $\mho$ if it such that:

(1) $\vartheta \vartriangle \theta \in Z, \forall \theta, \vartheta \in Z$,
(2) If $\vartheta \vartriangle \theta \in Z$ and $\theta \in Z$, then $\vartheta \in Z$.

Definition 2.3 (*Maji et al.* [9]) Suppos that $(\mho, \vartriangle, \nu)$ is a $(\delta - A)$ and $\varphi \neq Z \subseteq \mho$. The set Z is called a Delta Par-Idel ($\bar{\delta}$ – ideal) of $\mho$ if it such that:

(1) $\nu \in Z$,
(2) If $\vartheta \in Z$ and $\vartheta \in \mho$, then $\vartheta \vartriangle \theta \in Z$.

Definition 2.4 (*Zadeh* [1]) A mapping $Q : \pi \rightarrow [0, 1]$ is a fuzzy set (FS) of π, where $\pi \neq \varnothing$.

Definition 2.5
(*Mahmood and Hassan* [10]) Suppose that $O : \mho \rightarrow [0, 1]$ is a (FS). The set $N(O, t)$ is defined by $N(O, t) = \{\vartheta \in \mho | O(\vartheta) \geq t\}$, where $t \in [0, 1]$ and $0 \leq O(\vartheta) \leq 1$.

Definition 2.6 (*Dar* [11]) Let $\mho \neq \varnothing$, then $B = (\vartheta, O^-, O^+)$ is called bipolar fuzzy set (BFS) where $O^- : \mho \rightarrow [-1, 0]$ and $O^+ : \mho \rightarrow [0, 1]$ are mappings. Assume $B = (\vartheta, O^-, O^+)$ is a (BFS) and $(\lambda, \gamma) \in [-1, 0] \times [0, 1]$. The sets $B_\lambda^- = \{\vartheta \in \mho | O^-(\vartheta) \leq \lambda\}$ and $B_\gamma^+ = \{\vartheta \in \mho | O^+(\vartheta) \geq \gamma\}$ are called negative $\lambda-$ cut and positive $\gamma-$ cut, respectively.

Remark 2.7 (*Zhang* [12]) Assume that $B = (\vartheta, O^-, O^+)$ is (BFS) in $\mho$, we defin $\mathfrak{I}^-$ and $\mathfrak{I}^+$ by $\mathfrak{I}^- = -1 - inf\{O^-(\vartheta); \vartheta \in \mho\}$ and $\mathfrak{I}^+ = -1 - sup\{O^+(\vartheta); \vartheta \in \mho\}$.

Definition 2.8 (*Zhang* [12]) Assume that $B = (\vartheta, O^-, O^+)$ is a (BFS) of $\mho$. This term $B_{(\hbar,\bar{\lambda})} = (\vartheta, O_\hbar{}^-, O_{\bar{\lambda}}{}^+)$ is called a translation of $B = (\vartheta, O^-, O^+)$, whenever $(\hbar, \bar{\lambda}) \in [\mathfrak{I}^-, 0] \times [0, \mathfrak{I}^+]$ with $O_\hbar{}^-(\vartheta) : \mho \rightarrow [-1, 0]$ & $O_{\bar{\lambda}}{}^+(\vartheta) : \mho \rightarrow [0, 1]$ are mappings organized as: $O_\hbar^-(\vartheta) = O^-(\vartheta) + \hbar$ and $O_{\bar{\lambda}}^+(\vartheta) = O^+(\vartheta) + \bar{\lambda}, \forall \vartheta \in \mho$.

3 New Classes of Delta-Algebra with Their Applications

Some associations of ideals in algebraic semigroups and (FL) are shown and discussed in this part.

Definition 3.1 Assume that $\nu \in \mho \neq \varnothing$, we say $(\mho, \vartriangle, \triangleright, \nu)$ is a Delta-semigroup (for short, $\delta - SG$) if $\vartriangle$: $\mho \times \mho \rightarrow \mho$ and $\triangleright$; $\mho \times \mho \rightarrow \mho$ such that:

(1) $(\mho, \vartriangle, \nu)$ is a $(\delta - A)$,
(2) $(\mho, \triangleright)$ is a semigroup,
(3) $\vartheta \triangleright (\theta \vartriangle \sigma) = (\vartheta \triangleright \theta) \vartriangle (\vartheta \triangleright \sigma)$ and $(\theta \vartriangle \sigma) \triangleright \vartheta = (\theta \triangleright \vartheta) \vartriangle (\sigma \triangleright \vartheta), \forall \theta, \vartheta, \sigma \in \mho$

Example 3.2 Let $\{A_i | 1 \le i \le n\}$ be a set of chemical elements and $\{A_i{}^V | 1 \le i \le n\}$ be a set of their valency. Also, let $\mho = \{A_{\alpha_i}{}^V | 1 \le i \le n\}$, where $A_{\alpha_i}{}^V < A_{\alpha_{i+1}}{}^V$, for all $1 \le i \le n-1$. For example, if {Br, He, O, Cu} is a set of chemical elements, then $\mho = \{\mathrm{He}^V, \mathrm{O}^V, \mathrm{Cu}^V, \mathrm{Br}^V\} = \{0, 2, 4, 7\}$, see Table 1.

Define ($\rhd : \mho \times \mho \to \mho$) as $\rhd(x, y) = x \rhd y = \mathrm{He}^V, \forall x, y \in \mho$. Hence ($\mho$,$\lhd$) is a semigroup.

Also, define (Δ: $\mho \times \mho \to \mho$) by $\Delta\ (x, y) = x\ \Delta\ y =$

$$\begin{cases} xy, if x = \mathrm{He}^V, \\ x + y, if y = \mathrm{He}^V, \\ \mathrm{He}^V, if x = y, \\ \mathrm{Cu}^V, if \mathrm{He}^V \ne x \ne y \ne \mathrm{He}^V \end{cases}$$

Therefore, we can consider Table 2.

Then ($\mho$, Δ, $\rhd$, He^V) is a ($\delta - SG$).

Definition 3.3 Suppose that $\varphi \ne L \subseteq \mho$ and ($\mho$, Δ $\rhd$, ν) is a ($\delta - SG$). The set L is called a $\delta -$ subsemigroup ($\delta - SSG$) of $\mho$, if $\vartheta\ \Delta\ \theta \in L, \vartheta \rhd \theta \in L\ \forall \vartheta, \theta \in L$.

Definition 3.4 Suppose that $\varphi \ne L \subseteq \mho$ and ($\mho$, Δ, $\rhd$, ν) is a ($\delta - SG$). The set L is called a $\delta -$ ideal semigroup ($\delta - ISG$) of $\mho$, if

(i) L is an $\delta -$ ideal of $\mho$,
(ii) If $\vartheta \in L\ \ \&\ \theta \in \mho$, then $\vartheta\ \Delta\ \theta \in L\ \ \&\ \theta \rhd \vartheta \in L$.

Definition 3.5 Suppose that $\varphi \ne L \subseteq \mho$ and ($\mho$ Δ, $\rhd$, ν) is a ($\delta - SG$). The set L is called a $\overline{\delta} -$ ideal semigroup ($\overline{\delta} - ISG$) of $\mho$, if

(i) L is an $\overline{\delta} -$ ideal of $\mho$,
(ii) If $\vartheta \in L\ \&\ \theta \in \mho$, then $\vartheta\ \Delta\ \theta \in L \&\ \theta \rhd \vartheta \in L$.

Table 1 Valency of four elements

Element	Hydrogen	Oxygen	Copper	Bromine
Symbol	He	O	Cu	Br
Valency	0	2	4	7

Table 2 ($\mho$, Δ, He^V) is a ($\delta - A$)

(A)				(B)				
Δ	0	2	4	Δ	He^V	O^V	Cu^V	Br^V
0	0	0	0	He^V	He^V	He^V	He^V	He^V
2	2	0	4	O^V	O^V	He^V	Cu^V	Cu^V
4	4	4	0	Cu^V	Cu^V	Cu^V	He^V	Cu^V
7	7	4	4	Br^V	Br^V	Cu^V	Cu^V	He^V

Example 3.6 Assume that $(\mho, \triangle, \triangleright, \mathrm{He}^V)$ is a $(\delta - SG)$ in example (3.2) and let $L = \{\mathrm{He}^V, \mathrm{O}^V, \mathrm{Cu}^V\} \subseteq \mho = \{\mathrm{He}^V, \mathrm{O}^V, \mathrm{Cu}^V, \mathrm{Br}^V\}$. Hence L is $\bar{\delta}-$ ideal of $\mho$. Moreover, if $\vartheta \in L$ & $\theta \in \mho$, then $\vartheta \triangle \theta \in L$& $\theta \triangleright \vartheta \in L$. Therefore, L is a $(\bar{\delta} - ISG)$ of $\mho$.

Example 3.7 Assume that $(\mho, \triangle, \triangleright, \mathrm{He}^V)$ is a $(\delta - SG)$ in example (3.2) and let $w = \{\mathrm{O}^V, \mathrm{Br}^V\} \subseteq \mho = \{\mathrm{He}^V, \mathrm{O}^V, \mathrm{Cu}^V, \mathrm{Br}^V\}$. Thus $W = \{\mathrm{O}^V, \mathrm{Br}^V\}$ is not $(\delta - ISG)$ of $\mho$, since $\mathrm{O}^V, \mathrm{Br}^V \in W$, but $\mathrm{Br}^V \triangle \mathrm{O}^V = \mathrm{He}^V \notin W$. Therefore $W = \{\mathrm{O}^V, \mathrm{Br}^V\}$ is not $(\delta - ISG)$ of $\mho$.

Remark 3.8 Every $(\delta - ISG)$ does not have to be a $(\bar{\delta} - ISG)$. Take a look at the accompanying example.

Example 3.9 Assume that $(\mho, \triangle, \triangleright, \mathrm{He}^V)$ is a $(\delta - SG)$ in example (3.2) and let $J = \{\mathrm{He}^V, \mathrm{Br}^V\} \subseteq \mho = \{\mathrm{He}^V, \mathrm{O}^V, \mathrm{Cu}^V, \mathrm{Br}^V\}\}$. Hence $J = \{\mathrm{He}^V, \mathrm{Br}^V\}$ is $(\delta - ISG)$ of $\mho$. However, $J = \{\mathrm{He}^V, \mathrm{Br}^V\} \subseteq \mho$ is not $\delta-$ ideal of $\mho$, since $\mathrm{Br}^V \in J, \mathrm{Cu}^V \in \mho$, but $\mathrm{Br}^V \triangleright \mathrm{Cu}^V = \mathrm{He}^V \notin L$. Hence L is not $(\bar{\delta} - ISG)$ in $\mho$.

Definition 3.10 Let each of $\mho$ and $\mho'$ are $(\delta - SG)$. A mapping ξ: $(\mho, \triangle, \triangleright, \nu) \to (\mho', \triangle', \triangleright', \nu')$ is called a $\delta-$ semigroup homomorphism $(\delta - SGH)$ if $\xi(\vartheta \triangle \theta) = \xi(\vartheta)\triangle'\xi(\theta)$ and $\xi(\vartheta \triangleright \theta) = \xi(\vartheta)\triangleright'\xi(\theta)$, $\forall\vartheta, \theta \in \mho$. The sets $Ker\xi = \{\vartheta \in \mho | \xi(\vartheta) = \nu'\}$ and $Im\xi = \{\xi(\vartheta) \in \mho' | \vartheta \in \mho\}$ are called kernel and the image of ξ, respectively.

We will discuss some ideas about (FL).

Definition 3.11 Assume that O is a (FS) in $(\delta - A)$ $(\mho, \triangle, \nu)$. Then O is called a fuzzy $\delta-$ subalgebra $(F\delta - SA)$ of $\mho$ if it such that $\mathrm{O}(\vartheta \triangle \theta) \geq min\{\mathrm{O}(\vartheta), \mathrm{O}(\theta)\}$, $\forall\vartheta \triangle \theta \in \mho$.

Definition 3.12 Assume that O is a (FS) in $(\delta - SG)$ $(\mho, \triangle, \triangleright, \nu)$. Then O is called a fuzzy $\delta-$ subsemigroup $(F\delta - SSG)$ of $\mho$, if

(1) $\mathrm{O}(\vartheta \triangle \vartheta) \geq min\{\mathrm{O}(\vartheta), \mathrm{O}(\theta)\}$,
(2) $\mathrm{O}(\vartheta \triangleright \vartheta) \geq min\{\mathrm{O}(\vartheta), \mathrm{O}(\theta)\}$, $\forall\vartheta, \theta \in \mho$.

Definition 3.13 Assume that O is a (FS) in $(\delta - SG)$ $(\mho, \triangle, \triangleright, \nu)$. Then O is called a fuzzy $\delta-$ ideal semigroup $(F\delta - ISG)$ of $\mho$, if

(1) $\mathrm{O}(\vartheta \triangle \theta) \geq min\{\mathrm{O}(\vartheta), \mathrm{O}(\theta)\}$,
(2) $\mathrm{O}(\vartheta) \geq min\{\mathrm{O}(\vartheta \triangle \theta), \mathrm{O}(\theta)\}$,
(3) $\mathrm{O}(\vartheta\Box\theta) \geq min\{\mathrm{O}(\vartheta), \mathrm{O}(\theta)\}$, $\forall\vartheta, \theta \in \mho$.

Definition 3.14 Assume that O is a (FS) in $(\delta - SG)$ $(\mho, \triangle, \triangleright, \nu)$. Then O is called a fuzzy $\bar{\delta}$ ideal semigroup $(F\bar{\delta} - ISG)$ of $\mho$, if

(1) $\mathrm{O}(\nu) \geq \mathrm{O}(\vartheta)$ $\vartheta \in \mho$,
(2) $\mathrm{O}(\vartheta \triangle \theta) \geq min\{\mathrm{O}(\vartheta), \mathrm{O}(\theta)\}$,
(3) $\mathrm{O}(\vartheta \triangleright \theta) \geq min\{\mathrm{O}(\vartheta), \mathrm{O}(\theta)\}$, $\forall\vartheta, \theta \in \mho$.

Lemma 3.15 *If* O *is a* $(F\delta - ISG)$ *of* $\mho$, *then*

(1) $O(\nu) \geq O(\vartheta)\ \forall\vartheta \in \mho$,
(2) $O(\vartheta \vartriangle \theta) = O(\theta \vartriangle \vartheta)\}\ \forall\vartheta \neq \theta \in \mho - \{\nu\}$.

Proof

(1) Assume that O is a $(F\delta - ISG)$ of $\mho$, thus by [Definition 3.13 – (1)], we get $O(\vartheta \vartriangle \vartheta) \geq min\{O(\vartheta), O(\theta)\}$, for any $\forall\vartheta, \theta \in \Gamma$. Put $\vartheta = \theta$. Then $O(\vartheta \vartriangle \theta) \geq min\{O(\vartheta), O(\vartheta)\} = O(\vartheta)\ \forall\vartheta \in \mho \Rightarrow O(\nu) \geq O(\vartheta)\ \forall\vartheta \in \mho$.
(2) Assume that O is a $(F\delta - ISG)$ of $\mho$, thus by [Definition 2.2 – (4)], we consider $\vartheta \vartriangle \theta = \theta \vartriangle \vartheta\ \forall\vartheta \neq \vartheta \in \mho - \{\nu\}$, then $O(\vartheta \vartriangle \theta) = O(\theta \vartriangle \vartheta)\}$, $\forall\vartheta \neq \theta \in \mho - \{\nu\}$.

Remarks 3.16 If O is a $(F\bar{\delta} - ISG)$ of $\mho$, then $O(\vartheta \vartriangle \theta) = O(\theta \vartriangle \vartheta)\}$, $\forall\vartheta \neq \theta \in \mho - \{\nu\}$.

Definition 3.17 Assume that O is a (FS) in $\mho$ and $(\mho, \vartriangle, \triangleright, \nu)$ is a $(\delta - SG)$. Then O is called a bipolar fuzzy $\delta -$ subsemigroup $(BF\delta - SG)$ of $\mho$, if.

(1) $O^+(\vartheta \vartriangle \theta) \geq min\{O^+(\vartheta), O^+(\theta)\}$ and $O^-(\vartheta \vartriangle \theta) \leq max\{O^-(\vartheta), O^-(\theta)\}$,
(2) $O^+(\vartheta \triangleright \theta) \geq min\{O^+(\vartheta), O^+(\theta)\}$ and $O^-(\vartheta \triangleright \theta) \leq max\{O^-(\vartheta), O^-(\theta)\}$ $\forall\vartheta, \theta \in \mho$.

Example 3.18 Let $\mho = \{\nu, \vartheta, \theta, \tau\}$. Define $(\triangleright : \mho \times \mho \to \mho)$ by $\triangleright(\varpi, \mu) = \varpi \triangleright \mu = \nu, \forall\varpi, \mu \in \mho$. Hence $(\mho, \triangleright)$ is a semigroup. As well, determine $(\vartriangle)$ using the Table 3.

Then $(\mho, \vartriangle, \triangleright, \nu)$ is a$(\delta - SG)$. Define $B = (\vartheta, O^-, O^+)$ by $B = \{(\nu, -0.7, 0.8), (\vartheta, -0.6, 0.6), (\theta, -0.4, 0.5), (\tau, -0.3, 0.2)\}$. Therefore B is a $(BF\delta - SSG)$ of$\mho$.

Definition 3.19 Assume that O is a (FS) in $\mho$ and $(\mho, \vartriangle, \triangleright, \nu)$ is a $(\delta - SG)$. Then O is called a bipolar fuzzy $\delta -$ ideal semigroup $(BF\delta - ISG)$ of $\mho$, if

(1) 1)$O^+(\vartheta \vartriangle \theta) \geq min\{O^+(\vartheta), O^+(\theta)\}$ and $O^-(\vartheta \vartriangle \theta) \leq max\{O^-(\vartheta), O^-(\theta)\}$,
(2) $O^+(\vartheta) \geq min\{O^+(\vartheta \vartriangle \theta), O^+(\theta)\}$ and $O^-(\vartheta) \leq max\{O^-(\vartheta \vartriangle \theta), O^-(\theta)\}$,
(3) $O^+(\vartheta \triangleright \theta) \geq min\{O(\vartheta), O(\theta)\}$ and $O^-(\vartheta \triangleright \theta) \leq max\{O^-(\vartheta), O^-(\theta)\}\ \forall\vartheta, \theta \in \mho$.

Definition 3.20 Assume that O is a (FS) in $\mho$ and $(\mho, \vartriangle, \triangleright, \nu)$ is a $(\delta - SG)$. Then O is called a bipolar fuzzy $\bar{\delta}$ ideal semigroup $(BF\bar{\delta} - ISG)$ of $\mho$, if.

Table 3 $(\mho, \vartriangle, \nu)$ is a $(\delta - A)$

$\vartriangle$	ν	ϑ	θ	τ
ν	ν	ν	ν	ν
ϑ	ϑ	ν	τ	τ
θ	θ	τ	ν	τ
τ	τ	τ	τ	ν

(1) $O^+(\nu) \geq O^+(\vartheta)$ and $O^-(\nu) \leq O^-(\vartheta)$.
(2) $O^+(\vartheta \triangle \theta) \geq min\{O^+(\vartheta), O^+(\theta)\}$ and $O^-(\vartheta \triangle \theta) \leq max\{O^-(\vartheta), O^-(\theta)\}$,
(3) $O^+(\vartheta \rhd \theta) \geq min\{O^+(\vartheta), O^+(\theta)\}$ and $O^-(\vartheta \rhd \theta) \leq max\{O^-(\vartheta), O^-(\theta)\}$, $\forall \vartheta, \theta \in \mho$.

Proposition 3.21 *Suppose that* $B = (\vartheta, O^-, O^+)$ *is a* $(BF\delta - ISG)$. *Then* $O^+(\nu) \geq O^+(\vartheta)$ & $O^-(\nu) \leq O^-(\vartheta)$.

Proof since $B = (\vartheta, O^-, O^+)$ is a $(BF\delta - ISG)$, then from [definition (3.19)–(1)], we have $O^+(\vartheta \triangle \theta) \geq min\{O^+(\vartheta), O^+(\theta)\}$ and $O^-(\vartheta \triangle \theta) \leq max\{O(\vartheta), O(\theta)\}$ put $\vartheta = \theta$. Then $O^+(\vartheta \triangle \vartheta) \geq min\{O^+(\vartheta), O^+(\vartheta)\} = O^+(\vartheta)$ and $O^-(\vartheta \triangle \vartheta) \leq max\{O^-(\vartheta), O^-(\vartheta)\} = O^-(\vartheta)$. But, $\vartheta \triangle \vartheta = \nu, \forall \vartheta \in \mho$. Hence $O^+(\nu) \geq O^+(\vartheta)$ and $O^-(\nu) \leq O^-(\vartheta)$.

Lemma 3.22 *Suppose that* $(\mho, \triangle, \lhd, \nu)$ *is a* $(\delta - SG)$ *and* $B = (\vartheta, O^-, O^+)$ *is a* $(BF\delta - ISG)$ $\mho$. *Then B is* $(BF\overline{\delta} - ISG)$ *of* $\mho$.

Proof Suppose that $B = (\vartheta, O^-, O^+)$ is a $BF\delta - ISG)$ $\mho$, hence from [Proposition (3.21)], we get $O^+(\nu) \geq O^+(\vartheta)$ and $O^-(\nu) \leq O^-(\vartheta)$. Also, from [definition (3.19) – (1)] we obtain $O^+(\vartheta \triangle \theta) \geq min\{O^+(\vartheta), O^+(\theta)\}$ and $O^-(\vartheta \triangle \theta) \leq max\{O(\vartheta), O(\theta)\}, \forall \vartheta, \theta \in \mho$ and by [Definition (3.19) – (3)] we have $O^+(\vartheta \lhd \theta) \geq min\{O(\vartheta), O(\theta)\}$ and $O^-(\vartheta \lhd \theta) \leq max\{O^-(\vartheta), O^-(\theta)\} \forall \vartheta, \theta \in \Gamma$. Then B is $(BF\delta - ISG)$ of $\mho$.

Remarks 3.23

1. From Lemma 3.22 we consider that every $(BF\delta - ISG)$ is $(BF\overline{\delta} - ISG)$ however in general that not true in classical sets, see Example 3.9.
2. Every $(BF\overline{\delta} - ISG)$ does not have to be a $(BF\delta - ISG)$. Take a look at the accompanying example.

Example 3.24 Let $\{A_i | 1 \leq i \leq n\}$ be a set of chemical elements and $\{{A_i}^{AN} | 1 \leq i \leq n\}$ be a set of their atomic numbers. Also, let$\mho = \{{A_{\alpha_i}}^{AN} | 1 \leq i \leq n\}$, where${A_{\alpha_i}}^{AN} < {A_{\alpha_{i+1}}}^{AN}$, for all$1 \leq i \leq n-1$. For example, if $\{K, Mg, Na\}$ is a set of chemical elements, then$\mho = \{Na^{AN}, Mg^{AN}, K^{AN}\} = \{11, 12, 19\}$, see Table 4.

Define $(\rhd : \mho \times \mho \to \mho)$ as $\rhd(x, y) = x \rhd y = Na^{AN}, \forall x, y \in \mho$. Hence $(\mho, \rhd)$ is a semigroup. As well, determine $(\triangle)$ using the Table 5.

Then $(\mho, \triangle, \rhd, Na^{AN})$ is a $(\delta - SG)$. Define $B = (\vartheta, O^-, O^+)$ by $B = \{({A_{\alpha_i}}^{AN}, -(1/({A_{\alpha_i}}^{AN})^2), (1/({A_{\alpha_i}}^{AN})^2)) | 1 \leq i \leq 3\}$. Hence B is a $(BF\overline{\delta} - ISG)$ of $\mho$. But, it is not $(BF\delta - ISG)$, since $O^+(Mg^{AN}) < min\{O^+(Mg^{AN} \triangle K^{AN}), O^+(K^{AN})\}$.

Table 4 Atomic numbers of three chemical elements

Element	Sodium	Magnesium	Potassium
Symbol	Na	Mg	*K*
Atomic number	11	12	19

Table 5 $(\mho, \Delta, Na^{AN})$ is a $(\delta - A)$

(A)				(B)			
Δ	11	2	19	Δ	Na^{AN}	Mg^{AN}	K^{AN}
11	11	11	11	Na^{AN}	Na^{AN}	Na^{AN}	Na^{AN}
12	12	11	12	Mg^{AN}	Mg^{AN}	Na^{AN}	Mg^{AN}
19	19	12	11	K^{AN}	K^{AN}	Mg^{AN}	Na^{AN}

Proposition 3.25 *Suppose that* $(\mho, \Delta, \lhd, \nu)$ *is a* $(\delta - SG)$ *and* $B = (\vartheta, O^-, O^+)$ *is a* $(BF\delta - ISG)$ *of* $\mho$. *If* $\theta < \vartheta$, *then* $O^+(\vartheta) \geq O^+(\theta)$ *and* $O^-(\vartheta) \leq O^-(\theta)$.

Proof Let $B = (\vartheta, O^-, O^+)$ be a $(BF\delta - ISG)$, thus from [Definition (3.19) − (2)], we get $O^+(\vartheta) \geq min\{O^+(\vartheta \ \Delta \ \theta), O^+(\theta)\}$ and $O^-(\vartheta) \leq max\{\vartheta^-(\vartheta \ \Delta \ \theta), O^-(\theta)\}$. Now, assume $\theta < \vartheta$, this implies $\vartheta \ \Delta \ \theta = \nu$. Therefore $O^+(\vartheta) \geq min\{O^+(\vartheta \ \Delta \ \theta), O^+(\theta)\}$=$min\{O^+(\nu), O^+(\theta)\}$ and $O^-(\vartheta) \leq max\{O^-(\vartheta \ \Delta \ \theta), O^-(\theta)\} = max\{O^-(\nu), O^-(\theta)\}$. From [Proposition (3.21)] we get $O^+(\vartheta) \geq min\{O^+(\nu), O^+(\theta)\} = O^+(\theta)$ and $O^-(\vartheta) \leq max\{O^-(\nu), O^-(\theta)\} = O^-(\theta)$. Then $O^+(\vartheta) \geq O^+(\theta)$ and $O^-(\vartheta) \leq O^-(\theta)$.

Proposition 3.26 *Suppose that* $(\mho, \Delta, \lhd, \nu)$ *is a* $(\delta - SG)$ *and* $B = (\vartheta, O^-, O^+)$ *is a* $(BF\delta - ISG)$ *of* $\mho$. *If* $\tau < \vartheta \ \Delta \ \theta$, *then* $O^+(\vartheta \ \Delta \ \theta) \geq O^+(\tau)$ *and* $O^-(\vartheta \ \Delta \ \theta) \leq O^-(\tau)$.

Proof Assume $B = (\vartheta, O^-, O^+)$ is a$(BF\delta - ISG)$, thus from [Definition(3.16) − (2)], we get $O^+(\vartheta \ \Delta \ \theta) \geq min\{O^+((\vartheta \ \Delta \ \theta) \ \Delta \ \tau), O^+(\tau)\}$ and $O^-(\vartheta \ \Delta \ \theta) \leq max\{O^-((\vartheta \ \Delta \ \theta) \ \Delta \ \tau), O^-(\tau)\}$ Now, let$\tau < \vartheta \ \Delta \ \theta$, therefore$(\vartheta \ \Delta \ \theta) \ \Delta \ \tau = \nu$. And hence we consider that $O^+(\vartheta \ \Delta \ \theta) \geq min\{O^+(\nu), O^+(\tau)\}$ and$O^-(\vartheta \ \Delta \ \theta) \leq max\{O^-(\nu), O^-(\tau)\}$. From [Proposition(3.21)] we get $\min\{O^+(\nu), O^+(\tau)\} = O^+(\tau)$ and $\max\{O^-(\nu), O^-(\tau)\}$ =$O^-(\tau)$. Then $O^+(\vartheta \ \Delta \ \theta) \geq O^+(\tau)$ and$O^-(\vartheta \ \Delta \ \theta) \leq O^-(\tau)$.

Remarks 3.27

(1) Assume that $(\mho, \Delta, \lhd, \nu)$ is a $(\delta - SG)$ and $B = (\vartheta, O^-, O^+)$ is a $(BF\delta - ISG)$ of $\mho$. Then from [Definition (3.20) − (1), (2)], we get $B = (\vartheta, O^-, O^+)$ is $(BF\delta - SG)$ of $\mho$.

(2) Assume that $(\mho, \Delta, \lhd, \nu)$ is a $(\delta - SG)$ and $B = (\vartheta, O^-, O^+)$ is a $(BF\bar{\delta} - ISG)$ of $\mho$. Then from [Definition (3.19) − (1), (2)], we get $B = (\vartheta, O^-, O^+)$ is $(BF\delta - SG)$ of $\mho$.

Proposition 3.28 *Assume that* $(\mho, \Delta, \lhd, \nu)$ *is e a* $(\delta - SG)$ *and* $B = (\vartheta, O^-, O^+)$ *is a* $(BF\bar{\delta} - ISG)$ *of* $\mho$. *Then*

(1) $W_1 = \{\vartheta \in \mho | O^+(\vartheta) = O^+(\nu)\}$ *is* $(\bar{\delta} - ISG)$ of $\mho$.
(2) $W_2 = \{\vartheta \in \mho | O^-(\vartheta) = O^-(\nu)\}$ *is* $(\bar{\delta} - ISG)$ of$\mho$.

Proof

(1) Assume $(\mho, \triangle, \triangleleft, \nu)$ is a $(\delta - SG)$ and $B = (\vartheta, O^-, O^+)$ is a $(BF\bar{\delta} - ISG)$ of $\mho$, since $\nu \in \mho$, thus $O^+(\nu) = O^+(\nu)$, so $\nu \in W_1$. Also, let $\vartheta, \theta \in W_1$, then$O^+(\vartheta) = O^+(\nu)$ and $O^+(\theta) = O^+(\nu)$. Now, from [Definition (3.20) – (2)], we consider that $O^+(\vartheta \triangle \theta) \geq min\{O^+((\vartheta), O^+(\theta)\} = min\{O^+(\nu), O^+(\nu)\} = O^+(\nu)$, then $O^+(\vartheta \triangle \theta) \geq O^+(\nu)$, but $O^+(\nu) \geq O^+(\vartheta \triangle \theta)$,[since $\vartheta \triangle \theta \in \mho$]. So $O^+(\vartheta \triangle \theta) = O^+(\nu)$ and hence $\vartheta \triangle \theta \in W_1$. Also, let $\vartheta, \theta \in W_1$, then$O^+(\vartheta) = O^+(\nu)$ and $O^-(\vartheta) = O^-(\nu)$. Now, by [definition 3.20-(2)], we have $O^+(\vartheta \triangleleft \theta) \geq min\{O^+((\vartheta), O^+(\theta)\} = min\{O^+(\nu), O^+(\nu)\} = O^+(\nu)$, then $O^+(\vartheta \triangleleft \theta) \geq O^+(\nu)$, but $O^+(\nu) \geq O^+(\vartheta \triangleleft \theta)$ [since $\vartheta \triangleleft \theta \in \mho$]. So $O^+(\vartheta \triangleleft \theta) = O^+(\nu)$ and hence $\vartheta \triangleleft \theta \in W_1$. Then $W_1 = \{\vartheta \in \mho | O^+(\vartheta) = O^+(\nu)\}$ is $(\bar{\delta} - ISG)$ of $\mho$.

(2) Assume $(\mho, \triangle, \triangleleft, \nu)$ is a $(\delta - SG)$ and $B = (\vartheta, O^-, O^+)$ is a $(BF\bar{\delta} - ISG)$ of $\mho$, since $\nu \in \mho$, thus $O^-(\nu) = O^-(\nu)$, so $\nu \in W_1$. Also, let $\vartheta, \theta \in W_2$, then$O^-(\vartheta) = O^-(\nu)$ and $O^-(\theta) = O^-(\nu)$. Now, from [Definition (3.20) – (2)], we have $O^-(\vartheta \triangle \theta) \leq max\{\{O^-((\vartheta), O^-(\theta)\} = max\{O^-(\nu), O^-(\nu)\} = O^-(\nu)$, then $O^-(\vartheta \triangle \theta) \leq O^-(\nu)$, but $O^-(\nu) \leq O^-(\vartheta \triangle \theta)$,[since $\vartheta \triangle \theta \in \mho$]. So $O^-(\vartheta \triangle \theta) = O^-(\nu)$ and hence $\vartheta \triangle \theta \in W_2$. Also, let $\vartheta, \theta \in W_2$, then$O^-(\vartheta) = O^-(\nu)$ and $O^-(\vartheta) = O^-(\nu)$. Now, by [definition 3.20-(2)], we have $O^-(\vartheta \triangleleft \theta) \leq max\{\{O^-((\vartheta), O^+(\theta)\} = max\{O^-(\nu), O^-(\nu)\} = O^-(\nu)$, then $O^-(\vartheta \triangleleft \theta) \leq O^-(\nu)$, but $O^-(\nu) \leq O^-(\vartheta \triangleleft \theta)$ [since $\vartheta \triangleleft \theta \in \mho$]. So $O^-(\vartheta \triangleleft \theta) = O^-(\nu)$ and hence $\vartheta \triangleleft \theta \in W_2$. Then $W_2 = \{\vartheta \in \mho | O^-(\vartheta) = O^-(\nu)\}$ is $(\bar{\delta} - ISG)$ of $\mho$.

Corollary 3.29 *Assume that* $(\mho, \triangle, \triangleleft, \nu)$ *is a* $(\delta - SG)$ *and* $B = (\vartheta, O^-, O^+)$ *is a* $(BF\delta - ISG)$ *of* $\mho$. *Then*

(1) $W_1 = \{\vartheta \in \mho | O^+(\vartheta) = O^+(\nu)\}$ *is* $(\bar{\delta} - ISG)$ of $\mho$.
(2) $W_2 = \{\vartheta \in \mho | O^-(\vartheta) = O^-(\nu)\}$ *is* $(\bar{\delta} - ISG)$ of $\mho$.

Proof Assume that $(\mho, \triangle, \triangleleft, \nu)$ is a $(\delta - SG)$ and $B = (\vartheta, O^-, O^+)$ is a $(BF\delta - ISG)$ of $\mho$. Then from [Lemma (3.22)], we get $B = (\vartheta, O^-, O^+)$ is a $(BF\bar{\delta} - ISG)$ of $\mho$ and hence from [Proposition (3.28)], we obtain (1) and (2) are held.

Proposition 3.30 *Assume that* $(\mho, \triangle, \triangleleft, \nu)$ *is a* $(\delta - SG)$ *and* $B = (\vartheta, O^-, O^+)$ *is a* $(BF\delta - ISG)$ *of* $\mho$. *Then*

(1) $W_1 = \{\vartheta \in \mho | O^+(\vartheta) = O^+(\nu)\}$ *is* $(\delta - ISG)$ of $\mho$.
(2) $W_2 = \{\vartheta \in \mho | O^-(\vartheta) = O^-(\nu)\}$ *is* $(\delta - ISG)$ of $\mho$.

Proof

(1) Assume $(\mho, \triangle, \triangleleft, \nu)$ is a $(\delta - SG)$ and $B = (\vartheta, O^-, O^+)$ is a $(BF\delta - ISG)$ of $\mho$, since $\nu \in \mho$, thus $O^+(\nu) = O^+(\nu)$, so $\nu \in W_1$ and hence $W_1 \neq \varphi$, let $\vartheta, \theta \in W_1$, then$O^+(\vartheta) = O^+(\nu)$ and $O^+(\theta) = O^+(\nu)$. Now, from [Definition (3.19) – (1)], we get $O^+(\vartheta \triangle \theta) \geq min\{O^+((\vartheta), O^+(\theta)\} = min\{O^+(\nu), O^+(\nu)\} = O^+(\nu)$, then $O^+(\vartheta \triangle \theta) \geq O^+(\nu)$, but $O^+(\nu) \geq$

$O^+(\vartheta \triangle \theta)$, [since $\vartheta \triangle \theta \in \mho$]. So $O^+(\vartheta \triangle \theta) = O^+(\nu)$ and hence $\vartheta \triangle \theta \in W_1$. Now let $\vartheta \triangle \theta \in W_1$ and $\theta \in W_1$, then$O^+(\vartheta \triangle \theta) = O^+(\nu)$ and $O^+(\nu) = O^+(\nu)$. From [Definition (3.19) – (2)], we have $O^+(\vartheta) \geq min\{O^+(\vartheta \triangle \theta), O^+(\theta)\} = min\{O^+(\nu), O^+(\nu)\} = O^+(\nu) \Rightarrow O^+(\vartheta) \geq O^+(\nu)$....(1) from [Proposition (3.21)] $O^+(\nu) \geq O^+(\vartheta)$..........(2) thus from (1) and (2) $\Rightarrow O^+(\vartheta) = O^+(\nu)$ So, $\vartheta \in W_1$. Also, let $\vartheta, \theta \in W_1$, then$O^+(\vartheta) = O^+(\nu)$ and $O^+(\vartheta) = O^+(\nu)$. Now, from [Definition (3.19) – (3)], we have $O^+(\vartheta \lhd \theta) \geq min\{O^+((\vartheta), O^+(\theta)\} = min\{O^+(\nu), O^+(\nu)\} = O^+(\nu)$, then $O^+(\vartheta \lhd \theta) \geq O^+(\nu)$, but $O^+(\nu) \geq O^+(\vartheta \lhd \theta)$ [since $\vartheta \lhd \theta \in \mho$]. So $O^+(\vartheta \lhd \theta) = O^+(\nu)$ and hence $\vartheta \lhd \theta \in W_1$. Therefore, $W_1 = \{\vartheta \in \mho | O^+(\vartheta) = O^+(\nu)\}$ is $(\delta - ISG)$ of $\mho$.

(2) Assume $(\mho, \triangle, \lhd, \nu)$ is a $(\delta - SG)$ and $B = (\vartheta, O^-, O^+)$ is a $(BF\delta - ISG)$ of $\mho$, since $\nu \in \mho$, thus $O^-(\nu) = O^-(\nu)$, so $\nu \in W_2$ and hence $W_2 \neq \varphi$,let let $\vartheta, \theta \in W_2$, then$O^-(\vartheta) = O^-(\nu)$ and $O^-(\theta) = O^-(\nu)$. Now, from [Definition (3.16) – (i)], we get $O^-(\vartheta \triangle \theta) \leq max\{\{O^-((\vartheta), O^-(\theta)\} = max\{O^-(\nu), O^-(\nu)\} = O^-(\nu)$, then $O^-(\vartheta \triangle \theta) \leq O^-(\nu)$, but $O^-(\nu) \leq O^-(\vartheta \triangle \theta)$, [since $\vartheta \triangle \theta \in \mho$]. So $O^-(\vartheta \triangle \theta) = O^-(\nu)$ and hence $\vartheta \triangle \theta \in W_2$. Now, let $\vartheta \circ \theta \in W_2$ and $\theta \in W_2$, then $O^+(\vartheta \circ \theta) = O^+(\nu)$ and $O^-(\nu) = O^-(\nu)$. From [Definition (3.19) – (2)], we get $O^+(\vartheta) \leq max\{O^+(\vartheta \circ \theta), O^+(\theta)\} = max\{O^+(\nu), O^+(\nu)\} = O^-(\nu) \Rightarrow O^-(\vartheta) \leq O^-(\nu)$....(1) from [Proposition (3.21)] $O^-(\nu) \geq O^-(\vartheta)$..........(2) thus from (1) and (2), we get $O^-(\vartheta) = O^-(\nu)$. So, $\vartheta \in W_2$. Also, let $\vartheta, \theta \in W_2$, then$O^-(\vartheta) = O^-(\nu)$ and $O^-(\vartheta) = O^-(\nu)$. Now, from [Definition (3.19) – (3)], we have $O^-(\vartheta \lhd \theta) \leq \max\{O^-((\vartheta), O^+(\theta)\} = max\{O^-(\nu), O^-(\nu)\} = O^-(\nu)$, then $O^-(\vartheta \lhd \theta) \leq O^-(\nu)$, but $O^-(\nu) \leq O^-(\vartheta \lhd \theta)$ [since $\vartheta \lhd \theta \in \mho$]. So $O^-(\vartheta \lhd \theta) = O^-(\nu)$ and hence $\vartheta \lhd \theta \in W_2$. Therefore, $W_2 = \{\vartheta \in \mho | O^-(\vartheta) = O^-(\nu)\}$ is $(\delta - ISG)$ of $\mho$.

Theorem 3.31 *Assume that* $(\mho, \triangle, \lhd, \nu)$ *is a* $(\delta - SG)$ *and* $B = (\vartheta, O^-, O^+)$ *is a* $(BF\delta - ISG)$ *of* $\mho$. *The items are then identical:*

(1) $B = (\vartheta, O^-, O^+)$ *is a* $(BF\delta - ISG)$ *of* $\mho$,
(2) $B = (\vartheta, O^-, O^+)$ *such that the following:*
(i) *If* $B^-_\lambda = \{\vartheta \in \mho | O^-(\vartheta) \leq \lambda\} \neq \varphi, \forall \lambda \in [-1, 0]$, *then* B^-_λ *is* $(\delta - ISG)$ *of* $\mho$,
(ii) *If* $B^+_\gamma = \{\vartheta \in \mho | O^+(\vartheta) \geq \gamma\} \neq \varphi, \forall \lambda \in [0, 1]$, *then* B^+_γ *is* $(\delta - ISG)$ *of* $\mho$.

Proof (1) $\Rightarrow$ [(2) – (i)] Suppose that $\lambda \in [-1, 0]$ satisfies$B^-_\lambda \neq \varphi$, and$\vartheta, \theta \in B^-_\lambda$, thus $O^-(\vartheta) \leq \lambda \& O^-(\theta) \leq \lambda$. From [Definition(3.19) – (1)], we get $O^-(\vartheta \triangle \theta) \leq max\{O^-(\vartheta), O^-(\theta)\} \leq \max\{\lambda, \lambda\} = \lambda$, then$\vartheta \triangle \theta \in B^-_\lambda$. Also, if $\vartheta \triangle \theta \in B^-_\lambda$ $\& \theta \in B^-_\lambda$, hence $O^-(\vartheta \triangle \theta) \leq \lambda \& O^-(\theta) \leq \lambda$. From [Definition(3.19) – (2)], we get $O^-(\vartheta) \leq max\{O^-(\vartheta \triangle \theta), O^-(\theta)\} \leq \max\{\lambda, \lambda\} = \lambda$, then$\vartheta \in B^-_\lambda$. Also, let $\vartheta, \theta \in B^-_\lambda \Rightarrow O^-(\vartheta) = O^-(\nu)$ and$O^-(\theta) = O^-(\nu)$. Now, by [definition 3.19-(3)], we have $O^-(\vartheta \lhd \theta) \leq max\{O^-(\vartheta), O^-(\theta)\} = max\{O^-(\nu), O^-(\nu)\} = O^-(\nu)$, then$O^-(\vartheta \lhd \theta) \leq O^-(\nu)$, but$O^-(\nu) \leq O^-(\vartheta \lhd \theta)$, then $O^-(\vartheta \lhd \theta) = O^-(\nu)$ hence $\vartheta \lhd \theta \in B^-_\lambda$. Hence B^-_λ is $(\delta - ISG)$ of $\mho$.

(1) $\Rightarrow$ [(2) $-$ (*ii*)], assume $\gamma \in [0,1]$ and $B_\gamma^+ \neq \varphi$, if $\vartheta, \theta \in B_\gamma^+$, then $O^+(\vartheta) \geq \gamma$ & $O^+(\theta) \geq \gamma$. From [Definition (3.19) $-$ (1)], we get $O^+(\vartheta \vartriangle \theta) \geq min\{O^+(\vartheta), O^+(\theta)\} \leq \max\{\gamma, \gamma\} = \gamma$, then $\vartheta \vartriangle \theta \in B_\gamma^+$. Now, let $\vartheta \vartriangle \theta \in B_\gamma^+$ and $\theta \in B_\gamma^+$, then $O^+(\vartheta \vartriangle \theta) \leq \gamma$ and $O^+(\theta) \geq \gamma$. From [Definition (3.19) $-$ (2)], we have $O^+(\vartheta \geq min\{O^+(\vartheta \vartriangle \theta), O^+(\theta)\} \geq \min\{\gamma, \gamma\} = \gamma$, then $\vartheta \in B_\gamma^+$. Now, if $\vartheta, \theta \in B_\gamma^+ \Rightarrow O^+(\vartheta) = O^+(\nu)$ & $O^+(\theta) = O^+(\nu)$. From [Definition (3.19) $-$ (3)], we obtain $O^+(\vartheta \lhd \theta) \geq min\{O^+(\vartheta), O^+(\theta)\} = min\{O^+(\nu), O^+(\nu)\} = O^+(\nu)$, then $O^+(\vartheta \lhd \theta) \geq O^+(\nu)$, however $O^+(\nu) \geq O^+(\vartheta \lhd \theta)$, then $O^+(\vartheta \lhd \theta) = O^+(\nu)$ hence $\vartheta \lhd \theta \in B_\gamma^+$. Hence B_γ^+ is $(\delta - ISG)$ of $\mho$.

(2) $\Rightarrow$ (1) suppose that $O^+(\vartheta \vartriangle \theta) < min\{O^+(\vartheta), O^+(\theta)\}$, for some $\vartheta, \theta \in \mho$, and let $\gamma_1 = \frac{1}{2}(O^+(\vartheta \vartriangle \theta) + min\{O^+(\vartheta), O^+(\theta)\})$, thus $O^+(\vartheta \vartriangle \theta) < \gamma_1 < min\{O^+(\vartheta), O^+(\theta)\}$, thus $\vartheta, \theta \in B_{\gamma_1}^+$, but $\theta \vartriangle \vartheta \notin B_{\gamma_1}^+$ and this contradiction. Hence $O^+(\theta \vartriangle \vartheta) \geq min\{O^+(\vartheta), O^+(\theta)\}$, $\forall \vartheta, \theta \in \mho$. If $O^+(\vartheta) < min\{O^+(\vartheta \vartriangle \theta), O^+(\theta)\}$ for some $\vartheta, \theta \in \mho$, let $\gamma_2 = \frac{1}{2}(O^+(\vartheta) + min\{O^+(\vartheta \vartriangle \theta), O^+(\theta)\})$, thus we consider that $O^+(\vartheta) < \gamma_2 < min\{O^+(\vartheta \vartriangle \theta), O^+(\theta)\}$, thus $\vartheta \vartriangle \theta \in B_{\gamma_2}^+$ and $\theta \in B_{\gamma_2}^+$, but $\vartheta \in B_{\gamma_2}^+$ and this contradiction. Hence $O^+(\vartheta) \geq min\{O^+(\vartheta \vartriangle \theta), O^+(\theta)\}$, for all $\vartheta, \theta \in \mho$. Also, if $O^+(\vartheta \lhd \theta) < min\{O^+(\vartheta), O^+(\theta)\}$, for some $\vartheta, \theta \in \mho$, and let $\gamma_3 = \frac{1}{2}(O^+(\vartheta \lhd \theta) + min\{O^+(\vartheta), O^+(\theta)\})$, then $O^+(\vartheta \lhd \theta) < \gamma_3 < min\{O^+(\vartheta), O^+(\theta)\}$, thus $\vartheta, \theta \in B_{\gamma_3}^+$, but $\vartheta \lhd \theta \notin B_{\gamma_3}^+$ and this contradiction. Then $O^+(\vartheta \lhd \theta) \geq min\{O^+(\vartheta), O^+(\theta)\}$, for all $\vartheta, \theta \in \mho$. If $O^+(\vartheta) < min\{O^+(\vartheta \lhd \theta), O^+(\theta)\}$ for some $\vartheta, \theta \in \mho$, let $\gamma_4 = \frac{1}{2}(O^+(\vartheta) + min\{O^+(\vartheta \lhd \theta), O^+(\theta)\})$, thus we consider that $O^+(\vartheta) < \gamma_{\gamma_4} < min\{O^+(\vartheta \lhd \theta), O^+(\theta)\}$, thus $\vartheta \lhd \theta \in B_{\gamma_4}^+$ and $\theta \in B_{\gamma_4}^+$, but $\vartheta \in B_{\gamma_4}^+$ and this contradiction. Thus $O^+(\vartheta) \geq min\{O^+(\vartheta \lhd \theta), O^+(\theta)\}$,for all $\vartheta, \theta \in \mho$. If $O^-(\vartheta) > max\{O^-(\vartheta \vartriangle \theta), O^-(\theta)\}$ for some $\vartheta, \theta \in \mho$, let $\lambda_5 = \frac{1}{2}(O^-(\vartheta) + min\{O^-(\vartheta \vartriangle \theta), O^-(\theta)\})$, then $O^-(\vartheta) > \lambda_5 > max\{O^-(\vartheta \vartriangle \vartheta), O^-(\theta)\}$, thus $\vartheta \vartriangle \theta \in B_{\lambda_2}^-$ and $\theta \in B_{\lambda_2}^-$, but $\vartheta \in B_{\lambda_2}^-$ and this contradiction. Then $O^-(\vartheta) \leq max\{O^-(\vartheta \vartriangle \theta), O^-(\theta)\}$, for all $\vartheta, \theta \in \mho$. Also, if $O^-(\vartheta) > max\{O^-(\vartheta \lhd \theta), O^-(\theta)\}$ for some $\vartheta, \theta \in \mho$, let $\lambda_5 = \frac{1}{2}(O^-(\vartheta) + min\{O^-(\vartheta \lhd \theta), O^-(\theta)\})$, then $O^-(\vartheta) > \lambda_5 > max\{O^-(\vartheta \lhd \vartheta), O^-(\theta)\}$, thus $\vartheta \lhd \theta \in B_{\lambda_5}^-$ and $\theta \in B_{\lambda_5}^-$, but $\vartheta \in B_{\lambda_5}^-$ and this contradiction. Then $O^-(\vartheta) \leq max\{O^-(\vartheta \lhd \theta), O^-(\theta)\}$, for all $\vartheta, \theta \in \mho$. Then $B = (\vartheta, O^-, O^+)$ is *a* $(BF\delta - ISG)$ of δ.

4 The Translations of Bipolar Fuzzy (Delta/Delta Par)–Ideals Semigroups

The applications of Delta – semigroup homomorphism image and translations are found in this part Also, the product of $(BF\delta - ISGs)$ & $(BF\bar{\delta} - ISGs)$ are studied.

Theorem 4.1 *Supppose that* $(\mho, \vartriangle, \lhd, \nu)$ *is a* $(\delta - SG)$, $B = (\vartheta, O^-, O^+)$ *is a* (BFS) *in* $\mho$ *and* $B_{(\tau,\delta)} = (\vartheta, {}_\tau O^-, {}_\delta O^+)$ *is a translation of* $B = (\vartheta, O^-, O^+)$. *The items are then identical:*

(1) $B_{(\tau,\delta)} = (\vartheta, {}_\tau O^-, {}_\delta O^+)$ *is* $(BF\delta - SSG)$ *of* $\mho$,

(2) $B = (\vartheta, O^-, O^+)$ *is* $(BF\delta - SSG)$ *of* $\mho$.

Proof **(1) → (2):** Assume $B_{(\tau,\delta)} = (\vartheta, {}_\tau O^-, {}_\delta O^+)$ is $(BF\delta - SSG)$ of$\mho$, for some $\tau, \delta \in [\overset{-}{\leftarrow}, 0] \times [0, \overset{+}{\leftarrow}]$. Then $max\{{}_\tau O^-(\vartheta), {}_\tau O^-(\theta)\} = max\{O^-(\vartheta) + \tau, O^-(\theta) + \tau\} = max\{O^-(\vartheta), O^-(\theta)\} + \tau \Rightarrow O^-(\vartheta \triangle \theta) \le max\{O^-(\vartheta), O^-(\theta)\}, \forall \vartheta, \theta \in \mho$, and $O^+(\vartheta \triangle \theta) + \delta = {}_\delta O^+(\vartheta \triangle \theta) \ge min\{{}_\delta O^+(\vartheta), {}_\delta O^+(\theta)\} = min\{O^+(\omega) + \tau, O^+(\theta) + \tau\} = min\{O^+(\vartheta), O^+(\theta)\} + \delta \Rightarrow O^+(\vartheta \triangle \theta) \ge \min\{O^+(\vartheta), O^+(\theta)\}, \forall \vartheta, \theta \in \mho$. Therefore $B = (\vartheta, O^-, O^+)$ is $(BF\delta - SSG)$ of$\mho$.

(2)→ (1): Assume $B = (\vartheta, O^-, O^+)$ is $(BF\delta - SSG)$ of$\mho$. Then ${}_\tau O^-(\vartheta \triangle \theta) = max\{{}_\tau O^-(\vartheta), {}_\tau O^-(\theta)\}, \forall \vartheta, \theta \in \mho$, and ${}_\delta O^+(\vartheta \triangle \theta) = O^+(\vartheta \triangle \theta) + \delta \ge min\{O^+(\vartheta), O^+(\theta)\} + \delta = min\{O^+(\vartheta) + \tau, O^+(\theta) + \tau\} = min\{{}_\delta O^+(\vartheta), {}_\delta O^+(\theta)\}$, $\forall \vartheta, \theta \in \mho \Rightarrow B_{(\tau,\delta)} = (\vartheta, {}_\tau O^-, {}_\delta O^+)$ is $(BF\delta - SSG)$ of$\mho$.

Definition 4.2 Assume $T:(\mho, \triangle, \lhd, \nu) \to (\mho', \triangle', \lhd', \nu)$ is a $\delta -$ semigroup homomorphism $(\delta - SGH)$ and $(\mho, \triangle, \lhd, \nu)$ is a (BFS) of $\mho$. If $\mho = \mho'$. Let $B_T = (\vartheta, O_T{}^-, O_T{}^+)$ be (BFS) in $\mho$, where $O_T{}^-(\vartheta) = O^-(T(\vartheta))$ and $O_T{}^+(\vartheta) = O^+(T(\vartheta)), \forall \vartheta \in \mho$.

Theorem 4.3 *Assume*$T:(\mho, \triangle, \lhd, \nu) \to (\mho', \triangle', \lhd', \nu)$ *is a* $(\delta - SGH)$ *and onto. If* $B = (\vartheta', O^-, O^+)$ *is a* $(BF\delta - ISG)$ *of* $\mho'$. *Then* $B_T = (\vartheta, O_T{}^-, O_T{}^+)$ *is a* $(BF\delta - ISG)$ *of* $\mho$.

Proof Assume $B = (\vartheta', O^-, O^+)$ is a $(BF\delta - ISG)$ of $\mho'$, let $\vartheta, \theta \in \mho$. We have $O_T{}^+(\vartheta \triangle \theta) = O^+(T(\vartheta \triangle \theta)) = O^+(T(\vartheta)\triangle' T(\theta)) \ge min\{O^+(T(\vartheta)), O^+(T(\theta))\} = min\{O_T{}^+(\vartheta), O^+{}_T(\theta)\}$. Furthermore, we can consider that $O_T{}^-(\vartheta \triangle \theta) = O^-(T(\vartheta \triangle \theta)) = O^+(T(\vartheta)\triangle' T(\theta)) \le \max\{O^-(T(\vartheta)), O^-(T(\theta))\} = max\{O_T{}^-(\vartheta), O^-{}_T(\theta)\}$. Moreover, if $\vartheta \in \mho$ with $\forall \theta' \in \mho', \exists \theta \in \mho \ni T(\theta) = \theta'$ [because T is onto]. Hence $O_T^+(\vartheta) = O^+(T(\vartheta)) \ge \min\{O^+(T(\vartheta \triangle \theta)), O^+(T(\theta))\} = \{O^+(T\vartheta)\triangle'\theta'), O^+(\theta')\} = min\{O^+(T(\vartheta)\triangle' T(\theta)), O^+(T(\theta))\} = min\{O_T{}^+(\vartheta \triangle \theta), O_T{}^+(\theta)\}$, and $O_T{}^-(\vartheta) = O^-(T(\vartheta)) \le max\{O^-(T(\vartheta \triangle \theta)), O^-(T(\theta))\} = \{O^-(T\vartheta)\triangle'\theta'), O^-(\theta')\} = max\{O^-(T(\vartheta)\triangle' T(\theta)), O^-(T(\theta))\} = max\{O_T{}^-(\vartheta \triangle \theta), O_T{}^-(\theta)\}$. Now, let $\vartheta, \theta \in \Gamma$. Then

$$O_T{}^+(\vartheta \lhd \theta) = O^+(T(\vartheta \lhd \theta)) = O^+\big(T(\vartheta)\lhd' T(\theta)\big)$$

$$\ge \min\{O^+(\mathrm{T}(\vartheta))O^+(\mathrm{T}(\theta))\},$$

$$= \min\big\{O_T{}^+(\vartheta), O^+{}_T(\theta)\big\}.$$

And $O_T{}^-(\vartheta \lhd \theta) = O^-(T(\vartheta \lhd \theta)) = O^+\big(T(\vartheta)\lhd' T(\theta)\big)$

$$\le \max\big\{O^-(\mathrm{T}(\vartheta)), O^-(\mathrm{T}(\theta))\big\}$$

$$\leq max\{O_T^-(\vartheta), O^-{}_T(\theta)\}.$$

Also, let $\vartheta \in \mho$ and for any $\theta' \in \mho'$, $\exists\theta \in \mho \ni T(\theta) = \theta'$. (since T is onto). Then

$$\begin{aligned} O_T^+(\vartheta) &= O^+(T(\vartheta)) \geq min\{O^+(T(\vartheta \lhd \theta)), O^+(T(\theta))\} \\ &= \{O^+(T\vartheta)\lhd'\theta'), O^+(\theta')\} \\ &= min\{O^+(T(\vartheta)\lhd' T(\theta)), O^+(T(\theta))\} \\ &= min\{O_T^+(\vartheta \lhd \theta), O_T^+(\theta)\}. \end{aligned}$$

Also,

$$\begin{aligned} O_T^-(\vartheta) &= O^-(T(\vartheta)) \leq max\{O^-(T(\vartheta \lhd \theta)), O^-(T(\theta))\} \\ &= \{O^-(T\vartheta)\lhd'\theta'), O^-(\theta')\} \\ &= max\{O^-(T(\vartheta)\lhd' T(\theta)), O^-(T(\theta))\} \\ &= max\{O_T^-(\vartheta \lhd \theta), O_T^-(\theta)\}. \end{aligned}$$

Hence $B_T = (\vartheta, O_T^-, O_T^+)$ is a $(BF\delta - ISG)$ of $\mho$.

Corollary 4.4 *Assume* $T{:}(\mho, \Delta, \lhd, \nu) \to (\mho', \Delta', \lhd', \nu)$ *is a* $(\delta - SGH)$ *and onto. If* $B = (\vartheta', O^-, O^+)$ *is a* $(BF\delta - ISG)$ *of* $\mho'$ *Then* $B_T = (\vartheta, O_T^-, O_T^+)$ *is a* $(BF\bar{\delta} - ISG)$ *of* $\mho$.

Proof From [Theorem (4.3)] $B_T = (\vartheta, O_T^-, O_T^+)$ is a $(BF\delta - ISG)$ of $\mho'$. And hence from [Lemma (3.22)] we get $B_T = (\vartheta, O_T^-, O_T^+)$ is a $(BF\bar{\delta} - ISG)$ of $\mho$.

Definition 4.5 Assume that $B_1 = (\vartheta, O_1^-, O_1^+)$ and $B_2 = (\theta, O_2^-, O_2^+)$ are two $(BFSs)$ of $(\delta - SG)$ $\mho$. We define their product $B_1 \times B_2 = (\vartheta \times \theta, O_1^- \times O_2^-, O_1^+ \times O_2^+)$ as $O_1^+ \times O_2^+{:}\mho \times \mho \to [0, 1]$ where $(O_1^+ \times O_2^+)(\vartheta, \theta) = \min\{O_1^+(\vartheta), O_2^+(\theta)\}$, and $O_1^- \times O_2^-{:}\mho \times \mho \to [-1, 0]$ where $(O_1^- \times O_2^-)(\vartheta, \theta) = \max\{O_1^-(\vartheta), O_2^-(\theta)\}$.

Theorem 4.6 *Assume that* $(\mho, \Delta, \lhd, \nu)$ *is a* $\delta -$ *semigroup. If* $B_1 = (\vartheta, O_1^-, O_1^+)$ *and* $B_2 = (\theta, O_2^-, O_2^+)$ *are* $(BF\bar{\delta} - ISGs)$ *of* $\mho$. *Then* $B_1 \times B_2$ *is* $(BF\bar{\delta} - ISG)$ *of* $\mho \times \mho$.

Proof Let $(\vartheta, \theta) \in \mho \times \mho$, then we get $(O_1^- \times O_2^-)(\nu, \nu) = max\{O_1^-(\nu), O_2^-(\nu)\}$

$$\leq max\{O_1^-(\vartheta), O_2^-(\theta)\}$$

$$= (O_1^- \times O_2^-)(\vartheta, \theta).$$

And $(O_1^+ \times O_2^+)(\nu, \nu) = min\{O_1^+(\nu), O_2^+(\nu)\}$

$$\leq min\{O_1^+(\vartheta), O_2^+(\theta)\}$$

$$= \left(O_1^+ \times O_2^+\right)(\vartheta, \theta).$$

Also, let $(\vartheta_1, \vartheta_2), (\theta_1, \theta_2) \in \mho \times \mho$. Then

$$\left(O_1^- \times O_2^-\right)(\vartheta_1 \vartriangle \theta_1, \vartheta_2 \vartriangle \theta_2) = max\{O_1^-(\vartheta_1 \vartriangle \theta_1), O_2^-(\vartheta_2 \vartriangle \theta_2)$$

$$\leq max\{max\{O_1^-(\vartheta_1 \vartriangle \theta_1), O_1^-(\theta_1)\}, \max\{O_2^-(\vartheta_2 \vartriangle \theta_2), O_2^-(\theta_2)\}\}$$

$$= \max\{\max\{O_1^-(\vartheta_1 \vartriangle \vartheta_1), O_2^-(\vartheta_2 \vartriangle \theta_2)\}, \max\{O_1^-(\vartheta_1), O_2^-(\theta_2)\}\}$$

$$= max\{\left(O_1^- \times O_2^-\right)(\vartheta_1 \vartriangle \theta_1, \vartheta_2 \vartriangle \theta_2), (O_1^- \times O_2^-)(\vartheta_1, \vartheta_2)\},$$

And $(O_1^+ \times O_2^+)(\vartheta_1 \vartriangle \theta_1, \vartheta_2 \vartriangle \theta_2) = \min\left\{O_1^+(\vartheta_1 \vartriangle \theta_1), O_2^+(\vartheta_2 \vartriangle \theta_2)\right\}$

$$\geq min\{\min\{O_1^+(\vartheta_1 \vartriangle \theta_1), O_1^+(\theta_1)\}, \min\{O_2^+(\vartheta_2 \vartriangle \theta_2), O_2^+(\theta_2)\}\}$$

$$= min\{min\{O_1^+(\vartheta_1 \vartriangle \theta_1), O_2^+, (\theta_1 \vartriangle \theta_2)\}, min\{O_1^+(\theta_1), O_2^+(\theta_2)\}\}$$

$$= \min\{(O_1^+ \times O_2^+)(\vartheta_1 \vartriangle \theta_1, \vartheta_2 \vartriangle \theta_2), \left(O_1^+ \times O_2^+\right)(\vartheta_1, \vartheta_2)\}$$

Let $(\vartheta_1, \vartheta_2), (\theta_1, \theta_2) \in \mho \times \mho$. Then we have

$$\left(O_1^- \times O_2^-\right)(\vartheta_1 \triangleleft \theta_1, \vartheta_2 \triangleleft \theta_2) = max\{O_1^-(\vartheta_1 \triangleleft \theta_1), O_2^-(\vartheta_2 \triangleleft \theta_2)$$

$$\leq max\{max\{O_1^-(\vartheta_1 \triangleleft \theta_1), O_1^-(\theta_1)\}, \max\{(O_2^-(\vartheta_2 \triangleleft \theta_2), O_2^-(\theta_2)\}\}$$

$$= max\{max\{O_1^-(\vartheta_1 \triangleleft \vartheta_1), O_2^-(\vartheta_2 \triangleleft \theta_2)\}, max\{O_1^-(\vartheta_1), O_2^-(\theta_2)\}\}$$

$$= max\{(O_1^- \times O_2^-)(\vartheta_1 \triangleleft \theta_1, \vartheta_2 \triangleleft \theta_2), (O_1^- \times O_2^-)(\vartheta_1, \vartheta_2)\}.$$

Furthermore, $(O_1^+ \times O_2^+)(\vartheta_1 \triangleleft \theta_1, \vartheta_2 \triangleleft \theta_2) = \min\{O_1^+(\vartheta_1 \triangleleft \theta_1), O_2^+(\vartheta_2 \triangleleft \theta_2)\}$

$$\geq min\{min\{O_1^+(\vartheta_1 \triangleleft \theta_1), O_1^+(\theta_1)\}, \min\{O_2^+(\vartheta_2 \triangleleft \theta_2), O_2^+(\theta_2)\}\}$$

$$= min\{min\{O_1^+(\vartheta_1 \triangleleft \theta_1), O_2^+(\theta_1 \triangleleft \theta_2)\}, min\{O_1^+(\theta_1), O_2^+(\theta_2)\}\}$$

$$= min\{(O_1^+ \times O_2^+)(\vartheta_1 \triangleleft \theta_1, \vartheta_2 \triangleleft \theta_2), (O_1^+ \times O_2^+)(\vartheta_1, \vartheta_2)\}.$$

Hence $B_1 \times B_2$ is $(BF\overline{\delta} - ISG)$ of $\mho \times \mho$.

5 Conclusion and Future Works

We have been studying bipolar fuzzy ideals in algebraic semigroups. In this paper, we addressed some (FL) notions with their implementations using the the atomic numbers and valency of the chemical elements, as well as demonstrate additional applications of new (FL) techniques to research algebraic logic problems. In future work, we will provide more conceptions and investigate their applications using neutrosophic sets rather than fuzzy sets, and we will compare them.

References

1. Zadeh, L.A.: Fuzzy sets. Inf. Control. **8**, 338–353 (1965)
2. Mahmood, S., Hassan, A.N.: New class of algebraic fuzzy systems using cubic soft sets with their applications. IOP Conf. Ser.: Mater. Sci. Eng. **928**, 042019 (2020). https://doi.org/10.1088/1757-899X/928/4/042019
3. Noori, S.M., Ahmad, A.G., Khalil, S.: New class of doubt bipolar fuzzy sub measure algebra. Comput. Model. Eng. Sci. **135**(1), 293–300 (2023)
4. Khalil, S., Sharqi, M.S.: Some applications in decision-making using cosine maps and the relevance of the Pythagorean fuzzy. Eng. Appl. Artif. Intell. ScienceDirect **122**, 106089 (2023)
5. Jaber, L., Mahmood, S.: New category of equivalence classes of intuitionistic fuzzy delta-algebras with their applications. Smart Innov. Syst. Technol. **302**, 651–663 (2022). https://doi.org/10.1007/978-981-19-2541-2_54
6. Khalil, S., Hameed, F.: Applications of fuzzy ρ-ideals in ρ-algbras. Soft. Comput. (2020). https://doi.org/10.1007/s00500-020-04773-3
7. Lee, K.M.: Bipolar-valued fuzzy sets and their operations. In: Proceeding of International Conference on Intelligent Technologies, Bangkok, Thailand, pp. 307–312 (2000)
8. Muhiuddin, G.: Bipolar fuzzy KU-subalgebras\ideals of KU-algebras. Ann. Fuzzy Math. Inform. **8**(3), 409–418 (2014)
9. Maji, P.K., Roy, A.R., Biswas, R.: Fuzzy soft sets. J. Fuzzy Math. **9**(3), 589–602 (2001)
10. Mahmood, S., Hassan, A.: The characterizations of δ-algebras with their Ideals. IOP Conf. Ser: J. Phys. **1999**, 012108 (2021). https://doi.org/10.1088/1742-6596/1999/1/012108
11. Dar, K.H.: On fuzzy d-algebras, Punjab University. J. Math. **37**, 61–76 (2005)
12. Zhang, W.: Bipolar fuzzy sets and relations: a computational framework for cognitive modeling and multiagent decision analysis. IEEE, pp. 305–309 (1994)